Explorations in College Algebra

EXPLORATIONS IN COLLEGE ALGEBRA

Second Edition

Linda Almgren Kime
Judy Clark
University of Massachusetts, Boston

in collaboration with

Norma M. Agras
Miami Dade Community College

Robert F. Almgren
University of Toronto

Linda Falstein
University of Massachusetts, Boston
Emeritus

Meg Hickey
Massachusetts College of Art

John A. Lutts
University of Massachusetts, Boston

Peg McPartland
Golden Gate University

Beverly K. Michael
University of Pittsburgh

Jeremiah V. Russell
University of Massachusetts, Boston
Boston Public Schools

Software Developed by

Hubert Hohn
Massachusetts College of Art

Funded by a National Science Foundation Grant

JOHN WILEY & SONS

New York Chichester Weinheim Brisbane Toronto Singapore

EXECUTIVE EDITOR	Ruth Baruth
FREELANCE DEVELOPMENTAL EDITOR	Anne Scanlan-Rohrer
MARKETING MANAGER	Julie Lindstrom
PRODUCTION EDITOR	Ken Santor
TEXT DESIGNER	Lee Goldstein
ILLUSTRATION EDITOR	Sigmund Malinowski
ELECTRONIC ILLUSTRATIONS	Radiant
COVER PHOTO	Jeff Hunter/The Image Bank
COVER DESIGNER	Dawn Stanley

This book was set in Times Roman by Progressive Information Technologies and printed and bound by Von Hoffman Press. The cover was printed by Lehigh Press.

This book is printed on acid-free paper. 

This project was sponsored, in part, by the
National Science Foundation
Opinions expressed are those of the authors and not necessarily those of the Foundation.

This material is based upon work supported by the National Science Foundation under Grant No. USE-9254117

The paper in this book was manufactured by a mill whose forest management programs include sustained yield harvesting of its timberlands. Sustained yield harvesting principles ensure that the amount of timber cut each year does not exceed the amount of new growth.

Library of Congress Cataloging in Publication Data:
Kime, Linda Almgren.
Explorations in college algebra/Linda Almgren Kime, Judy Clark; in collaboration with Norma M. Agras . . . [et al.].—2nd ed.
p. cm.
Includes bibliographical references and index.
ISBN 0-471-37194-7 (pbk.: acid-free paper)
1. Algebra. I. Clark, Judy, 1945-II. Title.

QA152.2.K55 2001
512.9—dc21 00-043379

Printed in the United States of America

10 9 8 7 6 5 4

To all of our students, who have been thoughtful critics
and enthusiastic participants in the development of these materials.

A Letter to the Students From a Student

Have you ever thought of math as fun? No, I'm not crazy. You might think of math as merely a bunch of numbers and variables and long hours of monotonous repetition, but this class makes math interesting.

Math will come alive right in front of you. This class, unlike many other math courses, is different in that it shows you practical applications of math to real life situations and gives meaning to math. The numbers and variables are symbolic representations of things in the world. Math will finally make sense.

This class is also different in that there will be lecture, but much of the class time will be spent *doing* math. I found this to be very helpful in that you're actually doing the math right there and then and not trying to decipher and make sense of your lecture notes at home where it's too late to ask questions.

At first things will take a little time to get used to. But watch out. Once you see the big picture of this class it can be very addictive. All you have to do is give it a chance to show you what it's all about. This class opens up your mind and helps you think critically and thoughtfully about math as you never have before. I hope you enjoy it as much as I have.

Sincerely
Philip Wan
Nursing/Pre-Med Major
University of Massachusetts, Boston

Preface

Introduction

This text grew from our desire to reshape the traditional college algebra course, to make it relevant and accessible to all of our students. Our goal was to shift the focus from learning a set of discrete mechanical rules to exploring how algebra is used in the social and physical sciences. These materials were developed in conjunction with faculty from a nationwide Consortium of schools and were initially funded by a grant from the National Science Foundation.

General Principles

The following general principles served as our guide.

- Algebraic procedures and concepts can be developed from the investigation of practical problems.
- Mathematical ideas can be represented and understood in multiple ways –through words, numbers, graphs, and symbols.
- Students can make connections among various representations and understand concepts more deeply by becoming actively engaged in thinking mathematically, and by communicating their ideas to others.
- Conceptual understanding needs to be accompanied by sufficient practice in skill building.

Content

Families of functions are used to model real-world behavior. After an introductory chapter on data, we first focus on linear and exponential functions, since these are the two most commonly used mathematical models. We then discuss power functions, and quadratic and other polynomial functions. Real data sets are used extensively throughout, motivating abstract concepts, and forming the basis of many of the exercises and examples. Most of the data sets are provided in electronic form on the web.

Special Features and Supplements

The materials were designed to accommodate a broad range of teaching and learning styles, in a variety of settings. There are a number of special features such as Explorations, data sets, readings, and course software. The instructor should feel free to pick and choose among them, incorporating what is appropriate for his or her course and possibly layering in new elements over time. The following list includes descriptions of these features and how they are referenced in the text. All the web resources may be found at **www.wiley.com/college/kimeclark.**

Exploring Mathematical Ideas

The text adopts a problem-solving approach, where problems lie on a continuum from practicing algebraic skills to answering open-ended non-routine questions. In addition to the exercises the following features offer ways to explore mathematical ideas in greater depth.

Explorations
Explorations provide open-ended problems designed to be used in parallel with reading the text. They are located at the end of each chapter and in two chapter length Extended Explorations.

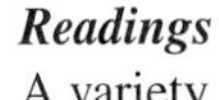

Readings
A variety of articles are available in the Appendix or on the Web (as indicated by a www under the reading icon).

Something to Think About
Provocative questions, posed throughout the text, can be used to generate class discussion or for independent inquiry.

60 Second Summaries
Short writing assignments in the Explorations and exercises offer students the opportunity to formulate hypotheses, make predictions, and summarize findings.

Using Technology

Technology is not required to teach this course. However, we provide interactive software for visualizing mathematical ideas along with resources for analyzing data using graphing calculators or spreadsheets.

Computers

Interactive software for MAC and PC
Programs for visualizing mathematical concepts, simulations, and practice in skill building are available on the Web. They may be used in classroom demonstrations, in a computer lab, or downloaded for student use at home.

Spreadsheet Files
Excel files containing all the major data sets used in the course are available on the Web.

Graphing Calculators

Graphing Calculator Manual
The manual includes basic instructions for using the TI-82 and TI-83 and is coordinated with the chapters in the text. It is available in hardcopy and on the Web.

Graph Link Files
Files for the TI-82 or TI-83 that contain all the major data sets used in the course are available on the Web.

Practice in Skill Building

Algebra Aerobics
Short collections of skill building practice problems (indicated by shading) are integrated throughout each chapter.

EGrade
EGrade is a browser-based program that contains a large bank of skill building questions and solutions. Each instructor has the ability to construct homework assignments, quizzes, and tests that will be automatically scored, recorded in a gradebook, and calculated into the class statistics. Students work on-line and receive immediate grading and feedback. For more information, visit **www.wiley.com/college/egrade.**

The Instructor's Manual provides support for using these features and contains a complete set of answers to the exercises. This manual is available in hard copy and is also available on a CD that contains the interactive software, the graph link and Excel data files, and the Graphing Calculator Manual as well.

The College Algebra Course

Many students enroll in college algebra as preparation for future mathematics or other quantitatively based courses. For many others, however, college algebra is their last mathematics course, often taken to meet a college requirement. These materials are intended to be appropriate for both sets of students.

In order to decide which topics to include in the text, we did an extensive survey of faculty teaching college algebra. One of the rather surprising conclusions we reached is that there is no canonical college algebra course. What one campus calls college algebra, another might label algebra for college students, math modeling, precalculus for non-math majors, or even quantitative reasoning.

In order to help the instructor design a course appropriate for your campus, see the flow chart on p. xiii, just before the Table of Contents.

Changes from the First Edition

In the second edition the number and types of exercises at the end of each chapter have been expanded and grouped by section. Certain chapters from the first edition have been merged or rearranged. The former Chapters 1 and 2 on data and functions have been merged and condensed to form the new Chapter 1. The former Chapters 3 and 4 on rates of change and linear functions have been merged to form the new Chapter 2. The former Chapter 12 (now Chapter 6) on logarithmic functions and exponential functions base e has been moved to directly follow the new Chapter 5 on exponential functions (former Chapter 8). All of the discussion of polynomial and quadratic functions now occurs in a single chapter (new Chapter 8: Quadratic and Other Polynomial Functions).

The two case study chapters (formerly Chapters 5 and 10) are now called "Extended Explorations" to reflect how they can be used. The Extended Exploration "Looking for Links: A Case Study in Education and Income" follows the new Chapter 2 on rates of change and linear functions. The Extended Exploration on Mathematics of Motion was moved after the chapter on quadratic and other polynomial functions.

Acknowledgments

We wish to express our appreciation to all those who have helped and supported us in this collaborative endeavor. We are grateful for the support of the National Science Foundation, whose funding made this project possible, and for the generous help from our program officers, Elizabeth Teles and Marjorie Enneking. We wish to thank members of our original Advisory Board, Deborah Hughes Hallett, and Philip and Phylis Morrison, who have continued to provide encouragement and invaluable advice.

The text could not have been produced without the generous and ongoing support we received from students, faculty, staff, and administrators at University of Massachusetts Boston. Our thanks especially to the administrators: Chancellor Sherry Penney, Vice Provosts Theresa Mortimer and Patricia Davidson, and Dean Christine Armett-Kibel, who have supported us throughout each phase of our work. We are grateful to the following of our UMass colleagues for their contributions to the text: Mark Powlack, Bob Seeley, George Lukas, Rachel Skvirsky, Ron Etter, Lowell Schwartz, Joan Lukas, Karen Callaghan, Randy Albelda, Brenda Cherry, Paul Foster, Brian Butler, and Jie Chen.

We will sorely miss Bob Lee's humor, pragmatism, and invaluable advice on how to adapt our materials for use with graphing calculators.

A text designed around the application of real world data would have been impossible if not for the long and exacting hours put in by Maria Naylor, Myrna Kustin and David Wilson, researching data sources around the globe.

We also wish to thank our terrific student teaching assistants: John Koveos, Irene Blach, Philip Wan, Kristen Demopoulos, Tony Horne, Cathy Briggs and Arlene Russo. They excelled as students and as teachers.

We are deeply indebted to Justin Gross, Nicole Kent Perez, Kristin Clark, David Hruby, Anthony Beckwith, Anu Karna, Ann Ostberg, Dean Hickerson, Neil Wigley, Peter Renz, Madalyn Stone and Dick Cluster for their gracious and incisive editorial help.

Particular thanks goes to our wonderful editor, Ruth Baruth, who has borne with us now through four versions of *Explorations*. She has long since evolved from editor to friend. Wiley is truly lucky to have such a gifted and dedicated acquisitions editor. Anne Scanlan-Rohrer provided us with the gracious, but firm, oversight necessary to shepherd a second edition to press. Ken Santor has kept us on track in production. Dawn Stanley, Sigmund Malinowski, Merillat Saint-Amand, Mary Johenk, and many others at Wiley have been extraordinarily helpful in dealing with the myriad of endless details.

One of the joys of this project has been working with so many dedicated faculty who are searching for new ways to reach out to students. These faculty, teaching assistants, and students all offered incredible support and encouragement, and a wealth of helpful suggestions for both the first and second editions. In particular our heartfelt thanks to: Sandi Athanassiou and all the wonderful TA's at University of Missouri-Columbia; Peggy Tibbs and John Watson, Arkansas Technical University; Josie Hamer, Robert Hoburg, and Bruce King, past and present faculty at Western Connecticut State University; Judy Stubblefield, Garden City Community College; Lida McDowell, Jan Davis, and Jeff Stuart, University of Southern Mississippi; Ann Steen, Santa Fe Community College; Leah Griffith, Rio Hondo College; Mark Mills, Central College; Tina Bond from Pensacola Junior College; and Curtis Card from Black Hills State University. All of them offered invaluable advice.

We would like to thank the following additional reviewers who helped shape the second edition:

Mailing/Telephone Survey: Rick Anderson, North Dakota State College of Science; Nathan Borchelt, Clayton College and State University; Shay Cardell, Central Arizona College; Klaus Fischer, George Mason University; Nora Franzova, Harford Community College; Joel K. Haack, University of Northern Iowa; Roy Hardy, Truett-McConnell College; Steve Hatfield, Brenau University; Judith M. Holbrook, Yavapai Community College; Charles Johnson, South Georgia College; Ashok Kumar, Valdosta State University; Nancy Leigh, Western Illinois University; Jane Lybrand, University of South Carolina at Aiken; James P. Marshall, Illinois College; Mary Marshall, Illinois College; Mary Ann Matras, East Strousburg University; Jeanie McGehee, University of Central Arkansas; David Olsen, Le Tourneau University (Houston Campus); Hari Pulapaka, Valdosta State University; Matt Seely, Salish Kootenai University; Karl Shaffer, De Anza College; Dona Sherrill, Arkansas Technical University; Mary H. Stephens, Clayton College and State University; Gerald White, Western Illinois University; Sharon Wilson, University of Tulsa.

Phone Interviews with Anne Scanlan-Rohrer: Ana DeArmas, Palm Beach Community College; Jim Graziose, Palm Beach Community College; Jeremy Haefner, University of Colorado, Colorado Springs; Elizabeth Hodes, Santa Barbara City College; Lynne Ipina, University of Wyoming; Paula Maida, Western Connecticut State University; Mary Pohl, Montana State University; Linda Schmidt, Greenville Tech; Judith Smalling, St. Petersburg Junior College; Ann Sitomer, Portland Community College.

Written Reviews: Rowan Lindley, Westchester Community College; Mark Mills, Central College; Ann Sitomer, Portland Community College; Rose Tan, Westchester Community College; Sheela Whelan, Westchester Community College.

Our families couldn't help but become caught up in this time-consuming endeavor. Linda's husband, Milford, and her son Kristian were invaluable scientific and, more importantly, emotional resources. They both offered unending encouragement and sympathetic shoulders. Judy's husband, Gerry, became our Consortium lawyer, and her daughters, Rachel, Caroline, and Kristin provided support, understanding, laughter, and whatever was needed. All our family members ran errands, made dinners, listened to our concerns and gave us the time and space to work on the text. Our love and thanks.

Finally, we wish to thank all of our students. Without them, this book would not have been written.

Covering the Contents

The following flow chart suggests some alternate paths through the chapters that have worked successfully for others.

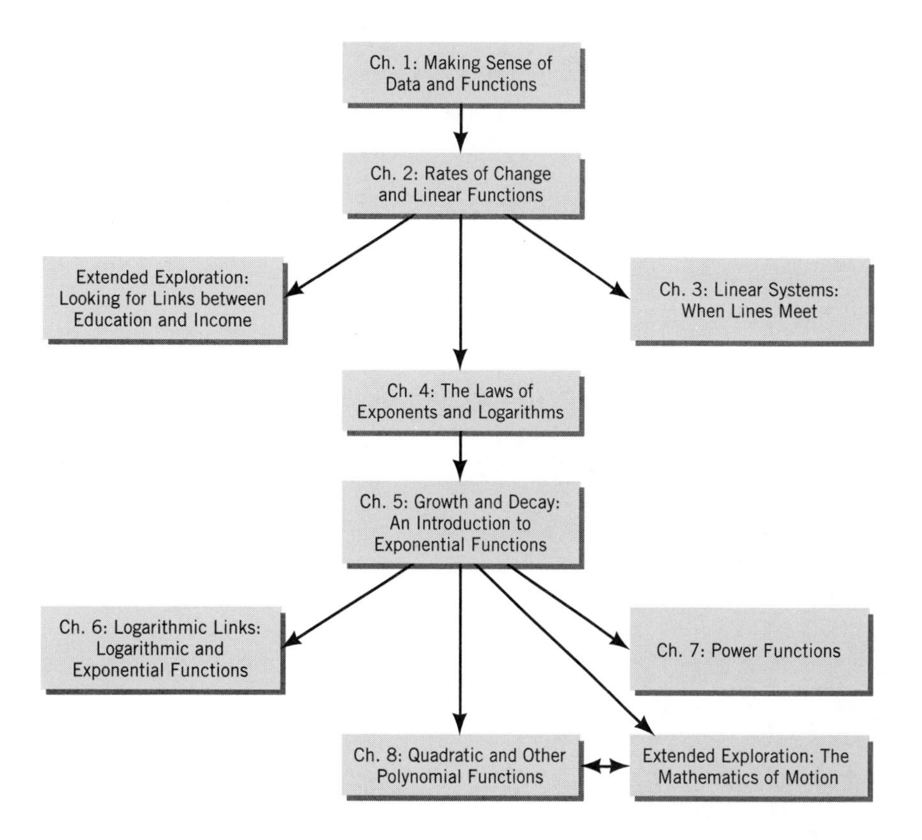

The straight vertical path through Chapters 1, 2, 4, 5, and 8, covering linear, exponential, and polynomial functions, indicates the core content of the text. You may choose to cover these chapters in depth, spending time on the explorations, readings and student discussions, writing, and presentations. Or you may pick up the pace and include as many of the other chapters and Extended Explorations as is appropriate for your department's needs.

Table of Contents

Making Sense of Data and Functions

Overview

How would you describe the distribution of ages in the United States? How has the population of the United States changed over time? In this chapter we investigate how to answer these questions by examining patterns in data. We explore how to use graphs to visualize the shape of single-variable data and to show changes in two-variable data. We introduce functions, a fundamental concept in mathematics, and show how to represent them using graphs, tables, and equations.

After reading this chapter, you should be able to:

- use graphs, tables, and numerical descriptors to represent single variable data
- construct a "60 second summary"
- understand the definition of a function
- represent functions with tables, graphs, equations, and function notation
- identify independent and dependent variables and the domain and range of a function

1.1 AN INTRODUCTION TO SINGLE VARIABLE DATA

This course starts with you. How would you describe yourself to others? Are you a 5-foot 6-inch, black, 26-year-old, female studying engineering? Or perhaps you are a 5-foot 10-inch, Chinese, 18-year-old, male English major, sharing an apartment with two friends. In statistical terms, characteristics such as height, race, age, and major that vary from person to person are called *variables*. Information collected about a variable is called *data*.[1]

Some variables, such as age, height, or number of people in your household, can be represented by a number and a unit of measure (such as 18 years, 6 feet, or 3 people). These are called *quantitative variables*. For other variables, such as gender or college major, we use categories (such as male and female or physics, English, and psychology) to classify information. These are called *qualitative variables*.

Exploration 1.1 provides an opportunity to collect your own data and to think about issues related to classifying and interpreting data.

Many of the controversies in the social sciences have centered on how particular variables are defined and measured. For nearly two centuries, the categories used by the U.S. Census Bureau to classify race and ethnicity have been the subject of debate. For example, Hispanic used to be considered a racial classification. It is now considered an ethnic classification, since Hispanics can be black, or white, or any other race.

In this section we will see how numerical descriptors, tables, and graphs can be used to describe patterns in single variable data. We start by looking at a small data set of student ages and then examine a large data set of the ages of the U.S. population. The search for ways to identify, summarize, and describe patterns in data can be thought of as a search for a "60 Second Summary." If you had to describe the U.S. population in terms of age in 60 seconds, what would you say? This section prepares you to construct such a summary.

Describing Data

To use data, we need to be able to describe them. We introduce some tools for describing single variable data by examining the ages of students in a college algebra class at the University of Massachusetts at Boston. The students' ages (in years) in ascending order are

18, 19, 20, 20, 20, 21, 22, 23, 23, 23, 26, 27, 28, 29, 30, 30, 33, 36, 39, 41, 46

Numerical Descriptors: Measures of Central Tendency

Numerical descriptors such as the mean and the median are often used to summarize data. They describe the "center of the data" and are called "measures of central tendency."

The mean. The *mean* is the arithmetic average,[2] with which you are probably already familiar.

[1]*Data* is the plural of the Latin word *datum* (meaning "something given"). Hence one datum, two data.

[2]The word "average" has an interesting derivation according to Klein's etymological dictionary. It comes originally from the Arabic word "awariyan," which means merchandise damaged by seawater. The idea being debated was that if your ships arrived with water-damaged merchandise, should you have to bear all the losses yourself or should they be spread around, or "averaged," among all the other merchants? The words *avería* in Spanish, *avaría* in Italian, and *avarie* in French still mean "damage."

> The *mean* of a list of numbers is their sum divided by the number of terms in the list.

The mean age of students in the college algebra class can be found by adding together all of the 21 observed ages and dividing the resulting sum by 21, the total number of terms in the list. Adding together the ages gives

Read the excerpt from *CHANCE News*, an electronic magazine, to see how the mean is sometimes incorrectly used in the popular press.

$$18 + 19 + 20 + 20 + 20 + 21 + 22 + 23 + 23 + 23 + 26 + 27 + 28 + 29 + 30 + 30 + 33 + 36 + 39 + 41 + 46 = 574$$

So the mean is

$$574/21 \approx 27.33$$

To describe this process in general, we can represent the age of each student by a letter with a numerical subscript. For a class with n students, the ages can be represented as $a_1, a_2, a_3, \ldots, a_n$, where a_1 represents the age of the first student, a_2 the age of the second student, and so on.

The mean of n values $a_1, a_2, a_3, \ldots, a_n$ is found by adding these values and dividing the sum by n, the number of values:

$$\text{mean} = \frac{a_1 + a_2 + a_3 + \cdots + a_n}{n}$$

The median. The *median* locates the "middle" of a numerically ordered list.

> The *median* separates a numerically ordered list of numbers into two parts, with half the numbers at or below the median and half at or above the median. If the number of observations is odd, the median is the middle number in the ordered list. If the number of observations is even, the median is the mean of the two middle numbers in the ordered list.

In the previous ordered list of ages, there are 21 numbers. Since the list has an odd number of terms, the median is the middle, or 11th, number, which is 26 years:

18, 19, 20, 20, 20, 21, 22, 23, 23, 23, **26,** 27, 28, 29, 30, 30, 33, 36, 39, 41, 46

$$\text{median} = 26$$

See the reading "The Median Isn't the Message" to find out how an understanding of the median gave renewed hope to the renowned scientist, Stephen J. Gould, when he was diagnosed with cancer.

If one of the 23-year-old people dropped out of the class, the list would contain the ages of only 20 students. Since that list now has an even number of terms, the median would be the mean of the middle two numbers, in this case the 10th and 11th numbers. So the median would be the mean of 26 and 27:

18, 19, 20, 20, 20, 21, 22, 23, 23, **26, 27,** 28, 29, 30, 30, 33, 36, 39, 41, 46

$$\text{median} = (26 + 27)/2 = 26.5$$

Note that neither the mean nor the median needs to be one of the numbers in the original list.

The significance of the median is that it divides the number of entries in the data set into two equal halves. If the median age in a large urban housing project is 17, then half the population is 17 or under. Hence issues of day care, recreation, supervision of minors, and kindergarten through 12th-grade education should be very high priorities with the management. The other half of the population is, of course, 17 or older. The mean, however, could be very different from the median. For instance, if the adults in that same project included a fairly large proportion of elders, the mean age could be 35; by itself this number would give no indication that a majority of the residents were minors.

If someone tells you that in his town "all of the children are above average," you might be skeptical. (This is called the "Lake Wobegon effect.") But could most (more than half) of the children be above average? Explain.

The major disadvantage of the median is that it is unchanged by the redistribution of values above and below the median. For example, as long as the median income is larger than the poverty level, it will remain the same even if all poor people suddenly increase their incomes up to that level and everyone else's income remains the same.

The mean is the most commonly cited statistic in the news media. One advantage of the mean is that it can be used for calculations relating to the whole data set. Suppose a corporation wants to open a new factory similar to its other factories. If the managers know the mean cost of wages and benefits for employees, they can make an estimate of what it will cost to employ the number of workers needed to run the new factory:

$$\text{total employee cost} = (\text{mean cost for employees}) \cdot (\text{number of employees})$$

The major disadvantage of the mean is that it is affected by extreme values in the data set called *outliers.* John Schwartz comments in a *Washington Post* article ("Mean statistics: When is the average best?" December 6, 1995) that if Bill Gates, billionaire founder of Microsoft, were to move into a town with 10,000 people without any income, the mean income would be more than a million dollars. This might suggest that the town is full of millionaires, which is obviously far from the truth. The median income, however, is still $0.

Numerical Descriptors: Frequency and Relative Frequency

A first step in searching for patterns in single variable data is to examine the overall shape of the data or how data are distributed over a range of values. When data are compressed into equal-sized intervals, certain features of the data may be revealed. Let's return to the data on age (in years) of students in a college algebra class. The raw data are

18, 19, 20, 20, 20, 21, 22, 23, 23, 23, 26, 27, 28, 29, 30, 30, 33, 36, 39, 41, 46

We can construct equal-sized intervals by choosing an interval size and counting the number of people whose ages fall into each interval. For example, in Table 1.1, we chose an interval size of 3 years, started our first interval at 18 years, and then determined there are five people in the 18- to 20-year interval (those who were 18, 19, or 20 years old). The number of observations in each interval is called the *frequency* or *frequency count.*

Table 1.1
Ages of Students in Three-Year Intervals

Age Interval	Frequency Count
18–20	5
21–23	5
24–26	1
27–29	3
30–32	2
33–35	1
36–38	1
39–41	2
42–44	0
45–47	1
Total	**21**

> The number of observations that fall within a given interval (or assume a given value) is called the *frequency count.*

To get a sense of the number of students in each age interval relative to the total count, we compare the frequency count in each interval to the total number of students in the class. Such a comparison is called a *relative frequency* and is expressed as either a decimal or a percentage.

The relative frequency for each age interval in our data set can be found by dividing the number of people in each age interval by the total number of people in the class. For example, there were 5 people in the first age interval of 18 to 20 years and the total number of students in the class was 21. To express the relative frequency of 18- to 20-year-olds as a percentage, we

divide 5 by 21 $5/21 \approx 0.24$

convert to % by multiplying by 100% $(0.24)(100\%) = 24\%$

> The fraction or percentage of the total number of observations that falls within a given interval (or assumes a given value) is called the *relative frequency*.

Table 1.2 is the age data table with an additional column for the relative frequency (expressed as a percentage) of students in each age interval.

Table 1.2
Student Ages with Relative Frequency

Age Interval	Frequency Count	Relative Frequency (%)*
18–20	5	24
21–23	5	24
24–26	1	5
27–29	3	14
30–32	2	10
33–35	1	5
36–38	1	5
39–41	2	10
42–44	0	0
45–47	1	5
Total	**21**	**100**

*Since we rounded off each of the percents, the percentages actually add to 102%. But 21 students represents 100% of our class.

Percentage of what? Relative frequencies are important because they can be used to compare data sets that are not the same size. For example, if we only know that 5 students in this one section are 18- to 20-year-olds and 1235 students in the college are 18- to 20-year-olds, then we cannot make a comparison between these sets of data. But if we know that 5 out of 21 students (or about 24%) of the class are 18- to 20-year-olds and 1235 out of 7251 students (or about 17%) of all the students in the college are 18- to 20-year-olds, then the numbers are seen relative to the size of the class and the population of the college. We now know that, compared to the college as a whole, this particular class has a larger percentage of 18- to 20-year-olds.

Visualizing Single Variable Data

Humans are visual creatures. By converting a data table into a picture, we can recognize patterns that may be hard to discern from a list of numbers. There are many different ways to display data. The kind of graph or chart chosen can influence which aspects of the data emerge and which remain hidden.

See the program "F1: Histograms".

A *histogram* shows how data for a quantitative variable are distributed over a range of values and thus can reveal patterns in the "shape of the data." The horizontal axis is usually marked off in equal-sized intervals to facilitate comparisons between intervals. The intervals on the vertical axis should also be of uniform size.

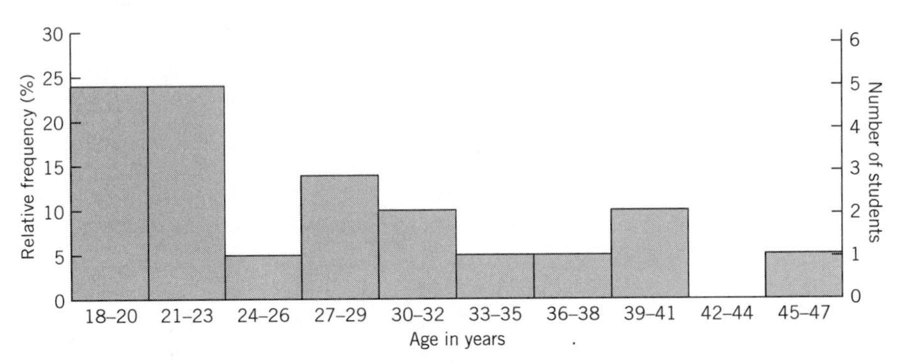

Figure 1.1 Distribution of ages of students in a college algebra class.
Source: Class survey, University of Massachusetts, Boston, 1997.

The horizontal axis of the histogram in Figure 1.1 represents the age in years of the students in the class. The vertical axis on the left side of the graph shows the relative frequency and the vertical axis on the right side shows the frequency count for each age interval. Because the shape of the frequency and relative frequency histograms are exactly the same, we can use the same graph to represent both the frequency and relative frequency distributions for a set of data.

The pie chart shown in Figure 1.2 contains the same information as the histogram in Figure 1.1. Pie charts are a visual display of the contribution of each category to the total. They can be constructed from frequency or relative frequency counts.

What are some trade-offs in using pie charts versus histograms?

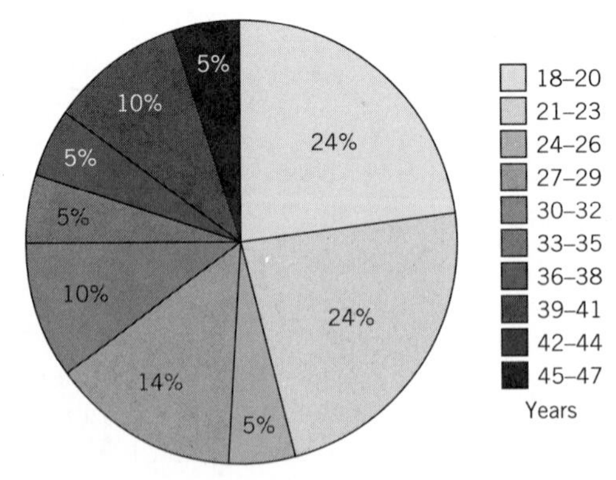

Since we rounded off each of the percentages, they add up to 102% rather than 100%

Figure 1.2 Ages of students in a college algebra class.
Source: Class survey, University of Massachusetts, Boston, 1997.

An Important Aside: What a Good Graph Should Include

When you encounter a graph in an article or you produce one for class, there are three elements that should always be present:

1. An informative title that succinctly describes the graph.
2. Clearly labeled axes (or legend) including the units of measurement (e.g., whether age is being measured in years or months).
3. The source of the data, cited in the data table, in the text, or on the graph.

An Introduction to Algebra Aerobics

In most sections in the text there are short "Algebra Aerobics" workouts with answers in the back of the book. They are intended to give you practice with the mechanical skills introduced in the section and to review other skills we assume you have learned in previous courses. In general, the problems shouldn't take you more than a few minutes to do. We recommend that you work out these practice problems and then check your solutions in the back of this book. The Algebra Aerobics are numbered according to the section of the chapter in which they occur.

Algebra Aerobics 1.1

1. **a.** Fill in Table 1.3.

Table 1.3

Age	Frequency Count	Relative Frequency (%)
1–20		38
21–40	35	
41–60	28	
61–80		
Total	137	

b. Calculate the percentage of the population that is over 40 years old.

2. Use Table 1.3 to create a histogram and pie chart.
3. From the histogram in Figure 1.3, create a frequency distribution table. Assume that the total number of people represented by the histogram is 1352. (*Hint:* Estimate the relative frequencies from the graph and then calculate the frequency count in each interval.)

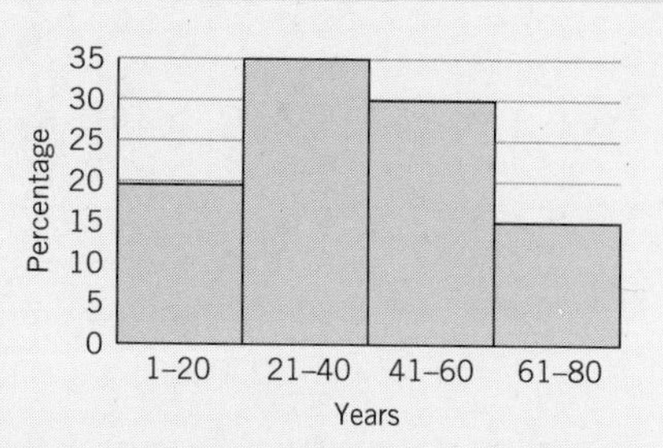

Figure 1.3 Distribution of ages (in years).

4. Calculate the mean and median for the following data:
   a. \$475, \$250, \$300, \$450, \$275, \$300, \$6000, \$400, \$300
   b. 0.4, 0.3, 0.3, 0.7, 1.2, 0.5, 0.9, 0.4
5. Explain why the mean may be a misleading numerical summary of the data in 4(a).
6. A \$200 TV set goes on sale for 10% off. You have a card that reduces the sale price by another 10%.
   a. How much do you pay for the TV set?
   b. How much did you save buying it during the sale and using your discount card?
   c. By what total percentage has the TV set been reduced from the original price?

1.2 WRITING ABOUT DATA

To communicate effectively, we need to be able to describe our results clearly and succinctly. In this section we describe a series of steps that are helpful when constructing a "60 second summary." Such summaries can be used to describe data, support a conjecture, make a prediction, or present a solution to a problem. Whether you are constructing a 60 second summary to present to your class or writing an executive summary for your boss, you can help your ideas take shape by putting them down on paper.

Constructing a "60 Second Summary"

How would you describe in 60 seconds the distribution of ages of the U.S. population?

1. Collect relevant information A first step is to collect relevant information. The web site for the U.S. Bureau of the Census (*www.census.gov*) provided the data for Table 1.4 that are graphed in Figure 1.4. This table and graph show the distribution of ages in 5-year intervals for the U.S. population in 1999.

Table 1.4
Ages of June 1, 1999, U.S. Population in Five-Year Intervals

Age Group (years)	Frequency Count (in 1000s)	Relative Frequency (%)
Under 5	18,920	6.9
5–9	19,964	7.3
10–14	19,516	7.2
15–19	19,741	7.2
20–24	18,023	6.6
25–29	18,268	6.7
30–34	19,783	7.3
35–39	22,556	8.3
40–44	22,255	8.2
45–49	19,309	7.1
50–54	16,398	6.0
55–59	12,843	4.7
60–64	10,513	3.9
65–69	9,457	3.5
70–74	8,781	3.2
75–79	7,324	2.7
80–84	4,814	1.8
85–89	2,621	1.0
90–94	1,147	0.4
95–99	342	0.1
100 and over	60	0.0
Total	**272,635**	**100.0**

Figure 1.4 U.S. Population, 1999—interval size 5.

2. Search for patterns: represent data in multiple ways The data we choose and the way we represent them will affect our conclusions. We compressed the data from Table 1.4 into 10-year intervals in Table 1.5 and then constructed the histogram shown in Figure 1.5. The histogram with 10-year intervals reveals patterns in the shape of the data that are not obvious in the histogram with 5-year intervals. Yet, other information is hidden. The histogram with 5-year intervals shows a sharp rise in percentage of people in the 35- to 39-year interval compared to the 30- to 34-year interval that is hidden in the histogram with 10-year intervals. The size of the interval affects what type of information is revealed or hidden.

Table 1.5
Ages of June 1, 1999, U.S. Population in Ten-Year Intervals

Age (years)	Frequency Count (in 1000s)	Relative Frequency (%)
Under 10	38,884	14.3
10–19	39,257	14.4
20–29	36,291	13.3
30–39	42,339	15.5
40–49	41,564	15.2
50–59	29,241	10.7
60–69	19,970	7.3
70–79	16,105	5.9
80 and over	8,984	3.3
Total	**272,635**	**100.0**

Source: U.S. Bureau of the Census, *www.census.gov.*

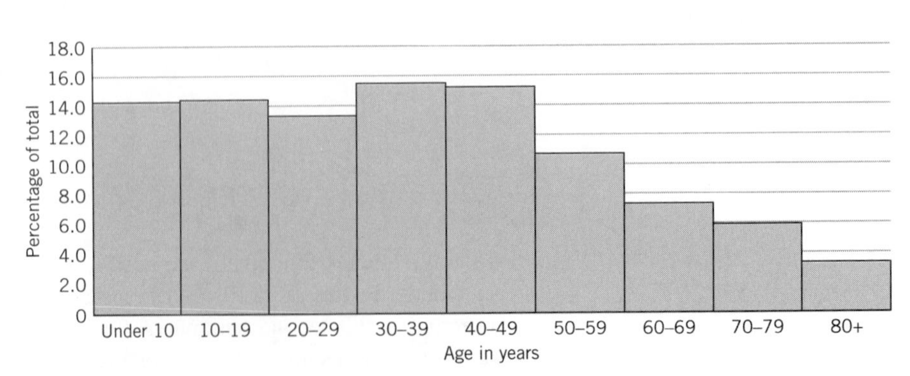

Figure 1.5 U.S. Population, 1999—interval size 10.

3. Make jottings Examine the graphs and data tables and jot down what you observe. Note any patterns or interesting trends. Here are some observations:

- Ages in 10-year intervals are distributed fairly evenly up to 30 years; there is a slight drop in the number of 20- to 29-year-olds.
- There is a bump on the 10-year interval between 30 and 49 years.
- Five-year interval data show a bump with a sharp rise in the number of people in the 35- to 39-year interval and a sharp drop in the number of people beginning with the 45- to 49-year interval.
- There is a steady decline in the number of people in each successive 10-year interval for those over 49 years and in each successive 5-year interval for those over 39 years.

4. Identify a key idea (out of possibly many ideas) This step forces you to reach some kind of summary statement about the data. A key idea is often used in the *topic* or *opening* sentence of a paragraph summarizing the data. The following is a key idea about the ages of the U.S. population in 1999.

> As seen from the 1999 data, there was a steady decline in the percentage of people in each successive 10-year interval beginning with the 10- to 19-year interval that was interrupted by a surge roughly between the ages of 30 and 49 years.

5. Provide supporting evidence Collect evidence that supports, expands, or modifies your key idea. Your jottings may give enough information or you may have to go back to tables and graphs to find more information. Sometimes you may need to find additional data to support your key idea.

- Approximately 31% of the people were between 30 and 49 years.
- The intervals with the highest percentage of people are from 30 to 39 and 40 to 49-years.

6. Put ideas together Construct a few sentences defining and defending your key idea. You may wish to start with a topic sentence containing your key idea, followed by supporting evidence that might include a graph or table. You will probably want to weave back and forth among the steps in order to refine or modify your ideas. If you are writing a paragraph, you could finish with a concluding sentence that raises a question about the data or with a speculation that requires more research.

A 60 Second Summary of
Patterns in the Distribution of Ages in 1999 U.S. Population

As seen from the 1999 data, there was a general decline in the percentage of the U.S. population in each successive 10-year interval that was interrupted by a surge roughly between the ages of 30 and 49 years. (See Figure 1.5.) From newborns to those 30 years old, there was a fairly even distribution in the percentage of people in each 10-year interval. Between the ages of 30 and 49 years, the percentage of people in each 10-year interval reached its highest values. For those 50 years of age or older, there was a steady decline in the percentage of people in each successive 10-year interval. This overall wave pattern may be due in part to the "baby boom," the surge of births after World War II.

Algebra Aerobics 1.2

1. Construct a topic sentence that describes some aspect of the accompanying graphs.

   a. Ten Fastest-Growing States, Percent Population Change: 1990 to 1997.

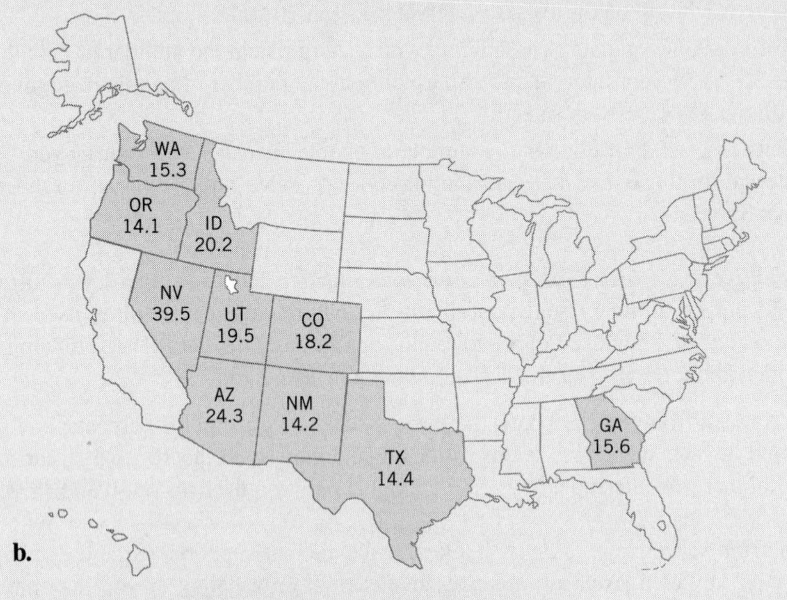

   b.

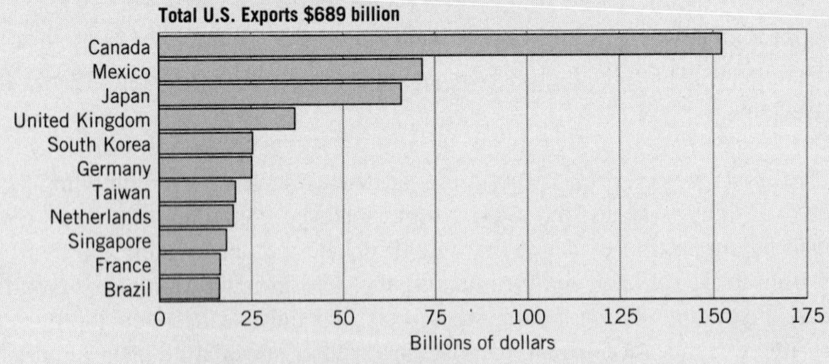

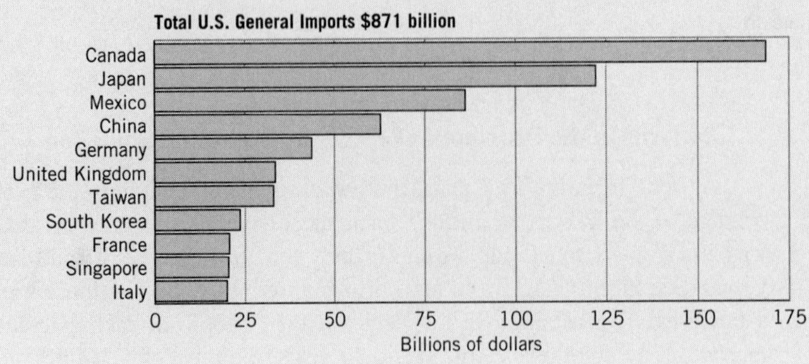

(continued on next page)

Algebra Aerobics 1.2 (continued)

c.

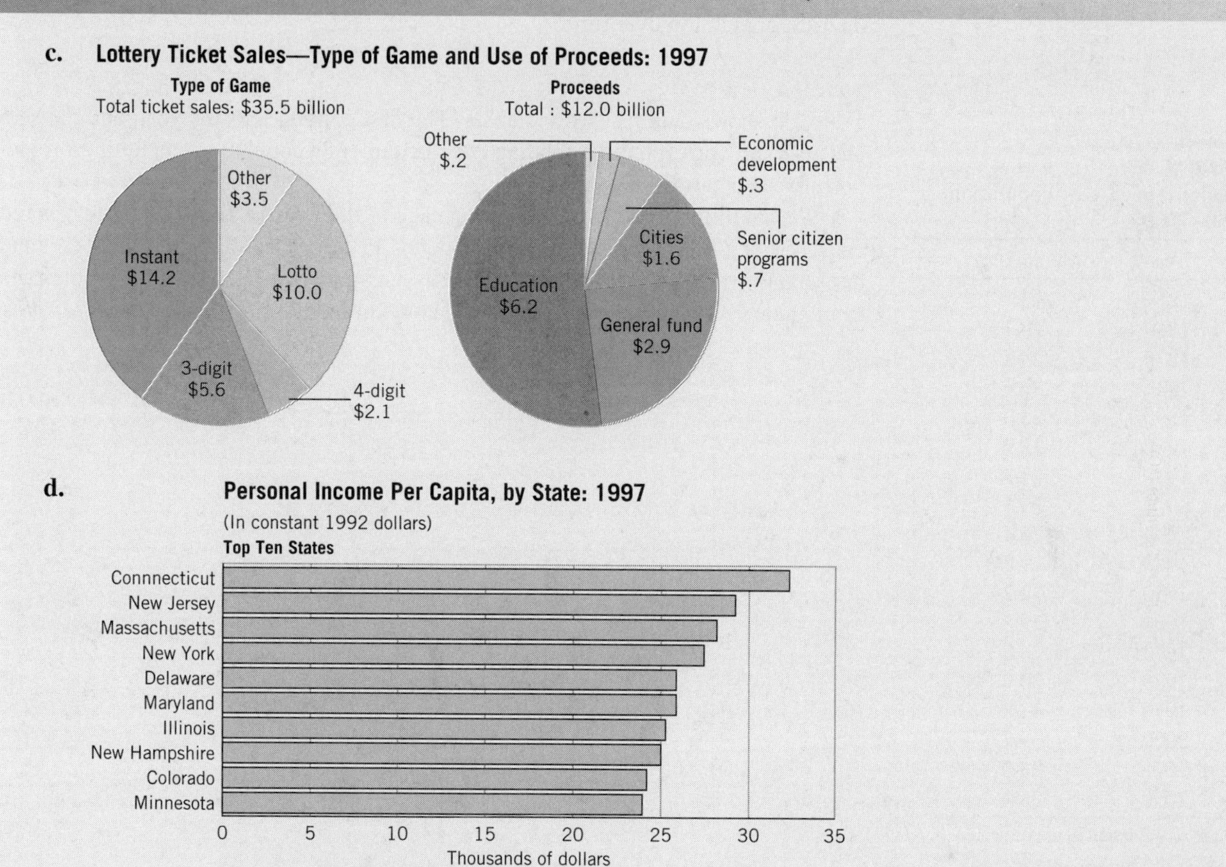

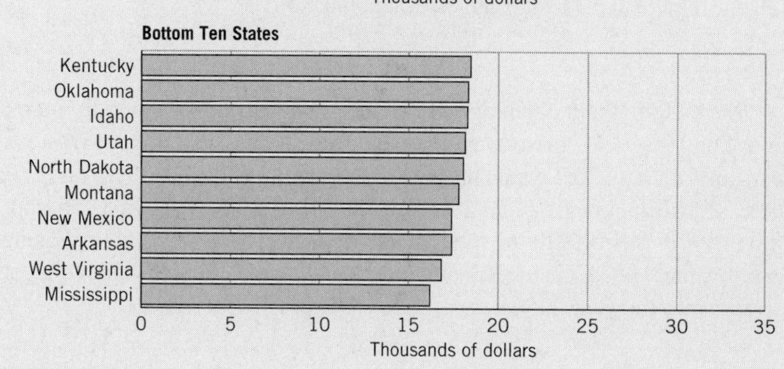

Source for graphs: U. S. Bureau of Census. *Statistical Abstract of the United States. 1998.*

1.3 VISUALIZING TWO VARIABLE DATA

We can obtain important but static snapshots of patterns in single variable data from images such as histograms and pie charts or from numbers such as the mean and median. By looking at two variable data, we can learn how change in one variable affects change in another. How does the weight of a child determine the amount of medication prescribed by a pediatrician? How does median age or income change over time?

Example 1

USMEDAGE

Construct a graph and a 60 second summary of the changes in the median age of the U.S. population over time.

Solution

Table 1.6 shows data for two variables. The first column lists the year, and the second lists the corresponding median age for the U.S. population, including projections for the years 2000 to 2050.

We can think of the two numbers in each of the rows in Table 1.6 as an ordered pair. The first row, for example, corresponds to the ordered pair (1850, 18.9) and the second row corresponds to (1860, 19.4). Figure 1.6 shows a plot of the data from Table 1.6. A plot of data points as shown in Figure 1.6 is called a *scatter plot.*

Table 1.6
Median Age of U.S. Population, 1850–2050*

Year	Median Age
1850	18.9
1860	19.4
1870	20.2
1880	20.9
1890	22.0
1900	22.9
1910	24.1
1920	25.3
1930	26.4
1940	29.0
1950	30.2
1960	29.5
1970	28.0
1980	30.0
1990	32.8
2000	35.7
2025	38.0
2050	38.1

*Data for 2000–2050 are projected.
Source: U.S. Bureau of the Census, *Statistical Abstract of the United States,* 1998.

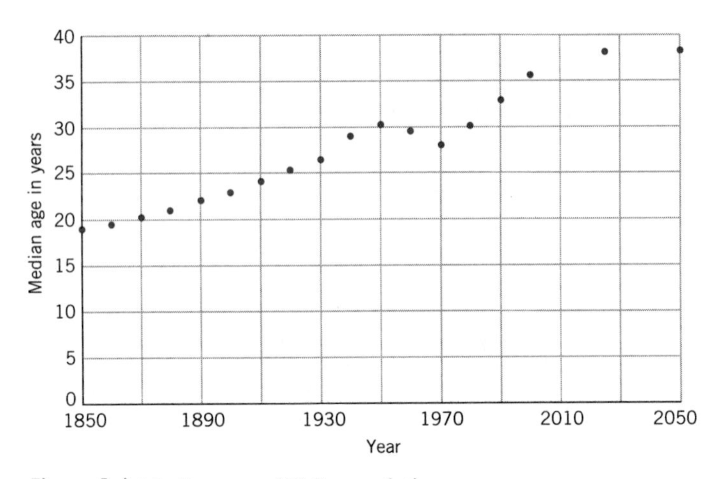

Figure 1.6 Median age of U.S. population.

What are some of the trade-offs in using the median instead of the mean age to describe changes over time?

Since 1850, the median age of the U.S. population has generally increased over time. For 100 years, there was a steady increase in median age, rising from 18.9 years in 1850 to 30.2 years in 1950. Between 1950 and 1970, the median age decreased, dropping from 30.2 years in 1950 to 29.5 years in 1960 and down to 28.0 in 1970. Since 1970 the median age has again been steadily increasing. Projections by the U.S. Census Bureau indicate that the median age will continue to increase into the year 2050.

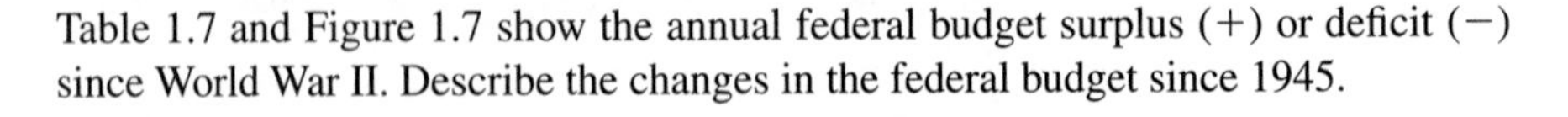

Example 2

Table 1.7 and Figure 1.7 show the annual federal budget surplus (+) or deficit (−) since World War II. Describe the changes in the federal budget since 1945.

Solution

The federal budget was fairly balanced (with little deficit or surplus) from 1950 to 1970. From 1971 to 1997, the federal budget ran an annual deficit, which generally was getting larger until 1992. If we look at Figure 1.7, we notice the graph reaches a minimum in 1992, when the annual deficit was almost *300 billion dollars.* From 1992 to 1997, the deficit steadily decreased, and in 1998 and 1999, there was a relatively large surplus. In Figure 1.7 we notice that the graph reaches a maximum in 1999, when there was a surplus of about *123 billion dollars.*

When the annual U.S. deficit reached zero, did that mean the federal debt was zero?

Table 1.7
Federal Budget: Surplus (+) or Deficit (−)

Year	Billions of Dollars	Year	Billions of Dollars
1945	− 47.6	1983	− 207.8
1950	− 3.1	1984	− 185.4
1955	− 3.0	1985	− 212.3
1960	+ 0.3	1986	− 221.2
1965	− 1.4	1987	− 149.8
1970	− 2.8	1988	− 155.2
1971	− 23.0	1989	− 152.5
1972	− 23.4	1990	− 221.2
1973	− 14.9	1991	− 269.4
1974	− 6.1	1992	− 290.4
1975	− 53.2	1993	− 255.1
1976	− 73.7	1994	− 203.1
1977	− 53.7	1995	− 163.9
1978	− 59.2	1996	− 107.5
1979	− 40.7	1997	− 21.9
1980	− 73.8	1998	+ 66.2
1981	− 79.0	1999	+ 122.7
1982	− 128.0		

Source: U.S. Bureau of the Census, *Statistical Abstract of the United States,* 1998 and *www.census.gov.*

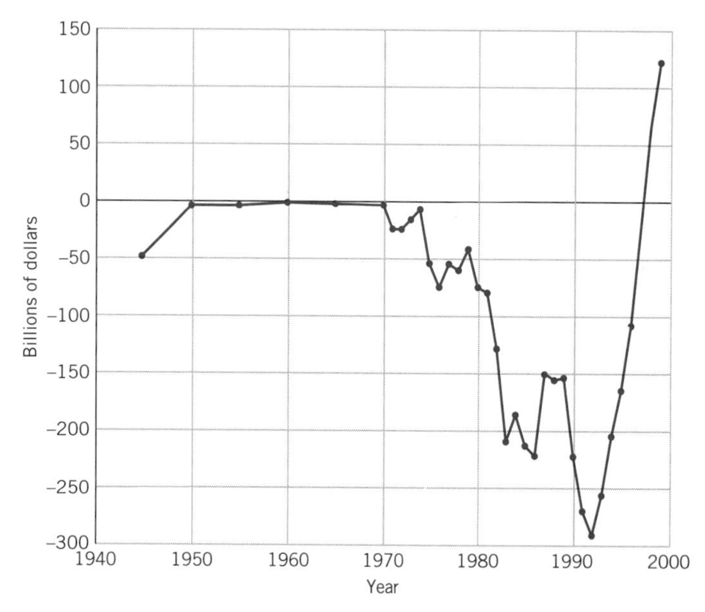

Figure 1.7 Annual federal budget surplus or deficit in billions of dollars.

1.4 INTERPRETING EQUATIONS AND THEIR GRAPHS

The relationship between two variables, such as time and median age, can be described by data tables, words, or graphs. Relationships can also be described by equations. For example, if you earn \$10 an hour at your job, then the relationship between your wages and the number of hours you work can be described with the equation

$$\text{wages} = \$10 \times \text{number of hours worked}$$

An equation gives us a rule that tells us how to determine the value of one variable when we know the value of another variable. In this case, our rule tells us to multiply the number of hours worked by \$10 to find our wages. The number of hours you work may *vary,* and thus your wages will also vary. It is common to use single letters as variable names. If w is used for the variable *wages* and h is used for the variable *hours,* then we can write the equation as

$$w = 10h$$

Variables can also represent quantities that are not associated with real objects or events. The following equation defines a relationship between two quantities, which are named by the abstract variables x and y:

$$y = x^2 + 2x - 3$$

In the equation above, if we know the value of x, then our rule tells us what to do to that value of x to find the value of y. For example, if $x = 1$, to find the value for y, the rule tells us to square 1, add 2 times 1 to this result, and then subtract 3. So, if $x = 1$, then $y = 0$:

$$\begin{aligned}(1)^2 + 2(1) - 3 &= 1 + 2 - 3 \\ &= 0\end{aligned}$$

We can think of equations as mathematical sentences. In the equation $y = x^2 + 2x - 3$, some combinations of values of x and y make the sentence true, such as $x = 1$ and $y = 0$. Other combinations make it false, for example $x = 0$ and $y = 5$. Pairs of values for x and y that make the sentence true are called *solutions* to the equation. We can express these solutions as ordered pairs of the form (x, y). Thus (1, 0) would be a solution to $y = x^2 + 2x - 3$, while (0, 5) would not be a solution, since $0^2 + 2(0) - 3 \neq 5$.

> The *solutions* of an equation in two variables x and y are the ordered pairs (x, y) that make the equation a true statement.

By substituting various values for x and finding the associated values for y, we can generate a table of solutions. There are infinitely many possible solutions since we could substitute any number for x and find a corresponding y. The rows in Table 1.8 represent ordered pairs that are solutions to the equation. Each of the ordered pairs can be plotted as a point to help us determine the graph of the equation, or we can use technology to graph the equation (see Figure 1.8).

Table 1.8

x	y
−4	5
−3	0
−2	−3
−1	−4
0	−3
1	0
2	5
3	12

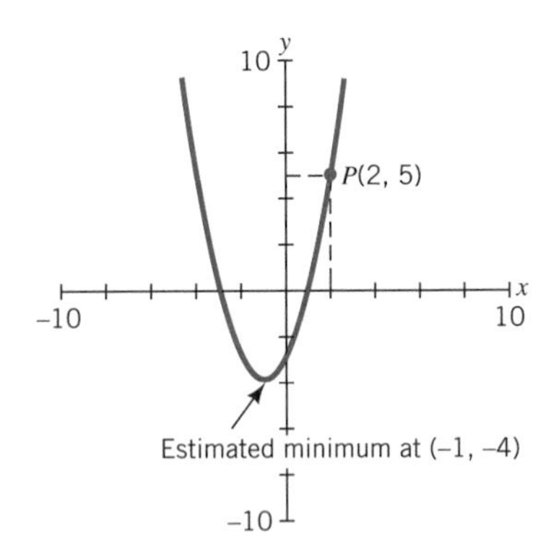

Figure 1.8 Graph of $y = x^2 + 2x - 3$.

In Figure 1.8, point ***P*** is located by moving 2 units to the right of the origin, and then moving up 5 units. The coordinates of point ***P*** are written as **(2, 5).** The *horizontal coordinate* is 2 and the *vertical coordinate* is 5. In this case, the horizontal axis is called the ***x**-axis* and the vertical axis is called the ***y**-axis.* Hence, the quantity on the horizontal axis is called the *x-coordinate* and the quantity on the vertical axis is called the *y-coordinate.* Each point on the graph in Figure 1.8 corresponds to an ordered pair in the form $(\boldsymbol{x}, \boldsymbol{y})$ that makes the equation a true statement.

> The *graph* of an equation in two variables is the set of points corresponding to the ordered pairs that are solutions to the equation.

We can "read" the graph of an equation in two variables from left to right and ask what effect changes in x have on y. In Figure 1.8, initially as x increases, y decreases, until it reaches an estimated minimum value of -4 when x is about -1. (In later chapters we'll learn how to determine this minimum value exactly.) When $x > -1$, as x increases, y now increases. Figure 1.8 shows only a small section of the graph of all the solutions to the equation. Both "arms" of the graph extend infinitely upward, so there is no maximum value for y.

Algebra Aerobics 1.4

1. Table 1.9 and the scatter plot in Figure 1.9 show median household income over time in constant 1996 dollars.[3]

Table 1.9
Median Household Income in Constant 1996 Dollars

Year	Median Household Income
1986	$35,642
1987	$35,994
1988	$36,108
1989	$36,575
1990	$35,945
1991	$34,705
1992	$34,261
1993	$33,922
1994	$34,158
1995	$35,082
1996	$35,492

Source: U.S. Bureau of the Census, *Statistical Abstract of the United States,* 1998.

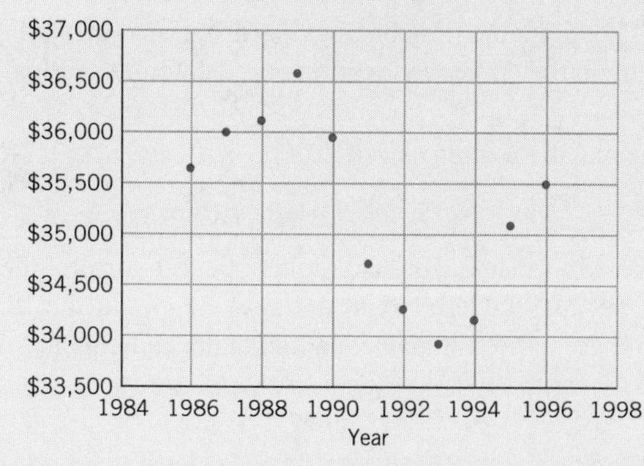

Figure 1.9 Median household income in 1996 dollars.

1. **a.** What is the maximum value for the median household income during this period? In what year does this occur? What are the coordinates of this point in Figure 1.9?

   b. What is the minimum value for the median household income? In what year does this occur? What are the coordinates of this point?

   c. Describe the changes in median household income since 1986.

2. **a.** Describe in your own words how to compute the value for y given a value for x using the following equation:

$$y = 3x^2 - x + 1$$

   b. Which of the following ordered pairs represent solutions to the equation?

$$(0, 0), (0, 1), (1, 0), (-1, 2), (-2, 3), (-1, 0)$$

   c. Use $x = 0, \pm 1, \pm 2, \pm 3$ to generate a small table of values that represent solutions to the equation.

3. Repeat the directions in 2(a), (b), and (c) using the equation $y = (x - 1)^2$.

[3]"Constant dollars" is a measure used by economists to compare incomes and other variables in terms of purchasing power, eliminating the effects of inflation. To say the median income in 1986 was $35,642 in "constant 1996 dollars" means that the median income in 1986 could buy an amount of goods and services that would cost $35,642 to buy in 1996. The actual median income in 1986 (measured in what economists call "current dollars") was much lower. Income corrected for inflation is sometimes called "real" income.

1.5 AN INTRODUCTION TO FUNCTIONS

What is a Function?

When we speak informally of one quantity being a function of some other quantity, we mean that one depends on the other. For example, someone may say that what they wear is a function of where they are going, meaning that what they wear is determined by where they are going. Or someone may claim that how well a car runs is a function of how well it is maintained, meaning that how well a car runs depends on how well it is maintained.

In mathematics, the word *function* has a more precise meaning. A function is a relationship between two quantities. When one quantity uniquely determines a second quantity, then the second quantity is a function of the first. We can think of the first quantity as the *input* and the second quantity as the *output* of the function.

EXPLORE

Exploration 1.2 will help you develop an intuitive sense of functions.

Median age and the federal debt are functions of time since each year determines a unique (one and only one) value of median age or the federal debt. The equation $y = x^2 + 2x - 3$ defines y as a function of x since each value of x we substitute in the equation determines a unique value of y.

> A variable y is a *function* of a variable x if each value of x determines a unique value of y.

Representing Functions with Words, Tables, Graphs, and Equations

We can think of a function as defining a "rule" that determines one and only one value of y for each value of x. The rule can be described using words, tables, graphs, or equations.

- Using words, the rule may say "y is five percent of x."
- Using a table, the rule may say "find the value of x in the left column of the table and read the corresponding value of y from the right column."
- Using a graph, it may say "go along the horizontal axis a distance x, then go up or down until you meet a point on the curve, and read y from the point's vertical coordinate."
- Using an equation, the rule may be written as $y = 0.05x$.

Example 1

The sales tax rate in Illinois is 8%; that is, for each dollar spent in a store, the law says that you should pay a tax of 8 cents or 0.08 dollars.[4] Represent the Illinois sales tax as a function of purchase price using an equation, table, and graph.

Solution

Equation

We can write this as an equation where T represents the amount of sales tax and P represents the price of your purchase (both measured in dollars) as $T = 0.08P$. Our rule for determining a value for T given a value for P says, "Take the given value of

[4] We have rounded off the sales tax to simplify calculations. The 1996 sales tax in Illinois was 7.75% and in the city of Chicago it was 8.75%.

Table 1.10
Illinois Sales Tax

P (purchase price in \$)	T (sales tax in \$)
0	0.00
1	0.08
2	0.16
3	0.24
4	0.32
5	0.40
6	0.48
7	0.56
8	0.64
9	0.72
10	0.80

Figure 1.10 Illinois sales tax.

P and multiply it by 0.08; the result is the corresponding value of T." The equation represents T as a function of P, since for each value of P the equation determines a unique value (one and only one value) of T.

When using this equation to calculate the sales tax, T, it is important to note that the purchase price, P, is restricted to dollar amounts greater than or equal to zero.

Table

We can use this formula to make a table of values for T determined by the many different values of P. Such tables were once posted next to many cash registers. Table 1.10 is one such table.

Graph

We can use the points in Table 1.10 to create a graph of the function, as shown in Figure 1.10. Table 1.10 only shows the sales tax for a few selected purchase prices, but we assume that we could have used any positive dollar amount for P. We connected the points on the scatter plot to suggest the many possible intermediate price values. For example, if $P = \$2.50$, then $T = \$0.20$.

Equations like the Illinois tax equation, used to describe real world situations, are called *mathematical models.* Such models offer a compact, often simplified description of what may be a complex situation. For example, our model of the sales tax ignored the fact that the tax should be rounded to the nearest cent.

When Is a Relationship Not a Function?

Not all relationships between quantities describe functions. In the following set of ordered pairs in the form (x, y), y is not a function of x:

$$(1, 2), (1, 3), (2, 4), (3, 5)$$

When $x = 1$, y can be 2 or 3. Thus, there is not one and only one value of y for each value of x.

Example 2

Consider the set of data in Table 1.11 at the top of the next page. The first column shows the year of the Olympics, T. The second column shows the record distance, R, in feet, for the men's Olympic 16-pound shot put. Is R a function of T? Is T a function of R?

Table 1.11
Olympic Shot Put

Year, T	Record Distance in Feet Thrown, R
1960	65
1964	67
1968	67
1972	70
1976	70
1980	70
1984	70
1988	74
1992	71
1996	71

Source: Reprinted with permission from the Universal Press Syndicate. From *The 1996 Universal Almanac.* 1996 data from http://sports.yahoo.com/oly/lgns/960726/fmshotmed.html.

Table 1.12
Olympic Shot Put

Record Distance in Feet Thrown, R	Year, T
65	1960
67	1964
67	1968
70	1972
70	1976
70	1980
70	1984
74	1988
71	1992
71	1996

Solution

R* is a function of *T

To determine if R can be described as a function of T, we need to find out if each value of T (the first variable or input) determines one and only one value for R (the second variable or output). For each T, there is one and only one R. Thus R, the record shot put distance, can be described as a function of the year of the Olympics, T. Note that different inputs (such as 1964 and 1968) can have the same output (67) and the relationship can still be a function.

1964 ↘ 67
1968 ↗

T* is NOT a function of *R

To determine if T can be described as a function of R, we need to find out if each value of R (now the first variable or input) determines one and only one value for T (the second variable or output). Table 1.12 shows this new pairing where R is thought of as the first variable or input and T as the second variable or output.

The year, T, is *not* a function of R, the record distance. Some values of R such as 67 give more than one value for T. For example, 67 would have two corresponding values for T, 1964 and 1968, and this violates the condition of one and only one output for each input.

67 ↗ 1964
67 ↘ 1968

What other R's would have more than one T value? For each input of a function, there should be one and only one output.

How to tell if a graph represents a function: The vertical line test. The mathematical convention is for the first variable or input of a function to be represented on the horizontal axis and the second or output on the vertical axis. For a graph to represent a function, each value on the horizontal axis must be associated with one and only one value on the vertical axis. If you can draw a vertical line that intersects a graph in more than one point, then the graph does not represent a function.

The graph in Figure 1.11 represents y as a function of x. For each value of x, there is only one corresponding value of y. No vertical line intersects the curve in more than one point.

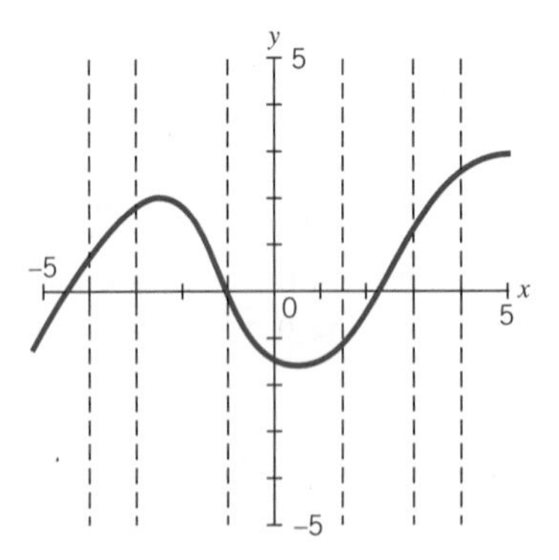

Figure 1.11 The graph represents y as a function of x since there is no vertical line that intersects the curve at more than one point.

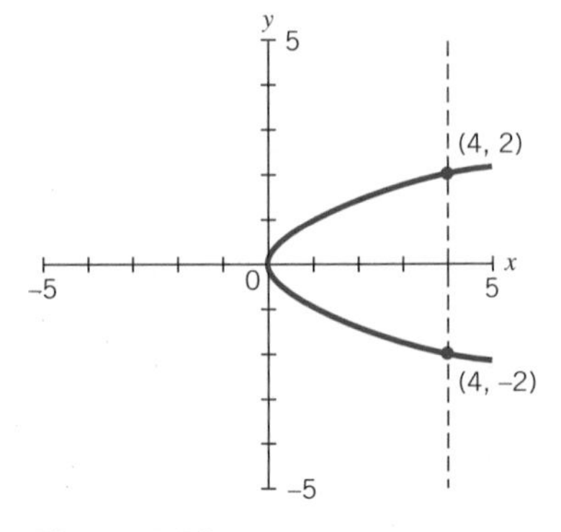

Figure 1.12 The graph does not represent y as a function of x since there is at least one vertical line that intersects this curve at more than one point.

The graph in Figure 1.12 does *not* represent a function. One can draw a vertical line (an infinite number in fact) that intersects the graph in more than one point. We show in Figure 1.12 a vertical line that intersects the graph at (4, 2) and (4, −2). That means that the value $x = 4$ does not determine one and only one value of y. It corresponds to y values of both 2 and − 2.

Algebra Aerobics 1.5

1. Refer to the graph in Figure 1.13.

a. Is y a function of x?

b. What seems to happen to the graph when the value of x approaches zero?

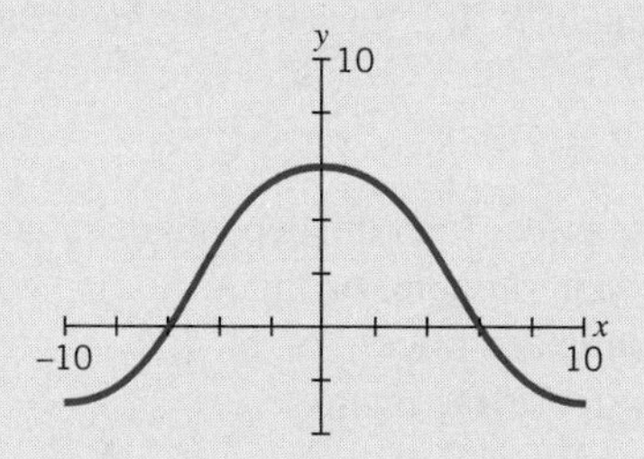

Figure 1.13 Graph of an abstract relationship between x and y.

2. Which of the graphs in Figure 1.14 represent functions and which do not? Why?

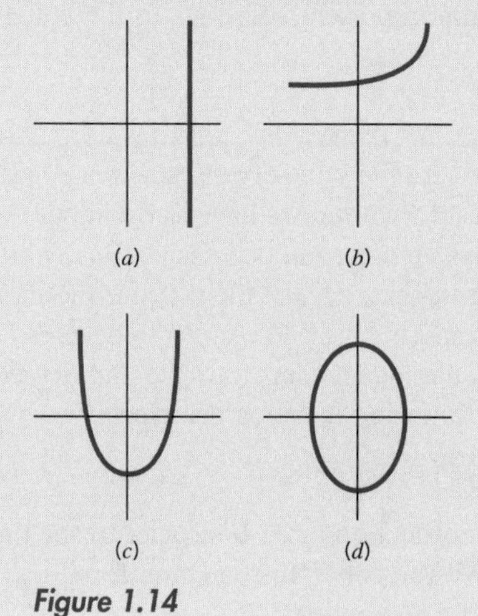

Figure 1.14

3. Consider the chart in Figure 1.15. Is weight a function of height? Is height a function of weight? Explain your answer.

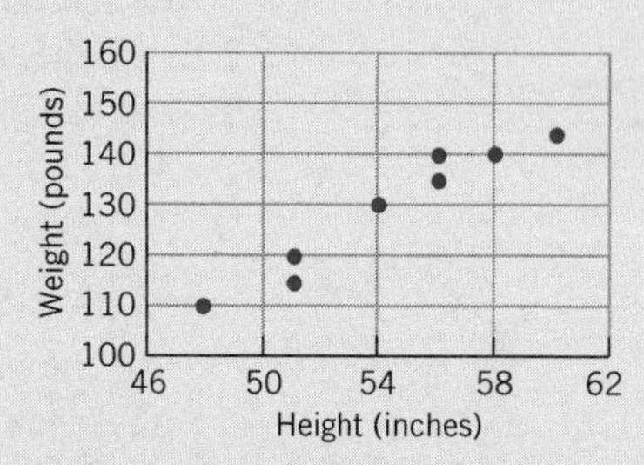

Figure 1.15 Graph of weight versus height.

4. Consider Table 1.13:

a. Is D a function of Y?

b. Is Y a function of D?

Table 1.13

Y	D
1992	\$2.50
1993	\$2.70
1994	\$2.40
1995	− \$0.50
1996	\$0.70
1997	\$2.70

1.6 THE LANGUAGE OF FUNCTIONS

Independent and Dependent Variables

Since a function is a rule that assigns to each input a unique output, we think of the output as being dependent on the input. We call the input the *independent variable* and the output the *dependent variable.* The dependent variable is a *function* of the independent variable if each value of the independent variable determines one and only one value of the dependent variable.

Solutions to equations that represent functions are written as ordered pairs in the form

(independent variable, dependent variable)

or equivalently,

(input, output)

Example 1

In the sales tax example in Section 1.5, the equation $T = 0.08P$ gives the Illinois sales tax, T, as a function of purchase price, P. The sales tax, T, in this case depends on the purchase price, P. So T is the dependent variable and P is the independent variable. But this equation also gives us P as a function of T; that is, each value of T corresponds to one and only one value of P. It is easier to see the relationship if we solve for P in terms of T, to get

$$P = T/0.08$$

Now, we are thinking of the purchase price, P, as the dependent variable, or output, and the sales tax, T, as the independent variable, or input. So, if you tell me how much tax you paid, I can find the purchase price. Yet, in practical terms it makes more sense to think of the purchase price as the input.

Example 2

In the shot put example of Section 1.5, we showed how R, the record distance in feet, could be described as a function of the year, T, but T could NOT be described as a function of R. The record, R, depends on the year, T, but T does not depend on R. In this case, we need to choose T as the independent variable and R as the dependent variable. The ordered pairs that represent solutions to the equation would be written as (T, R).

Example 3

Sometimes the choice as to which variable will serve as the independent variable is arbitrary or may be based on what information we have as our input.

In most countries, temperature is reported in degrees Celsius, while in the United States it is more common to use degrees Fahrenheit. If our thermometer records the temperature in degrees Celsius and we want to convert to degrees Fahrenheit, we would treat degrees Celsius, C, as the independent variable and degrees Fahrenheit, F, as the dependent variable. The following equation describes F as a function of C:

$$F = \tfrac{9}{5}C + 32$$

This equation also describes C as a function of F. It's easier to see the relationship if we solve for C in terms of F. We can solve this equation for C by doing the following calculations on each side of the equation:

$$F = \tfrac{9}{5}C + 32$$

Subtract 32 $\quad F - 32 = \tfrac{9}{5}C$

multiply by $\tfrac{5}{9}$ $\quad \tfrac{5}{9}(F - 32) = C$

thus $\quad C = \tfrac{5}{9}(F - 32)$

When we describe C as a function of F, then we are thinking of C as the dependent variable or output and F as the independent variable or input.

Domain and Range

A function is often only defined for certain values of the independent variable. The set of all possible values of the independent variable (or input numbers) is called the *domain* of the function. Also, the dependent variable often only takes on certain values. The set of values of the dependent variable (or output numbers) is called the *range* of the function.

> The *domain* of a function is the set of possible values of the independent variable. The *range* is the set of corresponding values of the dependent variable.

Example 4

In the Illinois sales tax example of Section 1.5, we used the equation

$$T = 0.08P$$

to represent the Illinois sales tax, T, as a function of the purchase price, P. Since negative values for P are meaningless, we noted that P is restricted to dollar amounts greater than or equal to zero. In theory there is no upper limit on prices, so we assume P has no maximum amount. In this example,

domain is all values of P greater than or equal to 0, where P is expressed in dollars and cents

What are the corresponding values for the tax T? The values for T in our model will also always be nonnegative. As long as there is no maximum value for P, there will be no maximum value for T. So,

range is all values of T greater than or equal to 0, where T is expressed in dollars and cents

---

Example 5

Consider the equation $T = 0.08P$ in the abstract, where the equation describes a relationship between two abstract variables, P and T. What are the domain and range of this function? Create a table of values for this function and graph it.

Solution

Since P no longer represents purchase price, P is not measured in dollars and cents. The values for P, the independent variable, are now all real numbers, positive, negative, or zero.[5] Values for the dependent variable T can also be any real number:

domain is all real numbers
range is all real numbers

Table 1.14 shows a small set of solutions. Figure 1.16 displays a graph of the function, which is a line that extends indefinitely in both directions.

[5] Recall that real numbers are numbers that can be written as signed decimal expressions, for example, 1, -3, 20.12, $\pi = 3.14159\ldots$, $\frac{1}{3} = 0.33333\ldots$. The real numbers split into two distinct categories: rational and irrational. Rational numbers are numbers whose decimal expansions eventually contain a fixed pattern that infinitely repeats or that terminates. For example, $\frac{1}{4} = 0.25$, $\frac{2}{3} = 0.66666\ldots$, and $-\frac{2}{11} = -0.1818181818\ldots$ are all rational numbers. Irrational numbers are numbers whose decimal patterns do not terminate and do not contain an infinitely repeating pattern. Some examples are $\sqrt{2} = 1.414213\ldots$ and $\pi = 3.14159\ldots$.

Table 1.14

P	T
−10	−0.80
−5	−0.40
0	0.00
5	0.40
10	0.80

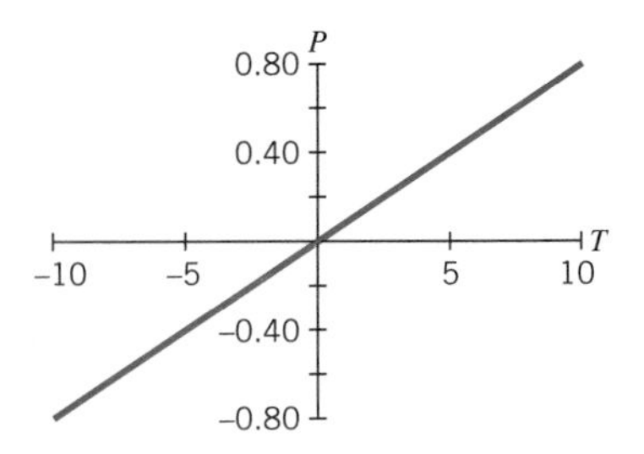

Figure 1.16 Graph of $T = 0.08P$.

Example 6

The equation $y = x^2 + 2x - 3$ describes y as a function of x since for each value of x there is one and only one corresponding value of y. What are the domain and range of this function?

Solution

There are no restrictions on the values for x; thus

domain is all real numbers

From the data in Table 1.15 and the graph in Figure 1.17 we can get an idea of the maximum or minimum values of this function. The graph of $y = x^2 + 2x - 3$ is U-shaped with an estimated minimum value at the point $(-1, -4)$. Thus there is a restriction on the values for y or the numbers that are outputs. Values for y are restricted to those greater than or equal to the value of the function at this minimum point, which is estimated at -4. In later chapters we will learn how to find this minimum value using algebraic formulas. There does not appear to be a maximum value since the arms of each side of the graph can extend upward indefinitely:

estimated range is all real numbers greater than or equal to -4

Table 1.15

x	y
−4	5
−3	0
−2	−3
−1	−4
0	−3
1	0
2	5

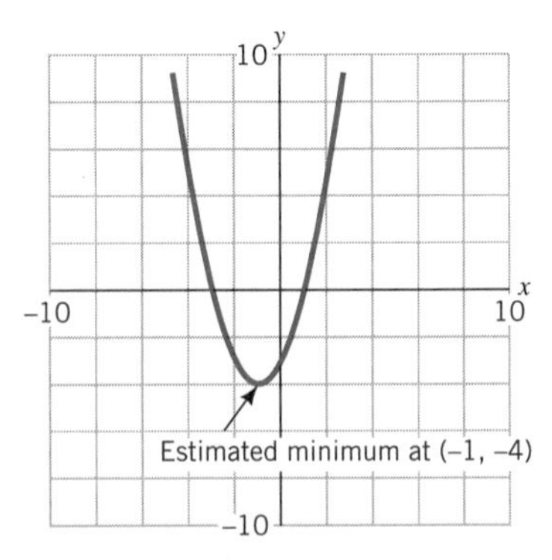

Figure 1.17 Graph of $y = x^2 + 2x - 3$.

Example 7a

The graph in Figure 1.18 shows the water level of the tides in Pensacola, Florida, over a 24-hour period. Are the Pensacola tides a function of the time of day? If so, identify the independent and dependent variables. What are the domain and range of this function?

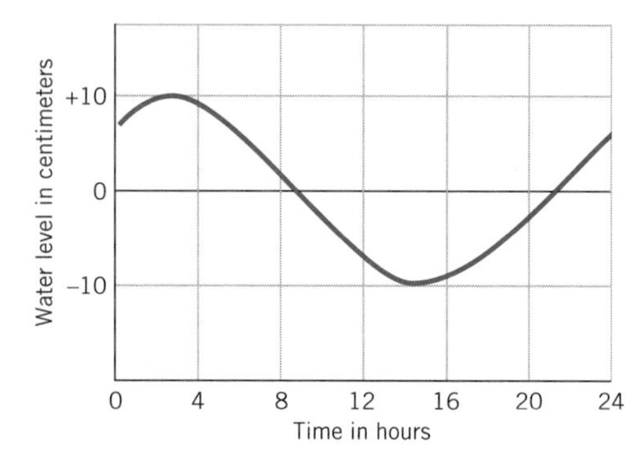

Figure 1.18 Diurnal tides in a 24-hour period in Pensacola, Florida.
Source: Adapted from Fig. 8-2, in *Oceanography: An Introduction to the Planet Oceanus,* by Paul R. Pinet. Copyright © 1992 by West Publishing Company, St. Paul, MN. All rights reserved.

Solution
The Pensacola tides are a function of the time of day since the graph passes the vertical line test. The independent variable is time, and the dependent variable is water level. The domain is from 0 to 24 hours, and the range is from about -10 to $+10$ centimeters.

Example 7b

Is the time of day a function of the Pensacola tides? Justify your answer.

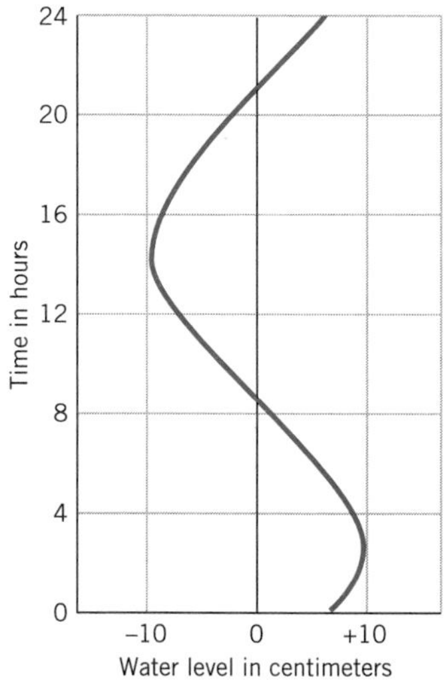

Solution
Time of day is NOT a function of the Pensacola tides. Figure 1.19 also shows the water level of the Pensacola tides, now with water level on the horizontal axis and time on the vertical axis. The graph fails the vertical line test. For instance, if you draw a vertical line corresponding to a water level of 0 centimeters (see the black line on the graph), the line crosses the graph in two different places (at about 9 and 21 hours). There is not one and only one hour associated with a water level of 0 centimeters. Therefore, the time of day is not a function of water level.

Figure 1.19 The time of day is not a function of the tide.

When specifying the domain and range of a function, we need to consider whether the function is undefined for any values. For example, for the function

$$y = 1/x$$

the expression $1/x$ is undefined when $x = 0$. We can use any other real number as a value for x except zero. Thus the domain is restricted to all real numbers except $x = 0$:

domain is all real numbers except 0

To find the range, we need to determine those numbers that are possible values for y. Sometimes it is easier to find the values that are not possible. In this case, it is not possible for y to equal zero. Why? Our rule says to take 1 and divide by x, but it is impossible to divide 1 by a real number in our domain and get zero as a result. Thus,

range is all real numbers except 0

From the graph of a function we can often get an idea of any restricted values for the domain and range. Using the values given in Table 1.16 and the graph of $y = 1/x$ in Figure 1.20, examine the behavior of the function when x approaches zero from the right and from the left side of the horizontal axis.

We can see that the function appears to "blow up" (takes arbitrarily large values) when x is near zero, splitting the graph into two pieces: one when $x > 0$ and one when $x < 0$.

Table 1.16

x	$y = 1/x$
-5	$-\frac{1}{5}$
-1	-1
$-\frac{1}{10}$	-10
$-\frac{1}{100}$	-100
0	Not defined
$\frac{1}{100}$	100
$\frac{1}{10}$	10
1	1
5	$\frac{1}{5}$

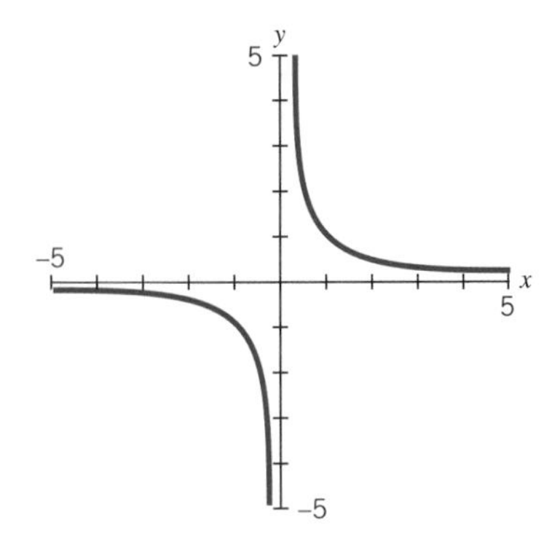

Figure 1.20 Graph of the function $y = 1/x$.

When $x > 0$, $y > 0$. As x increases, the values for y decrease, approaching but never reaching zero. For example, the following points are solutions to the equation $y = 1/x$ and hence lie on the graph: $(1, 1)$, $(10, \frac{1}{10})$, $(100, \frac{1}{100})$.

As x approaches 0 from the right, the solutions $(1, 1)$, $(\frac{1}{10}, 10)$, $(\frac{1}{100}, 100)$ suggest that the values for y get larger and larger. As x stays positive and gets closer and closer to 0, the y values get larger and larger. A comparable situation exists for $x < 0$. The function $y = 1/x$ has no maximum or minimum value. The range is all real numbers except 0.

Algebra Aerobics 1.6a

1. **a.** Write an equation for computing a 15% tip in a restaurant. Does your equation represent a function? If so, what are your choices for the independent and dependent variables? What are reasonable choices for the domain and range?
   b. How much would the equation suggest you tip for an \$8 meal?
   c. Compute a 15% tip on a total check of \$26.42.
   d. Think of the function for generating 15% tips as an abstract mathematical function having nothing to do with meal prices. What are the domain and range?
2. If we let D stand for ampicillin dosage expressed in milligrams and W stand for a child's weight in kilograms, then the equation

$$D = 50W$$

gives a rule for finding the safe maximum daily drug dosage of ampicillin (used to treat respiratory infections) for children who weigh less than 10 kilograms (about 22 pounds).[6]

   a. What are logical choices for the independent and dependent variables?
   b. Does the equation represent a function? Why?
   c. What is a reasonable domain? Range?
   d. Generate a small table and graph of the function.
   e. Think of the function $D = 50W$ for ampicillin dosage as an abstract mathematical equation. What is the domain? The range? How will the table and graph reflect these changes?
3. Table 1.17 lists the pulse rate, P, and the respiration rate, R, taken at different times for one individual.
   a. Based on this table, is R a function of P?
   b. Based on this table, is P a function of R?

Table 1.17

P (beats/min)	R (breaths/min)
75	12
95	17
75	14
130	20
110	19

[6]Information extracted from Anna M. Curren and Laurie D. Munday, *Math for Meds: Dosages and Solutions,* 6th ed. (San Diego: W. I. Publications, 1990), p. 198.

Putting Equations into "Function Form"

In Section 1.4, we defined the solutions of an equation such as

$$3x^2 - 2y = 5x$$

as the values for x and y that make the equation true. For example, $x = 1$ and $y = -1$ are a solution since they make it a true statement:

Given	$3x^2 - 2y = 5x$
If $x = 1$ and $y = -1$, then	$3(1)^2 - 2(-1) = 5(1)$
	$3 + 2 = 5$
results in the true statement	$5 = 5$

There are in fact an infinite number of solutions for this equation. But it is often difficult to find solutions to equations in a form such as $3x^2 - 2y = 5x$ by anything other than trial and error.

Not all equations in two variables represent functions. But when an equation represents a function, it is useful to solve the equation for the dependent variable in terms of the independent variable. This is called putting the equation in *function form.* Once an equation is in function form, we can easily generate many solutions by picking different values for the independent variable and using the functional rule to compute the corresponding values of the dependent variable. Many graphing calculators and computer graphing programs accept only equations in function form as input.

To put an equation into function form, we first need to identify the independent and the dependent variables. Sometimes the choice may be obvious, other times not. There may be more than one possible correct choice. For now, to practice the mechanics of solving for one variable in terms of another, we adopt the mathematical convention that x represents the independent and y the dependent variable.

To put each of the following equations into function form, we need to solve for y, the dependent variable, in terms of x, the independent variable. We want

$$y = \text{some rule involving } x$$

We start by putting all the terms involving y on one side of the equation and everything else on the other side.

Example 8

Analyze the equation $4x - 3y = 6$. Can it be put in function form? What do the symbols tell us to do? Generate a table and graph of solutions to the equation. Decide whether or not the equation represents a function. If it does, what are the domain and range?

Solution

First, try to put the equation into function form:

Given the equation	$4x - 3y = 6$
subtract $4x$ from both sides	$-3y = 6 - 4x$
divide both sides by -3	$\dfrac{-3y}{-3} = \dfrac{6 - 4x}{-3}$
simplify	$y = \dfrac{6}{-3} + \dfrac{-4x}{-3}$
simplify and rearrange terms	$y = \frac{4}{3}x - 2$

Table 1.18

x	y
−6	−10
−3	−6
0	−2
3	2
6	6

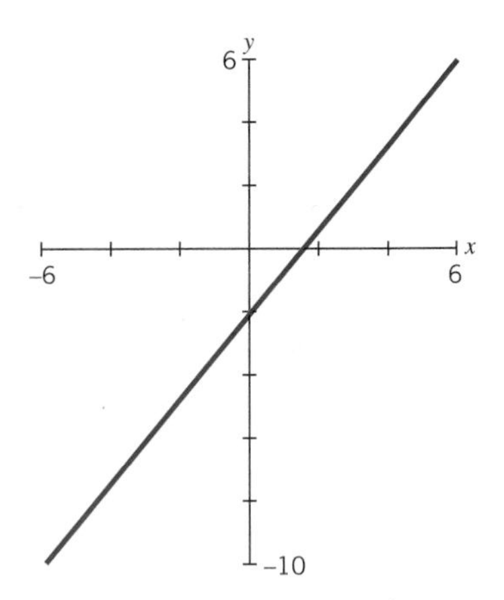

Figure 1.21 Graph of $y = \frac{4}{3}x - 2$.

We now have an expression for y in terms of x. We can generate a small table of solutions (Table 1.18) of the equation and graph the results (Figure 1.21). Does the equation $y = \frac{4}{3}x - 2$ represent y as a function of x? That is, given a particular value for x, does the equation determine one and only one value for y? This is not always an easy question to answer. The symbols in this equation tell us what to do. They say:

To find y, take the value of x, multiply it by $\frac{4}{3}$, and then subtract 2.

Starting with a particular value for x, each step in this process determines one and only one number, resulting in a unique corresponding value for y. So y is indeed a function of x. The graph also passes the vertical line test.

What is the domain, the set of possible values for x? In most cases throughout this text, we implicitly assume that the domain (and range) are restricted to real numbers. Are there any other constraints on the values that x may assume? The answer is no! So,

domain is all real numbers

What is the range, the corresponding set of possible values for y? Since the line in Figure 1.21 extends indefinitely in both directions, there are no maximum or minimum values for y. This implies that

domain is all real numbers

Example 9

Analyze the equation $y^2 - x = 0$. Can it be put into function form? Generate a table and graph of solutions to the equation. Decide whether or not the equation represents a function.

Solution

Put the equation in function form:

Given the equation	$y^2 - x = 0$
add x to both sides	$y^2 = x$

To solve this equation, we need to take the *square root* of both sides. An equation such as

$$y^2 = 4$$

Table 1.19

x	y
0	0
1	1 or -1
2	$\sqrt{2}$ or $-\sqrt{2}$
4	2 or -2
9	3 or -3

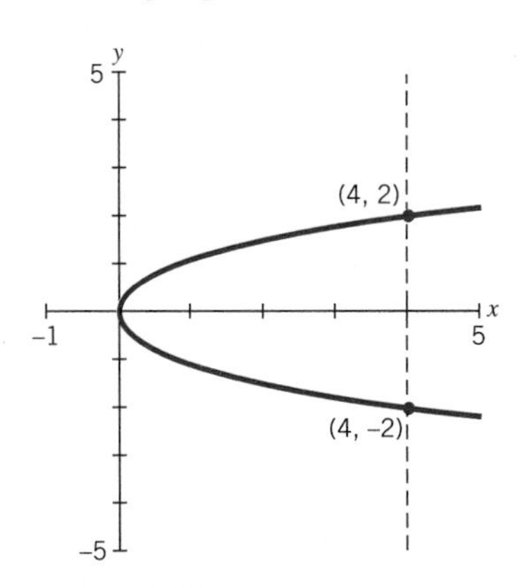

Figure 1.22 Graph of the equation $y^2 = x$.

has two solutions,

$$y = \sqrt{4} \text{ or } 2 \quad \text{and} \quad y = -\sqrt{4} \text{ or } -2$$

since both $2^2 = 4$ and $(-2)^2 = 4$. We can represent these two solutions as

$$y = \pm 2$$

So, to put

$$y^2 = x$$

into function form, we must take the square root of both sides of the equation. This gives us two solutions,

$$y = \pm\sqrt{x}$$

If we only consider real numbers, then we must assume that $x \geq 0$. If $x > 0$, then each value of x is associated with two different y values, namely $+\sqrt{x}$ and $-\sqrt{x}$. (If $x = 0$, then there exists one and only one value for y, namely zero.) So the equation $y^2 = x$ does *not* represent a function unless we restrict ourselves to a trivial domain of only zero.

Table 1.19 and the graph of the equation in Figure 1.22 confirm that $y^2 = x$ is not a function. Table 1.19 shows that two y values are associated with each x value (except for $x = 0$), and the graph of $y^2 = x$ in Figure 1.22 does not pass the vertical-line test. In this example, the equation $y = \pm\sqrt{x}$ is in "function form" but it does not represent a function.

Function Notation

When we want to indicate that one variable, T, is a function of another variable, P, we use the notation

$$T = f(P)$$

which is read "T is equal to f of P." Note that the use of the parentheses in $f(P)$ does *not* mean f times P. The notation $f(P)$ means the function f evaluated at P. We use the letter f to represent the relationship between T and P, but we could use any letter to represent this relationship.

A function, f, is a rule that tells us what to do to a value of an independent variable to produce a unique value of the dependent variable. The rule could be defined using an equation, words, graphs, or tables. If f is defined by the equation $T = 0.08P$, then

$$f(P) = T \quad \text{where} \quad T = 0.08P \qquad \text{or} \qquad f(P) = 0.08P$$

We must now be careful to distinguish among three different symbols for:

the independent variable, P
the dependent variable, T
the function name, f

Function notation is particularly useful when a function is being evaluated at a specific point. For example, instead of saying "the value for T when $P = 10$," we write simply "$f(10)$." Then we have

$$\begin{aligned} f(10) &= (0.08)(10) \\ &= 0.8 \end{aligned}$$

Similarly

$$\begin{aligned} f(0.5) &= (0.08)(0.5) \\ &= 0.04 \end{aligned} \qquad \begin{aligned} f(\tfrac{7}{4}) &= (0.08)(\tfrac{7}{4}) \\ &= 0.14 \end{aligned} \qquad \begin{aligned} f(6.5) &= (0.08)(6.5) \\ &= 0.52 \end{aligned}$$

The function notation $f(x)$ emphasizes the choice of one variable as an independent input and the other as a dependent variable or output. We encounter this notation regularly in mathematics texts. The $f(x)$ notation is used less frequently in social science and science texts. This may stem from either the reluctance to classify one variable as depending upon another or the desire to be able to retain the flexibility of an equation that can be rewritten in many different forms.

Example 10

In Example 8 we put the equation $4x - 3y = 6$ into function form as $y = (4/3)x - 2$ and determined that the equation did represent y as a function of x. Rewrite this equation using function notation.

Solution

We could name the function F and write $y = F(x)$ where $y = (4/3)x - 2$, or more simply just $F(x) = (4/3)x - 2$.

Example 11

In Algebra Aerobics 1.6(a), problem 2, we determined that the equation $D = 50W$ described D, the ampicillin dosage, as a function of W, a child's weight. Thus we could name the function h, and write $D = h(W)$ where $D = 50W$, or just $h(W) = 50W$. The following is a small table of values for the function.

Table 1.20

W	$h(W)$
0	0
2	100
5	250
8	400

What is:

a. $h(2)$? **b.** $h(8)$?

Solution

a. The expression $h(2)$ means to evaluate $h(W)$ when $W = 2$. Using the table we have $h(2) = 100$.

b. Similarly when $W = 8$, we have $h(8) = 400$.

Example 12

a. If $g(x) = \dfrac{1}{x - 1}$, what is the domain?

Evaluate the following expressions:

b. $g(0)$ **c.** $g(2)$ **d.** $g(-2)$

Solution

a. A fraction $\dfrac{a}{b}$ is undefined when $b = 0$. For the function $g(x) = 1/(x - 1)$, we want to find out what value of x will make the denominator equal to 0. When $x = 1$, the denominator will be equal to 0, since $1 - 1 = 0$. Thus,

$$\text{domain} = \text{all real numbers except } 1$$

b. Notice that in our formula, x stands for the independent variable or the input. For $g(0)$, the input is 0; so

$$g(0) = \frac{1}{0 - 1} = \frac{1}{-1} = -1$$

c.

$$g(2) = \frac{1}{2 - 1} = \frac{1}{1} = 1$$

d.

$$g(-2) = \frac{1}{-2 - 1} = \frac{1}{-3} = -\frac{1}{3}$$

Algebra Aerobics 1.6b

1. Consider the function $f(x) = x^2 - 5x + 6$. Find $f(0)$, $f(1)$, and $f(-3)$.

In problems 2–6, solve for y in terms of x. Determine if y is a function of x. If it is, rewrite using $f(x)$ notation. If y is a function of x, determine the domain.

2. $2(x - 1) - 3(y + 5) = 10$
3. $x^2 + 2x - 3y + 4 = 0$
4. $7x - 2y = 5$
5. $2xy = 6$
6. $x/2 + y/3 = 1$
7. From the graph in Figure 1.23, estimate $f(-4)$, $f(-1)$, $f(0)$, and $f(3)$. Find two approximate values for x such that $f(x) = 0$.
8. From Table 1.21 find $f(0)$ and $f(20)$. Find two values of x for which $f(x) = 10$. Explain why $f(x)$ is a function.

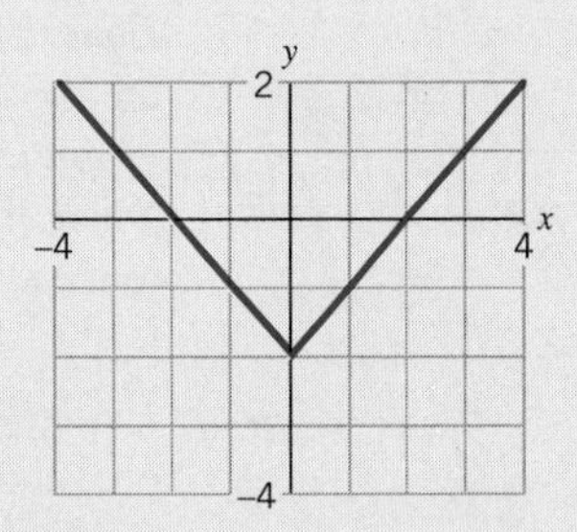

Figure 1.23 Graph of $f(x)$.

Table 1.21

x	$f(x)$
0	20
10	10
20	0
30	10
40	20

What Lies Ahead

The rest of this text is dedicated to building a library of functions that can serve as mathematical models for real world data sets. We will study the properties of these functions to enable us to better understand how they can be used to describe the world around us.

Chapter Summary

Variables and Data. Characteristics such as height, race, age, and college major that vary from person to person are called *variables.* Information collected about variables is called *data.*

Variables can be classified as *quantitative variables* or *qualitative variables.* Quantitative variables, such as age, height, or number of people in your household, can be represented by a number and a unit of measurement (e.g., 18 years, 6 feet, or 3 people). Qualitative variables, such as gender or college major, require us to construct categories (such as male and female or physics, English, and psychology).

Single Variable Data: Numerical Descriptors. The *mean* of a list of numbers is their sum divided by the number of terms in the list. The mean is often referred to as the "average."

The *median* separates a numerically ordered list of numbers into two parts, with half the numbers at or below the median and half at or above the median. If the number of observations is odd, the median is the middle number in the ordered list. If the number of observations is even, the median is the mean of the two middle numbers in the ordered list.

Outliers are extreme, nontypical values in a data set.

The *relative frequency* of any value or interval of values is the fraction (often expressed as a percentage) of all the data in the sample having that value or lying in that interval.

Equations in Two Variables. The *solutions* of an equation in two variables x and y are the ordered pairs of the form (x, y) that make the equation a true statement. The *graph* of an equation is the set of points corresponding to the ordered-pair solutions where the first coordinate corresponds to values on the horizontal axis and the second to values on the vertical axis.

Functions. A variable y is a *function* of a variable x if each value of x determines one and only one value of y. Functions can be represented with words, tables, graphs, and equations. Solutions of equations are written as ordered pairs in the form

(independent variable, dependent variable)

or equivalently as

(input, output)

By convention, the graph of a function has the independent variable on the horizontal axis and the dependent variable on the vertical axis. A graph does not represent a function if it fails the vertical line test. That is, if you can draw a vertical line that crosses the graph two or more times, the graph does not represent a function.

A formula or equation used to represent a real-world phenomenon is called a *mathematical model.*

The *domain* of a function is the set of all possible values of the independent variable. The *range* is the set of corresponding values of the dependent variable.

If an equation represents a function, then solving the equation for the dependent variable in terms of the independent variable is called putting the equation in *function form.* We write

$$y = f(x)$$

to indicate that y is a function of x. A function, f, is a rule that tells us what to do to a value of an independent variable to produce one and only one value of the dependent variable. The rule could be defined using an equation, words, a graph, or a table.

The $f(x)$ notation is particularly useful when a function is being evaluated at a specific point. For example, if $f(P) = T$ is defined by the equation $T = 0.08P$, instead of saying "the value for T when $P = 10$," we simply write "$f(10)$." Then we have

$$\begin{aligned} f(10) &= (0.08)(10) \\ &= 0.8 \end{aligned}$$

EXERCISES

Exercises for Section 1.1

1. **a.** Compute the mean and median for the list: 5, 18, 22, 46, 80, 105, 110.
   b. Change one of the entries in the list in part (a) so that the median stays the same and the mean increases.

2. Suppose that a church congregation has 100 members, each of whom donates 10% of his or her income to the church. The church collected $250,000 last year from its members.
   a. What was the mean contribution of its members?
   b. What was the mean income of its members?
   c. Can you predict the median income of its members? Explain your answer.

3. Suppose that annual salaries in a certain corporation are as follows:

Level I (30 employees)	$18,000
Level II (8 employees)	$36,000
Level III (2 employees)	$80,000

   Find the mean and median annual salary. Suppose that an advertisement is placed in the newspaper giving the average annual salary of employees in this corporation as a way to attract applicants. Why would this be a misleading indicator of salary expectations?

4. Suppose that your grades on your first four exams were 78, 92, 60, and 85%. What would be the lowest possible average that your last two exams could have so that your grade in the class, based on the average of the six exams, is at least 82%?

5. Read Stephen Jay Gould's article "The Median Isn't the Message" and explain how an understanding of statistics brought hope to a cancer victim.

6. **a.** On the first quiz (worth 25 points) given in a section of college algebra, one person received a score of 16, two people got 18, one got 21, three got 22, one got 23, and one got 25. What were the mean and median of the quiz scores for this group of students?
   b. On the second quiz (again worth 25 points), the scores for eight students were 16, 17, 18, 20, 22, 23, 25, and 25.

i. If the mean of the scores for the nine students was 21, then what was the missing score?

ii. If the median of the scores was 22, then what are possible scores for the missing ninth student?

7. Herb Caen, a Pulitzer Prize winning columnist for the *San Francisco Chronicle,* remarked that a person moving from state A to state B could raise the average IQ in both states. Is he right? Explain.

8. Why do you think most researchers use median rather than mean income when studying "typical" households?

9. According to the 1992 U.S. Census, the median net worth of American families was \$52,200. The mean net worth was \$220,300. How could there be such a wide discrepancy?

10. Read the *CHANCE News* article and explain why the author was concerned.

11. The Greek letter Σ (called sigma) is used to represent the sum of all of the terms of a certain group. Thus $a_1 + a_2 + a_3 \cdots + a_n$ can be written as

$$\sum_{i=1}^{n} a_i$$

which means to add together all of the values of a_i from a_1 to a_n.

$$\sum_{i=1}^{n} a_i = a_1 + a_2 \cdots + a_n$$

a. Using Σ notation, write an algebraic expression for the mean of the five numbers x_1, x_2, x_3, x_4, x_5.

b. Using Σ notation, write an algebraic expression for the mean of n numbers $t_1, t_2, t_3, \ldots, t_n$.

c. Evaluate the following sum:

$$\sum_{k=1}^{5} 2k$$

Ages of Students

Age Interval	Frequency Count
15–19	2
20–24	8
25–29	4
30–34	3
35–39	2
40–44	1
45–49	1
Total	21

12. The table to the left gives the ages of students in a mathematics class.

a. Use this information to estimate the mean age of the students in the class. Show your work.

b. What is the largest value the actual mean could have? The smallest? Why?

13. (Graphing calculator or spreadsheet recommended) Use the table below to generate an estimate of the mean age of the U.S. population. Show your work.

Ages of U.S. Population in 1999 in Ten-Year Intervals

Age (years)	Population (thousands)
Under 10	38,884
10–19	39,257
20–29	36,291
30–39	42,339
40–49	41,564
50–59	29,241
60–69	19,970
70–79	16,105
80 and over	8,984
Total	**272,635**

Source: U.S. Bureau of the Census, *www.census.gov.*

14. An article titled "Venerable Elders" (*The Economist,* July 24, 1999) reported that 'both Democratic and Republican images are selective snapshots of a reality in which the median net worth of households headed by Americans aged 65 or over is around double the national average—but in which a tenth of such households are also living in poverty."

What additional statistics would be useful in order to form an opinion on whether elderly Americans are wealthy or poor compared to Americans as a whole?

Salary (in thousands)	Number of Graduates Receiving Salary
11–15	2
16–20	5
21–25	20
26–30	10
31–35	6
36–40	1

15. Given here is a table of salaries taken from a survey of recent graduates (with bachelor degrees) from a well-known university in Pittsburgh.

a. How many graduates were surveyed?

b. Is this quantitative or qualitative data? Explain.

c. What is the relative frequency of people having a salary between $26,000 and $30,000?

d. Create a histogram of the data.

16. The accompanying bar chart shows the predictions of the U.S. Census Bureau about the future racial composition of American society.

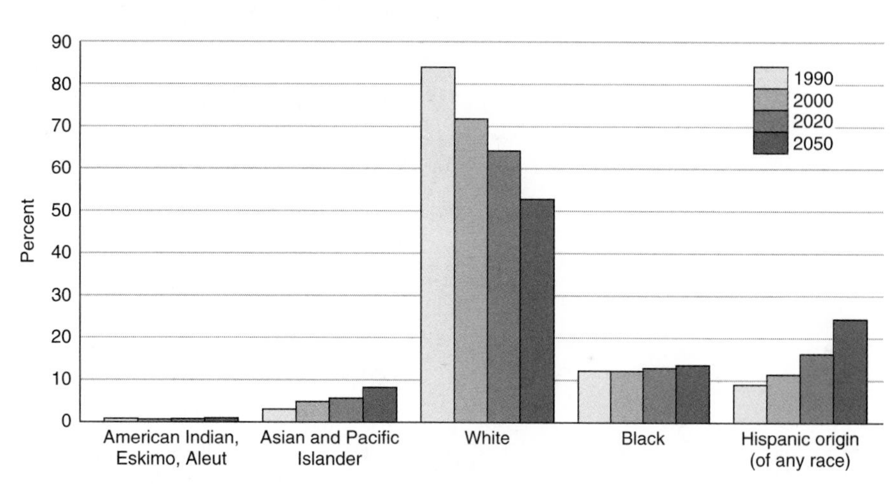

Percent distribution of the population, by race and Hispanic origin: 1990–2050.
Source: U.S. Bureau of the Census, Current Population Reports, P25-1095, in *The American Almanac 1993–1994: Statistical Abstract of the United States* (Austin, TX: The Reference Press).

a. Estimate the following percentages:

i. Asian and Pacific Islanders in the year 2050

ii. Combined white and black population in the year 2020

iii. Non-Hispanic population in the year 1990

b. The U.S. Bureau of the Census has projected that there will be approximately 392,031,000 people in the United States in the year 2050. Approximately how many people will be of Hispanic origin in the year 2050?

c. Write a topic sentence describing the overall trend.

17. On the top of page 34 is a pie chart of America's spending patterns at the end of 1992.

a. In what single category did Americans spend the largest percentage of their income? What was this percentage?

b. If an American family had an income of $30,000, according to this chart, how much of it would be spent on food and beverages?

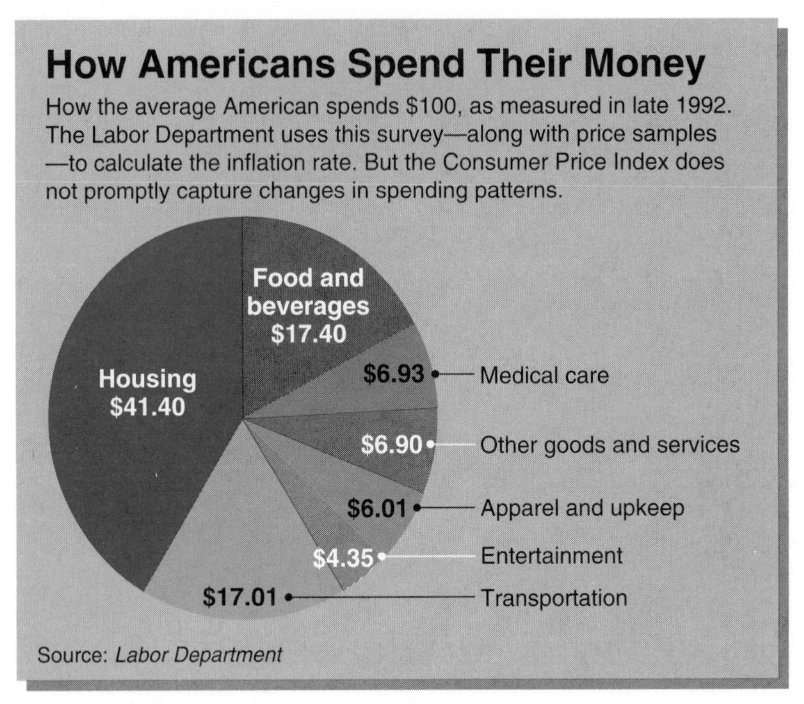

Source: New York Times, January 20, 1994. Copyright © 1994 by The New York Times Company. Reprinted by permission.

c. If you were to write a newspaper article to accompany this pie chart, what would your opening topic sentence be?

18. Attendance at a stadium for the last 30 games of a college baseball team is listed as follows:

5072	3582	2504	4834	2456	3956
2341	2478	3602	5435	3903	4535
1980	1784	1493	3674	4593	5108
1376	978	2035	1239	2456	5189
3654	3845	673	2745	3768	5227

Create a histogram that can be used to illustrate these data. Decide how large the intervals should be to illustrate the data well but not be overly detailed.

The program "F4: Measures of Central Tendency" in *FAM 1000 Census Graphs* can help you understand mean and median and their relationship to histograms.

19. Estimate the mean and median from the given histogram.

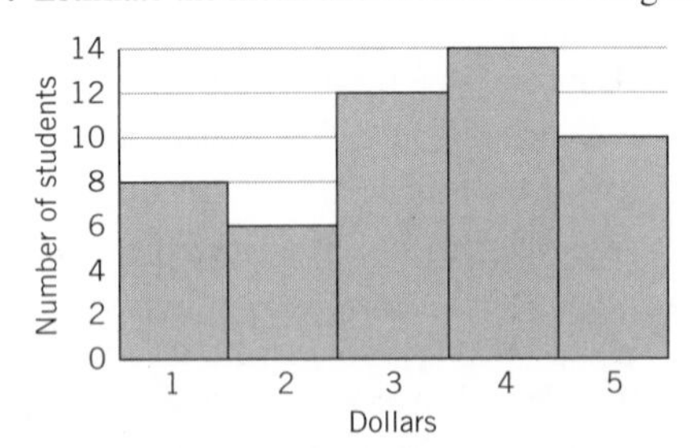

Weekly allowance of junior high school students.

20. Choose a paragraph of text from any source (perhaps even this textbook) and construct a histogram of the number of letters in the words. If the same word appears more than once, count it as many times as it appears. You will have to make some reasonable decisions about what to do with numbers, abbreviations, and contractions.

Compute the mean and median of word length from your graph. Indicate how you would expect the graph to be different if you used:

a. A children's book **b.** A work of literature **c.** A medical textbook

Exercises for Section 1.2

21. The following graph and table show the distribution of ages in the U.S. population in 1991 and projections for 2050. (See Excel or graph link file PROJAGES) Write a paragraph about these projected changes in the distribution of ages.

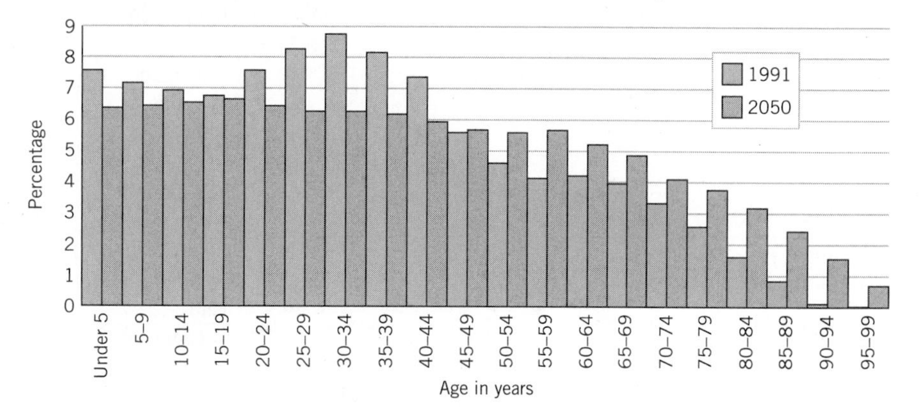

Distribution of ages of U.S. population in 1991 and 2050 (projected). *Source:* U.S. Bureau of the Census, Current Population Reports, P25-1095 and P25-1104, in *The American Almanac 1993–1994 and 1996–1997: Statistical Abstract of the United States* (Austin, TX: The Reference Press).

Distribution of Ages of U.S. Population in 1991 and 2050 (Projected)

	1991		2050			1991		2050	
Age	Total (in 1000s)	%	Total (in 1000s)	%	Age	Total (in 1000s)	%	Total (in 1000s)	%
Under 5	19,222	8	25,382	6	60–64	10,582	4	20,553	5
5–9	18,237	7	25,222	6	65–69	10,037	4	18,859	5
10–14	17,671	7	25,650	7	70–74	8,242	3	15,769	4
15–19	17,205	7	25,897	7	75–79	6,279	2	14,510	4
20–24	19,194	8	25,313	6	80–84	4,035	2	12,078	3
25–29	20,718	8	24,659	6	85–89	2,090	1	9,283	2
30–34	22,159	9	24,803	6	90–94	812	0	5,779	1
35–39	20,518	8	24,316	6	95–99	214	0	2,623	1
40–44	18,754	7	23,423	6	100 and over	44	0	1,208	0
45–49	14,094	6	22,266	6	**Total**	**252,177**	**100**	**392,031**	**100**
50–54	11,645	5	22,071	6	**Median age**	**33**		**39**	
55–59	10,423	4	22,367	6					

22. Open up the program "F1: Histograms" in *FAM1000 Census Graphs* in the course software. The 1999 U.S. Census data about 1000 randomly selected U.S. individuals and their families are imbedded in this program. You can use it to create histograms for education, age, and different measures of income. Try using different interval sizes to see what patterns emerge. Decide on one variable (say education) and then compare the histograms of this variable for different groups of people. For example, you could compare education histograms for men and women or for people living in two different regions of the country. Pick a comparison that you think is interesting and, if possible, print out your histograms. Then write a 60 second summary describing your observations.

23. The table at the top of the next page (and related Excel and graph link file MEDICAL) shows the number of community hospital beds and personnel per 1000 people for states in the Northeast and the Midwest in 1993. (The majority of all hospitals in the United States are community hospitals. This category excludes long-term, psychiatric, tuberculosis, and federal hospitals.)

a. What are the mean and median for the number of hospital beds per 1000 people for the Northeast? For the Midwest?

b. Write a summary paragraph that describes some aspect of the information in the table.

Northeast			
State	Beds (per 1000)	Occupancy Rate (%)	Personnel (per 1000)
Maine	4.4	68.0	18.5
New Hampshire	3.4	63.7	13.8
Vermont	1.9	64.2	7.0
Massachusetts	21.1	71.5	107.8
Rhode Island	3.0	73.3	14.7
Connecticut	9.2	74.4	44.8
New York	77.4	82.8	328.7
New Jersey	31.1	77.0	121.0
Pennsylvania	53.4	72.6	230.3
Mean	—	**71.9**	**98.5**
Median	—	**72.6**	**44.8**

Midwest			
State	Beds (per 1000)	Occupancy Rate (%)	Personnel (per 1000)
Ohio	41.1	60.5	176.2
Indiana	21.3	58.7	90.6
Illinois	44.1	63.5	180.0
Michigan	30.9	64.7	140.9
Wisconsin	17.7	63.4	65.5
Minnesota	18.4	66.0	55.0
Iowa	13.4	57.9	44.1
Missouri	23.6	58.9	95.9
North Dakota	4.4	64.2	12.2
South Dakota	4.3	60.6	11.4
Nebraska	8.4	55.2	25.7
Kansas	11.3	54.2	36.3
Mean	—	**60.7**	**77.8**
Median	—	**60.6**	**60.3**

Source: U.S. Bureau of the Census, Current Population Reports, P25-1104, in *The American Almanac 1996–1997: Statistical Abstract of the United States* (Austin, TX: The Reference Press).

24. *The New York Times* published the accompanying piece about the public's attitude toward spending the nation's budget surplus. Are the results contradictory? Describe the results in a 60 second summary.

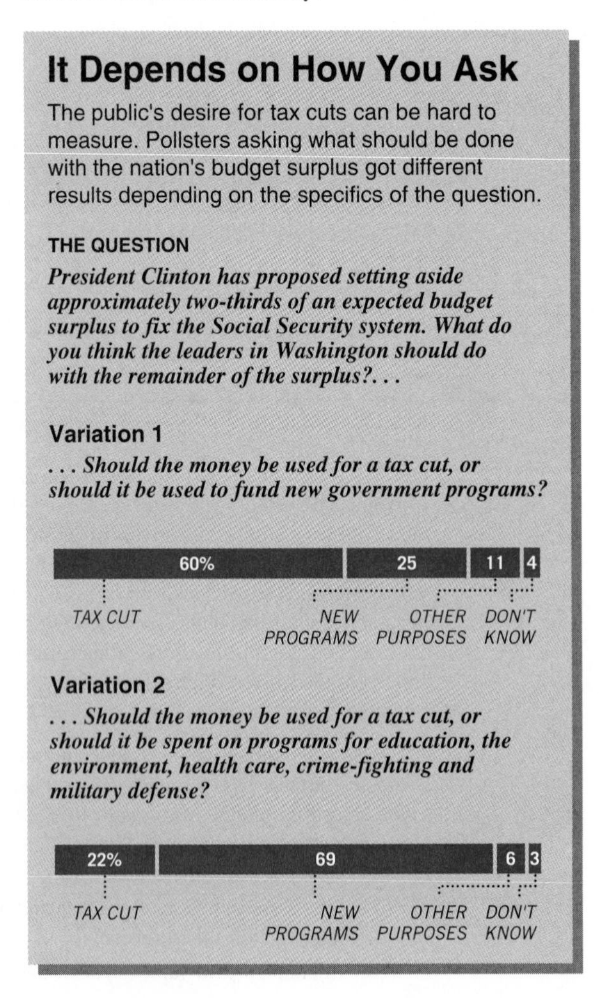

Source: From a nationwide Pew Research Center Poll; results published in *The New York Times,* January 30, 2000.

25. The following table and graphs (and related Excel and graph link file USCHINA) contain information about the populations of the United States and China. Write a 60 second summary comparing the two populations.

Age Distribution of the 1998 Population (in 1000s) of China and the United States

	China		United States	
Age Interval	Total	Percent	Total	Percent
0–4	97,396	7.87	19,020	7.04
5–9	111,211	8.99	19,912	7.37
10–14	110,638	8.94	19,184	7.10
15–19	98,012	7.92	19,473	7.20
20–24	102,095	8.25	17,768	6.57
25–29	127,336	10.29	18,680	6.91
30–34	121,072	9.79	20,209	7.48
35–39	82,084	6.64	22,638	8.38
40–44	91,118	7.37	21,891	8.10
45–49	75,137	6.07	18,855	6.98
50–54	53,743	4.34	15,728	5.82
55–59	44,169	3.57	12,408	4.59
60–64	40,913	3.31	10,256	3.79
65–69	33,124	2.68	9,575	3.54
70–74	24,084	1.95	8,781	3.25
75–79	14,322	1.16	7,195	2.66
80–84	7,127	0.58	4,712	1.74
85 and over	3,335	0.27	4,006	1.48
Total	**1,236,915**	**100.00**	**270,290**	**100.00**
Median age			**35.5**	
Mean age			**36.4**	

Source: U.S. Bureau of the Census, *www.census.gov.*

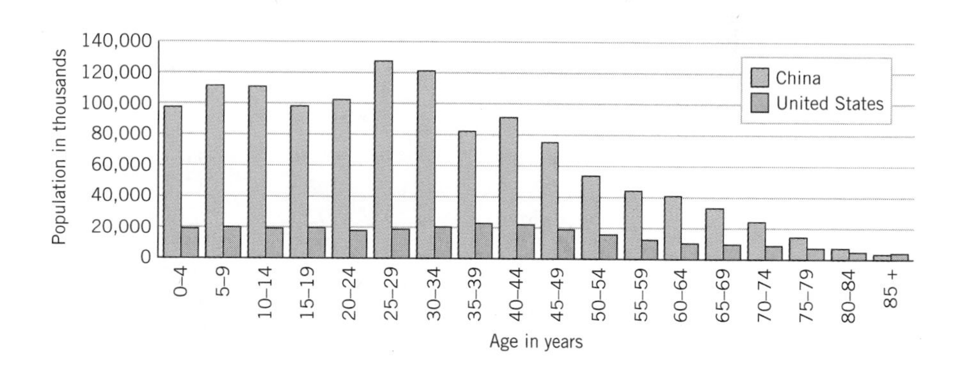

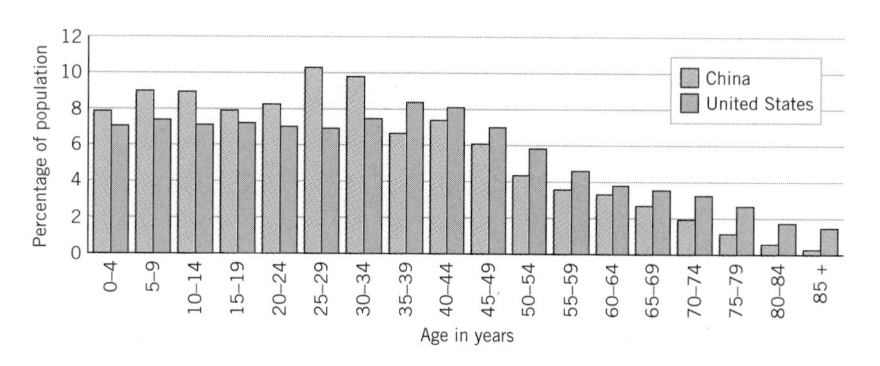

26. Read *The New York Times* OP-ED article, "A Fragmented War on Cancer" by Hamilton Jordan, President Jimmy Carter's chief of staff. Jordan claims that we are on the verge of a cancer epidemic.

a. Use what you have learned in Exercise 25 about the distribution of ages over time in the United States to refute his claim. Are there other arguments that refute his claim?

b. Read *The New York Times* letter to the editor by William M. London, Director of Public Health, American Council on Science and Health. London argues that Hamilton Jordan's assertions are misleading. What questions are raised by the arguments of William London? What additional data would you need to evaluate his arguments?

c. Write a paragraph refuting Hamilton Jordan's claim that we are on the verge of a cancer epidemic.

27. Consider the data given in the table below describing wolves captured and radio-collared in Montana and southern British Columbia, Canada, in 1993.

a. Look at the variables: sex, weight, and age. Identify each variable as quantitative or qualitative.

b. Calculate the mean and median weights of female wolves and of male wolves.

c. Discuss which measurement of central tendency (mean or median) might be more appropriate for describing the data in part (b).

d. Estimate the mean age of the females, of the males, and of the total population.

e. Write a 60 second summary of this wolf population.

Wolf Pack	Capture Date	Wolf Number	Ear tag Number	Sex	Weight (lb)	Age (years)
Spruce Creek	6/14	9318	102103	M	100	2–3
	6/16	9381	104105	F	73	1–2
North Camas	6/2	9375	none	F	77	4–5
	6/4	9378	100101	F	72	3–4
	6/4	9376	9596	F	66	2–3
	6/4	9377	9798	F	63	1–2
	6/9	9379	none	F	59	?
	6/11	9380	none	F	59	1
South Camas	5/27	9474	9091	F	61	1
	5/29	9317	9293	M	95	2–3
	5/30	8756	2627	F	77	6
Murphy Lake	6/22	1718	1718	F	73	3
	6/23	2627	2627	F	21	Pup
	6/23	3637	3637	M	96	4–5
	10/15	2223	2223	F	62	Pup
Sawtooth	2/26	8808	8808	M	122	5
	9/22	2829	2829	F	50	Pup
	9/25	3031	3031	F	53	Pup
Ninemile	8/27	4243	4243	F	73	1

Source: 1993 Annual Report of the Montana Interagency Wolf Working Group, prepared by the U.S. Fish and Wildlife Service.

28. Examine the accompanying graphs and tables showing production and consumption of conventional or primary power sources and nuclear electricity generation.

World's Major Producers of Primary Energy, 1996

Source: Energy Information Administration, International Energy Database; quadrillion Btu

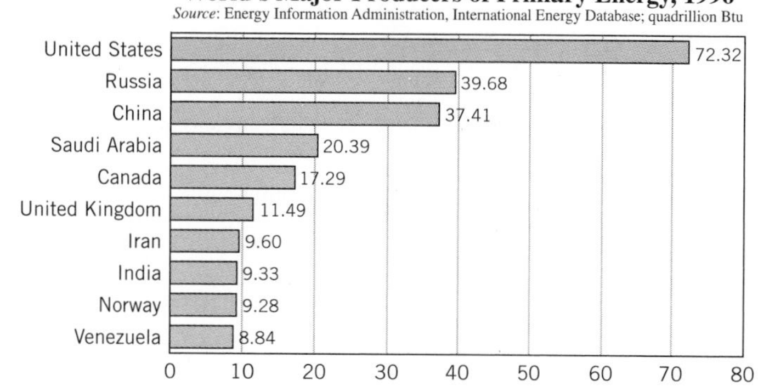

World's Major Consumers of Primary Energy, 1996

Source: Energy Information Administration, International Energy Database; quadrillion Btu

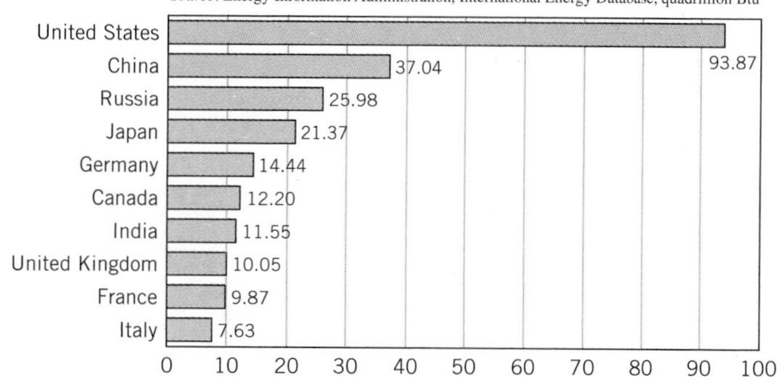

Nuclear Electricity Generation by Selected Country, March 1998

Source: Energy Information Administration, U.S. Dept. of Energy, *Monthly Energy Review*, June 1998

(Billions Kilowatt-hours E = estimate)

Country	Value	Country	Value	Country	Value	Country	Value
Argentina	0.7	France	34.7^E	Lithuania	1.3	Sweden	7.3
Belgium	3.7	Germany	14.0	Mexico	0.9	Switzerland	2.4
Brazil	0.4	Hungary	1.1	Netherlands	0.4	Taiwan	3.4
Bulgaria	2.2	India	1.0^E	Russia	11.1	Ukraine	7.2
Canada	7.2	Japan	27.3	South Africa	1.4	United Kingdom	10.1^E
Finland	2.0	Korea, South	6.7	Spain	4.6	United States	55.6^E

Nations Most Reliant on Nuclear Energy, 1997

Source: International Atomic Energy Agency, May 1998

(Nuclear electricity generation as % of total)

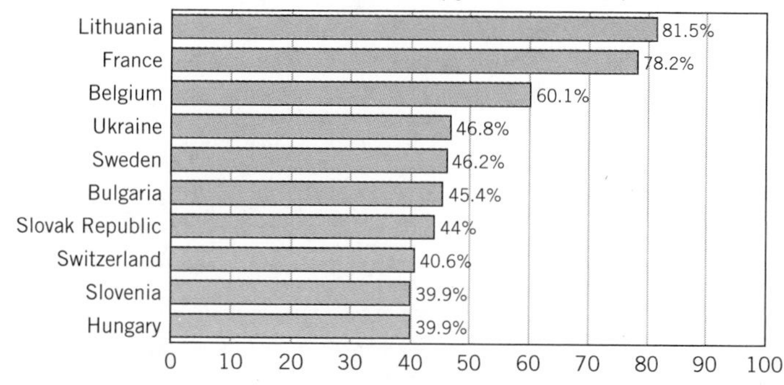

Source: The World Almanac and Book of Facts 1999.

a. Does the United States use more conventional power than it produces? If so, what percentage does it have to get from other sources?

b. Lithuania and France both generate about four-fifths of their electricity in nuclear plants. How many billions of kilowatt-hours of energy does France have to get from other sources? How many billions of kilowatt-hours of energy does Lithuania use altogether?

c. Does Sweden generate more or less nuclear electricity than Lithuania? Explain your answer using numerical measures.

d. Notice that the United States generates the most kilowatts of nuclear electricity. How is it possible that the United States is not shown as a major user of nuclear power?

e. Write a 60 second summary comparing the production and consumption of energy of one of the countries in these graphs and tables.

29. Population pyramids are a type of chart used to depict the overall age structure of a society.

a. Using the accompanying population pyramids of the United States, estimate the number of:

i. males that were between the ages 35 and 39 years in 1997

ii. females that were between the ages 55 and 59 years in 1997

iii. males 85 years and older in the year 2050; females 85 years and older in the year 2050

iv. all males and females between the ages of 0 and 9 years in the year 2050.

b. Write a 60 second summary describing the predicted shift in overall age structure in the United States between 1997 and 2050.

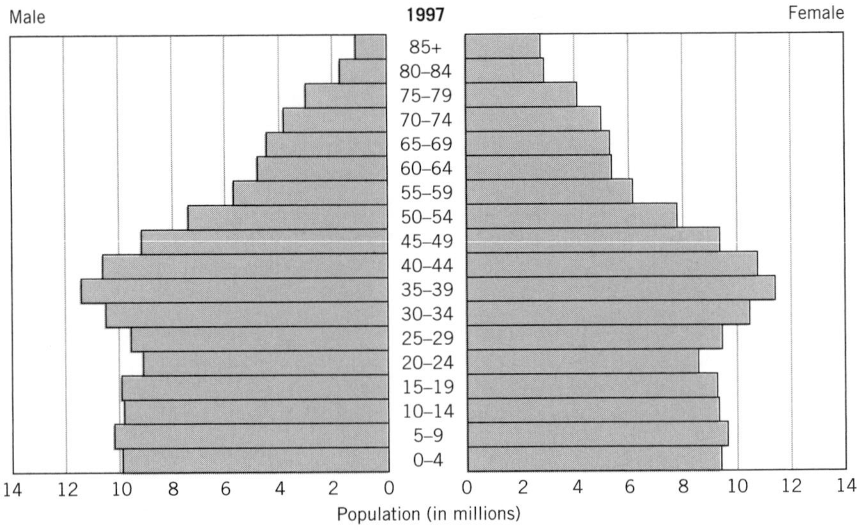

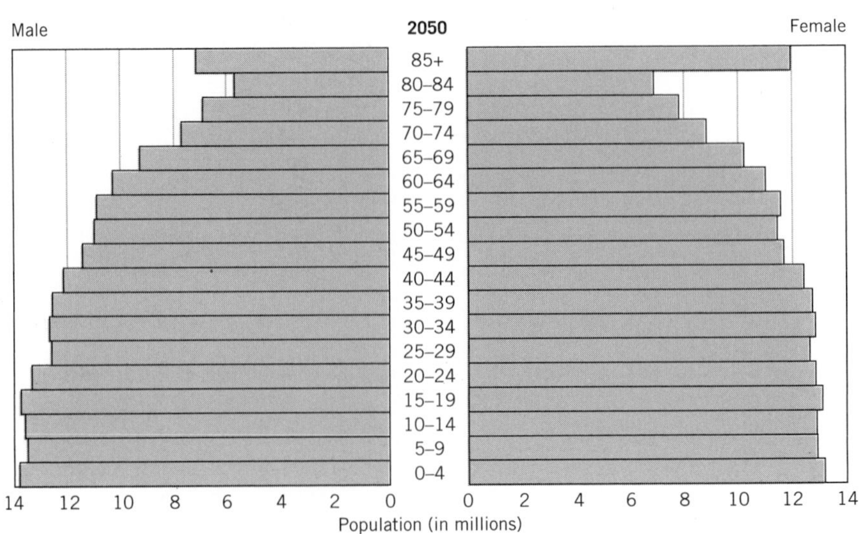

Population pyramid summary for the United States for 1997 and 2050.
Source: U.S. Census Bureau, International Data Base, *www.census.gov.*

30. The accompanying population pyramids show the age structure in Ghana, a developing country in Africa, for 1997. The previous exercise contains population pyramids for the United States, an industrialized nation, for 1997. Construct a paragraph describing at least three major differences in the distribution of ages in these two countries in 1997.

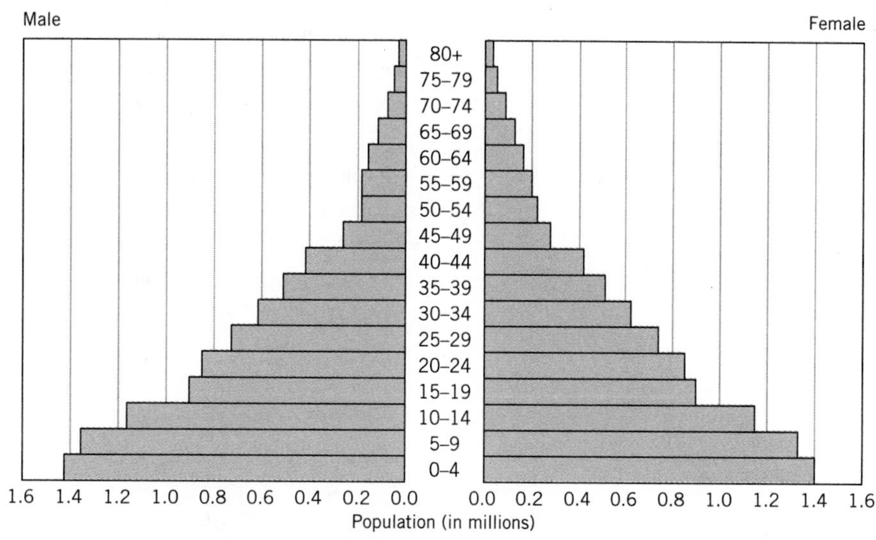

Population pyramid summary for Ghana, 1997.
Source: U.S. Census Bureau, International Data Base, *www.census.gov.*

31. Would you predict the mean or the median age to be greater for the United States? Why? Do you think your prediction holds true for most communities within the United States? What predictions would you make for other countries? Check your predictions with data from the U.S. Census at *www.census.gov.* Write a 60 second summary describing the reasons for your predictions and whether the data you found support or refute your predictions.

32. Make a prediction about the distribution of income for males and females in the United States. Check your predictions using the course software "F1: Histograms" in *FAM 1000 Census Graphs* and/or using data from the U.S. Census at *www.census.gov.* Write a paragraph describing your results.

Exercises for Section 1.3

33. Consider the accompanying graph of U.S. military sales to foreign governments:

a. During what years did sales increase?

b. During what years did sales decrease?

c. Estimate the maximum value for sales.

d. Estimate the minimum value for sales.

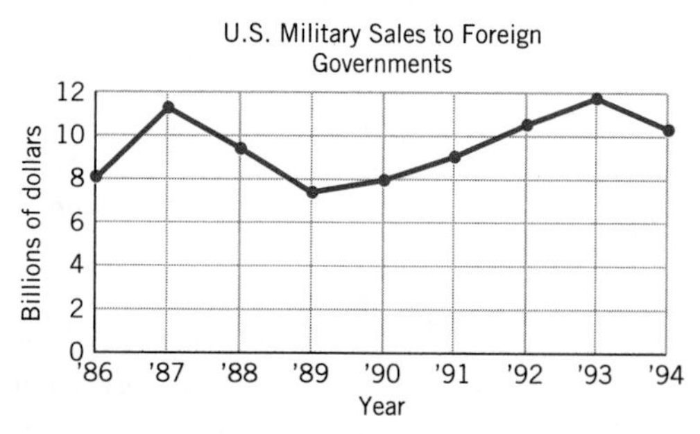

Source: U.S. Bureau of the Census, in *The American Almanac 1996–1997: Statistical Abstract of the United States,* 1996.

34. Sketch a plausible graph for each of the following and label the axes.

a. The amount of snow on your backyard each day from December 1 to March 1.

b. The temperature during a 24-hour period in your home town during one day in July.

c. The amount of water inside your fishing boat if your boat leaks a little and your fishing partner bails out water every once in a while.

d. The total hours of daylight each day of the year.

e. The temperature of a cup of hot coffee left to stand.

f. The temperature of an ice-cold drink left to stand.

35. The accompanying graph shows the 24-hour temperature cycle of a normal man. The man was confined to bed to minimize temperature fluctuations caused by activity.

a. Estimate the man's maximum temperature.

b. Estimate his minimum temperature.

c. Give a short general description of the 24-hour temperature cycle.

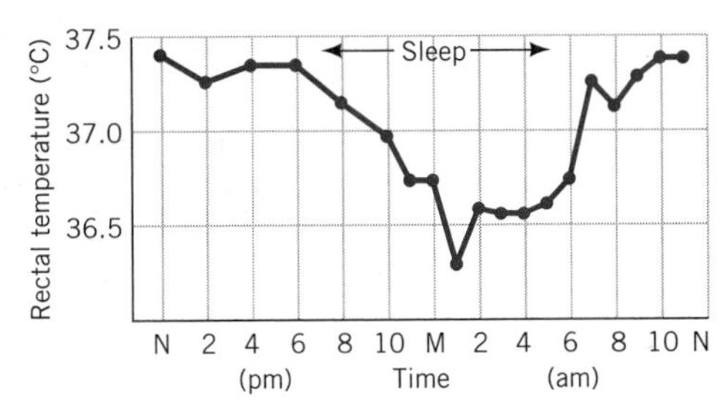

Source: V. B. Mountcastle, *Medical Physiology,* vol. 2, 14th ed. (St. Louis: Mosby-Year Book).

36. Consider the accompanying chart of juvenile arrests for murder.

a. During what intervals did the number of arrests show a decrease?

b. Approximate the number of juvenile arrests for murder in 1988 and 1993.

c. Estimate the ratio of the number of arrests in 1993 to the number of arrests in 1988.

d. What can be said generally about the number of juveniles arrested for murder based on this chart?

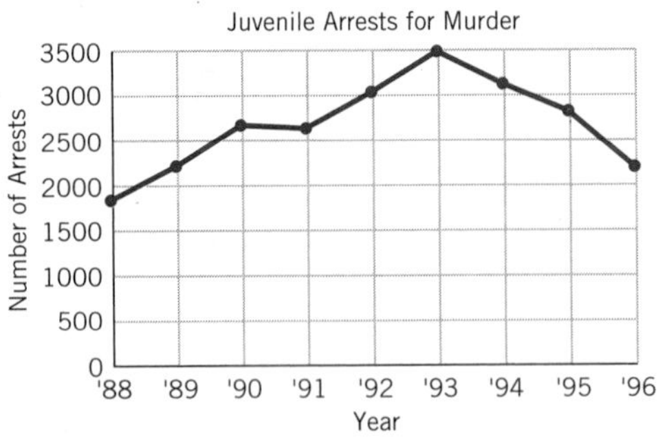

Source: U.S. Bureau of the Census, in *Statistical Abstract of the United States,* 1998.

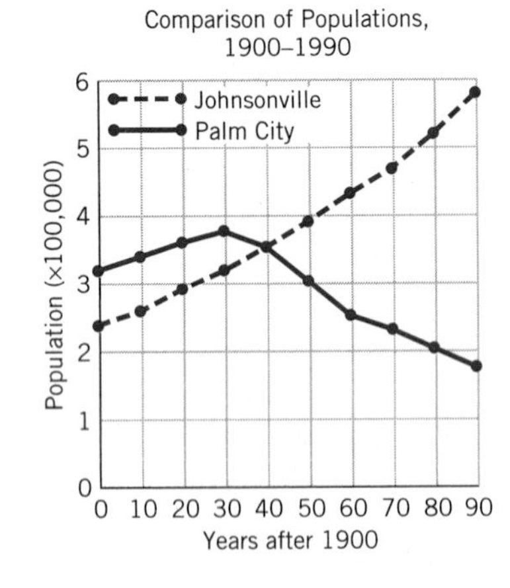

37. Examine the accompanying graph that shows the populations of two towns.

a. What is the range of population size for Johnsonville? For Palm City?

b. During what years did the population of Palm City increase?

c. During what years did the population of Palm City decrease?

d. When were the populations equal?

Exercises for Section 1.4

38. The following three graphs describe two cars, A and B.

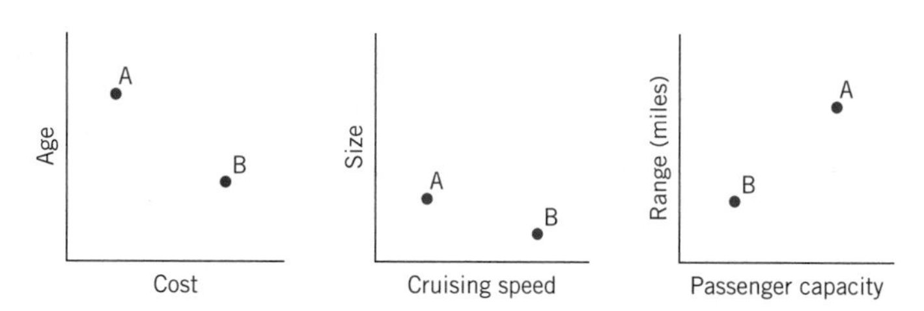

For parts (a)–(d), decide whether each statement below is true or false. Explain your reasoning.

a. The newer car is more expensive.

b. The slower car is larger.

c. The larger car is newer.

d. The less expensive car carries more passengers.

e. State two other facts you can derive from the graphs.

f. Which car would you buy? Why?

39. For parts (a)–(d) use the following equation: $y = \dfrac{x+1}{x-1}$.

a. Describe in words how to find the value for y given a value for x.

b. Find the ordered pair that represents a solution to the equation when the value of x is 5.

c. Find the ordered pair that represents a solution to the equation when the value of y is 3.

d. Is there an ordered-pair solution to the equation when the value of x is 1? If so, find it; if not, explain why.

40. For parts (a)–(d) use the following equation: $y = \dfrac{1}{x+1}$.

a. Describe in words how to find the value for y given a value for x.

b. Find the ordered pair that represents a solution to the equation when the value of x is zero.

c. Find the ordered pair that represents a solution to the equation when the value of y is 4.

d. Is there an ordered-pair solution to the equation when the value of x is -1? If so, find it; if not, explain why.

41. For parts (a)–(d) use the following equation: $y = -2x^2$

a. If $x = 0$, find the value of y.

b. If x is greater than zero, what can you say about the value of y?

c. If x is a negative number, what can you say about the value of y?

d. Can you find an ordered pair that represents a solution to the equation when y is greater than zero? If so, find it; if not, explain why.

42. Find the ordered pairs that represent solutions to each of the following equations when $x = 0$, $x = 3$, and $x = -2$:

a. $y = 2x^2 + 5x$

b. $y = -x^2 + 1$

c. $y = x^3 + x^2$

d. $y = 3(x-2)(x-1)$

43. Given the four points $(-1, 3)$, $(1, 0)$, $(2, 3)$, and $(1, 2)$, for each of the following equations, identify which points are solutions for that equation:

a. $y = 2x + 5$

b. $y = x^2 - 1$

c. $y = x^2 - x + 1$

d. $y = \dfrac{4}{x+1}$

Exercises for Section 1.5

44. Write a formula to express each of the following sentences:

a. The sale price is 20% off the original price. Use S for sale price and P for original price to express S as a function of P.

b. The time in Paris is 6 hours ahead of New York. Use P for Paris time and N for New York time to express P as a function of N. How would you adjust your formula if P comes out greater than 12?

c. For temperatures above 0°F the windchill effect can be estimated by subtracting two-thirds of the wind speed in miles per hour from the outside temperature. Use C for the effective windchill temperature, W for wind speed, and T for the actual outside temperature to write an equation expressing C in terms of W and T.

45. The basement of a large department store features discounted merchandise. Their policy is to reduce the previous month's price of the item by 10% each month for 5 months.

a. Use S_1 for sale price the first month and P for original price to express S_1 as a function of P. What is the price of a \$100 garment on sale for the first month?

b. Use S_2 for sale price the second month and P for original price to express S_2 as a function of P. What is the price of a \$100 garment on sale for the second month?

c. Use S_3 for sale price the third month and P for original price to express S_3 as a function of P. What is the price of a \$100 garment on sale for the third month?

d. Use S_5 for sale price the fifth month and P for original price to express S_5 as a function of P. What is the final price of a \$100 garment on sale for the fifth month? By what total percentage has the garment now been reduced from its original price?

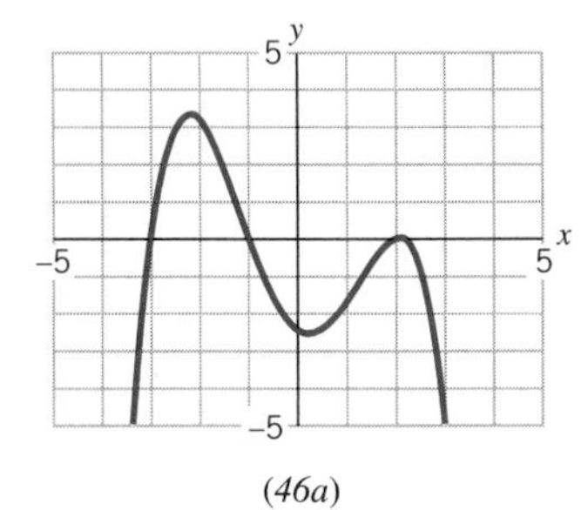

(46a)

46. a. Use the accompanying graph to estimate the location of the ordered pair where the maximum value of the function is found.

b. Use the accompanying graph to estimate the location of the ordered pair where the minimum value of the function is found.

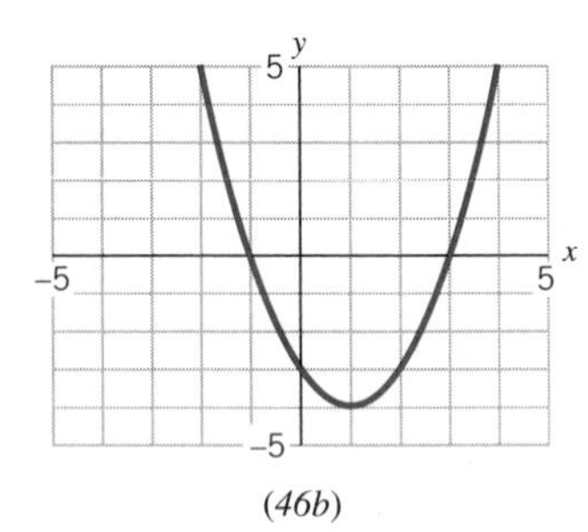

(46b)

47. In the accompanying graph, is y a function of x? Explain your answer.

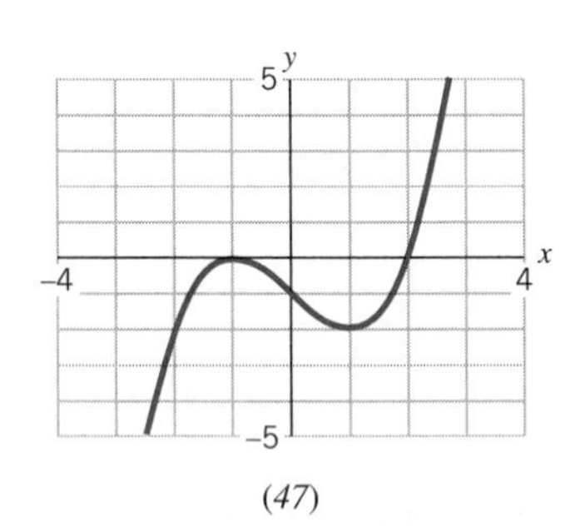

(47)

48. a. Which of the accompanying graphs describe functions?

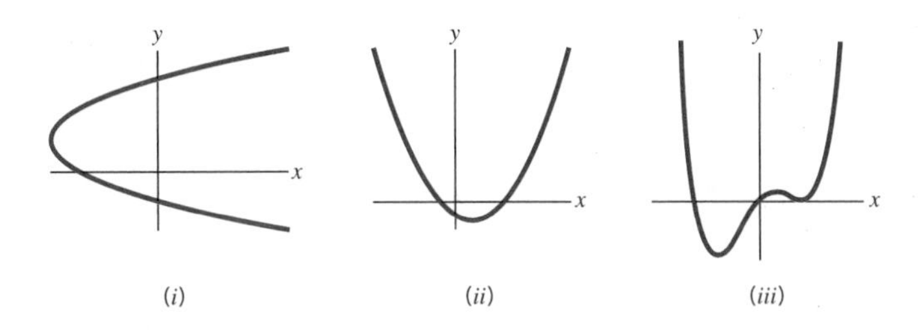

(i) (ii) (iii)

b. Explain your answers in part (a).

49. Consider the accompanying table, listing the weight (W) and height (H) of five individuals. Based on this table, is height a function of weight? Is weight a function of height? Justify your answers.

Weight W (lbs)	Height H (in.)
120	54
120	55
125	58
130	60
135	56

(49)

50. Determine whether y is a function of x in each of the following equations. If the equation does not define a function, find a value of x that is associated with two different y values.

a. $y = x^2 + 1$

b. $y = 3x - 2$

c. $y = 5$

d. $y^2 = x$

51. a. Which of the accompanying graphs describe a function?

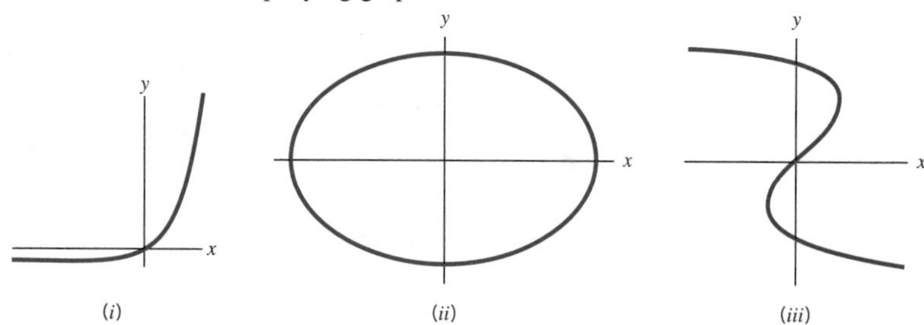

b. Explain your answers in part (a).

52. Which of the following tables describe a function? Explain your answers.

a.

Input Value	−2	−1	0	1	2
Output Value	−8	−1	0	1	8

b.

Input Value	0	1	2	1	0
Output Value	−4	−2	0	2	4

c.

Input Value	10	7	4	7	10
Output Value	3	6	9	12	15

d.

Input Value	0	3	9	12	15
Output Value	3	3	3	3	3

53. The parcel post rate schedule as of January 10, 1999, is shown here for different U.S. zones and weights of parcels. Say whether the following statements are true or not and explain your answers.

Parcel Post Rate Schedule

Weight Up to But	Rates by Zones						
Not Exceeding (lbs)	1 and 2	3	4	5	6	7	8
2	$3.15	$3.15	$3.15	$3.15	$3.15	$3.15	$3.15
3	3.59	3.90	4.25	4.25	4.25	4.25	4.25
4	3.73	4.16	4.91	5.35	5.35	5.35	5.35
5	3.86	4.39	5.33	6.45	6.45	6.45	6.45
6	3.99	4.62	5.71	7.10	7.40	7.60	8.15
7	4.11	4.82	6.07	7.72	8.35	8.75	9.85
8	4.24	5.01	6.38	8.26	9.30	9.90	11.55
9	4.33	5.19	6.71	8.76	10.25	11.05	13.25
10	4.45	5.36	6.99	9.23	10.92	12.20	14.95
11	4.54	5.53	7.27	9.66	11.47	13.30	16.10
12	4.64	5.68	7.53	10.06	11.97	14.30	17.35
13	4.73	5.81	7.77	10.44	12.44	15.17	18.65
14	4.82	5.97	8.01	10.80	12.89	15.74	19.90
15	4.90	6.10	8.24	11.13	13.31	16.28	21.15
16	4.98	6.23	8.45	11.45	13.70	16.77	21.85
17	5.07	6.34	8.66	11.74	14.08	17.25	22.49
18	5.14	6.46	8.85	12.02	14.42	17.69	23.10
19	5.23	6.58	9.04	12.29	14.76	18.12	23.67
20	5.29	6.68	9.20	12.54	15.07	18.52	24.21
21	5.36	6.80	9.37	12.79	15.38	18.90	24.72
22	5.43	6.89	9.54	13.02	15.66	19.26	25.21
23	5.50	7.01	9.71	13.23	15.93	19.60	25.67
24	5.55	7.10	9.85	13.45	16.19	19.94	26.12
25	5.62	7.19	10.01	13.64	16.44	20.24	26.54

Note: Inter BMC/ASF ZIP codes only, machinable parcels, no discount, no surcharge.
Source: The World Almanac and Book of Facts, 1999.

a. Within zones 1 and 2 the cost of mailing a parcel not exceeding 25 pounds is a function of weight.

b. The weight of a parcel is a function of the cost to mail it.

c. For parcels up to 2 pounds the cost of mailing is a function of zone.

d. The zone of a parcel is not a function of its weight.

54. For each of the following tables find a function formula which takes the x values and produces the y values.

a.

x	y
0	0
1	3
2	6
3	9
4	12

b.

x	y
0	−2
1	1
2	4
3	7
4	10

c.

x	y
0	0
1	−1
2	−4
3	−9
4	−16

55. a. Find an equation that represents the relationship between x and y in each of the accompanying tables.

x	y
0	5
1	6
2	7
3	8
4	9

x	y
0	1
1	2
2	5
3	10
4	17

x	y
0	3
1	3
2	3
3	3
4	3

b. Which of your equations represent y as a function of x? Justify your answers.

Exercises for Section 1.6

56. In Section 1.3 we examined the annual federal budget *surplus* or *deficit*. The federal *debt* takes into account the cumulative effect of all the deficits and surpluses for each year together with any interest or payback of principal. The accompanying table and graph show the accumulated gross federal debt from 1945 to 1999. (See related Excel or graph link file FEDDEBT.)

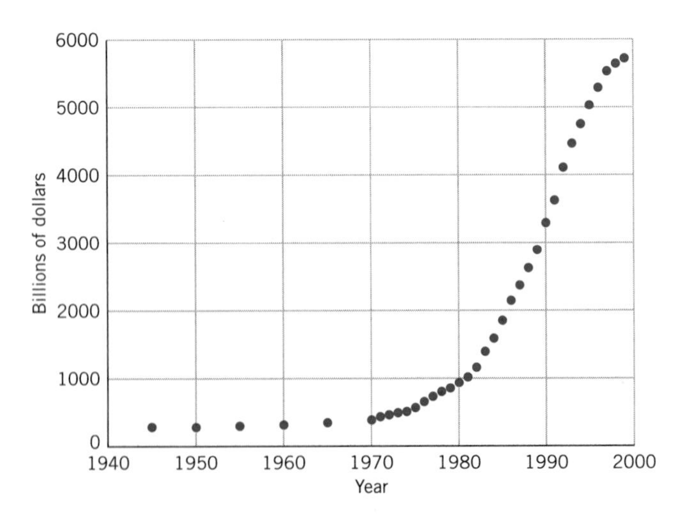

a. Identify which sets of data would be logical choices for the independent variable and dependent variable.

Accumulated Gross Federal Debt

Year	Billions of $	Year	Billions of $	
1945	260	1983	1371	
1950	257	1984	1564	
1955	274	1985	1827	
1960	291	1986	2120	
1965	322	1987	2346	
1970	361	1988	2601	
1971	408	1989	2868	
1972	436	1990	3266	
1973	466	1991	3599	
1974	484	1992	4082	
1975	541	1993	4436	
1976	629	1994	4721	
1977	706	1995	5001	
1978	777	1996	5260	
1979	829	1997	5502	
1980	909	1998	5614	
1981	994	1999	5691	(est)
1982	1137			

Source: U.S. Bureau of the Census, *Statistical Abstract of the United States,* 1998, and the Department of the Treasury, Bureau of the Public Debt On-Line, *www.publicdebt.treas.gov.*

b. Verify whether or not your choice for dependent variable is a function of the independent variable.

If one variable is a function of the other, then:

c. Identify the intervals on which the function is increasing or decreasing.

d. Identify any maximum or minimum values of the function.

57. The accompanying table shows the progress of national regulations in controlling carbon monoxide emissions.

a. Identify which sets of data would be logical choices for the independent variable and dependent variable.

b. Verify whether or not your choice for dependent variable is a function of the independent variable.

If one variable is a function of the other, then:

c. Express in words how the dependent variable relates to the independent variable.

d. Generate a graph.

e. Identify the intervals on which the function is increasing or decreasing.

f. Identify any maximum or minimum values of the function.

National Ambient Air Pollutant Concentrations, 1985–1996

Year	Carbon Monoxide* (ppm)
1985	6.97
1987	6.69
1988	6.38
1989	6.34
1990	5.87
1991	5.55
1992	5.18
1993	4.88
1994	5.10
1995	4.50
1996	4.20

*Air quality standard is 0 parts per million (ppm).

Source: U.S. Bureau of the Census, *Statistical Abstract of the United States,* 1998.

58. The formula

$$A = 25W$$

where A is ampicillin dosage in milligrams and W is the child's weight in kilograms, represents the *minimum* effective pediatric daily drug dosage of ampicillin as a function of a child's weight. (You may recall that $D = 50W$ represented the *maximum* recommended dosage in Algebra Aerobics 1.6a. Both formulas apply only for children weighing up to 10 kilograms.)

a. Identify which variables would be logical choices for the independent and dependent variable.

b. Verify whether or not your choice for dependent variable is a function of the independent variable.

If one variable is a function of the other, then:

c. Express in words how the dependent variable relates to the independent variable.

d. Generate a table of values and a graph.

e. Identify the intervals on which the function is increasing or decreasing.

f. Identify any maximum or minimum values of the function.

Y	P
1990	\$1.4
1991	\$2.3
1992	\$0
1993	− \$0.5
1994	\$1.4
1995	\$1.2

59. Consider the accompanying table.

a. Is P a function of Y?

b. What is the domain? What is the range?

c. What is the maximum value of P? In what year did this occur?

d. During what intervals was P increasing? Decreasing?

e. Now consider P as the independent variable and Y as the dependent variable. Is Y a function of P?

60. Assume that for persons who earn less than \$20,000 a year, income tax is 16% of their income.

a. Generate a formula that describes income tax in terms of income for people earning less than \$20,000 a year.

b. What are you treating as the independent variable? The dependent variable?

c. Does your formula represent a function? Explain.

d. If it is a function, what is the domain? The range?

61. Suppose that the price of gasoline is \$1.24 per gallon.

a. Generate a formula that describes the cost of buying gas as a function of the amount of gasoline purchased.

b. What is the independent variable? The dependent variable?

c. Does your formula represent a function? Explain.

d. If it is a function, what is the domain? The range?

e. Generate a small table of values and a graph.

62. For each equation below, write an equivalent equation that expresses z in terms of t. Is z a function of t? Why or why not?

a. $3t - 5z = 10$ **b.** $12t^2 - 4z = 0$ **c.** $2(t - 4) - (z + 1) = 0$

63. The cost of driving a car to work is estimated to be \$2.00 in tolls plus 32 cents per mile. Write an equation for computing the total cost C of driving M miles to work. Does your equation represent a function? What is the independent variable? What is the dependent variable? Generate a table of values and then graph the equation.

64. For each equation below, write an equivalent equation that expresses y in terms of x.

a. $3x + 5x - y = 3y$ **c.** $3x(5 - x) = x - y$

b. $x(x - 1) + y = 2x - 5$ **d.** $2(y - 1) = y + 5x(x + 1)$

65. If $f(x) = x^2 - x + 2$, find:

a. $f(2)$ **b.** $f(-1)$ **c.** $f(0)$ **d.** $f(-5)$

66. If $g(x) = 2x + 3$, evaluate $g(0)$, $g(1)$, and $g(-1)$.

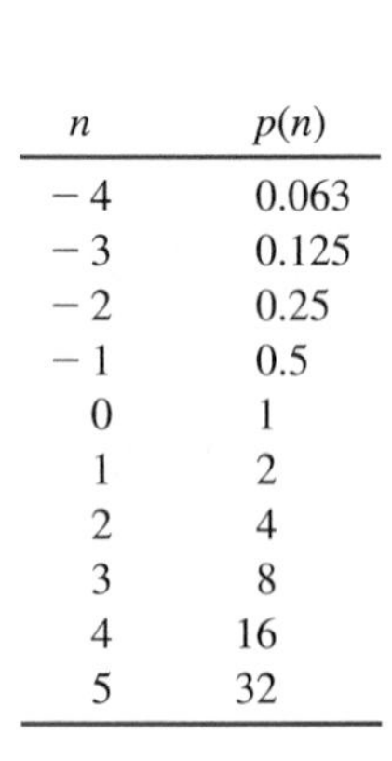

n	$p(n)$
−4	0.063
−3	0.125
−2	0.25
−1	0.5
0	1
1	2
2	4
3	8
4	16
5	32

67. Look at the accompanying table.

a. Find $p(-4)$, $p(5)$, and $p(1)$.

b. For what value(s) of n does $p(n) = 2$?

(68)

68. Consider the function $y = f(x)$ graphed in the accompanying figure.

a. Find $f(-3), f(0), f(1)$, and $f(2.5)$.

b. Find two values of x such that $f(x) = 0$.

69. From the accompanying graph of $y = f(x)$:

a. Find $f(-2), f(-1), f(0)$, and $f(1)$.

b. Find two values of x for which $f(x) = -3$.

c. Estimate the range of f. Assume that the arms of the graph extend upward indefinitely.

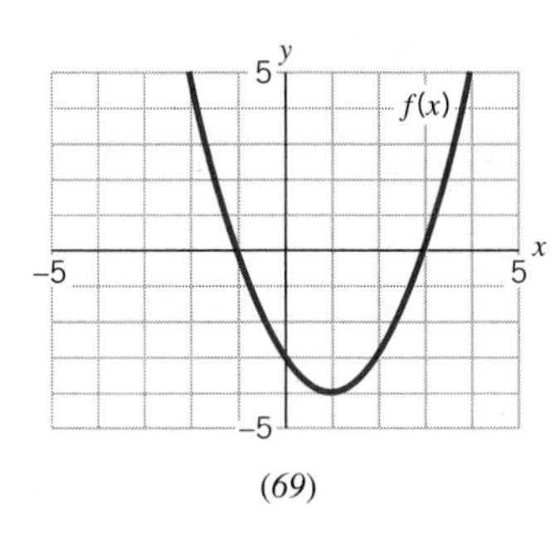

(69)

70. Find $f(3)$, if it exists, for each of the following functions:

a. $f(x) = (x - 3)^2$ **b.** $f(x) = \dfrac{1}{x}$ **c.** $f(x) = \dfrac{x + 1}{x - 3}$ **d.** $f(x) = \dfrac{2x}{x - 1}$

Determine the domain for each function.

71. If $f(x) = (2x - 1)^2$, evaluate $f(0), f(1)$, and $f(-2)$.

72. Find the domain for each of the following functions:

a. $f(x) = 300.4 + 3.2x$ **c.** $j(x) = \dfrac{1}{x + 1}$ **e.** $f(x) = x^2 + 3$

b. $g(x) = \dfrac{5 - 2x}{2}$ **d.** $k(x) = 3$

73. Given $f(x) = 1 - 0.5x$ and $g(x) = x^2 + 1$, evaluate:

a. $f(0) + g(0)$ **b.** $f(-2) - g(-3)$ **c.** $g(2)/f(1)$ **d.** $f(3) \cdot g(3)$

Exploration 1.1

Collecting, Representing, and Analyzing Data

Objectives

- explore issues related to collecting data
- learn techniques for organizing and graphing data using a computer (with a spreadsheet program) or a graphing calculator
- describe and analyze the overall shape of single-variable data using frequency and relative frequency histograms
- use the mean and median to represent single-variable data

Material/Equipment

- class questionnaire
- measuring tapes in centimeters and inches
- optional measuring devices: eye chart, flexibility tester, measuring device for blood pressure
- computer with spreadsheet program or graphing calculator with statistical plotting features
- data from class questionnaire or other small data set either as spreadsheet or graph link file
- overhead projector and projection panel for computer or graphing calculator
- transparencies for printing or drawing graphs for overhead projector (optional)

Related Readings

"U.S. Government Definitions of Census Terms" (In Anthology of Readings)
"Health Measurements" (On the web at www.wiley.com/college/kimeclark)

Related Software 

"F1: Histograms," in *FAM 1000 Census Graphs*

Procedure

This exploration may take two class periods.

Day One

In a Small Group or with a Partner

1. Pick (or your instructor will assign you) one of the undefined variables on the questionnaire. Spend about 15 minutes coming up with a workable definition and a strategy for measuring that variable. Be sure there is a way in which a number or single letter can be used to record each individual's response on the questionnaire.
2. Consult the reading "Health Measurements" if you decide to collect health data.

Class Discussion

After all of the definitions are recorded on the board, discuss your definition with the class. Is it clear? Does everyone in the class fall into one of the categories of your definition? Can anyone think of someone who might not fit into any of the categories? Modify the definition until all can agree on some wording. As a class, decide on the final version of the questionnaire. Write down the final definitions in your class notebook.

In a Small Group or with a Partner

Help each other when necessary to take measurements and fill out the entire questionnaire. Questionnaires remain anonymous, and you can leave blank any question you can't or don't want to answer. Hand in your questionnaire to your instructor by the end of class.

Exploration-Linked Homework

Read "U.S. Government Definitions of Census Terms" for a glimpse into the federal government's definitions of the variables you defined in class. How do the "class" definitions and the "official" ones differ?

Day Two

Class Demonstration

1. If you haven't used a spreadsheet or graphing calculator before you'll need a basic technical introduction. (Note: if you are using a TI-82 or TI-83 graphing calculator there are basic instructions in the Graphing Calculator Manual.) Then you'll need an electronic version of the data set from which you will choose one variable for the whole class to study (e.g., age from the class data).
   a. If you're using a spreadsheet:
   - Copy the column with the data onto a new spreadsheet and graph the data. What does this graph tell you about the data?
   - Sort the data and plot it again. Is this graph any better at conveying information about the data?

   b. If you're using a graphing calculator:
   - Discuss window sizes, changing interval sizes and statistical plot procedures.
2. Select an interval size and then construct a frequency histogram and a relative frequency histogram. If possible, label one of these carefully and print it out. If you have access to a laser printer, you can print onto an overhead transparency.
3. Calculate the mean and median using software functions or graphing calculator functions.

In a Small Group or with a Partner

Choose another variable from your data. Pick an interval size, and then generate both a frequency histogram and a relative frequency histogram. If possible, make copies of the histograms for both your partner and yourself. Calculate the mean and median.

Discussion/Analysis

With your partner(s), analyze and jot down patterns that emerge from the data. What are some limitations of the data? What other questions are raised and how might they be resolved? In your notebook, record jottings for a 60 second summary that would describe your results.

Exploration-Linked Homework

Prepare a verbal 60 second summary to give to the class. If possible, use an overhead projector with a transparency of your histogram or a projector linked to your graphing calculator. If not, bring in a paper copy of your histogram. Construct a written 60 second summary. (See Section 1.2 for some writing suggestions.)

ALGEBRA CLASS QUESTIONNAIRE

(You may leave any category blank)

1. Age (in years)
2. Sex (female = 1, male = 2)
3. Your height (in inches)
4. Distance from your navel to the floor (in centimeters)
5. Estimate your average travel time to school (in minutes)
6. What is your most frequent mode of transportation to school?
   (F = by foot, C = car, P = public transportation, B = bike, O = other)

The following variables will be defined in class. We will discuss ways of coding possible responses and then use the results to record our personal data.

7. The number of people in your household
8. Your employment status
9. Your ethnic classification
10. Your attitude toward mathematics

Health Data

11. Your pulse rate before jumping (beats per minute)
12. Your pulse rate after jumping for 1 minute (beats per minute)
13. Blood pressure: systolic (mm Hg)
14. Blood pressure: diastolic (mm Hg)
15. Flexibility (in inches)
16. Vision, left eye
17. Vision, right eye

Other Data

Exploration 1.2

Picturing Functions

Objective

- develop an intuitive understanding of functions

Material/Equipment

none required

Procedure

Part I

Class Discussion

Bridget, the 6-year-old daughter of a professor at the University of Pittsburgh, loves playing with her rubber duckie in the bath at night. Her mother drew the accompanying graph for her math class. It shows the water level (measured directly over the drain) in Bridget's tub as a function of time.

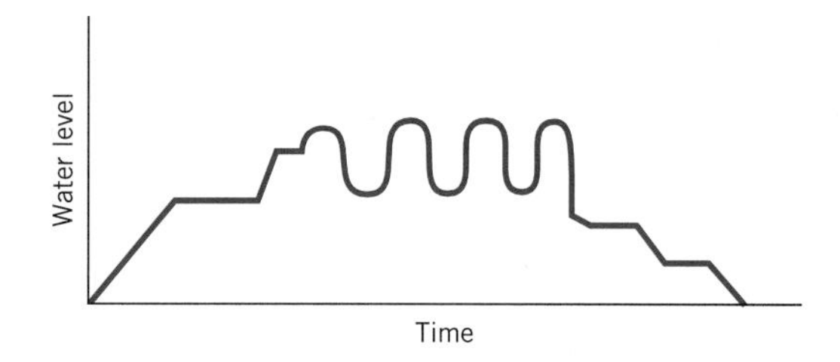

Pick out the time period during which:

- The tub is being filled
- Bridget is entering the tub
- She is playing with her rubber duckie
- She leaves the tub
- The tub is being drained

With a Partner

Create your own graph of a function that tells a story. Be as inventive as possible. Some students have drawn functions that showed the decibel levels during a phone conversation of a boy and girl friend, number of hours spent doing homework during 1 week, and amount of money in their pocket during the week.

Class Discussion

Draw your graph on the blackboard and tell its story to the class.

Part II

With a Partner

Generate a plausible graph for each of the following:

1. Time spent driving to work as a function of the amount of snow on the road. (Note: The first inch or so may not make any difference; the domain may be only up to about a foot of snow since after that you may not be able to get to work.)
2. The hours of daylight as a function of the time of year.
3. The temperature of the coffee in your cup as a function of time.
4. The distance that a cannonball (or javelin or baseball) travels as a function of the angle of elevation at which is it is launched. The maximum distance is attained for angles of around 45° from the horizontal.
5. Assume that you leave your home walking at a normal pace, realize you have forgotten your homework and run home, and then run even faster to school. You sit for a while in a classroom and then walk leisurely home. Now plot your distance from home as a function of time.

Bonus Question

Assume that water is pouring at a constant rate into each of the containers shown. The height of water in the container is a function of the volume of liquid. Sketch a graph of this function for each container.

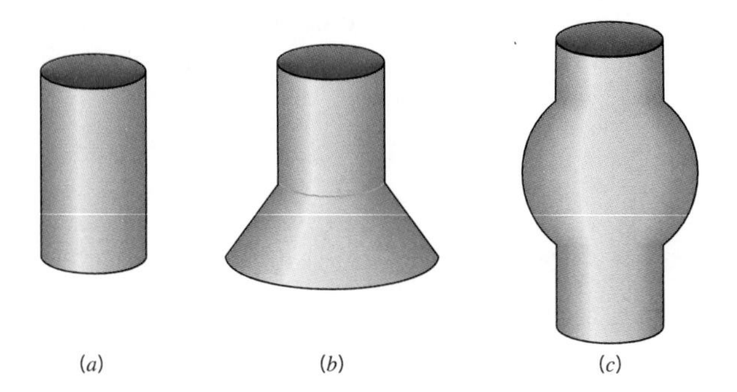

Discussion/Analysis

Are your graphs similar to those generated by the rest of the class? Can you agree as a class as to the basic shape of each of the graphs? Are there instances in which the graphs could look quite different?

Exploration 1.3

Deducing Formulas to Describe Data

Objectives

- find and describe patterns in data
- deduce functional formulas from data tables
- extend patterns using functional formulas

Material/Equipment

None required

Procedure

Class discussion

1. Examine data tables (a) and (b). In each table, look for a pattern in terms of how y changes when x changes. Explain in your own words how to find y in terms of x.

a.

x	y
0	0.0
1	0.5
2	1.0
3	1.5
4	2.0

b.

x	y
0	5
1	8
2	11
3	14
4	17

2. Assuming that the pattern continues indefinitely, use the rule you have found to extend the data table to include negative numbers for x.
3. Check your extended data tables. Did you find only one value for y given a particular value for x?
4. Use a formula to describe the pattern that you have found. Do you think this formula describes a function? Explain.

On Your Own

1. For each of the data tables (c) to (h), explain in your own words how to find y in terms of x. Then extend each table using the rules you have found.

c.

x	y
0	0
1	1
2	4
3	9
4	16

d.

x	y
0	0
1	1
2	8
3	27
4	64

e.

x	y
0	0
1	2
2	12
3	36
4	80

(*Hint:* For table (e) think about some combination of data in tables (c) and (d).

f.

x	y
-2	0
0	10
5	35
10	60
100	510

g.

x	y
0	-1
1	0
2	3
3	8
4	15

h.

x	y
0	3
10	8
20	13
30	18
100	53

2. For each table, construct a formula to describe the pattern you have found.

Discussion/Analysis

With a Partner

Compare your results. Do the formulas that you have found describe functions? Explain.

Class Discussion

Does the rest of the class agree with your results? Remember that formulas that look different may give the same results.

Exploration-Linked Homework

1. a. For data tables (i) and (j), explain in your own words how to find y in terms of x. Using the rules you have found, extend the data tables to include negative numbers.

i.

x	y
-10	10.0
0	0.0
3	0.9
8	6.4
10	10.0

j.

x	y
0	-3
1	1
2	5
3	9
4	13

b. For each table, find a formula to describe the pattern you have found. Does your formula describe a function? Explain.

2. Make up a functional formula, generate a data table, and bring the data table on a separate piece of paper to class. The class will be asked to find your rule and express it as a formula.

k.

x	y

Rates of Change and Linear Functions

Overview

How does the U.S. population change over time? How do children's heights change as they age? Average rates of change provide a tool for measuring how change in one variable affects a second variable. When average rates of change are constant, the relationship is linear. By judiciously choosing the endpoints we use to calculate the average rate of change, and by manipulating graphs and descriptions, we can distort the impression data convey.

After reading this chapter you should be able to:

- calculate and interpret average rates of change
- understand how representations of data can be biased
- recognize that a constant rate of change denotes a linear relationship
- construct a linear function given a table, graph, or description
- derive by hand a linear model for a set of data

2.1 AVERAGE RATES OF CHANGE

In Chapter 1, we looked at how change in one variable could affect change in a second variable. In this section we'll examine how to measure that change.

Describing Change in the U.S. Population over Time

See Excel or graph link file USPOP.

We can think of the U.S. population as a function of time. Table 2.1 and Figure 2.1 are two representations of that function. They show the changes in the size of the U.S. population since 1790, the year the U.S. government conducted its first decennial census. Time, as usual, is the independent variable and population size is the dependent variable.

Table 2.1
Population of the United States: 1790–2000

Year	In Millions
1790	3.9
1800	5.3
1810	7.2
1820	9.6
1830	12.9
1840	17.1
1850	23.2
1860	31.4
1870	39.8
1880	50.2
1890	62.9
1900	76.0
1910	92.0
1920	105.7
1930	122.8
1940	131.7
1950	151.9
1960	180.0
1970	204.0
1980	227.2
1990	249.4
2000	274.6 (est.)

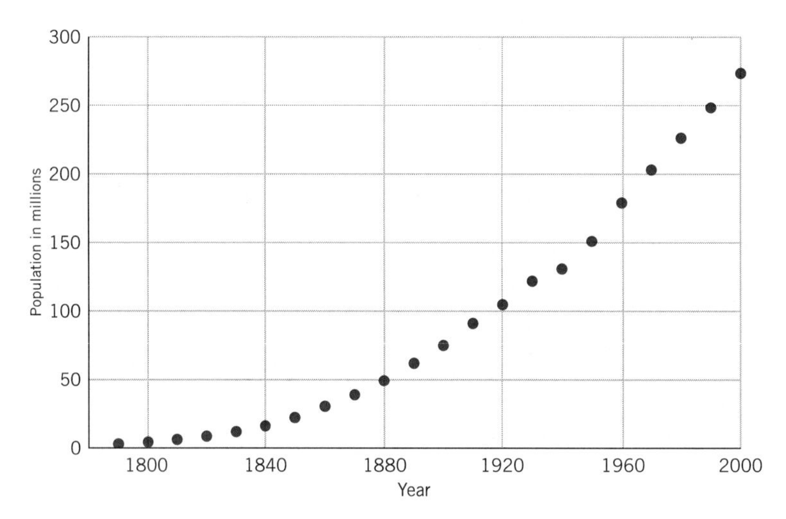

Figure 2.1 Population of the United States.
Source: U. S. Bureau of the Census, Current Population Reports, *Statistical Abstract of the United States,* 1998.

See "C1: U.S. Population" in *Rates of Change.*

Figure 2.1 clearly shows that the size of the U.S. population has been growing over the last two centuries, and growing at what looks like an increasingly rapid rate. How can the change in population over time be described quantitatively? One way is to pick two points on the graph of the data and calculate how much the population has changed during the time period between them.

Suppose we look at the change in the population between 1900 and 1990. In 1900 the population was 76.0 million; by 1990 the population had grown to 249.4 million. How much did the population increase? Since it rose from 76.0 million to 249.4 million, it increased by the difference between these two values.

$$\begin{aligned}\text{change in population} &= (249.4 - 76.0) \text{ million people}\\ &= 173.4 \text{ million people}\end{aligned}$$

In Figure 2.2, we have drawn two parallel horizontal lines through the points (1900, 76.0) and (1990, 249.4) across to the vertical population axis. The 173.4 million population change is represented by the difference in the height of the two points or the change in the variable on the vertical axis.

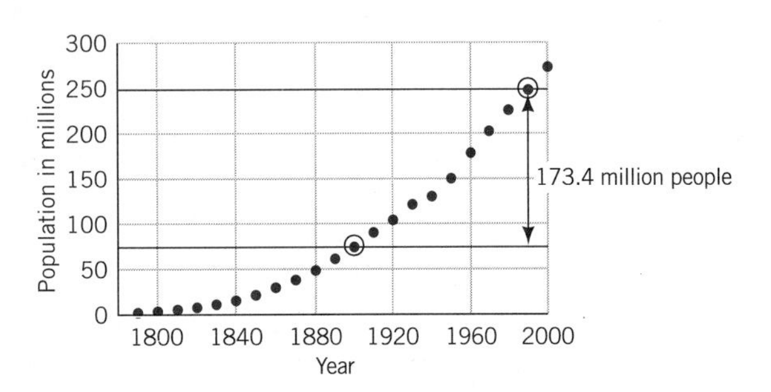

Figure 2.2 Population change: 173.4 million people.

Knowing that the population changed by 173.4 million tells us nothing about how rapid the change was; this change clearly represents much more dramatic growth if it happened over 20 years than if it happened over 200 years. In this case, the length of time over which the change in population occurred is

$$\begin{aligned}\text{change in years} &= (1990 - 1900) \text{ years}\\ &= 90 \text{ years}\end{aligned}$$

In Figure 2.3 we have drawn two parallel vertical lines from the points (1900, 76.0) and (1990, 249.4) down to the horizontal time axis. The 90-year change is represented by the horizontal difference in the positions of the two points or the change in the variable on the horizontal axis.

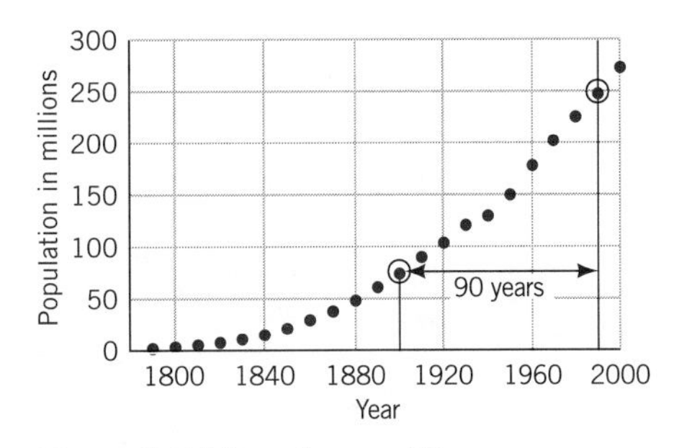

Figure 2.3 Time change: 90 years.

If a line segment is drawn connecting the two points, it forms the hypotenuse of the right triangle sketched in Figure 2.4. The length of the horizontal section of the triangle represents a change of 90 years, and the length of the vertical section of the triangle represents a change of 173.4 million in population size.

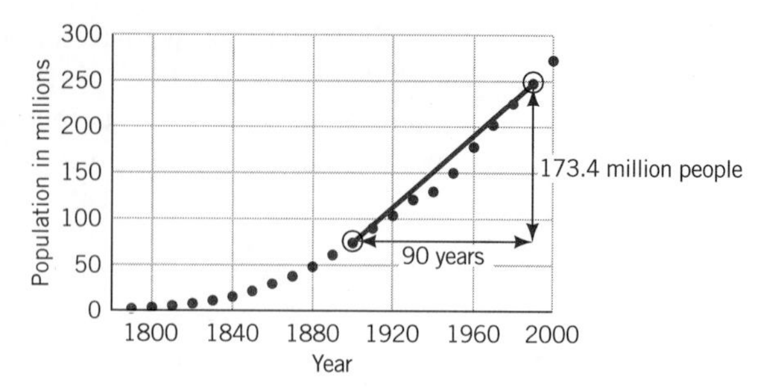

Figure 2.4 Population change over 90 years.

To find the *average rate of change* in population per year from 1900 to 1990, divide the change in the population by the change in years:

$$\begin{aligned}\text{average rate of change} &= \frac{\text{change in population}}{\text{change in years}}\\ &= \frac{173.4 \text{ million people}}{90 \text{ years}}\\ &\approx 1.93 \text{ million people/year}\end{aligned}$$

In the phrase "million people/year" the slash sign represents division and is read as "per." So our calculation shows that "on average" the population grew at a rate of 1.93 million people per year from 1900 to 1990.

Defining the Average Rate of Change

The notion of the average rate of change can be used to describe the change in any variable with respect to another. If you have a graph that represents a plot of data points of the form (x, y), then the average rate of change between any two points is the change in the y value divided by the change in the x value.

> The *average rate of change* of y with respect to $x = \dfrac{\text{change in } y}{\text{change in } x}$

If the variables represent real world quantities which have units of measure (e.g., millions of people or years), then the average rate of change should be represented in terms of the appropriate units:

$$\text{units of the average rate of change} = \frac{\text{units of } y}{\text{units of } x}$$

For example, the units might be dollars/year (read as dollars per year) or pounds/person (read as pounds per person).

Example 1

Between 1850 and 1950 the median age in the United States rose from 18.9 to 30.2, but by 1970 it had dropped to 28.0.

a. Calculate the average rate of change in the median age between 1850 and 1950.

b. Compare your answer in part (a) to the average rate of change between 1950 and 1970.

Solution

a. Between 1850 and 1950,

$$\begin{aligned}\text{average rate of change} &= \frac{\text{change in median age}}{\text{change in years}}\\ &= \frac{(30.2 - 18.9)\text{ years}}{(1950 - 1850)\text{ years}} = \frac{11.3\text{ years}}{100\text{ years}}\\ &= 0.113\text{ year/year}\end{aligned}$$

The units are a little confusing. But the results mean that between 1850 and 1950 the median age increased by 0.113 years each year.

b. Between 1950 and 1970,

$$\begin{aligned}\text{average rate of change} &= \frac{\text{change in median age}}{\text{change in years}}\\ &= \frac{(28.0 - 30.2)\text{ years}}{(1970 - 1950)\text{ years}} = \frac{-2.2\text{ years}}{20\text{ years}}\\ &= -0.110\text{ year/year}\end{aligned}$$

Note that since the median age dropped in value between 1950 and 1970, the average rate is negative. In particular, the median age decreased by 0.110 year each year.

Limitations of the Average Rate of Change

The average rate of change is an average. Average rates of change have the same limitations as any *average.* Although the average rate of change of the U.S. population from 1900 to 1990 was 1.93 million people/year, it is highly unlikely that in each year the population grew by exactly 1.93 million. The number 1.93 million people/year is, as the name states, an average. Similarly, if the arithmetic average, or *mean,* height of students in your class is 67 inches, you wouldn't expect every student to be 67 inches tall.

The average rate of change depends on the end points. If the data points do not all lie on a straight line, the average rate of change varies for different intervals. For instance, the average rate of change in population for the time interval 1840 to 1940 is 1.15 million people/year and from 1880 to 1980 is 1.77 million people/year. (See Table 2.2. Note: here we abbreviate "million people" as "million.") You can see on the graphs that the line segment is much steeper from 1880 to 1980 than from 1840 to 1940 (Figures 2.5 and 2.6). Different intervals give different impressions of the rate of change in the U.S. population, so it is important to state which end points are used.

Table 2.2

Time Interval	Change in Time	Change in Population	Average Rate of Change
1840–1940	100 yr	$131.7 - 17.1 = 114.6$ million	$\dfrac{114.6 \text{ million}}{100 \text{ yr}} \approx 1.15$ million/yr
1880–1980	100 yr	$227.2 - 50.2 = 177.0$ million	$\dfrac{177.0 \text{ million}}{100 \text{ yr}} \approx 1.77$ million/yr

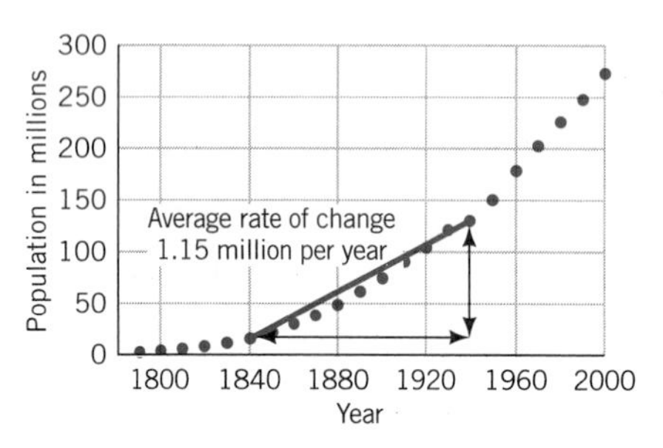

Figure 2.5 Average rate of change: 1840–1940.

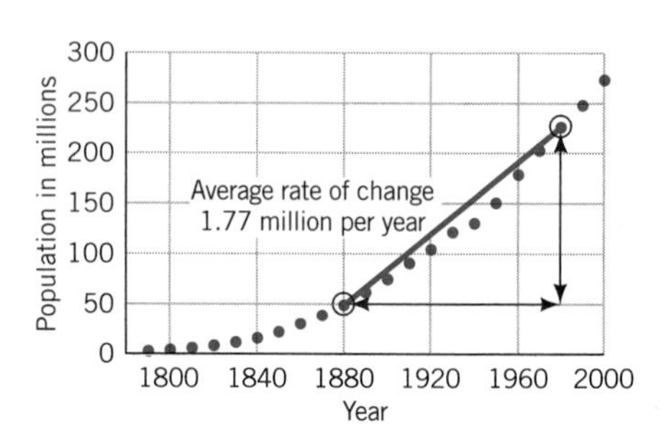

Figure 2.6 Average rate of change: 1880–1980.

The average rate of change does not reflect all the fluctuations in population size that may occur between the end points. For more specific information, the average rate of change can be calculated for smaller intervals.

Algebra Aerobics 2.1

1. Suppose your weight five years ago was 135 pounds and your weight today is 143 pounds. Find the average rate of change in your weight with respect to time.
2. The U.S. general imports totaled \$581 trillion in 1983 and \$871 trillion in 1997. Find the average rate of change from 1983 to 1997 in the value of general imports.
3. Table 2.3 indicates the number of deaths in motor vehicle accidents in the United States.

Table 2.3
Annual Deaths in Motor Vehicle Accidents (thousands)

1970	1980	1990	1996
114.6	105.7	92.0	93.9

Find the average rate of change:

a. From 1970 to 1980

b. From 1980 to 1996

2.2 CHANGE IN THE AVERAGE RATE OF CHANGE

We can obtain an even better sense of patterns in the U.S. population if we look at how the average rate of change varies over time. One way to do this is to pick a fixed interval for time and then to calculate the average rate of change for each successive time period. Since we have the U.S. population data in 10-year intervals, we can calculate the average rate of change for each successive decade. The third column in Table 2.4 shows the results of these calculations. Each entry represents the average population growth *per year* (the average annual rate of change) during the previous decade. A few of these calculations are worked out in the last column of the table.

Table 2.4
Average Annual Rates of Change of U.S Population: 1790–2000

Year	Population (millions)	Average Annual Rate for Prior Decade (millions/yr)	Sample Calculations
1790	3.9	Data not available	
1800	5.3	0.14	0.14 = (5.3 − 3.9)/(1800 − 1790)
1810	7.2	0.19	
1820	9.6	0.24	
1830	12.9	0.33	
1840	17.1	0.42	0.42 = (17.1 − 12.9)/(1840 − 1830)
1850	23.2	0.61	
1860	31.4	0.82	
1870	39.8	0.84	
1880	50.2	1.04	
1890	62.9	1.27	
1900	76.0	1.31	
1910	92.0	1.60	
1920	105.7	1.37	
1930	122.8	1.71	
1940	131.7	0.89	0.89 = (131.7 − 122.8)/(1940 − 1930)
1950	151.9	2.02	
1960	180.0	2.81	
1970	204.0	2.40	
1980	227.2	2.32	
1990	249.4	2.22	
2000	274.6 (est)	2.52	

What is happening to the average rate of change over time? Start at the top of the third column and scan down the numbers. Notice that until 1910 the average rate of change increases every year. Not only is the population growing every decade, but it is growing at an increasing rate. It's like a car that is not only moving forward but also accelerating. A feature that was not so obvious in the original data is now evident: in the intervals 1910 to 1920, 1930 to 1940, and 1960 to 1990 we see an increasing population but a decreasing rate of growth. It's like a car decelerating—it is still moving forward but it is slowing down.

The graph in Figure 2.7 with years on the horizontal axis and average rates of change on the vertical axis shows more clearly how the average rate of change fluctuates over time. The first point, corresponding to the year 1800, shows an average rate of change of 0.14 million people/year for the decade 1790 to 1800. The rate 1.71, corresponding to the year 1930, means that from 1920 to 1930 the population was increasing at a rate of 1.71 million people/year.

What might be some reasons for the slowdown in population growth from the 1960s through the 1980s?

Take the time to interpret a few points; this graph is somewhat more abstract than the previous ones. The pattern of growth was fairly steady up until about 1910. Why did it change? A possible explanation for the slowdown in the decade prior to 1920 might be World War I and the 1918 flu epidemic, which by 1920 had killed nearly 20,000,000 people, including about 500,000 Americans.

In Figure 2.7, the steepest decline in the average rate of change is between 1930 and 1940. One obvious suspect for the big slowdown in population growth in the 1930s is the Great Depression. Look back at Figure 2.1, the original graph that shows the overall growth in the U.S. population. The decrease in the average rate of

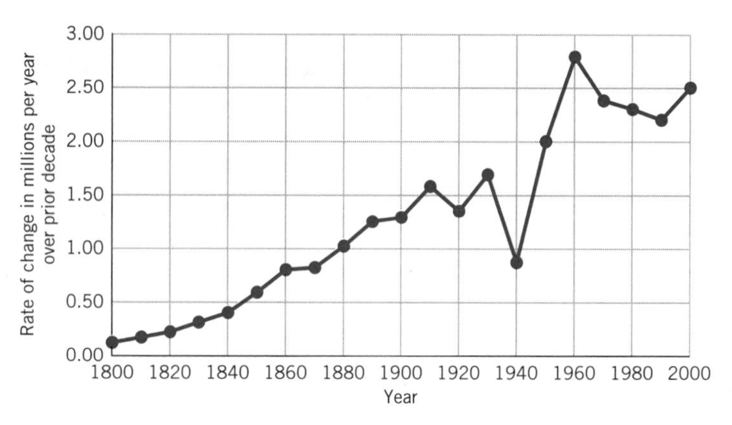

Figure 2.7 Average rates of change in the U.S. population by decade.

change in the 1930s is large enough to show up in our original graph as a visible slowdown in population growth.

The average rate of change increases again between 1940 and 1960, then drops off from the 1960s through the 1980s. The rate increases once more in the 1990s. This latest surge in the growth rate is attributed partially to the "baby boom echo" (the result of baby boomers having children) and to a rise in birth rates and immigration.

Algebra Aerobics 2.2

1. Table 2.5 and Figure 2.8 show estimates for world population between 1800 and 2050.
   a. Fill in the third column of the table by calculating the average annual rate of change.

Table 2.5
World Population

Year	Total Population (millions)	Average Rate of Change
1800	980	n.a.
1850	1260	
1900	1650	
1950	2520	
2000	6060	
2050	8910 (est)	

Source: United National Population Division of the United Nations, *www.popin.org*.

   b. Graph the average annual rate of change versus time.
   c. During what period was the average annual rate of change the largest?
   d. Describe in general terms what happened to the average rate of change between 1800 and 2050.

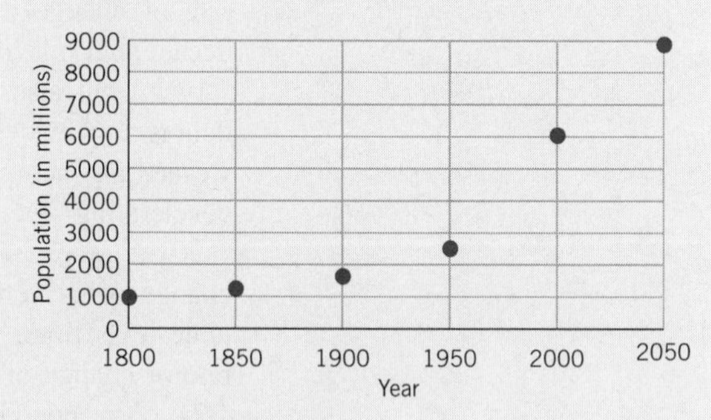

Figure 2.8 World population.

2.3 THE AVERAGE RATE OF CHANGE IS A SLOPE

On a graph, the average rate of change is the *slope* of the line connecting two points. The slope is an indicator of the steepness of the line.

If (x_1, y_1) and (x_2, y_2) are two points, then the change in y equals $y_2 - y_1$ (Figure 2.9). This difference is often denoted by Δy, read as "delta y", where Δ is the Greek letter capital D (think of D as representing difference): $\Delta y = y_2 - y_1$. Simi-

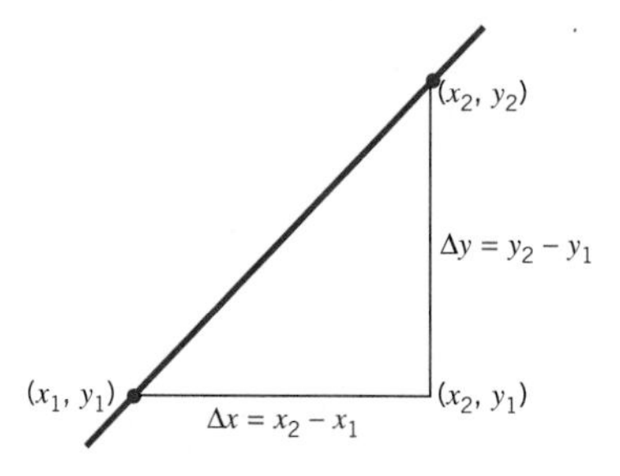

Figure 2.9 Slope = $\Delta y/\Delta x$.

The reading "Slopes" describes many of the practical applications of slopes, from cowboy boots to handicap ramps.

larly, the change in x (delta x) can be represented by $\Delta x = x_2 - x_1$. Then

$$\text{slope} = \frac{y_2 - y_1}{x_2 - x_1} = \frac{\Delta y}{\Delta x} = \frac{\text{change in } y}{\text{change in } x}$$

The average rate of change represents a *slope*. Given two points (x_1, y_1) and (x_2, y_2),

$$\text{average rate of change} = \frac{\text{change in } y}{\text{change in } x} = \frac{\Delta y}{\Delta x} = \frac{y_2 - y_1}{x_2 - x_1} = \text{slope}$$

A note about calculating slopes: it doesn't matter which point is first. Given two points, (x_1, y_1) and (x_2, y_2), it doesn't matter which one we use as the first point when we calculate the slope. In other words we can calculate the slope between (x_1, y_1) and (x_2, y_2) as

$$\frac{y_2 - y_1}{x_2 - x_1} \quad \text{or as} \quad \frac{y_1 - y_2}{x_1 - x_2}$$

Both calculations result in the same value.

We can show both forms are equivalent.

$$\text{slope} = \frac{y_2 - y_1}{x_2 - x_1}$$

multiply by $\frac{-1}{-1}$ $\quad = \frac{-1}{-1} \cdot \frac{y_2 - y_1}{x_2 - x_1}$

simplify $\quad = \frac{-y_2 + y_1}{-x_2 + x_1}$

rearrange terms $\quad = \frac{y_1 - y_2}{x_1 - x_2}$

In calculating the slope, we do need to be consistent in the order in which the coordinates appear in the numerator and the denominator. If y_1 is the first term in the numerator, then x_1 must be the first term in the denominator.

Example 1

Given the two points $(-2, -6)$ and $(7, 12)$, we can calculate the slope of the line passing through them. Treating $(-2, -6)$ as (x_1, y_1) and $(7, 12)$ as (x_2, y_2) (Figure 2.10), then

$$\text{slope} = \frac{y_2 - y_1}{x_2 - x_1} = \frac{12 - (-6)}{7 - (-2)} = \frac{18}{9} = 2$$

Or, we could have used -6 and -2 as the first terms in the numerator and denominator, respectively:

$$\text{slope} = \frac{y_1 - y_2}{x_1 - x_2} = \frac{-6 - 12}{-2 - 7} = \frac{-18}{-9} = 2$$

Either way we obtain the same answer.

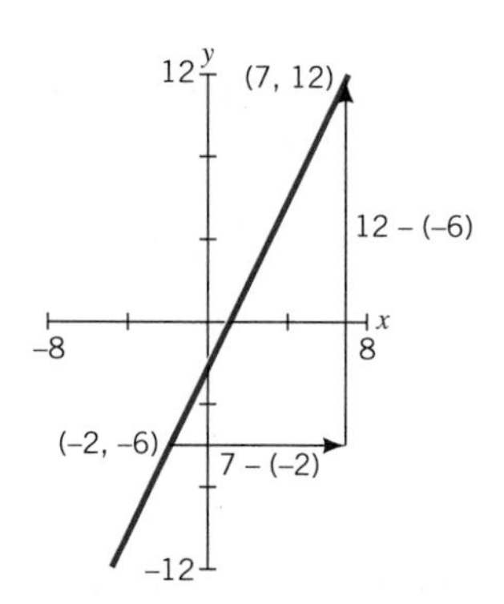

Figure 2.10 $(y_1 - y_2)/(x_1 - x_2) = (y_2 - y_1)/(x_2 - x_1)$.

Example 2

Construct a graph of the data in Table 2.6. Calculate the average rate of change of the percentage of the U.S. population living in rural areas from 1850 to 1940 and interpret the result.

Table 2.6
Percentage of U.S. Population Living in Rural Areas

Year	%	Year	%	Year	%	Year	%	Year	%
1850	84.7	1880	71.8	1910	54.3	1940	43.5	1970	26.4
1860	80.2	1890	64.9	1920	48.8	1950	36.0	1980	26.3
1870	74.3	1900	60.3	1930	43.8	1960	30.1	1990	24.8

Source: U.S. Bureau of the Census. Historical Statistics—Colonial Times to 1970, *Statistical Abstract of the United States,* 1993, 1998.

Solution

The data are graphed in Figure 2.11.

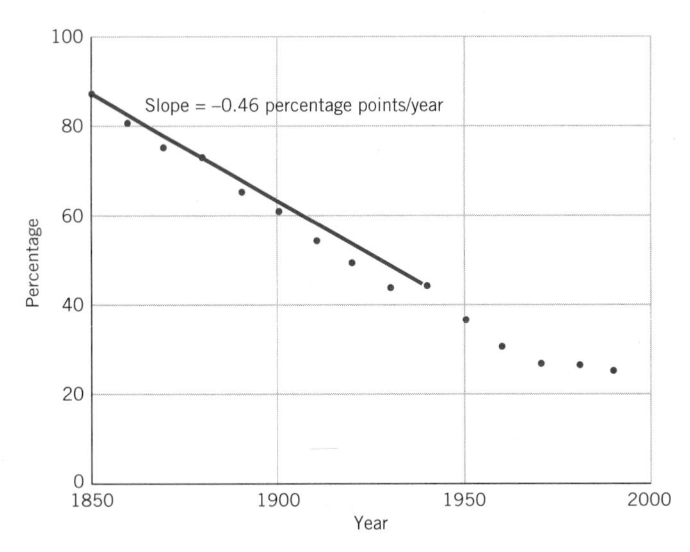

Figure 2.11 Percentage of the U.S. population living in rural areas.

A line drawn through the two end points (1850, 84.7%) and (1940, 43.5%) slopes down, since the percentage of the population that is rural is *declining.* Here the slope, or average rate of change, is negative:

$$\text{average rate of change} = \frac{\text{change in percent of population that is rural}}{\text{change in years}}$$

$$= \frac{43.5\% - 84.7\%}{1940 - 1850}$$

$$= \frac{-41.2\%}{90 \text{ yr}}$$

$$\approx -0.46\%/\text{yr}$$

Were 1850 and 1940 good choices for end points, or would another time period have given an average rate of change (or slope) more typical of all the data? What kind of social and economic implications does a population shift of this magnitude have on society?

This means that the percentage of the population that lived in rural areas decreased on the average by 0.46 percentage point (a little under one-half of a percent) each year from 1850 to 1940. That may not seem like very much on an annual basis, but in less than a century the rural population went from being the large majority (84.7%) to less than half of the U.S. population (43.5%). By 1990 the rural population represented less than 25% of the total population.

> If the slope, or rate of change, of y with respect to x is *positive,* then the graph of the relationship rises up when read from left to right. This means that as x increases in value, y also increases in value.
>
> If the slope is *negative,* the graph falls when read from left to right. As x increases, y decreases.
>
> If the slope is *zero,* the graph is flat. As x increases, there is no change in y.

Example 3

Given Table 2.7 of civil disturbances over time, plot and then connect the points, and (without doing any calculations) indicate on the graph when the average rate of change between adjacent data points is positive (+), negative (−), or zero (0).

Table 2.7
Civil Disturbances in U.S. Cities

	Period	Number of Disturbances
1968	Jan.–Mar.	6
	Apr.–June	46
	July–Sept.	25
	Oct.–Dec.	3
1969	Jan.–Mar.	5
	Apr.–June	27
	July–Sept.	19
	Oct.–Dec.	6
1970	Jan.–Mar.	26
	Apr.–June	24
	July–Sept.	20
	Oct.–Dec.	6
1971	Jan.–Mar.	12
	Apr.–June	21
	July–Sept.	5
	Oct.–Dec.	1
1972	Jan.–Mar.	3
	Apr.–June	8
	July–Sept.	5
	Oct.–Dec.	5

Solution

The data are graphed in Figure 2.12. Each line segment is labeled +, −, or 0, indicating whether the average rate of change between adjacent points is positive, negative, or zero. The largest positive average rate of change, or steepest upward slope, seems to be between the January to March and April to June counts in 1968. The largest negative average rate of change, or steepest downward slope, appears later in the same year (1968) between the July to September and October to December counts.

Civil disturbances between 1968 and 1972 occurred in cycles: The largest numbers occurred in the summer months and the smallest in the winter months. The peaks decrease over time. What was happening in America that might correlate with the peaks? This was a tumultuous period in our history. Many previously silent factions of society were finding their voices. Recall that in April 1968 Martin Luther King was assassinated and that in January 1973 the last American troops were withdrawn from Vietnam.

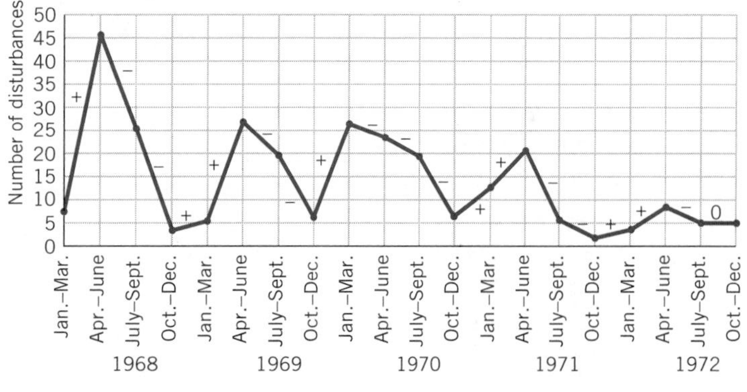

Figure 2.12 Civil disturbances: 1968–1972.
Source: D. S. Moore and G. P. McCabe, *Introduction to the Practice of Statistics.*

Algebra Aerobics 2.3

1. a. Plot each pair of points and then calculate the slope of the line that passes through them.
   i. (4, 1) and (8, 11)   iii. (0, − 3) and (− 5, − 1)
   ii. (− 3, 6) and (2, 6)

   b. Recalculate the slopes in part (a), reversing the order of the points. Check that your answers are the same.

2. Specify the intervals on the graph in Figure 2.13 for which the average rate of change between adjacent data points is approximately zero.

3. Specify the intervals on the graph in Figure 2.14 for which the average rate of change between adjacent data points appears positive, negative, or zero.

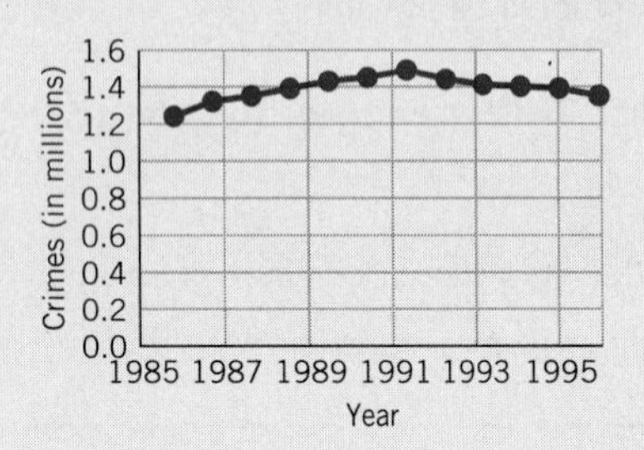

Figure 2.13 Violent crimes in the United States.

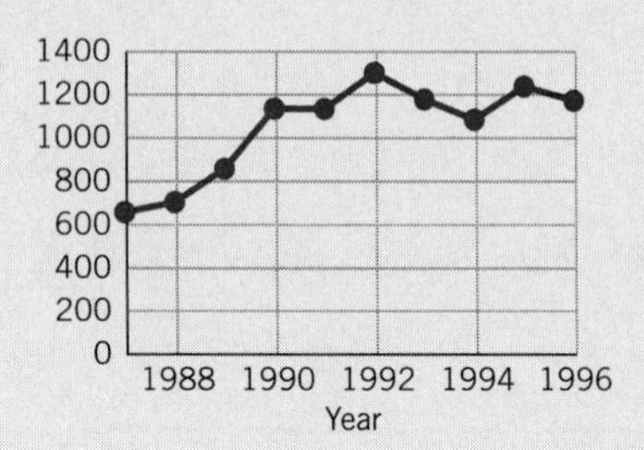

Figure 2.14 Tornadoes: 1987–1996.

2.4 PUTTING A SLANT ON DATA

Whenever anyone summarizes a set of data, choices are being made. One choice may not be more "correct" than another. But these choices can convey, either accidentally or on purpose, very different impressions.

Slanting the Slope: Choosing Different End Points

Within the same data set, one choice of end points may paint a rosy picture, while another choice may portray a more pessimistic outcome.

Example 1

The data in Table 2.8 and the scatter plot in Figure 2.15 show median family income during the 1980s and 1990s. How could we use the information to make a case that families are better off? Worse off?

Table 2.8
Median Household Income (constant 1996 $)

Year	Income	Year	Income
1986	$35,642	1992	$34,261
1987	$35,994	1993	$33,922
1988	$36,108	1994	$34,158
1989	$36,575	1995	$35,082
1990	$35,945	1996	$35,492
1991	$34,705		

Source: U.S. Bureau of the Census, *Statistical Abstract of the United States,* 1998.

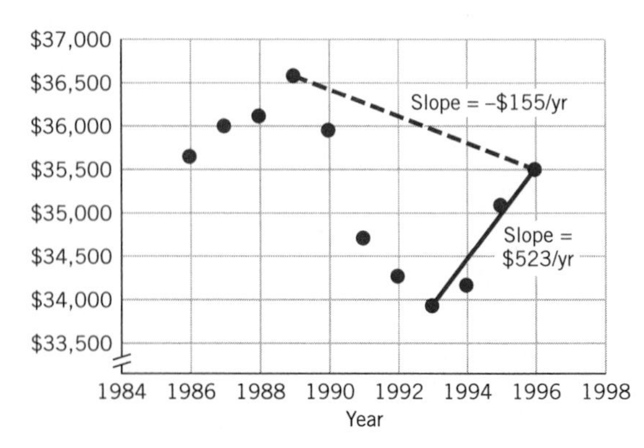

Figure 2.15 Median household income (1996 dollars).

Solution

To make an optimistic case that households are better off, we could choose as end points (1993, \$33922) and (1996, \$35492). Then

$$
\begin{aligned}
\text{average rate of change} &= \frac{\text{change in median income}}{\text{change in years}} \\
&= \frac{\$35{,}492 - \$33{,}922}{1996 - 1993} \\
&= \frac{\$1570}{3 \text{ yr}} \\
&\approx \$523/\text{yr}
\end{aligned}
$$

So between 1993 and 1996 the median household income *increased* on average by \$523 per year. We can see this reflected in Figure 2.15 in the positive slope of the line connecting (1993, \$33922) and (1996, \$35492).

A gloomy interpretation of the same data could be constructed merely by changing the left-hand end point to (1989, \$36575). The average rate of change from 1989 to 1996 is

$$
\begin{aligned}
\text{average rate of change} &= \frac{\text{change in median income}}{\text{change in years}} \\
&= \frac{\$35{,}492 - \$36{,}575}{1996 - 1989} \\
&= \frac{-\$1083}{7 \text{ yr}} \\
&\approx -\$155/\text{yr}
\end{aligned}
$$

So between 1989 and 1996 the median household income *decreased* on average by \$155 per year! Figure 2.15 shows the negative slope of the dotted line connecting (1989, \$36575) and (1996, \$35492). Both average rates of change are correct, but they certainly give very different impressions of the well-being of American households.

Slanting the Data with Words and Graphs

If we wrap data in suggestive vocabulary and shape graphs to support a particular viewpoint, we can influence the interpretation of information. In Washington, D.C., this would be referred to as "putting a spin on the data."

Take a close look at the following three examples. Each contains exactly the same underlying facts: the same average rate of change calculation and a graph with a plot of the same two data points (1990, 249.4) and (2000, 274.6) representing the U.S. population (in millions) in 1990 and in 2000.

See "C4: Distortion by Clipping and Scaling" in *Rates of Change.*

1. The U.S. population increased by *only* 2.52 million/year between 1990 and 2000 (Figure 2.16).

   Stretching the scale of the horizontal axis relative to the vertical axis makes the slope of the line look almost flat and hence minimizes the impression of change.

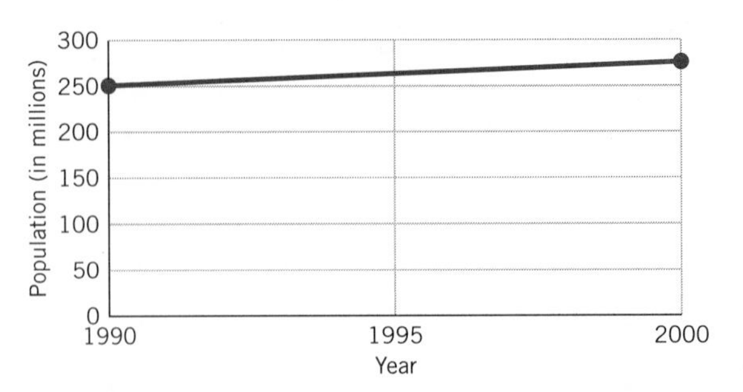

Figure 2.16 "Modest" growth in the U.S. population.

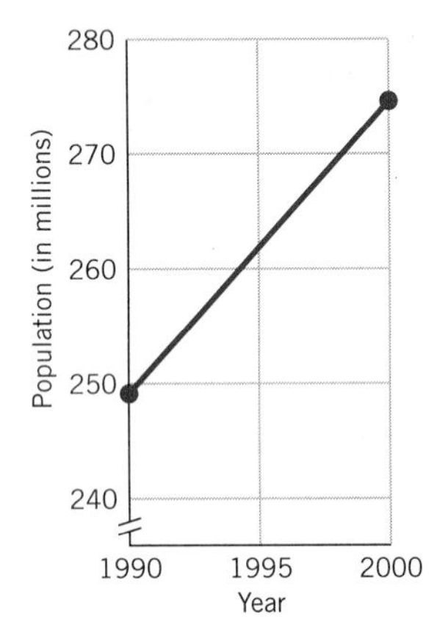

Figure 2.17 "Explosive" growth in the U.S. population.

2. The U.S. population had an *explosive* growth of over 2.52 million/year between 1990 and 2000 (Figure 2.17).

   Cropping the vertical axis (which now starts at 240 instead of 0) and stretching the scale of the vertical axis relative to the horizontal axis makes the slope of the line look steeper and emphasizes the impression of dramatic change.
3. The U.S. population grew at a *reasonable* rate of 2.52 million/year during the 1990s. (Figure 2.18).

   Visually the steepness of the line here seems to lie roughly halfway between the previous two graphs. *In fact, the slope of 2.52 million/year is precisely the same for all three graphs.*

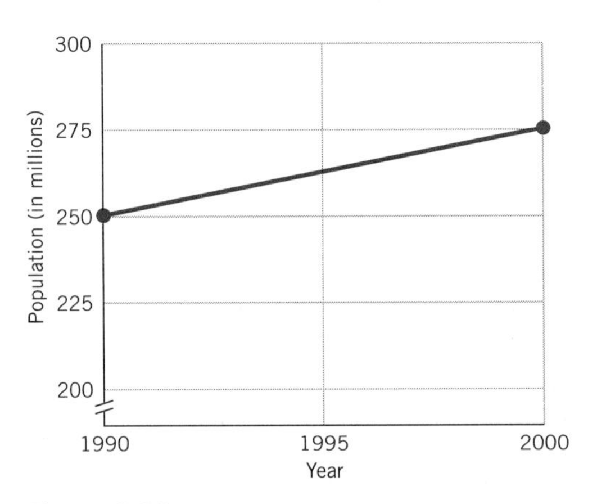

Figure 2.18 "Reasonable" growth in the U.S. population.

How could you decide upon a "fair" interpretation of the data? You might try to put the data in context by asking: How does the growth between 1990 and 2000 compare to other decades in the history of the United States? How does it compare to growth in other countries at the same time? Was this rate of growth easily accommodated or did it strain national resources and overload the infrastructure?

Exploration 2.1 gives you a chance to put your own "spin" on data.

Although the three examples are exaggerated, a statistical claim is never completely free of bias. For every statistic that is quoted, others have been left out. This does not mean that you should discount all statistics. However, you do need to become an educated consumer, constantly asking common sense questions and then coming to your own conclusions.

Algebra Aerobics 2.4

1. Use the data in Table 2.9 on immigration into the United States between 1901 and 1990 to answer the following questions:
   a. What two end points might you use to argue that the immigration level has declined slightly?
   b. What two end points might you use to argue that immigration levels have increased dramatically?

Table 2.9
Immigration: 1901–1990

Period	Number of Immigrants (thousands)
1901–1910	8795
1911–1920	5736
1921–1930	4107
1931–1940	528
1941–1950	1035
1951–1960	2515
1961–1970	3322
1971–1980	4493
1981–1990	7338

2.5 WHEN RATES OF CHANGE ARE CONSTANT

In our examples so far, the average rates of change have varied. Now we will examine the special case when the average rate of change remains constant.

What If the U.S. Population Had Grown at a Constant Rate? A Hypothetical Example

In Section 2.2 we calculated the average rate of change in the population between 1790 and 1800 as 0.14 million people/year. What if the average rate of change had remained constant? What if in every decade after 1790 the U.S. population had grown at exactly the same rate, namely at 0.14 million people/year? That would mean that starting with a population estimated at 3,900,000 in 1790, the population would have grown by exactly 140,000 people each year, or 1.4 million people each decade. The slopes of all the little line segments connecting adjacent population data points would be identical, namely 0.14 million people/year. The graph would be a straight line, indicating a constant average rate of change.

On the graph of the actual U.S. population data, the slopes of the line segments connecting adjacent points are increasing, so the graph curves upward. Table 2.10 and Figure 2.19 on page 72 compare the actual and hypothetical results.

This hypothetical example represents a linear function. Any function with a constant average rate of change is linear, and its graph will be a straight line. Note that when the average rate of change is constant, we often drop the word "average" and just say "rate of change."

Experiment with varying the average velocities and then setting them all constant in "C3: Average Velocity and Distance" in *Rates of Change.*

> A linear function has a constant rate of change. Its graph is a straight line.

A Real Example of a Constant Rate of Change

According to the standardized growth and development charts used by many American pediatricians, the median weight for girls during their first 6 months of life increases at an almost constant rate. Starting at 7.0 pounds at birth, the female median weight increases by approximately 1.5 pounds/month. Thus, in creating a

Table 2.10
U. S. Population

Year	Actual Population (Millions)	Population in Millions if Average Rate of Change Remained Constant at 0.14 million/yr
1790	3.9	3.9
1800	5.3	5.3
1810	7.2	6.7
1820	9.6	8.1
1830	12.9	9.5
1840	17.1	10.9
1850	23.2	12.3
1860	31.4	13.7
1870	39.8	15.1
1880	50.2	16.5
1890	62.9	17.9
1900	76.0	19.3
1910	92.0	20.7
1920	105.7	22.1
1930	122.8	23.5
1940	131.7	24.9
1950	151.9	26.3
1960	180.0	27.7
1970	204.0	29.1
1980	227.2	30.5
1990	249.4	31.9
2000	274.6	33.3

Source: U.S. Bureau of the Census, *Statistical Abstract of the United States,* 1998.

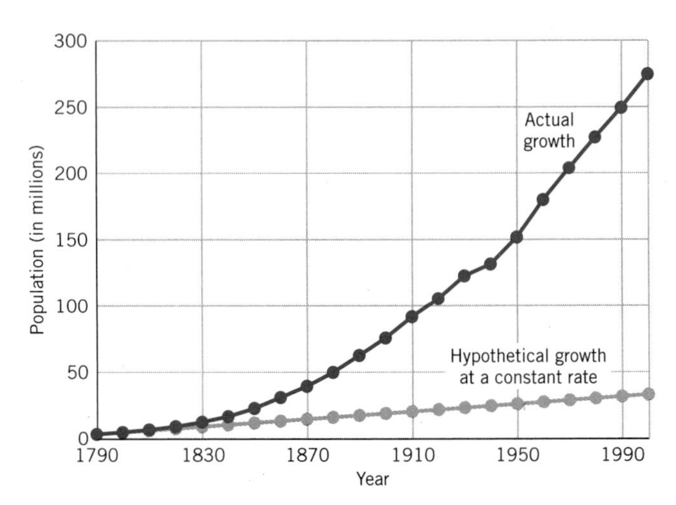

Figure 2.19 U.S. population: a hypothetical example.

model, we will use 1.5 pounds/month as the rate of change of girls' weight from birth to 6 months.

Since each age corresponds to a unique median weight, the median weight for baby girls is a function of age. The function is linear because the rate of change is constant. Table 2.11 and Figure 2.20 show two representations of the function. Note that the domain is between 0 and 6 months.

Table 2.11
Median Weight for Girls

Age (months)	Weight (lbs)
0	7.0
1	8.5
2	10.0
3	11.5
4	13.0
5	14.5
6	16.0

Source: Data derived from the Ross Growth and Development Program, Ross Laboratories, Columbus, OH.[1]

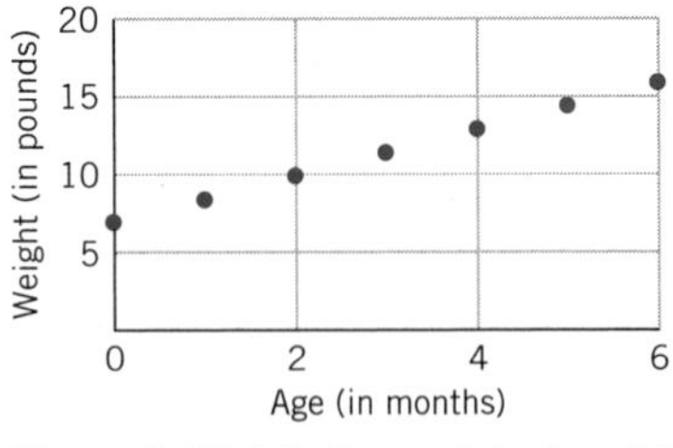

Figure 2.20 Median weight for girls.

The rate of change, 1.5 pounds/month, means that as age increases by 1 month, weight increases by 1.5 pounds. On the graph, if we move 1 unit (1 month) to the

[1]Ross Laboratories adapted data from P. V. V. Hamill, T. A. Drizd, C. L. Johnson, R. B. Reed, A. F. Roche, and W. M. Moore, "Physical Growth: National Center for Health Statistics Percentiles," *Am J Clin Nutr* 32:607–629, 1979. Data from the Fels Longitudinal Study, Wright State University School of Medicine; Yellow Springs, OH.

right, then we need to move up 1.5 units (1.5 pounds) to find the next data point. Just as in the hypothetical case of the constant U.S. population growth, the graph shows the points lying on a straight line.

Finding an Equation to Model the Relationship Between Female Infant Weight and Age

Can we find an equation, a *mathematical model,* that gives female infant weight as a function of age? Using Table 2.11, we can think of the 7.0 lb weight at birth (when age = 0) as our base value. In Figure 2.20, the base value represents the intercept on the vertical axis, so the graph crosses the weight axis at 7 pounds. The median weight for girls starts at the base value of 7.0 pounds at birth and increases by 1.5 pounds every month, which is the average rate of change. Table 2.12 shows the pattern in the monthly weight gain.

Table 2.12
Pattern in Median Weight for Girls

Age (months)	Weight (lb)	Pattern	Generalized Expression
0	7.0	7.0 + 1.5(0)	= 7.0 + 1.5(age in months)
1	8.5	7.0 + 1.5(1)	= 7.0 + 1.5(age in months)
2	10.0	7.0 + 1.5(2)	= 7.0 + 1.5(age in months)
3	11.5	7.0 + 1.5(3)	= 7.0 + 1.5(age in months)
4	13.0	7.0 + 1.5(4)	= 7.0 + 1.5(age in months)
5	14.5	7.0 + 1.5(5)	= 7.0 + 1.5(age in months)
6	16.0	7.0 + 1.5(6)	= 7.0 + 1.5(age in months)

An equation for this relationship between weight and age is

$$\text{median weight for girls} = 7.0 + 1.5 \cdot (\text{age in months})$$

where the weight is in pounds. If we let W represent the median weight in pounds and A the age in months, we can write this equation more compactly as

$$W = 7.0 + 1.5A$$

If the median birth weight for baby boys is the same as for baby girls but boys put on weight at a faster rate, which numbers in the model would change and which would stay the same? What would you expect to be different about the graph?

This equation can be used as a *model* of the relationship between age and median weight for baby girls in their first 6 months of life.

Since our equation represents quantities in the real world, each term in the equation has units attached to it. W (which represents weight) and 7.0 are in pounds, 1.5 is in pounds per month, and A (which represents age) is in months. The rules for canceling units are the same as the rules for canceling numbers in fractions. The units of the term $1.5A$ are

$$\left(\frac{\text{lb}}{\cancel{\text{month}}}\right)\cancel{\text{month}} = \text{lb}$$

The units of the right-hand side and left-hand side of our equation must match, and they do. That is, in the equation

$$W = 7.0 + 1.5A$$

the units are

$$\text{lb} = \text{lb} + \left(\frac{\text{lb}}{\cancel{\text{month}}}\right)\cancel{\text{month}}$$

$$= \text{lb} + \text{lb}$$

$$\text{lb} = \text{lb}$$

Our equation $W = 7.0 + 1.5A$ defines W as a function of A, since for each value of A (age) the equation determines a unique value of W (weight). If we name the function f, then $f(A) = W$.

Looking at this function in the abstract. If we treat $f(A) = 7.0 + 1.5A$ as an abstract function, then we can substitute any positive or negative value for A. The graph would be a line extending infinitely to the left and to the right. The domain (and corresponding range) would be all real numbers.

If we treat $f(A) = 7.0 + 1.5A$ as a model of the relationship between age and height, then the model is valid only for values of A between 0 and 6 months. So the domain of our model is $0 \le A \le 6$. Negative values of age are meaningless and the data for the median weights for girls older than 6 months will deviate from this equation. We would certainly not expect a female to continue to gain 1.5 pounds/month for the rest of her life! The range of the model would be from a birth weight of 7 pounds to 16 pounds, the median weight at 6 months.

We derived the function using only ages that were integer values (whole numbers of months), but the function can be extended to noninteger values of A between 0 and 6. This is called *interpolation.* For instance, we can use the function $f(A) = 7.0 + 1.5A$ to predict the median weight for girls who are 2.5 months old.

When $A = 2.5$, then

$$\begin{aligned} f(2.5) &= 7.0 + 1.5(2.5) \\ &= 7.0 + 3.75 \\ &= 10.75 \text{ lb} \end{aligned}$$

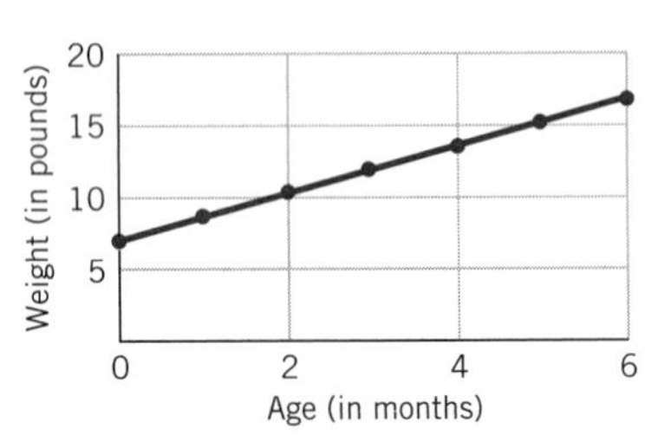

Figure 2.21 Median weight for girls.

The point (2.5, 10.75) is a *solution of,* or *satisfies,* the equation $W = 7.0 + 1.5A$.

The original data set contained seven *discrete* points, whereas Figure 2.21 shows the continuous function $f(A) = 7.0 + 1.5A$, where A can be *any* real number from 0 to 6.

Algebra Aerobics 2.5

1. Determine which, if any, of the following points satisfy the abstract equation $W = 7.0 + 1.5A$.
   a. (2, 10) **b.** (5, 13.5) **c.** (− 1.3, 5.05)
2. From Figure 2.21, estimate the weight W of a baby girl who is 4.5 months old. Then, use the equation $W = 7.0 + 1.5A$ to calculate the corresponding value for W. How close is your estimate?
3. From the same graph, estimate the age of a baby girl who weighs 11 pounds. Then use the equation to calculate the value for A.
4. Select any two points of the form (A, W) from Table 2.11 that satisfy the equation $W = 7.0 + 1.5A$. Use these points to verify that the rate of change between them is 1.5.
5. Find two points that satisfy the equation $W = 7 + 1.5A$ that are *not* in Table 2.11. Try using a negative value for A, then find the corresponding value for W. Then try a very large positive value for A and find the corresponding value for W.
6. If you calculate the slope using the two points you generated in Problem 5, what do you think your answer will be? Now do the calculations. Were you right?
7. **a.** If $C = 15P + 10$ describes the relationship between the number of persons (P) in a dining party and the total cost in dollars (C) of their meals, what is the unit of measure for 15? For 10?
   b. The equation $W = 7.0 + 1.5A$ (modeling weight as a function of age) expressed in units of measure only is
   $$\text{lb} = \text{lb} + \left(\frac{\text{lb}}{\text{month}}\right) \text{months}$$
   Express $C = 15P + 10$ in units of measure only.
8. The program "L4: Finding 2 Points on a Line" in *Linear Functions* will give you practice in finding solutions to equations.

2.6 LINEAR FUNCTIONS

The General Linear Equation

In the equation

$$W = 7.0 + 1.5A$$

modelling the relationship between median weight of female infants and age, the number 7.0 (the weight in pounds at birth) is the base value, or vertical intercept. The number 1.5 (in pounds/month) is the rate of change, or slope.

This equation is in the form

dependent variable = base value + rate of change · independent variable

The traditional mathematical choice for the independent variable is x, and for the dependent variable is y. So the general linear equation is often written in the form

$$y = b + m \cdot x$$

Any equation that can be written in this form represents a *linear function.* The equation represents a function, since any value of x will determine one and only one corresponding y value. The function is called linear because its graph is a straight line (see Figure 2.22).

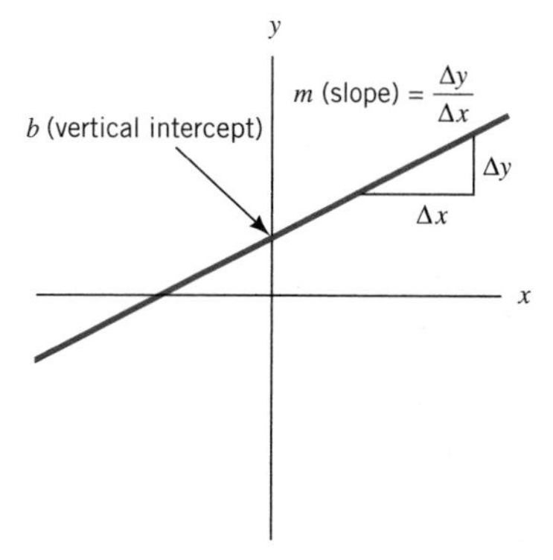

Figure 2.22 Graph of $y = b + mx$.

> A function $y = f(x)$ is called *linear* if it can be represented by an equation of the form
>
> $$y = b + mx$$
>
> Its graph is a straight line where
>
> m is the *slope*, or *rate of change* of y with respect to x
> b is the *vertical intercept,* or *base value,* the value of y when x is zero

Explorations 2.2a and b (along with "L1: *m* & *b* Sliders" in *Linear Functions*) allow you to examine the effects of *m* and *b* on the graph of a linear function.

Linear functions can also be written in the form

$$y = mx + b$$

Social scientists tend to prefer the $y = b + mx$ format because placing the b term first suggests starting at a base value (when $x = 0$) and then adding on multiples of m. Both forms describe equivalent equations and are used interchangeably.

Note: You do need to be careful about the order of the terms when identifying the slope (m) and vertical intercept (b). For example:

In the linear function $y = 15 - 2x$, we have $b = 15$ and $m = -2$.
In the linear function $y = 15x - 2$, we have $b = -2$ and $m = 15$.

Algebra Aerobics 2.6a

1. Identify the slope, m, and the vertical intercept, b, of the line with the given equation:
   a. $y = 5x + 3$
   b. $y = 5 + 3x$
   c. $y = 5x$
   d. $y = 3$
   e. $f(x) = 7.0 - x$
   f. $h(x) = -11x + 10$
   g. $y = 1 - (2/3)x$
   h. $2y + 6 = 10x$
2. If $f(x) = 50 - 25x$:
   a. Why does $f(x)$ describe a linear function?
   b. Evaluate $f(0)$ and $f(2)$.
   c. Use your answers in part (b) to verify that the slope is -25.

Finding the Graph of a Linear Function

The linear function

$$h(A) = 30.0 + 2.5A$$

where A = age (in years) and $h(A)$ = median height (in inches) can be used to model the relationship between median age and height for children between the ages of 2 and 12 years.[2]

To graph this equation, we could evaluate the function at two points, plot the points, and draw a line connecting them. For example, at 2 and 12 years the predicted median heights are respectively

$$h(2) = 30.0 + 2.5(2)$$
$$= 35.0 \text{ inches}$$
$$h(12) = 30.0 + 2.5(12)$$
$$= 60.0 \text{ inches}$$

The points (2, 35.0) and (12, 60.0) can be used to determine the graph in Figure 2.23.

Table 2.13 and Figure 2.23 show a number of points that satisfy the equation $h(A) = 30.0 + 2.5A$. Since our model is only valid for ages between 2 and 12 years, we must restrict the domain of the model such that $2 \leq A \leq 12$. So on the graph only the section of the line between the dotted lines is relevant to our model. Note that the vertical intercept, the point (0, 30.0), lies outside the domain.

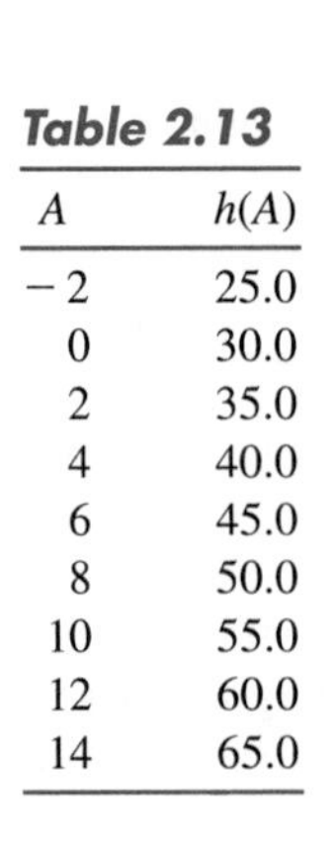

Table 2.13

A	$h(A)$
−2	25.0
0	30.0
2	35.0
4	40.0
6	45.0
8	50.0
10	55.0
12	60.0
14	65.0

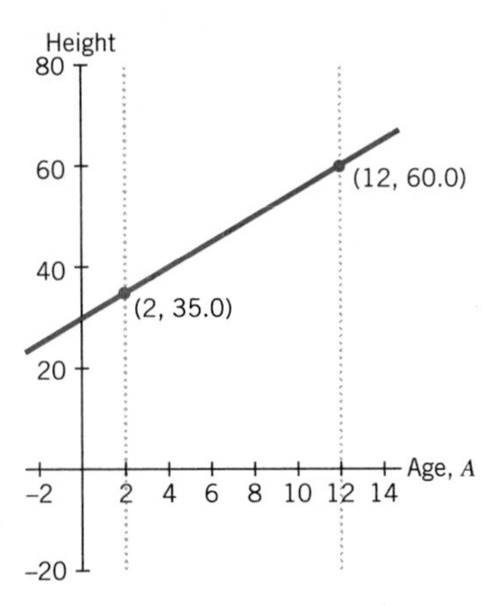

Figure 2.23 Graph of $h(A) = 30.0 + 2.5A$. Section between dotted lines indicates region relevant to model.

Finding the Equation of a Linear Function

Any linear function can be described by an equation of the form

$$y = b + mx$$

After identifying the independent (x) and dependent (y) variables, we can determine the equation for the function by finding the values for m (the slope or rate of change) and b (the base or vertical intercept).

[2]Adapted from R. E. Behrman and V. C. Vaughan, III (eds.), W. E. Nelson (sr. ed.), *Nelson Textbook of Pediatrics,* 12th ed. (Philadelphia: W.B. Saunders, 1983) p. 19.

Finding the equation from a graph

Example 1

Find the equation of the linear function graphed in Figure 2.24.

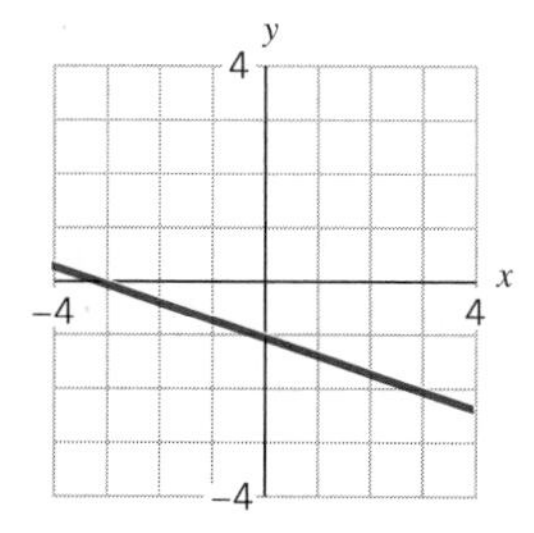

Figure 2.24 Graph of a linear function.

Solution

We can use any two points on the graph to calculate m, the slope. If, for example, we take $(-3, 0)$ and $(3, -2)$, then

$$\text{slope} = \frac{\text{change in } y}{\text{change in } x} = \frac{-2-0}{3-(-3)} = \frac{-2}{6} = \frac{-1}{3}$$

From the graph we see that the y intercept is at -1. So

$$b = -1$$

Hence the equation is

$$y = -1 - \tfrac{1}{3}x$$

Finding the equation from words

When making the transition from words to an equation, it's important first to identify the independent and dependent variables. Then you can dissect the sentences for information about the slope or points that satisfy the equation.

Example 2

A long distance phone carrier charges a monthly base fee of \$5.00 plus \$.07 per minute for long distance calls. Construct an equation to model your monthly phone bill.

Solution

Your phone bill depends upon the number of minutes you talk. So it would make sense to let the independent variable, say M, be the number of minutes you talk, and the dependent variable, say P, be your phone bill. The number \$5.00 represents the value of P when $M = 0$. Hence the point (0, \$5.00) must satisfy your equation, so \$5.00 is the vertical intercept (or b). The number \$.07 is the rate of change of your phone bill with respect to time. Since the rate is constant, the relationship is linear, so \$.07 is the slope (or m). Hence the linear equation is

$$P = 5.00 + 0.07M$$

Example 3

For tax purposes, the \$1,200 computer you purchased today is allowed to be *straight line depreciated* (or assumed to decrease linearly in value) until it is assumed to be worth \$0 at the end of 5 years. Construct an equation to model its worth over time.

Solution

Let the independent variable N be the number of years from today, and let the dependent variable W be the worth of the computer. Then we know two points that must satisfy our linear model: when $N = 0$, $W = \$1{,}200$ and when $N = 5$, $W = \$0$. The first point represents (0, \$1200) or the vertical intercept, so $b = 1200$.

We can find the slope, m, by calculating the average rate of change between the two points (0, \$1200) and (5, \$0).

$$m = \frac{1200 - 0}{0 - 5} = \frac{1200}{-5} = -\$240/\text{yr}$$

So the linear equation is

$$W = 1200 - 240N$$

Example 4

Harvard Pilgrim Healthcare, an HMO in Massachusetts, distributes the following recommended weight formula for men: "Give yourself 106 lb for the first 5 ft, plus 6 lb for every inch over 5 ft tall." Construct an equation that models the relationship.

Solution

If we think of recommended weight as a function of height, then we would treat height as the independent variable and weight as the dependent variable. We could let h represent height in inches, but since we are concerned only with heights over 5 feet, we instead let h = height in inches in *excess* of 5 feet and w = weight in pounds. Then according to the recommendations, when $h = 0$ inches (for a 5-foot male), $w = 106$ pounds. So the w-intercept, or base value, is 106 pounds. For each additional inch in height, the recommendations permit an additional 6 pounds. The rate of change of weight with respect to height is a constant 6 pounds/inch. The linear function $f(h) = w$, where

$$w = 106 + 6h$$

gives us a mathematical model for the relationship. The variable h = height in inches above 5 feet, and w = weight in pounds. Note that w and 106 are in pounds, 6 is in pounds/inch, and h is in inches.

In our model we need to think about realistic values for the domain. The recommendations are for men 5 feet (or 60 inches) and taller. Our estimate for a reasonable maximum male height is 7 feet (or 84 inches). Hence the constraints on h, the possible values for the number of inches in excess of 5 feet, would be $0 \leq h \leq 24$ inches.

Since the recommended weight increases with height, the range of the model is from $f(0) = 106$ pounds to $f(24) = 106 + 6(24) = 250$ pounds.

Table 2.14 and Figure 2.25 are both representations of this function.

According to the HMO recommendations, what region of the grid on Figure 2.25 represents men who are overweight? What region represents men who are underweight?

Table 2.14

Height, h, (inches in excess of 5 ft)	Weight, w (lb)
0	106
4	130
8	154
12	178
16	202
20	226
24	250

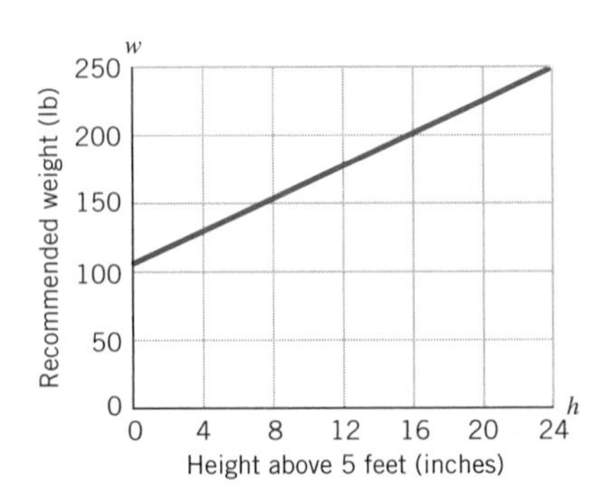

Figure 2.25 Recommended weight for males between 5 and 7 feet tall.

Finding the equation from a table

Example 5

a. Determine if the data in Table 2.15 represent a linear relationship between values of blood alcohol concentration and number of drinks consumed for a 160-pound person. (One drink is defined as 5 oz of wine, 1.25 oz of 80-proof liquor, or 12 oz of beer.)

b. If the relationship is linear, determine the corresponding equation.

Table 2.15

D No. of Drinks	A Blood Alcohol Concentration
2	0.047
4	0.094
6	0.141
10	0.235

Solution

a. We can generate a third column in the table that represents the average rate of change between consecutive points (see Table 2.16). Since the average rate of change of A with respect to D remains constant at 0.0235, these data represent a linear relationship.

Table 2.16

D	A	Average Rate of Change
2	0.047	n.a.
4	0.094	$\frac{0.094 - 0.047}{4 - 2} = \frac{0.047}{2} = 0.0235$
6	0.141	$\frac{0.141 - 0.094}{6 - 4} = \frac{0.047}{2} = 0.0235$
10	0.235	$\frac{0.235 - 0.141}{10 - 6} = \frac{0.094}{4} = 0.0235$

b. The rate of change is the slope, so the corresponding linear equation will be of the form

$$A = b + 0.0235D \qquad (1)$$

To find b, we can substitute any of the original paired values for D and A, for example, (4, 0.094), into Equation (1) above to get

$$0.094 = b + (0.0235 \cdot 4)$$

$$0.094 = b + 0.094$$

$$0.094 - 0.094 = b$$

$$0 = b$$

So when D, the number of drinks, is 0, A, the blood alcohol concentration, is 0, which makes sense. So the final equation is

$$A = 0 + 0.0235D$$

or just $$A = 0.0235D$$

Algebra Aerobics 2.6b

For Problems 1 and 2, find an equation, make an appropriate table, and sketch the graph of:

1. A line with slope 1.2 and vertical intercept -4.
2. A line with slope -400 and vertical intercept 300 (be sure to think about axis scales).
3. Write an equation for the line graphed in Figure 2.26.

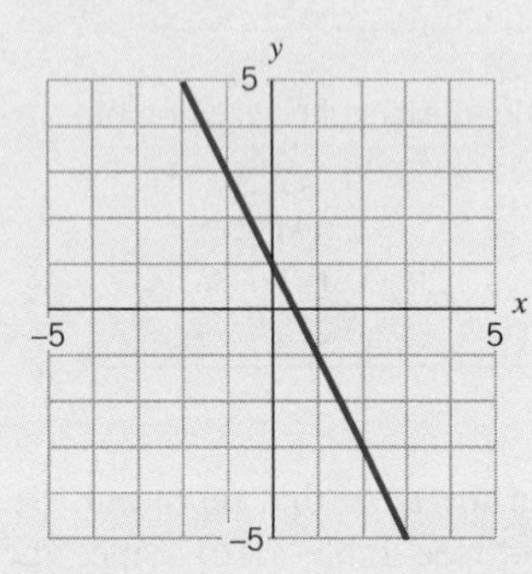

Figure 2.26 Graph of a linear function.

4. **a.** Find an equation to represent the current salary after x years of employment if the starting salary is \$12,000 with annual increases of \$3000.
   b. Create a small table of values and sketch a graph.
5. **a.** Plot the following data.

Years of Education	Hourly Wage
8	\$5.30
10	\$8.50
13	\$13.30

   b. Is the relationship between hourly wage and years of education linear? Why or why not?
   c. If it is linear, construct a linear equation to model it.
6. The program "L3: Finding a Line Through 2 Points" in *Linear Functions* will give you practice in this skill.

2.7 SPECIAL CASES

Direct Proportionality

The simplest possible relationship between two variables is when one is equal to a constant times the other. For instance, in the previous example $A = 0.0235D$; blood alcohol concentration A equals a constant, 0.0235, times D, the number of drinks. We say that *A is directly proportional to D* or that *A varies directly with D*.

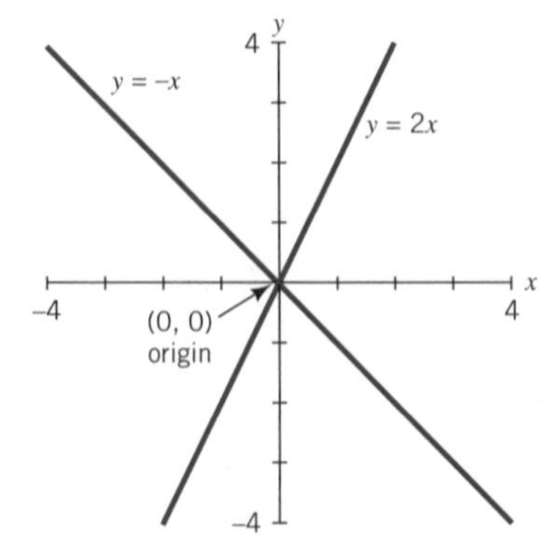

Figure 2.27 Graphs of two relationships in which y is directly proportional to x. Note that both graphs are lines that go through the origin.

How to recognize direct proportionality. Linear functions in the form

$$y = mx$$

describe a relationship where y is directly proportional to x. Notice that we could write the equation $y = mx$ as

$$y = 0 + mx$$

which would be in the general linear form $y = b + mx$, where $b = 0$. Direct proportionality is a special case of the linear function where b, the vertical intercept, is 0. The x-coordinate of the vertical intercept is already 0. So if two variables are directly proportional to each other, their graph will be a straight line that passes through the point (0, 0), the origin. Figure 2.27 shows the graphs of two relationships in which y is directly proportional to x: namely, $y = 2x$ and $y = -x$.

If y is directly proportional to x, is x directly proportional to y?

> If $y = mx$, where $m \neq 0$, the relationship between x and y is described by saying that
>
> *y is directly proportional to x*
>
> or that
>
> *y varies directly with x.*
>
> The graph is a straight line through the origin.

Example 1

Suppose you go on a road trip, driving at a constant speed of 60 miles per hour. After 1 hour, you will have traveled 60 miles; after 2 hours, you will have traveled 120 miles; and so on. The distance d (in miles) you have traveled varies directly with the time t (in hours) according to the formula $d = 60t$.

Example 2

In Table 2.17, C, the total cost of gasoline, is directly proportional to G, the amount of gasoline bought. We could describe the relationship with the equation $C = 1.50G$.

Table 2.17

Gasoline Bought, G (gal)	Total Cost of Gasoline, C (dollars)
1	1.50
2	3.00
3	4.50
4	6.00
5	7.50

One of the most important attributes of a directly proportional relationship, $y = mx$, is that when x is multiplied by a constant, y is multiplied by the same constant. When x doubles in value (is multiplied by 2), y doubles in value. When x triples in value, y triples in value. In the blood alcohol equation, $A = 0.0235D$, when the value of D doubles from 2 to 4, the value of A doubles from 0.047 to $0.094 = 2(0.047)$. When D doubles again from 4 to 8, A also doubles from 0.094 to $0.188 = 2(0.094)$.

Functional notation is useful in helping to understand direct proportionality. If we write $f(x) = mx$ and evaluate f at a particular value of x, say 4, then,

$$f(4) = m \cdot 4$$

If we

multiply 4 by 2, then $\quad f(2 \cdot 4) = m(2 \cdot 4) = 2(m \cdot 4) = 2f(4)$

multiply 4 by 3, then $\quad f(3 \cdot 4) = m(3 \cdot 4) = 3(m \cdot 4) = 3f(4)$

multiply 4 by a constant k, then $\quad f(k \cdot 4) = m(k \cdot 4) = k(m \cdot 4) = kf(4)$

So, in general, if we evaluate f at some arbitrary value x_1, then, if we

multiply x_1 by 2, then $\quad f(2x_1) = m(2x_1) = 2(mx_1) = 2f(x_1)$

multiply x_1 by 3, then $\quad f(3x_1) = m(3x_1) = 3(mx_1) = 3f(x_1)$

multiply x_1 by a constant k, then $\quad f(kx_1) = m(kx_1) = k(mx_1) = kf(x_1)$

So, multiplying x_1 by any constant k means that $f(x_1)$, the value of the function corresponding to x_1, is also multiplied by k.

Example 3

Converting from one currency into another can be an example of direct proportionality. For example, on August 5, 1999, you could buy 1.47 Canadian dollars for 1 U.S. dollar. The linear function

$$C_1 = 1.47d$$

converts U.S. dollars, d, to Canadian dollars, C_1. The number of Canadian dollars you receive is directly proportional to the number of U.S. dollars you exchange.

But suppose the Exchange Bureau decides to charge a service fee of \$2.00 (two U.S. dollars) each time you exchange currencies. Now the linear function

$$C_2 = 1.47(d - 2.00)$$
$$= 1.47d - 2.94$$

represents converting d U.S. dollars after first subtracting off the \$2.00 service fee. The number of Canadian dollars you receive is now no longer directly proportional to the number of U.S. dollars you exchange. For example, if you double the number of U.S. dollars from say \$100 to \$200, the number of Canadian dollars you receive will not double. It will go from $1.47(100) - 2.94 = 144.06$ to $1.47(200) - 2.94 = 291.06$.

Algebra Aerobics 2.7a

1. Give an equation and draw the graph of the line that passes through the origin and has the given slope.
   a. $m = -1$ **b.** $m = 0.5$
2. For each of Tables 2.18 and 2.19, determine whether the variables vary directly with each other. Represent each relationship with an equation.
   a. ***Table 2.18***

x	y
−2	6
−1	3
0	0
1	−3
2	−6

   b. ***Table 2.19***

x	y
0	5
1	8
2	11
3	14
4	17

3. In February 1996, the exchange rate was \$1.00 U.S. to 1.45 Deutschmarks (German marks).
   a. Find a linear function that converts U.S. dollars to German marks.
   b. Find a linear function that converts U.S. dollars to German marks with a service fee of \$2.50.
   c. Which function represents a directly proportional relationship and why?

Horizontal and Vertical Lines

The slope, m, of any horizontal line is 0. So the general form for the equation of a horizontal line is

$$y = b + 0x$$

or just

$$y = b$$

For example, Table 2.20 and Figure 2.28 show points that satisfy the equation of the horizontal line $y = 1$. If we calculate the slope between any two points in the table, for example $(-2, 1)$ and $(2, 1)$, we get

$$\text{slope} = \frac{1 - 1}{-2 - 2} = \frac{0}{-4} = 0$$

For a vertical line the slope, m, is undefined, so we can't use the standard $y = b + mx$ format. The graph of a vertical line (as in Figure 2.29) fails the vertical line test, so y is not a function of x. However, every point on a vertical line does have the

Table 2.20

x	y
−4	1
−2	1
0	1
2	1
4	1

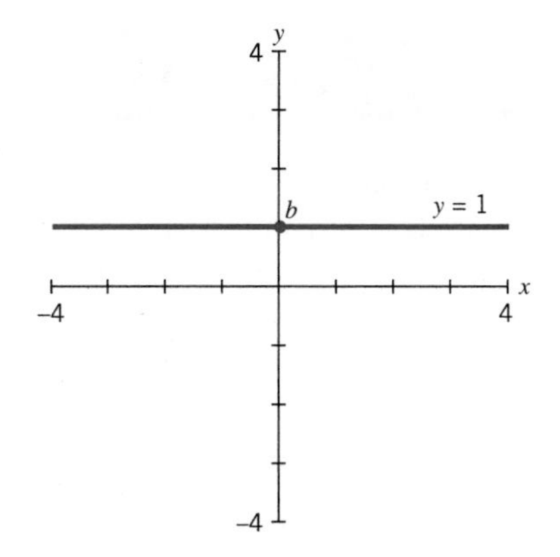

Figure 2.28 Graph of the horizontal line $y = 1$

same horizontal coordinate, which equals the coordinate of the horizontal intercept. Therefore the general equation for a vertical line is of the form

$$x = \mathrm{c} \quad \text{where c is a constant (the horizontal intercept).}$$

For example, Table 2.21 and Figure 2.29 show points that satisfy the equation of the vertical line $x = 1$.

Table 2.21

x	y
1	−4
1	−2
1	0
1	2
1	4

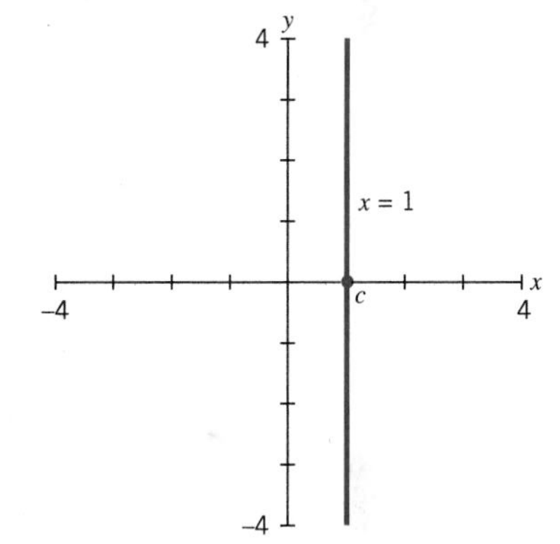

Figure 2.29 Graph of the vertical line $x = 1$

Note that if we tried to calculate the slope between two points, say $(1, -4)$ and $(1, 2)$, on the vertical line $x = 1$ we would get

$$\text{slope} = \frac{-4-2}{1-1} = \frac{-6}{0} \quad \text{which is undefined}$$

> The general equation of a *horizontal line* is
>
> $$y = b$$
>
> where b is a constant (the vertical intercept) and the slope is 0.
>
> The general equation of a *vertical line* is
>
> $$x = c$$
>
> where c is a constant (the horizontal intercept) and the slope is undefined.

Describe the equation for any line perpendicular to the horizontal line $y = b$.

Parallel and Perpendicular Lines

Parallel lines have the same slopes. So if the two equations $y = b_1 + m_1x$ and $y = b_2 + m_2x$ describe two parallel lines, then $m_1 = m_2$. For example, the two lines $y = 2.0 - 0.5x$ and $y = -1.0 - 0.5x$ each have a slope of -0.5 and thus are parallel (see Figure 2.30).

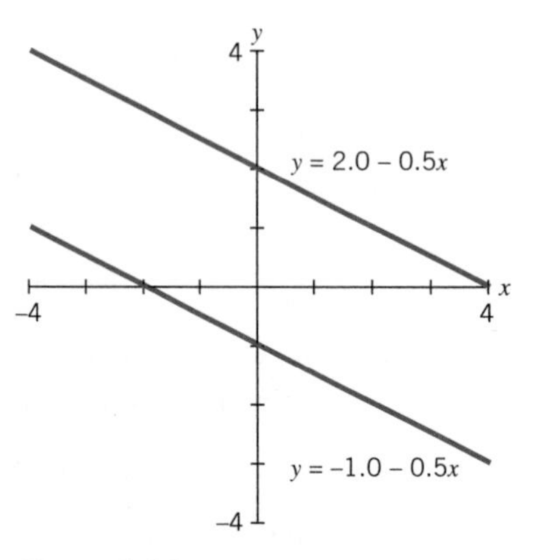

Figure 2.30 Two parallel lines with the same slope.

Figure 2.31 Two perpendicular lines with slopes that are negative reciprocals.

Two lines are perpendicular if their slopes are negative reciprocals. If $y = b_1 + m_1x$ and $y = b_2 + m_2x$ describe two perpendicular lines, then $m_1 = -1/m_2$. For example, in Figure 2.31 the two lines $y = 3 - 2x$ and $y = -2 + \frac{1}{2}x$ have slopes of -2 and $\frac{1}{2}$ respectively. Since -2 is the negative reciprocal of $\frac{1}{2}$ (i.e., $-\frac{1}{\left(\frac{1}{2}\right)} = -1 \div \frac{1}{2} = -1 \cdot \frac{2}{1} = -2$), the two lines are perpendicular.

Why does this relationship hold for perpendicular lines? Consider a line whose slope is given by v/h. Now imagine rotating the line 90 degrees clockwise to generate a second line perpendicular to the first (Figure 2.32).

What would the slope of this new line be?

The positive vertical change, v, becomes a positive horizontal change. The positive horizontal change, h, becomes a negative vertical change.

The slope of the original line is v/h, and the slope of the line rotated 90 degrees clockwise is $-h/v$. Note that $-h/v = -1/(v/h)$, which is the original slope inverted and multiplied by -1.

In general, the slope of a perpendicular line is the negative reciprocal of the slope of the original line. If the slope of a line is m_1, then the slope, m_2, of a line perpendicular to it is $-1/m_1$.

This is true for any pair of perpendicular lines for which slopes exist. It does not work for horizontal and vertical lines since vertical lines have undefined slopes.

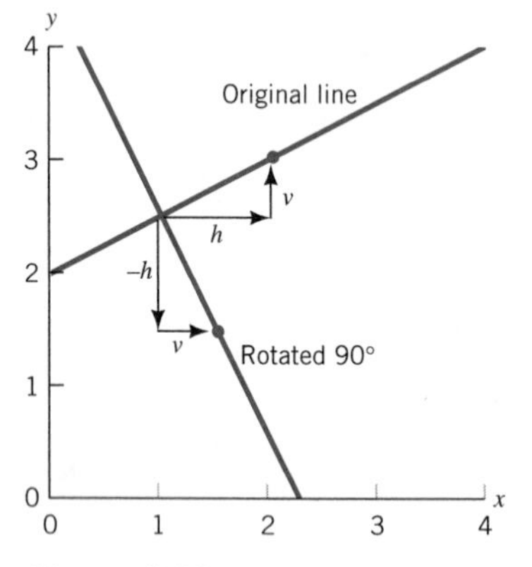

Figure 2.32 Perpendicular lines $m_2 = -1/m_1$.

> Parallel lines have the same slope.
>
> Perpendicular lines have slopes that are negative reciprocals of each other.

Algebra Aerobics 2.7b

1. In each case, find an equation for the horizontal line that passes through the given point.
   a. $(3, -5)$ **c.** $(-3, 5)$
   b. $(5, -3)$ **d.** $(-5, 3)$
2. In each case, find an equation for the vertical line that passes through the given point.
   a. $(3, -5)$ **c.** $(-3, 5)$
   b. $(5, -3)$ **d.** $(-5, 3)$
3. Construct the equation of the line that passes through the points
   a. $(0, -7)$, $(3, -7)$, $(-1, -7)$, and $(350, -7)$
   b. $(-4.3, 0)$, $(-4.3, 8)$, $(-4.3, -1000)$, and $(-4.3, 280)$
4. Find the equation of the line that is parallel to $y = 4 - x$ and that passes through the origin.
5. Find the equation of the line that is parallel to $W = 358.9C + 2500$ and passes through the point where $c = 4$ and $w = 1000$.
6. Find the slope of a line perpendicular to each of the following:
   a. $y = 4 - 3x$ **c.** $y = 3.1x - 5.8$
   b. $y = x$ **d.** $y = -\frac{3}{5}x + 1$
7. **a.** Find an equation for the line that is perpendicular to $y = 2x - 4$ and passes through $(3, -5)$. Graph both lines on the same coordinate system.
   b. Find the equations of two other lines that are perpendicular to $y = 2x - 4$ but do not pass through the point $(3, -5)$.
   c. How do the three lines from parts (a) and (b) that are perpendicular to $y = 2x - 4$ relate to each other?

2.8 FINDING LINEAR MODELS FOR DATA

According to Edward Tufte in *Data Analysis of Politics and Policy,* "Fitting lines to relationships is the major tool of data analysis." When we work with real two variable data with an underlying linear relationship, the data points will rarely fall exactly in a straight line. However, we can model the trends in the data with a linear equation.

Linear relationships are also of particular importance, not because most relationships are linear, but because straight lines are easily drawn and analyzed. A human can fit a straight line by eye to a scatter plot almost as well as a computer. This paramount convenience of linear equations as well as their relative ease of manipulation and interpretation means that lines are often used as first approximations to patterns in data.

Fitting a Line to Data

Children's heights were measured monthly over several years as a part of a study of nutrition in developing countries. Table 2.22 and Figure 2.33 show data collected on the mean heights of 161 children in Kalama, Egypt.

Although the data points do not lie exactly on a straight line, the overall pattern seems clearly linear. Rather than generating a line through two of the data points, try eyeballing a line that approximates all the points. A ruler or a piece of black thread laid down through the dots will give you a pretty accurate fit.

In the Extended Exploration on education and income following this chapter, we will see that there are ways to use technology to find a "best fit" line. Figure 2.34 shows a line that approximates the data points. This line does not necessarily pass through any of the original points.

What is an equation for this linear model? Estimating the coordinates of two points *on the line,* say (20, 77.5) and (26, 81.5), we can calculate the slope, m, or

See Excel or graph link file KALAMA and the program "R5: Age-Height Data: Kalama, Egypt" in *Linear Regression.*

Table 2.22
Mean Height of Kalama Children

Age (months)	Height (cm)
18	76.1
19	77.0
20	78.1
21	78.2
22	78.8
23	79.7
24	79.9
25	81.1
26	81.2
27	81.8
28	82.8
29	83.5

Source: D. S. Moore and G. P. McCabe, *Introduction to the Practice of Statistics.* Copyright © 1989 by W.H. Freeman and Company. Used with permission.

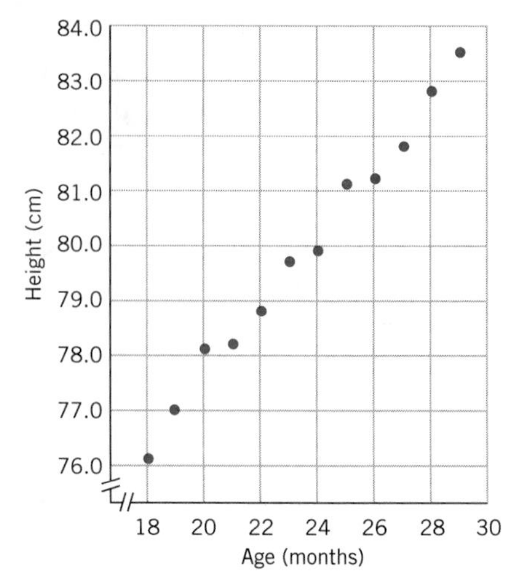

Figure 2.33 Mean height of children in Kalama, Egypt.

rate of change, as

$$m = \frac{(81.5 - 77.5) \text{ cm}}{(26 - 20) \text{ months}}$$

$$= \frac{4.0 \text{ cm}}{6 \text{ months}}$$

$$\approx 0.67 \text{ cm/month}$$

So our model predicts that for each additional month between 18 and 29 months, an "average" child will grow about 0.67 centimeter.

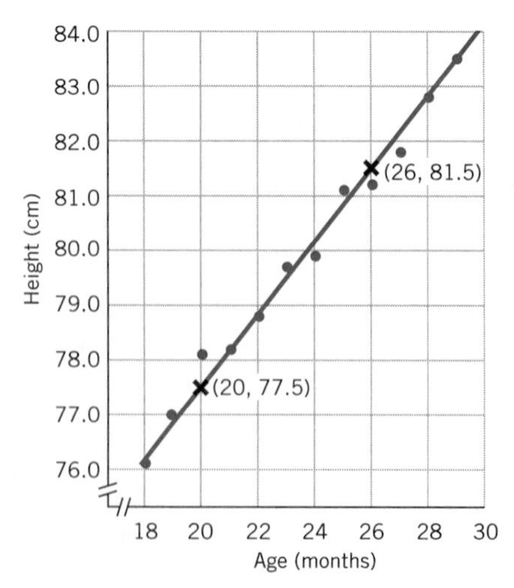

Figure 2.34 Estimated coordinates of two points on the line.

Now we know our linear equation is of the form

$$H = b + 0.67A \tag{1}$$

where A = age in months and H = mean height in centimeters.

How can we find b, the vertical intercept? We have to resist the temptation to estimate b directly from the graph using the coordinates where we would project that our line crosses the height axis. As is frequently the case in social science graphs, both the horizontal and the vertical axes are cropped. Because the horizontal axis is cropped, we can't read the vertical intercept off the graph. We'll have to calculate it.

Since the line passes through (20, 77.5) we:

substitute (20, 77.5) in Equation (1)	$77.5 = b + (0.67)(20)$
simplify	$77.5 = b + 13.4$
solve for b	$b = 64.1$

Our linear equation modeling the Kalama data is

$$H = 64.1 + 0.67A$$

where A = age in months and H = height in centimeters. It offers a compact summary of the Kalama data.

What is the domain of this model? The data were collected on children aged 18 to 29 months. We don't know its predictive value outside these ages, so

$$\text{domain is values of } A \text{ where } 18 \leq A \leq 29$$

Note that although the H-intercept is necessary to write the equation for the line, it lies outside of the set of values for which our model applies.

Compare Figure 2.35 to Figure 2.34. They both show graphs of the same equation $H = 64.1 + 0.67A$. In Figure 2.34 both axes are cropped, while Figure 2.35 includes the origin (0, 0). On Figure 2.35 the vertical intercept is now visible and the dotted lines indicate the region that applies to our model. So, a word of warning when reading graphs: Always look carefully to see if the axes have been cropped.

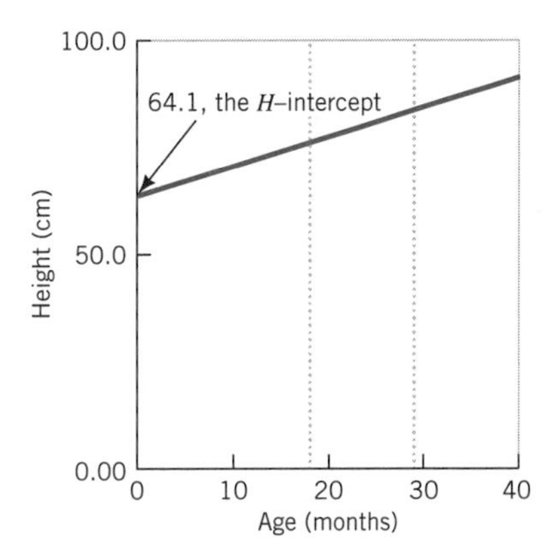

Figure 2.35 Graph of $H = 64.1 + 0.67A$ that includes the origin (0, 0). Dotted lines show the region that models Kalama data.

Fitting a Line to a "Cloud" of Data Points

The Kalama data set was fairly clean, because for each month we were given a single value for the average height instead of all the individual heights. For many data sets there may be many y values associated with one x value. In these cases the scatter plots look more like clouds of data, as in Figure 2.36.

Figure 2.36 shows mean SAT verbal scores vs. percentage of seniors taking the SAT in 1989 for each state in the United States. Each point is labeled with an abbreviation of the state's name. The point labeled WA near the middle of the graph with approximate coordinates (40, 450) means that in 1989, 40% of seniors in the state of Washington took the SAT test and their mean verbal score was 450. There appears to be a general downward trend. As the percentage of students taking the test increases, the mean SAT verbal score decreases. We can sketch a best fit line through the data (as shown on the scatter plot), estimate its slope, and determine its equation.

As before we can estimate the coordinates of two points on the line, say (10, 480) and (60, 430), and use them to calculate the slope, m, or rate of change:

$$m = \frac{(480 - 430)\text{ SAT points}}{(10 - 60)\%} = \frac{50 \text{ SAT points}}{-50\%} = -1.0 \text{ SAT points/\%}$$

Read the Anthology article "Verbal SAT Scores." What very different conclusions might you draw looking at the SAT scores as a single variable as opposed to looking at their relationship to the percentage taking the SAT?

On average, for each additional 1% of seniors taking the SAT verbal test, our model predicts that the mean verbal score was roughly 1 point lower. Equivalently, when the percentage of seniors taking the test increased by 10, the mean SAT verbal score dropped on average by 10 points.

Let S = mean SAT verbal score and P = percentage of students taking the SAT. Substituting -1 for the slope gives us a linear equation of the form

$$S = b - P$$

We estimated that the line passes through (60, 430), so we can substitute the values 60 and 430 for P and S in the equation of the line to solve for b:

$$430 = b - 60$$

$$490 = b$$

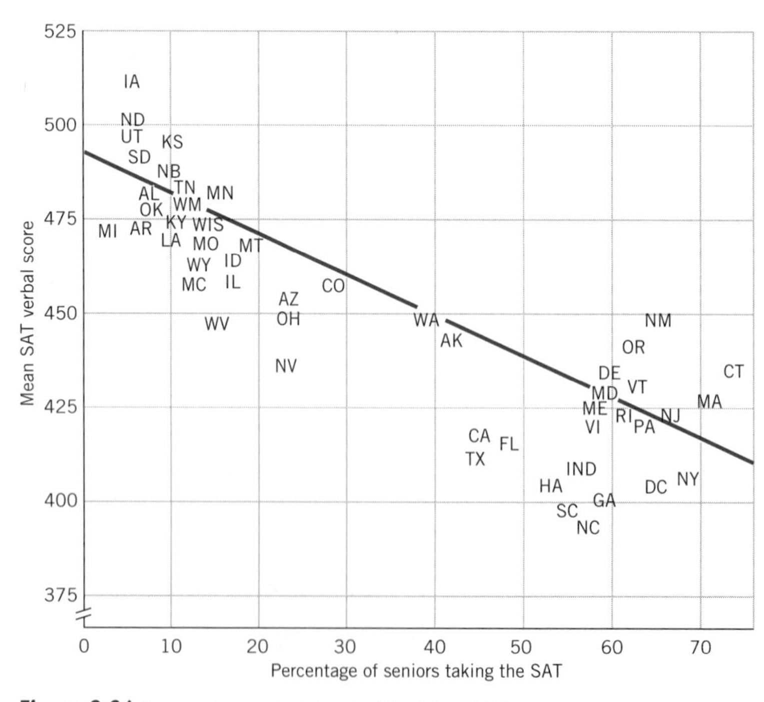

Figure 2.36 Percentage of seniors taking the SAT.

Source: "Verbal SAT Scores," *Quantitative Examples,* J. Truxal (ed.), NLA Monograph Series, Suny Research Foundation, 1991. Note: State abbreviations are taken from source.

The vertical axis is cropped in Figure 2.36, but since the x-axis is not we can still see the real vertical intercept and confirm that 490 is a reasonable value. Substituting this value for b, we get

$$S = 490 - P$$

which is a linear function, where P = percentage of seniors taking the SAT verbal test and S = mean SAT verbal score. Given the plotted data, a reasonable domain for P might be $5\% \leq P \leq 75\%$.

Reinitializing the Independent Variable

Question How can we find an equation that models the trend in smoking in the United States?

One way to answer this would be to look on the Internet for information about tobacco consumption. The web site for the National Center for Health Statistics provided the data reproduced in Table 2.23 and graphed in Figure 2.37. They show a decline in the percentage of adult smokers between 1965 and 1995.

The relationship appears fairly linear. So the equation of a best fit line would provide a compact description of the relationship. Suppose we let x = year and y = percentage of adults who smoke. Since the horizontal axis is cropped and starts at the year 1965, the real vertical intercept would occur 1965 units to the left, at 0 A.D.! If you drew a big enough graph, you'd find that the vertical intercept would occur at approximately (0, 1230). A nonsensical *extrapolation* or extension of the model outside its known values would say that in 0 A.D., 1230% of the adult population smoked. So a useful strategy is instead to define x = number of years *since* 1965. Table 2.24 and Figure 2.38 show these new values for x.

Table 2.23

Year	Percentage of Adults Who Smoke
1965	42.4
1966	42.6
1970	37.4
1974	37.1
1978	34.1
1979	33.5
1980	33.2
1983	32.1
1985	30.1
1987	28.8
1988	28.1
1990	25.5
1991	25.6
1992	26.5
1993	25.0
1994	25.5
1995	24.7

Source: National Health Interview Surveys, National Center for Health Statistics, *www.fedstats.gov.*

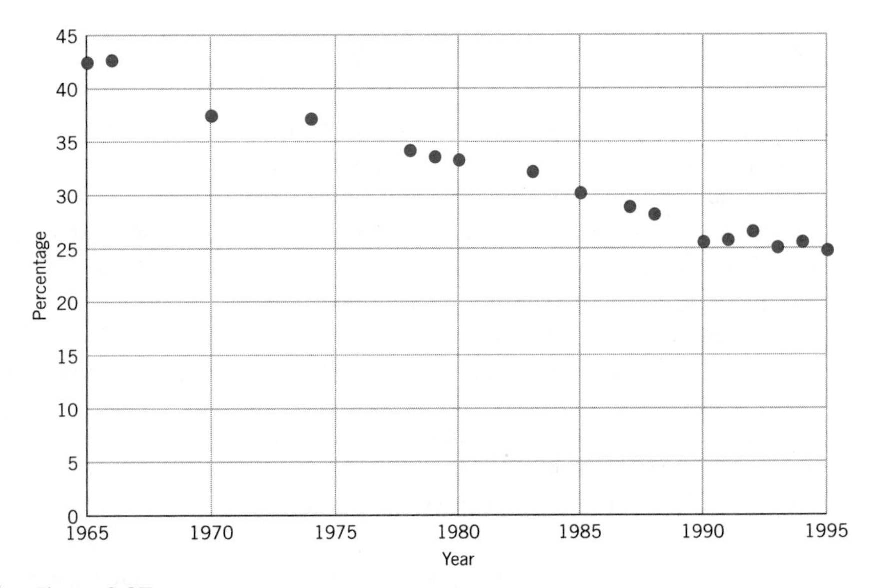

Figure 2.37 Percentage of adults who smoke.

Table 2.24

Year	No. of Years Since 1965	Percentage of Adults Who Smoke
1965	0	42.4
1966	1	42.6
1970	5	37.4
1974	9	37.1
1978	13	34.1
1979	14	33.5
1980	15	33.2
1983	18	32.1
1985	20	30.1
1987	22	28.8
1988	23	28.1
1990	25	25.6
1991	26	25.7
1992	27	26.5
1993	28	25.0
1994	29	25.5
1995	30	24.7

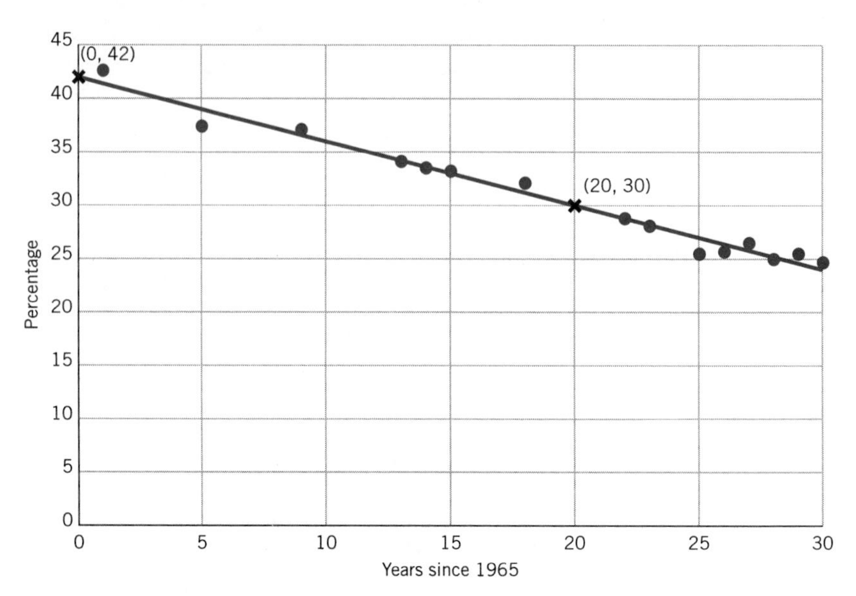

Figure 2.38 Percentage of adults who smoke.

Estimating two points on the line at (0, 42) and (20, 30) gives us

$$b = 42 \quad \text{and} \quad m = \frac{42 - 30}{0 - 20} = \frac{12}{-20} = -0.6$$

so the equation would be

$$y = 42 - 0.6x$$

where x = number of years since 1965 and y = percentage of adult smokers. The domain is $0 \leq x \leq 30$. This model says that, starting in 1965 when about 42% of U.S. adults smoked, on average, the percentage of the adult population that smokes has declined by 0.6 percentage points a year for 30 years.

Algebra Aerobics 2.8

1. Figure 2.39 shows the number of students in the United States who were 35 or older between the years of 1990 and 1998.
   a. Sketch a line through the data that best approximates their pattern.
   b. Estimate the coordinates of two points on your line and use them to find the slope.
   c. If x = number of years *since* 1980 and y = number (in thousands) of U.S. students 35 or older, what would the coordinates of your two points in part (b) be in terms of x and y?
   d. Construct a linear equation using the x and y of part (c).
   e. What does your model tell you about older students in the United States?

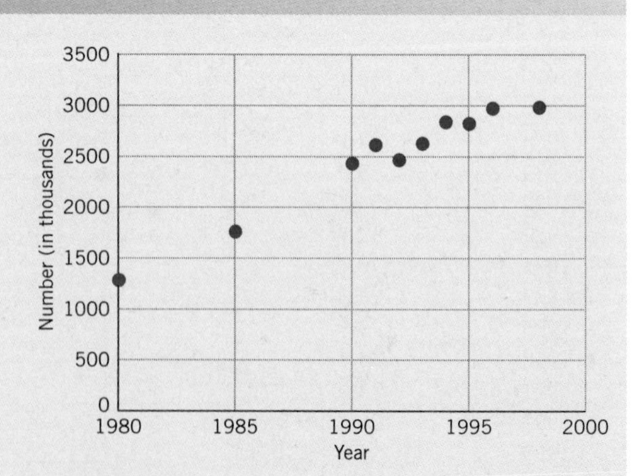

Figure 2.39 U.S. population 35 years old and above enrolled in school.
Source: U.S. National Center for Education Statistics, *Digest of Education Statistics,* annual.

Chapter Summary

The Average Rate of Change. The average rate of change of y with respect to $x = \dfrac{\text{change in } y}{\text{change in } x}$.

If the variables represent real world quantities that have units of measure (e.g., dollars or years), then the average rate of change should be represented in terms of the appropriate units:

$$\text{units of average rate of change} = \frac{\text{units of } y}{\text{units of } x}$$

For example, the units might be dollars/year (read as dollars per year) or pounds/person (read as pounds per person).

On a graph, the average rate of change between two points is the slope of the straight line connecting the two points. Given two points (x_1, y_1) and (x_2, y_2),

$$\text{average rate of change} = \frac{\text{change in } y}{\text{change in } x} = \frac{\Delta y}{\Delta x} = \frac{y_2 - y_1}{x_2 - x_1} = \text{slope}$$

where $\Delta y = y_2 - y_1$ and $\Delta x = x_2 - x_1$.

If the slope, or average rate of change, of y with respect to x is *positive,* then the graph of the relationship rises up when read from left to right. This means that as x increases in value, y increases in value. If the slope is *negative,* the graph falls when read from left to right. As x increases, y decreases. If the slope is *zero,* the graph is flat. As x increases, there is no change in y.

Linear Functions. A *linear function* has a constant average rate of change. It can be described by an equation of the form

$$y = b + mx \quad \text{or equivalently} \quad y = mx + b$$

Its graph is a straight line where b is the *vertical intercept,* or the value of y when x is 0, and m is the *slope,* or *rate of change* of y with respect to x. (When the average rate of change is constant, we usually drop the word "average" and just refer to the "rate of change.")

The following are special cases of lines:

1. *Direct proportionality:* y is said to be *directly proportional to* x or to *vary directly with* x if there is a *real number* m, such that

   $$y = mx$$

   This equation represents a linear function in which the y-intercept is 0, so the graph passes through the origin.

2. *Horizontal lines:* A horizontal line has an equation of the form $y = b$. Its slope is 0.

3. *Vertical lines:* A vertical line has an equation of the form $x = c$. Its slope is undefined.

4. *Parallel lines:* Two lines are parallel if they have the same slope. For example, $y = 5 + 2x$ and $y = -7 + 2x$ are parallel since they both have a slope of 2.

5. *Perpendicular lines:* Two lines are perpendicular if their slopes are negative reciprocals. For example, the lines $y = 5 + 2x$ and $y = 8 - \frac{1}{2}x$ are perpendicular since $-\frac{1}{2}$ is the negative reciprocal of 2.

For a data set whose graph exhibits a linear pattern, we can visually position a line to fit the data. The equation of this line offers an approximate but compact description of the data.

Lines are of special importance because they are easily drawn and manipulated. Lines offer the simplest choice for a rough description of patterns in data.

EXERCISES

Exercises for Section 2.1

Year	Score
1967	466
1970	460
1975	434
1980	424
1985	431
1990	424
1992	423
1993	424
1994	423

1. The accompanying table indicates average male verbal SAT scores for the years 1967 to 1994, before the scores were "recentered."

   Find the average rate of change:

   a. From 1967 to 1994 **b.** From 1992 to 1994

2. **a.** In 1990, aerospace industry net profits in the United States were $4.49 billion, whereas in 1992, the aerospace industry showed a net loss (negative profit) of $1.84 billion. Find the average annual rate of change in net profits from 1990 to 1992.

   b. In 1997, aerospace industry net profits were $7.22 billion. Find the average rate of change in net profits

   i. From 1990 to 1997

   ii. From 1992 to 1997

3. According to the U.S. Bureau of the Census, in elementary and secondary schools, in the academic year ending in 1985, there were 631,983 microcomputers being used for student instruction, or 62.7 students per microcomputer. In the academic year ending in 1998, there were 8,049,875 microcomputers being used, or 6.4 students per microcomputer.

   Find the average rate of change from 1985 to 1998 in:

   a. The number of microcomputers being used

   b. The number of students per microcomputer

Year	Percentage of Persons 25 Years Old and Over Completing 4 or More Years of College
1940	4.6
1998	24.4

Source: U.S. Bureau of the Census, Current Population Reports, P-60 Series, *The Statistical Abstract of the United States,* 1998.

4. Using the information in the accompanying table on college completion:

   a. Plot the data, labeling both axes and the coordinates of the points.

   b. Calculate the average rate of change (in percentage points per year).

   c. Write a topic sentence summarizing what you think is the central idea to be drawn from these data.

5. Use the information in the accompanying table to answer the following questions:

Percentage of Persons 25 Years Old and Over Who Have Completed 4 Years of High School or More

	1940	1998
All	24.5	82.8
White	26.1	83.7
Black	7.3	76.0
Asian/Pacific Islander	22.6	85.0

Source: U.S. Bureau of the Census, Current Population Reports, P-60 Series, *The Statistical Abstract of the United States,* 1998.

   a. What has been the average rate of change (of percentage points per year) of completion of 4 years of high school from 1940 to 1998 for whites? For blacks? For Asian/Pacific Islanders? For all?

b. If these rates continue, what percentages of whites, of blacks, of Asian/Pacific Islanders, and of all will have finished 4 years of high school in the year 2000?

c. Write a 60 second summary describing the key elements in the high school completion data. Include rates of change and possible projections for the near future.

d. If these rates continue, in what year would 100% of whites have completed 4 years of high school or more? In what year 100% of blacks? 100% of Asian/Pacific Islanders? Do these projections make sense?

6. The accompanying data and graph show U.S. consumption and export of cigarettes.

Cigarette Market

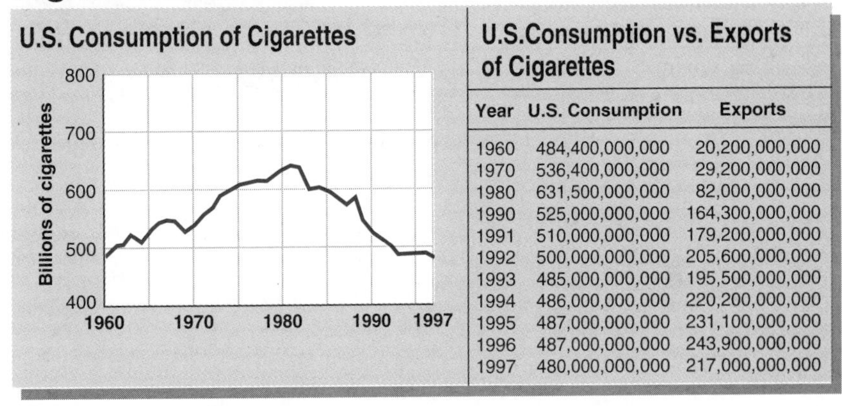

U.S.Consumption vs. Exports of Cigarettes

Year	U.S. Consumption	Exports
1960	484,400,000,000	20,200,000,000
1970	536,400,000,000	29,200,000,000
1980	631,500,000,000	82,000,000,000
1990	525,000,000,000	164,300,000,000
1991	510,000,000,000	179,200,000,000
1992	500,000,000,000	205,600,000,000
1993	485,000,000,000	195,500,000,000
1994	486,000,000,000	220,200,000,000
1995	487,000,000,000	231,100,000,000
1996	487,000,000,000	243,900,000,000
1997	480,000,000,000	217,000,000,000

Source: U.S. Agriculture Department.

a. Calculate the average rates of change in U.S. cigarette consumption from 1960 to 1980, from 1980 to 1997, and from 1960 to 1997.

b. Compute the average rates of change for the same time periods for cigarette exports.

c. Sketch what the graph of exports would look like from 1960 to 1997.

d. The total number of cigarettes consumed in the United States in 1960 is very close to the number consumed in 1997. Does that mean smoking is as popular in current times as it was in 1960? Explain your answer.

e. Write a paragraph summarizing what the data tell you about the consumption and exports of the cigarette industry since 1960, using average rates of change.

7. Use the accompanying table on life expectancy to answer the following questions. (See also Excel or graph link file LIFEXPEC.)

Average Number of Years of Life Expectancy in the United States by Race and Sex Since 1900

Life Expectancy at Birth by Year	White Males	White Females	Black Males	Black Females
1900	46.6	48.7	32.5	33.5
1950	66.5	72.2	58.9	62.7
1960	67.4	74.1	60.7	65.9
1970	68.0	75.6	60.0	68.3
1980	70.7	78.1	63.8	72.5
1990	72.7	79.4	64.5	73.6
2000 (est)	74.2	80.5	64.6	74.7

Source: U.S. National Center for Health Statistics, *The Statistical Abstract of the United States,* 1995, 1997, 1998.

a. What group had the highest life expectancy in 1900? In 2000? What group had the lowest life expectancy in 1900? In 2000?

b. Which group had the largest average rate of change in life expectancy between 1900 and 2000?

c. Write a short summary of the patterns in U.S. life expectancy over the last century using average rates of change to support your points.

8. The data and chart given show the changing uses of domestic TVs and the hours spent watching them.

Movie Box-Office Gross and Attendance (in millions)

Year	Box office gross	Attendance/ admissions
1950	$1,379.0	3,017.5
1960	984.4	1,304.5
1970	1,429.2	920.6
1980	2,748.5	1,021.5
1981	2,965.6	1,067.0
1982	3,452.7	1,175.4
1983	3,766.0	1,196.9
1984	4,030.6	1,199.1
1985	3,749.4	1,056.1
1986	3,778.0	1,017.2
1987	4,252.9	1,088.5
1988	4,458.4	1,084.8
1989	5,033.4	1,262.8
1990	5,021.8	1,188.6
1991	4,803.2	1,140.6
1992	4,871.0	1,173.2
1993	5,154.2	1,244.0
1994	5,396.2	1,291.7
1995	5,493.5	1,262.6
1996	5,911.5	1,338.6
1997	6,365.9	1,387.7

Source: Wall Street Journal Almanac, 1999.

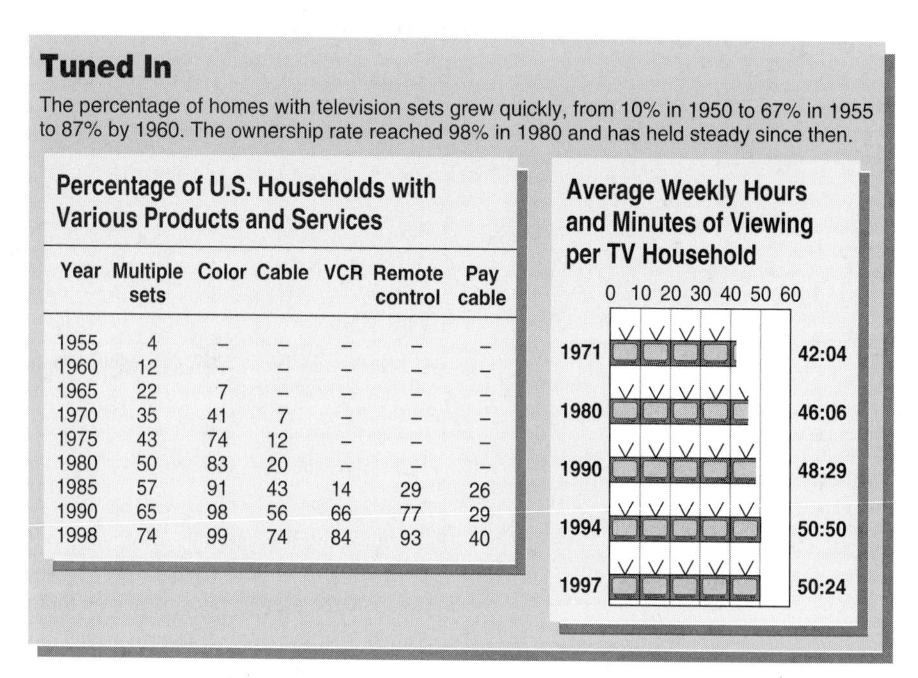

Tuned In

The percentage of homes with television sets grew quickly, from 10% in 1950 to 67% in 1955 to 87% by 1960. The ownership rate reached 98% in 1980 and has held steady since then.

Percentage of U.S. Households with Various Products and Services

Year	Multiple sets	Color	Cable	VCR	Remote control	Pay cable
1955	4	–	–	–	–	–
1960	12	–	–	–	–	–
1965	22	7	–	–	–	–
1970	35	41	7	–	–	–
1975	43	74	12	–	–	–
1980	50	83	20	–	–	–
1985	57	91	43	14	29	26
1990	65	98	56	66	77	29
1998	74	99	74	84	93	40

Source: Nielsen Media Research.

a. What is the average rate of change between 1985 and 1998 of the percentage of households with cable, VCR, and pay cable? Which of these appears to have the greatest potential for further growth?

b. Since network TV, cable, VCR, and pay cable all are watched on a TV set, what do the data imply about the popularity of network TV?

c. Another side of the screen entertainment industry is represented by the movie data given on box office gross and movie attendance. From 1950 to 1970 what was the average rate of change of movie attendance? What factors do you think caused this change?

d. i. What was the average rate of change of movie attendance from 1970 to 1997?

ii. If the U.S. population was 204.0 million in 1970 and 267.6 million in 1997, what was the movie attendance per person in the United States for each of those years?

iii. What was the average rate of change from 1970 to 1997 of movie attendance per person?

e. From the information given, does it appear that the range of entertainment available on the domestic TV has affected the growth of big-screen entertainment at movie theaters? Write a paragraph on your conclusions.

Exercises for Section 2.2

9. The accompanying table and graph illustrate the average annual salary of professional baseball players from 1990 to 1996.

Year	Salary (millions)	Average Rate of Change over Prior Year
1990	\$0.60	n.a.
1991	\$0.85	
1992	\$1.03	
1993	\$1.08	
1994	\$1.17	
1995	\$1.11	
1996	\$1.12	

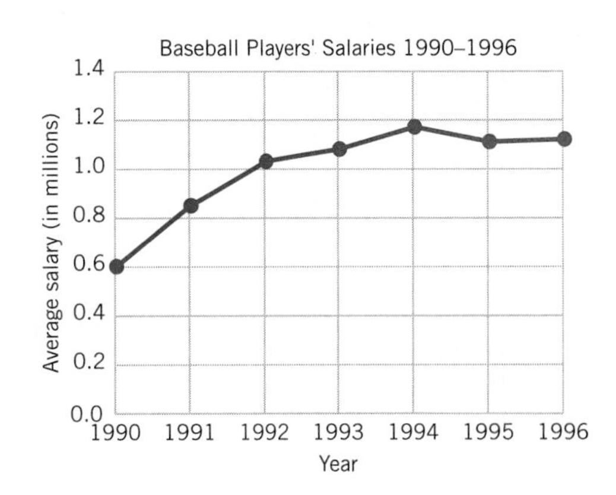

a. Fill in the third column in the table.

b. During which 1-year interval was the average rate of change the smallest in absolute size?

c. During which 1-year interval was the average rate of change the largest?

d. Write a paragraph describing the change in professional baseball players' salaries between 1990 and 1996.

10. The accompanying table indicates the number of juvenile arrests (in thousands) in the United States for aggravated assaults.

Year	Arrests (thousands)	Annual Average Rate of Change over prior 5 years
1980	33.5	n.a.
1985	36.8	
1990	54.5	
1995	68.5	

a. Fill in the third column in the table by calculating the annual average rate of change.

b. Graph the annual average rate of change vs. time.

c. During what period was the annual average rate of change the largest?

d. Describe the growth of aggravated assault cases during these years by referring to the annual average rates of change.

11. The accompanying table and graph show the change in median age of the U.S. population from 1850 through the present and projected into the next century. (See also Excel or graph link file USMEDAGE.)

a. Using the given data, specify the longest time period over which the median age has been increasing. Calculate the average rate of change between the two end points of this time period.

b. Calculate the average rate of change between 1850 and 1900, 1900 and 1950, and 1950 and 2000 and the projected average rate of change between 2000 and 2050. Write a paragraph comparing these results.

c. Does an increase in median age necessarily mean that more people are living longer? What changes in society could make this change in median age possible?

Median Age of U.S. Population: 1850–2050

Year	Median Age
1850	18.9
1860	19.4
1870	20.2
1880	20.9
1890	22.0
1900	22.9
1910	24.1
1920	25.3
1930	26.4
1940	29.0
1950	30.2
1960	29.5
1970	28.0
1980	30.0
1990	32.8
2000	35.7
2025	38.0
2050	38.1

Note: Data for 2000–2050 are projected.
Source: U.S. Bureau of the Census, *www.census.gov.*

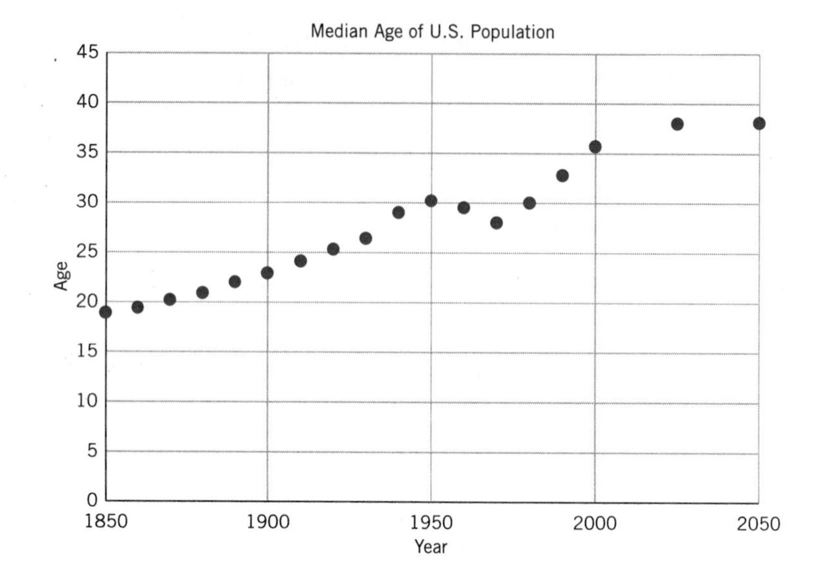

A graphing calculator or computer is helpful for the following parts:

d. Generate a partial third column for the table, starting in 1860 and ending in 2000, that for each listed year contains the average rate of change of median age over the previous decade.

e. Plot average rate of change vs. years (columns 3 and 1 in your table).

f. Identify periods with negative average rates of change and suggest reasons for the declining rates.

12. The accompanying table and graph are both representations of the accumulated gross debt of the federal government as a function of time. (See also Excel and graph link file FEDDEBT.)

Year	Debt (billions of dollars)	Year	Debt (billions of dollars)
1945	260	1983	1,371
1950	257	1984	1,564
1955	274	1985	1,827
1960	291	1986	2,120
1965	322	1987	2,346
1970	361	1988	2,601
1971	408	1989	2,868
1972	436	1990	3,266
1973	466	1991	3,599
1974	484	1992	4,082
1975	541	1993	4,436
1976	629	1994	4,721
1977	706	1995	5,001
1978	777	1996	5,260
1979	829	1997	5,502
1980	909	1998	5,614
1981	994	1999	5,691 (est.)
1982	1,137		

Source: U.S. Bureau of the Census, *Statistical Abstract of the United States,* 1998 and the Department of the Treasury, Bureau of the Public Debt On-Line at www.publicdebt.treas.gov.

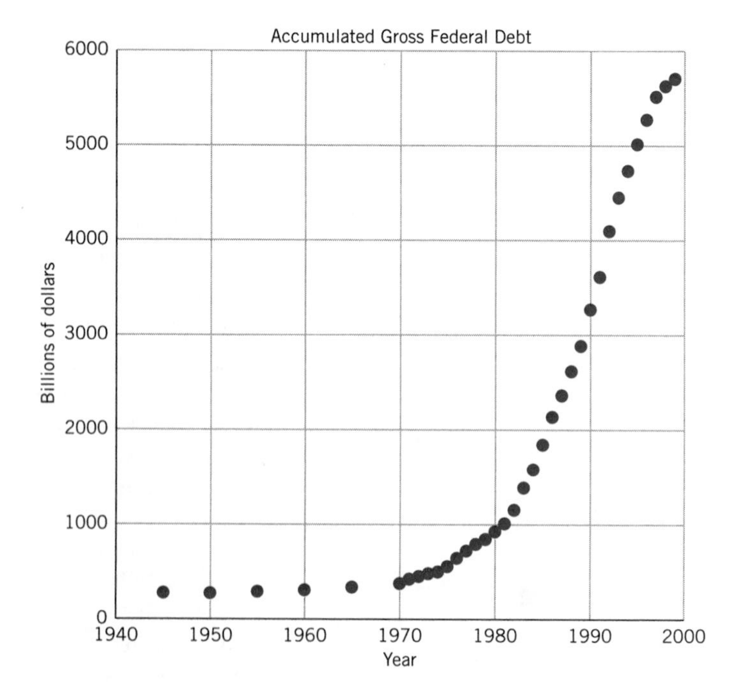

A graphing calculator is helpful for the following tasks:

a. Generate a third column for the table containing the average annual rate of change of the debt during the previous interval. Write a paragraph describing what these numbers represent.

b. Graph the average rates of change from your third column vs. time. Write a paragraph describing what your graph and the data show.

13. The accompanying data and graphs give a picture of the two major methods of news communication in the United States. (See also Excel or graph link files NEWPRINT and ONAIRTV.)

Year	Newspapers (thousands of copies printed)	Number of Newspapers Published	Year	Newspapers (thousands of copies printed)	Number of Newspapers Published
1915	28,777	2580	1960	58,882	1763
1920	27,791	2042	1965	60,358	1751
1925	33,739	2008	1970	62,108	1748
1930	39,589	1942	1975	60,655	1756
1935	38,156	1950	1980	62,202	1745
1940	41,132	1878	1985	62,766	1676
1945	48,384	1749	1990	62,324	1611
1950	53,829	1772	1995	58,200	1583
1955	56,147	1760			

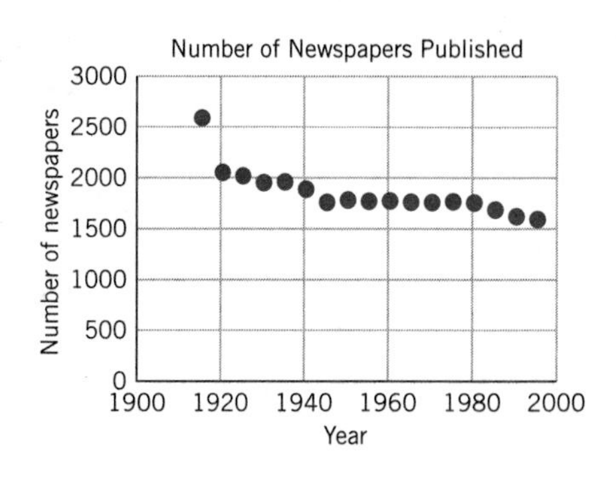

Year	Number of On-Air TV Stations
1950	98
1955	411
1960	515
1965	569
1970	677
1975	706
1980	734
1985	883
1990	1092
1995	1532

Source: U.S. Bureau of the Census, *www.census.gov.*

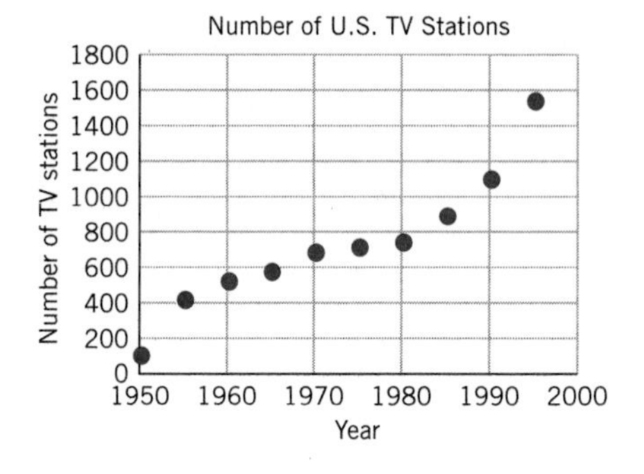

a. Use the U.S. population numbers from Table 2.1 to calculate and compare the number of copies of newspapers *per person* in 1920 and in 1990.

b. Create a table that displays the annual average rate of change in TV stations for each 5-year period since 1950. Create a similar table that displays the annual average rate of change in newspapers published for the same period. Graph the results. (Use of a graphing calculator or computer is recommended.)

c. If new TV stations continue to come into existence at the same rate as from 1990 to 1995, how many will there be by the year 2005? Do you think this is likely to be a reasonable projection or is it overly large or small judging from past rates of growth? Explain.

d. What trends do you see in the dissemination of news as reflected in these data?

Exercises for Section 2.3

14. Plot each pair of points and calculate the slope of the line that passes through them.

a. $(3, 5)$ and $(8, 15)$
b. $(-1, 4)$ and $(7, 0)$
c. $(5, 9)$ and $(-5, 9)$
d. $(-2, 6)$ and $(2, -6)$
e. $(-4, -3)$ and $(2, -3)$

15. Find the value of t if m is the slope of the line that passes through the given points:

a. $(3, t)$ and $(-2, 1)$, $m = -4$
b. $(5, 6)$ and $(t, 9)$, $m = \frac{2}{3}$

16. a. Find the value of x so that the slope of the line through $(x, 5)$ and $(4, 2)$ is $\frac{1}{3}$.
b. Find the value of y so that the slope of the line through $(1, -3)$ and $(-4, y)$ is -2.
c. Find the value of y so that the slope of the line through $(-2, 3)$ and $(5, y)$ is 0.

17. Points that lie on the same line are said to be *collinear*. Determine if the following points are collinear:

a. $(2, 3)$, $(4, 7)$, and $(8, 15)$
b. $(-3, 1)$, $(2, 4)$, and $(7, 8)$

18. Specify the intervals on the accompanying graph for which the slope of the line segment between adjacent data points appears positive. For which does it appear negative? For which zero?

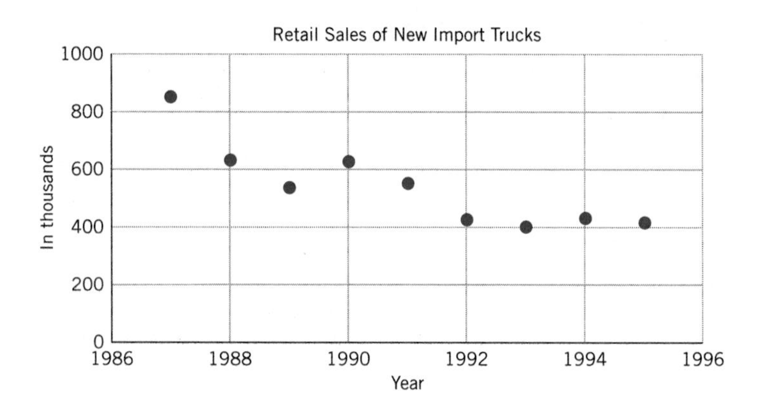

19. Given the accompanying graph of a function, specify the intervals over which:

a. The *function* is positive, negative, or zero
b. The *rate of change between any two points in the interval* is positive, is negative, or is zero

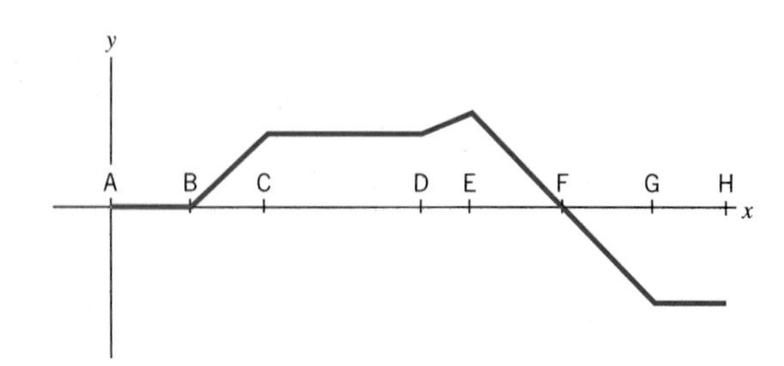

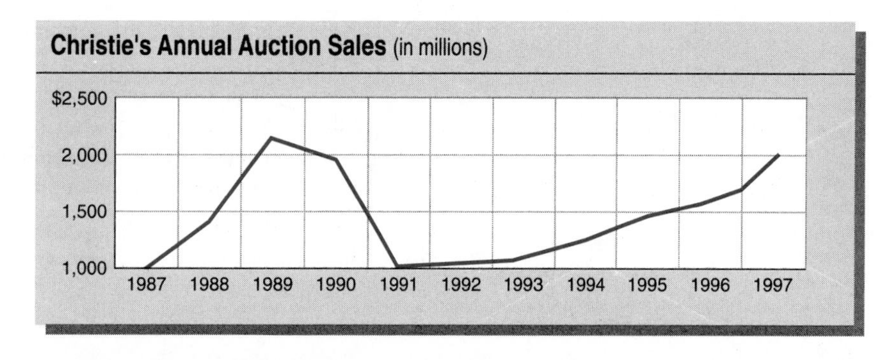

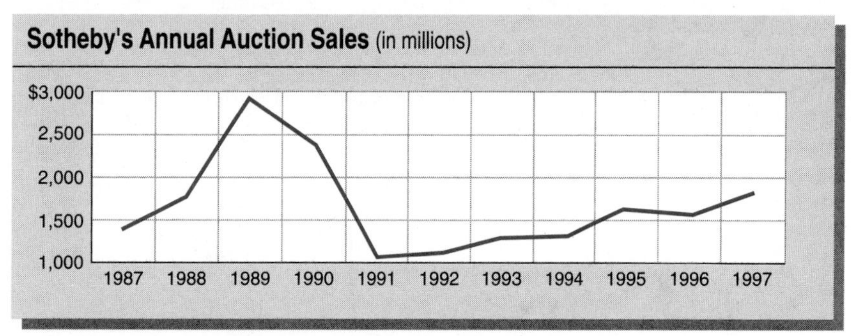

Sources: Sotheby's Holdings Inc. and *Christie's International PLC*

20. Examine the given graphs of Christie's and Sotheby's auction sales showing their annual sales in the art market.

a. Identify a period of negative slope on the Christie's graph. Describe in words what negative slope means in this art market. Estimate the slope, giving units.

b. Identify a period of positive slope on the Sotheby's graph. Describe in words what positive slope means in this art market. Estimate the slope, giving units.

c. Write a paragraph about these art markets using numerical descriptions or comparisons and mentioning both discouraging and encouraging features you see.

21. Read the Anthology article "Slopes" and describe two practical applications of slopes, one of which is from your own experience.

Exercises for Section 2.4

22. Examine the data given on women in the U.S. military forces.

Women in Uniform: Female Active-Duty Military Personnel

Year	Total	Army	Navy	Marine Corps	Air Force
1965	30,610	12,326	7,862	1,581	8,841
1970	41,479	16,724	8,683	2,418	13,654
1975	96,868	42,295	21,174	3,186	30,213
1980	171,418	69,338	34,980	6,706	60,394
1985	211,606	79,247	52,603	9,695	70,061
1990	227,018	83,621	59,907	9,356	74,134
1991	221,138	80,306	59,391	9,005	72,436
1992	210,048	73,430	59,305	8,524	68,789
1993	203,506	71,328	57,601	7,845	66,732
1994	199,688	69,878	55,825	7,671	66,314
1995	196,116	68,046	55,830	8,093	64,147
1996	197,693	69,623	54,692	8,564	64,814
1997	200,526	72,827	52,578	9,286	65,835

Source: U.S. Defense Department.

a. Make the case with graphs and numbers that women are a growing presence in the U.S. military.

b. Make the case with graphs and numbers that women are a declining presence in the U.S. military.

c. Write a paragraph that gives a fair picture of the changing presence of women in the military using appropriate statistics to make your points. What additional data would be helpful?

23. The accompanying table and graph show the number of new AIDS cases reported in Florida from 1986 to 1997.

Year	Number of New AIDS Cases
1986	1,031
1987	1,633
1988	2,650
1989	3,448
1990	4,018
1991	5,471
1992	5,086
1993	10,958
1994	8,617
1997	6,098

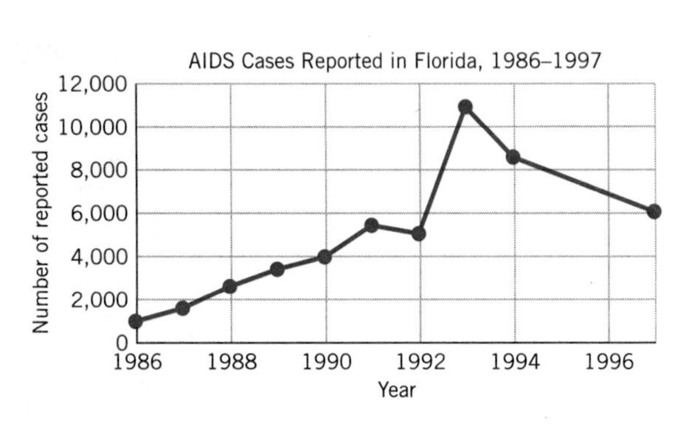

a. Find something encouraging to say about these data by using numerical evidence, including average rates of change.

b. Find numerical support for something discouraging to say about the data.

c. How might we explain the enormous jump in new AIDS cases reported from 1992 to 1993 and the drop-off the following year? Is there anything potentially misleading about the data?

24. The accompanying graphs are an approximation of the costs of doctors' bills and Medicare between 1963 and 1975.

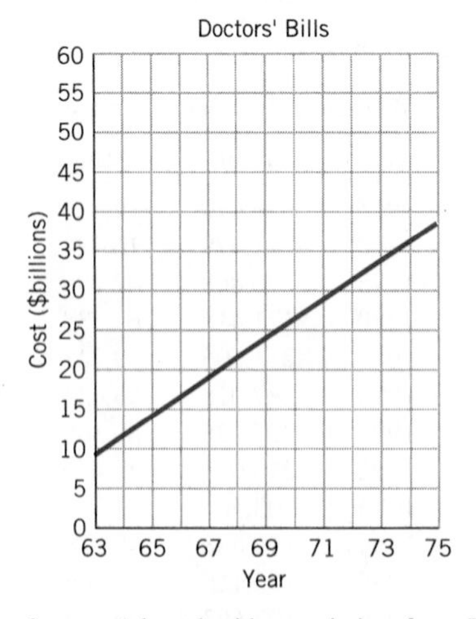

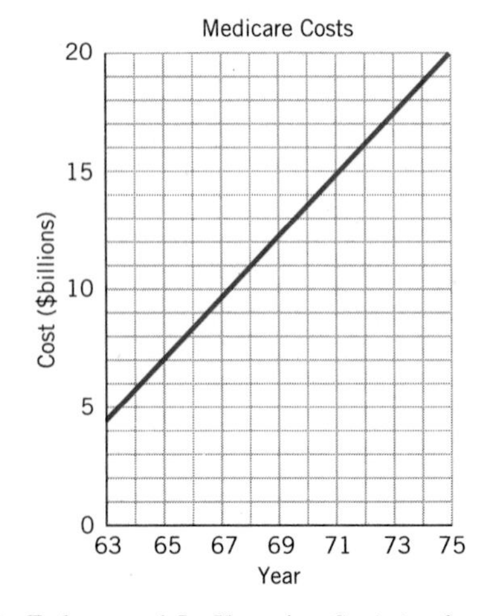

Source: Adapted with permission from L. Ferleger and L. Horowitz, *Statistics for Social Change* (Boston, South End Press, 1980).

a. At first glance, which *appears* to have been growing at a faster average annual rate of change: doctors' bills or Medicare costs? Why?

b. Which actually grew at a faster average annual rate of change? How can you tell?

25. The accompanying graphs show the same data on the income of black people as a percentage of the income of white people over a 25-year period. Describe the impression each graph gives and how the axes have been altered in each graph to convey these impressions.

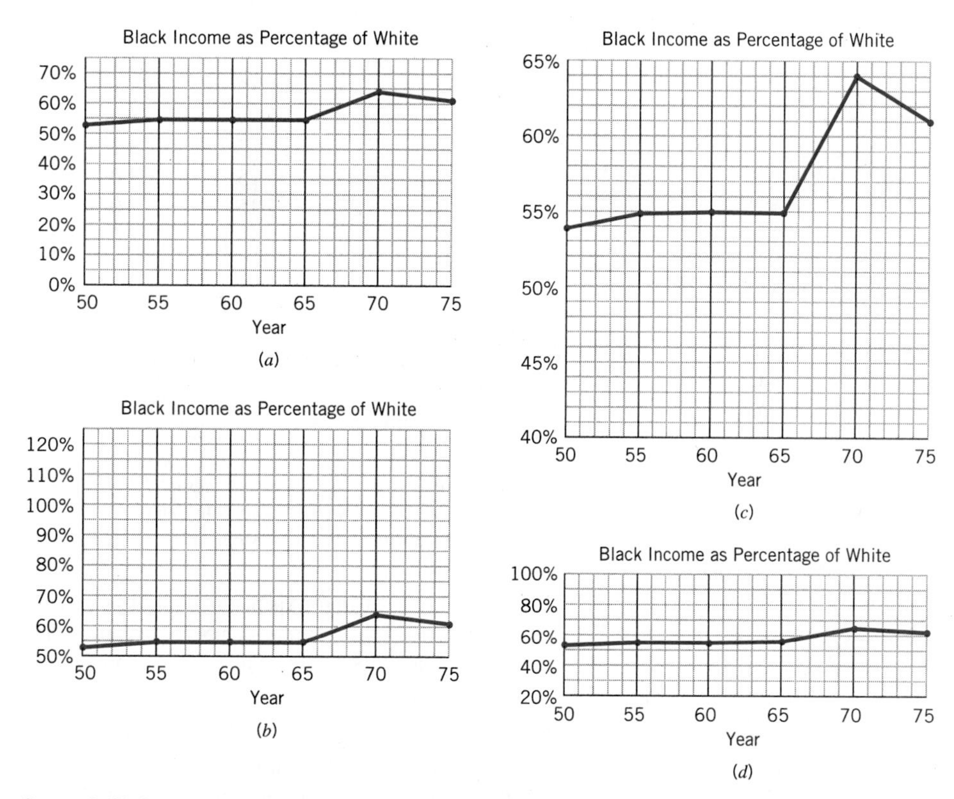

Source: L. Ferleger and L. Horowitz, *Statistics for Social Change* (Boston, South End Press, 1980).

26. Laws to regulate environmental pollution in America are a very recent phenomenon, with the first federal regulations appearing in the 1950s. The Clean Air Act, passed in 1963 and amended in 1970, established for the first time uniform national air pollution standards. The act placed national limits and a timetable on three classes of automotive pollutants: hydrocarbons (HC), carbon monoxide (CO), and nitrogen oxides (NO). All new cars for each particular year were restricted from exceeding these standards. The accompanying tables on the next page show the results. (See also Excel or graph link files POLLUTE for emission standards per mile and EMISSION for total annual automobile emission estimates.)

a. Make a convincing argument that the Clean Air Act was a success.

b. Make a convincing argument that the Clean Air Act was a failure. (Hint: How is it possible that the emissions per vehicle mile were down but the total amount of emissions did not improve?)

National Autmobile Emission Control Standards

Model Year Applicable	HC (grams/ mile)	CO (grams/ mile)	NO_2* (grams/ mile)
Pre-1968	8.7	87	4.4
1968	6.2	51	n.r.[†]
1970	4.1	34	n.r.
1972	3.0	28	n.r.
1973	3.0	28	3.1
1975	1.5	15	3.1
1975 (C)[‡]	0.9	9	2.0
1976	1.5	15	3.1
1977	1.5	15	2.0
1978	1.5	15	2.0
1979	1.5	15	2.0
1980	0.41	7	2.0
1981 and beyond	0.41	3.4	1.0

*NO_2, nitrogen dioxide, is a form of nitrogen oxide.

[†]No requirement.

[‡]California standards.

Source: P. Portney (ed.), *Current Issues in U.S. Environmental Policy* (Baltimore, MD: Johns Hopkins University Press, 1978), p. 76.

National Automobile Emissions Estimates (million metric tons per year)

Year	HC	CO	NO_2
1970	28.3	102.6	19.9
1971	27.8	103.1	20.6
1972	28.3	104.4	21.6
1973	28.4	103.5	22.4
1974	27.1	99.6	21.8
1975	25.3	97.2	20.9
1976	27.0	102.9	22.5
1977	27.1	102.4	23.4
1978	27.8	102.1	23.3

Source: Environmental Quality—1980: The Eleventh Annual Report of the Council on Environmental Quality, p. 170.

27. Open "L6: Changing Axis Scales" in Linear Functions. Generate a line in the upper left-hand box. The same line will appear graphed in the three other boxes but with the axes scaled differently. Describe how the axes are rescaled in order to create such different impressions.

28. The accompanying graph shows the number of Nobel Prizes awarded in science for various countries between 1901 and 1974. It contains accurate information but gives the impression that the number of prize winners declined drastically in the 1970s, which was not the case. What flaw in the construction of the graph leads to this impression?

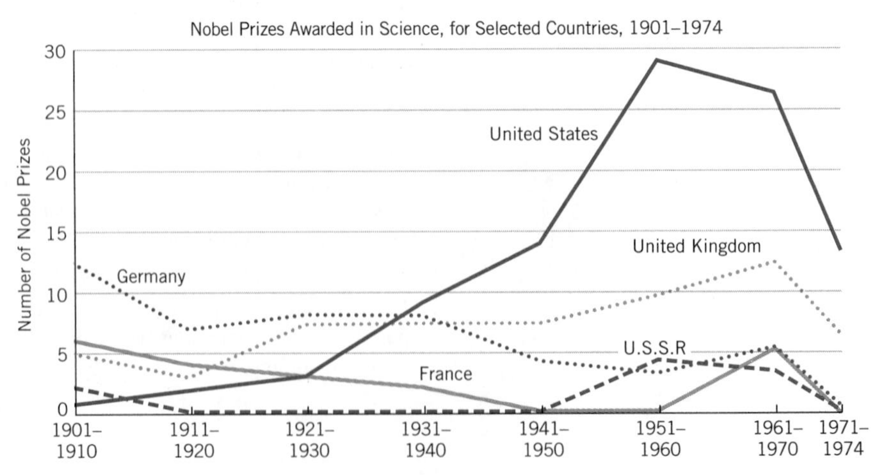

Source: E. R. Tufte, *The Visual Display of Quantitative Information,* (Cheshire, Connecticut: Graphics Press, 1983).

Exercises for Section 2.5

29. Consider the equation $E = 5000 + 1000n$.

a. Find the value of E for $n = 0, 1, 20$.

b. Express your answers to part (a) as points with coordinates (n, E).

30. Consider the equation $G = 12{,}000 + 800n$.

a. Find the value of G for $n = 0, 1, 20$.

b. Express your answers to part (a) as points with coordinates (n, G).

31. Determine if any of the following points satisfy one or both of the equations in questions 29 and 30:

a. (5000, 0) **b.** (15, 24000) **c.** (35, 40000)

32. Suppose the equations $E = 5000 + 1000n$ and $G = 12{,}000 + 800n$ give the total cost of operating an electrical (E) vs. a gas (G) heating/cooling system in a home for n years.

a. Find the cost of heating a home using electricity for 10 years.

b. Find the cost of heating a home using gas for 10 years.

c. Find the initial (or installation) cost for each system.

d. Determine how many years it will take before \$40,000 has been spent in heating/cooling a home that uses:

i. Electricity **ii.** Gas

33. If the equation $E = 5000 + 1000n$ gives the total cost of heating/cooling a home after n years, rewrite the equation using only units of measure.

34. a. If $S(x) = 20{,}000 + 1000x$ describes the annual salary in dollars for a person who has worked for x years for the Acme Corporation, then what is the unit of measure for 20,000? For 1000?

b. Rewrite $S(x)$ as an equation using only units of measure.

c. Evaluate $S(x)$ for x values of 0, 5, and 10 years.

d. How many years will it take for a person to earn an annual salary of \$43,000?

35. The following represent linear equations written only using units of measure. In each case supply the missing unit.

a. inches = inches + (inches/hour) (?)

b. miles = miles + (?) (gallons)

c. calories = calories + (?) (grams of fat)

Exercises for Section 2.6

You might wish to hone your mechanical skills with three programs in *Linear Functions:* "L1: Finding m & b," "L3: Finding a Line through 2 Points," and "L4: Finding 2 Points on a Line." They offer practice in predicting values for m and b, generating linear equations and finding corresponding solutions.

36. Identify the slopes and the vertical intercepts of the lines with the given equations:

a. $y = 3 + 5x$ **c.** $y = 4$ **e.** $f(E) = 10{,}000 + 3000E$

b. $f(t) = -t$ **d.** $Q = 35t - 10$

37. For each of the following, find the slope and the vertical intercept. Sketch a graph.

a. $y = 0.4x - 20$ **b.** $P = 4000 - 200C$

38. Construct an equation and sketch the graph of the line with the given slope, m, and vertical intercept, b.

a. $m = 3, b = -2$ **b.** $m = -\frac{3}{4}, b = 1$ **c.** $m = 0, b = 50$

In Exercises 39 to 41 find an equation, generate a small table of solutions, and sketch the graph.

39. A line that has a vertical intercept of -2 and a slope of 3.

40. A line that crosses the vertical axis at 3.0 and has a rate of change of -2.5.

41. A line that has a vertical intercept of 1.5 and a slope of 0.

42. Estimate b (the y-intercept) and m (the slope) for each of the accompanying graphs. Then, for each graph, write the corresponding linear function. Be sure to note the scales on the axes.

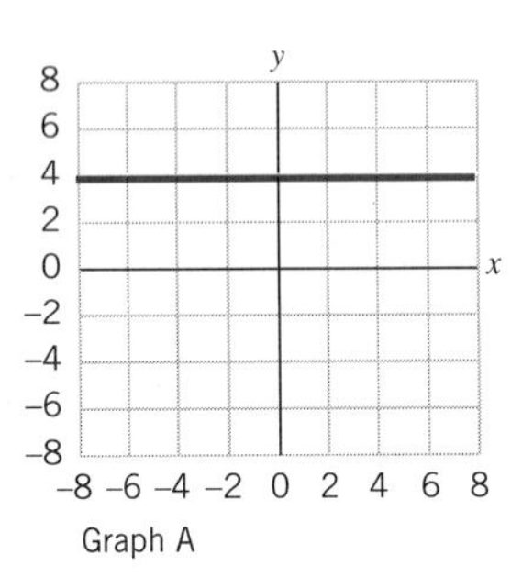

Graph A

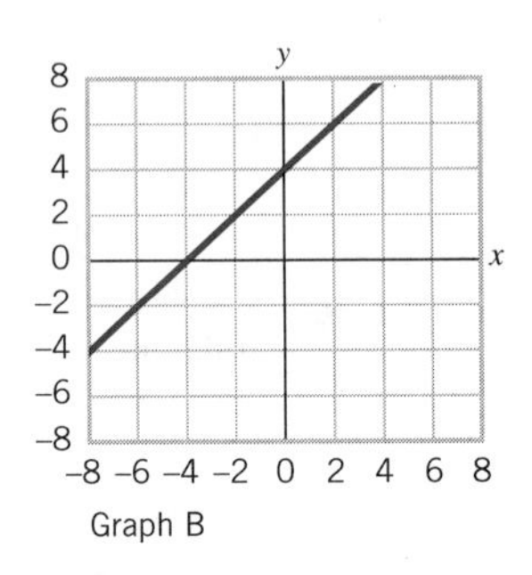

Graph B

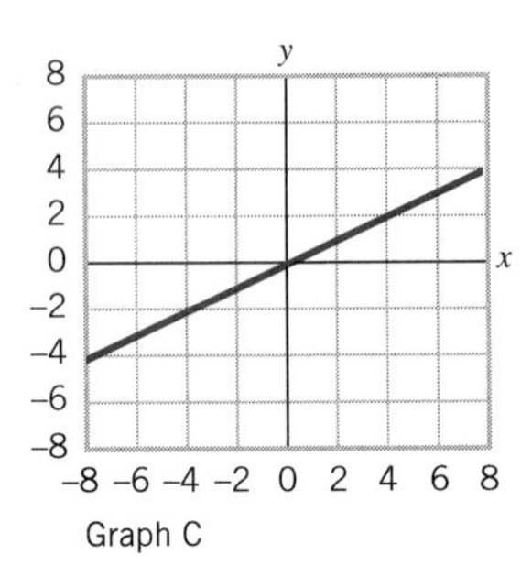

Graph C

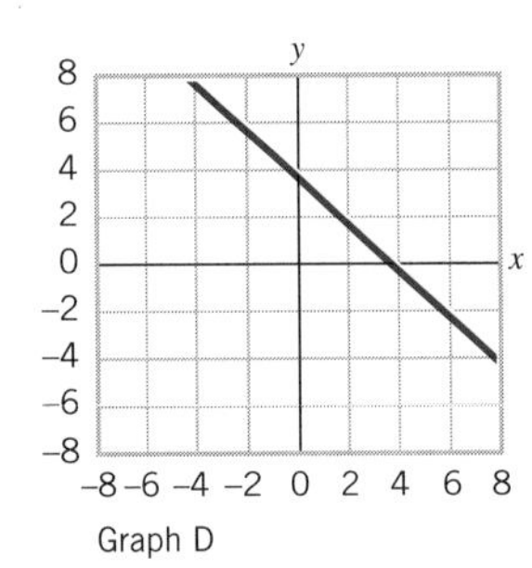

Graph D

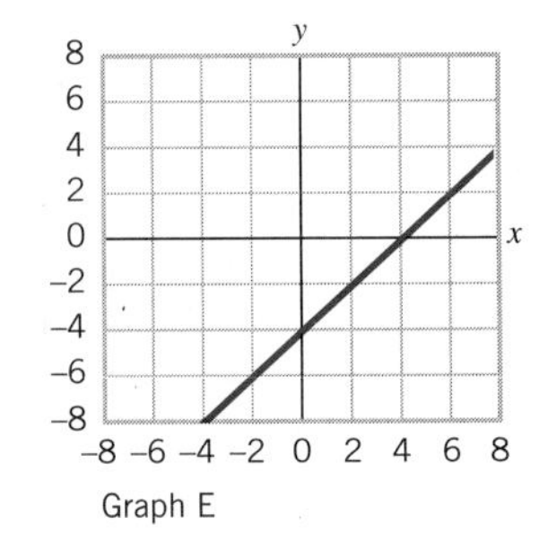

Graph E

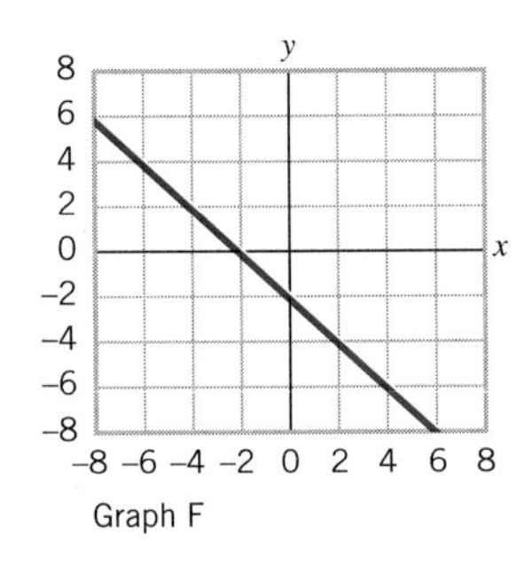

Graph F

43. Write an equation for the line through $(-2, 3)$ that has slope:

a. 5 **b.** $-\frac{3}{4}$ **c.** 0

44. Write an equation for the line through (0, 50) that has slope:

a. -20 **b.** 5.1 **c.** 0

45. Calculate the slope and write an equation for the linear function represented by each of the given tables.

a.

x	y
2	7.6
4	5.1

b.

A	W
5	12
7	16

46. Construct a linear equation that satisfies each of the following properties:

a. A negative slope and a positive y-intercept

b. A positive slope and a vertical intercept of -10.3

c. A constant rate of change of $1300/year

47. Find an equation to represent the cost of attending college classes if application and registration fees are $150 and classes cost $120 per credit.

48. a. Write an equation that describes the total cost to produce x items if the startup cost is $200,000 and the production cost per item is $15.

b. Why is the total cost per item less if the item is produced in large quantities?

49. Your bank charges you a \$2.50 monthly maintenance fee on your checking account and an additional \$.10 for each check you cash. Write an equation to describe your monthly checking account costs.

50. If a town starts with a population of 63,500 that declines by 700 people each year, construct an equation to model its population size over time. How long would it take for the population to drop to 53,000?

51. A teacher's union has negotiated a uniform salary increase for each year of service up to 20 years. If a teacher started at \$26,000 and 4 years later had a salary of \$32,000:

a. What was the annual increase?

b. What function would describe the teacher's salary over time?

c. What would be the domain for the function?

52. Your favorite aunt put money in a savings account for you. The account earns simple interest, that is, it increases by a fixed amount each year. After 2 years your account has \$8,250 in it and after 5 years it has \$9,375.

a. Construct an equation to model the amount of money in your account.

b. How much did your aunt put in initially?

c. How much will you have after 10 years?

53. The equation $K = 4F - 160$ models the relationship between F, the temperature in degrees Fahrenheit, and K, the number of cricket chirps per minute for the snow tree cricket.

a. Assuming F is the independent variable and K is the dependent variable, identify the slope and vertical intercept in the above equation.

b. Identify the units for K, 4, F, and -160.

c. What is a reasonable domain for this model?

d. Generate a small table of points that satisfy the equation. Be sure to choose realistic values for F from the domain of your model.

e. Calculate the slope directly from two data points. Is this value what you expected? Why?

f. Graph the equation, indicating the domain.

54. The equation $F = 32 + \frac{9}{5}C$ describes the relationship between degrees Fahrenheit, F, and degrees Celsius, C.

a. Assuming C is the independent variable and F is the dependent variable, what are the values of the slope and the vertical intercept?

b. Identify the units for F, 32, $\frac{9}{5}$, and C.

c. Generate a small table of points that satisfy the equation.

d. Calculate the slope directly from two data points. Is this value what you expected? Why?

e. Graph the equation.

f. At what temperature are the Celsius and Fahrenheit temperatures equal? Show your work.

55. You read in the newspaper that the river is polluted with 285 parts per million (ppm) of a toxic substance, and local officials estimate they can reduce the pollution by 15 ppm each year.

a. Derive an equation that represents the amount of pollution, P, as a function of time, t.

b. The article states the river will not be safe for swimming until pollution is reduced to 40 ppm. If the cleanup proceeds as estimated, in how many years will it be safe to swim in the river?

56. The women's recommended weight formula from Harvard Pilgrim Healthcare says: "Give yourself 100 lb for the first 5 ft plus 5 lb for every inch over 5 ft tall."

a. Find a mathematical model for this relationship. Be sure you clearly identify your variables.

b. Specify a reasonable domain for the function and then graph it.

c. Use your model to calculate the recommended weight for a woman 5 feet, 4 inches tall; for one 5 feet, 8 inches tall.

57. In 1973 a math professor bought her house in Cambridge, Massachusetts, for $20,000. The value of the house has risen steadily so that in 1999 real estate agents tell her the house is now worth $250,000.

a. Find a formula to represent these facts about the value of the house, V, as a function of time, t.

b. If she retires in 2010, what does your formula predict her house will be worth then?

c. If she is now 57, and the house continues to gain value at the same rate, how old will she be when her house is worth half a million dollars?

58. The y-axis, the x-axis, the line $x = 6$, and the line $y = 12$ determine the four sides of a 6-by-12 rectangle in the first quadrant of the xy plane. Imagine that this rectangle is a pool table. There are pockets at the four corners and at the points (0, 6) and (6, 6) in the middle of each of the longer sides. When a ball bounces off one of the sides of the table, it obeys the "pool rule": The slope of the path after the bounce is the negative of the slope before the bounce. (*Hint:* It helps to sketch the pool table on a piece of graph paper first.)

a. Your pool ball is at (3, 8). You hit it toward the y-axis, along the line with slope 2.

i. Where does it hit the y-axis?

ii. If the ball is hit hard enough, where does it hit the side of the table next? And after that? And after that?

iii. Show that the ball ultimately returns to (3, 8). Would it do this if the slope had been different from 2? What is special about the slope 2 for this table?

b. A ball at (3, 8) is hit toward the y-axis, and bounces off it at $(0, \frac{16}{3})$. Does it end up in one of the pockets? If so, what are the coordinates of that pocket?

c. Your pool ball is at (2, 9). You want to shoot it into the pocket at (6, 0). Unfortunately, there is another ball at (4, 4.5) that may be in the way.

i. Can you shoot directly into the pocket at (6, 0)?

ii. You want to get around the other ball by bouncing yours off the y-axis. If you hit the y-axis at (0, 7), do you end up in the pocket? Where do you hit the line $x = 6$?

iii. If bouncing off the y-axis at (0, 7) didn't work, perhaps there is some point $(0, b)$ on the y-axis from which the ball would bounce into the pocket at (6, 0). Try to find that point.

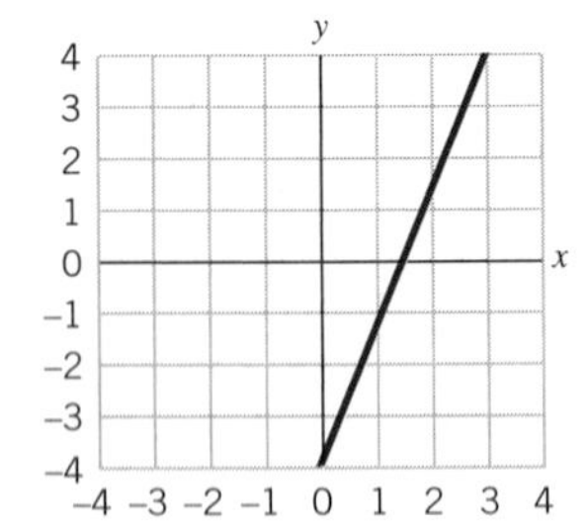

59. Find the equation of the line shown on the accompanying graph. Use this equation to create two new graphs, taking care to label the scales on your new axes. For one of your graphs, choose scales that make the line appear steeper than in the original graph. For your second graph, choose scales that make the line appear less steep than in the original graph.

60. The exchange rate a bank gave for French francs in June 1999 was 6 F for $1 U.S., and they also charged a constant fee of $5 per transaction. In France when changing francs to British pounds, £, the rate was 0.27£ for 1 F, with a transaction fee of 25 F.

a. Write a general equation for how many francs you got when changing dollars. Use F for French money and D for the dollars being exchanged. Make a graph of F vs. D.

b. Would it have made any sense to exchange $10 for francs? Explain.

c. Find a general expression for the percentage of the total francs converted from dollars that the bank kept for the transaction fee.

d. Write a general equation for how many pounds you got when changing francs. Use P for British pounds and F for the francs being exchanged. Make a graph of P vs. F.

e. If you changed dollars to francs, then changed the francs into pounds, find an expression for the number of pounds, P, you got for dollars, D. Make a graph of P vs. D.

f. Find out the current exchange rates for dollars, francs and pounds.

61. Suppose that

i. for 8 years of education, the mean salary for women is approximately $6800;

ii. for 12 years of education, the mean salary for women is approximately $11,600; and

iii. for 16 years of education, the mean salary for women is approximately $16,400.

a. Plot this information on a graph.

b. What sort of relationship does this information suggest between mean salary for women and education? Justify your answer.

c. Generate an equation that could be used to model the data from the limited information given (letting E = years of education and S = mean salary). Show your work.

62. a. Fill in the third column in the Tables 1 and 2.

Table 1

t	d	Average Rate of Change
0	400	n.a.
1	370	
2	340	
3	310	
4	280	
5	250	

Table 2

t	d	Average Rate of Change
0	1.2	n.a.
1	2.1	
2	3.2	
3	4.1	
4	5.2	
5	6.1	

b. In either table, is d a linear function of t? If so, construct a linear equation relating d and t for that table.

Relationship between Salinity and Freezing Point

Salinity (ppt)	Freezing Point (°C)
0	0.00
5	− 0.27
10	− 0.54
15	− 0.81
20	− 1.08
25	− 1.35

Source: Data adapted from P. R. Pinet, *Oceanography: An Introduction to the Planet Oceanus* (St. Paul, MN: West Publishing Company, 1992), p. 522.

63. Adding minerals or organic compounds to water lowers its freezing point. Antifreeze for car radiators contains glycol (an organic compound) for this purpose. The accompanying table shows the effect of salinity (dissolved salts) on the freezing point of water. Salinity is measured in the number of grams of salts dissolved in 1000 grams of water. So our units for salinity are in parts per thousand, abbreviated as *ppt.*

Is the relationship between the freezing point and salinity linear? If so, construct an equation that models the relationship. If not, explain why.

64. The accompanying data show average rounded values for blood alcohol concentration (BAC) for people of different weights, according to how many drinks (5 oz wine, 1.25 oz 80-proof liquor, or 12 oz beer) they have consumed.

	Blood Alcohol Concentration for Selected Weights					
Number of Drinks	100 lb	120 lb	140 lb	160 lb	180 lb	200 lb
2	0.075	0.063	0.054	0.047	0.042	0.038
4	0.150	0.125	0.107	0.094	0.083	0.075
6	0.225	0.188	0.161	0.141	0.125	0.113
8	0.300	0.250	0.214	0.188	0.180	0.167
10	0.375	0.313	0.268	0.235	0.208	0.188

a. Examine the data on BAC for a 100-pound person. Is this linear data? If so, find a formula to express blood alcohol concentration, A, as a function of the number of drinks, D, for a 100-pound person.

b. Examine the data on BAC for a 140-pound person. Is this linear data? If it's not precisely linear, what might be a reasonable estimate for the average rate of change of blood alcohol concentration, A, with respect to number of drinks, D? Find a formula to estimate blood alcohol concentration, A, as a function of number of drinks, D, for a 140-pound person. Can you make any general conclusions about BAC as a function of number of drinks for all of the weight categories?

c. Examine the data on BAC for people who drink two drinks. Is this linear data? If so, find a formula to express blood alcohol concentration, A, as a function of weight, W, for people who drink two drinks. Can you make any general conclusions about BAC as a function of weight for any particular number of drinks?

Exercises for Section 2.7

65. For each of the following linear functions, determine the independent and dependent variables. Write an equation.

a. Sales tax is 6.5% of the purchase price.

b. The height of a tree is directly proportional to the amount of sunlight it receives.

c. The average salary for full-time employees of American domestic industries has been growing at an annual rate of \$1300/year since 1985, when the average salary was \$25,000.

66. On the scale of a map 1 inch represents a distance of 35 miles.

a. What is the distance between two points that are 4.5 inches apart on the map?

b. Construct an equation that converts inches on the map to miles in the real world.

t (hours)	d (miles)
0	0
1	5
2	10
3	15
4	20

67. Find a function that represents the relationship between distance, d, and time, t, of a moving object using the data in the accompanying table. Is d directly proportional to t? Which is a more likely choice for the object, a person jogging or a moving car?

68. Determine which of the following variables (w, y, or z) vary directly with x:

x	w	y	z
0	1	0	0
1	2	2.5	$-\frac{1}{3}$
2	5	5	$-\frac{2}{3}$
3	10	7.5	-1
4	17	10	$-\frac{4}{3}$

69. The accompanying figure shows worldwide shipments of 5.25-inch computer disk drives between 1983 and 1993. Construct the equation of a horizontal line that would be a reasonable model for these data.

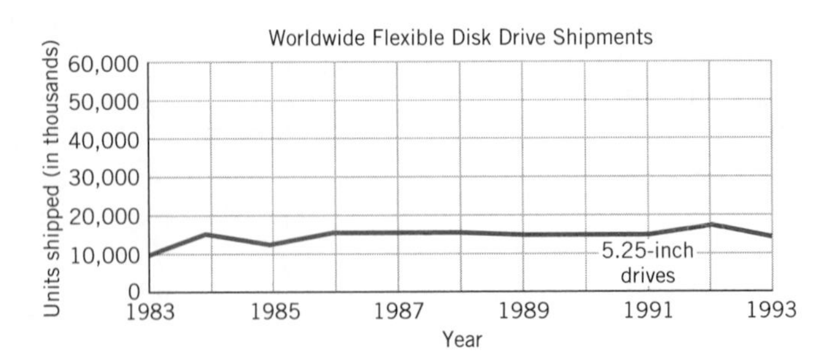

70. An employee for an aeronautical corporation has a starting salary of \$25,000/year. After working there for 10 years and not receiving any raises, he decides to seek employment elsewhere. Graph the employee's salary as a function of time for the time he was employed with this corporation. What is the domain? What is the range?

71. For each of the given points write equations for three lines such that one of the three lines is horizontal, one is vertical, and one has slope 2.

a. $(1, -4)$ **b.** $(2, 0)$ **c.** $(8, 50)$

72. Consider the function $f(x) = 4$.

a. What is $f(0)$? $f(30)$? $f(-12.6)$?

b. Describe the graph of this function.

c. Describe the slope of this function's graph.

73. A football player who weighs 175 pounds is instructed at the end of spring training that he has to put on 30 pounds before reporting for fall training.

a. If fall training begins 3 months later, at what (monthly) rate must he gain weight?

b. Suppose that he eats a lot and takes several nutritional supplements in order to gain weight, but due to his metabolism he still weighs 175 pounds throughout the summer and at the beginning of fall training. Sketch a graph of his weight vs. time for those 3 months.

c. Describe the graph you drew in part (b). If you had drawn a graph of the player's weight versus time for the three months by plotting daily (date, weight), how might this differ from your graph in part (b)?

74. a. Write an equation for the line parallel to $y = 2 + 4x$ that passes through the point $(3, 7)$.

b. Find an equation for the line perpendicular to $y = 2 + 4x$ that passes through the point $(3, 7)$.

75. a. Write an equation for the line parallel to $y = 4 - x$ that passes through the point $(3, 7)$.

b. Find an equation for the line perpendicular to $y = 4 - x$ that passes through the point $(3, 7)$.

76. Construct the equations of two lines that:

a. Are parallel to each other

b. Intersect at the same point on the y-axis

c. Both go through the origin

d. Are perpendicular to each other

77. For each of the accompanying graphs compare the m values and then the b values. You don't need to do any calculations or determine the actual equations. Using just the graphs determine if the lines have the same slope. Are the slopes both positive or both negative, or is one negative and one positive? Do they have the same y-intercept?

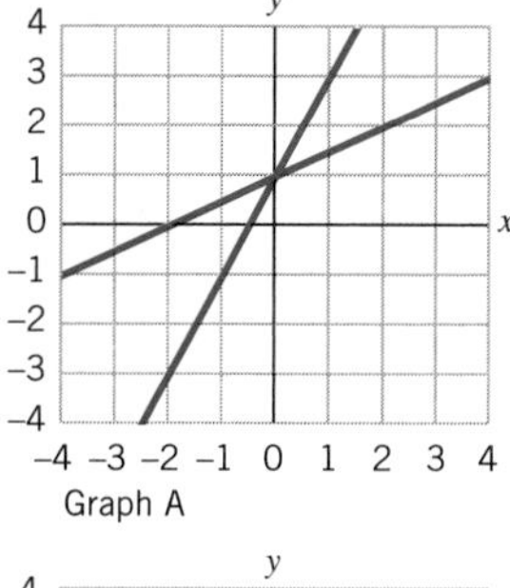

Graph A

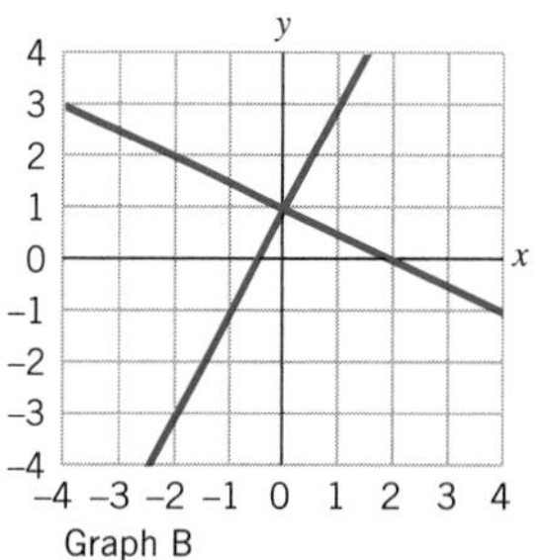

Graph B

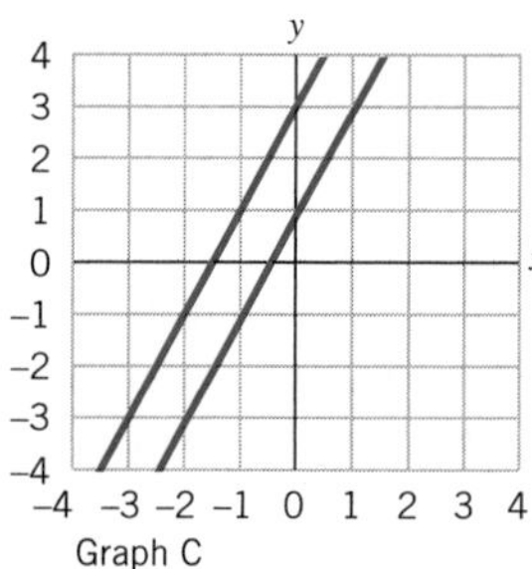

Graph C

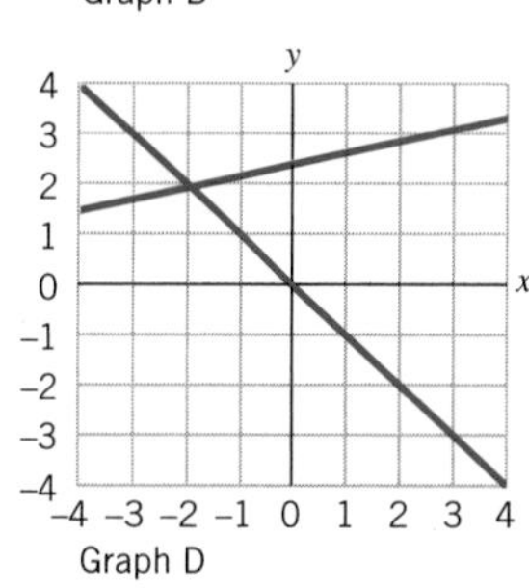

Graph D

78. Find the equation of the line in the form $y = mx + b$ for each of the following sets of conditions. Show your work.

a. Slope is \$1400/year and line passes through the point (10 yr, \$12,000).

b. Line is parallel to $2y - 7x = y + 4$ and passes through the point $(-1, 2)$.

c. Equation is $1.48x - 2.00y + 4.36 = 0$.

d. Line is horizontal and passes through (1.0, 7.2).

e. Line is vertical and passes through (275, 1029).

f. Line is perpendicular to $y = -2x + 7$ and passes through (5, 2).

79. In the equation $Ax + By = C$:

a. Solve for y so as to rewrite the equation in the form $y = mx + b$.

b. Identify the slope.

c. What is the slope of any line parallel to $Ax + By = C$?

d. What is the slope of any line perpendicular to $Ax + By = C$?

80. Use the results of Exercise 79, parts (c) and (d) to find the slope of any line that is (i) parallel and (ii) perpendicular to the given line.

a. $5x + 8y = 37$ **b.** $7x + 16y = -14$ **c.** $30x + 47y = 0$

Exercises for Section 2.8

81. The percentage of dentistry degrees awarded to women in the United States between 1970 and 1995 is shown in the accompanying table and graph.

Percentage of Dentistry Degrees Awarded to Women

1970	1975	1980	1985	1990	1995
0.9	3.1	13.3	20.7	30.9	36.4

Source: U.S. National Center for Education Statistics, "Digest of Education Statistics," annual, *Statistical Abstract of the United States*, 1998.

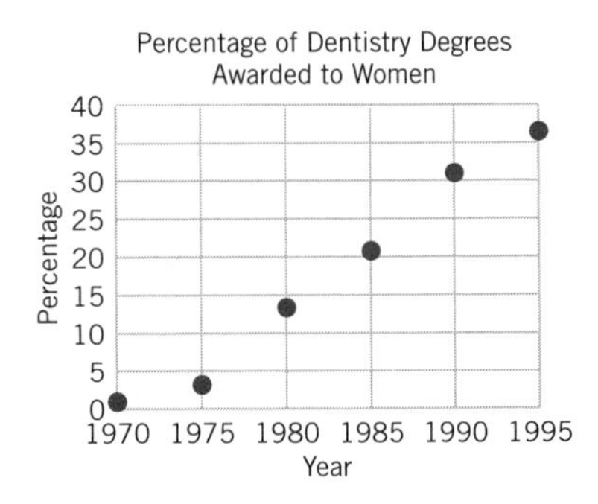

The data show that since 1970 the percentage of dentists who are women has been rising.

a. Sketch a line to represent the data points. What is the rate of change of percentage of dentistry degrees awarded to women according to your line?

b. According to your estimate of the rate of increase, when will 100% of dentistry degrees be awarded to women?

c. Since it seems extremely unlikely that 100% of dentistry degrees will ever be granted to women, comment on what is likely to happen to the rate of growth of women's degrees in dentistry; sketch a likely graph for the continuation of the data into the next century.

82. The accompanying table and graph show the mortality rates (in deaths per 1000) for male and female infants in the United States from 1980 to 1997.

a. Sketch a line through the graph of the data for the female infant mortality rates. Does the line seem to be a reasonable model for the data? What is the approximate slope of the line through these points? Show your work.

b. Sketch a line through the male mortality rates. Is it a reasonable approximation? Estimate its slope. Show your work.

c. List at least two important conclusions from the data set.

Year	Male	Female
1980	13.9	11.2
1981	13.1	10.7
1982	12.8	10.2
1983	12.3	10.0
1984	11.9	9.6
1985	11.9	9.3
1986	11.5	9.1
1987	11.2	8.9
1988	11.0	8.9
1989	10.8	8.8
1990	10.3	8.1
1991	10.0	7.8
1992	9.4	7.6
1993	9.3	7.4
1994	8.8	7.2
1995	8.3	6.8
1996	8.0	6.6
1997	8.0	6.5

Source: Centers for Disease Control and Prevention, National Vital Statistics Reports, *www.cdc.gov.*

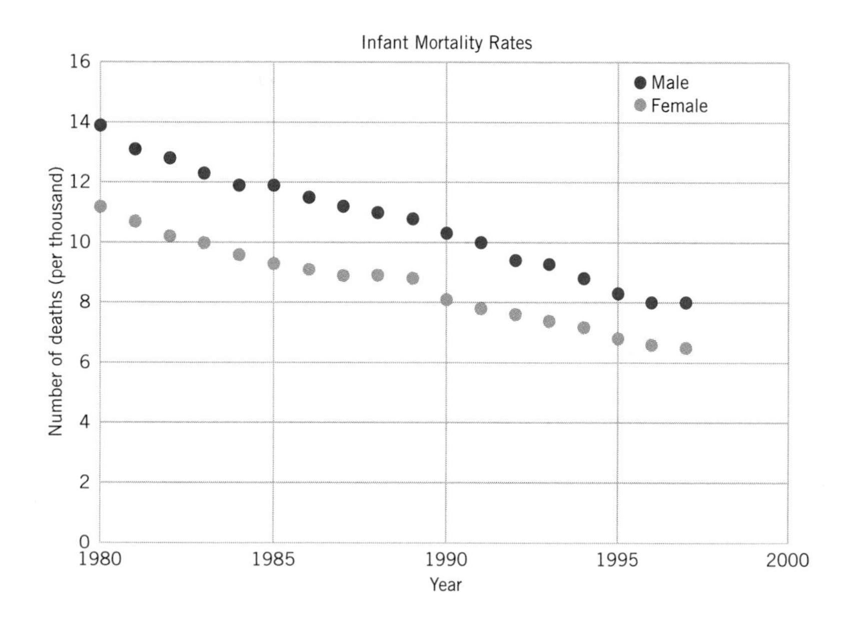

83. The accompanying scatter plot shows the relationship between literacy rate (the percentage of the population that can read and write) and infant mortality rate (infant deaths per 1000 live births) for 91 countries. The raw data are contained in the Excel or graph link file NATIONS and are described at the end of the Excel file. (You might wish to identify the outlier, the country with about a 20% literacy rate and a low infant mortality rate of about 40 per 1000 live births.) Construct a linear model. Show all your work and clearly identify the variables and units. Interpret your results.

84. The accompanying table and graph show data for the men's Olympic 16-pound shot put.

Olympic Shot Put

Year	Feet Thrown
1900	46
1904	49
1908	48
1912	50
1920	49
1924	49
1928	52
1932	53
1936	53
1948	56
1952	57
1956	60
1960	65
1964	67
1968	67
1972	70
1976	70
1980	70
1984	70
1988	74
1992	71
1996	71

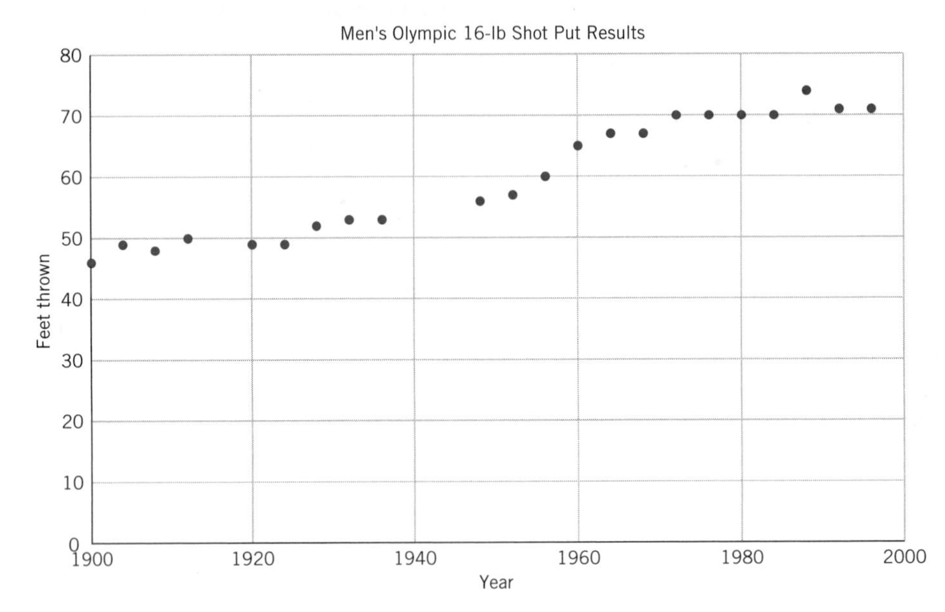

Source: Reprinted with permission from *The 1996 Universal Almanac,* Universal Press Syndicate. (The data points are missing for 1916 and the early 1940s since the Olympics were not held during World Wars I and II.)

a. Draw a line approximating the data and find its equation. Show your work and interpret your results. For ease of calculation you may wish to think of 1900 as year 0 and let your x coordinate measure the number of years since 1900.

b. If the shot put results continued to change at the same rate, in what year would you predict that the winner will put the shot a distance of 80 feet? Does this seem like a realistic estimate? Why or why not?

85. The accompanying table compares the trends in home computer ownership in the United States, Japan, and Europe. It would be a typical part of a presentation by an outside information specialist to a high-technology corporation to help in formulating a business strategy.

Home Computer Ownership

	1996	1997	1998	1999	2000	2001
United States						
Home computers in use (millions)	33.1	37.7	42.3	46.9	51.4	56.2
Household penetration (%)	32	36	40	44	47	51
Japan						
Home computers in use (millions)	5.4	6.9	8.5	10.1	12.1	14.8
Household penetration (%)	13	16	20	24	28	34
Europe						
Home computers in use (millions)	22.6	26.1	30.0	34.3	39.0	44.3
Household penetration (%)	15	17	20	22	25	28

Numbers for 2000 and 2001 are projections.

a. On a single graph, plot the number of home computers (in millions) in use in the United States, Japan, and Europe between 1996 and 2001.

b. Does the growth in number of home computers seem roughly linear for each of these countries? If so, construct a linear model for:

i. The United States **ii.** Japan **iii.** Europe

In each case, interpret the average rate of change of your model.

iv. If the current average rates of change continue, what would be the number of home computers in use in 2005 in the United States? In Japan? In Europe?

c. On a single graph, plot the household penetration (in %) of computers in the United States, Japan, and Europe between 1996 and 2001.

d. Does the growth in household penetration seem roughly linear for each of these countries? If so, construct a linear model for:

i. The United States **ii.** Japan **iii.** Europe

Again in each case, interpret the average rate of change of your model.

iv. If the current average rates of change continue, what would be the percentage of household penetration in 2005 in the United States? In Japan? In Europe?

e. Write a 60 second summary describing the growth in home computer ownership.

86. The accompanying graph shows the relationship between the age a woman has her first child and her chance of getting breast cancer (relative to a childless woman).

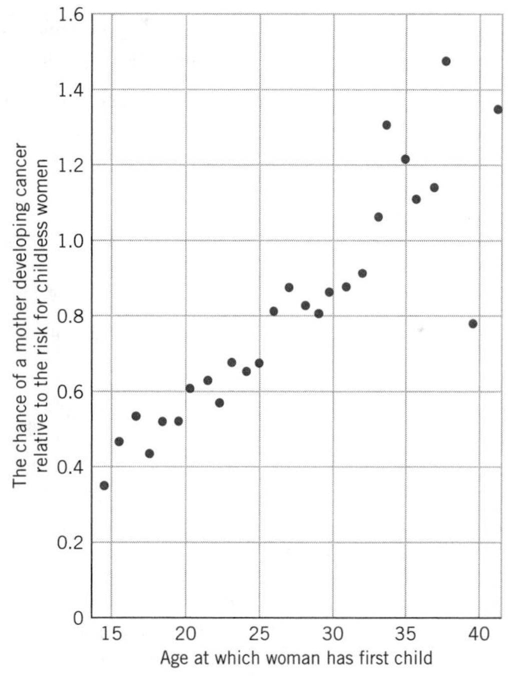

Source: J. Cairns, *Cancer: Science and Society* (W. H. Freeman and Company: San Francisco, 1978), p. 49.

a. If a woman has her first child at age 18, approximately what is her risk of developing cancer relative to a woman who has never born a child?

b. At roughly what age is the chance the same that a woman will develop breast cancer whether or not she has a child?

c. If a woman is beyond the age you specified in part (b), is she more or less likely to develop breast cancer than a childless woman?

d. Sketch a line that looks like a best fit to the data, estimate the coordinates of two points on the line, and use them to calculate the slope.

e. Interpret the slope in this context.

f. Construct a linear model for these data, identifying your independent and dependent variables.

87. The given data show that health care is becoming more expensive and taking a bigger share of the U.S. gross domestic product (GDP). The GDP is the market value of all goods and services that have been bought for final use.

Year	1960	1965	1970	1975	1980	1985	1990	1995
U.S. health care costs as a percentage of GDP	5.1	5.7	7.1	8.0	8.9	10.2	12.2	13.6
Amount per person, $	141	202	341	582	1052	1733	2691	3633

Source: U.S. Health Care Financing Administration.

a. Graph health care costs as percentage of GDP vs. year with time on the horizontal axis. Measure time in years since 1960. Draw a straight line by eye that appears to be the closest fit to the data. Figure out the slope of your line and write an equation for H, health care percent of the GDP, as a function of t, year since 1960.

b. What does your formula predict for health care as a percentage of GDP for the year 2010?

c. Why do you think the health care cost per person has gone up so much more dramatically than the health care percentage of the GDP?

88. a. From the accompanying chart showing sport utility vehicle (SUV) sales, estimate what the rate of increase of sales has been from 1991 to 1997.

Sport Utility Vehicle Sales in the U.S., 1988-97

Source: American Automobile Manufacturers Assn.

In 1988, 960,852 sport utility vehicles (SUVs) were sold in the United States, accounting for 6.3% of all sales of light vehicles (SUVs, minivans, vans, pickup trucks, and trucks under 14,000 lbs.). By 1997, sales of SUVs in the U.S. increased to 2,435,301, accounting for 16.1% of total light vehicle sales.

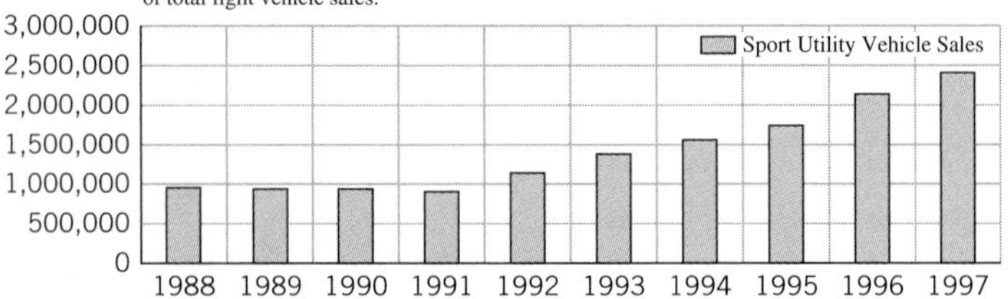

Source: The World Almanac & Book of Facts, 1999.

b. Estimate a linear formula to represent sales in years since 1991.

c. If the popularity of SUVs continues to grow at the same rate, how many would be sold in 2005?

Exploration 2.1

Having It Your Way

Objective

- construct arguments supporting opposing points of view from the same data

Material/Equipment

- excerpts from the *Fact Book* of the University of Southern Mississippi, from the *Student Statistical Portrait* of the University of Massachusetts, Boston, or from the equivalent for the student body at your institution.
- computer with spreadsheet program and printer or graphing calculator with projection system (optional)
- graph paper and/or overhead transparencies

Procedure

Working in Small Groups

Examine the data and graphs from the *Fact Book* of the University of Southern Mississippi or the *Student Statistical Portrait* of the University of Massachusetts, Boston. Explore how you would use the data to construct arguments that support at least two different points of view. Decide on the arguments you are going to make and divide up tasks among your team members.

Rules of the Game

- Your arguments needn't be lengthy, but you need to use graphs and numbers to support your position. You may only use legitimate numbers, but you are free to pick and choose only those that support your case. If you construct your own graphs, you may of course use whatever scaling you wish on the axes.
- For any data that represent a time series, as part of your argument, pick two appropriate end points and calculate the associated average rate of change.
- Use "loaded" vocabulary (e.g., "surged ahead," "declined drastically"). This is your chance to be outrageously biased, write absurdly flamboyant prose, and commit egregious sins of omission.
- Decide as a group how to present your results to the class. Some students enjoy realistic "role playing" in their presentations and have added creative touches such as mock protesters complete with picket signs.

Suggested Topics

Your instructor might ask your group to construct one or both sides of the arguments on one topic. If you're using data from your own institution, answer the questions provided by your instructor.

Using the* Fact Book *from the University of Southern Mississippi

1. Use the data on enrollment by ethnic group and gender from "10-Year Trend" to support each of the following cases:

a. You are a student activist trying to convince the Board of Trustees that they have not done enough to increase diversity of the student body at USM.

b. You are the Affirmative Action Officer arguing that your office has done a good job of increasing the diversity of the student body at USM.

2. Use data on "Retention of First-Time Entering Freshmen" and "10-Year Trend" to support each of the following cases:

a. You are a student trustee lobbying the state legislature for more money for USM.

b. You are a taxpayer writing an editorial to the local paper arguing that USM does not need more money.

3. Use the data from the "10-Year Trend" and "Degrees Awarded" to support each of the following:

a. You are the president of the university arguing at a press conference that the university is growing and hence needs increased funding.

b. You are a student making a speech before the Student Senate that the university is shrinking and hence the university needs more resources to reverse this trend.

Using the* Student Statistical Portrait *from the University of Massachusetts, Boston

1. You are the Dean of the College of Management. Use the data on "SAT Scores of New Freshmen by College/Program" to make the case that:

a. The freshmen admitted to the College of Management are not as prepared as the students in the College of Arts and Sciences and therefore you need more resources to support the freshmen in your program.

b. The freshmen admitted to the College of Management are better prepared than the students in the College of Arts and Sciences and therefore you would like to expand your program.

2. You are an Associate Provost lobbying the state legislature. Use the data on "Undergraduate Admissions" to present a convincing argument that:

a. UMass is becoming less desirable as an institution for undergraduates and so more funds are needed to strengthen the undergraduate program.

b. UMass is becoming more desirable as an institution for undergraduates and so more funds are needed to support the undergraduate program.

3. Use the data on "Distribution of High School Rank and SAT Scores" to support each of the following:

a. You are the president of the student body, arguing that UMass/Boston is becoming a more elite institution and hence is turning its back on its urban mission.

b. You are the head of the Honors Society at UMass/Boston, writing a letter to the student newspaper proclaiming that the university is lowering its academic standards and hence is in danger of compromising its academic credibility.

Exploration-Linked Homework

With your partner or group prepare a short class presentation of your arguments, using, if possible, overhead transparencies or a projection panel. Then write individual 60 second summaries to hand in.

Exploration 2.2A

Looking at Lines

Objective

- find patterns in the graphs of linear equations in the form $y = mx + b$

Equipment

- computer with course software "L1: m & b Sliders" in *Linear Functions*

Procedure

In each part try working first in pairs, comparing your observations and taking notes. Your instructor may then wish to bring the whole class back together to discuss everyone's results.

Part I: Exploring the Effect of m and b on the Graph of $y = mx + b$

Open the program *Linear Functions* and click on the button "L1: m & b Sliders."

1. What is the effect of m on the graph of the equation?

   Fix a value for b. Construct four graphs with the same value for b but with different values for m. Continue to vary m, jotting down your observations about the effect on the line when m is positive, negative, or equal to zero. Do you think your conclusions work for values of m that are not on the slider?

   Choose a new value for b and repeat your experiment. Are your observations still valid? Compare your observations with those of your partner.

2. What is the effect of b on the graph of the equation?

   Fix a value for m. Construct four graphs with the same value for m but with different values for b. What is the effect on the graph of changing b? Record your observations. Would your conclusions still hold for values of b that are not on the slider?

   Choose a new value for m and repeat your experiment. Are your observations still valid? Compare your observations with those of your partner.

3. Write a 60 second summary on the effect of m and b on the graph of $y = mx + b$.

Part II: Constructing Lines under Certain Constraints

1. Construct the following sets of lines still using "L1: m & b Sliders." Be sure to write down the equations for the lines you construct. What generalizations can you make about the lines in each case? Are the slopes of the y-intercepts of the lines related in some way?

   Construct any line. Then construct another line that has a steeper slope, and then construct one that has a shallower slope.

   Construct three *parallel lines.*

   Construct three lines with the *same y-intercept,* the point where the line crosses the y-axis.

   Construct a pair of lines that are *horizontal.*

   Construct a pair of lines that go *through the origin.*

   Construct a pair of lines that are *perpendicular* to each other.

2. Write a 60 second summary of what you have learned about the equations of lines.

Exploration 2.2B

Looking at Lines with a Graphing Calculator

Objective

- find patterns in the graphs of linear equations in the form $y = mx + b$

Material/Equipment

- graphing calculator (instructions for the TI-82 and the TI-83 are available in the Graphing Calculator Workbook)

Procedure

Getting Started

Set your calculator to the integer window setting. For the TI-82 or TI-83 do the following:

```
WINDOW FORMAT
Xmin=-47
Xmax=47
Xscl=10
Ymin=-31
Ymax=31
Yscl=10
```

1. Press ZOOM, select [6:ZStandard].
2. Press ZOOM, select [8:ZInteger], ENTER.
3. Press WINDOW to see whether the settings are the same as the duplicated screen image.

Working in Pairs

In each part try working first in pairs, comparing your observations and taking notes. Your instructor may then wish to bring the whole class back together to discuss everyone's results.

Part I: Exploring the Effect of m and b on the Graph of $y = mx + b$

1. What is the effect of m on the graph of the equation $y = mx$?

 a. Case 1: $m > 0$.

 Enter the following functions into your calculator and then sketch the graphs by hand. To get started, try $m = 1, 2, 5$. Try a few other values of m where $m > 0$.

$$Y1 = x$$

$$Y2 = 2x$$

$$Y3 = 5x$$

$$Y4 = \cdots$$

$$Y5 =$$

$$Y6 =$$

Compare your observations with your partner. In your notebook describe the effect of multiplying x by a positive value for m in the equation $y = mx$.

b. Case 2: $m < 0$.

Begin by comparing the graphs of the lines when $m = 1$ and $m = -1$. Then experiment with other negative values for m and compare the graphs of the equations.

$$Y1 = x$$

$$Y2 = -x$$

$$Y3 = \cdots$$

$$Y4 =$$

$$Y5 =$$

$$Y6 =$$

Alter your description in Part 1(a) to describe the effect of multiplying x by any real number m for $y = mx$ (remember to also explore what happens when $m = 0$).

2. What is the effect of b on the graph of an equation, $y = mx + b$?

a. Enter the following into your calculator and then sketch the graphs by hand. To get started, try $m = 1$ and $b = 0, 20, -20$. Try other values for b as well.

$$Y1 = x$$

$$Y2 = x + 20$$

$$Y3 = x - 20$$

$$Y4 = \cdots$$

$$Y5 =$$

$$Y6 =$$

b. Discuss with your partner the effect of adding any number b to x for $y = x + b$. (*Hint:* Use "trace" to find where the graph crosses the y-axis.) Record your comments in your notebook.

c. Choose another value for m and repeat the exercise. Are your observations still valid?

3. Write a 60 second summary on the effect of m and b on the graph of $y = mx + b$.

Part II: Constructing Lines under Certain Constraints

1. Construct the following sets of lines using your graphing calculator. Be sure to write down the equations for the lines you construct. What generalizations can you make about the lines in each case? Are the slopes of the y-intercepts of the lines related in any way?

Construct any line. Then construct another line that has a steeper slope, then one that has a shallower slope.

Construct three *parallel lines.*

Construct three lines with the *same y-intercept.*

Construct a pair of lines that are *horizontal.*

Construct a pair of lines that go *through the origin.*

Construct a pair of lines that are *perpendicular* to each other.

2. Write a 60 second summary of what you have learned about the equations of lines.

An Extended Exploration: Looking for Links Between Education and Income

Overview

Robert Reich, former U.S. Secretary of Labor for the Clinton administration, stated that "learning is the key to earning." Is this true? Is there a relationship between education and income?

In this extended exploration, we use a large data set from the Bureau of the U.S. Census to examine ways in which education and income may be related. We use technology to fit lines to data and explore how to interpret the resulting linear models, called regression lines. You can explore further by finding evidence to support or disprove conjectures, examining questions raised by the analysis, and posing your own questions.

In this exploration, you will:

- analyze U.S. Census data
- use regression lines to summarize data
- make conjectures about the relationship between education and income in the United States
- examine the distinction between correlation and causation

USING U.S. CENSUS DATA

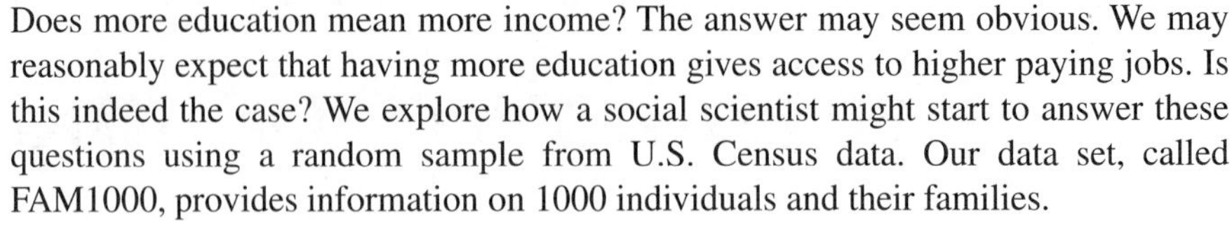

Does more education mean more income? The answer may seem obvious. We may reasonably expect that having more education gives access to higher paying jobs. Is this indeed the case? We explore how a social scientist might start to answer these questions using a random sample from U.S. Census data. Our data set, called FAM1000, provides information on 1000 individuals and their families.

Read the article "Who Collects Data and Why" for an overview of the types of data collected by institutions other than the federal government and the purposes to which the information is put.

The Bureau of the Census, as mandated by the Constitution, conducts a nationwide census every 10 years. To collect up-to-date information, the Census Bureau also conducts a monthly survey of American households for the Bureau of Labor Statistics called the Current Population Survey, or CPS. The CPS is the largest survey taken between census years. The CPS is based on data collected each month from approximately 50,000 households. Questions are asked about race, education, housing, number of people in the household, income, and employment status.[1] The March surveys are the most extensive. Our sample of census data FAM1000, was extracted from the March 1999 Current Population Survey. (See *www.bls.census.gov/cps*) It contains information about 1000 individuals randomly chosen from those who worked at least 1 week in 1998.

You can use the FAM1000 data and the related software, called *FAM1000 Census Graphs,* to follow the discussion in the text and/or conduct your own case study. The full FAM1000 data set is in the Excel file, FAM1000, and condensed versions are in the graph link files, FAM1000 A, B, C and D. These data files and related software can be downloaded from the web at *www.wiley.com/college/kimeclark* and are also available on the Instructor's CD. The software provides easy to use interactive tools for analyzing the FAM1000 data. Instructions for using graph link files can be found in the Graphing Calculator Manual.

Table 2 shows all of the information for 52 of the 1000 individuals in FAM1000 and Table 1 is a *data dictionary* with short definitions for each data category. Think of the data as a large array of rows and columns of facts. Each row represents all the information obtained from one particular respondent about his or her family. Each column contains the coded answers of all the respondents to one particular question. Try deciphering the information below that comes from the first row of our data array in Table 2.

region	cencity	famsize	faminc	marstat	sex	race	reorgn	age	occup	educ	wkswork	hrswork	pwages	ptotinc	yrft
2	2	1	$24,800	1	1	1	8	55	4	10	52	40	$24,000	$24,800	1

Referring to the data dictionary, we learn that the respondent lives in the Midwest, in the suburbs, by herself (family size of 1), with a family income of $24,800 in 1998. The respondent is not married, is female, white, non-Hispanic, and 55 years old. She had a 10th-grade education, worked 52 weeks a year in 1998, 40 hours a week and earned $24,000 in personal wages and salary, had a personal total income of $24,800 (which equals her family income since she lives by herself), and considers herself a year-round full-time worker.

[1]The results of the survey are used to estimate numerous economic and demographic variables, such as the size of the labor force, the employment rate, and income and education levels. The results are widely quoted in the popular press and are published monthly in *The Monthly Labor Review* and *Employment and Earnings,* irregularly in the *Current Population Reports* and *Special Labor Force Reports,* and yearly in the *Statistical Abstract of the United States* and *The Economic Report of the President.*

Table 1
Data Dictionary for March 1999 Current Population Survey

Variable	Definition	Unit of Measurement (Code and Allowable Range)
region	Census region	1 = Northeast 2 = Midwest 3 = South 4 = West
famsize	Family size	Range is 1 to 39
faminc	Family income	Range is −$10,000 to $600,000
marstat	Marital status	0 = Presently married 1 = Presently not married
sex	Sex	0 = Male 1 = Female
race	Race of respondent	1 = White 2 = Black 3 = American Indian, Aleut Eskimo 4 = Asian or Pacific Islander 5 = Other
reorgn	Ethnic origin of respondent (Hispanic or other)	1 = Mexican American 2 = Chicano 3 = Mexican (Mexicano) 4 = Puerto Rican 5 = Cuban 6 = Central or South American 7 = Other Spanish 8 = All other 9 = Don't know 10 = Not available
age	Age	Range is 16 to 90
occup	Occupation group of respondent	0 = Not in universe 1 = Executive, administrative managerial 2 = Professional specialty 3 = Technicians and related support 4 = Sales 5 = Administrative support, including clerical 6 = Private household service 7 = Protective service 8 = Other service occupations 9 = Precision production, craft and repair 10 = Machine operators, assemblers and inspectors 11 = Transportation and materials moving equipment 12 = Handlers, equipment cleaners, helpers, etc. 13 = Farming, forestry, and fishing 14 = Armed forces

Variable	Definition	Unit of Measurement (Code and Allowable Range)
cencity	Residence location	1 = Central city 2 = Suburban 3 = Rural 4 = Not identified
educ	Years of education	1 = Less than first grade 4 = First, second, third, or fourth grade 6 = Fifth or sixth grade 8 = Seventh or eighth grade 9 = Ninth grade 10 = Tenth grade 11 = Eleventh grade 12 = Twelfth grade, no diploma HG = High school graduate (diploma or equivalent) SC = college, but no degree AO = Associate's degree in occupation/vocation program AP = Associate's degree in academic program BD = Bachelor's degree MD = Master's degree (M.A., M.S., M.S.W., M.B.A.) PD = Professional school degree (M.D., D.D.S., D.V.M., L.L.B., J.D.) PH = Doctoral degree (Ph.D., Ed.D.)

Note: In the FAM 1000 data, substitutions of numbers for symbols were made in order to plot points. For example, we replaced HG (high school graduate) with 12, representing 12 years of school. SC (some college but no degree), we estimate as 14, or 2 years more than high school. Similar estimates were made for the higher education categories.

Variable	Definition	Unit of Measurement (Code and Allowable Range)
wkswork	Weeks worked in 1998 (even for a few hours)	Range is 1 to 52
hrswork	Usual hours worked	Range is 1 to 99 per week in 1998
pwages	Total personal wages	Range is $0 to $500,000
ptotinc	Personal total income	Range is − $10,000 to $600,000
yrft	Employed full-time year-round	1 = full-time all year-round 2 = part-time all year-round 3 = full-time part of the year 4 = part-time part of the year

Source: Adapted from the U.S. Census Bureau, *Current Population Survey,* March 1999, by Jie Chen, Computing Services, University of Massachusetts, Boston.

Table 2
FAM 1000 Data Set

region	cencity	famsize	faminc	marstat	sex	race	reorgn	age	occup	educ	wkswork	hrswork	pwages	ptotinc	yrft
2	2	1	$24,800	1	1	1	8	55	4	10	52	40	$24,000	$24,800	1
1	2	3	$52,024	1	0	1	8	21	9	12	52	40	$14,000	$14,000	1
3	1	2	$184,823	0	0	1	8	66	0	20	30	35	$59,540	$137,491	3
2	2	1	$33,610	1	1	1	8	67	1	14	43	37	$12,000	$33,610	3
3	2	3	$31,200	0	1	1	8	45	9	12	28	30	$0	− $2,000	4
3	4	2	$51,400	0	0	1	8	51	4	12	52	40	$0	$51,400	1
2	2	5	$164,877	0	1	2	8	40	2	18	52	40	$43,000	$43,000	1
4	2	3	$55,143	0	1	1	8	49	8	12	51	40	$19,652	$19,665	1
4	2	3	$116,650	0	0	1	8	34	1	12	52	45	$42,000	$42,450	1
3	4	2	$16,269	1	0	2	8	18	4	10	8	45	$3,000	$3,000	3
4	2	4	$59,451	1	1	1	8	78	6	12	51	40	$7,000	$13,501	1
1	2	3	$32,007	0	0	1	6	40	11	12	52	40	$24,000	$24,007	1
1	2	3	$49,761	0	1	1	8	49	5	14	17	6	$510	$2,435	4
1	2	4	$120,700	0	0	1	8	39	1	16	52	45	$100,000	$104,350	1
4	3	4	$124,500	0	0	4	8	38	4	16	52	48	$68,000	$69,750	1
3	4	2	$44,500	0	1	1	8	19	5	14	52	38	$16,500	$16,500	1
1	4	3	$50,690	1	1	1	8	62	8	12	24	40	$9,080	$18,693	3
3	3	3	$12,000	0	0	1	8	19	7	11	40	40	$9,000	$9,000	3
4	2	1	$21,000	1	0	1	8	27	3	16	52	50	$21,000	$21,000	1
3	2	4	$84,172	0	1	2	8	44	4	16	52	40	$13,318	$19,144	1
3	3	2	$58,600	0	0	1	8	52	1	14	52	40	$32,400	$32,550	1
2	4	5	$150,408	0	1	1	8	34	2	16	52	24	$16,000	$53,127	2
2	3	4	$68,094	0	1	1	8	35	5	16	52	40	$20,000	$20,047	1
3	2	6	$304,981	0	1	1	8	31	2	16	40	5	$0	− $1,875	4
4	1	2	$140,536	0	1	1	8	58	8	16	40	40	$0	$27,846	3
2	2	5	$82,616	0	1	1	8	42	5	16	52	25	$14,000	$14,808	2

Table 2 *(continued)*
FAM 1000 Data Set

region	cencity	famsize	faminc	marstat	sex	race	reorgn	age	occup	educ	wkswork	hrswork	pwages	ptotinc	yrft
4	2	1	$9,550	1	0	1	3	29	8	9	52	40	$12,050	$9,550	1
2	1	4	$32,555	0	0	1	8	26	9	12	52	20	$0	$31,200	2
2	3	3	$34,600	0	1	1	8	27	8	12	52	40	$19,000	$19,000	1
3	1	2	$31,210	0	1	1	6	24	2	16	40	20	$7,000	$7,000	4
3	2	3	$47,755	1	0	1	8	18	12	12	52	25	$8,416	$9,094	2
4	1	3	$60,400	0	1	1	2	39	4	12	52	30	$24,000	$24,000	2
2	2	3	$82,200	0	0	1	8	47	9	12	52	40	$42,000	$42,000	1
3	3	4	$41,698	0	1	1	8	47	5	16	52	40	$20,000	$20,349	1
4	4	1	$401	1	0	1	8	59	13	12	52	60	$0	$401	1
1	2	3	$64,200	0	1	2	8	36	5	12	52	38	$21,000	$21,000	1
2	2	2	$65,608	0	0	1	8	35	9	16	50	45	$3,600	$35,608	1
2	3	2	$62,093	0	1	1	8	53	1	16	52	40	$40,000	$42,084	1
2	2	3	$391,016	1	0	1	8	22	2	14	30	8	$8,700	$10,500	4
3	2	2	$61,714	0	1	1	8	53	5	14	52	40	$32,000	$33,057	1
3	1	4	$110,456	0	1	1	8	42	5	12	52	50	$40,000	$40,000	1
4	4	3	$130,240	0	1	1	8	39	0	14	3	12	$240	$240	4
4	2	3	$118,777	0	0	1	8	61	0	18	15	25	$25,000	$99,277	4
3	1	5	$28,000	0	1	2	8	33	8	14	52	60	$0	$19,000	1
1	3	4	$52,710	0	1	1	8	37	1	12	52	40	$20,400	$20,448	1
2	1	2	$62,500	0	1	1	8	34	5	12	52	40	$24,000	$24,800	1
3	2	2	$47,000	0	0	1	8	39	2	14	52	60	$17,000	$17,000	1
1	2	1	$13,000	1	0	1	8	53	1	14	52	50	$13,000	$13,000	1
4	4	4	$31,040	0	0	1	8	27	9	14	52	60	$30,000	$30,000	1
1	2	1	$25,078	1	1	1	8	50	10	12	52	48	$25,000	$25,078	1
4	3	4	$63,080	0	0	1	8	55	2	16	52	40	$60,000	$61,540	1
4	1	4	$88,200	0	1	1	3	32	5	14	52	40	$53,000	$53,000	1

Source: From the U.S. Census Bureau, *Current Population Survey,* March 1999.
Data can be accessed through the Census Bureau website: *http://www.bls.census.gov/cps/cpsmain.htm.*

Going Further: How Good Are the Data?

How does the U.S. Bureau of the Census define years of educational attainment? What are some of the ways in which income is defined by the U.S. Bureau of the Census? What are some of the issues raised by these definitions?

What groups of people may be undercounted in the U.S. Current Population Survey? What are some of the current controversies about how the U.S. Bureau of the Census collects the census data?

The following readings give a closer look at data collection and analysis. You may also want to search for current newspaper articles on controversies surrounding the Year 2000 Census.

On the web at *www.bls.census.gov/cps and www.wiley.com/college/kimeclark.*

- M. Maier, "Wealth, Income, and Poverty," from *The Data Game: Controversies in Social Science Statistics.*
- T. Roman, "U.S. Government Data Collection."
- "Money Income" and "Educational Attainment," *Population Profile of the United States,* 1998; Bureau of the Census, Current Population Reports, 1998.
- "Questions to Raise When Analyzing Data."

In the Anthology of Readings

- T. Roman, "U.S. Government Definition of Census Terms."

SUMMARIZING THE DATA: REGRESSION LINES

What Is the Relationship between Education and Income?

In the physical sciences, the relationship among variables is often quite direct; if you hang a weight on a spring, it is clear, even if the exact relationship is not known, that the amount the spring stretches is definitely dependent on the heaviness of the weight. Further, it is reasonably clear that the weight is the *only* important variable; the temperature and the phase of the moon, for example, can safely be neglected.

In the social and life sciences it is usually difficult to tell whether one variable truly depends on another. For example, it is certainly plausible that a person's income depends in part on how much formal education he or she has had, since we may suspect that having more education gives access to higher paying jobs, but many other factors also play a role. Some of these factors, such as the person's age or type of work, are measured in the FAM1000 data set; others, such as family background or good luck, may not have been measured or even be measurable. Despite this complexity, we attempt to determine as much as we can by first looking at the relationship between income and education alone.

We begin to explore this question with a *scatter plot* of education and personal wages from the FAM1000 data set. If we hypothesize that income depends on education, then the convention is to graph education on the horizontal axis. Each ordered pair of data values gives a point with coordinates of the form

(education, personal wages)

Take a moment to examine the scatter plot in Figure 1. Each point refers to a particular respondent and has two coordinates. The first coordinate gives the years of education and the second gives the personal wages. For example, the two points near the top of the graph represent individuals who make $150,000 in personal wages. One of these people has 16 years of education and the other has attained at least 20 years of education. The coordinates for each of these points are

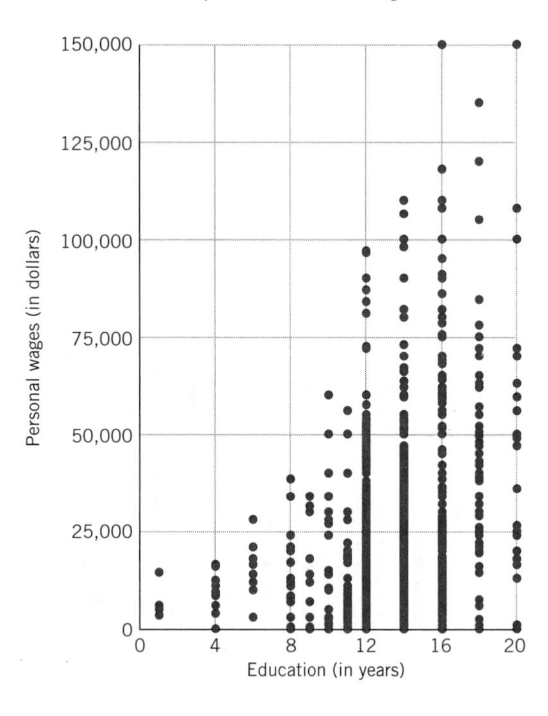

Figure 1 Personal wages versus education. In attempting to pick a reasonable scale to display personal wages on this graph, the vertical axis was cropped at $150,000, which meant excluding from the display the points for thirteen individuals who each earned more than $150,000 in wages.

(16, 150000) and (20, 150000), respectively. We can refer back to the original data set to find out more information about these points, as well as outliers that are not shown on the graph.

How might we think of the relationship between these two variables? Clearly, personal income is not a function of education in the mathematical sense since people who have the same amount of education earn widely different amounts. The scatter plot obviously fails the vertical line test.

But suppose that, to form a simple description of these data, we were to insist on finding a simple functional description. And suppose we insisted that this simple relationship be a linear function. In Chapter 2, we informally fit linear functions to data. A formal mathematical procedure called *regression analysis* lets us determine what linear function is the "best" approximation to the data; the resulting "best fit" line is called a *regression line* and is similar in spirit to reporting only the mean of a set of single variable data, rather than the entire data set. "It is often a useful and powerful method of summarizing a set of data."[2]

WWW.

If you are interested in a standard technique for generating regression lines, a summary of the method used in the course software is provided in the reading "Linear Regression Summary."

We can measure how well a line represents a data set by adding up the squares of the vertical distances between the line and the data points. The regression line is the line that makes this sum as small as possible. The calculations necessary to compute this line are tedious, although not difficult, and are easily carried out by computer software and graphing calculators. The *FAM1000 Census Graphs* program can be used to find regression lines. The various techniques for determining regression lines are beyond the scope of this course.

In Figure 2, we show the FAM1000 data set, along with a regression line determined from the data points. The equation of the line is

$$\text{pwages}_{\text{all data}} = -30{,}420 + 4480 \cdot \text{yrs. of educ.}$$

Since personal wages are in dollars, the units for the term 30,420 must be dollars, and the units for 4480 must be dollars per year.

[2]E. Tufte, *Data Analysis of Politics and Policy* (Upper Saddle River, NJ: Prentice-Hall, 1974), p. 65.

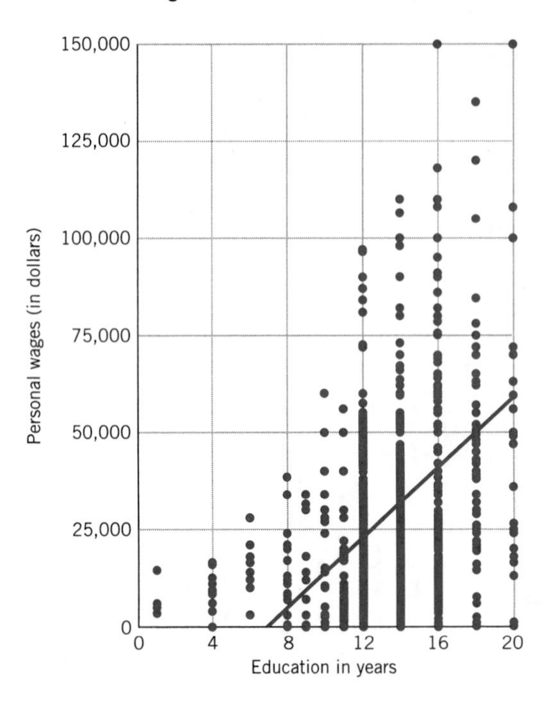

Figure 2 Regression line for personal wages vs. education.

This line is certainly more concise than the original set of 1000 data points. Looking at the graph, you may judge with your eyes to what extent the line is a good description of the original data set. Note from the equation that the vertical intercept, − \$30,420, of this line is *negative*; according to this regression line, people with less than seven years of education earn *negative* wages, even though all wages in the original data set are positive. The linear model is clearly inaccurate for those people with less than seven years of education, so we should limit the domain of our model to values greater than 7 years of education. The number 4480 represents the slope of the regression line, or the average rate of change of personal wages with respect to years of education. Thus, this model predicts that for each additional year of education, individual personal wages increased by \$4480.

The data points are widely scattered about the line, for reasons that are clearly not captured by the linear model. Some of this scatter is simply a result of randomness and the fact that each data point is a different individual; however, some of the scatter may result from other variables such as age, which a more sophisticated analysis could include.

We emphasize that, although we can construct an approximate linear model for any data set, this does not mean that we really believe that the data are truly represented by a linear relationship. In the same way, we may report the mean to summarize a set of data, without believing that the data values are all the same number. In both cases, there are features of the original data set that may or may not be important and that we do not report.

Another way to look for structure in this data set is to eliminate clutter by grouping together all people with the same years of education and to plot the *mean* personal wages of each group. The resulting graph (Figure 3) is sometimes called the *graph of averages.* Note that the vertical scale now only goes up to \$100,000.

This plot is less cluttered. For each year of education, only the single mean income point has been graphed. For instance, the point corresponding to 12 years of

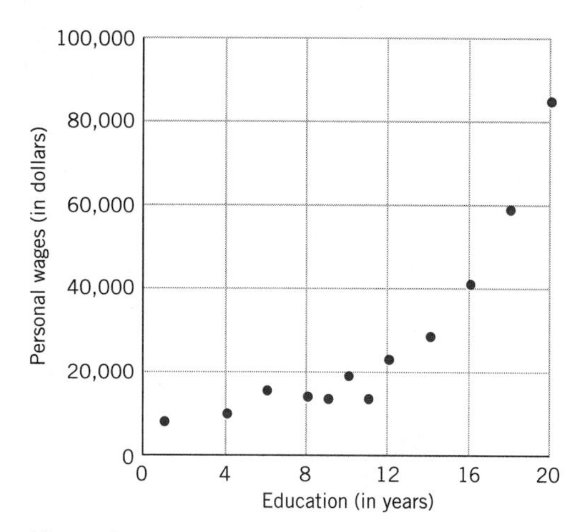

Figure 3 Mean personal wages vs. education.

education has a vertical value of approximately \$22,000; hence the mean of the personal wages of everyone in the FAM1000 data with 12 years of education is about \$22,000. The pattern is clearer: An upward trend to the right is more obvious in this graph.

Every time we construct a simplified representation of an original data set, we should ask ourselves what information has been suppressed. For the graph of averages, we have suppressed the spread of data in the vertical direction. We also do not see how many different people are represented by a single data point at each year of education; for example, there are only *four* people with 1 year of education but almost two hundred with 16 years, and each of these sets is represented by a single point in the graph of averages.

In the graph of averages, we can see some additional features of the data that have not been evident in either the full scatter plot or the equation of the straight-line model. The graph is fairly flat below 12 years; perhaps if you do not graduate from high school, it does not matter very much how many years you went to school. Starting at 12 years it slopes upward, perhaps representing the payoff from college and graduate education. All of these observations are completely suppressed by the simple linear model.

We can fit a line to the graph of averages by the same method of linear regression. The equation of this straight line is

$$\text{pwages}_{\text{mean data}} = -11{,}530 + 3580 \cdot \text{yrs. of educ.}$$

Here, − 11,530 is the vertical intercept and 3580 is the slope or rate of change of personal wages with respect to education. Again, since the vertical intercept is negative, we would ignore the left-end segment of the line. This model predicts that for each additional year of education the *mean* personal wage increases by \$3580. Note that this linear model predicts *mean personal wages for the group,* not personal wages for an individual.

Figure 4 shows two regression lines: one represents the fit to the averages and the other represents the fit to all of the data. Both of these straight lines are reasonable answers to the question, "What straight line best describes the relationship between education and income?" and the difference between them indicates the uncertainty in answering such a question. We may argue that the benefit in income of each year of education is \$3580, or \$4480, or somewhere between these values.

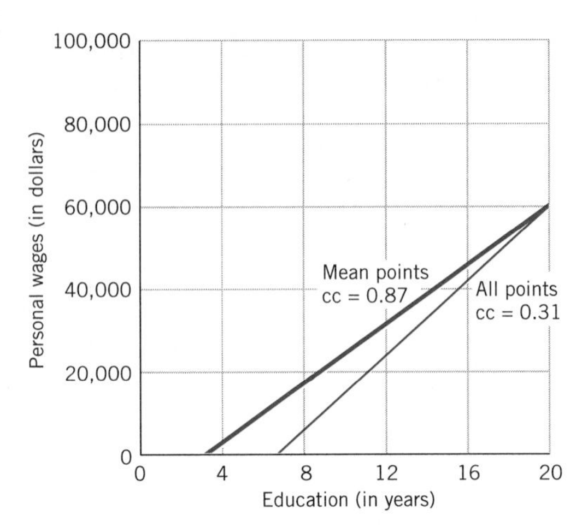

Figure 4 Regression lines for personal wages vs. education.

The programs R1-R4 and R7 in *Linear Regression* can help you visualize the links among scatter plots, best fit lines, and correlation coefficients.

Regression Line: How Good a Fit?

Once we have determined a line that approximates our data, we must ask, "How good a fit is our regression line?" To help answer this, statisticians calculate a quantity called the *correlation coefficient.* This number is computed by most regression software and graphing calculators, and we have included it on our graphs, labeled "cc."[3] The correlation coefficient is always between -1 (negative association and no scatter; the data points fit exactly on a line with a negative slope) and 1 (positive association and no scatter; the data points fit exactly on a line with a positive slope). The closer the absolute value of the correlation coefficient is to 1, the better the fit and the stronger the linear association between the variables.

Remember that the absolute value of a number, N, is expressed in symbols as $|N|$. It means the distance from zero without taking into account the sign. If $N \geq 0$, then $|N| = N$. If $N < 0$, then $|N| = -N$. Thus $|0.89| = 0.89$ and $|-0.89| = 0.89$.

The reading "*The Correlation Coefficient*" will tell you how to calculate the correlation coefficient and give you an intuitive sense of how the size of the correlation coefficient helps to answer the question, "How good a fit to the data is the regression line?"

A small correlation coefficient (with absolute value close to zero) indicates that the variables do not depend linearly on each other. This may be because there is no relationship between them, or because there is a relationship that is something more complicated than linear. In future chapters we discuss many possible nonlinear functional relationships.

There is no absolute answer to the question of when a correlation coefficient is "good enough" to say that the linear regression line is a good fit to the data. A fit to the graph of averages generally gives a higher correlation coefficient than a fit to the original data set because the scatter has been smoothed out. When in doubt, plot all the data along with the linear model and use your best judgment. The correlation coefficient is only a tool that may help you decide among different possible models or interpretations.

On Your Own: Interpret the Correlation Coefficient

Draw two scatter plots of data, one with a correlation coefficient of $+1$ and the other with a correlation coefficient of -1. In each case explain what the correlation coefficient tells you about the relationship between the two variables.

Now draw a scatter plot that suggests zero correlation between two variables.

[3]We use the label cc for the correlation coefficient in the text and software in order to minimize confusion. In a statistics course the correlation coefficient is usually referred to as *Pearson's r* or just *r*.

INTERPRETING REGRESSION LINES: CORRELATION VS. CAUSATION

One is tempted to conclude that increased education *causes* increased income. This may be true, but the model we have used does not offer conclusive proof. This model can show how strong or weak a relationship exists between variables but does not answer the question, "Why are the variables related?" We need to be cautious in how we interpret our findings.

Regression lines show *correlation,* not *causation.* We say that two events are correlated when there is a statistical link. If we find a regression line with a correlation coefficient that is close to 1 in absolute value, a strong relationship is suggested. In our previous example, education is positively correlated with personal wages. If education increases, personal wages increase. Yet this does not prove that education causes an increase in personal wages. The reverse might be true; that is, an increase in personal wages might cause an increase in education. The correlation may be due to other factors altogether. It might occur purely by chance or be jointly caused by yet another variable. Perhaps both educational opportunities and income levels are strongly affected by parental education or a history of family wealth. Thus a third variable, such as parental socioeconomic status, may better account for both more education and higher income. We call this type of variable that may be affecting the results a *hidden variable.*

Figure 5 shows a clear correlation between the number of radios and the proportion of insane people in England between 1924 and 1937. (People were required to have a license to own a radio.)

Are you convinced that radios cause insanity? Or are both variables just increasing with the years? We tend to accept as reasonable the argument that an increase in education causes an increase in personal wages, because the results seem intuitively possible and they match our preconceptions. But we balk when asked to believe that an increase in radios causes an increase in insanity. Yet the

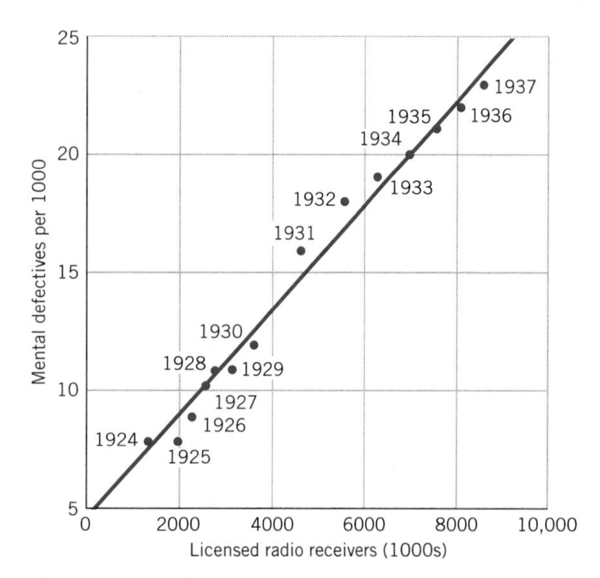

Figure 5 A curious correlation?

Source: From E. A. Tufte, *Data Analysis for Politics and Policy,* p. 90. Copyright © 1974 by Prentice Hall, Inc., Upper Saddle River, NJ. Reprinted with permission.

arguments are based on the same sort of statistical reasoning. The flaw in the reasoning is that statistics can only show that events occur together or are correlated, but *statistics can never prove that one event causes another.* Any time you are tempted to jump to the conclusion that one event causes another because they are correlated, think about the radios in England!

NEXT STEPS: RAISING MORE QUESTIONS

When a strong link is found between variables, often the next step is to raise questions whose answers may provide more insight into the nature of the relationship. How can the evidence be strengthened? If we use other income measures, such as total personal income or total family income, will the relationship between education and income still hold? Are there other variables that affect the relationship?

Does Income Depend on Age?

We started our exploration by looking at how income depended on education, because it seemed natural that more education might lead to more income. But it is equally plausible that a person's income might depend on his or her age. People may generally earn more as they advance through their working careers, but their incomes may drop when they eventually retire. We can examine the FAM1000 data to look for evidence to support this hypothesis.

It's hard to see much when we plot all the data points. But the plot of mean personal wages vs. age in Figure 6 seems to suggest that up until about age 50, as age increases, mean personal wages increase. After age 50, as people move into middle age and retirement, mean personal wages tend to decrease. So age does seem to affect personal wages, in a way that is roughly consistent with our intuition. But the relationship appears to be nonlinear, and so linear regression may not be a very effective tool to explore this dependence.

The FAM1000 data set contains internal relationships that are not obvious on a first analysis. We might for example also investigate the relationship between education and age. Age may be acting as a hidden variable influencing the relationship between education and income.

There are a few simple ways to attempt to minimize the effect of age. For example, we can restrict our analysis to individuals who are all roughly the same

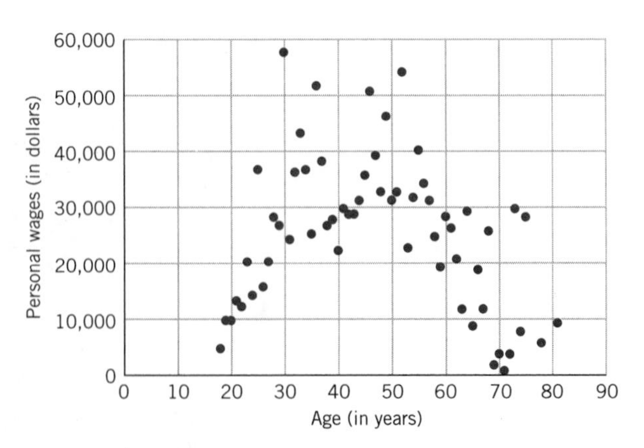

Figure 6 Mean personal wages vs. age.

age. This sample still would include a very diverse collection of people. More sophisticated strategies involve statistical techniques such as *multivariable analysis,* a topic beyond the scope of this course.

Does Income Depend on Gender?

We can continue to look for simple relationships in the FAM1000 data set by using some of the other variables to sort the data in different ways. For example, we can look at whether the relationship between income and education is different for men and women. To do this, we compute the mean personal wages for each year of education for men and women separately. By plotting the results, we begin to see some patterns.

Figures 7 and 8 indicate that in general, personal wages for both men and women increase as education increases.

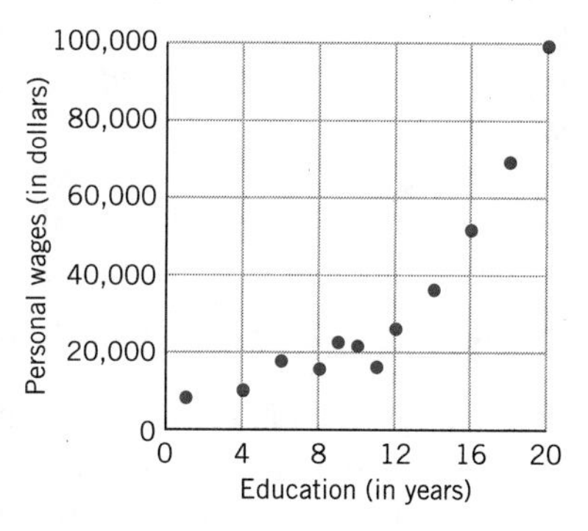

Figure 7 Mean personal wages vs. education for men.

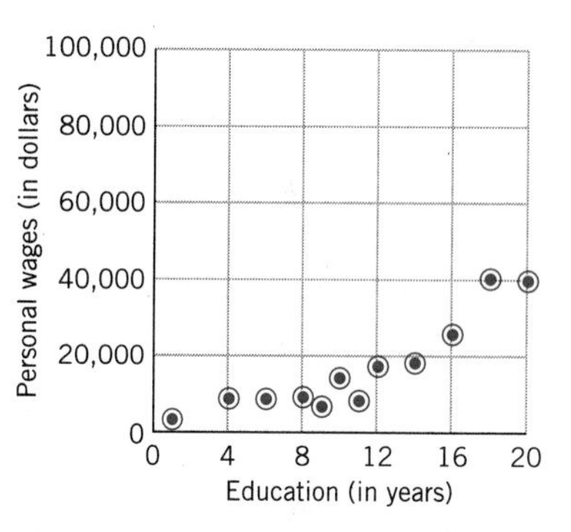

Figure 8 Mean personal wages vs. education for women.

If we put the data for men and women on the same graph (Figure 9), it is easier to make comparisons. We can see in Figure 9 that the mean personal wages of men are consistently higher than the mean personal wages of women. We can also examine the best fit lines for mean personal wages vs. education for men and for women shown in Figure 10.

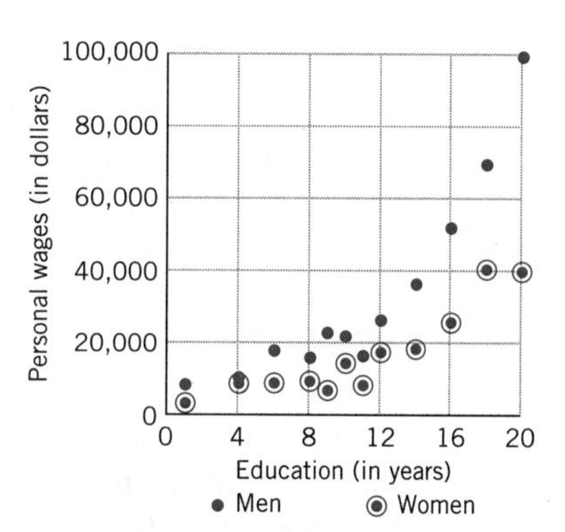

Figure 9 Mean personal wages vs. education for women and for men.

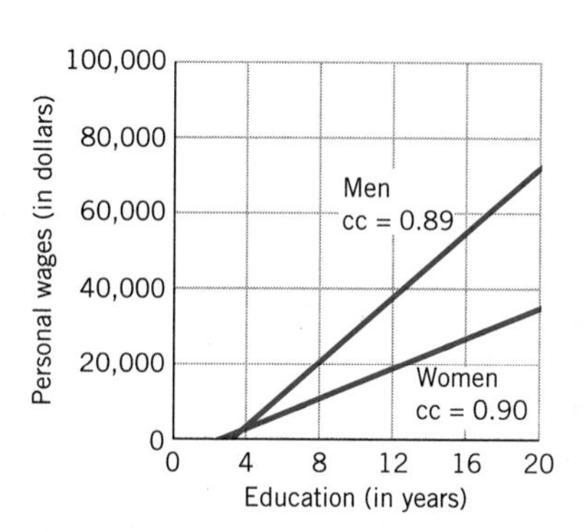

Figure 10 Regression lines for mean personal wages vs. education for women and for men.

The linear model for mean personal wages for men is given by

$$W_{men} = -12{,}850 + 4290E$$

where E = years of education and W_{men} = mean personal wages for men. The correlation coefficient is 0.89. The rate of change of mean personal wages with respect to education is approximately $4290/year. For males in this set, the mean personal wage increases by roughly $4290 each additional year of education.

For women the comparable linear model is

$$W_{women} = -4360 + 1980E$$

where E = years of education and W_{women} = mean personal wages for women. The correlation coefficient is 0.90. As you might predict from the relative status of men and women in the U.S. work force, the rate of change for women is much lower. For women in this sample, the model predicts that the mean personal wages increase by only $1980/year for each additional year of education. In Figure 10, we can see that the regression line for men is steeper than the one for women when plotted on the same grid. In addition, as we can see in Figure 9, the mean personal wage for any particular number of years of education is consistently lower for women than for men. Although the vertical intercept of the regression line for men's personal wages is below that for women, within a realistic domain the regression line for men lies above that for women. The disparity in persomal wages between men and women seems to increase with the level of education.

Going Deeper: Asking More Questions

The regression lines graphed in Figure 10 estimate mean personal wages for *all* the women and men in FAM1000. What other variables have we ignored that we may want to consider in a more refined analysis of the impact of gender on personal income? We could consider type of job or amount of work or other things we think may be important.

- *Type of job*
  Do women and men make the same salaries when they hold the same types of jobs? We could compare only people within the same profession and ask whether the same level of education corresponds to the same level of personal wages for women as for men. There are many more questions, such as: Are there more men than women in higher paying professions? Do men and women have the same access to higher paying jobs, given the same level of education?

- *Amount of work:*
  Typically part-time jobs pay less than full-time jobs, and more women hold part-time jobs than men. In addition, there are usually more women than men who are unemployed. We can examine the personal wages of only those who are working full time and determine whether it is still true that given equal amounts of education, women get paid less on average than men.

EXPLORING ON YOUR OWN

Your journey into exploratory data analysis is just beginning. You now have some tools to further examine the complex relationship between education and income. You may want to explore answers to the questions raised above, or explore answers to your own questions and conjectures. For example, what other variables besides age do you think affect income? What other variables besides gender may affect the relationship between education and income?

Working with Partners

You may want to work with a partner. You can discuss questions that may be worth pursuing and you can help each other interpret and analyze the findings. In addition, you can compare two regression lines more easily by using two computers or two graphing calculators.

Generating Conjectures

One way to start is to generate conjectures about the effects of other variables on the relationship between education and income. (See the Data Dictionary in Table 1 for variables included in the FAM1000 Census data.)

See Excel file FAM1000.

Or you may want to start by generating conjectures about how different income measures may affect the relationship between education vs. income. The FAM1000 census data include the following measures of income: personal wages, personal total income, and family income.

You can then generate and compare regression lines using the procedures described below.

Procedures for Finding Regression Lines

1. Finding regression lines:

a. *If using a computer:*

Open "F3: Regression with Multiple Subsets" in *FAM1000 Census Graphs.* This program allows you to find regression lines for education versus income for different income variables and for different groups of people. Select (by clicking on the appropriate box) one of the four income variables: personal wages, personal wages per hour, personal total income, or family income. Then select at least two regression lines that it would make sense to compare (e.g., men vs. women, white vs. nonwhite, two or more regions of the country). You should do some browsing through the various regression line options to pick those that are the most interesting. Print out your regression lines (on overhead transparencies if possible).

b. *If using graphing calculators and graph link files:*

See graph link files FAM1000A, B, C, and D.

FAM1000 graph link files A, B, C, and D contain data for generating the graph of averages and associated regression lines for several income variables as a function of years of education. The Graphing Calculator Manual contains descriptions and hardcopy of the files, as well as instructions for downloading and transferring them to T1-82 and 83 calculators. Decide on at least two regression lines which are interesting to compare.

2. For each of the regression lines you choose, work together with your partner to record the following information in your notebooks:

The equation of the regression line.
What the variables represent.
A reasonable domain.
The subset of the data the line represents (men? nonwhites?).
The correlation coefficient.
Whether or not the line is a good fit and why.
The slope.
Interpretation of the slope (e.g., for each additional year of education, average total personal wages rise by such and such an amount).

Discussion/Analysis

With your partner, explore ways of comparing the two regression lines. What do the correlation coefficients tell you about the strength of these relationships? How do the two slopes compare? Can you say that one group is better off? Is that group better off no matter how many years of education? What factors are hidden or not taken into account?

Were your original conjectures supported by your findings? What additional evidence could be used to support your interpretation of your findings? Are your findings surprising in any way? If so, why?

Prepare a 60 second summary of your results. Discuss with your partner how to present your findings. What are the limitations of the data? What are the strengths and weaknesses of your analysis? What factors are hidden or not taken into account? What questions are raised?

You may wish to continue researching questions raised by your analysis by returning to the data or referring to additional sources such as the related readings at *www.wiley.com/college/kimeclark* or the current population survey web site at *www.bls.census.gov/cps*

EXERCISES

1. **a.** Evaluate each of the following:

$$|0.65| \qquad |-0.68| \qquad |-0.07| \qquad |0.70|$$

b. List the absolute values in part (a) in ascending order from the smallest to the largest.

2. The figures below show regression lines and corresponding correlation coefficients for four different scatter plots. Which of the lines describes the strongest linear relationship between the variables? Which of the lines describes the weakest linear relationship?

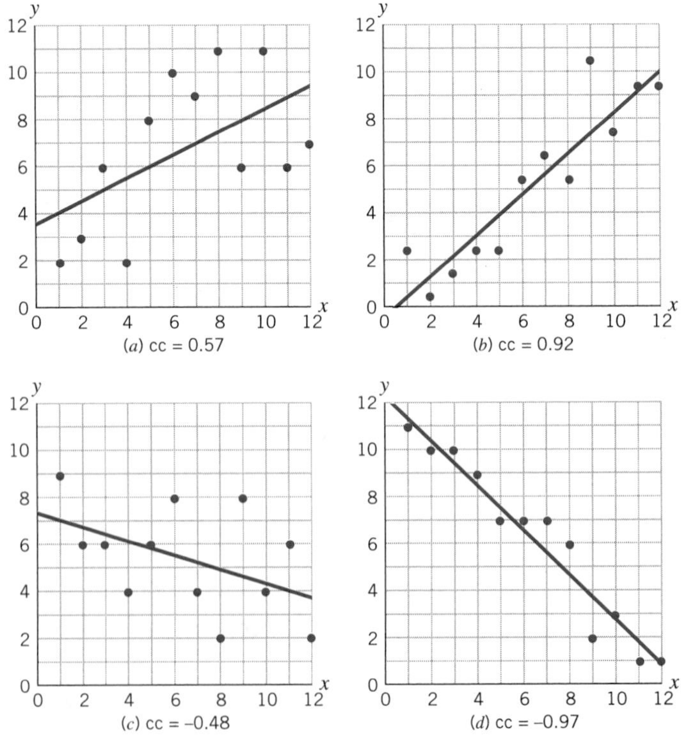

Four regression lines with their correlation coefficients.

3. The following equation represents the best fit regression line for mean personal wages versus years of education for the 299 people in the FAM1000 data who live in the southern region of the United States:

$$W = -16{,}520 + 4210E$$

where W = mean personal wages and E = years of education; cc = 0.88.

a. Identify the slope of the regression line, the vertical intercept, and the correlation coefficient.

b. What does the slope mean in this context?

c. By what amount does this regression line predict that mean personal wages for those who live in the South change for 1 additional year of education? For 10 additional years of education?

4. a. Use a straight edge to fit a line through the data in the accompanying graph. Draw it, and find its equation. Interpret the slope.

b. Enter the data from the graph into a spreadsheet or graphing calculator. Find the equation for this regression line. Graph it.

c. Compare your estimated line of best fit with the actual one.

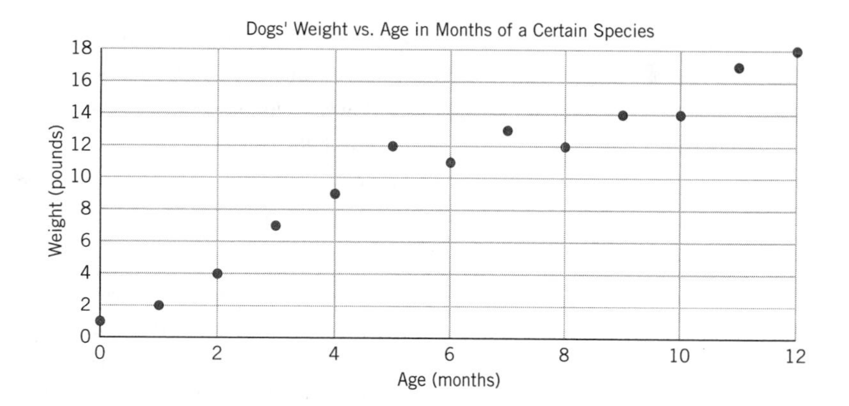

In Exercises 5 to 8, the data analyzed are from the FAM1000 data files and the income measure is personal total income. Personal total income includes personal wages and other sources of income, such as interest and dividends on investments. The numbers in the equations are rounded to the nearest 10. You can generate your own regression lines using the FAM1000 Excel file or the software programs in *FAM1000 Census Graphs* (if you have access to a computer) or the graph link files FAM1000A, FAM1000B, FAM1000C, or FAM1000D (if you have access to a graphing calculator). The graph link files are described in the Graphing Calculator Manual.

5. The accompanying graph and regression line show mean personal total income versus years of education.

a. What is the slope of the regression line?

b. Interpret the slope in this context.

c. By what amount does this regression line predict that mean personal total income changes for 1 additional year of education? For 10 additional years of education?

d. What features of the data are not well described by the regression line?

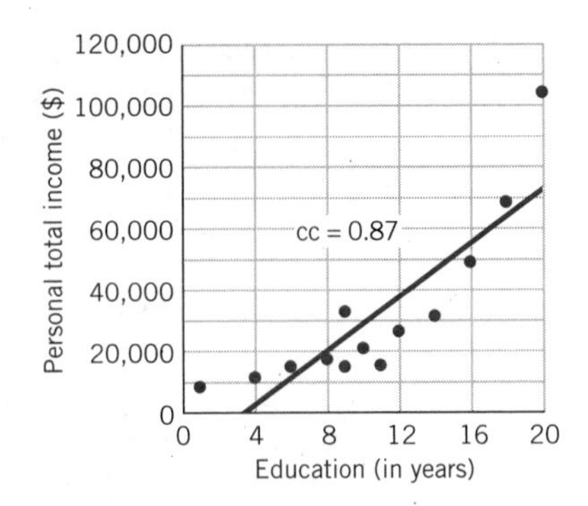

Personal total income = −14,960 + 4370 • educ yrs

6. The following equation represents a best fit regression line for mean personal total income of white males vs. years of education:

$$\text{personal total income} = -17{,}280 + 5240 \cdot \text{years of education}$$

The correlation coefficient is 0.89 and the sample size is 460 white males.

a. What is the rate of change of mean personal total income with respect to years of education?

b. Generate a small table with three points that lie on this regression line. Use two of these points to calculate the slope of the regression line.

c. How does this slope relate to your answer to part (a)?

d. Sketch the graph.

7. From the FAM1000 data, the best fit regression line for mean personal total income of white females vs. years of education is

$$\text{personal total income} = -3400 + 2100 \cdot \text{years of education}$$

The correlation coefficient is 0.87 and the sample size is 383 white females.

a. Interpret the number 2100 in this equation.

b. Generate a small table with three points that lie on this regression line. Use two of these points to calculate the slope of the regression line.

c. How does this slope relate to your answer to part (a)?

d. Sketch the graph.

e. Describe some differences between mean personal total income versus education for white females and for white males (see Exercise 6). What are some of the limitations of the model in making this comparison?

8. From the FAM1000 data, the best fit regression line for mean personal total income for the Census category of nonwhite females vs. years of education is

$$\text{personal total income} = -8270 + 2760 \cdot \text{years of education}$$

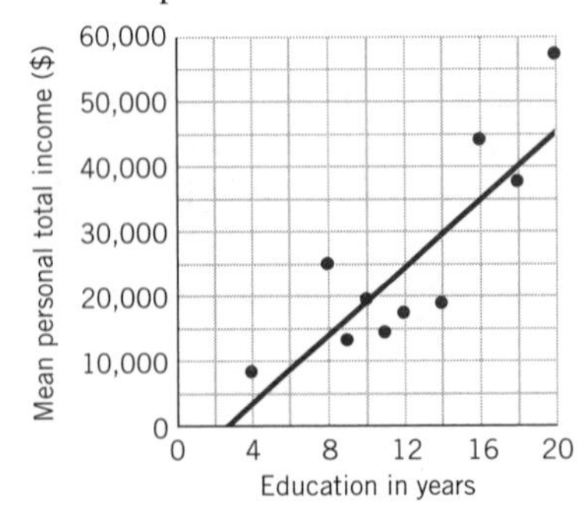

Personal total income =
−8,270 + 2760 • educ yrs
Non-white women

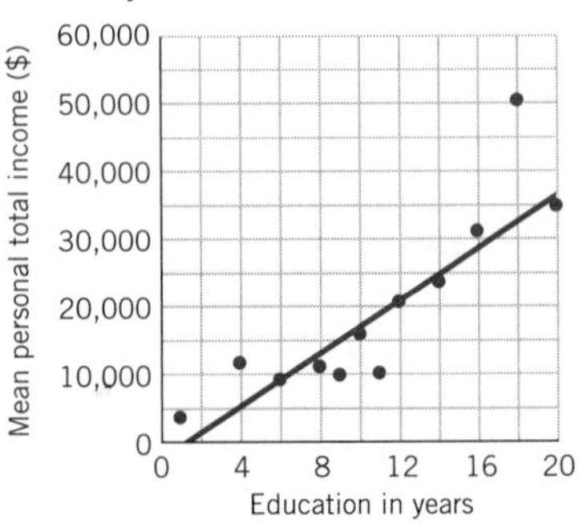

Personal total income =
−3,400 + 2100 • educ yrs
White women

The correlation coefficient is 0.85 and the sample size is 82 nonwhite females. Exercise 7 gives the best fit regression line and cc for mean personal total income for white females vs. years of education. The accompanying figures show the graph of averages for the mean personal total income for white and nonwhite females.

Write a 60 second summary comparing the relationship between education and personal total income for white and nonwhite females.

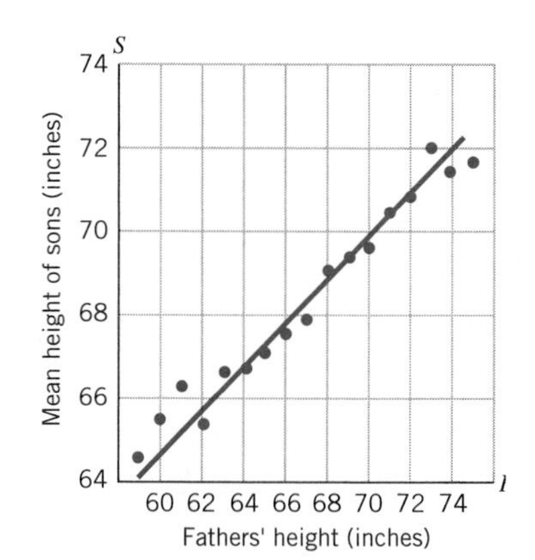

From Snedecor and Cochran, *Statistical Methods*, 8th ed. By permission of the Iowa State University Press. Copyright © 1967.

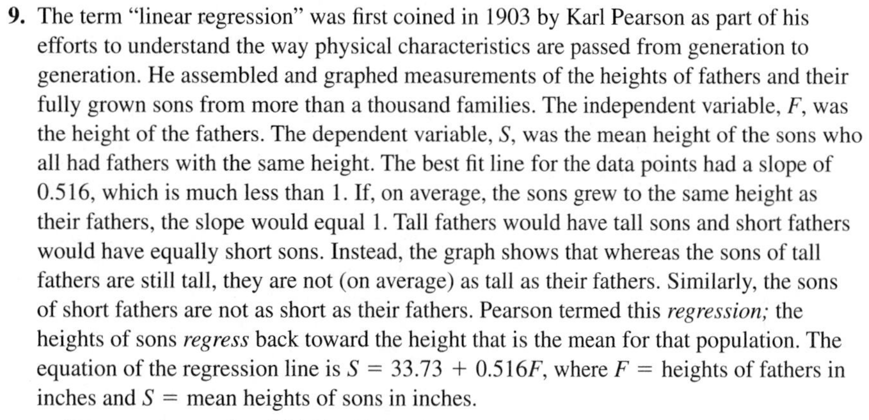

9. The term "linear regression" was first coined in 1903 by Karl Pearson as part of his efforts to understand the way physical characteristics are passed from generation to generation. He assembled and graphed measurements of the heights of fathers and their fully grown sons from more than a thousand families. The independent variable, F, was the height of the fathers. The dependent variable, S, was the mean height of the sons who all had fathers with the same height. The best fit line for the data points had a slope of 0.516, which is much less than 1. If, on average, the sons grew to the same height as their fathers, the slope would equal 1. Tall fathers would have tall sons and short fathers would have equally short sons. Instead, the graph shows that whereas the sons of tall fathers are still tall, they are not (on average) as tall as their fathers. Similarly, the sons of short fathers are not as short as their fathers. Pearson termed this *regression;* the heights of sons *regress* back toward the height that is the mean for that population. The equation of the regression line is $S = 33.73 + 0.516F$, where F = heights of fathers in inches and S = mean heights of sons in inches.

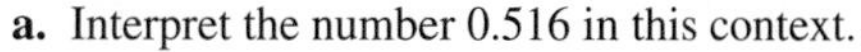

a. Interpret the number 0.516 in this context.

b. Use the regression line to predict the mean height of sons whose fathers are 64 inches tall and of those whose fathers are 73 inches tall.

c. Predict the height of a son who has the same height as his father.

d. If there were over 1000 families, why are there only 17 data points on this graph?

10. Read the excerpt from the book *Performing Arts—The Economic Dilemma.* Describe what the regression line tells you about the relationship between attendance per concert and the number of concerts for a major orchestra.

See Excel or graph link file EDUCOSTS.

11. (Optional use of graphing calculator or computer.)

The data in the table and the accompanying graph give the mean annual cost for tuition and fees at public and private 4-year colleges in the United States since 1985.

Annual Cost for Tuition and Fees at Public and Private Four-year Colleges in the United States

Year	Cost for Public Education ($)	Cost for Private Education ($)
1985	1,386	6,843
1987	1,651	8,118
1988	1,726	8,771
1989	1,846	9,451
1990	2,035	10,348
1991	2,159	11,379
1992	2,410	12,192
1993	2,604	13,055
1994	2,820	13,874
1995	2,977	14,357
1996	3,151	15,605
1997	3,321	16,531

Source: U.S. Bureau of the Census, *Statistical Abstract of the United States,* 1998.

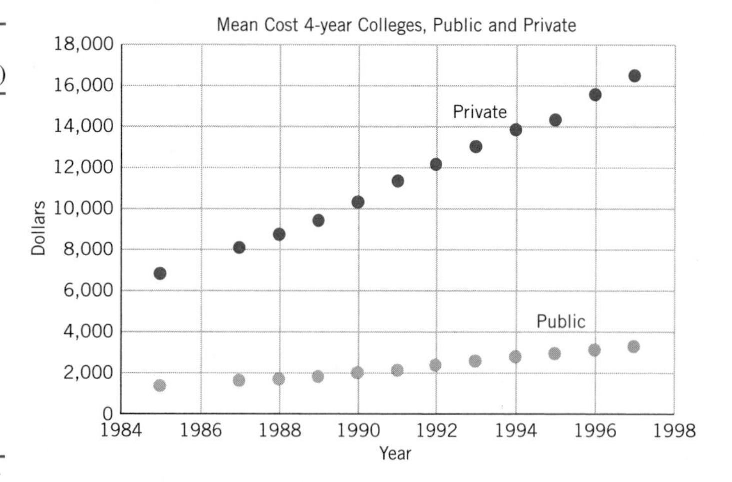

It is clear that the cost of education is going up, but public education is still less expensive than private. The graph suggests that costs of both public and private education vs. time can be roughly represented as straight lines.

a. By hand, sketch two lines to represent the data. Calculate the rate of change of education cost per year for public and for private education by estimating the coordinates of two points that lie on the line, and then estimating the slope.

b. Construct an equation for each of your lines in part (a). If you are using technology, generate two regression lines and compare the equations with the ones you constructed.

c. If the costs continue to rise at the same rates for both sorts of schools, what would be the respective costs for public and private education in the year 2010? Does this seem plausible to you? Why or why not?

12. The charts shown demonstrate the rapid worldwide growth in the number of home computers that are going on-line to the Internet.

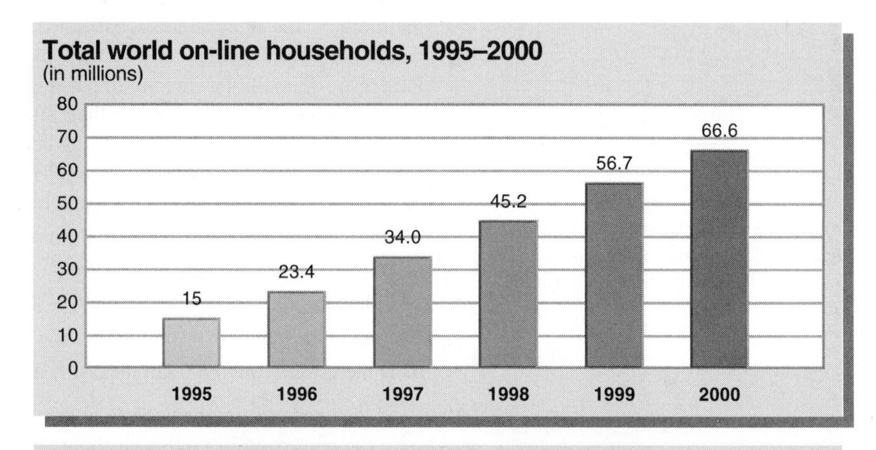

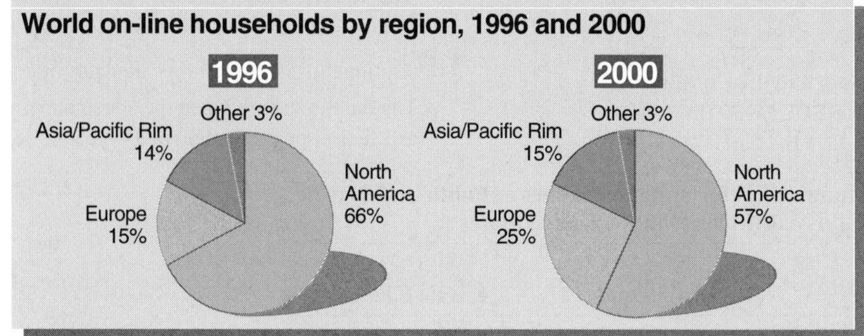

Source: Wall Street Journal Almanac, 1999.

a. Estimate or use technology to compute the regression line for the number of millions of on-line households vs. years since 1995.

b. If this rate of growth is sustained, how many on-line households will there be worldwide in 2010?

c. From the pie charts identify the region where the greatest relative on-line growth has occurred.

d. Do you think the total number of on-line households in North America has declined from 1996 to 2000? From the data given in the charts estimate the number of on-line households in North America in 1996 and in 2000.

See Excel or graph link file KALAMA.

13. (Requires graphing calculator or computer.)

In Chapter 2, we generated by hand a linear model for the mean heights of a group of 161 children in Kalama, Egypt.

Mean Height of Kalama Children

Age (months)	Height (cm)	Age (months)	Height (cm)
18	76.1	24	79.9
19	77.0	25	81.1
20	78.1	26	81.2
21	78.2	27	81.8
22	78.8	28	82.8
23	79.7	29	83.5

Source: D. S. Moore and G. P. McCabe, *Introduction to the Practice of Statistics.* Copyright © 1989 by W.H. Freeman and Company. Used with permission.

a. Use technology to find the best fit regression line.

b. Identify the variables for your model, specify the domain, and interpret the slope and vertical intercept in this context. How good a fit is your line?

c. Use your line to predict the mean height for children 26.5 months old.

See Excel or graph link file CALORIES.

14. (Optional use of graphing calculator or computer.)

The accompanying table shows the calories per minute burned by a 154-pound person moving at speeds from 2.5 to 12 miles/hour (mph). (*Note:* A fast walk is about 5 mph; faster than that is considered jogging or slow running.) Marathons, about 26 miles long, are now run in slightly over 2 hours, so that top distance runners are approaching a speed of 13 mph.

Speed (mph)	Calories per Minute
2.5	3.0
3.0	3.7
3.5	4.2
3.8	4.9
4.0	5.5
4.5	7.0
5.0	8.3
5.5	10.1
6.0	12.0
7.0	14.0
8.0	15.6
9.0	17.5
10.0	19.6
11.0	21.7
12.0	24.5

a. Plot the data.

b. Does it look as if the relationship between speed and calories per minute is linear? If so, generate a linear model. Identify the variables and a reasonable domain for the model, and interpret the slope and vertical intercept. How well does your line fit the data?

c. Describe in your own words what the model tells you about the relationship between speed and calories per minute.

See Excel or graph link file SMOKERS.

15. (Optional use of graphing calculator or computer.)

The accompanying table shows (for years 1965 to 1995 and for people 18 and over) the total percentage of cigarette smokers, the percentage of males that are smokers, and the percentage of females that are smokers.

Percentage of Smokers

Year	Total Population 18 and Older	All Males	All Females
1965	42.4	51.9	33.9
1974	37.1	43.1	32.1
1979	33.5	37.5	29.9
1983	32.1	35.1	29.5
1985	30.1	32.6	27.9
1987	28.8	31.2	26.5
1988	28.1	30.8	25.7
1990	25.5	28.4	22.8
1991	25.6	28.1	23.5
1992	26.5	28.6	24.6
1993	25.0	27.7	22.5
1994	25.5	28.2	23.1
1995	24.7	27.0	22.6

Source: U.S. Bureau of the Census, *Statistical Abstract of the United States,* 1998.

a. By hand, draw a scatter plot of the percentage of all current smokers 18 and older vs. time.

i. Calculate the average rate of change from 1965 to 1995.

ii. Calculate the average rate of change from 1990 to 1995.

Be sure to specify the units in each case.

b. On your graph, sketch an approximate regression line. By estimating coordinates of points on your regression line, calculate the average rate of change of the percentage of total smokers with respect to time.

c. Using a calculator or computer, generate a regression line for the percentage of all smokers 18 and older as a function of time. (You may wish to set 1965 as year 0.) Record the equation and the correlation coefficient. How good a fit is this regression line to the data? Compare the rate of change for your hand-generated regression line to the rate of change for the technology-generated regression line.

d. Generate and record regression lines (and their associated correlation coefficients if you are using technology) for the percentages of both males and females that are smokers as functions of time.

e. Write a summary paragraph using the results from your graphs and calculations to describe the trends in smoking from 1965 to 1995. Would you expect this overall trend to continue? Why?

See Excel or graph link file SKIJUMP.

16. (Optional use of graphing calculator or computer.)

The accompanying table is a list of U.S. ski jumping records set on the 90-meter jumping hill at Howelsen Hill, Steamboat Springs, Colorado. Note that the record was broken more than once in the years 1950 and 1963.

Year	Distance (ft)	Ski Jumper
1916	192	Ragnar Omtvedt
1917	203	Henry Hall
1950	301	Gordon Wten
1950	305	Merrill Barber
1950	307	Art Devlin
1951	316	Ansten Samuelstuen
1963	318	Gene Kotlarek
1963	322	Gene Kotlarek
1978	354	Jim Denny

Source: Tread of Pioneers Museum, Steamboat Springs, CO.

a. Plot the data on a scatter plot. (You may want to define 1916 to be year 0.)

b. Do the data seem to be linear? Explain your answer.

c. Assuming the data are linear, construct a best fit line (by hand or using technology). Carefully identify your variables, and discuss the meaning of the slope and the vertical intercept within the context of this problem. How good a fit is your line?

d. Use your linear model to predict the record ski jump in the year 2000. Does your estimate seem reasonable?

See Excel or graph link file NATIONS.

17. (Requires computer or TI-83 graphing calculator.)

Examine the Excel file NATIONS and the data dictionary at the end of the file. Which pairs of variables would you suspect to be linearly related? Explore the data to find out which variables you want to study. Using either the Excel or graph link version of NATIONS, choose one pair, and generate a best fit regression line. Interpret your results in a 60 second summary. What are your hypotheses as to why these variables may be related? What hidden factors might be influencing the relationship?

See Excel or graph link file SHOTPUT.

18. (Requires graphing calculator or computer.)

In the exercises in Chapter 2, we saw the table and graph given here for the men's Olympic 16-pound shot put.

Olympic Shot Put

Year	Feet Thrown	Year	Feet Thrown
1900	46	1956	60
1904	49	1960	65
1908	48	1964	67
1912	50	1968	67
1920	49	1972	70
1924	49	1976	70
1928	52	1980	70
1932	53	1984	70
1936	53	1988	74
1948	56	1992	71
1952	57	1996	71

Source: From *The 1996 Universal Almanac,* Universal Press Syndicate. Reprinted with permission. (The data points are missing for 1916 and the early 1940s since the Olympics were not held during World Wars I and II.) 1996 data from *http://sports.yahoo.com/oly/lgns/960726/tfmshotmed.html.*

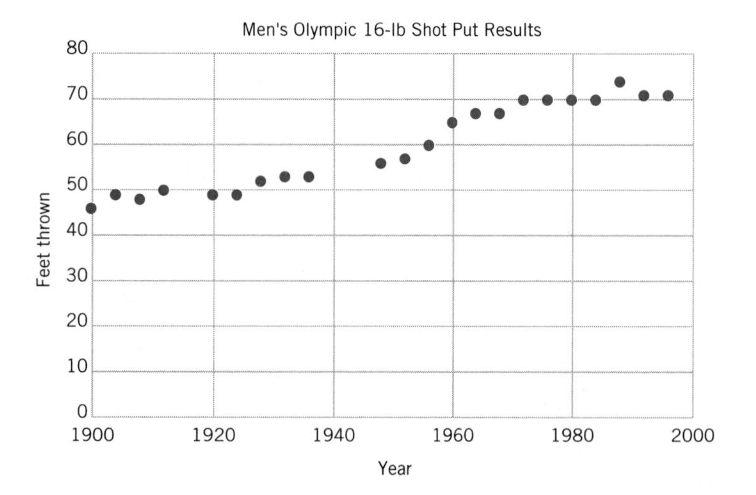

a. Using a graphing calculator or computer, find the equation of the best fit line and, if possible, its correlation coefficient. When you enter the data, you may want to set 1900 as year 0. Is the line a good fit?

b. Describe in your own words your linear model for the data, interpreting each of the terms in your equation. Be sure to specify the units.

See Excel or graph link file FARMPOP.

19. (Requires graphing calculator or computer.)

The accompanying table contains data on the farm population for each decade from 1880 to 1990. Farm population consists of all persons living on farms in rural areas.

Farm Population: 1880–1990

Year	Population (thousands)	Percentage of Total U.S. Population
1880	21,973	43.8
1890	24,771	42.3
1900	29,875	41.9
1910	32,077	34.9
1920	31,974	30.1
1930	30,529	24.9
1940	30,547	23.2
1950	23,048	15.3
1960	15,635	8.7
1970	9,712	4.8
1980	6,051	2.7
1990	4,591	1.9

Source: For 1880–1970, U.S. Bureau of Census, *Historical Statistics of the United States: Colonial Times to 1970.* For 1970 on, U.S. Bureau of Census, *Statistical Abstract of the United States*, 1996.

a. Make scatter plots of the farm population in absolute numbers over time and in percentage of total population over time.

b. Is a linear model appropriate for describing either of the two scatter plots? If so, generate a regression line and interpret your equation.

c. If you selected a different time period, would you find a stronger association between farm population and time? Confirm your prediction using a graphing calculator or function graphing program.

20. The decline of the purchasing power of the U.S. dollar is described here in data from the U.S. Bureau of Labor Statistics. Each of the two tables uses as a standard the amount you could purchase with $1.00 in 1983; in years before 1983 a dollar bought more of the same goods; in years after it bought less. The purchasing power is rounded to the nearest cent.

Purchasing Power of 1983 Dollar in 5-Year Intervals: 1950–1995

	1950	1955	1960	1965	1970	1975	1980	1985	1990	1995
Purchasing power	$4.15	$3.73	$3.37	$3.16	$2.57	$1.86	$1.21	$0.93	$0.77	$0.66

Purchasing Power of 1983 Dollar in 1-Year Intervals: 1983–1997

	1983	1984	1985	1986	1987	1988	1989	1990	1991	1992	1993	1994	1995	1996	1997
Purchasing power	$1.00	$0.96	$0.93	$0.91	$0.88	$0.85	$0.81	$0.77	$0.73	$0.71	$0.69	$0.67	$0.66	$0.64	$0.62

a. Graph the data for the 5-year intervals between 1950 and 1995 and for the 1-year intervals between 1983 and 1997. Estimate or use technology to calculate a regression line for each graph.

b. How does the rate of decline for the 1950-to-1995 period compare to the one for the period since 1983? If the purchasing power continues to decline as it has from 1983, what will a dollar buy in 2010?

c. According to the 1983-to-1997 formula, when will the dollar purchase nothing? Comment on the likelihood of this.

See Excel or graph link file MARATHON.

21. (Optional use of graphing calculator or computer.)

The accompanying table and graph show the winning running time in minutes for women in the Boston Marathon.

Women's Boston Marathon Winning Times

Year	Time (min)	Year	Time (min)
1972	190	1987	145
1973	186	1988	145
1974	167	1989	145
1975	162	1990	145
1976	167	1991	144
1977	166	1992	144
1978	165	1993	145
1979	155	1994	142
1980	154	1995	145
1981	147	1996	147
1982	150	1997	146
1983	143	1998	143
1984	149	1999	143
1985	154	2000	146
1986	145		

Source: From *The 1993 Universal Almanac,* Universal Press Syndicate. Reprinted with permission. 1994–1999 data from *http://www.bostonmarathon.org.*

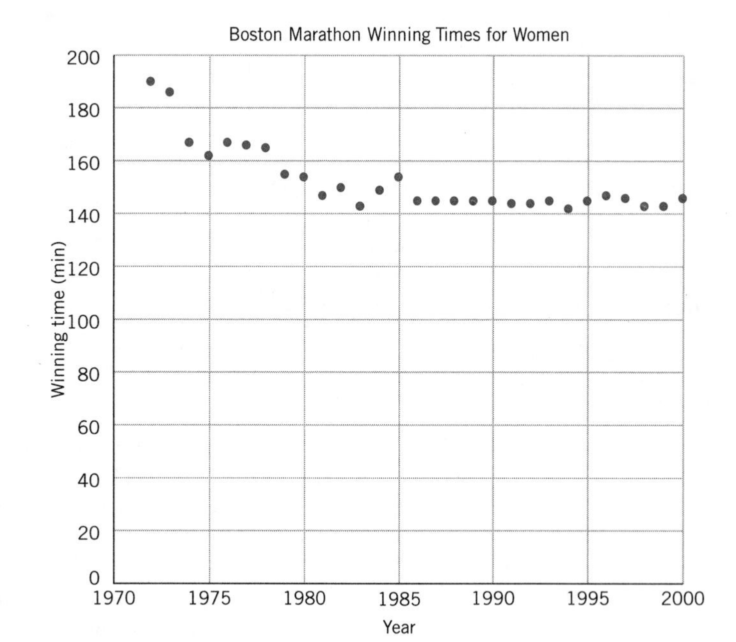

a. Sketch a line that approximates the data by hand. Set 1972 as year 0 and generate an equation of this line. Interpret the slope of your line in this context. Using technology, generate a regression line and compare its equation to the equation you generated.

b. If the marathon times continued to change at the rate in your linear model, predict the winning running time for the women's marathon in 2010. Does that seem reasonable? If not, why not?

c. The graph seems to flatten out after about 1981. Construct a second regression line for the data from 1981 on. If you're using technology, what is the correlation coefficient for this line? What would this line predict for the winning running time for the women's marathon in 2010? Does this estimate seem more realistic than your previous estimate?

d. Write a short paragraph summarizing the trends in the Boston Marathon winning times for women.

See Excel or graph link file MENSMILE.

Record Times for Men's Mile

Year	Time (min)	Year	Time (min)
1913	4.24	1958	3.91
1915	4.21	1962	3.91
1923	4.17	1964	3.90
1931	4.15	1965	3.89
1933	4.13	1966	3.86
1934	4.11	1967	3.85
1937	4.11	1975	3.85
1942	4.10	1975	3.82
1942	4.10	1979	3.82
1942	4.08	1980	3.81
1943	4.04	1981	3.81
1944	4.03	1981	3.81
1945	4.02	1981	3.79
1954	3.99	1985	3.77
1954	3.97	1993	3.74
1957	3.95		

Source: Data extracted from the website at *http://www.uta.fi/~csmipe/sports/eng/mwr.html.*

22. (Optional use of graphing calculator or computer.)

The accompanying table and graph show the world record times for the men's mile. Note that several times the standing world record was broken more than once during a year.

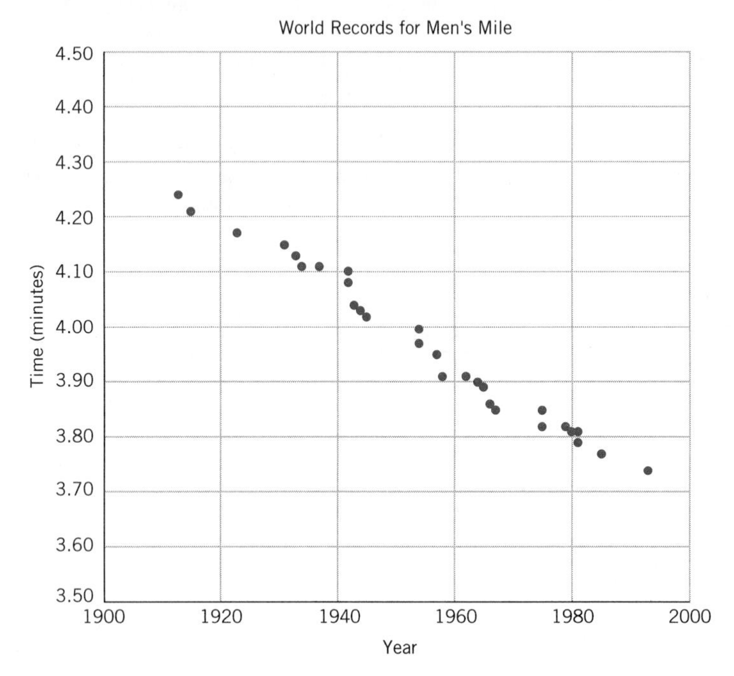

a. Generate a line that approximates the data (by hand, with a graphing calculator or a function graphing program). You may wish to set 1910 or 1913 as year 0. If you are using technology, specify the correlation coefficient. Interpret the slope of your line in this context.

b. If the world record times continued to change at the rate specified in your linear model, predict the record time for the men's mile in 2010. Does your prediction seem reasonable? If not, why not?

c. In what year would your linear model predict the world record to be 0 minutes? Since this is impossible, what do you think is a reasonable domain for your function? Describe what you think would happen in the years after those contained in your domain.

d. Write a short paragraph summarizing the trends in the world record times for the men's mile.

23. The temperature at which water boils is affected by the difference in atmospheric pressure at different altitudes above sea level. The classic cookbook *The Joy of Cooking* by Irma S. Rombauer and Marion Rombauer Becker gives the data in the accompanying table (rounded to the nearest degree) on the boiling temperature of water at different altitudes above sea level.

Boiling Temperature of Water

Altitude (ft above sea level)	Boiling Temperature °F	°C
0	212	100
2,000	208	98
5,000	203	95
7,500	198	92
10,000	194	90
15,000	185	85
30,000	158	70

a. Use the accompanying table for the following:

i. Plot boiling temperature in degrees Centigrade, °C, vs. the altitude. Find a formula to describe the boiling temperature of water, in °C, as a function of altitude.

ii. Plot boiling temperature in degrees Fahrenheit, °F, vs. the altitude. Find a formula to describe the boiling temperature of water, in °F, as a function of altitude.

iii. According to your formula, what is the temperature at which water will boil where you live? Can you verify this? What other factors could influence the temperature at which water will boil?

b. The highest point in the United States is Mount McKinley in Alaska, at 20,320 feet above sea level; the lowest point is Death Valley in California, at 285 feet below sea level. You can think of distances below sea level as negative altitudes from sea level. At what temperature in degrees Fahrenheit will water boil in each of these locations according to your formula?

c. A healthy human has a normal temperature of around 98.6°F. A certain amount of water in the human body is necessary for life; at what altitude does your formula predict that the water in the body will boil?

d. A scientist who develops a formula for observed phenomena is interested in testing the limits of its accuracy. It is interesting to ask, and then test, whether there is an altitude at which water can be made to boil at 0°C, the freezing point of water at sea level. At what altitude would your formula predict this would happen? Does this seem reasonable? (Note that airplane cabins are pressurized to near sea-level atmospheric pressure conditions to avoid unhealthy conditions resulting from high altitude.)

See Excel or graph link file LONGJUMP.

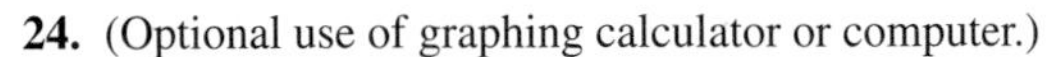

24. (Optional use of graphing calculator or computer.)

The accompanying table and graph show the world distance records for the women's long jump. Several times a new long-jump record was set more than once during a given year.

Women's Long-Jump Records

Year	Distance (m)	Year	Distance (m)
1954	6.28	1976	6.99
1955	6.28	1978	7.07
1955	6.31	1978	7.09
1956	6.35	1982	7.15
1956	6.35	1982	7.20
1960	6.40	1983	7.21
1961	6.42	1983	7.27
1961	6.48	1983	7.43
1962	6.53	1985	7.44
1964	6.70	1986	7.45
1964	6.76	1986	7.45
1968	6.82	1987	7.45
1970	6.84	1988	7.45
1976	6.92	1988	7.52

Source: Data extracted from the website at *http://www.uta.fi/~csmipe/sports/eng/mwr.html.*

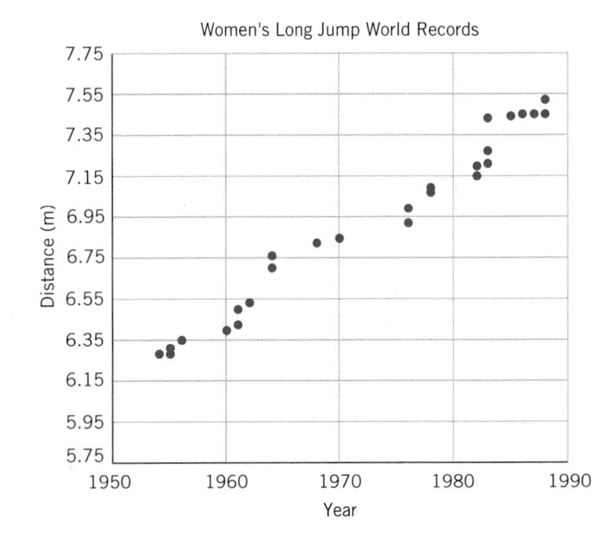

a. Generate a line that approximates the data (by hand or with a graphing calculator or function graphing program). You may wish to set 1950 or 1954 as year 0. If you are using technology, specify the correlation coefficient. Interpret the slope of your line in this context.

b. If the world record distances continued to change at the rate described in your linear model, predict the world record distance for the women's long jump in the year 2000.

c. What would your model predict for the record in 1943? How does this compare with the actual 1943 record of 6.25 meters? What do you think would be a reasonable domain for your model? What do you think the data would look like for years outside your specified domain?

See Excel or graph link file MOTOR.

Motor Vehicle Registrations

Year	Vehicles (millions)
1945	31.0
1950	49.2
1955	62.7
1960	73.9
1965	90.4
1970	108.4
1975	132.9
1980	155.8
1985	170.2
1990	188.8
1995	201.5

Source: U.S. Federal Highway Administration, *Highway Statistics,* annual.

25. (Optional use of graphing calculator or computer.)

The accompanying table and graph show the increasing number of motor vehicle registrations (cars and trucks) in the United States.

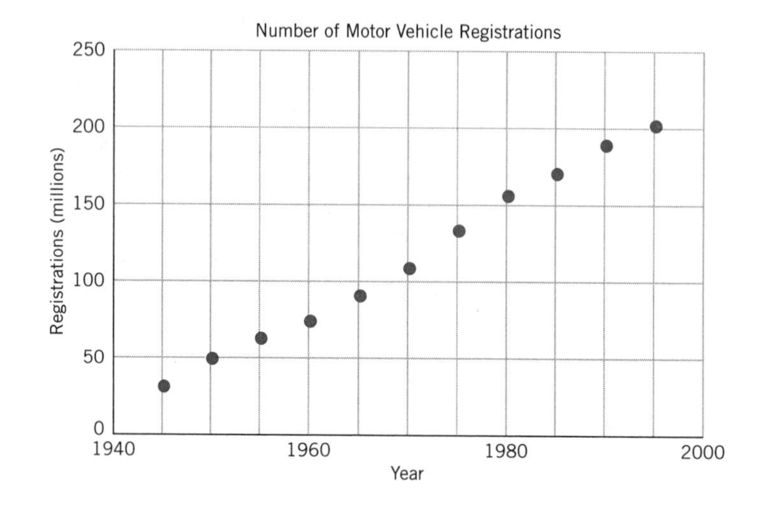

a. Using 1945 as a base year, find a linear equation that would be a reasonable model for the data.

b. Interpret the slope of your line in this context.

c. If this rate of growth continues, how many motor vehicles will be registered in the United States in 2010?

26. Examine the following data on U.S. union membership from 1930 to 1997:

U.S. Union Membership, 1930–97

Year	Labor Force* (thousands)	Union Members† (thousands)	Percentage of Labor Force	Year	Labor Force* (thousands)	Union Members† (thousands)	Percentage of Labor Force
1930	29,424	3,401	11.6	1986	96,903	16,975	17.5
1935	27,053	3,584	13.2	1987	99,303	16,913	17.0
1940	32,376	8,717	26.9	1988	101,407	17,002	16.8
1945	40,394	14,322	35.5	1989	103,480	16,960	16.4
1950	45,222	14,267	31.5	1990	103,905	16,740	16.1
1955	50,675	16,802	33.2	1991	102,786	16,568	16.1
1960	54,234	17,049	31.4	1992	103,688	16,390	15.8
1965	60,815	17,299	28.4	1993	105,067	16,598	15.8
1970	70,920	19,381	27.3	1994	107,989	16,748	15.5
1975	76,945	19,611	25.5	1995	110,038	16,360	14.9
1980	90,564	19,843	21.9	1996	111,960	16,269	14.5
1985	94,521	16,996	18.0	1997	114,533	16,110	14.1

*Does not include agricultural employment; from 1985, does not include self-employed or unemployed persons.

†From 1930 to 1980, includes dues-paying members of traditional trade unions, regardless of employment status; from 1985, includes members of employee associations that engage in collective bargaining with employers.

Source: Bureau of Labor Statistics, U.S. Dept. of Labor. *The World Almanac & Book of Facts,* 1999.

a. Graph the percentage of labor force in unions vs. time from 1985 to 1997. Measuring time in years since 1985, find a linear regression formula for these data either using a graphing calculator or a computer or estimating by hand.

b. When does your formula predict that only 10% of the labor force will be unionized?

c. What other data would you want to look at to understand why union membership is declining?

27. If a study shows that smoking and lung cancer have a high positive correlation, does this mean that smoking causes lung cancer? Explain your answer.

28. Parental income has been found to have a high positive correlation with academic success. What are two different ways you could interpret this finding?

29. How did a Princeton professor's statistical analysis influence a decision by a federal judge in Philadelphia to give a Pennsylvania senate seat to a losing Republican candidate? See the reading *His Stats Can Oust a Senator or Price a Bordeaux.*

Chapter 3

When Lines Meet: Linear Systems

Overview

When is solar heating cheaper than conventional heating? Will you pay more tax under a flat tax or a graduated tax plan? In this chapter we use a collection of two or more equations (called a system of equations) to compare different heating systems and different tax plans.

After reading this chapter, you should be able to:

- construct, graph, and interpret systems of linear equations
- find a solution for a system of two linear equations
- use systems of linear equations to model some social and physical phenomena
- construct, graph, and interpret piecewise linear functions

3.1 AN ECONOMIC COMPARISON OF SOLAR VS. CONVENTIONAL HEATING SYSTEMS

On a planet with limited fuel resources, heating decisions involve both monetary and ecological considerations. Typical costs for three different kinds of heating systems for a three-bedroom housing unit are given in Table 3.1.

Table 3.1
Typical Costs for Three Heating Systems

Type of System	Installation Cost ($)	Operation Cost/Year ($/yr)
Electric	5,000	1,100
Gas	12,000	700
Solar	30,000	150

Solar heating is clearly the most costly to install and the least expensive to run. Electric heating, conversely, is the cheapest to install and the most expensive to run. By converting this information to equation form, we can find out when the solar system begins to pay back the initially higher cost. If no allowance is made for inflation or changes in fuel price,[1] the general equation for the total cost, C, is

$$C = \text{installation cost} + (\text{annual operating cost})(\text{years of operation})$$

If we let n equal the number of years of operation and use the data from Table 3.1, we can construct the following linear equations:

$$C_{\text{electric}} = 5000 + 1100n$$
$$C_{\text{gas}} = 12{,}000 + 700n$$
$$C_{\text{solar}} = 30{,}000 + 150n$$

Together they form a *system of linear equations.* Table 3.2 gives the cost data at 5-year intervals, and Figure 3.1 shows the costs over a 40-year period for the three heating systems.

Table 3.2
Heating System Total Costs

Year	Electric ($)	Gas ($)	Solar ($)
0	5,000	12,000	30,000
5	10,500	15,500	30,750
10	16,000	19,000	31,500
15	21,500	22,500	32,250
20	27,000	26,000	33,000
25	32,500	29,500	33,750
30	38,000	33,000	34,500
35	43,500	36,500	35,250
40	49,000	40,000	36,000

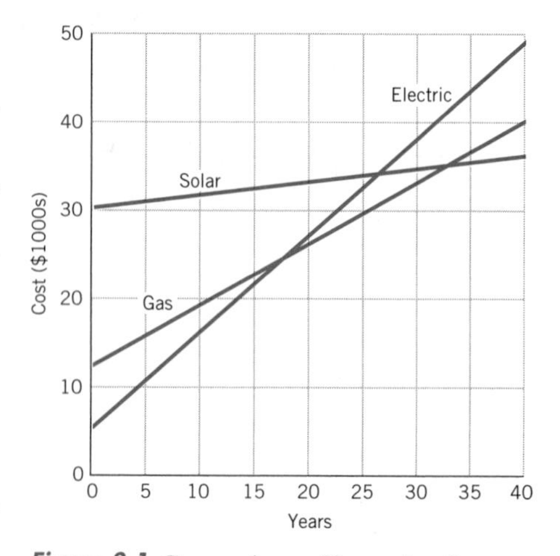

Figure 3.1 Comparison of home heating costs.

[1]A more sophisticated model might include many other factors, such as interest, repair costs, the cost of depleting fuel resources, risks of generating nuclear power, or what economists call opportunity costs.

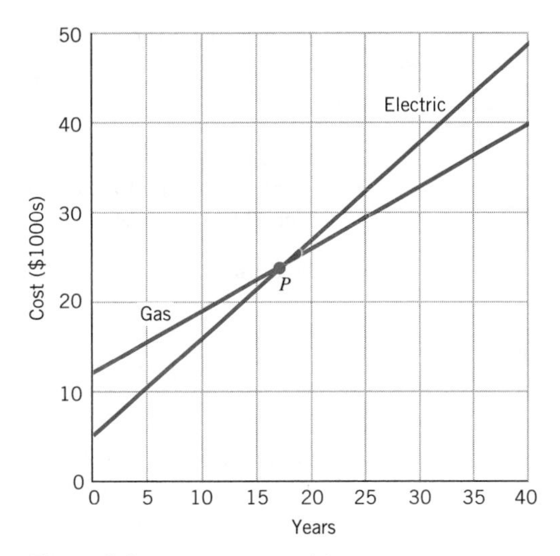

Figure 3.2 Gas versus electric.

Let's compare the costs for gas and electric heat. Figure 3.2 shows the graphs of the equations for these two heating systems. The point of intersection, P, shows where the lines predict the *same* total cost for both gas and electricity, given a certain number of years of operation. From the graph, we can estimate the coordinates of the point P:

$$P = \text{(number of years of operation, cost)}$$

$$\approx (17, \$24000)$$

The total cost of operation is about \$24,000 for each type of energy after about 17 years of operation. We can compare the relative costs of each system to the left and right of the point of intersection. Figure 3.2 shows that gas is less expensive than electricity to the right of the intersection point and more expensive than electricity to the left of the intersection point. Using the equations, we can find exact values for the coordinates of the intersection point.

When the gas and electric lines intersect, the coordinates satisfy both equations. At the point of intersection, the total cost of electric heat, C_{electric}, equals the total cost of gas heat, C_{gas}. Thus, the two expressions for the total cost can be set equal to each other to find exact values for the coordinates of the intersection point:

$$C_{\text{electric}} = 5000 + 1100n \qquad (1)$$

$$C_{\text{gas}} = 12{,}000 + 700n \qquad (2)$$

Set (1) equal to (2) $\qquad C_{\text{electric}} = C_{\text{gas}}$

substitute $\qquad 5000 + 1100n = 12{,}000 + 700n$

subtract 5000 from each side $\qquad 1100n = 7000 + 700n$

subtract $700n$ from each side $\qquad 400n = 7000$

divide each side by 400 $\qquad n = 17.5$ years

When $n = 17.5$ years, the total cost for electric or gas heating is the same. The total cost can be found by substituting this value for n in Equation (1) or (2):

Substitute 17.5 for n in Equation (1)

$$C_{\text{electric}} = 5000 + 1100(17.5)$$
$$= 5000 + 19{,}250$$
$$C_{\text{electric}} = \$24{,}250$$

Since we claim that the pair of values (17.5, 24250) satisfies both equations, we need to check, when $n = 17.5$ years, that C_{gas} is also \$24,250:

Substitute 17.5 for n in Equation (2)

$$C_{\text{gas}} = 12{,}000 + 700(17.5)$$
$$= 12{,}000 + 12{,}250$$
$$C_{\text{gas}} = \$24{,}250$$

The coordinates (17.5, 24250) satisfy both equations. When $n = 17.5$ years, then $C_{\text{electric}} = C_{\text{gas}} = \$24{,}250$. The point (17.5, 24250) is called a *solution* to the system of these two equations. After 17.5 years, a total of \$24,250 has been spent on heat for either an electric or a gas heating system. Electricity is a more economical fuel supply for 17.5 years, then gas becomes cheaper.

The intersection points for electric vs. solar heating and gas vs. solar heating can be estimated using the graphs in Figures 3.3 and 3.4. You will be asked to find more accurate solutions in the homework exercises.

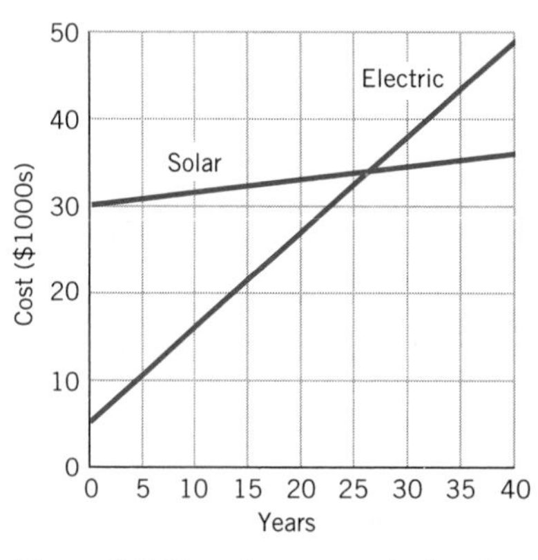

Figure 3.3 Electric versus solar heating.

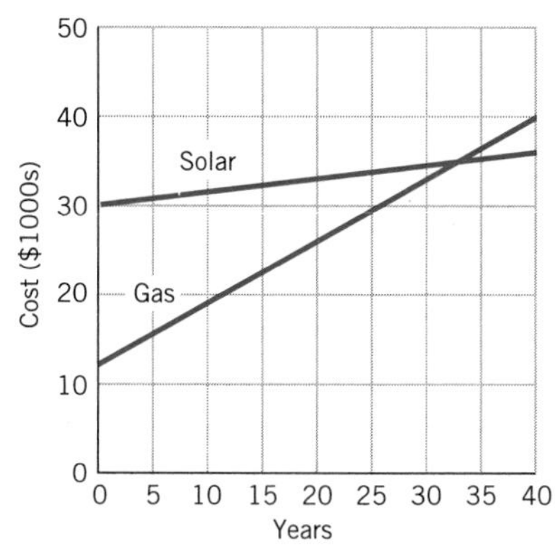

Figure 3.4 Gas versus solar heating.

Algebra Aerobics 3.1

1. Our model tells us that electric systems are cheaper than gas for 17.5 years of operation. Using Figure 3.1, estimate the interval over which gas is the cheapest of the three heating systems. When does solar heating become the cheapest system compared to the other two heating systems?
2. For the following system of equations:

$$4x + 3y = 9$$
$$5x + 2y = 13$$

   a. Determine whether $(3, -1)$ is a solution.

   b. Show why $(1, 4)$ is not a solution for this system.

3. **a.** Show that the following equations are equivalent:

$$4x = 6 + 3y$$
$$12x - 9y = 18$$

   b. How many solutions are there for the system of equations in part (a)?

3.2 FINDING SOLUTIONS TO SYSTEMS OF LINEAR EQUATIONS

In the heating example, we found a solution for a system of two equations by finding the point where the graphs of the two equations intersected. At the point of intersection, the equations share the same value for the independent variable and the same value for the dependent variable.

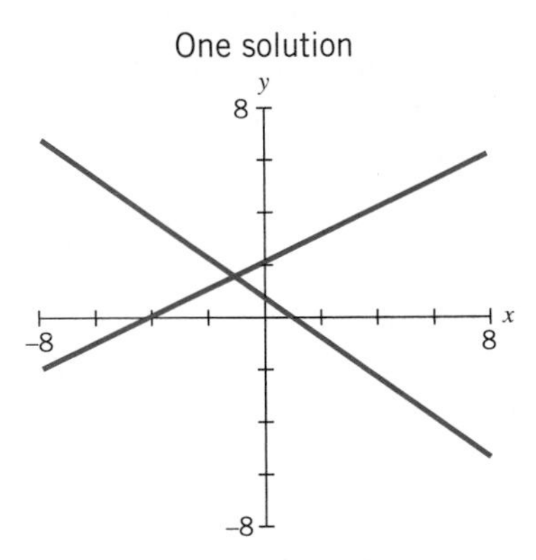

Figure 3.5 Lines intersect at a single point.

> A pair of real numbers is a *solution* to a system of linear equations in two variables if and only if the pair of numbers is a solution to each equation in the system.
>
> On the graph of a system of linear equations in two variables, a solution is a point at which all the lines intersect.

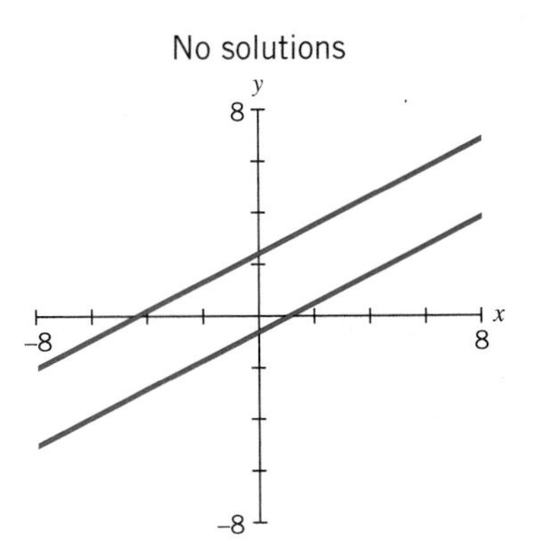

Figure 3.6 Parallel lines never intersect.

We begin analyzing a system of equations by finding out if there is a solution, that is, if the graphs of the equations intersect.

Visualizing Solutions to Systems of Linear Equations

For a single linear equation, the graph of its solutions is a line and every point on the line is a solution. This equation has an unlimited number of solutions. In a system of two linear equations, a solution must satisfy both equations. We can easily visualize what might happen. If we graph two different straight lines and the lines intersect (Figure 3.5), there is only one solution—at the intersection point. If the lines are parallel, they never intersect and there are no solutions (Figure 3.6).

If the lines are not parallel, the coordinates of the point of intersection can be estimated by inspecting the graph. For example, in Figure 3.5, the two lines appear to cross near the point $(-1, 1.5)$.

How else might the graphs of two lines be related? We may find that the two equations represent the same line. Consider, for example, the following two equations:

$$y = 0.5x + 2 \quad (1)$$

$$3y = 1.5x + 6 \quad (2)$$

If we multiply each side of Equation (1) by 3, we obtain Equation (2). The two equations are *equivalent,* since any solution of one equation is also a solution of the other equation. Both equations represent the same line. There are an infinite number of points on that line, and they are all solutions for both equations (Figure 3.7).

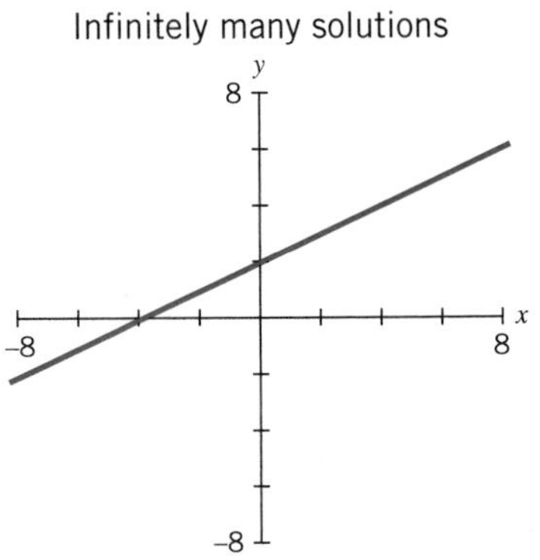

Infinitely many solutions

Figure 3.7 The two equations represent the same line. Every point on the line is a solution for both equations.

Using Equations to Find Solutions

A system of two linear equations can be solved in several ways. The form of the equations usually determines the most efficient method. The methods we examine here all give the same final answer. In each case, a solution consists of a pair of numbers (or a set of pairs of numbers), which we write in the form (x, y).

Substitution Method

When at least one of the equations is in function form, $y = mx + b$, or can easily be put in function form, we can use the substitution method. The point of intersection represents a solution, and at the point of intersection both equations have the same value for x and the same value for y. Thus, to find the solution, we can substitute the expression for y from one equation into the other equation. The steps for this strategy are:

1. Create one equation in one unknown by substituting the expression for y of one equation into the other equation. Solve the resulting equation to find the value for x.
2. Substitute this value for x in one of the two original equations to find the corresponding value for y.
3. Check your results in the other original equation.

It is a good idea to graph the equations to verify that the solution you have found gives the coordinates of the intersection point.

Example 1

Find the solution for the following system of equations:

$$y = 2x + 8 \quad (1)$$

$$y = -3x - 7 \quad (2)$$

Solution

This is the simplest case of substitution, when both equations are in function form. Since y is written in terms of x in both equations, we can set the expressions for y equal to each other or substitute the expression for y in Equation (1) into Equation (2),

$$2x + 8 = -3x - 7$$

then solve for x

$$5x = -15$$

$$x = -3$$

The two lines cross when $x = -3$. To find the y value at the point of intersection, we can substitute -3 for x in either of the two original equations. Using Equation (1):

$$y = 2x + 8 \quad (1)$$

Substitute -3 for x $\quad y = 2(-3) + 8$

multiply $\quad = -6 + 8$

$$y = 2$$

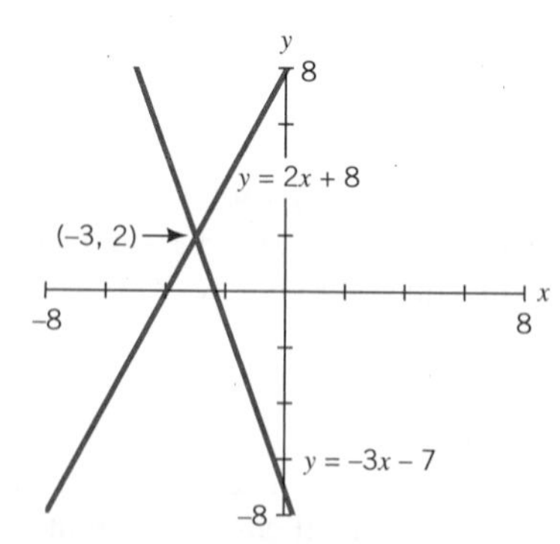

Figure 3.8 Graphs of $y = 2x + 8$ and $y = -3x - 7$.

We can check that the pair of values $(-3, 2)$ does indeed work in *both* equations. Using Equation (2), we verify that when $x = -3$, then $y = 2$:

$$y = -3x - 7 \quad (2)$$

Substitute -3 for x $\quad y = -3(-3) - 7$

simplify $\quad = 9 - 7$

$$y = 2$$

So $x = -3$ and $y = 2$ is the solution to the system. In Figure 3.8 we get visual confirmation that $(-3, 2)$ represents the intersection point of the two lines and thus is a solution to the system.

Example 2

Find the point (if any) where the graphs of the following two linear equations intersect:

$$6x + 7y = 25 \qquad (1)$$

$$y = 15 + 2x \qquad (2)$$

Solution

In Equation (1) substitute the expression for y from Equation (2):

$$6x + 7(15 + 2x) = 25$$

Simplify $\qquad 6x + 105 + 14x = 25$

$$20x = -80$$

$$x = -4$$

We can use one of the original equations to find the value for y when $x = -4$. Using Equation (2):

$$y = 15 + 2x \qquad (2)$$

Substitute -4 for x $\qquad y = 15 + 2(-4)$

multiply $\qquad y = 15 - 8$

$$y = 7$$

We can check that $(-4, 7)$ satisfies Equation (1) by substituting -4 for x and 7 for y and verifying that the result is a true statement:

$$6x + 7y = 25 \qquad (1)$$

$$6(-4) + 7(7) = 25$$

$$-24 + 49 = 25$$

True statement $\qquad 25 = 25$

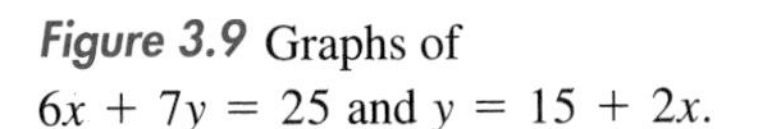

Figure 3.9 Graphs of $6x + 7y = 25$ and $y = 15 + 2x$.

Figure 3.9 shows a graph of the two equations and the intersection point at $(-4, 7)$.

Algebra Aerobics 3.2a

1. Solve the following systems of equations:

a. $y = x + 4$
$y = -2x + 7$

b. $y = -1700 + 2100x$
$y = 4700 + 1300x$

c. $F = C$
$F = 32 + \frac{9}{5}C$

(Part (c) was a question on the T V program, "Who Wants To Be a Millionaire?")

2. Solve the following systems of equations:

a. $y = x + 3$
$5y - 2x = 21$

b. $z = 3w + 1$
$9w + 4z = 11$

c. $x = 2y - 5$
$4y - 3x = 9$

d. $r - 2s = 5$
$3r - 10s = 13$

Elimination Method

Another method called *elimination* is convenient when neither equation can be represented in a function form that is easy to use. The strategy is to eliminate one variable by adding the equations together. Elimination occurs if the coefficients of one variable are opposites of each other (e.g., $2x$ and $-2x$, or $-4y$ and $4y$). After the equations are added the coefficient of that variable will be zero.

Example 3

The following system can be solved by adding the equations and thus eliminating the x terms:

$$2y + 3x = -6 \qquad (1)$$

$$5y - 3x = 27 \qquad (2)$$

Solution

Add Equation (1) and	$2y + 3x = -6$
Equation (2) to	$\underline{5y - 3x = 27}$
eliminate the x terms	$7y + 0x = 21$
	$7y = 21$
divide by 7	$y = 3$

Now we can solve for x by substituting $y = 3$ in one of the original equations. Using Equation (1):

Let $y = 3$	$2(3) + 3x = -6$
multiply	$6 + 3x = -6$
add -6 to both sides	$3x = -12$
divide by 3	$x = -4$

We can check this solution in the other equation. Using Equation (2):

	$5y - 3x = 27$
Let $x = -4$ and $y = 3$	$5(3) - 3(-4) = 27$
multiply	$15 + 12 = 27$
true statement	$27 = 27$

The solution to the system is $x = -4$ and $y = 3$, or $(-4, 3)$.

In this case our work was simplified since the coefficients of x (namely 3 and -3) were opposites of each other, so the sum of the x terms was 0. When neither variable has coefficients in different equations that are opposites of each other, the strategy is to:

1. Find equivalent equations for one or both equations such that when the equations are added together, one variable is eliminated since the coefficient of that variable is now zero.
2. Solve the new equation for the remaining variable.

3. Substitute the value found in step 2 into one of the original equations to determine a value for the eliminated variable.
4. Check your answer in the other original equation.

Example 4

Solve the following system of equations:

$$4x - 3y = 13 \qquad (1)$$

$$3x - 5y = -4 \qquad (2)$$

Solution

In this system we can find equivalent equations such that the coefficients of one of the variables are opposites of each other. There are many ways to accomplish this. First choose which variable is to be eliminated. Sometimes the computation required to eliminate one variable is easier than the computation involved to eliminate the other. As in this case, it usually doesn't matter. If we choose to eliminate y, then we need to find out how to make the coefficients of y opposites of each other. An easy way is to multiply both sides of Equation (1) by 5 and both sides of Equation (2) by -3. The result is a system of equations [Equations (3) and (4)] that is equivalent to our original system, but now the coefficients of y are 15 and -15:

Multiply each side of Equation (1) by 5	$5(4x - 3y) = 5(13)$	
simplify	$20x - 15y = 65$	(3)
multiply each side of Equation (2) by -3	$-3(3x - 5y) = -3(-4)$	
simplify	$-9x + 15y = 12$	(4)

then add Equations (3) and (4)	$20x - 15y = 65$
to eliminate the y terms	$-9x + 15y = 12$
	$11x = 77$
divide by 11	$x = 7$

Solve Equation (1) for y when $x = 7$:

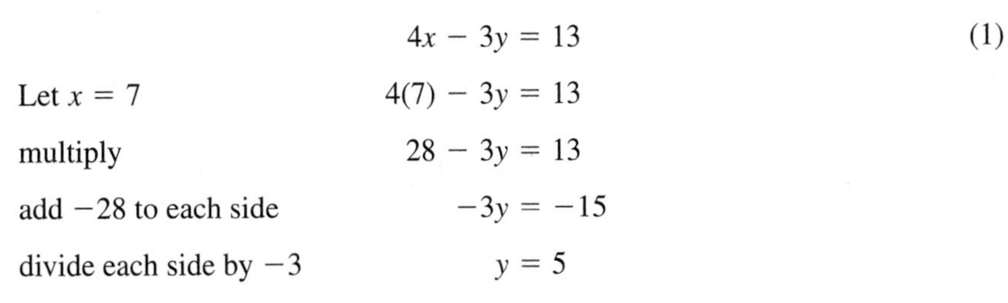

	$4x - 3y = 13$	(1)
Let $x = 7$	$4(7) - 3y = 13$	
multiply	$28 - 3y = 13$	
add -28 to each side	$-3y = -15$	
divide each side by -3	$y = 5$	

Use Equation (2) to check whether $x = 7$ and $y = 5$ is a solution:

	$3x - 5y = -4$	(2)
Let $x = 7$ and $y = 5$	$3(7) - 5(5) = -4$	
multiply	$21 - 25 = -4$	
true statement	$-4 = -4$	

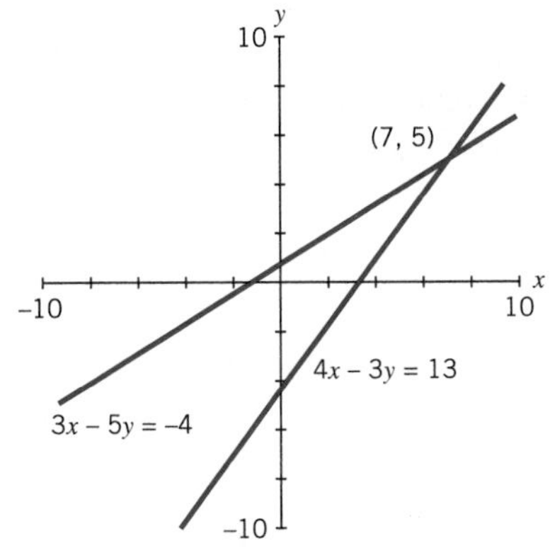

Figure 3.10 Graphs of $4x - 3y = 13$ and $3x - 5y = -4$.

Figure 3.10 shows the graphs of the two equations and the intersection point at (7, 5).

Special Cases: How Can You Tell If There Is No Unique Intersection Point?

Example 5

A system with no solution

Solve the following system of two linear equations:

$$y = 20{,}000 + 1500x \qquad (1)$$

$$2y - 3000x = 50{,}000 \qquad (2)$$

If we assume the two lines intersect, then the y values for Equations (1) and (2) are the same at some point. We can try to find this value for y by substituting the expression for y from Equation (1) into Equation (2):

$$2(20{,}000 + 1500x) - 3000x = 50{,}000$$

Simplify $\quad 40{,}000 + 3000x - 3000x = 50{,}000$

false statement $\quad 40{,}000 = 50{,}000$ (???)

What could this possibly mean? Where did we go wrong? If we return to the original set of equations and solve Equation (2) for y in terms of x, we can see why there is no solution for this system of equations:

$$2y - 3000x = 50{,}000 \qquad (2)$$

Add $3000x$ to both sides $\quad 2y = 50{,}000 + 3000x$

divide by 2 $\quad y = 25{,}000 + 1500x \qquad (3)$

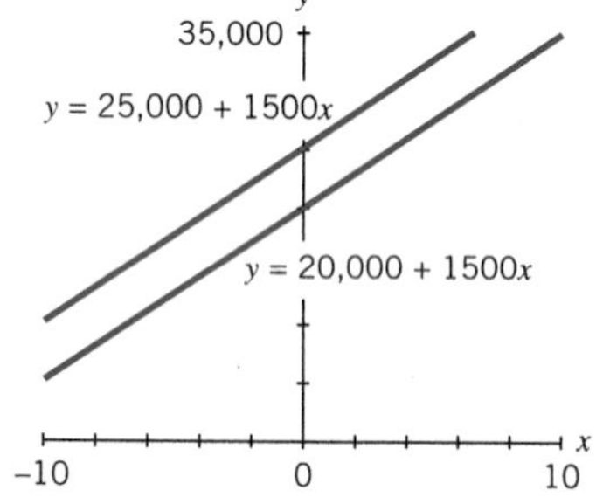

Figure 3.11 Graphs of $y = 25{,}000 + 1500x$ and $y = 20{,}000 + 1500x$.

Now examine Equations (1) and (3). Equation (3) is a rewritten form of the original Equation (2):

$$y = 20{,}000 + 1500x \qquad (1)$$

$$y = 25{,}000 + 1500x \qquad (3)$$

We can see that we have two parallel lines, both with the same slope of 1500 but different y intercepts (20,000 and 25,000, respectively). So there is no intersection point for these lines (see Figure 3.11). There is no value of x that yields the same value of y in both equations, because the value of y in Equation (3) must always be 5000 greater than the corresponding y value in Equation (1). Our initial premise, that the two y values were equal at some point, allowed us to substitute one y expression for another. However, analyzing these equations allows us to discover that our premise was incorrect.

Example 6

A system with infinitely many solutions

Solve the following system:

$$45x = -y + 33 \qquad (1)$$

$$2y + 90x = 66 \qquad (2)$$

As always, there are multiple ways of solving the system. Let's put both equations in function form:

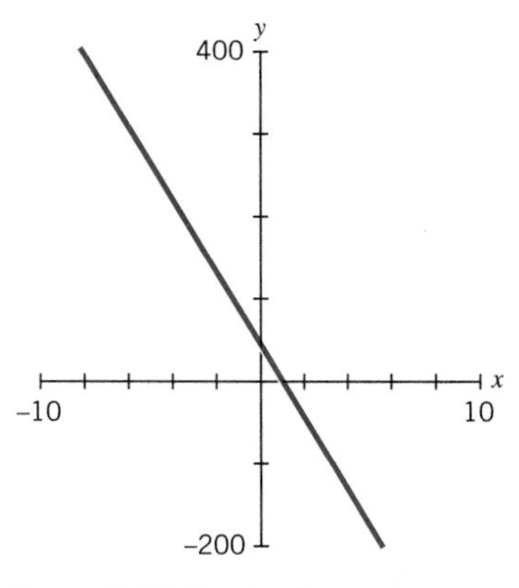

Figure 3.12 Graph of $y = -45x + 33$.

Solve Equation (1) for y	$45x = -y + 33$	(1)
add y to both sides	$y + 45x = 33$	
add $-45x$ to both sides	$y = -45x + 33$	
Solve Equation (2) for y	$2y + 90x = 66$	(2)
add $-90x$ to both sides	$2y = -90x + 66$	
divide by 2	$y = -45x + 33$	

We can now see that the original equations are both equivalent to $y = -45x + 33$. Thus, the two equations represent the same line. Every point on the line determined by Equation (1) is also on the line determined by Equation (2), so there are infinitely many solutions. In Figure 3.12 the graphs of the two equations are identical, and every point on the line is a solution to the system.

Algebra Aerobics 3.2b

1. Solve each system of equations using the method you think is most efficient.
   a. $2y - 5x = -1$, $3y + 5x = 11$
   b. $3x + 2y = 16$, $2x - 3y = -11$
   c. $t = 3r - 4$, $4t + 6 = 7r$
   d. $z = 2000 + 0.4(x - 10{,}000)$, $z = 800 + 0.2x$
2. Solve each system of equations and check your answers by graphing each system.
   a. $5y + 30x = 20$, $y = -6x + 4$
   b. $y = 1500 + 350x$, $2y = 700x + 3500$
   c. $y = 2x + 4$, $y = -x + 4$
3. Construct a system of two linear equations in which there is no solution for the system.
4. Solve the following system of equations:
   $$y = x + 4$$
   $$\frac{x}{2} + \frac{y}{3} = 3$$

3.3 INTERSECTION POINTS REPRESENTING EQUILIBRIUM

When systems of equations are used to model phenomena, an intersection point often represents a state of equilibrium. In this section we examine how economists use and interpret intersection points.

Supply and Demand Curves

In economics, a graph of two lines or curves is often used to model the relationship between supply and demand.[2] The horizontal axis represents the quantity of a product (measured perhaps in thousands of units per month) and the vertical axis represents the market price of the product (perhaps in dollars per unit).

[2]Note that economists use the term "supply and demand *curves*" even though the graphs are frequently drawn as straight lines.

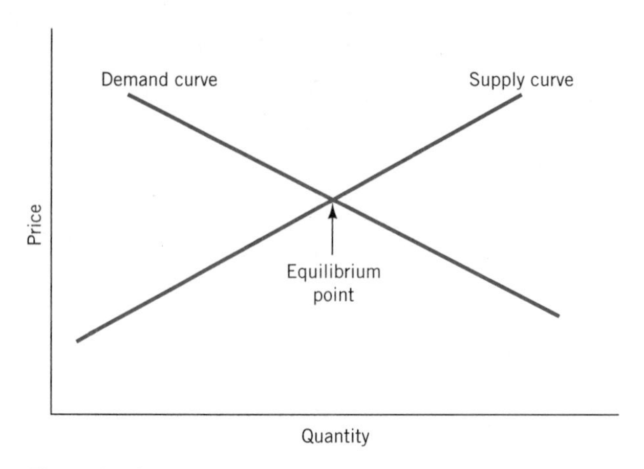

Figure 3.13 Supply and demand curves.

The demand curve represents the point of view of the consumer. It shows the relationship between the market price of a product and the quantity demanded of that product. It typically has a negative slope. When the price of a product is high, consumers are reluctant to buy it, so the quantity demanded is low. As the price of the product decreases, consumers are more willing to buy, so the quantity demanded increases. Economists assume that the quantity demanded depends on the price; that is, price is considered the independent variable that determines demand.[3] But you should note they usually graph price on the vertical axis.

The supply curve represents the producer's point of view. It shows the relationship between market prices and the amounts that producers are willing to supply. It typically has a positive slope. When the market price of a product is low, the profit is smaller, and producers will supply a smaller quantity of the product. If the price rises, a better profit can be made and producers will make larger quantities of the product to sell.

The intersection of the two curves represents the point at which supply equals demand (Figure 3.13). At the price indicated by the intersection point, the quantity demanded by consumers equals the quantity supplied by producers. The intersection point is called the *equilibrium point,* and the price at that point is called the *equilibrium price.* At any lower price, demand would exceed supply; at any higher price, supply would exceed demand.

Questions. What if the overall demand increases? What if consumers want more of the product and are willing to pay a higher price to get it? What happens to the equilibrium point?

Discussion. The graph in Figure 3.14 shows what happens when overall demand increases. A *shift* of the whole demand curve to the right means that more will be bought at each price. Pick a particular price, P, and trace that price horizontally to the original demand curve to locate the quantity consumers have previously demanded, Q_{old}, at that price. If overall demand has increased, then that same price P is now associated with a higher quantity, Q_{new}. Since this reasoning holds for *any*

[3]For this analysis, economists also assume "other things being equal." See P. Samuelson, *Economics: An Introductory Analysis* (New York: McGraw Hill, 1964), pp. 58ff.

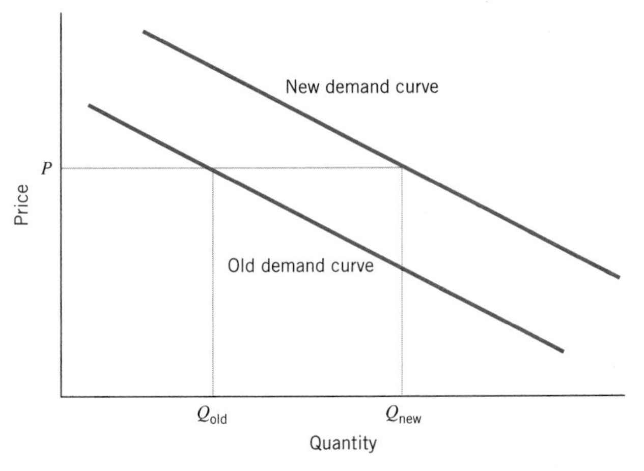

Figure 3.14 An increase in demand shifts the demand curve to the right.

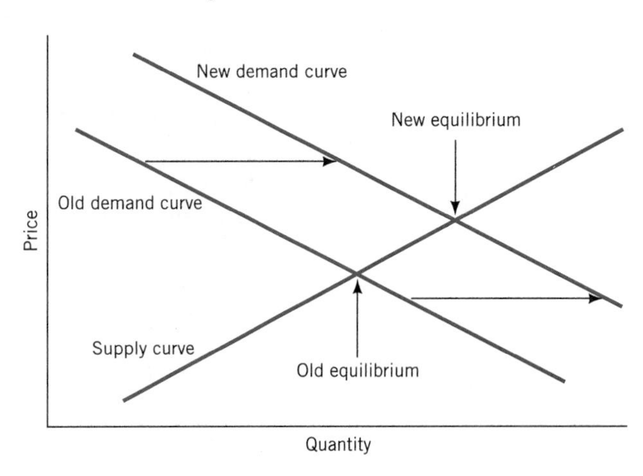

Figure 3.15 An increase in demand shifts the equilibrium point to the right and up.

price P, all points on the original demand curve shift to the right, forming a new demand curve.

The supply curve has not changed. But the shift in the demand curve moves the equilibrium point up and to the right (Figure 3.15). The new equilibrium point represents a larger quantity and a higher price. In other words, an increase in overall consumer demand causes producers to make a larger quantity that sells at a higher price.

3.4 GRADUATED VS. FLAT INCOME TAX: USING PIECEWISE LINEAR FUNCTIONS

Simple Tax Models

Income taxes may be based on either a flat or a graduated tax rate. With a flat tax rate, no matter what the income level, everyone is taxed at the same percentage. Flat taxes are often said to be unfair to those with lower incomes, because paying a fixed percentage of income is considered more burdensome to someone with a low income than to someone with a high income. A graduated tax rate means that people with higher incomes are taxed at a higher rate. Such a tax is called *progressive* and is generally less popular with those who have high incomes. Whenever the issue appears on the ballot, the pros and cons of the graduated vs. the flat tax rate are hotly debated in the news media and paid political broadcasting. Of the 42 states with an income tax, 35 had a graduated income tax in 1994.

The *New York Times* article *"How a Flat Tax Would Work for You and for Them"* discusses the trade-offs in using a flat tax.

For the taxpayer there are two primary questions in comparing the effect of flat and graduated tax schemes. For what income level will the taxes be the same under both plans? And, given a certain income level, how will taxes differ under the two plans?

Taxes are influenced by many factors, such as filing status, exemptions, and deductions. For our comparisons of flat and graduated income tax plans, we examine one filing status and assume that exemptions and deductions have already been subtracted from income.

Table 3.3
Taxes under Flat Tax Plan

Income after Deductions ($)	Taxes at Flat Tax Rate of 15% ($)
0	0
10,000	1500
20,000	3000
30,000	4500
40,000	6000
50,000	7500

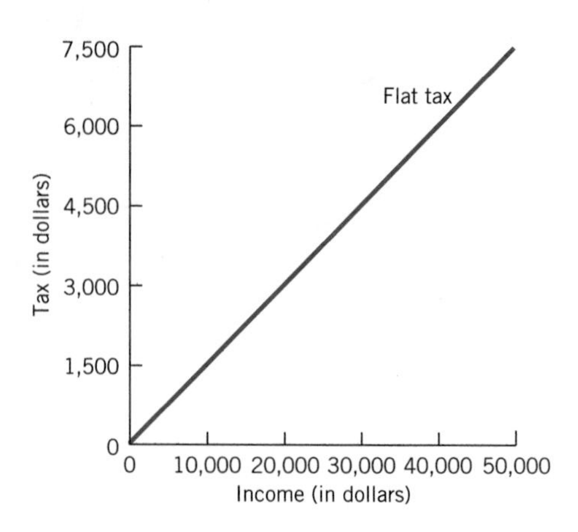

Figure 3.16 Flat tax at a rate of 15%.

A flat tax model. Flat taxes are uniquely determined by income. So we can use income, i, as the independent variable, flat taxes, T_F, as the dependent variable, and write T_F as a function, f, of i. If the flat tax rate is 15% (or 0.15 in decimal form), then

$$f(i) = T_F = 0.15i$$

This flat tax plan is represented in Table 3.3 and Figure 3.16.

A graduated tax model: a piecewise linear function. Under a graduated income tax, the tax rate changes for different portions of the income after deductions. Let's consider a graduated tax where the first $10,000 of income is taxed at a 10% rate and any income over $10,000 is taxed at a 20% rate. For example, if your income after deductions is $30,000, then your taxes under this plan are

$$\begin{aligned}\text{graduated tax} &= (10\% \text{ of } \$10{,}000) + (20\% \text{ of income over } \$10{,}000)\\ &= 0.10(\$10{,}000) + 0.20(\$30{,}000 - \$10{,}000)\\ &= 0.10(\$10{,}000) + 0.20(\$20{,}000)\\ &= \$1000 + \$4000\\ &= \$5000\end{aligned}$$

In general, if an income is over $10,000, then under this plan,

$$\begin{aligned}\text{graduated tax} &= 0.10(\$10{,}000) + 0.20(i - \$10{,}000)\\ &= \$1000 + 0.20(i - \$10{,}000)\end{aligned}$$

This graduated tax plan is represented in Table 3.4 and Figure 3.17.

The graph of the graduated tax is the result of piecing together two different line segments that represent the two different formulas used to find taxes. The short segment represents taxes for low incomes between $0 and $10,000 and the longer, steeper segment represents taxes for higher incomes that are greater than $10,000.

To find the algebraic expression for the graduated tax, we need to consider that the formula used for finding taxes is different for different levels of income. Functions that use different formulas for different intervals of the domain are said to be

Table 3.4
Taxes under Graduated Tax Plan

Income after Deductions ($)	Taxes ($)
0	0
5,000	500
10,000	1000
20,000	1000 + 2000 = 3000
30,000	1000 + 4000 = 5000
40,000	1000 + 6000 = 7000
50,000	1000 + 8000 = 9000

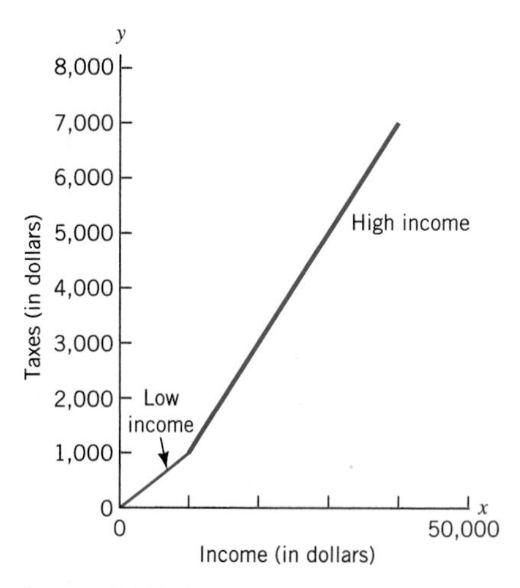

Figure 3.17 Graduated tax.

piecewise defined. Since each income determines a unique tax, we can define the graduated tax T_G as a piecewise function, g, of income i:

$$g(i) = T_G = \begin{cases} 0.10i & \text{for } i \leq \$10{,}000 \\ 1000 + 0.20(i - 10{,}000) & \text{for } i > \$10{,}000 \end{cases}$$

The value of the independent variable i for income determines which formula to use to evaluate the function. This function is called a *piecewise linear function,* since each piece consists of a different linear formula. The formula for incomes equal to or below \$10,000 is different from the formula for incomes above \$10,000.

To find $g(\$8000)$, the value of T_G when $i = \$8000$, use the upper formula in the definition since income, i, in this case is less than \$10,000:

For $i \leq \$10{,}000$ $\qquad g(i) = 0.10i$

substituting \$8000 for i $\qquad g(\$8000) = 0.10 \cdot (\$8000)$

$$= \$800$$

To find $g(\$40{,}000)$, we use the lower formula in the definition, since in this case income is greater than \$10,000:

For $i > \$10{,}000$ $\qquad g(i) = \$1000 + 0.20(i - \$10{,}000)$

substituting \$40,000 for i $\qquad g(\$40{,}000) = \$1000 + 0.20(\$40{,}000 - \$10{,}000)$

$$= \$1000 + 0.20(\$30{,}000)$$

$$= \$1000 + \$6000$$

$$= \$7000$$

Comparing the two tax models. In Figure 3.18 we compare the different tax plans by plotting the flat and graduated tax equations on the same graph.

The intersection points indicate the incomes at which the amount of tax is the same under both plans. From the graph, we can estimate the coordinates of the two points as (0, 0) and (20000, 3000). That is, under both plans with \$0 income you pay \$0 taxes, and with approximately \$20,000 in income you would pay approximately \$3000 in taxes.

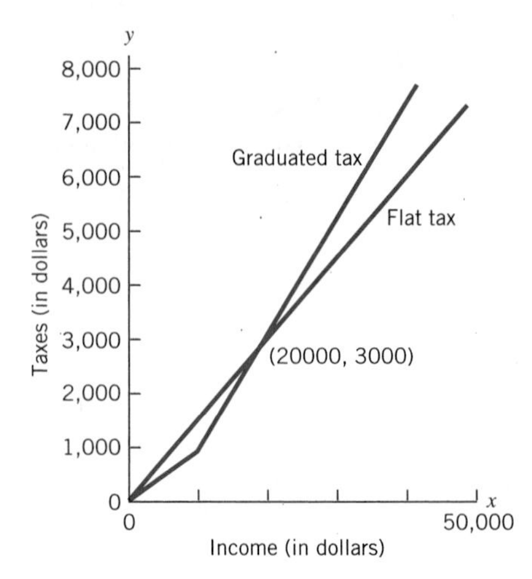

Figure 3.18 Flat tax vs. graduated tax.

Individual voters want to know what impact these different plans will have on their taxes. From the graph in Figure 3.18, we can see that to the left of the intersection point at (20000, 3000), the flat tax is *greater* than the graduated tax for the same income. To the right of this intersection point, the flat tax is *less* than the graduated tax for the same income. So for incomes *less than* \$20,000, taxes are *greater* under the flat tax plan, and for incomes *greater than* \$20,000, taxes will be *less* under the flat tax plan.

To find more accurate values for the coordinates of the point(s) of intersection, we need to set $T_F = T_G$. We know that $T_F = 0.15i$. Which of the two expressions do we use for T_G? The answer depends upon what value of income, i, we consider. For $i \leq \$10,000$, we have $T_G = 0.10i$.

If	$T_F = T_G$
and $i \leq \$10,000$, then	$0.15i = 0.10i$
This can only happen when	$i = 0$

If $i = 0$, both T_F and T_G equal 0; therefore one intersection point is indeed (0, 0).

For $i > \$10,000$, we use $T_G = \$1000 + 0.20(i - \$10,000)$ and again set $T_F = T_G$.

If	$T_F = T_G$
and $i > \$10,000$, then	$0.15i = \$1000 + 0.20(i - \$10,000)$
apply distributive property	$0.15i = \$1000 + 0.20i - (0.20)(\$10,000)$
multiply	$0.15i = \$1000 + 0.20i - \$2000$
combine terms	$0.15i = -\$1000 + 0.20i$
add $-0.20i$ to each side	$-0.05i = -\$1000$
divide by -0.05	$i = -\$1000/(-0.05)$
	$i = \$20,000$

We have assumed that deductions were the same under both plans. If we drop this assumption, how could we make a flat tax plan less burdensome for people with low incomes?

So each plan results in the same tax for an income of \$20,000. How much tax is required? We can substitute \$20,000 for i into either the flat tax formula or the graduated tax formula for incomes over \$10,000 and solve for the tax. Given the flat tax function,

$$T_F = 0.15i$$

if $i = \$20{,}000$, then $\quad T_F = (0.15)(\$20{,}000)$

$$= \$3000$$

We can check to make sure that when $i = \$20{,}000$, the graduated tax, T_G, will also be \$3000:

If $i > \$10{,}000$, then $\quad T_G = \$1000 + 0.20(i - \$10{,}000)$

so, if $i = \$20{,}000$, then $\quad T_G = \$1000 + 0.20(\$20{,}000 - \$10{,}000)$

perform operations $\quad = \$3000$

The other intersection point is, as we estimated, (\$20000, \$3000).

Algebra Aerobics 3.4a

1. In our simple tax models, the amount of flat tax is a function f of income i where $f(i) = 0.15i$ and the amount of graduated tax is described by the piecewise linear function

$$g(i) = \begin{cases} 0.10i & \text{for } i \leq \$10{,}000 \\ 1000 + 0.20(i - 10{,}000) & \text{for } i > \$10{,}000 \end{cases}$$

   a. Use Figure 3.18 to predict which tax is larger:
   i. $f(\$5000)$, $g(\$5000)$
   ii. $f(\$15{,}000)$, $g(\$15{,}000)$
   iii. $f(\$40{,}000)$, $g(\$40{,}000)$

   b. Now use the equations defining f and g to check your predictions in part (a).

2. Construct a new graduated tax function, if:
   a. The tax is 5% on the first \$50,000 of income and 8% on any income in excess of \$50,000.
   b. The tax is 6% on the first \$30,000 of income and 9% on any income in excess of \$30,000.

The Case of Massachusetts

The state of Massachusetts currently has a flat tax of 5.95% on earned income, but there has been an ongoing debate about whether or not to change to a graduated income tax. In 1994 Massachusetts voters considered a proposal called Proposition 7. Proposition 7 would have replaced the flat tax rate with graduated income tax rates (called marginal rates) as shown in Table 3.5.

Table 3.5
Massachusetts Graduated Income Tax Proposal

	Marginal Rates		
Filing Status	5.5%	8.8%	9.8%
Married/joint	< \$81,000	\$81,000–\$150,000	\$150,000+
Married/separate	< \$40,500	\$40,500–\$ 75,000	\$75,000+
Single	< \$50,200	\$50,200–\$ 90,000	\$90,000+
Head of household	< \$60,100	\$60,100–\$120,000	\$120,000+

Source: Office of the Secretary of State, Michael J. Connolly, Boston, MA. 1994

EXPLORE

In Exploration 3.1 you can analyze the impact of Proposition 7 on people using different filing statuses.

Questions. For what income and filing status would the taxes be equal under both plans? Who will pay less tax and who will pay more under the graduated income tax plan? We analyze the tax for a single person and leave the analyses of the other filing categories for you to do.

Discussion. The proposed graduated income tax is designed to tax at higher rates only that portion of the individual's income that exceeds a certain threshold. For example, for those who file as single people, the graduated tax rate means that earned income under \$50,200 would be taxed at a rate of 5.5%. Any income between \$50,200 and \$90,000 would be taxed at 8.8%, and any income over \$90,000 would be taxed at 9.8%.

The tax for a single person earning \$100,000 would be the sum of three different dollar amounts.

$$\begin{aligned} 5.5\% \text{ on the first } \$50{,}200 &= (0.055)(\$50{,}200) \\ &= \$2761 \end{aligned}$$

$$\begin{aligned} &8.8\% \text{ on the next } \$39{,}800 \text{ (the portion of income between } \$50{,}200 \text{ and } \$90{,}000) \\ &\qquad = (0.088)(\$39{,}800) \\ &\qquad \approx \$3502 \end{aligned}$$

$$\begin{aligned} 9.8\% \text{ on the remaining } \$10{,}000 &= (0.098)(\$10{,}000) \\ &= \$980 \end{aligned}$$

The total graduated tax would be \$2761 + \$3502 + \$980 = \$7243. Under the flat tax rate the same individual pays 5.95% of \$100,000 = 0.0595(\$100,000), or \$5950.

Table 3.6 shows the differences between the flat tax and the proposed graduated tax plan for single people at several different income levels. We can write T_F, flat taxes for single people, as a linear function, f, of income, i, where

$$f(i) = T_F = 0.0595i$$

Table 3.6
Massachusetts Taxes: Flat Rate vs. Graduated Rate for Single People

Income after Exemptions and Deductions (\$)	Current Flat Tax at 5.95% (\$)	Graduated Tax Under Proposition 7 (\$)
0	0	0
25,000	1488	1375
50,000	2975	2750
75,000	4463	4943
100,000	5950	7243

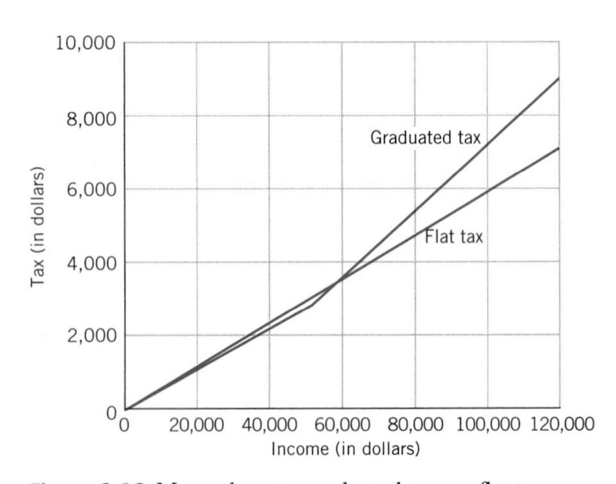

Figure 3.19 Massachusetts graduated tax vs. flat tax.

We can write T_G, graduated tax for single people, as a piecewise linear function, g, of income i, where

$$g(i) = T_G = \begin{cases} 0.055i & \text{for } 0 \leq i < \$50{,}200 \\ \$2761 + 0.088(i - \$50{,}200) & \text{for } \$50{,}200 \leq i \leq \$90{,}000 \\ \$6263 + 0.098(i - \$90{,}000) & \text{for } i > \$90{,}000 \end{cases}$$

Note that \$2761 in the second formula of the definition is the tax on the first \$50,200 of income (5.5% of \$50,200), and \$6263 in the third formula of the definition is the sum of the taxes on the first \$50,200 and the next \$39,800 of income (5.5% of \$50,200 + 8.8% of \$39,800 ≈ \$2761 + \$3502 = \$6263). The flat tax and the graduated income tax for single people are compared in Figure 3.19.

For what income would single people pay the same tax under both plans? An intersection point on the graph indicates when taxes are equal. One intersection point occurs at (0, 0). That makes sense since under either plan if you have zero income, you pay zero taxes. The second intersection point is at approximately (58000, 3500). That means for an income of approximately \$58,000, the taxes are the same and are approximately \$3500. In the Algebra Aerobics you are asked to find more accurate values for this point of intersection and to describe what happens to the right and to the left of the intersection point.

Algebra Aerobics 3.4b

1. **a.** From the graph in Figure 3.19 we estimated the coordinates of one of the points of intersection to be approximately (58000, 3500). Use the equations in the text for the flat tax and proposed graduated tax in Massachusetts to find more accurate values for that intersection point.

   b. **i)** How does the flat tax compare to the graduated tax for incomes to the left of this intersection point?

   ii) What happens to the right of this intersection point?

2. Use the graph in Figure 3.19 to predict which tax is larger for each of the following incomes and by approximately how much. Then use the equations defining f and g to check your predictions.

   a. \$30,000

   b. \$60,000

   c. \$120,000

Chapter Summary

A pair of real numbers is a *solution* to a system of linear equations in two variables if and only if the pair of numbers is a solution of each equation.

On the following graphs of systems of linear equations, a solution appears as an intersection point. In a system of two linear equations, if the lines intersect once, there is one solution. If the lines are parallel, they never intersect, and there are no solutions. If the two equations are equivalent, they represent the same line and there are an infinite number of solutions.

The coordinates of the point of intersection can be estimated from a graph of the system of equations. Using the equations in the system, exact values for the point(s) of intersection can be found by using the methods of *substitution* or *elimination*.

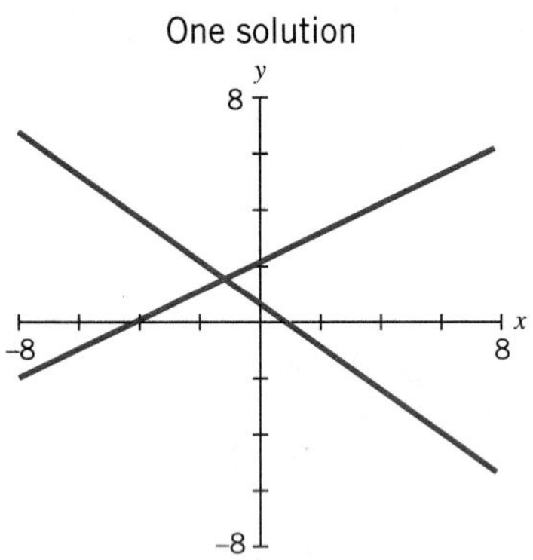

Lines intersect at a single point.

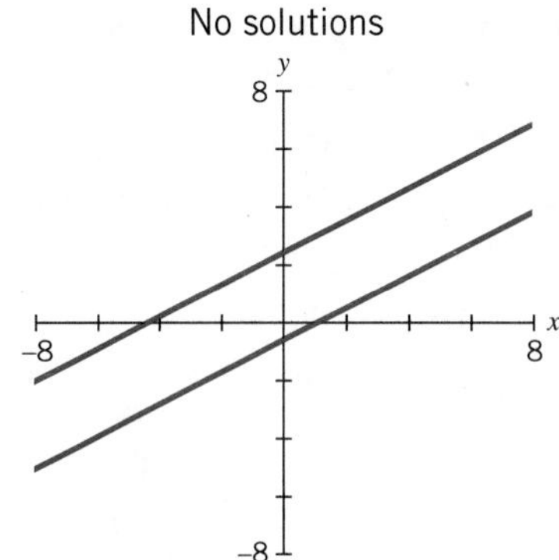

Parallel lines never intersect.

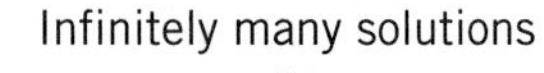

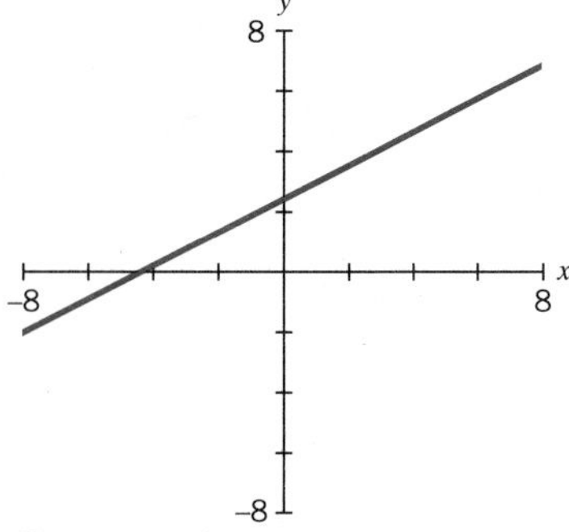

The two equations represent the same line. Every point on the line satisfies both equations.

Functions that use different formulas for different intervals of the domain are said to be *piecewise defined.* Some functions, like the graduated income tax, can be constructed out of pieces of several different linear functions. They are called *piecewise linear* functions.

EXERCISES

Exercises for Section 3.1

1. Explain what is meant by "a solution to a system of equations."

2. a. Determine whether $(-2, 3)$ is a solution for the following system of equations:

$$3x + y = -3$$
$$x + 2y = 4$$

   b. Explain why $(3, -2)$ is not a solution for the system in part (a).

3. a. Determine whether $(5, -10)$ is a solution for the following system of equations:

$$4x - 3y = 50$$
$$2x + 2y = 5$$

   b. Explain why $(-10, 5)$ is not a solution for the system in part (a).

4. Estimate the solution of the system of linear equations graphed in the accompanying figure.

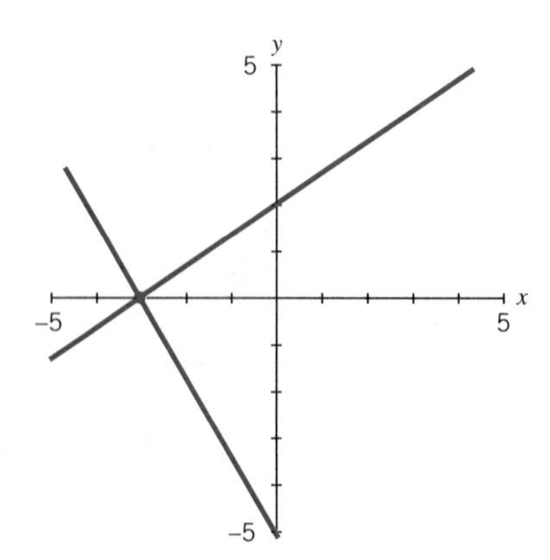

5. a. Estimate the coordinates of the point of intersection on the accompanying graph.

b. Describe what happens to the population of Johnsonville in relation to the population of Palm City to the right and to the left of this intersection point.

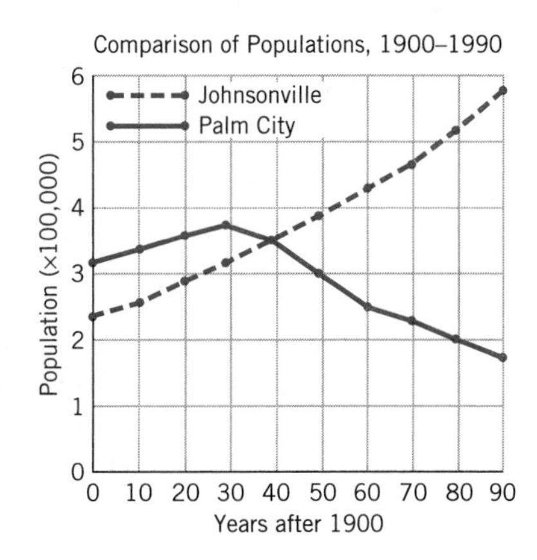

6. In the text the following cost equations were given for gas and solar heating:

$$C_{\text{gas}} = 12{,}000 + 700n$$

$$C_{\text{solar}} = 30{,}000 + 150n$$

where n represents the number of years since installation and the cost represents the total accumulated costs up to and including year n.

a. Sketch the graph of this system of equations.

b. What do the coefficients 700 and 150 represent on the graph, and what do they represent in terms of heating costs?

c. What do the constant terms 12,000 and 30,000 represent on the graph? What does the difference between 12,000 and 30,000 say about the costs of gas vs. solar heating?

d. Label the point on the graph where gas and solar heating costs are equal. Make a visual estimate of the coordinates, and interpret what the coordinates mean in terms of heating costs.

e. Use the equations to find a better estimate for the intersection point. To simplify the computations, you may want to round values to whole numbers. Show your work.

f. When is the total cost of solar heating more expensive? When is gas heating more expensive?

7. Answer the questions in Exercise 6 (with suitable changes in wording) for the cost equations for electric and solar heating:

$$C_{\text{electric}} = 5000 + 1100n$$

$$C_{\text{solar}} = 30{,}000 + 150n$$

8. Consider the following job offers. At Acme Corporation, you are offered a starting salary of \$20,000 per year, with raises of \$2500 annually. At Boca Corporation, you are offered \$25,000 to start and raises of \$2000 annually.

a. Find an equation to represent your salary, S_A, after n years of employment with Acme. Assume n is an integer.

b. Find an equation to represent your salary, S_B, after n years of employment with Boca. Assume n is an integer.

c. Create a table of values showing your salary at each of these corporations for values of n up to 12 years.

d. At what year of employment would the two corporations pay you the same salary?

Exercises for Section 3.2

9. a. Solve the following system algebraically:

$$S = 20{,}000 + 2500n$$

$$S = 25{,}000 + 2000n$$

b. Graph the system in part (a) and locate the solution on your graph. Check your estimate with your answer in part (a).

10. Predict the number of solutions to each of the following systems. Give reasons for your answer. You don't need to find any actual solutions.

a. $y = 20{,}000 + 700x$
$y = 15{,}000 + 800x$

b. $y = 20{,}000 + 700x$
$y = 15{,}000 + 700x$

c. $y = 20{,}000 + 700x$
$y = 20{,}000 + 800x$

11. a. Solve the following system algebraically:

$$x + 3y = 6$$

$$5x + 3y = -6$$

b. Graph the system of equations in part (a) and locate the solution to the system on your graph. Check your estimate with your answers in part (a).

12. Calculate the solution(s), if any, to each of the following systems of equations. Use any method you like.

a. $y = -11 - 2x$
$y = 13 - 2x$

b. $t = -3 + 4w$
$-12w + 3t + 9 = 0$

c. $y = 2200x - 700$
$y = 1300x + 4700$

d. $3x = 5y$
$4y - 3x = -3$

In some of the following examples you may wish to round off your answers:

e. $y = 2200x - 1800$
$y = 1300x - 4700$

f. $y = 4.2 - 1.62x$
$1.48x - 2y + 4.36 = 0$

g. $4r + 5s = 10$
$2r - 4s = -3$

h. $2x + 3y = 13$
$3x + 5y = 21$

i. $xy = 1$
$x^2y + 3x = 2$
(A nonlinear system! *Hint:* Solve $xy = 1$ for y and use substitution.)

13. Assume you have $2000 to invest for 1 year. You can make a safe investment that yields 4% interest a year or a risky investment that yields 8% a year. If you want to combine safe and risky investments to make $100 a year, how much of the $2000 should you invest at the 4% interest? How much at the 8% interest? (*Hint:* Set up a system of two equations in two variables, where one equation represents the total amount of money you have to invest and the other equation represents the total amount of money you want to make on your investments.)

14. Two investments in high technology companies total $1000. If one investment earns 10% annual interest and the other earns 20%, find the amount of each investment if the total interest earned is $140 for the year.

15. Solve the following systems:

a. $\dfrac{x}{3} + \dfrac{y}{2} = 1$
$x - y = \dfrac{4}{3}$

b. $\dfrac{x}{4} + y = 9$
$y = \dfrac{x}{2}$

16. For each of the following systems of equations, describe the graph of the system and determine if (i) there is no solution, (ii) there is an infinite number of solutions, and (iii) there is exactly one solution:

a. $2x + 5y = -10$
$y = -0.4x - 2$

b. $3x + 4y = 5$
$3x - 2y = 5$

c. $2x - y = 5$
$6x - 3y = 4$

17. If $y = b + mx$, solve for values for m and b by constructing two linear equations in m and b for the given sets of ordered pairs.

a. When $x = 2$, $y = -2$ and when $x = -3$, $y = 13$

b. When $x = 10$, $y = 38$ and when $x = 1.5$, $y = -4.5$

18. The following are formulas predicting future raises for four different groups of union employees. N represents the number of years from the date of contract. Each equation represents the salary that will be earned after N years.

Group A: Salary $= 30{,}000 + 1500N$

Group B: Salary $= 30{,}000 + 1800N$

Group C: Salary $= 27{,}000 + 1500N$

Group D: Salary $= 21{,}000 + 2100N$

a. Will group A ever earn more per year than group B? Explain.

b. Will group C ever catch up to group A? Explain.

c. Which group will be making the highest yearly salary in 5 years? How much will that salary be?

d. Will group D ever catch up to group C? If so, after how many years and at what salary?

e. How much total salary would an individual in each group have earned 3 years after the contract?

19. Explain what is meant by "two equivalent equations." Give an example of two equivalent equations.

20. In 1997 AT&T offered two long-distance calling plans. The "One Rate" plan charged a flat rate of \$0.15 per minute. The "One Rate Plus" plan charged a service fee of \$4.95 a month plus \$0.10 per minute.

a. Consider the monthly telephone cost as a function of the number of minutes of phone use that month. Construct a cost function (covering a 1-month period) for each calling plan.

b. Sketch both functions on the same grid.

c. Use your graph to estimate when the costs will be the same under both plans.

d. Calculate when the costs will be equal, label the corresponding point on the graph, and interpret your answer.

21. Solve the following system of three equations in three variables:

$$2x + 3y - z = 11 \quad (1)$$

$$5x - 2y + 3z = 35 \quad (2)$$

$$x - 5y + 4z = 18 \quad (3)$$

Hint: You can apply the same basic strategy of elimination as we did in solving two equations in two variables and systematically eliminate variables until you end up with one equation in one variable.

a. Use Equations (1) and (2) to eliminate one variable, creating a new equation (4) in two variables.

b. Use Equations (1) and (3) to eliminate the same variable as in part (a). You should end up with a new equation (5) that has the same variables as Equation (4).

c. Equations (4) and (5) derived in parts (a) and (b) represent a system of two equations in two variables. Solve the system.

d. Find the corresponding value for the variable eliminated in part (a).

e. Check your work by making sure your solution works in all three original equations.

22. a. Using the strategy described in Exercise 21, solve the following system:

$$2a - 3b + c = 4.5 \quad (1)$$

$$a - 2b + 2c = 0 \quad (2)$$

$$3a - b + 2c = 0.5 \quad (3)$$

b. How many distinct equations would you need in order to solve a system with five variables? With n variables?

23. a. Construct a system of linear equations in two variables that has no solution.

b. Construct a system of linear equations in two variables that has exactly one solution.

c. Solve the system of equations you constructed in part (b) by using two different algebraic strategies and by graphing the system of equations. Do your answers all make sense? Why or why not?

24. Nenuphar wants to invest a total of $30,000 into two savings accounts, one paying 6% simple interest and the other paying 9% simple interest (a more risky investment). If after 1 year she wants the total interest from both accounts to be $2100, how much should she invest in each account?

25. When will the following system of equations have no solution? Justify your answer.

$$y = m_1x + b_1$$

$$y = m_2x + b_2$$

26. Although total solar energy powered home energy systems are quite expensive to install, passive solar systems are much more economical. Many passive solar features can be incorporated at the time of construction with little additional cost. These features enable energy costs to be one-half to one-third of costs in conventional homes.

For the case study Esperanza del Sol, here is the cost analysis:[4]

Cost of installation of a conventional energy system is $10,000.
Additional cost to install passive solar features: $150
Annual energy costs for conventional home: $740
Annual energy costs for passive solar system: $540

a. Write the cost equation for the conventional system.

b. Write the cost equation for the passive solar system.

c. When would the passive solar system be exactly as costly as the conventional system?

d. After 5 years, what would be the total energy cost of the passive solar system? What would be the total cost of the conventional system?

27. a. Construct a system of linear equations where both of the following conditions are met:

- The coordinates of the point of intersection are (2, 5).
- One of the lines has a slope of -4 and the other line has a slope of 3.5.

b. Graph the system of equations you found in part (a). Verify that the coordinates of the point of intersection are the same as the coordinates specified in part (a).

[4]Adapted from *Buildings for a Sustainable America: Case Studies,* The American Solar Energy Society, Boulder, CO.

28. **a.** Construct a system of linear equations where both of the following conditions are met:

- The coordinates of the point of intersection are (60, 100).
- The two lines are perpendicular to each other and one of the lines has a slope of -4.

b. Graph the system of equations you found in part (a). Verify that the coordinates of the point of intersection are the same as the coordinates specified in part (a).

29. Worker Alpha can lay A bricks in an hour, worker Beta lays B bricks in an hour.

a. For a job requiring Q bricks to be laid, find an expression for how long it will take Alpha and Beta working together to do the job.

b. If Alpha lays 50 bricks/hour and Beta lays 70 bricks/hour but starts 2 hours later than Alpha, when will they have laid the same number of bricks?

c. Alpha and Beta both work 8 hours/day at the same pace as described in part (b). Alpha charges \$30/hour and a daily travel charge of \$12, Beta charges \$45/hour but no travel charge. Fill out the data chart given.

Alpha			Beta		
Time (hr)	Bricks Laid	Cost (\$)	Time (hr)	Bricks Laid	Cost (\$)
1	50	$12 + 30 = 42$	1	70	45
2	100	$12 + 2(30) = 72$	2		
3			3		
4			4		
5			5		
6			6		
7			7		
8			8		

On the same axes, graph the cost vs. bricks laid for Alpha and the cost vs. bricks laid for Beta. Estimate the cost and number of bricks for a job where Alpha and Beta lay the same number of bricks and get paid the same amount. For what size of job is it cheaper to hire Alpha?

d. Derive formulas for the number of bricks laid by Alpha and Beta as a function of hours worked, T_a and T_b, respectively, as well as the cost for both. What is the cost, number of bricks, and hours worked for Alpha and Beta for a job where Alpha and Beta lay the same number of bricks and each get paid the same amount?

30. A husband drives a heavily loaded truck that can only go 55 mph on a 650-mile turnpike trip. His wife leaves on the same trip 2 hours later in the family car averaging 70 mph. Recall that distance traveled = speed · time traveled.

a. Derive an expression for distance, D_{h}, the husband travels in t hours since he started.

b. How many hours has the wife been traveling if the husband has traveled t hours?

c. Derive an expression for distance, D_{w}, that the wife will have traveled while the husband has been traveling for t hours.

d. Graph distance vs. time for husband and wife on the same axes.

e. Calculate when and where the wife will overtake the husband.

f. Suppose the husband and wife wanted to arrive at a restaurant at the same time, and the restaurant is 325 miles from home. How much later should she leave, assuming he still travels at 55 mph and she at 70 mph?

31. Life-and-death travel problems are dealt with by air traffic computers and controllers who are trying to prevent intersections of planes traveling at various speeds in three-dimensional space. To get a taste of what is involved consider this situation: airplanes A and B are traveling at the same altitude on the paths shown on the position plot.

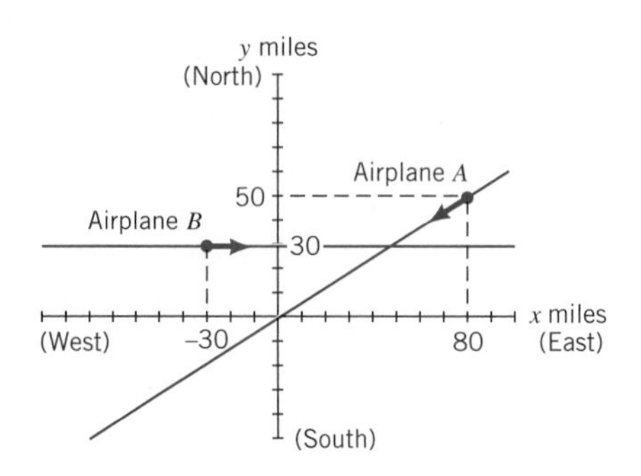

a. Find formulas to express positions of Airplanes A and B on x and y coordinates, using y_A and y_B to denote the North/South coordinates of Airplanes A and B, respectively.

b. What are the coordinates of the intersection of the airplane paths?

c. Airplane A travels at 2 miles/minute and Airplane B travels at 6 miles/minute. Clearly their paths will intersect if they each continue on the same course, but will they arrive at the intersection point at the same time? How far does plane A have to travel to the intersection point? How many minutes will it take to get there? How far does plane B have to travel to the intersection point? How many minutes will it take to get there? Recall the rule of Pythagoras for finding the hypotenuse of a right triangle: $A^2 + B^2 = C^2$, where C is the hypotenuse and A and B are the other sides. Is this a safe situation?

Exercises for Section 3.3

32. Assume that you are given the graph of a standard model for supply and demand as shown in Section 3.3 of the text.

a. What happens if the overall demand decreases; that is, at each price level consumers purchase a smaller quantity? Sketch a graph showing the supply curve and both the original and the changed demand curve. Label the graph. How has the demand curve changed? Describe the shift in the equilibrium point in terms of price and quantity.

b. What happens if the supply decreases; that is, at each price level manufacturers produce lower quantities? Sketch a graph showing the demand curve and both the original and the changed supply curve. Label the graph. How has the supply curve changed? Describe the shift in the equilibrium point in terms of price and quantity.

33. When studying populations (human or otherwise), the two primary factors affecting population size are the birth rate and the death rate. There is abundant evidence that, other things being equal, as the population density increases, the birth rate tends to decrease and the death rate tends to increase. (See E. O. Wilson and W. H. Bossert's *A Primer of Population Biology,* Sunderland, MA: Sinauer Associates, 1971, p. 104.)

a. Generate a rough sketch showing birth rate as a function of population density. Note the units for population density on the horizontal axis are the number of individuals for a given area. The units on the vertical axis represent a rate, such as the number of individuals per 1000. Now add to your graph a rough sketch of the relationship between death rate and population density. In both cases assume the relationship is linear.

b. At the intersection point of the two lines the growth of the population is zero. Why? *Note:* We are ignoring all other factors such as immigration.

The intersection point is called the *equilibrium point.* At this point the population is said to have stabilized, and the size of the population that corresponds to this point is called the *equilibrium number.*

c. What happens to the equilibrium point if the overall death rate decreases; that is, at each value for population density the death rate is lower? Sketch a graph showing the birth rate and both the original and the changed death rate. (Label the graph carefully.) Describe the shift in the equilibrium point.

d. What happens to the equilibrium point if the overall death rate increases? Analyze as in part (c).

34. Use the information in Exercise 33 to answer the following questions:

a. What if the overall birth rate increases (that is, if at each population density level the birth rate is higher)? Sketch a graph showing the death rate and both the original and the changed birth rate. Be sure to label the graph carefully. Describe the shift in the equilibrium point.

b. What happens if the overall birth rate decreases? Analyze as in part (a).

Exercises for Section 3.4

35. Find out what kind of income tax (if any) your state has. Is it flat or graduated? Construct and graph a function that describes your state's income tax for one filing status. Identify the income tax for various income levels.

36. a. Construct a graduated tax function where the tax is 10% on the first $30,000 of income, then 20% on any income in excess of $30,000.

b. Construct a flat tax function where the tax is 15% of income.

c. Calculate the tax for both the flat tax function from part (b) and the graduated tax function from part (a) for each of the following incomes: $10,000; $20,000; $30,000; $40,000; and $50,000.

d. Graph the graduated and flat tax functions on the same grid and estimate the coordinates of the points of intersection. Interpret the points of intersection.

37. Construct a piecewise linear function for each of the accompanying graphs.

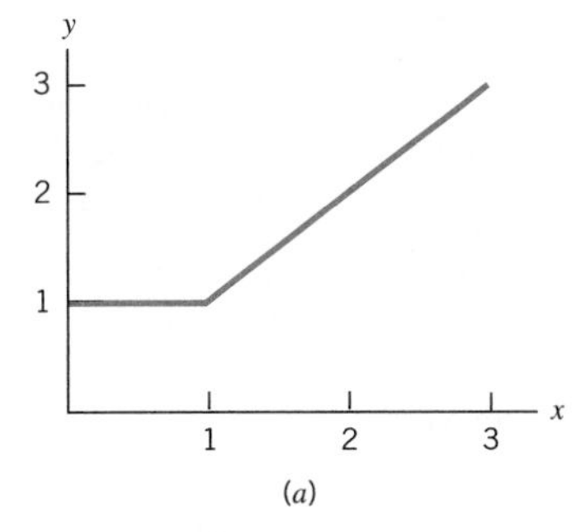

(*a*)

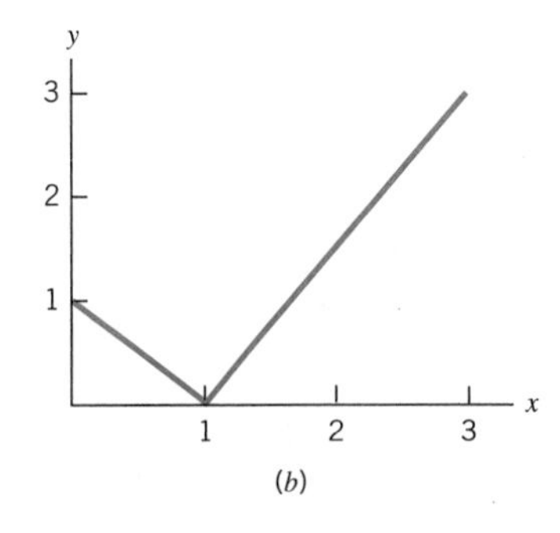

(*b*)

38. Missouri currently has a flat rate tax of 4% on income after deductions. This flat rate replaced a graduated income tax plan. The tax rate for single people under the graduated plan is shown in the accompanying table on the next page. For what income levels would a single person pay less tax under the current flat tax plan than under the graduated tax plan?

Missouri State Tax for a Single Person

Income after Deductions ($)	Marginal Tax Rate (%)
≤ 1000	1.50
1001–2000	2.00
2001–3000	2.50
3001–4000	3.00
4001–5000	3.50
5001–6000	4.00
6001–7000	4.50
7001–8000	5.00
8001–9000	5.50
> 9000	6.00

39. Consider the following function:

$$f(x) = \begin{cases} 2x + 1 & x \leq 0 \\ 3x & x > 0 \end{cases}$$

Evaluate $f(-10), f(-2), f(0), f(2)$, and $f(4)$.

40. Consider the following function:

$$g(x) = \begin{cases} 3 & x \leq 1 \\ 4 + 2x & x > 1 \end{cases}$$

Evaluate $g(-5)$, $g(-2)$, $g(0)$, $g(1)$, $g(1.1)$, $g(2)$, and $g(10)$.

41. Construct a small table of values and graph the following piecewise linear functions. In each case specify the domain.

a. $f(x) = \begin{cases} 5 & \text{for } x < 10 \\ -15 + 2x & \text{for } x \geq 10 \end{cases}$

b. $g(t) = \begin{cases} 1 - t & \text{for } -10 \leq t \leq 1 \\ t & \text{for } 1 < t < 10 \end{cases}$

42. The accompanying table, taken from a pediatric text, provides a set of formulas for the approximate "average" height and weight of normal infants and children.

Age	Weight in Kilograms	Weight in Pounds
At birth	3.25	7
3–12 months	(age in months + 9)/2	(age in months) + 11
1–5 years	2 · (age in years) + 8	5 · (age in years) + 17
6–12 years	3.5 · (age in years) − 2.5	7 · (age in years) + 5
Age	**Height in Centimeters**	**Height in Inches**
At birth	50	20
At 1 year	75	30
2–12 years	6 · (age in years) + 77	2.5 · (age in years) + 30

Source: R. E. Behrman and V. C. Vaughan (eds.), *Nelson Textbook of Pediatrics,* 12th ed. (Philadelphia: W.B. Saunders, 1983), p. 19.

For children from birth to 12 years of age, construct and graph a piecewise linear function for each of the following (assuming age is always the independent variable):

a. Weight in kilograms

b. Weight in pounds (How does this model compare to the model for female infants in Chapter 2, Section 5?)

c. Height in centimeters

d. Height in inches

Note that there are certain gaps in the table that need to be resolved in order to construct piecewise linear functions. For example you will need to decide which weight formula to use for a child who is 5½ years old and which height formula to use for a child who is 1½ years old.

43. Suppose a flat tax is 10% of income. Suppose a graduated tax is a fixed $1000 for any income ≤ $20,000 plus 20% of any income over $20,000.

a. Construct functions for the flat tax and the graduated tax.

b. Construct a small table of values:

Income	Flat Tax	Graduated Tax
$0		
$10,000		
$20,000		
$30,000		
$40,000		

c. Graph both tax plans and estimate any point(s) at which the two plans would be equal.

d. Use the function definitions to calculate any point(s) at which the two plans would be equal.

e. For what levels of income would the flat tax be more than the graduated tax? For what levels of income would the graduated tax be more than the flat tax?

f. Are there any conditions in which an individual might have negative income under either of the above plans; that is, the amount of taxes would exceed an individual's income? This is not as strange as it sounds. For example, many states impose a minimum corporate tax on a company, even if it is a small one-person operation with no income in that year.

44. You check around for the best deal on your prescription medicine. At your local pharmacy it costs $4.39 a bottle; by mail order catalog it costs $3.85 a bottle but there is a flat shipping charge of $4.00 for any size order; by a source found on the Internet it costs $3.99 a bottle and shipping costs $1.00 for each bottle, but for orders of 10 or more bottles it costs $3.79 a bottle, and handling is $2.50 per order plus shipping costs of $1.00 for each bottle.

a. Find a formula to express each of these costs if N is the number of bottles purchased and $C_{pharmacy}$, $C_{catalog}$, and $C_{internet}$ are the respective order costs.

b. Graph the costs for purchases up to 30 bottles at a time. Which is the cheapest source if you buy fewer than 10 bottles at a time? If you buy more than 10 at a time? Explain.

45. Heart health is a prime concern, because heart disease is the leading cause of death in the United States. Aerobic activities such as walking, jogging, and running are recommended for cardiovascular fitness, because they increase the heart's strength and stamina.

a. A typical training recommendation for a beginner is to walk at a moderate pace of about 3.5 miles/hour (or approximately 0.0583 miles/minute) for 20 minutes. Construct a function that describes the distance traveled *in miles*, $D_{beginner}$, as a function of time, T, *in minutes,* for someone maintaining this pace. Construct a small table of solutions and graph the function using a reasonable domain.

b. A more advanced training routine is to walk at a pace of 3.75 miles/hour (or 0.0625 miles/minute) for 10 minutes and then jog at 5.25 miles/hour (or 0.0875 miles/minute) for 10 minutes. Construct a piecewise linear function that gives the total distance, D_{advanced}, as a function of time, T, *in minutes.* Generate a small table of solutions and plot the graph of this function on your graph in part (a).

c. Do these two graphs intersect? If so, what do the intersection point(s) represent?

46. A graduated income tax is proposed in Borduria to replace an existing flat rate of 8% on all income. The new proposal is that persons will pay no tax on their first \$20,000 of income, 5% on income over \$20,000 and less than or equal to \$100,000, and 10% on their income over \$100,000. (*Note:* Borduria is a fictional totalitarian state in the Balkans that figures in the adventures of Tintin.)

a. Construct a table of values that shows how much tax persons will pay under both the existing 8% flat tax and the proposed new tax for each of the following incomes: \$0; \$20,000; \$50,000; \$100,000; \$150,000; \$200,000.

b. Construct a graph of tax dollars vs. income for the 8% flat tax.

c. On the same graph plot tax dollars vs. income for the proposed new graduated tax.

d. Construct a function that describes tax dollars under the existing 8% tax as a function of income.

e. Construct a piecewise function that describes tax dollars under the proposed new graduated tax rates as a function of income.

f. Use your graph to estimate income level for which the taxes are the same under both plans. What plan would benefit people with incomes below your estimate? What plan would benefit people with incomes above your estimate?

g. Use your equations to find the coordinates that represent the point at which the taxes are the same for both plans. Label this point on your graph.

h. If the median income in the state is \$27,000 and the mean income is \$35,000, do you think the new graduated tax would be voted in by the people? Explain your answer.

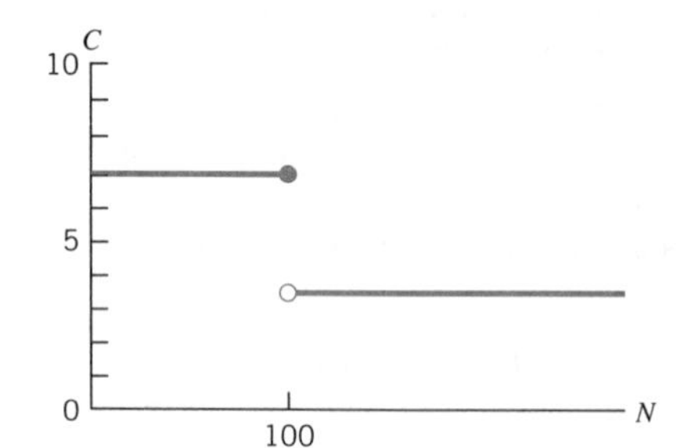

47. The pieces of a piecewise linear function may not always be "connected." The accompanying is a graph of a *step function,* called this since its graph looks like a series of steps. *Note:* When there might be confusion about whether an endpoint is included in a line, we draw a solid dot, ●, to indicate when the point is included and a hollow dot, ○, to indicate when it is not.

a. What does C equal when $N < 100$? When $N = 100$? When $N > 100$?

b. Write a piecewise linear function that represents this graph.

c. If we let N = number of photocopies of one page and C = cost per copy, then this function tells you what a national copy chain charges per copy. Describe its pricing policy.

d. If the page you are copying is part of a group of pages that are to be copied and collated (bound together), the price structure changes. The cost per copy for a page that will be collated is 7¢ per copy for the first 5000 copies and 3.5¢ per copy thereafter. Draw a new graph and construct a new step function to reflect this.

Challenge questions:

e. Using the information in part (c), construct and graph a piecewise linear function that describes, not the individual cost, but the *total* cost, T, of making N copies of a page.

f. Modify the function you constructed in part (e) to reflect the total cost of making N copies of a page that is part of a group that will be collated.

48. In Chapter 2, Exercise 57, you read of a math professor who purchased her house in Cambridge, MA for \$20,000 in 1973. Its assessed value has climbed at a steady rate so that it was worth \$250,000 as of 1999. Alas, one of her colleagues has not been so fortunate. He too bought a house in that same year for \$55,000. Not long after his family moved in, rumors began to circulate that the housing complex had been built on the site of a former toxic dump. Although never substantiated, the rumors adversely affected the value of his home, which has steadily decreased in value over the years and in 1999 was worth a meager \$23,000. In what year would the two homes have been assessed at the same value?

Exploration 3.1

Flat vs. Graduated Income Tax: Who Benefits?

Objectives

- compare the effects of different tax plans on individuals in different income brackets
- interpret intersection points

Material/Equipment

- spreadsheet or graphing calculator (optional). If using a graphing calculator, see examples on graphing piecewise functions in the Graphing Calculator Manual
- graph paper

Related Readings

"How a Flat Tax Would Work, for You and for Them." *New York Times,* Jan. 21, 1996.
"Flat Tax Goes From 'Snake Oil' to G.O.P. Tonic." *New York Times,* Nov. 14, 1999.

Procedure

If you need practice using the graphing calculator to graph systems of equations and find intersection points, see the Graphing Calculator Manual.

A variety of tax plans were debated in the 1996 and 2000 elections for president. (See related readings.) One plan recommended a flat tax of 19% on income after exemptions and deductions. In this exploration we examine who benefits from this flat tax as opposed to the current graduated tax plan.

The questions we explore are:

For what income will taxes be equal under both plans?

Who will benefit under the graduated income tax plan as compared to the flat tax plan?

Who will pay more taxes under the graduated plan as compared to the flat tax plan?

In a small group or with a partner

The accompanying table gives the 1999 federal graduated tax rates on income after deductions for single people.

1999 Tax Rate Schedules

Schedule X—Use if your filing status is **Single**

If the amount on Form 1040, line 39, is: *Over—*	*But not over—*	Enter on Form 1040, line 40	*of the amount over—*
$0	$25,750	- - - - - - - - - -15%	$0
25,750	62,450	$3,862.50 + 28%	25,750
62,450	130,250	14,138.50 + 31%	62,450
130,250	283,150	35,156.50 + 36%	130,250
283,150	- - - - - -	90,200.50 + 39.6%	283,150

1. Construct a function representing flat taxes of 19% where income, i, represents income after deductions.
2. Construct a piecewise linear function representing the graduated federal tax for single people in 1999, where income, i, represents income after deductions.
3. Graph your two functions on the same grid. Estimate from your graph any intersection points for the two functions.
4. Use your equations to calculate more accurate values for the points of intersection.
5. (Extra credit) Use your results to make changes in the tax plans. Decide on a different income for which taxes will be equal under both plans. You can use what you know about the distribution of income in the United States from the FAM 1000 data to make your decision. Alter one or both of the original functions such that both tax plans will generate the same tax given the income you have chosen.

Analysis

- Interpret your findings. Assume deductions are treated the same under both the flat tax plan and the graduated tax plan. What do the intersection points tell you about the differences between the tax plans?
- What information is useful in deciding on the merits of each of the plans?
- What if the amount of deductions that most people can take under the flat tax is less than the graduated tax plan? How will your analysis be affected?

Exploration-Linked Homework

Reporting Your Results

Take a stance for or against a flat federal income tax. Using supportive quantitative evidence, write a 60 second summary for a voters' pamphlet advocating your position. Present your arguments to the class.

Chapter 4

The Laws of Exponents and Logarithms: Measuring the Universe

Overview

Most of the examples we've studied so far have come from the social sciences. In order to delve into the physical and life sciences, we need to know about the tools that scientists use to compactly describe and compare the extremes in deep time and deep space. In this chapter we examine how to write, manipulate, and display the units and numbers of science.

After reading this chapter, you should be able to:

- write expressions in scientific notation
- convert between English and metric units
- simplify expressions using the rules of exponents
- compare numbers of widely differing sizes
- calculate logarithms base 10 and plot numbers on a logarithmic scale

4.1 MEASURING TIME AND SPACE: THE NUMBERS OF SCIENCE

On a daily basis we encounter quantities measured in tenths, tens, hundreds, or perhaps thousands. Finance or politics may bring us news of "1.2 billion people living in China" or "a federal debt of $6 trillion." In the physical sciences the range of numbers encountered is much larger. *Scientific notation* was developed to write compactly and to compare the sizes found in our universe, from the largest object we know, the observable universe, to the tiniest, the minuscule quarks oscillating inside the nucleus of an atom. We use examples from deep space and deep time to demonstrate powers of 10 and the use of scientific notation.

The Metric System and Powers of Ten

The international scientific community and most of the rest of the world uses the *metric system,* a system of measurements based on the meter (which is about 39.37 inches, a little over 3 feet). In daily life Americans have resisted converting to the metric system and still use the *English system* of inches, feet, and yards. In order to understand scientific measurements, we need to know how to convert between metric and English units. Table 4.1 shows the conversion for three standard units of length: the meter, the kilometer, and the centimeter. Most of the measurements in the examples in this and subsequent chapters will be in metric units. For a more complete conversion table, see the inside back cover.

Table 4.1
Conversions from Metric to English for Some Standard Units

Metric Unit	Abbreviation	In Meters	Equivalence in English Units	Informal Conversion
meter	m	1 m	3.28 ft	The width of a twin bed, a little more than a yard
kilometer	km	1000 m	0.62 mile	A casual 12-min walk, a little over a half a mile
centimeter	cm	0.01 m	0.39 in	The length of an ant, a little under a half an inch

Deep Space

For a real appreciation of the size of things in the universe, we highly recommend the video and related book by Philip and Phylis Morrison entitled *Powers of Ten: About the Relative Size of Things.*

The observable universe. Current measurements with the most advanced scientific instruments generate a best guess for the radius of the observable universe at about 100,000,000,000,000,000,000,000,000 meters, or "one hundred trillion trillion meters." Obviously we need a more convenient way to read, write, and say this number. In order to avoid writing a large number of zeros, exponents can be used as a shorthand:

- 10^{26} can be written as a 1 with 26 zeros after it.
- 10^{26} means: $10 \cdot 10 \cdot 10 \cdot \ldots \cdot 10$, the product of twenty-six 10s.
- 10^{26} is read as "ten to the twenty-sixth" or "ten to the twenty-sixth power."

So the estimated size of the radius of the observable universe is 10^{26} meters.

The Milky Way. Most of the objects we see in the night sky are the luminous balls of gas we call stars that are in our own galaxy, the Milky Way. There are approximately 100 billion (100,000,000,000 or 10^{11}) stars in the Milky Way. Most of these are thought to have planets circling them, though the first confirmation of a planet outside our solar system came only in 1993. The radius of the Milky Way is approximately 1,000,000,000,000,000,000,000 or 10^{21} meters.[1]

Our solar system. The approximate radius of our solar system is 1,000,000,000,000 or 10^{12} meters. In the metric system the prefix *tera-* means 10^{12}, so this distance is one *terameter.*

Our sun. The radius of our own sun, a modest star about halfway through its life cycle, is approximately 1,000,000,000, or 10^9 meters (1 billion meters). In the metric system, *giga-* is the prefix for 10^9, so the distance is one *gigameter.*

Earth. Our home, the third closest planet orbiting our sun, has a radius of somewhat less than 10,000,000 or 10^7 meters.

Us. Human beings are roughly in the middle of the scale of measurable objects in the universe. Human heights, including children's, vary from about one-third of a meter to 2 meters. In the wide scale of objects in the universe, a rough estimate for human height is 1 meter.

In order to continue the system of writing all sizes using powers of 10, we need a way to express 1 as a power of 10. Since $10^3 = 1000$, $10^2 = 100$, and $10^1 = 10$, a logical way to continue would be to say that $10^0 = 1$. Since reducing a power of 10 by 1 is equivalent to dividing by 10, the following calculations give justification for defining 10^0 as equal to 1.

$$10^2 = \frac{10^3}{10} = \frac{(10)(10)\cancel{(10)}}{\cancel{10}} = 100$$

$$10^1 = \frac{10^2}{10} = \frac{(10)\cancel{(10)}}{\cancel{10}} = 10$$

$$10^0 = \frac{10^1}{10} = \frac{10}{10} = 1$$

DNA molecules. A DNA strand provides genetic information for a human being. It is made up of a chain of building blocks called nucleotides. The chain is tightly coiled into a double helix, but stretched out it would measure about 0.01 meter in length. How does this DNA length translate into a power of 10? The number 0.01 or one hundredth equals $1/10^2$. We define $1/10^2$ to be 10^{-2}. So a DNA strand, uncoiled and measured lengthwise, is approximately 10^{-2} meters, or one *centimeter.*

By using negative exponents, we can continue to use powers of 10 to represent numbers less than 1. We want a system in which reducing the power by 1 remains equivalent to dividing by 10. So we define $10^2 = 100$, $10^1 = 10$, $10^0 = 1$, $10^{-1} = 1/10$, $10^{-2} = 1/10^2$, and so on. For any positive integer, n, we define

$$10^{-n} = \frac{1}{10^n}.$$

[1]The rough estimates for the sizes of objects in the universe in this section are taken from Timothy Ferris, *Coming of Age in the Milky Way* (New York: Doubleday, 1988).

Living cells. The average width of a living cell is

$$0.000\,01 = \frac{1}{100{,}000} = \frac{1}{10^5} = 10^{-5} \text{ meter}$$

Atoms. Atoms have an approximate average radius of

$$0.000\,000\,000\,1 = \frac{1}{10{,}000{,}000{,}000} = \frac{1}{10^{10}} = 10^{-10} \text{ meter}$$

Hydrogen atoms. The hydrogen atom, the smallest of the atoms, has an approximate radius of

$$0.000\,000\,000\,01 = \frac{1}{100{,}000{,}000{,}000} = \frac{1}{10^{11}} = 10^{-11} \text{ meter}$$

Summary of Powers of Ten

When n is a positive integer:

$$10^n = \underbrace{10 \cdot 10 \cdot 10 \; . \; . \; . \; 10}_{n \text{ factors}}$$ which can be written as 1 followed by n zeros.

$$10^{-n} = \frac{1}{10^n}$$ which can be written as a decimal point followed by $n - 1$ zeros and a 1

$$10^0 = 1$$

Multiplying by 10 is equivalent to moving the decimal point one place to the right. Dividing by 10 is equivalent to moving the decimal point one place to the left. The metric system is built on powers of ten. The prefix that is added to a basic metric unit of measure indicates the number of times the basic unit size has been multiplied or divided by ten. The metric prefixes for multiples or subdivisions by powers of 10 are given in Table 4.2. For example, millimeter is abbreviated mm; kilometer km; and megameter Mm. Abbreviations are usually not used for deci, deka, and hecto.

Table 4.2
Metric Prefixes for Powers of 10

atto-	a	10^{-18}	centi	c	10^{-2}	kilo-	k	10^3
femto-	f	10^{-15}	deci-		10^{-1}	mega-	M	10^6
pico-	p	10^{-12}	(unit)		10^0	giga-	G	10^9
nano-	n	10^{-9}	deka-		10^1	tera-	T	10^{12}
micro-	μ	10^{-6}	hecto-		10^2	peta-	P	10^{15}
milli-	m	10^{-3}				exa-	E	10^{18}

SCIENTIFIC NOTATION

In the previous examples we estimated the sizes of objects to the nearest power of 10 without worrying about the exact numbers. Getting the power of 10 correct is the first consideration when dealing with physical objects. You want to know whether something is closer to a thousand or to 10 billion meters in size. Once the power of 10 is known, the number can be refined to a greater level of accuracy.

For example, a more accurate measure of the radius of a hydrogen atom is 0.000 000 000 052 9 meter across. This can be written more compactly by using *scientific notation*; that is, we can rewrite it as a number times a power of 10.

In order to convert a nonzero number into scientific notation:

- Write down the first nonzero digit and all the digits following it. Put a decimal point right after the first digit. This new number is called the *coefficient.* In this case the coefficient is 5.29.
- Figure out what power of 10 is needed to convert the coefficient back to the original number. In this case the decimal place has to move 11 places to the left to get 0.000 000 000 052 9. This is equivalent to dividing 5.29 by 10 eleven times, or dividing it by 10^{11}:

$$0.000\,000\,000\,052\,9 = \frac{5.29}{10^{11}}$$

$$= 5.29\left(\frac{1}{10^{11}}\right)$$

$1/10^{11} = 10^{-11}$, so we get:

$$= 5.29 \cdot 10^{-11}$$

This number is now said to be in scientific notation.[2]

When a number is negative, then its coefficient is negative. Thus, the number $-0.000\,000\,000\,052\,9$ in scientific notation is $-5.29 \cdot 10^{-11}$. The coefficient is -5.29.

Although the coefficient, N, can be positive or negative, we insist that the coefficient stripped of its sign, called its *absolute value* and written $|N|$, be such that $1 \leq |N| < 10$.

Remember when calculating the absolute value of N, if $N \geq 0$, then $|N| = N$. For example, $|2| = 2$, $|11.57| = 11.57$, and $|0| = 0$. But if $N < 0$, then $|N| = -N$. When N is negative, $-N$ is actually positive. For example, $|-2| = -(-2) = 2$. The important fact is that no matter what the value of N, the absolute value of N, or $|N|$, will always be greater than or equal to zero.

> The expression $|N|$ is read "the absolute value of N."
>
> If $N \geq 0$, then $|N| = N$.
> If $N < 0$, then $|N| = -N$.
>
> So whether N is positive, negative, or zero, $|N| \geq 0$.

[2]Most calculators or computers automatically translate a number into scientific notation when it is too large or small to fit into the display. The notation is often slightly modified by using the letter E (short for "exponent") to replace the expression "times 10 to some power". So $3.0 \cdot 10^{26}$ may appear as 3.0E+26. The number after the E tells how many places and the sign (+ or −) indicates in which direction to move the decimal point of the coefficient.

Any nonzero number, positive or negative, can be written in scientific notation, that is, written as the product of a coefficient N multiplied by 10 to some power, where $1 \leq |N| < 10$. Thus 2,000,000 or 2 million or $2 \cdot 10^6$ are all correct representations of the same number. The one you choose to use depends on the context.

A number is in *scientific notation* if it is in the form

$$N \cdot 10^n$$

where $1 \leq |N| < 10$ and n is an integer (positive, negative, or zero).

Deep Time

The poem "Imagine" offers a creative look at the Big Bang.

The Big Bang. In 1929 the American astronomer Edwin Hubble published an astounding paper that claimed that the universe was expanding. Most astronomers and cosmologists now agree with his once controversial theory and believe that somewhere between 8 and 15 billion years ago the universe began an explosive expansion from an infinitesimally small point. This event is referred to as the "Big Bang" and the universe has been expanding ever since it occurred.[3]

Powers of 10 and scientific notation can be used to record the progress of the universe since the Big Bang.

Age of the universe. One current estimate places the age of our universe at about 15,000,000,000 years, or about 15 billion years. Using scientific notation, we would say the universe is $1.5 \cdot 10^{10}$ years old.

Age of Earth. Earth is believed to have been formed during the last third of the universe's existence—about 4,600,000,000, or 4.6 billion, years ago. In scientific notation the age is $4.6 \cdot 10^9$ years.

Age of Pangaea. About 200,000,000, or 200 million, years ago, all Earth's continents collided to form one giant land mass now referred to as Pangaea. Pangaea existed $2.0 \cdot 10^8$ years ago.

This metaphor is played out in Carl Sagan's video *Cosmos* and in his book *The Dragons of Eden*, which is excerpted in the Anthology of Readings.

Age of human life. ***Homo sapiens sapiens*** first walked on Earth about 100,000, or $1.0 \cdot 10^5$, years ago. In the life of the universe, this is almost nothing. If all of time, from the Big Bang to today, were scaled down into a single year, with the Big Bang on January 1, our early human ancestors would not appear until less then 4 minutes before midnight on December 31, New Year's Eve.

[3]Depending on its total mass, the universe will either expand forever or collapse back upon itself. However, cosmologists are unable to estimate the total mass of the universe, since they are in the embarrassing position of not being able to find about 90% of it. Scientists call this missing mass *dark matter,* which describes not only its invisibility but scientists' own mystification.

Algebra Aerobics 4.1

1. Express as a power of 10:
   a. 10,000,000,000 **b.** 0.000 000 000 000 01
2. Express in standard notation (without exponents):
   a. 10^{-8} **b.** 10^{13}
3. Express as a power of 10 and then in standard notation:
   a. A nanosecond in terms of seconds
   b. A decimeter in terms of meters
   c. A gigabyte in terms of bytes
4. Rewrite each measurement in meters, first using a power of 10 and then standard notation:
   a. 7 cm **b.** 9 mm **c.** 5 km
5. Avogadro's number is $6.02 \cdot 10^{23}$. A mole of any substance is defined to be Avogadro's number of particles of that substance. Express this number in standard notation.
6. The distance between Earth and its moon is 384,000,000 meters. Express this in scientific notation.
7. An angstrom (denoted by Å), a unit commonly used to measure the size of atoms, is 0.000 000 01 cm. Express its size using scientific notation.
8. The width of a DNA double helix is approximately 2 nanometers, or $2 \cdot 10^{-9}$ meters. Express the width in standard notation.
9. Express in standard notation:
   a. $-7.05 \cdot 10^{8}$ **b.** $-4.03 \cdot 10^{-5}$
10. Express in scientific notation:
    a. $-43{,}000{,}000$ **b.** $-0.000\ 008\ 3$

4.2 SIMPLIFYING EXPRESSIONS WITH POSITIVE INTEGER EXPONENTS

What can we say about the value of $(-1)^n$ when n is an even integer? When n is an odd integer?

No matter what the base, whether it is 10 or any other number, repeated multiplication leads to *exponentiation.* For example,

$$3 \cdot 3 \cdot 3 \cdot 3 = 3^4$$

Here 4 is the *exponent* of 3 and 3 is called the *base.* In general, if a is a real number and n is a positive integer, then we define a^n as an abbreviation for the multiplication of a as a factor n times.

> In the expression a^n, where a is any real number, the number a is called the *base* and n is called the *exponent* or *power.*
>
> If a is any nonzero real number and n is a positive integer, then
>
> $$a^n = \underbrace{a \cdot a \cdot a \cdots a}_{n \text{ factors}} \quad \text{(the product of } n \text{ factors of } a\text{)}$$

In this section we'll see how the rules for manipulating expressions with positive exponents make sense if we remember what the exponent tells us to do to the base. The rules are summarized in the following box. We first examine these rules when the exponents m and n are positive integers. In the following sections, we extend the rules to cases where m and n are any rational numbers, such as negative integers and fractions. With additional mathematics, the rules can be extended to include cases where m and n are any real numbers.

Exponent Rules

If a and b are any nonzero real numbers, then:

1. $a^n \cdot a^m = a^{(n+m)}$	**4.** $(ab)^n = a^n b^n$
2. $\dfrac{a^n}{a^m} = a^{(n-m)}$	**5.** $\left(\dfrac{a}{b}\right)^n = \dfrac{a^n}{b^n}$
3. $(a^m)^n = a^{(m \cdot n)}$	

We show below how Rules 1, 3, and 5 make sense and leave Rules 2 and 4 for you to justify in the exercises.

Rule 1: To justify this rule, think about the total number of times a is taken as a factor when a^n is multiplied by a^m:

$$a^n \cdot a^m = \underbrace{a \cdot a \cdot a \cdots a}_{n \text{ factors}} \cdot \underbrace{a \cdot a \cdots a}_{m \text{ factors}} = \underbrace{a \cdot a \cdot a \cdots a}_{n+m \text{ factors}} = a^{(n+m)}$$

Rule 3: First think about how many times we are taking a^m as a factor when we raise it to the nth power:

$$(a^m)^n = \underbrace{a^m \cdot a^m \cdots a^m}_{n \text{ factors of } a^m}$$

Use Rule 1:

$$= a^{\overbrace{(m+m+\ldots+m)}^{n \text{ terms}}}$$

Represent adding m n times as $m \cdot n$

$$= a^{(m \cdot n)}$$

Rule 5: Remember that the exponent n in the expression $(a/b)^n$ applies to the whole expression within the parentheses:

$$\left(\frac{a}{b}\right)^n = \underbrace{\left(\frac{a}{b}\right) \cdot \left(\frac{a}{b}\right) \cdots \left(\frac{a}{b}\right)}_{n \text{ factors of } a/b}$$

$$= \frac{\overbrace{a \cdot a \cdots a}^{n \text{ factors of } a}}{\underbrace{b \cdot b \cdots b}_{n \text{ factors of } b}} = \frac{a^n}{b^n}$$

Example 1

Deneb is 1600 light years from Earth. How far is Earth from Deneb when measured in miles?

Solution

The distance that light travels in 1 year, called a *light year,* is approximately 5.88 trillion miles.

Since $$1 \text{ light year} = 5{,}880{,}000{,}000{,}000 \text{ miles}$$

then the distance from Earth to Deneb is

$$\begin{aligned} 1600 \text{ light years} &= (1600)\cdot(5{,}880{,}000{,}000{,}000 \text{ miles}) \\ &= (1.6\cdot 10^3)\cdot(5.88\cdot 10^{12} \text{ miles}) \\ &= (1.6\cdot 5.88)\cdot(10^3\cdot 10^{12}) \text{ miles} \\ &\approx 9.4\cdot 10^{3+12} \text{ miles} \\ &\approx 9.4\cdot 10^{15} \text{ miles} \end{aligned}$$

Using ratios to compare sizes of objects.

In comparing two objects of about the same size, it is common to subtract one size from the other and say, for instance, that one person is 6 inches taller than another. This method of comparison is not effective for objects that have vastly different sizes. To say that the difference between the estimated radius of our solar system (1 terameter, or 1,000,000,000,000 meters) and the average size of a human (about 10^0, or 1, meter) is 1,000,000,000,000 − 1 = 999,999,999,999 meters is not particularly useful. In fact, since our measurement of the solar system certainly isn't accurate to within 1 meter, this difference is meaningless. As shown in the following example, a more useful method for comparing objects of wildly different sizes is to calculate the ratio of the two sizes.

Example 2

How many times larger is the sun than Earth?

Solution 1

The radius of the sun is approximately 10^9 meters and the radius of Earth is about 10^7 meters. One way to answer the question "How many *times* larger is the sun than Earth?" is to form the ratio of the two radii:

$$\begin{aligned} \frac{\text{radius of the sun}}{\text{radius of Earth}} &= \frac{10^9 \text{ m}}{10^7 \text{ m}} \\ &= \frac{10^9 \cancel{\text{m}}}{10^7 \cancel{\text{m}}} = 10^{9-7} = 10^2 \end{aligned}$$

The units cancel, so 10^2 is unitless. The radius of the sun is approximately 10^2, or 100, times larger than the radius of Earth.

Solution 2

Another way to answer the question is to compare the volumes of the two objects. The sun and Earth are both roughly spherical. The formula for the volume V of a sphere with radius r is

$$V = \tfrac{4}{3}\pi r^3$$

The radius of the sun is approximately 10^9 meters and the radius of Earth is about 10^7 meters. The ratio of the two volumes is

$$\begin{aligned}
\frac{\text{volume of the sun}}{\text{volume of Earth}} &= \frac{(4/3)\pi(10^9)^3\ \text{m}^3}{(4/3)\pi(10^7)^3\ \text{m}^3} \\
&= \frac{(10^9)^3}{(10^7)^3} \qquad (\textit{Note: } \tfrac{4}{3}\pi \text{ and m}^3 \text{ cancel.}) \\
&= \frac{10^{27}}{10^{21}} \\
&= 10^6
\end{aligned}$$

So while the radius of the sun is 100 times larger than the radius of Earth, the *volume* of the sun is approximately $10^6 = 1{,}000{,}000$, or 1 million, times larger than the volume of Earth!

Example 3

According to the U.S. Bureau of the Census, in 1999 the estimated gross federal debt was \$5.69 trillion and the estimated U.S. population was 274 million. What was the approximate federal debt *per person?*

Solution

$$\begin{aligned}
\frac{\text{federal debt}}{\text{U.S. population}} &= \frac{5.69 \cdot 10^{12}\ \text{dollars}}{2.74 \cdot 10^8\ \text{people}} \\
&= \left(\frac{5.69}{2.74}\right)\cdot\left(\frac{10^{12}}{10^8}\right)\frac{\text{dollars}}{\text{people}} \approx 2.08 \cdot 10^4\,\frac{\text{dollars}}{\text{people}}
\end{aligned}$$

So the federal debt amounted to about $\$2.08 \cdot 10^4$ or \$20,800 per person.

Example 4

Simplify and write as an expression with exponents:

$$7^3 \cdot 7^2 = 7^{3+2} = 7^5 \qquad (x^5)^3 = x^{5\cdot 3} = x^{15}$$

$$w^3 \cdot w^5 = w^{3+5} = w^8 \qquad (11^2)^4 = 11^{2\cdot 4} = 11^8$$

$$\frac{10^8}{10^3} = 10^{8-3} = 10^5 \qquad \frac{z^8}{z^3} = z^{8-3} = z^5$$

Example 5

Simplify:

$$(3a)^4 = 3^4a^4 = 81a^4$$

$$(-5x)^3 = (-5)^3x^3 = -125x^3$$

$$\left(\frac{2}{3}\right)^3 = \frac{2^3}{3^3} = \frac{8}{27}$$

$$\left(\frac{-2a}{3b}\right)^3 = \frac{(-2a)^3}{(3b)^3} = \frac{(-2)^3a^3}{3^3b^3} = \frac{-8a^3}{27b^3}$$

Common Errors

1. Don't confuse a term such as $-x^4$ with $(-x)^4$. For example, $-2^4 = -(2^4) = -16$. But $(-2)^4 = (-2)(-2)(-2)(-2) = +16$. You have to remember what is being raised to the power. In the expression -2^4, order of operations says to compute the power first, before applying the negation sign. In the expression $(-2)^4$, everything inside the parentheses is raised to the fourth power. So 2 is negated first, and then the result (-2) is raised to the fourth power.

   For example, $(-3a)^2$ means that everything in the parentheses is squared:

$$(-3a)^2 = (-3a)(-3a) = (-3)(-3)a \cdot a = 9a^2$$

   However, in $-3a^2$, the exponent 2 applies only to the base a.

2. It is important to remember that in order to combine the exponents of two multiplied terms, the terms *must have the same base.* For example,

$$(9300)^2 \cdot (9300)^6 = (9300)^8$$

   But an expression like $81^2 \cdot 47^6$, which means $(81 \cdot 81) \cdot (47 \cdot 47 \cdot 47 \cdot 47 \cdot 47 \cdot 47)$, cannot be simplified by adding exponents since *the bases are not equal.*

3. The sum of terms with the same base, such as $10^2 + 10^3$, *cannot* be simplified by adding exponents. Be careful to avoid this common error. The sum $10^2 + 10^3 \neq 10^5$. We can verify this by calculating the values of both sides:

$$10^2 + 10^3 = 100 + 1000$$
$$= 1100$$

   but

$$10^5 = 100{,}000$$

   A sum of terms with the *same exponents and bases* can be simplified, although not by adding powers. For example, $10^3 + 10^3 = 2 \cdot 10^3$. Why can these terms be combined?

   Use the distributive property to rewrite $10^3 + 10^3$ $\quad 10^3 + 10^3 = 10^3(1 + 1)$

   add $\quad = 10^3 \cdot 2$

   use the commutative property $\quad = 2 \cdot 10^3$

   For example, $7^3 \cdot 6^5$ and $7^3 + 6^5$ cannot be simplified since the bases 7 and 6 are different. $6^5 \cdot 6^5 \cdot 6^5 = 6^{15}$, but $6^5 + 6^5 + 6^5 = 3 \cdot 6^5$.

Algebra Aerobics 4.2a

1. Simplify where possible, leaving the answer in a form with exponents:
   a. $10^5 \cdot 10^7$
   b. $8^6 \cdot 8^{14}$
   c. $z^5 \cdot z^4$
   d. $5^5 \cdot 6^7$
   e. $7^3 + 7^3$

2. Simplify (if possible), leaving the answer in exponent form:
   a. $\dfrac{10^{15}}{10^7}$
   b. $\dfrac{8^6}{8^4}$
   c. $\dfrac{3^5}{3^4}$
   d. $\dfrac{5}{6^7}$

3. Simplify, leaving the answer in exponent form:
   a. $(10^4)^5$
   b. $(7^2)^3$
   c. $(x^4)^5$
   d. $(2x)^4$
   e. $(2a^4)^3$
   f. $(-2a)^3$

4. Simplify:
   a. $\left(\dfrac{-4}{5}\right)^2$
   b. $\left(\dfrac{-2x}{4y}\right)^3$
   c. $(-5)^2$
   d. -5^2

5. A high-density diskette has a storage capacity of about 1.44 megabytes ($1.44 \cdot 10^6$ bytes). If a hard drive has a capacity of 2 gigabytes ($2 \cdot 10^9$ bytes), how many diskettes would it take to equal the storage capacity of the hard drive?

Estimating Answers

By rounding off numbers and using scientific notation and the rules for exponents, we can often make quick estimates of answers to complicated calculations. In this age of calculators and computers, we need to be able to roughly estimate the size of an answer, in order to make sure our answer makes sense.

Example 6

Estimate the value of $\dfrac{(382{,}152)\cdot(490{,}572{,}261)}{(32{,}091)\cdot(1942)}$. Express your answer in both scientific and standard notation.

Solution

Round off each number $\quad \dfrac{(382{,}152)\cdot(490{,}572{,}261)}{(32{,}091)\cdot(1942)} \approx \dfrac{(400{,}000)\cdot(500{,}000{,}000)}{(30{,}000)\cdot(2000)}$

rewrite in scientific notation $\quad \approx \dfrac{(4\cdot 10^5)\cdot(5\cdot 10^8)}{(3\cdot 10^4)\cdot(2\cdot 10^3)}$

group the coefficients and the powers of 10 $\quad \approx \left(\dfrac{4\cdot 5}{3\cdot 2}\right)\cdot\left(\dfrac{10^5\cdot 10^8}{10^4\cdot 10^3}\right)$

simplify each expression $\quad \approx \dfrac{20}{6}\cdot\dfrac{10^{13}}{10^7}$

we get $\quad \approx 3.33\cdot 10^6$

or in standard notation $\quad \approx 3{,}330{,}000$

Using a calculator on the original problem, we get a more precise answer of 3,008,199.595.

Example 7

In 2000 the world population was approximately 6.073 billion people. There are roughly 57.9 million square miles of land on Earth, of which about 22% are favorable for agriculture. Estimate how many people per square mile of farmable land there were in 2000.

Solution

$$\begin{aligned}
\frac{\text{size of world population}}{\text{amount of farmable land}} &= \frac{6.073 \text{ billion people}}{22\% \text{ of } 57.9 \text{ million square miles}} \\
&= \frac{6.073\cdot 10^9 \text{ people}}{(0.22)\cdot(57.9)\cdot 10^6 \text{ mile}^2} \\
&\approx \frac{6\cdot 10^9 \text{ people}}{(0.2)\cdot 60\cdot 10^6 \text{ mile}^2} \\
&\approx \frac{6\cdot 10^9 \text{ people}}{12\cdot 10^6 \text{ mile}^2} \\
&\approx 0.5\cdot 10^3 \text{ people/mile}^2 \\
&\approx 500 \text{ people/mile}^2
\end{aligned}$$

So there are roughly 500 people/mile^2 of farmable land in the world. Using a calculator and rounding to the nearest whole number, we get 477 people/mile^2 of farmable land.

Algebra Aerobics 4.2b

1. Estimate the value of:
   a. $(0.000\ 297\ 6) \cdot (43{,}990{,}000)$
   b. $\dfrac{453{,}897 \cdot 2{,}390{,}702}{0.004\ 38}$
2. The radius of Jupiter, the largest of the planets in our solar system, is approximately $7.14 \cdot 10^4$ km. (If r is the radius of a sphere, the sphere's surface area equals $4\pi r^2$, and its volume equals $\frac{4}{3}\pi r^3$.)
   a. Estimate the surface area of Jupiter.
   b. Estimate the volume of Jupiter.
3. Only about three-sevenths of the land favorable for agriculture is actually being farmed. Using the facts in Example 7, estimate the number of people per square mile of farmable land that is being used. Should your estimate be larger or smaller than the ratio of people to farmable land? Explain.

4.3 SIMPLIFYING EXPRESSIONS WITH NEGATIVE INTEGER EXPONENTS

The definitions for raising any base, a, to the zero power or to a negative power follow a logic that is similar to the one that was used to define $10^0 = 1$ and $10^{-n} = 1/10^n$.

If a is a real number and n is a positive integer, the following definitions hold:

$$a^0 = 1 \qquad (a \neq 0)$$

$$a^{-n} = \frac{1}{a^n} \qquad (a \neq 0)$$

It is important to note that $a^1 = a$; thus $a^{-1} = 1/a^1 = 1/a$.

In the following examples, we show how to use the five rules for exponents when the exponents are negative integers or zero.

Example 1

Simplify $x^2 \cdot x^{-5}$.

Solution

Using Rule 1 for exponents,

$$x^2 \cdot x^{-5} = x^{2+(-5)} = x^{-3}$$

or we can simplify by first writing x^{-5} as $1/x^5$ and then using Rule 2 for exponents:

$$x^2 \cdot x^{-5} = x^2 \cdot \frac{1}{x^5} = \frac{x^2}{x^5} = x^{2-5} = x^{-3}$$

Example 2

Simplify. Express your answer with positive exponents.

a. $\dfrac{10^2}{10^6}$ **b.** $\dfrac{6^2}{6^{-7}}$ **c.** $\dfrac{(-5)^2}{(-5)^6}$ **d.** $\dfrac{x^{-2}}{x^4}$

Solution

Using Rule 2 for exponents:

a. $\dfrac{10^2}{10^6} = 10^{2-6} = 10^{-4} = \dfrac{1}{10^4}$

b. $\dfrac{6^2}{6^{-7}} = 6^{2-(-7)} = 6^9$

c. $\dfrac{(-5)^2}{(-5)^6} = (-5)^{2-6} = (-5)^{-4} = \dfrac{1}{(-5)^4} = \dfrac{1}{(-1)^4(5)^4} = \dfrac{1}{5^4}$

d. $\dfrac{x^{-2}}{x^4} = x^{-2-4} = x^{-6} = \dfrac{1}{x^6}$

Example 3

Simplify:

a. $(13^{-8})^3$ **b.** $(w^2)^{-7}$

Solution

Using Rule 3 for exponents,

a. $(13^{-8})^3 = 13^{(-8)\cdot 3} = 13^{-24}$ **b.** $(w^2)^{-7} = w^{2\cdot(-7)} = w^{-14}$

Example 4

Simplify $\dfrac{v^{-2}(w^5)^2}{(v^{-1})^4 w^{-3}}$.

Solution

Apply Rule 3 twice $\dfrac{v^{-2}(w^5)^2}{(v^{-1})^4 w^{-3}} = \dfrac{v^{-2}w^{10}}{v^{-4}w^{-3}}$

Apply Rule 2 twice $= v^{-2-(-4)}w^{10-(-3)}$

Simplify $= v^2 w^{13}$

The rule for applying negative powers is the same whether a is an integer or a fraction:

$$a^{-n} = 1/a^n \quad \text{where } a \neq 0$$

For example,

$$\left(\frac{1}{2}\right)^{-1} = \frac{1}{(1/2)^1} = 1 \div \left(\frac{1}{2}\right) = 1\cdot\left(\frac{2}{1}\right) = 2$$

In general,

$$\left(\frac{a}{b}\right)^{-n} = \frac{1}{(a/b)^n} = 1 \div \left(\frac{a}{b}\right)^n = 1\cdot\left(\frac{b}{a}\right)^n = \left(\frac{b}{a}\right)^n = \frac{b^n}{a^n}$$

Example 5

Simplify:

a. $\left(\dfrac{1}{2}\right)^{-11}\cdot\left(\dfrac{1}{2}\right)^{-2}$ **b.** $\left(\dfrac{a}{b}\right)^3\cdot\left(\dfrac{a}{b}\right)^{-5}$

Solution

a. Using Rule 1 for exponents and the definition of a^{-n};

$$\left(\frac{1}{2}\right)^{-11}\left(\frac{1}{2}\right)^{-2} = \left(\frac{1}{2}\right)^{-11+(-2)}$$

$$= \left(\frac{1}{2}\right)^{-13} = \left(\frac{2}{1}\right)^{13} = 2^{13} = 8192$$

b. Using Rules 1 and 5 for exponents and the definition of a^{-n};

$$\left(\frac{a}{b}\right)^{3}\cdot\left(\frac{a}{b}\right)^{-5} = \left(\frac{a}{b}\right)^{3+(-5)}$$

$$= \left(\frac{a}{b}\right)^{-2} = \left(\frac{b}{a}\right)^{2} = \frac{b^2}{a^2}$$

Algebra Aerobics 4.3

1. Simplify (if possible). Express with positive exponents.

   a. $10^5 \cdot 10^{-7}$ **b.** $\frac{11^6}{11^{-4}}$ **c.** $\frac{3^{-5}}{3^{-4}}$ **d.** $\frac{5^5}{6^7}$ **e.** $\frac{7^3}{7^3}$ **f.** $a^{-2} \cdot a^{-3}$

2. A typical TV signal, traveling at the speed of light, takes $3.3 \cdot 10^{-6}$ second to travel 1 kilometer. Estimate how long it would take the signal to travel across the United States (a distance of approximately 4300 kilometers).
3. Distribute and simplify $x^{-2}(x^5 + x^{-6})$.
4. Simplify:

   a. $(10^4)^{-5}$ **b.** $(7^{-2})^{-3}$ **c.** $(2a^3)^{-2}$ **d.** $\left(\frac{8}{x}\right)^{-2}$

5. Simplify:

   a. $\frac{t^{-3}t^0}{(t^{-4})^3}$ **b.** $\frac{v^{-3}w^7}{(v^{-2})^3 w^{-10}}$

Exploration 4.1 will help you understand the relative ages and sizes of objects in our universe and give you practice in scientific notation and unit conversion.

4.4 CONVERTING UNITS

Problems in science constantly require converting back and forth between different units of measure. In order to do so, we need to be comfortable with the laws of exponents and the basic metric and English units (see Table 4.1 or a more complete table on the inside back cover). The following unit conversion examples describe a strategy based upon *conversion factors.*

Converting Units within the Metric System

Example 1

Light travels at a speed of approximately $3.00 \cdot 10^5$ kilometers per second (km/sec). Describe the speed of light in meters per second (m/sec).

Solution

The prefix *kilo-* means thousand. One kilometer (km) is equal to 1000 or 10^3 meters (m):

$$1 \text{ km} = 10^3 \text{ m} \tag{1}$$

Dividing both sides of Equation (1) by 1 km, we can rewrite it as

$$1 = \frac{10^3 \text{ m}}{1 \text{ km}}$$

If instead we divide both sides of Equation (1) by 10^3 m, we get

$$\frac{1 \text{ km}}{10^3 \text{ m}} = 1$$

The ratios (10^3 m)/(1 km) and (1 km)/(10^3 m) are called *conversion factors,* because we can use them to convert between kilometers and meters.

When solving unit conversion problems, the crucial question is always; "What is the right conversion factor?" If units in km/sec are multiplied by units in meters per kilometer, we have

$$\frac{\cancel{\text{km}}}{\text{sec}} \cdot \frac{\text{m}}{\cancel{\text{km}}}$$

and the result is in meters per second. So multiplying the speed of light in km/sec by a conversion factor in m/km will give us the correct units of m/sec. Since a conversion factor always equals 1, we will not change the value of the original quantity by multiplying it by a conversion factor. In this case, we use the conversion factor of (10^3 m)/(1 km):

$$\begin{aligned} 3.00 \cdot 10^5 \text{ km/sec} &= 3.00 \cdot 10^5 \frac{\cancel{\text{km}}}{\text{sec}} \cdot \frac{10^3 \text{ m}}{1 \cancel{\text{km}}} \\ &= 3.00 \cdot 10^5 \cdot 10^3 \text{ m/sec} \\ &= 3.00 \cdot 10^8 \text{ m/sec} \end{aligned}$$

Hence light travels at approximately $3.00 \cdot 10^8$ m/sec.

Example 2

Check your answer in Example 1 by converting $3.00 \cdot 10^8$ m/sec back to km/sec.

Solution

Here we use the same strategy, but now we need to use the other conversion factor.

Multiplying $3.00 \cdot 10^8$ m/sec by (1 km)/(10^3 m) gives us

$$\begin{aligned} 3.00 \cdot 10^8 \frac{\cancel{\text{m}}}{\text{sec}} \cdot \frac{1 \text{ km}}{10^3 \cancel{\text{m}}} &= 3.00 \cdot \frac{10^8 \text{ km}}{10^3 \text{ sec}} \\ &= 3.00 \cdot 10^5 \text{ km/sec} \end{aligned}$$

which was the original value given for the speed of light.

Converting between the Metric and English Systems

Example 3

If light travels $3.00 \cdot 10^5$ km/sec, how many *miles* does light travel in a second?

Solution

The crucial question is: "What conversion factor should be used?" From Table 4.1 we know that

$$1 \text{ km} \approx 0.62 \text{ mile}$$

This equation can be rewritten in two ways:

$$1 \approx \frac{0.62 \text{ mile}}{1 \text{ km}} \quad \text{or} \quad 1 \approx \frac{1 \text{ km}}{0.62 \text{ mile}}$$

It produces two possible conversion factors:

$$\frac{0.62 \text{ mile}}{1 \text{ km}} \quad \text{and} \quad \frac{1 \text{ km}}{0.62 \text{ mile}}$$

Why is the conversion factor 1 km/0.62 miles not helpful in solving this problem?

Which one will convert kilometers to miles? We need one with kilometers in the denominator and miles in the numerator, namely (0.62 mile)/(1 km), so that km will cancel.

Multiplying $3.00 \cdot 10^5$ km/sec by the conversion factor (0.62 mile)/(1 km) gives us

$$3.00 \cdot 10^5 \frac{\cancel{\text{km}}}{\text{sec}} \cdot \frac{0.62 \text{ mile}}{1 \cancel{\text{km}}} = 1.86 \cdot 10^5 \text{ mile/sec}$$

So light travels at $1.86 \cdot 10^5$ or 186,000 miles/sec.

Using Multiple Conversion Factors

Example 4

Light travels at $3.00 \cdot 10^5$ km/sec. How many kilometers does light travel in one *year*?

Solution

Here our strategy is to use more than one conversion factor to convert from seconds to years. Use your calculator to perform the following calculations:

$$3.00 \cdot 10^5 \frac{\text{km}}{\cancel{\text{sec}}} \cdot \frac{60 \cancel{\text{sec}}}{1 \cancel{\text{min}}} \cdot \frac{60 \cancel{\text{min}}}{1 \cancel{\text{hr}}} \cdot \frac{24 \cancel{\text{hr}}}{1 \cancel{\text{day}}} \cdot \frac{365 \cancel{\text{days}}}{1 \text{ year}} = 94{,}608{,}000 \cdot 10^5 \text{ km/year}$$

$$\approx 9.46 \cdot 10^7 \cdot 10^5 \text{ km/year}$$

$$= 9.46 \cdot 10^{12} \text{ km/year}$$

So a light year, the distance light travels in one year, is approximately equal to $9.46 \cdot 10^{12}$ kilometers.

Algebra Aerobics 4.4

1. The mean distance from our sun to Jupiter is $7.8 \cdot 10^8$ kilometers. Express this distance in meters.
2. In Section 4.2 we said that a light year was about $5.88 \cdot 10^{12}$ miles. Verify that $9.46 \cdot 10^{12}$ kilometers $\approx 5.88 \cdot 10^{12}$ miles.
3. 1 Angstrom $= 10^{-8}$ cm. Express 1 Angstrom in meters.
4. If a road sign says the distance to Quebec is 218 km, what is the distance in miles?
5. The distance from Earth to the sun is about 93,000,000 miles. There are 5280 feet in a mile, and a dollar bill is approximately 6 inches long. Estimate how many dollar bills would have to be placed end to end to reach from Earth to the sun.

4.5 SIMPLIFYING EXPRESSIONS WITH FRACTIONAL EXPONENTS

So far we have derived rules for operating with expressions of the form a^n, where n is any integer. These rules can be extended to expressions of the form $a^{m/n}$, where the exponent is a fraction. We need first to consider what an expression such as $a^{m/n}$ means.

The expression m/n can also be written as $m \cdot (1/n)$ or $(1/n) \cdot m$. If the laws of exponents are consistent, then

$$a^{m/n} = (a^m)^{1/n} = (a^{1/n})^m$$

For example, if we apply Rule 3 for exponents to the expression $(a^{1/2})^2$, then the following should be true:

$$(a^{1/2})^2 = a^{(1/2)\cdot 2} = a^1 = a$$

Square Roots: Expressions of the Form $a^{1/2}$

The expression $a^{1/2}$ is called the *principal square root* (or just the *square root*) of a and is often written as $\sqrt{a}$. The symbol $\sqrt{\ }$ is called a radical. The principal square root of a is the *nonnegative* number b such that $b^2 = a$. Both the square of -2 and the square of 2 are equal to 4, but the notation $\sqrt{4}$ is defined as *only the positive root.* If both -2 and 2 are to be considered, we write $\pm\sqrt{4}$, which means "plus or minus the square root of 4."

In the real numbers, $\sqrt{a}$ is not defined when a is negative. For example, $\sqrt{-4}$ is undefined, since there is no real number b such that $b^2 = -4$.

> For $a \geq 0$,
>
> $$a^{1/2} = \sqrt{a}$$
>
> where $\sqrt{a}$ is the nonnegative number whose square is a.

For example, $25^{1/2} = \sqrt{25} = 5$, since $5^2 = 25$.

Estimating square roots. A number is called a *perfect square* if its square root is an integer. For example, 25 and 36 are both perfect squares since $25 = 5^2$ and $36 = 6^2$, so $\sqrt{25} = 5$ and $\sqrt{36} = 6$. If we don't know the square root of some number x and don't have a calculator handy, we can estimate the square root by bracketing it between perfect squares, a and b, for which we do know the square roots. If $0 \leq a < x < b, \leq$ then $\sqrt{a} < \sqrt{x} < \sqrt{b}$. For example, to estimate $\sqrt{10}$,

we know	$9 < 10 < 16$	where 9 and 16 are perfect squares
so	$\sqrt{9} < \sqrt{10} < \sqrt{16}$	
and	$3 < \sqrt{10} < 4$	

Therefore $\sqrt{10}$ lies somewhere between 3 and 4, probably closer to 3 because 10 is closer to 9. According to a calculator $\sqrt{10} \approx 3.16$.

nth Roots: Expressions of the Form $a^{1/n}$

The term $a^{1/n}$ denotes the nth root of a, often written as $\sqrt[n]{a}$. For $a \geq 0$, the nth root of a is the nonnegative number whose nth power is a.

$$8^{1/3} = \sqrt[3]{8} = 2 \quad \text{since } 2^3 = 8 \quad \text{(we call 2 the third or cube root of 8)}$$

$$16^{1/4} = \sqrt[4]{16} = 2 \quad \text{since } 2^4 = 16 \quad \text{(we call 2 the fourth root of 16)}$$

For $a < 0$, if n is odd, $\sqrt[n]{a}$ is the negative number whose n^{th} power is a. Note that if n is even, then $\sqrt[n]{a}$ is not a real number when $a < 0$.

$$(-8)^{1/3} = \sqrt[3]{-8} = -2 \quad \text{since } (-2)^3 = -8.$$

$$(-27)^{1/3} = \sqrt[3]{-27} = -3 \quad \text{since } (-3)^3 = -27.$$

$$(-16)^{1/4} = \sqrt[4]{-16} \text{ is not a real number.}$$

> If a is a real number and n is a positive integer
>
> $$a^{1/n} = \sqrt[n]{a}$$
>
> For $a \geq 0$,
>
> $\sqrt[n]{a}$ is the nonnegative number whose nth power is a.
>
> For $a < 0$,
>
> If n is odd, $\sqrt[n]{a}$ is the negative number whose nth power is a.
> If n is even, $\sqrt[n]{a}$ is not a real number.

Using a Calculator

Many calculators and spreadsheet programs have a square root function often labeled $\sqrt{\ }$ or perhaps "SQRT." You can also calculate square roots by raising a number to the $\frac{1}{2}$ or 0.5 power using the ^ key, as in 4 ^ .5. Try using a calculator to find $\sqrt{4}$ and $\sqrt{9}$. What happens on your calculator when you evaluate $\sqrt{-4}$?

In any but the simplest cases where the square root is immediately obvious, you will probably use the calculator. For example, use your calculator to find

$$8^{1/2} = \sqrt{8} \approx 2.8284$$

Double check the answer by verifying that $(2.8284)^2 \approx 8$.

If the nth root exists, you can find its value on a calculator. For example, to determine a fifth root, raise the number to the $\frac{1}{5}$ or the 0.2 power. So

$$3125^{1/5} = \sqrt[5]{3125} = 5$$

Double check your answer by verifying that $5^5 = 3125$.

Rules for Computations with Radicals

The following rules apply to computations with radicals, when $\sqrt[n]{a}$ and $\sqrt[n]{b}$ exist:

$$\sqrt[n]{ab} = \sqrt[n]{a} \cdot \sqrt[n]{b} \quad \text{and} \quad \sqrt[n]{a/b} = \sqrt[n]{a} \big/ \sqrt[n]{b} \; (b \neq 0).$$

Be aware that for most values of a, b, and n,

$$\sqrt[n]{a+b} \neq \sqrt[n]{a} + \sqrt[n]{b} \quad \text{and} \quad \sqrt[n]{a-b} \neq \sqrt[n]{a} - \sqrt[n]{b}$$

For example,

$$\sqrt{64} = \sqrt{16 \cdot 4} = \sqrt{16} \cdot \sqrt{4} = 4 \cdot 2 = 8$$

but $\quad \sqrt{25+16} = \sqrt{41} \approx 6.40 \quad$ while $\quad \sqrt{25} + \sqrt{16} = 5 + 4 = 9$

Example 1

Estimate $\sqrt{27}$.

Solution

We know $\quad 25 < 27 < 36$

therefore $\quad \sqrt{25} < \sqrt{27} < \sqrt{36}$

and $\quad 5 < \sqrt{27} < 6$

So $\sqrt{27}$ lies somewhere between 5 and 6. Would you expect $\sqrt{27}$ to be closer to 5 or to 6? Check your answer with a calculator.

Example 2

The function $S = \sqrt{30d}$ describes the relationship between S, the speed of a car in miles per hour, and d, the distance in feet a car skids after applying the brakes on a dry tar road. Estimate the speed of a car that leaves 40-foot long skid marks on a dry tar road and the speed that leaves 150-foot long skid marks.

Solution

If $d = 40$ feet, then $S = \sqrt{30 \cdot 40} = \sqrt{1200} \approx 35$, so the car was traveling at about 35 miles per hour.

If $d = 150$ feet, then $S = \sqrt{30 \cdot 150} = \sqrt{4500} \approx 67$, so the car was traveling at almost 70 miles per hour.

Example 3

If we solve for the radius, r, in the formula for the surface of a sphere, $S = 4\pi r^2$, we get

$$r = \sqrt{\frac{S}{4\pi}} = \sqrt{\frac{1}{4}} \cdot \sqrt{\frac{S}{\pi}} = \frac{1}{2}\sqrt{\frac{S}{\pi}}$$

If we assume that Earth has a spherical shape, and we know that its surface area is approximately 200,000,000 square miles, we can use this formula to estimate Earth's radius. Substituting for S, we get

$$\begin{aligned} r &= \frac{1}{2}\sqrt{\frac{200{,}000{,}000 \text{ miles}^2}{\pi}} \\ &\approx \frac{1}{2}\sqrt{63{,}661{,}977 \text{ miles}^2} \\ &\approx \frac{1}{2} \cdot 7979 \text{ miles} \approx 3989 \text{ miles} \end{aligned}$$

So Earth has a radius of about 4000 miles.

Example 4

Simplify:

a. $625^{1/4}$ **b.** $(-625)^{1/4}$ **c.** $125^{1/3}$ **d.** $(-125)^{1/3}$

Solution

a. $625^{1/4} = 5$ since $5^4 = 625$ **b.** $(-625)^{1/4}$ does not have a real number solution

c. $125^{1/3} = 5$ since $5^3 = 125$ **d.** $(-125)^{1/3} = -5$ since $(-5)^3 = -125$

Example 5

The volume of a sphere is given by the equation $V = \frac{4}{3}\pi r^3$. Rewrite the formula, solving for the radius as a function of the volume.

Given	$V = \frac{4}{3}\pi r^3$
multiply both sides by 3	$3V = 4\pi r^3$
divide by 4π	$\frac{3V}{4\pi} = r^3$
take the cube root and switch sides	$r = \sqrt[3]{\frac{3V}{4\pi}}$

Algebra Aerobics 4.5a

1. Evaluate each of the following without a calculator:
   a. $81^{1/2}$ b. $144^{1/2}$
2. Simplify and rewrite without radical signs:
   a. $\sqrt{9x}$ b. $\sqrt{\frac{x^2}{25}}$ $(x \geq 0)$
3. Use the formula in Example 2 to estimate the following:
   a. the speed of a car that leaves 60-foot long skid marks on a dry tar road
   b. the speed that leaves 200-foot long skid marks
4. Without a calculator, find two consecutive integers between which the given number lies:
   a. $\sqrt{29}$ b. $\sqrt{92}$
5. Evaluate each of the following without a calculator:
   a. $27^{1/3}$ b. $16^{1/4}$ c. $8^{-1/3}$
6. Use a calculator or spreadsheet to estimate the following:
   a. $\sqrt[4]{1295}$ b. $\sqrt[3]{372{,}783}$
7. Evaluate:
   a. $\sqrt[3]{-27}$ b. $(-10{,}000)^{1/4}$ c. $(-1000)^{1/3}$
8. Estimate the radius of a spherical balloon with a volume of 2 cubic feet.

Fractional Powers: Expressions of the Form $a^{m/n}$

EXPLORE

You may want to do Exploration 4.2 on Kepler's laws of planetary motion after reading this section.

What about $a^{m/n}$? We can write $a^{m/n}$ either as $(a^m)^{1/n}$ or $(a^{1/n})^m$. Writing it as $(a^m)^{1/n}$ means that we would first raise the base, a, to the mth power and then take the nth root of that. Writing it as $(a^{1/n})^m$ implies first finding the nth root of a and then raising that to the mth power. For example,

$$2^{3/2} = (2^3)^{1/2}$$
$$= (8)^{1/2}$$

using a calculator ≈ 2.8284

Equivalently,

$$2^{3/2} = (2^{1/2})^3$$
$$\approx (1.414)^3$$
$$\approx 2.8284$$

We could, of course, use a calculator to compute $2^{3/2}$ (or $2^{1.5}$) directly by raising 2 to the $\frac{3}{2}$ or 1.5 power.

If $a \geq 0$ and m and n are integers ($n \neq 0$), then,

$$a^{m/n} = \left(\sqrt[n]{a}\right)^m$$

or equivalently $= \sqrt[n]{a^m}$

Exponents expressed as ratios, of the form m/n, are called *rational exponents.* The set of laws for simplifying exponential expressions also holds for rational exponents.

We have dealt with integer exponents and rational (fractional) exponents in expressions such as $8^{2/3}$. Terms with irrational exponents, such as 2^{π}, are well defined but are beyond the scope of this course. You may recall that irrational real numbers are those that cannot be expressed as a ratio of integers. However, the same *rules* for exponents apply.

Example 6

Find the product of $(\sqrt{5}) \cdot (\sqrt[3]{5})$ leaving the answer in exponent form.

Solution

$(\sqrt{5}) \cdot (\sqrt[3]{5}) = 5^{1/2} \cdot 5^{1/3} = 5^{(1/2)+(1/3)} = 5^{5/6}$.

Example 7

According to McMahon and Bonner in *On Size and Life*,[4] common nails range from 1 to 6 inches in length. The weight varies even more, from 11 to 647 nails per pound. Longer nails are relatively thinner than shorter ones. A good approximation of the relationship between length and diameter is given by the equation

$$d = 0.07L^{2/3}$$

where d = diameter and L = length, both in inches. Estimate the diameters of nails that are 1, 3, and 6 inches long.

Solution

When $L = 1$ inch, $d = 0.07 \cdot (1)^{2/3} = 0.07 \cdot 1 = 0.07$ inches.
When $L = 3$ inches, then the diameter equals $0.07 \cdot (3)^{2/3} \approx 0.07 \cdot 2.08 \approx 0.15$ inches.
When $L = 6$ inches, then $d = 0.07 \cdot (6)^{2/3} \approx 0.07 \cdot 3.30 \approx 0.23$ inches.

Note that L and d are both in inches. Since $d = 0.07L^{2/3}$, then $0.07L^{2/3}$ must also be in inches. The units for $L^{2/3}$ are $(\text{in.})^{2/3}$! So the coefficient, 0.07, must have units of $(\text{in.})^{1/3}$, since $(\text{in.})^{1/3} \cdot (\text{in.})^{2/3} = (\text{in.})^1 = \text{in.}$ The coefficients of variables with fractional powers are often in strange units that are hard to figure out. Fortunately, knowing the actual units in such cases is less interesting and less important than knowing the power.

Algebra Aerobics 4.5b

1. Find the product expressed in exponent form:
   a. $\sqrt{2}\sqrt[3]{2}$ **b.** $\sqrt{5}\sqrt[4]{5}$ **c.** $\sqrt{3}\sqrt[3]{9}$
2. Find the quotient by first expressing in exponent form. Leave the answer in positive exponent form.
   a. $\dfrac{\sqrt{2}}{\sqrt[3]{2}}$ **b.** $\dfrac{2}{\sqrt[4]{2}}$ **c.** $\dfrac{\sqrt[4]{5}}{\sqrt[3]{5}}$
3. McMahon and Bonner give the relationship between chest circumference and body weight of adult primates as

$$c = 17.1m^{3/8}$$

where m = weight in kilograms and c = chest circumference in centimeters. Estimate the chest circumference of a:

a. 0.25-kilogram tamarin **b.** 25-kilogram baboon

4.6 ORDERS OF MAGNITUDE

Comparing Numbers of Widely Differing Sizes

We have seen that a useful method of comparing two objects of widely different sizes is to calculate the ratio rather than the difference of the sizes. The ratio can be estimated by computing *orders of magnitude,* the number of times we would have to multiply or divide by 10 to convert one size into the other. Each factor of 10 represents one order of magnitude.

For example, the radius of the observable universe is approximately 10^{26} meters and the radius of our solar system is approximately 10^{12} meters. To compare

[4] T. A. McMahon and J. Tyler Bonner, *On Size and Life* (New York: Scientific American Library, Scientific American Books, 1983).

the radius of the observable universe to the radius of our solar system, calculate the ratio

$$\frac{\text{radius of the universe}}{\text{radius of our solar system}} \approx \frac{10^{26}\text{ meters}}{10^{12}\text{ meters}}$$
$$\approx 10^{26-12}$$
$$\approx 10^{14}$$

The radius of the universe is roughly 10^{14} times larger than the radius of the solar system; that is, we would have to multiply the radius of our solar system by 10 fourteen times in order to obtain the radius of the universe. Since each factor of 10 is counted as a single order of magnitude, the radius of the universe is *14 orders of magnitude larger* than the radius of our solar system. Equivalently, we could say that the radius of our solar system is *14 orders of magnitude smaller* than the radius of the universe.

When something is one order of magnitude larger than a *reference object,* it is 10 times larger. You *multiply* the *reference size* by 10 to get the other size. If the object is two orders of magnitude larger, it is 100, or 10^2, times larger, so you would multiply the reference size by 100. If it is one order of magnitude smaller, it is 10 times smaller, so you would *divide* the reference size by 10. Two orders of magnitude smaller means the reference size is divided by 100, or 10^2.

The following examples demonstrate order of magnitude comparisons.

- We previously found that the radius of the sun is 100, or 10^2, times larger than the radius of Earth. This is equivalent to saying that the radius of the sun is 2 orders of magnitude larger than the radius of Earth. We also found that the volume of the sun is 10^6 times, or 6 orders of magnitude, larger than the volume of Earth.
- The radius of the sun, at 10^9 meters, is 20 orders of magnitude larger than the radius of a hydrogen atom, at 10^{-11} meter, since

$$\frac{\text{radius of sun}}{\text{radius of the hydrogen atom}} \approx \frac{10^{9}\text{ meters}}{10^{-11}\text{ meter}}$$
$$\approx 10^{9-(-11)}$$
$$\approx 10^{20}$$

  So the radius of the sun is 10^{20} times larger than the radius of the hydrogen atom.
- Surprisingly enough, the average width of a living cell is approximately three orders of magnitude *smaller* than one of the single strands of DNA it contains, if the DNA is uncoiled and measured lengthwise:

$$\frac{\text{length of DNA strand}}{\text{radius of the living cell}} \approx \frac{10^{-2}\text{ meter}}{10^{-5}\text{ meter}}$$
$$\approx 10^{-2-(-5)}$$
$$\approx 10^{-2+5}$$
$$\approx 10^{3}$$

The reading "Earthquake Magnitude Determination" describes how earthquake tremors are measured.

A Measurement Scale Based on Orders of Magnitude: The Richter Scale

The *Richter scale,* designed by the American Charles Richter in 1935, allows us to compare the magnitudes of earthquakes throughout the world. The Richter scale measures the maximum ground movement (tremors) as recorded on an instrument

Table 4.3
Description of the Richter Scale

Richter Scale Magnitude	Description
2.5	Generally not felt, but recorded on seismographs
3.5	Felt by many people
4.5	Felt by all; some slight local damage may occur
6	Considerable damage in ordinary buildings; a destructive earthquake
7	"Major" earthquake; most masonry and frame structures destroyed; ground badly cracked
8 and above	"Great" earthquake; a few per decade worldwide; almost total or total destruction; bridges collapse, major openings in ground, and tremors visible

called a seismograph. The sizes of earthquakes vary widely, so Richter designed the scale to measure order of magnitude differences. The scale ranges from less than 1 to over 8. Each increase of 1 unit on the Richter scale represents an increase of 10 times, or 1 order of magnitude, in the maximum tremor size of the earthquake. So an increase from 2.5 to 3.5 indicates a 10-fold increase in maximum tremor size. An increase of 2 units from 2.5 to 4.5 indicates an increase in maximum tremor size by a factor of 10^2, or 100.

Table 4.3 contains some typical values on the Richter scale along with a description of how humans near the center (called the *epicenter*) of an earthquake perceive its effects. There is no theoretical upper limit on the Richter scale, but the biggest earthquakes measured so far registered as 8.6 on the Richter scale (in Japan and Chile).[5]

Algebra Aerobics 4.6a

1. In 1987 Los Angeles had an earthquake that measured 5.9 on the Richter scale. In 1988 Armenia had an earthquake that measured 6.9 on the Richter scale. Compare the sizes of the two earthquakes using orders of magnitude.
2. In 1983 Hawaii had an earthquake that measured 6.6 on the Richter scale. Compare the size of this earthquake to the largest ever recorded, 8.6 on the Richter scale.
3. If my salary is $100,000 per year and you make an order of magnitude more, what is your salary? If Henry makes two orders of magnitude less money than I do, what is his salary?
4. For each of the following pairs, determine the order of magnitude difference:
   a. The radius of the sun (10^9 meters) and the radius of the Milky Way (10^{21} meters)
   b. The radius of a hydrogen atom (10^{-11} meter) and the radius of a proton (10^{-15} meter)

[5]The following table gives values on the Richter scale and the approximate number per year (worldwide):

Richter Scale Magnitude	Average Number per Year (Worldwide)
< 4	More than 1,000,000
4 to 5.9	A few thousand
6 to 6.9	120
7 to 7.9	18
≥ 8	1

Source: National Earthquake Information Center at *www.neic.cr.usgs.gov/neis/eqlists*

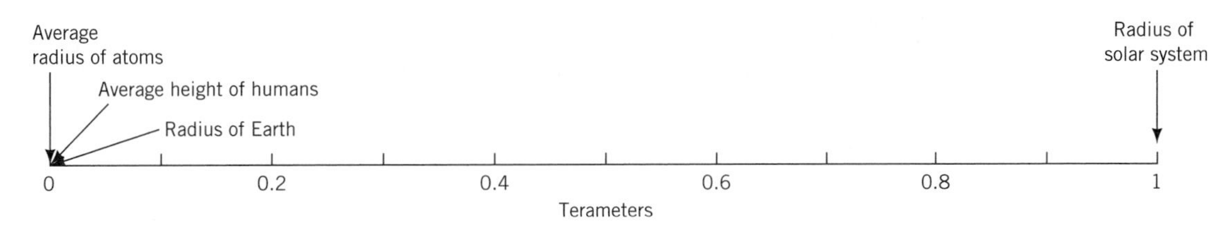

Figure 4.1 Sizes of various objects in the universe on a linear scale. (*Note:* One terameter = 10^{12} meters.)

Graphing Numbers of Widely Differing Sizes

Exploration 4.1 asks you to construct a graph using logarithmic scales on both axes.

If the sizes of various objects in our solar system are plotted on a standard linear axis, we get the uninformative picture shown in Figure 4.1. The largest value stands alone, and all the others are so small when measured in terameters that they all appear to be zero. When objects of widely different orders of magnitude are compared on a linear scale, the effect is similar to pointing out an ant in a picture of a baseball stadium.

A more effective way of plotting sizes with different orders of magnitude is to use an axis that has orders of magnitude (powers of 10) evenly spaced along it. This is called a *logarithmic* or *log scale.* The plot of the previous data graphed on a logarithmic scale is much more informative (Figure 4.2).

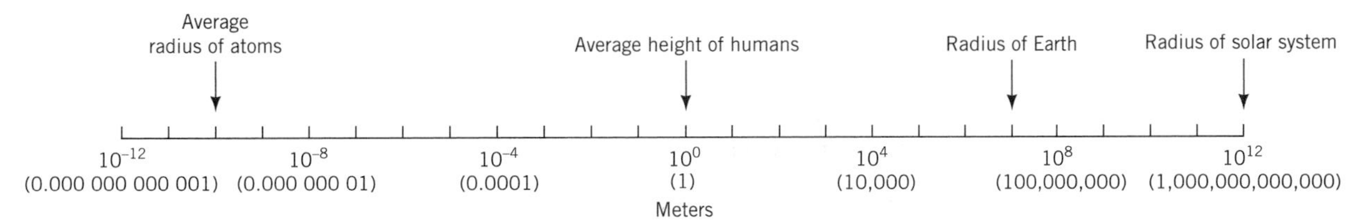

Figure 4.2 Sizes of various objects in the universe on an order-of-magnitude (or logarithmic) scale.

Graphing sizes on a log scale is extremely useful, but we have to read the scales very carefully. When we use a linear scale, each move of one unit to the right is equivalent to *adding* 1 unit to the number, and each move of k units to the right is equivalent to *adding* k units to the number (Figure 4.3).

When we use a log scale (see Figure 4.4), we need to remember that now one unit of length represents a change of one order of magnitude. Moving one unit to the right is equivalent to *multiplying by 10.* So moving from 10^4 to 10^5 is equivalent

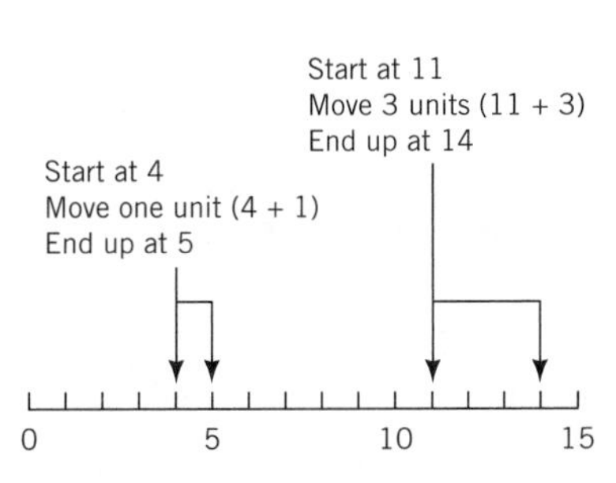

Figure 4.3 Linear scale.

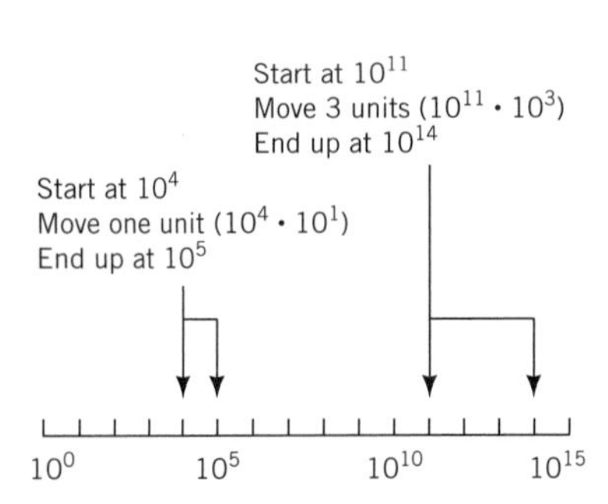

Figure 4.4 Order of magnitude (or logarithmic) scale.

to multiplying 10^4 by 10. Moving three units to the right is equivalent to *multiplying* the starting number by 10^3, or 1000. In effect, a linear scale is an "additive" scale and a logarithmic scale is a "multiplicative" scale.

Algebra Aerobics 4.6b

1. By rounding the number to the nearest power of 10, find the approximate location of each of the following on the logarithmic scale in Figure 4.2.
   a. The radius of the sun at approximately 1 billion meters.
   b. The radius of a virus at 0.000 000 7 meter.
2. a. Plot on Figure 4.2 an object whose radius is two orders of magnitude smaller than that of Earth.
   b. What object is plotted on Figure 4.2 that is six orders of magnitude larger than that of a virus?

4.7 LOGARITHMS BASE 10

Logarithms were originally used to help carry out calculations for astronomical problems. The logarithm and the ideas that underlie it have had many other important scientific and practical applications. We use logarithmic scales to display physical quantities such as sound intensities and earthquake strengths.

Finding the Logarithms of Powers of 10

When handling very large or very small numbers, we have seen that it is often easier to write the number using powers of 10. For example,

$$100{,}000 = 10^5$$

We refer to 10 as the base and 5 as the power or exponent. We say that

100,000 equals the base 10 to the 5th power

But we could also rephrase this as

5 is the power of the base 10 that is needed to produce 100,000

The more technical way to say this is

5 is the *logarithm* base 10 of 100,000

In symbols we write

$$5 = \log_{10} 100{,}000$$

So the expressions

$$100{,}000 = 10^5 \quad \text{and} \quad 5 = \log_{10} 100{,}000$$

are two ways of saying the same thing.

The key point to remember is that a logarithm is an exponent.

> The *logarithm base 10 of x* is the power to which we raise 10 in order to produce x:
>
> $$\log_{10} x = c \quad \text{means that} \quad 10^c = x$$

Example 1

Since $1{,}000{,}000{,}000 = 10^9$

then $\log_{10} 1{,}000{,}000{,}000 = 9$

and we say that the logarithm base 10 of 1,000,000,000 is 9. The logarithm of a number tells us the exponent, in this case of the base 10. Since here the logarithm is 9, that means that when we write 1,000,000,000 as a power of 10, the exponent is 9.

Example 2

Since $1 = 10^0$

then $\log_{10} 1 = 0$

and we say that the logarithm base 10 of 1 is 0. Since logarithms represent exponents, this says that when we write 1 as a power of 10, the exponent is 0.

Example 3

How do we calculate the logarithm base 10 of decimals such as 0.000 01?

Solution

Since $0.000\,01 = 10^{-5}$

then $\log_{10} 0.000\,01 = -5$

and we say that the logarithm base 10 of 0.000 01 is -5.

In the previous example, we found that the log (short for "logarithm") of a number can be negative. This makes sense if we think of logarithms as exponents, since exponents can be any real number. But we cannot take the log of a negative number or zero: that is, $\log_{10} x$ is not defined when $x \leq 0$. Why? If $\log_{10} x = c$, where $x \leq 0$, then $10^c = x$ (a number ≤ 0). But 10 to any power will never produce a number that is negative or zero, so $\log_{10} x$ is not defined if $x \leq 0$.

> $\log_{10} x$ is not defined when $x \leq 0$.

Table 4.4
Logarithms of Powers of 10

x	Exponential Notation	$\text{Log}_{10}\, x$
0.0001	10^{-4}	-4
0.001	10^{-3}	-3
0.01	10^{-2}	-2
0.1	10^{-1}	-1
1	10^{0}	0
10	10^{1}	1
100	10^{2}	2
1,000	10^{3}	3
10,000	10^{4}	4

Table 4.4 gives a sample set of values for x and their associated logarithms base 10. For each value of x, in order to find the logarithm base 10, we write the value as a power of 10, and the logarithm is just the exponent.

Most scientific calculators and spreadsheet programs have a LOG function that calculates common logarithms. Try using technology to double check some of the numbers in Table 4.4.

Logarithms base 10 are used frequently in our base 10 number system and are called *common logarithms.* We write $\log_{10} x$ as $\log x$.

> Logarithms base 10 are called *common logarithms.*
> $\log_{10} x$ is written as $\log x$.

Algebra Aerobics 4.7a

1. Without using a calculator, find the logarithm base 10 of:

   a. 10,000,000 **c.** 10,000 **e.** 1,000 **g.** 1

   b. 0.000 000 1 **d.** 0.0001 **f.** 0.001

   Now use a calculator to double check your answers.

2. Rewrite the following expressions in an equivalent form using powers of 10:

   a. $\log 100{,}000 = 5$ **c.** $\log 10 = 1$

   b. $\log 0.000\,000\,01 = -8$ **d.** $\log 0.01 = -2$

Finding the Logarithms of Numbers between 1 and 10

When a number such as 100 or 0.001 can be written as an integer power of 10, it's easy to find its logarithm base 10. Express the number as a power of 10, and the log is just the exponent. But what about the logarithms of other numbers, such as 2?

Example 4

Estimate log 2.

Solution

$$\log 2 = c \quad \text{is true if and only if} \quad 10^c = 2$$

So we need to find a number c such that 10 raised to that power will give us 2. Since

$$1 < 2 < 10$$

and

$$1 = 10^0 \qquad 2 = 10^c \qquad 10 = 10^1$$

then

$$10^0 < 10^c < 10^1$$

So we might suspect that

$$0 < c < 1$$

Using a calculator, we can estimate values of 10^c where c is between 0 and 1. Remember that $10^{0.1}$ (or $10^{1/10}$) means the 10th root of 10:

$$10^{0.1} \approx 1.258\,925 \qquad 10^{0.2} \approx 1.584\,893 \qquad 10^{0.3} \approx 1.995\,262 \qquad 10^{0.4} \approx 2.511\,886$$

So $10^{0.3}$ is very close to 2, whereas $10^{0.4}$ is larger than 2. Hence a good estimate is

$$\log 2 \approx 0.3$$

By trial and error we could come even closer to the actual value:

$$10^{0.31} \approx 2.041\,738 \qquad 10^{0.301} \approx 1.999\,862 \qquad 10^{0.3015} \approx 2.002\,166$$

Since 1.999 862 is closer to 2 than 2.002 166, we can improve our original estimate to

$$\log 2 \approx 0.301$$

One calculator gave

$$\log 2 \approx 0.301\,029\,996$$

so our estimates were pretty close.

Finding the Logarithm of Any Positive Number

How can we find the logarithms of positive numbers that are not between 1 and 10? Our knowledge of scientific notation can help. Recall that a positive number, c, written in scientific notation is of the form

$$c = N \cdot 10^n$$

where $1 \leq N < 10$ and n is an integer. We can put together what we know about finding $\log N$ and $\log 10^n$ to find $\log c$.

Example 5

What is log 2000?

Solution

Write 2000 in scientific notation	$2000 = 2 \cdot 10^3$
substitute $10^{0.301}$ for 2 (see Example 4 in this section)	$\approx 10^{0.301} \cdot 10^3$
combine powers	$\approx 10^{0.301+3}$
to get	$2000 \approx 10^{3.301}$
We can rewrite as a logarithm	$\log 2000 \approx 3.301$

A calculator gives $\log 2000 = 3.301\,029\,996$.

Example 6

Find log 0.000 02.

Solution

Write 0.000 02 in scientific notation	$0.000\,02 = 2 \cdot 10^{-5}$
substitute $10^{0.301}$ for 2 (see Example 4)	$\approx 10^{0.301} \cdot 10^{-5}$
combine powers	$\approx 10^{0.301-5}$
to get	$0.00002 \approx 10^{-4.699}$
We can rewrite as a logarithm	$\log 0.000\,02 \approx -4.699$

A calculator gives $-4.698\,970\,004$.

The logarithm of any positive number written in scientific notation equals the log of the coefficient plus the exponent of the power of 10. That is, if c written in scientific notation is

$$c = N \cdot 10^n$$

(where $c > 0$ and hence $N > 0$), then

$$\log c = (\log N) + n$$

Once you understand what logarithms represent, you can use a calculator to do the actual computation.

Using logarithms to write any positive number as a power of 10. We will see later that in many cases it is useful to write a number as a power of 10.

Example 7

Write 6,370,000 (the radius of Earth in meters) as a power of 10.

Solution

By definition log 6,370,000 is the number c such that $10^c = 6{,}370{,}000$. A calculator gives

$$\log_{10} 6{,}370{,}000 \approx 6.804$$

From the definition of logarithms we know that

$$6{,}370{,}000 \approx 10^{6.804}$$

Algebra Aerobics 4.7b

1. Estimate each of the following, and then check with a calculator:

   a. log 3 **b.** log 6 **c.** log 6.37

2. Use the answers from Problem 1 to estimate values for:

   a. log 3,000,000 **b.** log 0.006

   Then use a calculator to check your answers.

3. Write each of the following as a power of 10:

   a. 0.000 000 7 m (the radius of a virus)

   b. 780,000,000 km (the mean distance from our sun to Jupiter)

   c. 0.0042

   d. 5,400,000,000

Chapter Summary

The Numbers of Science. In describing the sizes of objects in the universe, the scientific community and most countries other than the United States use the metric system, a measurement system based on the meter and powers of ten. The following table shows the conversion into English units for three of the most common metric length measurements.

Metric Unit	Abbreviation	In Meters	Equivalence in English Units	Informal Conversion
meter	m	1 m	3.28 ft	The width of a twin bed: a little more than a yard
kilometer	km	1000 m	0.62 mile	A casual 12-min walk: a little over a half a mile
centimeter	cm	0.01 m	0.39 in	The length of an ant: a little under a half an inch

Powers of 10 are useful in describing the scale of objects. When n is a positive integer, we define

$$10^n = 10 \cdot 10 \cdot 10 \cdots 10$$ the product of n tens or 1 followed by n zeros

$$10^{-n} = \frac{1}{10^n}$$ 1 divided by 10 n times or a decimal point followed by $n - 1$ zeros and a 1

$$10^0 = 1$$

To write very small or large numbers compactly, we use scientific notation. A number is in *scientific notation* if it is in the form

$$N \cdot 10^n$$

where $1 \leq |N| < 10$ and n is an integer (positive, negative, or zero).

The expression $|N|$ denotes "the absolute value of N." No matter what the value of N, the absolute value of N will always be greater than or equal to 0. If $N \geq 0$, then $|N| = N$. If $N < 0$, then $|N| = -N$.

The Laws of Exponents. In the expression a^n, where a is any real number, the number a is called the *base* and n is called the *exponent* or *power*.
If a is any nonzero real number and n is a positive integer, then

$$a^n = \underbrace{a \cdot a \cdot a \cdots a}_{n \text{ factors}} \qquad \text{(the product of } n \text{ factors of } a\text{)}$$

$$a^0 = 1 \qquad \text{(for } a \neq 0\text{)}$$

$$a^{-n} = \frac{1}{a^n} \qquad \text{(for } a \neq 0\text{)}$$

If a and b are nonzero real numbers, and m and n are real numbers, then

$$a^m \cdot a^n = a^{(m+n)}$$

$$\frac{a^n}{a^m} = a^{(n-m)}$$

$$(a^m)^n = a^{(m \cdot n)}$$

$$(ab)^n = a^n b^n$$

$$\left(\frac{a}{b}\right)^n = \frac{a^n}{b^n}$$

Fractional Powers. If a is any nonnegative real number,

$a^{1/2} = \sqrt{a}$ where $\sqrt{a}$ is that nonnegative number whose square is a.

For example, $\sqrt{16} = 4$ since $4^2 = 16$.

If a is any real number and n is a positive integer, then

$$a^{1/n} = \sqrt[n]{a}$$

For $a \geq 0$,

$\sqrt[n]{a}$ is the nonnegative number whose n^{th} power is a.

For $a < 0$,

If n is odd, $\sqrt[n]{a}$ is the negative number whose n^{th} power is a.

If n is even, $\sqrt[n]{a}$ is not a real number.

For example, $8^{1/3} = 2$ since $2^3 = 8$, and $(-8)^{1/3} = -2$ since $(-2)^3 = -8$.

If a is any nonnegative real number and m and n are integers ($n \neq 0$), then

$$a^{m/n} = (a^m)^{1/n} = (a^{1/n})^m$$
$$= \sqrt[n]{a^m} = \left(\sqrt[n]{a}\right)^m$$

For example, $8^{2/3} = (8^{1/3})^2 = 2^2 = 4$ or equivalently $8^{2/3} = (8^2)^{1/3} = (64)^{1/3} = 4$.

We can also write $8^{2/3}$ as $\left(\sqrt[3]{8}\right)^2 = \sqrt[3]{8^2} = 4$.

Orders of Magnitude. When comparing two objects of widely different sizes, it is common to use *orders of magnitude.*

For example, to compare the radius of the universe to the radius of the solar system, calculate the ratio of the two sizes:

$$\frac{\text{radius of the universe}}{\text{radius of the solar system}} \approx \frac{10^{26}\ \cancel{m}}{10^{12}\ \cancel{m}}$$
$$\approx 10^{26-12}$$
$$\approx 10^{14}$$

The radius of the universe is 10^{14} times larger than the radius of the solar system, so it is 14 orders of magnitude larger than the solar system.

Each factor of 10 is counted as a single order of magnitude. When something is one order of magnitude larger than a reference object, it is 10 times larger.

If it is one order of magnitude smaller, it is 10 times smaller, or one-tenth the size.

Orders of magnitude or logarithmic scales are used to graph objects of widely differing sizes.

Logarithms. The *logarithm base 10 of x* is the power to which we raise 10 in order to produce x. So

$$\log_{10} x = c \quad \text{means that} \quad 10^c = x$$

We say c is the logarithm base 10 of x. For example, $\log_{10} 6{,}370{,}000 \approx 6.804$ means that $10^{6.804} \approx 6{,}370{,}000$.

Logarithms base 10 are called *common logarithms* and $\log_{10} x$ is written as $\log x$.

When $x \leq 0$, $\log x$ is not defined.

EXERCISES

Exercises for Section 4.1

1. Write each expression as a power of 10:

a. $10 \cdot 10 \cdot 10 \cdot 10 \cdot 10 \cdot 10$

b. $\frac{1}{10 \cdot 10 \cdot 10 \cdot 10 \cdot 10}$

c. one billion

d. one-thousandth

2. Express in standard decimal notation (without exponents):

a. 10^{-7} **b.** 10^{7}

3. Express each in meters, using powers of 10:

a. 10 cm **b.** 4 km **c.** 3 terameters **d.** 6 nanometers

4. Express as a power of 10:

a. 10,000,000,000,000

b. 0.000 000 000 001

c. $10 \cdot 10 \cdot 10 \cdot 10$

d. $\frac{1}{10 \cdot 10 \cdot 10 \cdot 10}$

e. one million

f. one-millionth

5. Write each of the following in scientific notation:

a. 0.000 29 **b.** 654.456 **c.** 720,000 **d.** 0.000 000 000 01

6. Why are the following expressions *not* in scientific notation? Rewrite each in scientific notation.

a. $25 \cdot 10^{4}$ **b.** $0.56 \cdot 10^{-3}$

7. Write each of the following in standard decimal form:

a. $7.23 \cdot 10^{5}$ **b.** $5.26 \cdot 10^{-4}$ **c.** $1.0 \cdot 10^{-3}$ **d.** $1.5 \cdot 10^{6}$

8. Express in scientific notation. (Refer to the chart in Exploration 4.1.)

a. The age of the observable universe

b. The size of the first living organism on Earth

c. The size of Earth

d. The age of Pangaea

e. The size of the first cells with a nucleus

9. Evaluate:

a. $|9|$ **b.** $|-9|$ **c.** $|-1000|$ **d.** $-|-1000|$

10. Hubble's Law states that galaxies are receding from one another at velocities directly proportional to the distances separating them. The accompanying graph illustrates that Hubble's Law holds true across the known universe. The plot includes 10 major clusters of galaxies. The boxed area at the lower left represents the galaxies observed by Hubble when he discovered the law. The easiest way to understand this graph is to think of Earth as being at the center of the universe (at 0 distance) and not moving (at 0 velocity). In other words, imagine Earth at the origin of the graph (a favorite fantasy of humans). Think of the horizontal axis as measuring the distance of a galaxy cluster from Earth and the vertical axis as measuring the velocity at which a galaxy cluster is moving away from Earth (the recession velocity). Then answer the following questions:

a. Identify the coordinates of two data points that lie on the regression line drawn on the graph. (*Hint:* For the horizontal coordinate it is easier to use such numbers as 1 or 2 with the units being billion of light years and for the vertical coordinate to use such numbers as 10 or 20 with units in thousands of kilometers per second.)

b. Use the coordinates of the points in part (a) to calculate the slope of the line. That slope is called the Hubble constant.

c. What does the slope mean in terms of distance from Earth and recession velocity?

d. Construct an equation for your line in the form $y = mx + b$. Show your work.

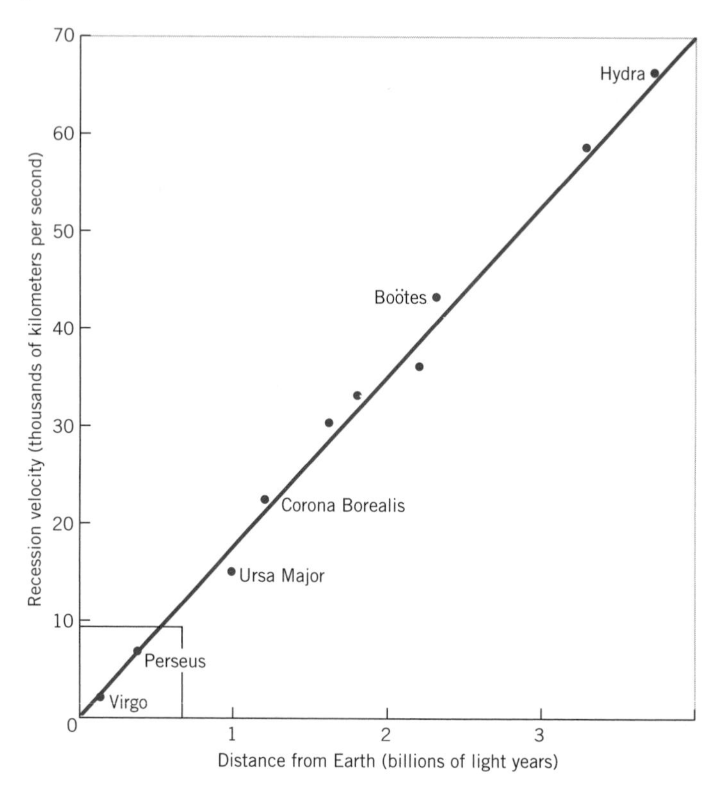

Source: T. Ferris, *Coming of Age in the Milky Way* (New York: William Morrow, 1988). Copyright © by Timothy Ferris. By permission of William Morrow & Company, Inc.

11. a. Generate a small table of values and plot the function $y = |x|$.

b. On the same graph, plot the function $y = |x - 2|$.

12. The accompanying amusing graph shows a roughly linear relationship between the "scientifically" calculated age of Earth and the year the calculation was published. For instance, in about 1935 Ellsworth calculated that Earth was about 2 billion years old. The age is plotted on the horizontal axis and the year the calculation was published on the vertical axis. The triangle on the horizontal coordinate represents the presently accepted age of Earth.

a. Who calculated that Earth was less than 1 billion years old? Give the coordinates of the points that give this information.

b. In about what year did scientists start putting the age of Earth at over a billion years? Give the coordinates that represent this point.

c. On your graph sketch an approximation of a best fit line for these points. Use two points on the line to calculate the slope of the line.

d. Interpret the slope of that line in terms of the year of calculation and the estimated age of Earth.

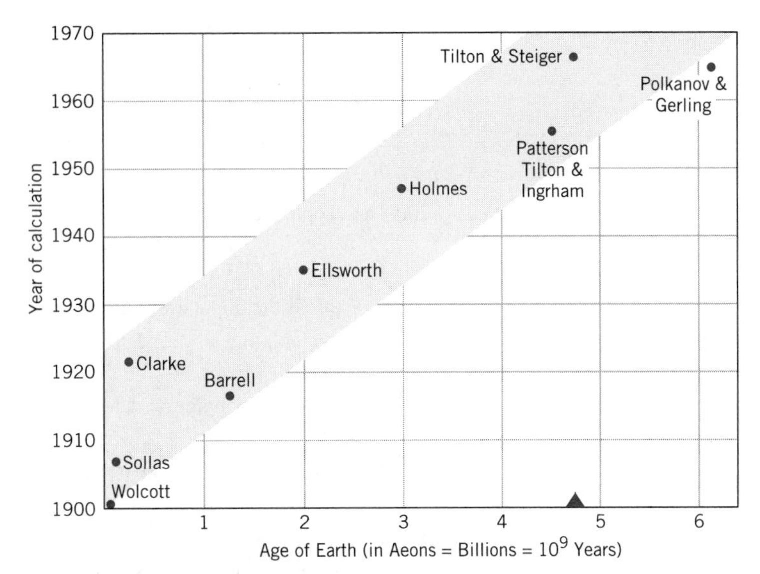

A graph of calculations for the age of Earth.
Source: American Scientist, Research Triangle Park, NC. Copyright © 1980.

Exercises for Section 4.2

13. Simplify, when possible, writing your answer as an expression with exponents:

a. $10^4 \cdot 10^3$ **d.** $x^5 \cdot x^{10}$ **g.** $\dfrac{z^7}{z^2}$ **j.** $4^5 \cdot (4^2)^3$

b. $10^4 + 10^3$ **e.** $(x^5)^{10}$ **h.** 256^0

c. $10^3 + 10^3$ **f.** $4^7 + 5^2$ **i.** $\dfrac{3^5 \cdot 3^2}{3^8}$

14. Simplify:

a. $(-1)^4$ **c.** $(a^4)^3$ **e.** $(2a^4)^3$ **g.** $(10a^2b^3)^3$

b. $-(1)^4$ **d.** $-(2a^2)^3$ **f.** $(-2a^4)^3$ **h.** $(2ab)^2 - 3a^2b^2$

15. Simplify:

a. $(-2a)^4$ **c.** $(-x^5)^3$ **e.** $(2x^4)^5$ **g.** $(50a^{10})^2$

b. $-2(a)^4$ **d.** $(-2ab^2)^3$ **f.** $(-4x^3)^2 + x^3(2x^3)$ **h.** $(3ab)^3 + ab$

16. Simplify and write your answer as an expression with positive exponents:

a. $-\left(\dfrac{3}{5}\right)^2$ **b.** $\left(\dfrac{-5a^3}{a^2}\right)^4$ **c.** $\left(\dfrac{10a^3}{5b}\right)^2$ **d.** $\left(\dfrac{-2x^3}{3y^2}\right)^3$

17. Simplify and write your answer as an expression with positive exponents:

a. $-\left(\dfrac{5}{8}\right)^2$ **b.** $\left(\dfrac{3x^3}{5y^2}\right)^3$ **c.** $\left(\dfrac{-10x^5}{2b^2}\right)^4$ **d.** $\left(\dfrac{-x^5}{x^2}\right)^3$

18. Evaluate and express your answer in standard decimal form:

a. $-2^4 + 2^2$ **d.** $-10^4 + 10^5$ **g.** $2 \cdot 10^3 + (-10)^3$

b. $-2^3 + (-4)^2$ **e.** $10^3 + 2^3$ **h.** $(1000)^0$

c. $2 \cdot 3^2 - 3(-2)^2$ **f.** $2 \cdot 10^3 + 10^3 + 10^2$

19. Express your answer as a power of 10 and in standard decimal form. (Refer to Table 4.2.)

a. How many times larger is a gigabyte of memory than a megabyte?

b. How many times farther is a kilometer than a dekameter?

c. How many times heavier is a kilogram than a milligram?

d. How many times longer is a microsecond than a nanosecond?

20. The People's Republic of China was estimated in 1999 to have about 1,247,000,000 people, and Monaco about 32,150. Monaco has an area of 0.75 miles^2, and China has an area of 3,704,000 miles^2.

a. Express the populations and geographic areas in scientific notation.

b. By calculating a ratio, determine how much larger China's population is than Monaco's.

c. What is the population density (the number of people per square mile) for each country?

d. Write a paragraph comparing and contrasting the population size and density for these two nations.

21. a. In 1999 Japan had a population of approximately 126.1 million people and a total land area of about 152.5 thousand square miles. What was the population density (the number of people per square mile)?

b. In 1999 the United States had a population of approximately 272.6 million people and a total land area of about 3620 thousand square miles. What was the population density of the United States?

c. Compare the population densities of Japan and the United States.

22. The distance that light travels in 1 year (a light year) is $5.88 \cdot 10^{12}$ miles. If a star is $2.4 \cdot 10^{8}$ light years from Earth, what is this distance in miles?

23. a. For any non-zero real number a, what can we say about the sign of the expression $(-a)^n$ when n is an even integer? What can we say about the sign of $(-a)^n$ when n is an odd integer?

b. What is the sign of the resulting number if a is a positive number? If a is a negative number? Explain your answer.

24. Round off the numbers and then estimate the values of the following expressions without a calculator. Show your work, writing your answers in scientific notation. Use a calculator to verify your answers.

a. $(2{,}968{,}001{,}000) \cdot (189{,}000)$

b. $(0.000\,079) \cdot (31{,}140{,}284{,}788)$

c. $\dfrac{4{,}083{,}693 \cdot 49{,}312}{213 \cdot 1945}$

25. Justify the following rule for exponents. If a and b are any nonzero real numbers, then

$$(ab)^n = a^n b^n$$

26. Justify the following rule for exponents. First consider the case of $n > m$ and then the case of $n < m$. If a and b are any nonzero real numbers, then

$$\frac{a^n}{a^m} = a^{(n-m)}$$

27. In 1998 the United Kingdom generated approximately 10.1 billion kilowatt-hours of nuclear energy with a population of about 59 million on 94,525 miles^2. In the same year the United States generated approximately 55.6 billion kilowatt-hours of nuclear energy with a population of about 270 million on 3,675,031 miles^2 (total area including land and water).

a. How many kilowatt-hours is the United Kingdom generating per person? How many kilowatt-hours are they generating per square mile? Express each in scientific notation.

b. How many kilowatt-hours is the United States generating per person? How many kilowatt-hours are we generating per square mile? Express each in scientific notation.

c. How much nuclear power is being generated in the United Kingdom per square mile relative to the United States?

d. Write a brief statement comparing the relative magnitude of generation of nuclear power per person in the United Kingdom and the United States.

Exercises for Section 4.3

28. Simplify and express your answer with positive exponents:

a. $(x^{-3})\cdot(x^{4})$ **c.** $(x^{2})^{-3}$ **e.** $(2n^{-2})^{-3}$

b. $(x^{-3})\cdot(x^{-2})$ **d.** $(n^{-2})^{-3}$ **f.** $n^{-4}(n^{5} - n^{2}) + n^{-3}(n - n^{4})$

29. Simplify where possible. Express your answer with positive exponents.

a. $\dfrac{2^{3}x^{4}}{2^{5}x^{8}}$ **b.** $\dfrac{x^{4}y^{7}}{x^{3}y^{-5}}$ **c.** $\dfrac{x^{-2}y}{xy^{3}}$ **d.** $\dfrac{(x+y)^{4}}{(x+y)^{-7}}$ **e.** $\dfrac{a^{-2}bc^{-5}}{(ab^{2})^{-3}c}$

30. Simplify where possible. Express your answer with positive exponents.

a. $(3\cdot 3^{8})^{-2}$ **c.** $2^{6} + 2^{6} + 2^{7} + 2^{-4}$ **e.** $10^{-5} + 5^{-2} + 10^{10}$

b. $x^{3}\cdot x^{-4}\cdot x^{12}$ **d.** $2x^{-3} + 3x(x^{-4})$

31. Evaluate and write the results using scientific notation:

a. $(2.3\cdot 10^{4})(2.0\cdot 10^{6})$ **c.** $\dfrac{8.19\cdot 10^{23}}{5.37\cdot 10^{12}}$ **e.** $(6.2\cdot 10^{52})^{3}$

b. $(3.7\cdot 10^{-5})(1.1\cdot 10^{8})$ **d.** $\dfrac{3.25\cdot 10^{8}}{6.29\cdot 10^{15}}$ **f.** $(5.1\cdot 10^{-11})^{2}$

32. Write each of the following in scientific notation:

a. $725\cdot 10^{23}$ **c.** $\dfrac{1}{725\cdot 10^{23}}$ **e.** $-725\cdot 10^{-23}$

b. $725\cdot 10^{-23}$ **d.** $-725\cdot 10^{23}$

33. A TV signal traveling at the speed of light takes about $8\cdot 10^{-5}$ seconds to travel 15 miles. How long would it take the signal to travel a distance of 3000 miles?

34. Round off the numbers and then estimate the values of the following expressions without a calculator. Show your work, writing your answers in scientific notation. Use a calculator to verify your answers.

a. $(0.000\,359)\cdot(0.000\,002\,47)$ **b.** $\dfrac{0.00000731\cdot 82{,}560}{1{,}891{,}000}$

35. Simplify and express your answer with positive exponents.

a. $\dfrac{x^{-2} - y^{-1}}{(xy^{2})^{-1}}$ **b.** $(5x^{-2}y^{-3})^{-2}$

36. The robot spacecraft NEAR (Near Earth Asteroid Rendezvous) is on a four-year mission through the inner solar system to study asteroids. Recently, scientists on Earth sent a radio message to its computers with instructions to fire its thruster rockets in order to make the craft go into orbit around Eros, a Manhattan-sized asteroid 160 million miles from Earth.

a. If radio messages travel at the speed of light, how long will it take for a message sent back from the NEAR spacecraft to reach the scientists informing them whether or not the maneuver was successful?

b. The near-Earth asteroid Cruithne is now known to be a companion, and an unusual one, of Earth. This asteroid shares Earth's orbit, its motion "choreographed" in such a way as to remain stable and avoid colliding with our planet. At its closest approach Cruithne gets to within 0.1 astronomical units of Earth (about 15 million kilometers).

The asteroid is currently about 0.3 astronomical units (45 million kilometers) from Earth. If NEAR were in orbit around Cruithne right now, how long would a radio signal transmitted from Earth take to reach the spacecraft?

Exercises for Section 4.4

37. For the following questions, make an estimate and then check your estimate by doing the calculations:

a. One foot is how many centimeters?

b. One foot is what part of a meter?

38. A football field is 100 yards long. How many meters is this? What part of a kilometer is this?

39. If a falling object accelerates at the rate of 9.8 meters per second every second, how many feet per second does it accelerate each second?

40. Convert the following to feet and express your answers in scientific notation:

a. The radius of the solar system is approximately 10^{12} meters.

b. The radius of a proton is approximately 10^{-15} meters.

41. The speed of light is approximately $1.86 \cdot 10^5$ miles/sec.

a. Write this number in decimal form and express your answer in words.

b. Convert the speed of light into meters per year. Show your work.

42. The average distance from Earth to the sun is about 150,000,000 km and the average distance from the planet Venus to the sun is about 108,000,000 km.

a. Express these distances in scientific notation.

b. Divide the distance from Venus to the sun by the distance from Earth to the sun and express your answer in scientific notation.

c. The distance from Earth to the sun is called 1 astronomical unit (1 A.U.) How many astronomical units is Venus from the sun?

d. Pluto is 5,900,000,000 km from the sun. How many astronomical units is it from the sun?

43. The distance from Earth to the sun is approximately 150 million kilometers. If the speed of light is $3.00 \cdot 10^5$ km/sec, how long does it take light from the sun to reach Earth? If a solar flare occurs right now, how long would it take for us to see it?

44. Estimate the number of heartbeats in a lifetime. Explain your method.

45. A nanosecond is 10^{-9} seconds. Modern computers can perform on the order of one operation every nanosecond. Approximately how many feet does an electrical signal moving at the speed of light travel in a computer in 1 nanosecond?

46. Since light takes time to travel, everything we see is from the past. When you look in the mirror, you see yourself not as you are, but as you were nanoseconds ago.

a. Suppose you look up tonight at the bright star Deneb. Deneb is 1600 light years away. When you look at Deneb, how old is the image you are seeing?

b. Even more disconcerting is the fact that what we see as simultaneous events do not necessarily occur simultaneously. Consider the two stars Betelgeuse and Rigel in the constellation Orion. Betelgeuse is 300 and Rigel 500 light years away. How many years apart were the images generated that we see simultaneously?

47. The National Institutes of Health guidelines suggest that adults over 20 should have a body mass index, or BMI, of under 25. This index is created according to the formula

$$\text{BMI} = \frac{\text{weight in kilograms}}{(\text{height in meters})^2}$$

a. Given that 1 kilogram = 2.2 pounds and 1 meter = 39.37 inches, calculate the body mass index of President Clinton, who is 6 feet 2 inches tall and in 1997 weighed 216 pounds. According to the guidelines, how would you describe his weight?

b. Most Americans don't use the metric system. So in order to make the BMI easier to use, convert the formula into an equivalent one using weight in pounds and height in inches. Check your new formula by using Clinton's weight and height and confirm that you get the same BMI.

c. The following excerpt from the article "America Fattens Up" (*The New York Times,* October 20, 1996) describes a very complicated process for determining your BMI:

> *To estimate your body mass index you first need to convert your weight into kilograms by multiplying your weight in pounds by 0.45. Next, find your height in inches. Multiply this number by .0254 to get meters. Multiply that number by itself and then divide the result into your weight in kilograms. Too complicated? Internet users can get an exact calculation at http://141.106.68.17/bsa.acgl.*

Can you do a better job of describing the process?

d. A letter to the editor from Brent Kigner, of Oneonta, N.Y., in response to *The New York Times* article says:

> *Math intimidates partly because it is often made unnecessarily daunting. Your article "American Fattens Up" convolutes the procedure for calculating the Body Mass Index so much that you suggest readers retreat to the Internet. In fact, the formula is simple: Multiply your weight in pounds by 703, then divide by the square of your height in inches. If the result is above 25, you weigh too much.*

Is Brent Kigner right?

48. Computer technology refers to the storage capacity for information with its own special units. Each minuscule electrical switch is called a "bit" and can be off or on. As the information capacity of computers has increased, the industry has developed some much larger units based on the bit:

1 byte = 8 bits

1 kilobit = 2^{10} bits, or 1024 bits (a kilobit is sometimes abbreviated Kbit)

1 kilobyte = 2^{10} bytes, or 1024 bytes (a kilobyte is sometimes abbreviated Kbyte)

1 megabit = 2^{20} bits, or 1,048,576 bits

1 megabyte = 2^{20} bytes, or 1,048,576 bytes

1 gigabyte = 2^{30} bytes, or 1,073,741,824 bytes

a. How many kilobytes are there in a megabyte? Express your answer as a power of 2 and in scientific notation.

b. How many bits are there in a gigabyte? Express your answer as a power of 2 and in scientific notation.

49. The article following talks about Planck's length, the smallest length or size anything can be in the universe, which is 10^{-35} meters (from G. Johnson, "How is the Universe Built? Grain by Grain," in the science section of the Dec. 7, 1999, *New York Times,* p. D1). Read the article and then answer the following questions:

a. How many kilometers is Planck's length?

b. How many miles is Planck's length?

c. If light travels at $3 \cdot 10^8$m/sec, how long will it take to cross a distance equivalent to Planck's length?

Slightly smaller than what Americans quaintly insist on calling half an inch, a centimeter (one-hundredth of a meter) is easy enough to see. Divide this small length into 10 equal slices and you are looking, or probably squinting, at a millimeter (one-thousandth, or 10 to the minus 3 meters). By the time you divide one of these tiny units into a thousand minuscule micrometers, you have far exceeded the limits of the finest bifocals. But in the mind's eye, let the cutting continue, chopping the micrometer into a thousand nanometers and the nanometers into a thousand picometers, and those in steps of a thousandfold into femtometers, attometers, zeptometers, and yoctometers. At this point, 10 to the minus 24 meters, about one-billionth the radius of a proton, the roster of Greek names runs out. But go ahead and keep dividing, again and again until you reach a length only one hundred-billionth as large as that tiny amount: 10 to the minus 35 meters. . . . You have finally hit rock bottom: a span called the Planck length, the shortest anything can get. According to recent developments in the quest to devise the "theory of everything," space is not an infinitely divisible continuum. It is not smooth but granular, and the Planck length gives the size of the smallest possible grains.

The time it takes for a light beam to zip across this ridiculously tiny distance . . . is called Planck time, the shortest possible tick of an imaginary clock.

Exercises for Section 4.5

50. Evaluate:

a. $\sqrt{10{,}000}$ **c.** $625^{1/2}$ **e.** $\left(\frac{1}{9}\right)^{1/2}$

b. $\sqrt{-25}$ **d.** $100^{1/2}$ **f.** $\left(\frac{625}{100}\right)^{1/2}$

51. Assume that all variables represent positive quantities and simplify.

a. $\sqrt{\frac{a^2b^4}{c^6}}$ **b.** $\sqrt{36x^4y}$ **c.** $\sqrt{\frac{49x}{y^6}}$ **d.** $\sqrt{\frac{x^4y^2}{100z^6}}$

52. Estimate the radius, r, of a circular region with an area, A, of 35 ft^2 (where $A = \pi r^2$).

53. Estimate the radius, r, of a sphere with surface area, S, of 20 m^2 (where $S = 4\pi r^2$).

54. Calculate the following:

a. $4^{1/2}$ **c.** $27^{1/3}$ **e.** $8^{2/3}$ **g.** $16^{1/4}$

b. $-4^{1/2}$ **d.** $-27^{1/3}$ **f.** $-8^{2/3}$ **h.** $16^{3/4}$

55. Calculate:

a. $\left(\frac{1}{100}\right)^{1/2}$ **b.** $25^{-1/2}$ **c.** $\left(\frac{9}{16}\right)^{-1/2}$ **d.** $\left(\frac{1}{1000}\right)^{1/3}$

56. Estimate the length of a side, s, of a cube with volume, V, of 6 cm^3 (where $V = s^3$).

57. Evaluate when $x = 2$:

a. $(-x)^2$ **c.** $x^{1/2}$ **e.** $x^{3/2}$

b. $-x^2$ **d.** $(-x)^{1/2}$ **f.** x^0

58. Simplify:

a. $\frac{6}{\sqrt{6}}$ **b.** $\frac{7}{\sqrt[3]{7}}$ **c.** $\frac{5}{\sqrt[3]{25}}$ **d.** $\frac{2}{\sqrt[4]{8}}$

59. Evaluate:

a. $27^{2/3}$ **b.** $16^{-3/4}$ **c.** $25^{-3/2}$ **d.** $81^{-3/4}$

60. Without using a calculator, find two *consecutive* integers such that one is smaller and one is larger than each of the following (for example, $3 < \sqrt{11} < 4$). Show your reasoning.

a. $\sqrt{13}$ **b.** $\sqrt{22}$ **c.** $\sqrt{40}$

Now verify your answers using a calculator.

61. Estimate the radius of a spherical balloon that has a volume of 4 ft^3.

62. *Constellation I.* Reduce each of the following expressions to the form $r^a \cdot n^b$; then plot the exponents as points with coordinates (a, b) on graph paper. Do you recognize the constellation?

a. $\dfrac{(r^5)^2}{n^6 \cdot r^{-1}}$ **c.** $\dfrac{(r \cdot n^5)^0 \cdot r \cdot n^3}{r^{-0.4} \cdot n^5}$ **e.** $\dfrac{r^{-6} \cdot r^{-13} \cdot r^7}{(r \cdot n^2)^3}$ **g.** $\dfrac{1}{r^{-0.3} \cdot n^{0.6}}$

b. $\dfrac{r^{10}}{r^{21} \cdot (n^{-2})^3}$ **d.** $\dfrac{n}{r^{0.8}}$ **f.** $(r^3 \cdot n^2)^3$

Constellation II. Reduce each of the following expressions to the form $u^a \cdot m^b$; then plot the exponents as points with coordinates (a, b) on graph paper. Do you recognize the constellation?

a. $\dfrac{(u^2)^2 \cdot m}{u^2 \cdot m^{-4}}$ **d.** $\dfrac{(um^2)^3 \cdot u^2}{(um)^4}$ **g.** $\dfrac{(mu)^0 \cdot (u^{10})^{-1} \cdot m^{1/4}}{(m^{-3} \cdot u^{-1/3})^3}$

b. $\dfrac{u^{-9/5} \cdot m^3}{(umu^2)^1 \cdot m^{-1}}$ **e.** $\dfrac{u^{-3/2} \cdot u^{-7/2} \cdot m^1 \cdot (m^3)^3}{(um)^2}$

c. $\dfrac{u^2 \cdot u^{-4}}{u^3 \cdot (m^{-2})^3}$ **f.** $\dfrac{1}{u^{12} \cdot m^{-9}}$

Constellation III. Reduce each of the following expressions to the form $p^a \cdot c^b$; then plot the exponents as points with coordinates (a, b) on graph paper. What constellation appears?

a. $\dfrac{p^7 \cdot p^{-3} \cdot (c^7)^0}{p^4}$ **c.** $\dfrac{p^{-5}}{p^{2.5} \cdot p^2 \cdot c^{-2}}$ **e.** $\dfrac{1}{(p^{0.5})^3 \cdot (c^{-13/9})^3}$

b. $\dfrac{(p^{-2})^{-3} \cdot c^4}{p^{-5} \cdot c^{7.5}}$ **d.** $\dfrac{(p^2 \cdot c^3)^2}{c^5}$

63. The breaking strength S (in pounds) of a three-strand manila rope is a function of its diameter, D (in inches). The relationship can be described by the equation $S = 1700D^{1.9}$. Calculate the breaking strength when D equals:

a. 1.5 in. **b.** 2.0 in.

64. If a rope is wound around a wooden pole, the pounds of frictional force, F, between the pole and the rope is a function of the number of turns, N, according to the equation $F = 14 \cdot 10^{0.70N}$. What is the frictional force when the number of turns is:

a. 0.5 **b.** 1 **c.** 3

Exercises for Section 4.6

65. Refer to the chart in Exploration 4.1.

a. How many orders of magnitude larger is the Milky Way than the first living organism on Earth?

b. How many orders of magnitude older is the Pleiades (a cluster of stars) than the first *Homo sapiens?*

66. Water boils (changes from a liquid to a gas) at 373°K (degrees Kelvin). The temperature of the core of the sun is 20 million degrees Kelvin. By how many orders of magnitude is the sun's core hotter than the boiling temperature of water?

67. In the December 1999 issue of the journal *Science,* two Harvard scientists describe a pair of "nanotweezers" they created which are capable of manipulating objects as small as one-50,000th of an inch wide. The scientists used the tweezers to grab and pull clusters of polystyrene molecules, which are of the same size as structures inside cells. A future use of these nanotweezers may be to grab and move components of biological cells.

a. Express one-50,000th of an inch in scientific notation.

b. Express the size of objects the tweezers are able to manipulate in meters.

c. The prefix nano refers to 9 subdivisions by 10, or a multiple of 10^{-9}. So a nanometer would be 10^{-9} meters. Is the name for the tweezers given by the inventors appropriate?

d. If not, how many orders of magnitude larger or smaller would the tweezers' ability to manipulate small objects have to be in order to grasp things of nanometer size?

68. Determine the order of magnitude difference in the sizes of the radii for:

a. The solar system (10^{12} meters) compared to Earth (10^{7} meters)

b. Protons (10^{-15} meter) compared to the Milky Way (10^{21} meters)

c. Atoms (10^{-10} meter) compared to neutrons (10^{-15} meter)

69. In order to compare the sizes of different objects, we need to use the same unit of measure.

a. Convert each of these to meters:

i. The radius of the moon is approximately 1,922,400 yards.

ii. The radius of Earth is approximately 6400 km.

iii. The radius of the sun is approximately 432,000 miles.

b. Determine the order of magnitude difference between:

i. The surface areas of the moon and Earth

ii. The volumes of the sun and the moon

70. The pH scale measures the hydrogen ion concentration in a liquid, which determines whether the substance is acidic or alkaline (One M equals $6.02 \cdot 10^{23}$ particles, such as atoms, ions, molecules, etc., per liter or 1 mole per liter.) A strong acid solution has a hydrogen ion concentration of 10^{-1} M. A strong alkali solution has a hydrogen ion concentration of 10^{-14} M. Pure water with a concentration of 10^{-7} M is neutral. The pH value is the power without the minus sign, so pure water has a pH of 7, acidic substances have a pH less than 7, and alkaline substances have a pH greater than 7.

a. Tap water has a pH of 5.8. Before the industrial age, rain water commonly had a pH of about 5. With the spread of modern industry, rain in northeastern United States and parts of Europe now has a pH of about 4, and in extreme cases the pH is about 2. Lemon juice has a pH of 2.1. If acid rain with a pH of 3 is discovered in an area, how much more acidic is it than preindustrial rain?

b. Blood has a pH of 7.4; wine has a pH of about 3.4. By what order of magnitude is wine more acidic than blood?

71. Which is an additive scale? Explain why. Which is a multiplicative or logarithmic scale? Explain why.

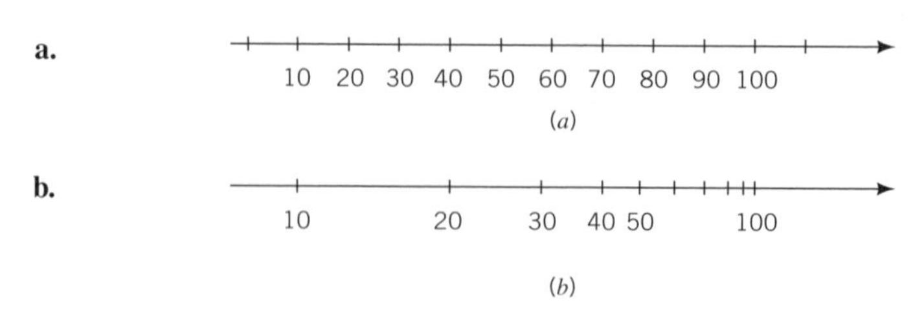

72. Round the following numbers to the nearest power of 10 and then find the approximate location on the logarithmic scale in Figure 4.2. (Refer to the chart in Exploration 4.1.)

a. The radius of the first atoms

b. The radius of the sun

73. (Refer to the chart in Exploration 4.1.) Plot on the logarithmic scale in Figure 4.2 an object whose radius is:

a. Five orders of magnitude larger than the radius of the first atoms

b. Twenty orders of magnitude smaller than the radius of the sun

74. a. Place the number 50 on the *additive* scale below:

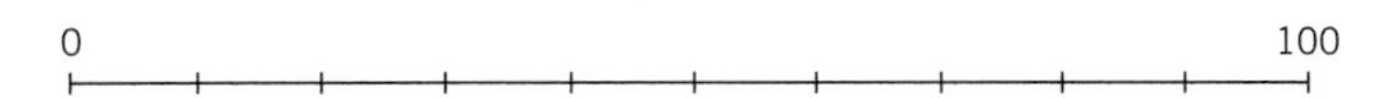

b. Place the number 50 on the *multiplicative* scale below: (*Hint:* Use a calculator to find the value of each tick mark or use logs (Section 4.7) to solve.)

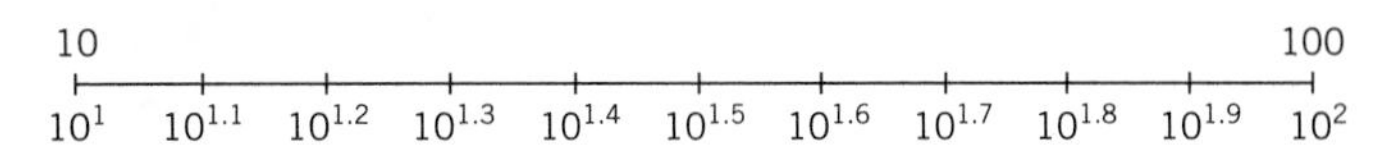

75. The coordinate system below uses a multiplicative scale. Position the point whose coordinates are (708, 25). See hint in Exercise 74(b).

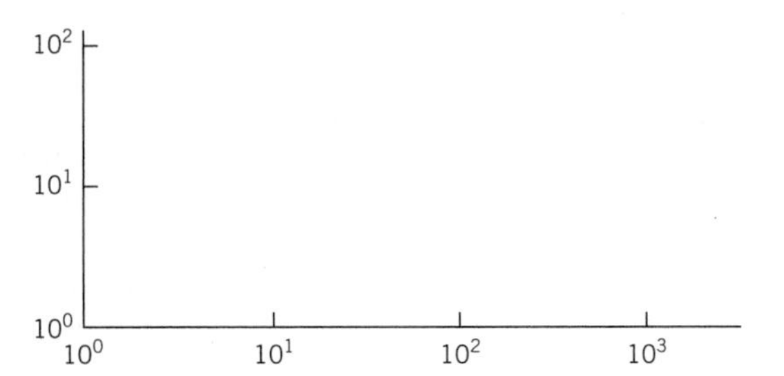

76. a. Read the chapter entitled "The Cosmic Calendar" from Carl Sagan's book *The Dragons of Eden.*

b. Carl Sagan tried to give meaning to the cosmic chronology by imagining the 15-billion-year lifetime of the universe compressed into the span of one calendar year. To get a more personal perspective, consider your date of birth as the time the Big Bang took place. Map the following five cosmic events onto your own life span:

i. The Big Bang
ii. Creation of Earth
iii. First life on Earth
iv. First *Homo sapiens*
v. American Revolution

Once you have done the necessary mathematical calculations and placed your results on either a chart or a timeline, form a topic sentence and write a playful paragraph about what you were supposedly doing when these cosmic events took place. Hand in your calculations along with your writing.

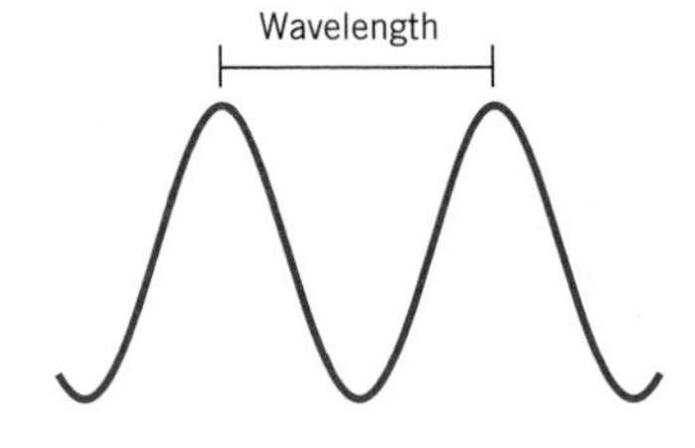

77. Light or electric and magnetic fields, such as X-rays and radio waves, can be thought of as radiation waves moving through space at the speed of light ($3.00 \cdot 10^5$ km, or 186,000 miles per second.) The distance between the crest of one wave and the crest of the next is called the wavelength. These radiation waves are referred to as *electromagnetic radiation.* We can draw the entire *electromagnetic spectrum* on an order of magnitude (or logarithmic) scale that shows the wavelengths in centimeters from cosmic rays to long wave radio waves.

a. Within the range of X-ray wavelengths, what is a reasonable size for an "average" wavelength of an X-ray?

b. What is a reasonable size for an "average" long wave radio wave?

c. How many orders of magnitude larger is an average radio wave than an average X-ray?

d. Very short wavelengths are often measured in Angstroms. One Angstrom, written 1 Å, equals 10^{-8} cm. What is the approximate range of wavelengths of ultraviolet rays in

Angstroms? (Sunscreen or sunglasses are commonly used to provide protection against ultraviolet light.)

e. What is the approximate range of values for wavelengths of infrared radiation in Angstroms? (Although human vision cannot detect infrared radiation, night vision scopes like those used by the military can pick up infrared signals emitted by warm bodies.)

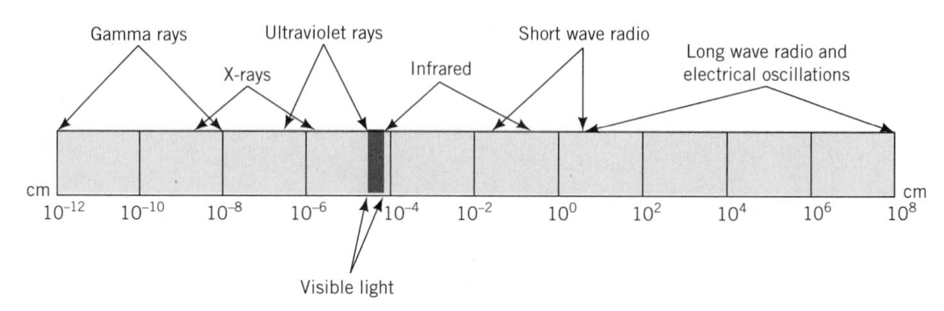

The Electromagnetic Spectrum.

78. Radio waves, sent from a broadcast station and picked up by the antenna of your radio, are a form of electromagnetic radiation (EM), as are microwaves, X-rays, and visible, infrared, and ultraviolet light. They all travel at the speed of light. Electromagnetic radiation can be thought of as oscillations like the vibrating strings of a violin or guitar or like ocean swells that have crests and troughs. (See image in Exercise 77.) The distance between the crest or peak of one wave and the next is called the wavelength. The number of times a wave crests per minute—or per second for fast oscillating waves—is called its frequency. Wavelength and frequency are inversely proportional: the longer the wavelength, the smaller the frequency, and vice versa, the faster the oscillation, the shorter the wavelength. For radio waves and other EM, the number of oscillations per second of a wave is measured in hertz, after the German scientist who first demonstrated that electrical waves could transmit information across space. One cycle or oscillation per second equals one *hertz*.

For the following exercise you may want to find an old radio or look on a stereo tuner at the AM and FM radio bands. You may see the notation kHz beside the AM and MHz beside the FM. AM radio waves oscillate at frequencies measured in the kilohertz range, while FM radio waves oscillate at frequencies measured in the megahertz range.

a. The Boston FM rock station WBCN transmits at 104 MHz. Write its frequency in *hertz* using scientific notation.

b. The Boston AM radio news station WBZ broadcasts at 1030 kHz. Write its frequency in *hertz* using scientific notation.

The wavelength λ (Greek lambda) and frequency μ (Greek mu) are related by the formula

$$\lambda = \frac{c}{\mu} \quad \text{where } c \text{ is tye speed of light in meters per second}$$

c. Estimate the wavelength of the WBCN radio transmission.

d. Estimate the wavelength of the WBZ-AM radio transmission.

e. Compare your answers in parts (c) and (d) in order of magnitude to the length of a football field (approximately 100 meters).

Exercises for Section 4.7

79. Rewrite in an equivalent form using logarithms:

a. $10^4 = 10{,}000$ **c.** $10^0 = 1$

b. $10^{-2} = 0.01$ **d.** $10^{-5} = 0.000\,01$

80. Use your calculator to evaluate to two decimal places:

a. $10^{0.4}$ **d.** $10^{0.7}$

b. $10^{0.5}$ **e.** $10^{0.8}$

c. $10^{0.6}$ **f.** $10^{0.9}$

81. Express the number 375 in the form 10^x.

82. Estimate the value of each of the following:

a. log 4000 **b.** log 5,000,000 **c.** log 0.000 8

83. Rewrite the following statements using logs:

a. $10^2 = 100$ **b.** $10^7 = 10{,}000{,}000$ **c.** $10^{-3} = 0.001$

Rewrite the following statements using exponents:

d. $\log 10 = 1$ **e.** $\log 10{,}000 = 4$ **f.** $\log 0.000\,1 = -4$

84. Do the following without a calculator:

a. Find the following values:

log 100

log 1000

log 10,000,000

What is happening to the values of $\log x$ as x gets larger?

b. Find the following values:

log 0.1

log 0.001

log 0.000 01

What is happening to the values of $\log x$ as x gets closer to 0?

c. What is log 0?

d. What is $\log(-10)$? What do you know about $\log x$ when x is any negative number?

85. Without using a calculator for each number in the form $\log x$, find some integers a and b such that $a < \log x < b$. Justify your answer. Then verify your answers with a calculator.

a. log 11 **b.** log 12,000 **c.** log 0.125

86. Use a calculator to determine the following logs. Double check each answer by writing down the equivalent expression using exponents, and then verify this equivalence using a calculator.

a. log 15 **b.** log 15,000 **c.** log 1.5

87. On a logarithmic scale, what would correspond to moving over:

a. 0.001 unit **b.** $\frac{1}{2}$ unit **c.** 2 units **d.** 10 units

88. The difference in the noise level of two sounds is measured in decibels, where decibels $= 10 \log(I_2 / I_1)$ and I_1 and I_2 are the intensities of the two sounds. Calculate the difference in noise level when $I_1 = 10^{-15}$ watts/cm^2 and $I_2 = 10^{-8}$ watts/cm^2.

89. The concentration of hydrogen ions in a water solution typically ranges from 10 M to 10^{-15} M. (One M equals $6.02 \cdot 10^{23}$ particles, such as atoms, ions, molecules, etc., per liter or 1 mole per liter.)[6] Because of this wide range, chemists use a logarithmic scale,

[6]You may recall from Algebra Aerobics 4.1 that $6.02 \cdot 10^{23}$ is called Avogadro's number. A mole of a substance is defined as Avogadro's number of particles of that substance. M is called a molar unit.

called the pH scale, to measure the concentration (see Exercise 70). The formal definition of pH is pH $= -\log[H^+]$, where $[H^+]$ denotes the concentration of hydrogen ions. Chemists use the symbol H^+ for hydrogen ions and the brackets [ ] mean "the concentration of."

a. Pure water at 25°C has a hydrogen ion concentration of 10^{-7} M. What is the pH?

b. In orange juice, $[H^+] \approx 1.4 \cdot 10^{-3}$ M. What is the pH?

c. Household ammonia has a pH of about 11.5. What is its $[H^+]$?

d. Does a higher pH indicate a lower or a higher concentration of hydrogen ions?

e. A solution with a pH > 7 is called basic; with a pH $= 7$ is called neutral; and with a pH < 7 is called acidic. Identify pure water, orange juice, and household ammonia as either acidic, neutral, or basic. Then plot their positions on the accompanying scale, which shows both the pH and the hydrogen ion concentration.

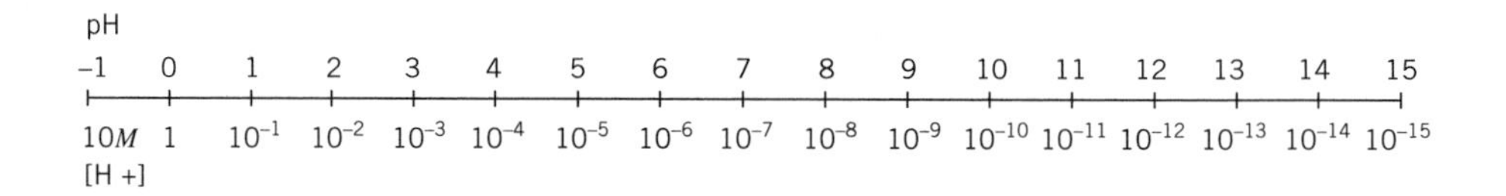

Exploration 4.1

The Scale and the Tale of the Universe

Objective

- gain an understanding of the relative sizes and relative ages of objects in the universe using scientific notation and unit conversions

Materials/Equipment

- tape, pins, paper, and string to generate a large wall graph (optional)
- enclosed worksheet and conversion table

Related Readings

- excerpts from *Powers of Ten* and "The Cosmic Calendar" from *The Dragons of Eden*

Related Videos

- *Powers of Ten* and "The Cosmic Calendar" in the PBS series *Cosmos*

Related Software

- "E1: Tale and Scale of the Universe" in *Exponential & Log Functions*

Procedure

Work in small groups. Each group should work on a separate subset of objects on the accompanying worksheet.

1. Convert the ages and sizes of objects so they can be compared. You can refer to the conversion table that shows equivalences between English and metric units (Table 4.1) and Table 4.2, which lists the meanings of the metric prefixes such as kilo-, giga-, and tera-. In addition, 1 light year $\approx 9.46 \cdot 10^{12}$ km.
2. Generate on the blackboard or on the wall a blank graph whose axes are marked off in orders of magnitude (integer powers of 10), with the units on the vertical axis representing age of object, ranging from 10^0 to 10^{11} years, and the units on the horizontal axis representing size of object, ranging from 10^{-12} to 10^{27} meters.
3. Each small group should plot the approximate coordinates of their selected objects (age in years, size in meters) on the graph. You might want to draw and label a small picture of your object to plot on your graph.

Discussion/Analysis

- Scan the plotted objects from left to right, looking only at relative sizes. Does your graph make sense in terms of what you know about the sizes of these objects?

- Now scan the plotted objects from top to bottom, only considering relative ages. Does your graph make sense in terms of what you know about the relative ages of these objects?
- Compare the largest and smallest objects in the table. By how many orders of magnitude do they differ? Compare the sizes of other pairs of objects in the table. Compare the ages of the oldest and youngest objects in the table. By how many orders of magnitude do they differ?

			In Scientific Notation	
Object	Age (in years)	Size (of radius)	Age (in years)	Size (in meters)
Observable universe	15 billion (?)	10^{26} meters		
Surtsey (Earth's newest land mass)	35 years	0.5 miles		
Pleiades (a galactic cluster)	100 million	32.6 light years		
First living organisms on Earth	3.5 billion	0.000 05 meters		
Pangaea (Earth's prehistoric supercontinent)	200 million	4500 miles		
First *Homo sapiens sapiens*	100 thousand	100 centimeters		
First *Tyrannosaurus rex*	200 million	20 feet		
Eukaryotes (first cells with nuclei)	2 billion	0.000 05 meters		
Earth	5 billion	6400 kilometers		
Milky Way galaxy	14 billion	50,000 light years		
First atoms	15 billion	0.000 000 000 1 meters		
Our sun	5 billion	1 gigameter		
Our solar system	5 billion	1 terameter		

Exploration 4.2

Patterns in the Positions and Motions of the Planets

Objective

- explore patterns in the positions and motions of the planets and discover Kepler's Law

Introduction and Procedure

Four hundred years ago, before Newton's laws of mechanics, Johannes Kepler discovered a law that relates the periods of planets with their average distances from the sun. (A period of a planet is the time it takes a planet to complete one orbit of the sun.) Kepler's strong belief that the solar system was governed by harmonious laws drove him to try to discover hidden patterns and correlations among the positions and motions of the planets. He used the trial-and-error method and continued his search for years.

At the time of his work, Kepler did not know the distance from the sun to each planet in terms of measures of distance such as the kilometer. But he was able to determine the distance from each planet to the sun in terms of the distance from Earth to the sun, now called the astronomical unit, or A.U. for short. One A.U. is the distance from Earth to the sun. The first column in the table below gives the average distance from the sun to each of the planets in astronomical units.

Patterns in the Positions and Motions of the Planets: Kepler's Discovery

Fill in the following table and look for the relationship that Kepler found:

Kepler's Third Law: The First Planet Table (Inner Planetary System)

Planet	Average Distance from Sun (A.U.)*	Cube of the Distance (A.U.^{3})	Orbital Period (years)	Square of the Orbital Period (years^2)
Mercury	0.3870		0.2408	
Venus	0.7232		0.6151	
Earth	1.0000		1.0000	
Mars	1.5233		1.8807	
Jupiter	5.2025		11.8619	
Saturn	9.5387		29.4557	

*1 A.U. $\approx 149.6 \cdot 10^6$ km; 1 year $\approx$ 365.26 days.

Source: Data from S. Parker and J. Pasachoff, *Encyclopedia of Astronomy,* 2nd ed. (New York: McGraw-Hill, 1993), Table 1, Elements of Planetary Orbits.

The planets Uranus, Neptune, and Pluto were discovered after Kepler made his discovery. Check to see whether the relationship you found above holds true for these three planets.

The Second Planet Table (Outer Planetary System)

Planet	Average Distance from Sun (A.U.)*	Cube of the Distance ($A.U.^3$)	Orbital Period (years)	Square of the Orbital Period ($years^2$)
Uranus	19.1911		84.0086	
Neptune	30.0601		164.7839	
Pluto	39.5254		248.5900	

Source: Data from S. Parker and J. Pasachoff, *Encyclopedia of Astronomy,* 2nd ed. (New York: McGraw-Hill, 1993), Table 1, Elements of Planetary Orbits.

Summary

- Express your results in words.
- Construct an equation showing the relationship between distance from the sun and orbital period. Solve the equation for distance from the sun. Then solve the equation for orbital period.
- Do your conclusions hold for all of the planets?

Growth and Decay: An Introduction to Exponential Functions

Overview

The simplest common function models are linear and exponential. Linear functions are used to represent quantities to which a fixed amount is added (or subtracted) during each time period. Exponential functions are used to represent quantities that are multiplied by a fixed amount during each time period. Exponential functions can model such diverse phenomena as bacteria growth, radioactive decay, compound interest rates, inflation, musical pitch, and family trees.

After reading this chapter you should be able to:

- recognize the properties of exponential functions
- understand the differences between exponential and linear growth
- use exponential functions to model growth and decay phenomena
- identify the graph of an exponential function on a semi-log plot

5.1 EXPONENTIAL GROWTH

The Growth of *E. coli* Bacteria

Measuring and predicting growth is of concern to population biologists, ecologists, demographers, economists, and politicians alike. By studying the growth of a population of the bacterium *E. coli,* we can construct a simple model that can be generalized to describe the growth of other phenomena such as cells, countries, or capital.

Bacteria are very tiny, single-celled organisms that are by far the most numerous organisms on Earth. One of the most frequently studied bacteria is *E. coli,* a rod-shaped bacterium approximately 10^{-6} meter (or 1 micrometer) long that inhabits the intestinal tracts of humans.[1] The cells of *E. coli* reproduce by a process called fission: The cell splits in half, forming two "daughter cells."

The program "E2: Exponential Growth & Decay" in *Exponential & Log Functions* offers a dramatic visualization of growth and decay.

The rate at which fission occurs depends on several conditions, in particular, available nutrients and temperature. Under ideal conditions *E. coli* can divide every 20 minutes. If we start with an initial population of 100 *E. coli* bacteria that doubles during every 20-minute time period, we generate the data in Table 5.1. The initial 100 bacteria double to become 200 bacteria at the end of the first time period, double again to become 400 at the end of the second time period, and keep on doubling to become 800, 1600, 3200, etc., at the ends of successive time periods. At the end of the twenty-fourth time period (at $24 \cdot 20$ minutes $= 480$ minutes, or 8 hours), the initial 100 bacteria in our model have grown to over 1.6 billion bacteria!

Because the numbers become astronomically large so quickly, we run into the problems we saw in Chapter 4 when graphing numbers of widely different sizes. Figure 5.1 shows a graph of the data in Table 5.1 for only the first 10 time periods. We can see from the graph that the relationship between number of bacteria and time is not linear. The number of bacteria seems to be increasing more and more rapidly over time.

Table 5.1
Growth of *E. coli* Bacteria

Time Periods (of 20 minutes each)	Number of *E. coli* Bacteria
0	100
1	200
2	400
3	800
4	1,600
5	3,200
6	6,400
7	12,800
8	25,600
9	51,200
10	102,400
11	204,800
12	409,600
13	819,200
14	1,638,400
15	3,276,800
16	6,553,600
17	13,107,200
18	26,214,400
19	52,428,800
20	104,857,600
21	209,715,200
22	419,430,400
23	838,860,800
24	1,677,721,600

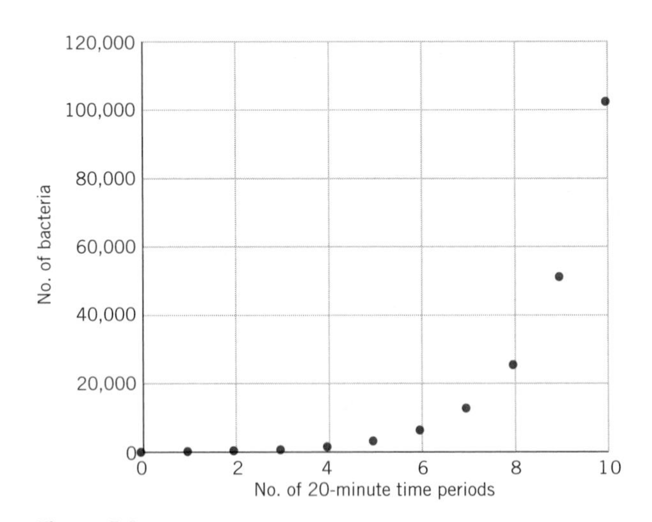

Figure 5.1 Growth of *E. coli* bacteria.

[1]Most types of *E. coli* are beneficial to humans, aiding in human digestion. A few types are lethal. You may have read about deaths resulting from people eating certain deadly strains of *E. coli* bacteria in undercooked hamburgers.

A mathematical model for E. coli *growth.* If we examine the values in Table 5.1, we see that the initial number of 100 bacteria is repeatedly doubled or multiplied by 2. If we record in a third column (Table 5.2) the number of times we multiply 2 times the original value of 100, we can begin to see a pattern emerge.

Table 5.2
Pattern in *E. coli* Growth

Number of Time Periods	Number of *E. coli* Bacteria	Generalized Expression
0	100	$100 = 100 \cdot 2^0$
1	200	$100 \cdot 2 = 100 \cdot 2^1$
2	400	$100 \cdot 2 \cdot 2 = 100 \cdot 2^2$
3	800	$100 \cdot 2 \cdot 2 \cdot 2 = 100 \cdot 2^3$
4	1,600	$100 \cdot 2 \cdot 2 \cdot 2 \cdot 2 = 100 \cdot 2^4$
5	3,200	$100 \cdot 2 \cdot 2 \cdot 2 \cdot 2 \cdot 2 = 100 \cdot 2^5$
6	6,400	$100 \cdot 2 \cdot 2 \cdot 2 \cdot 2 \cdot 2 \cdot 2 = 100 \cdot 2^6$
7	12,800	$100 \cdot 2 \cdot 2 \cdot 2 \cdot 2 \cdot 2 \cdot 2 \cdot 2 = 100 \cdot 2^7$
8	25,600	$100 \cdot 2 \cdot 2 \cdot 2 \cdot 2 \cdot 2 \cdot 2 \cdot 2 \cdot 2 = 100 \cdot 2^8$
9	51,200	$100 \cdot 2 \cdot 2 \cdot 2 \cdot 2 \cdot 2 \cdot 2 \cdot 2 \cdot 2 \cdot 2 = 100 \cdot 2^9$
10	102,400	$100 \cdot 2 \cdot 2 \cdot 2 \cdot 2 \cdot 2 \cdot 2 \cdot 2 \cdot 2 \cdot 2 \cdot 2 = 100 \cdot 2^{10}$

Remembering that 2^0 equals 1 by definition, we can describe the relationship by

$$\text{number of } E.\ coli \text{ bacteria} = 100 \cdot 2^{\text{no. of time periods}}$$

If we let N = number of bacteria and t = number of time periods, we can write the equation more compactly as

$$N = 100 \cdot 2^t$$

Since each value of t determines one and only one value for N, the equation represents N as a function of t. If we call the function f, we can write $N = f(t)$, where $N = 100 \cdot 2^t$. The domain of the function, as a model of bacteria growth, includes values of t between 0 and some unspecified positive number since in theory the number of bacteria could increase without bound. This function is called *exponential* since the independent variable, t, occurs in the exponent of the base 2. The number 100 is the *initial bacteria population.* We also call 2 the *growth factor,* or the multiple by which the population grows during each time period. So the bacteria population is doubling or, stated another way, is growing by 100% during each time period.

The General Exponential Growth Function

The *E. coli* growth equation

$$N = 100 \cdot 2^t$$

is in the form

$$\text{dependent variable} = (\text{initial population}) \cdot (\text{growth factor})^{\text{independent variable}}$$

Such an equation, with the independent variable in the exponent and a growth factor > 1, describes an *exponential growth function.*

An exponential growth function $y = f(x)$ can be represented by an equation of the form

$$y = Ca^x \qquad (a > 1)$$

where C = initial value of y (when $x = 0$) and a = *growth factor,* the amount by which y is multiplied when x increases by 1.

Example 1

If you started with 200 cells that tripled during every time period t, an equation to model its growth would be

$$N = 200 \cdot 3^t$$

Example 2

The equation

$$Q = (4 \cdot 10^6) \cdot 2.5^T$$

would represent an initial population of $4 \cdot 10^6$, or 4,000,000, that grew by a factor of 2.5 (that is, was multiplied by 2.5) each time period, T.

Algebra Aerobics 5.1a

1. Identify the initial population and the growth factor in each of the following exponential growth functions:
   a. $y = 350 \cdot 5^x$ **c.** $P = (7 \cdot 10^3) \cdot 4^t$
   b. $Q = 25{,}000 \cdot 1.5^t$
2. Write an equation for an exponential growth function where:
   a. The initial population is 3000 and the growth factor is 3.
   b. The initial population is $4 \cdot 10^7$ and the growth factor is 1.3.
   c. The initial population is 75 and the population quadruples during each time period.

Linear vs. Exponential Growth

If a quantity N depends linearly on another quantity t, we know that the equation relating the two quantities can be written in the form

$$N = \text{intercept} + (\text{slope}) \cdot t$$

Suppose the dependence is exponential. We have seen from the previous examples that the equation relating N and t can be written in the form

$$N = \text{intercept} \cdot (\text{growth factor})^t$$

Linear growth is intrinsically *additive*. Linear growth means that for each unit increase in the independent variable, we must *add* a fixed amount (the slope or rate of change) to the value of the dependent variable. For example, in the linear function $N = 100 + 2t$, each time we increase t by 1, we add 2 to the value of N. After t time periods, we would add $2t$ to the value of N. Assuming t is an integer, we can write this linear function as

$$N = 100 + \underbrace{2 + 2 + \ldots\ldots\ldots + 2}_{t \text{ times}}$$

Exponential growth is *multiplicative,* which means that for each unit increase in the independent variable, we *multiply* the value of the dependent variable by the growth factor. For example, in the exponential function $N = 100 \cdot 2^t$, each time we increase t by 1, we *multiply* the value of N by 2. After t time periods have elapsed, we multiply the value of N by 2^t. If t is an integer, we can write this exponential function as

$$N = 100 \cdot \underbrace{2 \cdot 2 \cdot \ldots\ldots\ldots\ldots \cdot 2}_{t \text{ times}}$$

The linear function involves repeated addition, the exponential function repeated multiplication. In each case the independent variable counts the number of repetitions.

Table 5.3 shows the additive nature of the linear function $N = 100 + 2t$ and Table 5.2 on page 235 shows the multiplicative nature of the exponential function $N = 100 \cdot 2^t$.

Table 5.3
Additive Pattern of a Linear Function

t	$N = 100 + 2t$	Generalized Expression
0	100	$100 = 100 + 2 \cdot 0$
1	102	$100 + 2 = 100 + 2 \cdot 1$
2	104	$100 + 2 + 2 = 100 + 2 \cdot 2$
3	106	$100 + 2 + 2 + 2 = 100 + 2 \cdot 3$
4	108	$100 + 2 + 2 + 2 + 2 = 100 + 2 \cdot 4$
5	110	$100 + 2 + 2 + 2 + 2 + 2 = 100 + 2 \cdot 5$
6	112	$100 + 2 + 2 + 2 + 2 + 2 + 2 = 100 + 2 \cdot 6$
7	114	$100 + 2 + 2 + 2 + 2 + 2 + 2 + 2 = 100 + 2 \cdot 7$
8	116	$100 + 2 + 2 + 2 + 2 + 2 + 2 + 2 + 2 = 100 + 2 \cdot 8$
9	118	$100 + 2 + 2 + 2 + 2 + 2 + 2 + 2 + 2 + 2 = 100 + 2 \cdot 9$
10	120	$100 + 2 + 2 + 2 + 2 + 2 + 2 + 2 + 2 + 2 + 2 = 100 + 2 \cdot 10$

Exponential growth is more rapid as time goes on, as can be seen in Table 5.4. For both the linear function $N = 100 + 2t$ and the exponential function $N = 100 \cdot 2^t$,

Table 5.4
Comparing Additive Pattern of Linear Growth to Multiplicative Pattern of Exponential Growth

Time, t	Linear Function, $N = 100 + 2t$ Pattern	Exponential Function, $N = 100 \cdot 2^t$ Pattern
0	$100 + 2 \cdot 0 = 100$	$100 \cdot 2^0 = 100$
1	$100 + 2 \cdot 1 = 102$	$100 \cdot 2^1 = 200$
2	$100 + 2 \cdot 2 = 104$	$100 \cdot 2^2 = 400$
3	$100 + 2 \cdot 3 = 106$	$100 \cdot 2^3 = 800$
4	$100 + 2 \cdot 4 = 108$	$100 \cdot 2^4 = 1{,}600$
5	$100 + 2 \cdot 5 = 110$	$100 \cdot 2^5 = 3{,}200$
6	$100 + 2 \cdot 6 = 112$	$100 \cdot 2^6 = 6{,}400$
7	$100 + 2 \cdot 7 = 114$	$100 \cdot 2^7 = 12{,}800$
8	$100 + 2 \cdot 8 = 116$	$100 \cdot 2^8 = 25{,}600$
9	$100 + 2 \cdot 9 = 118$	$100 \cdot 2^9 = 51{,}200$
10	$100 + 2 \cdot 10 = 120$	$100 \cdot 2^{10} = 102{,}400$

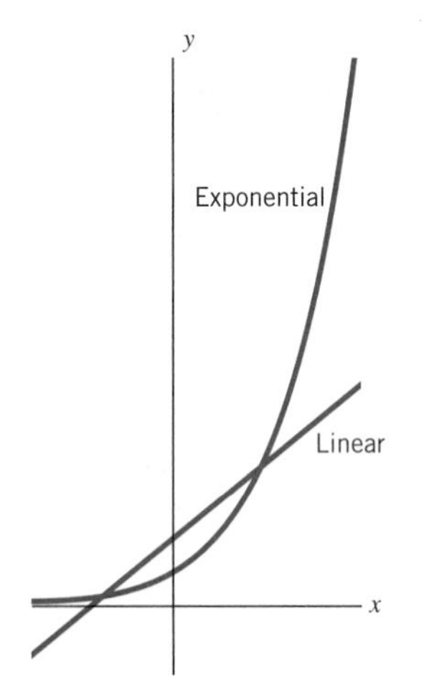

Figure 5.2 Graph comparing linear and exponential growth.

when $t = 0$, then $N = 100$. After 10 time periods, the values for N are strikingly different: 102,400 for the exponential function vs. 120 for the linear function. The initial value of 100 has been multiplied by 2 ten times in the exponential function to get $100 \cdot 2^{10}$. In the linear function, 2 has been added to 100 ten times, to get $100 + 2 \cdot 10$.

If we compare any linear growth function (whose graph will be a straight line with a positive slope) to any exponential growth function (whose graph will curve upward), it's not hard to see that sooner or later the exponential curve will permanently lie above the linear graph and continue to grow faster and faster (Figure 5.2). We say that the exponential function eventually dominates the linear function.

> Every exponential growth function will eventually dominate every linear growth function.

Algebra Aerobics 5.1b

1. Fill in the following table:

t	$N = 10 + 3t$	$N = 10 \cdot 3^t$
0		
1		
2		
3		
4		

2. Sketch a graph of the linear function $N = 10 + 3t$ for $0 \leq t \leq 4$. Is the slope constant? Justify your answer. What is the vertical intercept?
3. Sketch the graph of the exponential function $N = 10 \cdot 3^t$ for $0 \leq t \leq 4$. Compare this graph to that of the linear function $N = 10 + 3t$ in Problem 2.
4. Sketch the functions in Problems 2 and 3 on one coordinate plane for $0 \leq t \leq 2$. Compare the graphs.

Comparing the Average Rate of Change of Linear and Exponential Functions

Another way to compare linear and exponential functions is to examine average rates of change. Recall that if N is a function of t, then

$$\text{average rate of change} = \frac{\text{change in } N}{\text{change in } t} = \frac{\Delta N}{\Delta t}$$

Examine Table 5.5, which contains the average rate of change for both the linear and exponential function, where $\Delta t = 1$. For all linear functions, we know that the average rate of change is constant. For the linear function $N = 100 + 2t$, we can tell from the equation, Table 5.5, and Figure 5.3 that the (average) rate of change is constant at 2 units. Average rates of change for the exponential function $N = 100 \cdot 2^t$ are calculated in Table 5.5 and graphed in Figure 5.4. These suggest that the average rates of change of an exponential growth function grow *exponentially.*

Table 5.5
Comparing Average Rate of Change Calculations

	Linear Function		Exponential Function	
t	$N = 100 + 2t$, N	Average Rate of Change (between $t - 1$ and t)	$N = 100 \cdot 2^t$, N	Average Rate of Change (between $t - 1$ and t)
0	100	n.a.	100	n.a.
1	102	2	200	100
2	104	2	400	200
3	106	2	800	400
4	108	2	1,600	800
5	110	2	3,200	1,600
6	112	2	6,400	3,200
7	114	2	12,800	6,400
8	116	2	25,600	12,800
9	118	2	51,200	25,600
10	120	2	102,400	51,200

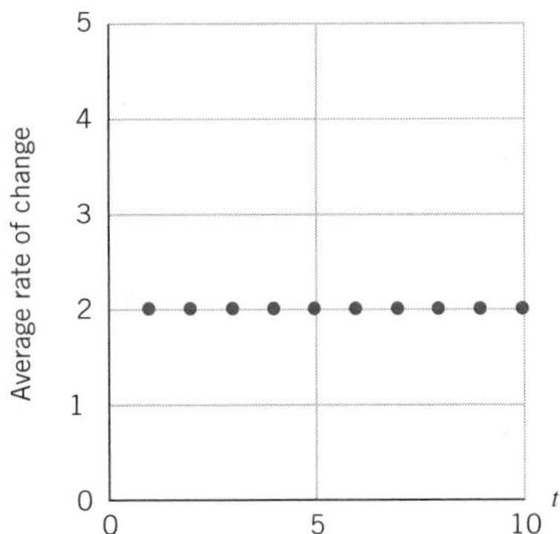

Figure 5.3 Graph of the average rates of change between several pairs of points on the linear function $N = 100 + 2t$.

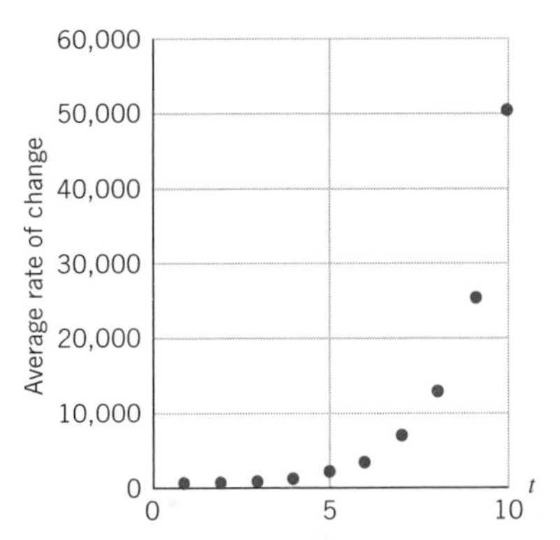

Figure 5.4 Graph of the average rates of change between several pairs of points on the exponential function $N = 100 \cdot 2^t$.

Looking at Real Growth Data for *E. coli* Bacteria

The idealized model for *E. coli* bacteria (in Table 5.1) showed phenomenal growth. It is not surprising that when we actually go to a biology laboratory and watch real bacteria grow, the growth is substantially slower. Conditions are rarely ideal; bacteria die, nutrients are used up, and temperatures may not be optimal. Table 5.6 and Figure 5.5 show some real data collected by a graduate student in biology. A solution was inoculated with an indeterminate amount of bacteria at time $t = 0$, and then the number of *E. coli* cells per milliliter was measured at regular time periods of 20 minutes each. For the first two time periods, even though there were probably millions of bacteria per milliliter, the amount was too small to measure.

Using a curve-fitting program to find a best fit exponential function to model the *E. coli* growth, we get

$$N = (1.37 \cdot 10^7) \cdot 1.5^t$$

where t represents the number of 20-minute time periods. This model tells us that there were approximately $1.37 \cdot 10^7$ (or 13.7 million) *E. coli* cells at $t = 0$, when

Table 5.6
Growth of Real *E. coli* for 12 Time Periods

Time Period (20 min each)	Number of Cells/ml
0	n.a.
1	n.a.
2	n.a.
3	$3.00 \cdot 10^7$
4	$7.00 \cdot 10^7$
5	$1.00 \cdot 10^8$
6	$1.80 \cdot 10^8$
7	$2.70 \cdot 10^8$
8	$4.25 \cdot 10^8$
9	$5.80 \cdot 10^8$
10	$8.00 \cdot 10^8$
11	$1.00 \cdot 10^9$
12	$1.16 \cdot 10^9$

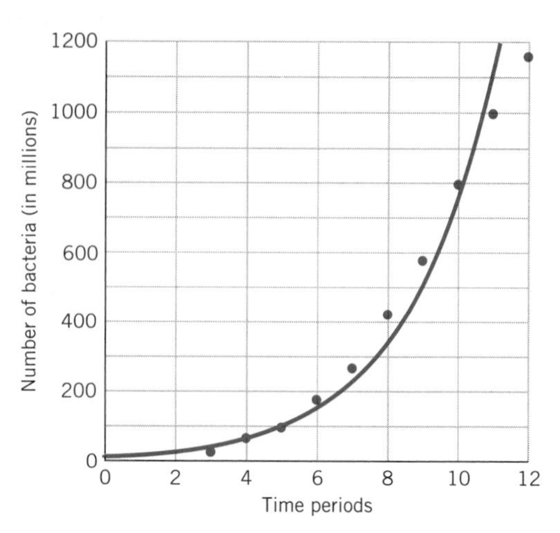

Figure 5.5 Real *E. coli* growth for 12 time periods.

the bacteria were initially placed in the broth. More importantly, it says that the growth factor was 1.5; that each time t increased by 1, N was multiplied by 1.5. Multiplying by 1.5 increases the value of N by 50%. We call 50%, the percentage increase over each unit of time, the *growth rate.* So in the real experiment, the number of bacteria actually increased by 50% each 20-minute time period, instead of doubling or increasing by 100%, as in our first idealized example.

Limitations of the Model

Clearly our model described the potential, not the actual, growth of *E. coli* bacteria. If *E. coli* bacteria really doubled every 20 minutes, then the offspring from a single cell could cover the entire surface of Earth with a layer a foot deep in fewer than 36 hours! The bacterial growth rate in the laboratory was half of that of the idealized example, but even this growth rate can't be sustained for long. At the slower rate of a 50% increase every 20 minutes, the offspring of a single cell would still cover the surface of Earth with a foot-deep layer in fewer than 3 days. Under most circumstances exponential growth is quickly restricted by environmental limits imposed by shortages of food, space, or oxygen, by predation, or by the accumulation of waste products.

So, while the growth may be exponential at first, it eventually slows down. Figure 5.6 shows the growth of *E. coli* over 24 time periods (or 8 hours), instead of the original 12 time periods (or 4 hours). The first part of the curve (up to $t = 12$) repeats the data points from Table 5.6 and Figure 5.5 with their exponential growth pattern. Then the growth slows down and the curve flattens out as N approaches a maximum population size, called the *carrying capacity,* indicated by the dotted line. Hence the growth rate, the percentage increase for each time period, must start decreasing. This S-shaped curve, typical of real population growth, is called a *sigmoid,* for sigma, the Greek letter S. Our exponential model is applicable to only a part of the growth curve, and thus limited its domain to values of t roughly between 0 and 12 time periods.

The arithmetic of exponentials leads us to the inevitable conclusion that in the long term, the *growth rate* of populations—whether they consist of bacteria or mosquitoes or humans—must approach zero.

See Excel or graph link file ECOLI.

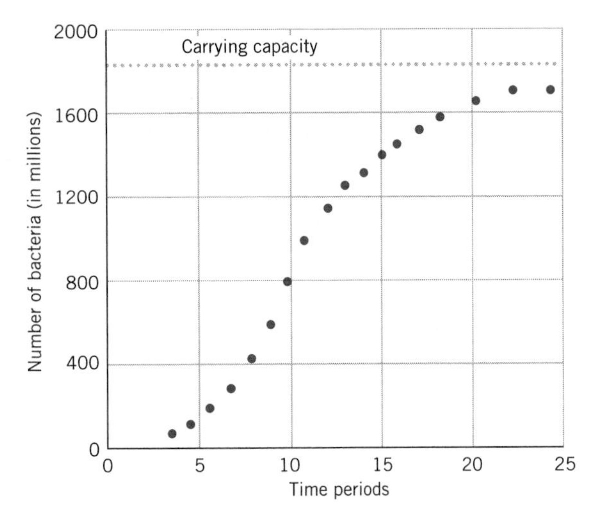

Figure 5.6 Plot of *E. coli* growth for 24 time periods, showing an initial exponential growth that tapers off.

5.2 EXPONENTIAL DECAY

The equation $y = Ca^x$, where $a > 1$, represents growth. Each time x increases by 1, the population is multiplied by a, a number greater than 1, so the population increases. But what if a were positive but less than 1 ($0 < a < 1$)? Then each time x increases by 1 and the population is multiplied by a, the population size would be reduced. Then we would have *exponential decay.*

The Decay of Iodine-131

We can use an exponential function to describe the decay of the radioactive isotope iodine-131. After being generated by a fission reaction, iodine-131 decays into a nontoxic, stable isotope. Every 8 days the amount of iodine-131 remaining is cut in half. For example, we show in Table 5.7 and Figure 5.7 how 160 milligrams (mg) of iodine-131 decays over 4 time periods (or $4 \cdot 8 = 32$ days).

Table 5.7

Time Periods (of 8 days each)	Amount of Iodine-131 (mg)		General Expression
0	160	=	$160 = 160\,(\frac{1}{2})^0$
1	80	=	$160(\frac{1}{2}) = 160\,(\frac{1}{2})^1$
2	40	=	$160(\frac{1}{2})(\frac{1}{2}) = 160\,(\frac{1}{2})^2$
3	20	=	$160(\frac{1}{2})(\frac{1}{2})(\frac{1}{2}) = 160\,(\frac{1}{2})^3$
4	10	=	$160(\frac{1}{2})(\frac{1}{2})(\frac{1}{2})(\frac{1}{2}) = 160\,(\frac{1}{2})^4$

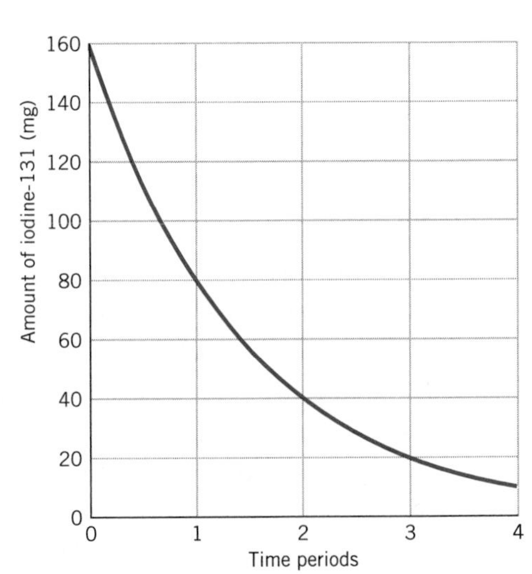

Figure 5.7 Exponential decay of iodine-131.

We can describe the decay of iodine-131 with the equation

$$Q = 160 \cdot \left(\tfrac{1}{2}\right)^T$$

where Q = quantity of iodine-131 (in mg) and T = number of time periods (of 8 days each). The number 160 is the initial amount of iodine-131 and $\frac{1}{2}$ is the *decay factor*. This equation is in the form

$$\text{dependent variable} = (\text{initial population}) \cdot (\text{decay factor})^{\textit{independent variable}}$$

This is the standard format for an exponential decay function.

> An *exponential decay function* $y = f(x)$ can be represented by an equation of the form
>
> $$y = Ca^x \qquad (0 < a < 1)$$
>
> where C = initial value of y (when $x = 0$) and a = *decay factor*, the amount by which y is multiplied when x increases by 1.

Example 1

If you started with 500 mg of iodine-131, the exponential decay function describing the amount left after T time periods would be

$$Q = 500 \cdot \left(\tfrac{1}{2}\right)^T$$

Example 2

The equation

$$N = (3 \cdot 10^4) \cdot (0.25)^T$$

represents an initial quantity of size $3 \cdot 10^4$ with a decay factor of 0.25 (or $\frac{1}{4}$). This would mean that at the end of each time period there would only be one-fourth as much as at the beginning of the time period.

Algebra Aerobics 5.2

1. Which of the following exponential functions represent growth and which represent decay?
   a. $y = 100 \cdot 3^x$
   b. $f(t) = 75 \cdot \left(\frac{2}{3}\right)^t$
   c. $w = 250 \cdot (0.95)^r$
   d. $g(r) = (2 \cdot 10^6) \cdot (1.15)^r$
   e. $y = (7 \cdot 10^9) \cdot (0.20)^z$
   f. $h(x) = 150 \cdot \left(\frac{5}{2}\right)^x$
2. Write an equation for an exponential decay function where:
   a. initial population is 2300; decay factor is $\frac{1}{3}$
   b. initial population is $3 \cdot 10^9$; decay factor is 0.35
   c. initial population is 375; population drops to one-tenth its previous size at the end of each time period
3. Does the exponential function $y = 12 \cdot (5)^{-x}$ represent growth or decay? (*Hint:* Rewrite the function in the standard form $y = Ca^x$.)

5.3 THE GRAPHS OF EXPONENTIAL FUNCTIONS

We have seen that:

> Exponential functions can be represented by equations of the form
>
> $$y = Ca^x \qquad (a > 0, a \neq 1)$$
>
> where C = initial value of y (when $x = 0$).
>
> If $a > 1$, the function represents growth and a is called the *growth factor.* If $0 < a < 1$, the function represents decay and a is called the *decay factor.*

The Effect of the Base *a*

Exploration 5.1 (along with "E3: $y = Ca^x$ Sliders" in *Exponential & Log Functions*) allows you to examine the effects of a and C on the graph of an exponential function.

When $a > 1$: Exponential growth. Given an exponential function $y = Ca^x$, each time x increases by 1, y is multiplied by a. When $a > 1$, the value of y will increase, so the function represents growth. The larger the value of a, the more rapid the growth. In Figure 5.8 we compare the functions

$$y = 100 \cdot 1.2^x \qquad y = 100 \cdot 1.3^x \qquad y = 100 \cdot 1.4^x$$

Each function has the same initial value of $C = 100$ but different growth factors of 1.2, 1.3, and 1.4, respectively. Note that since C and a are both positive, the value of y is always positive. So all the graphs lie above the x-axis.

What would the graph of the function $y = Ca^x$ look like if $a = 1$? What kind of function would you have?

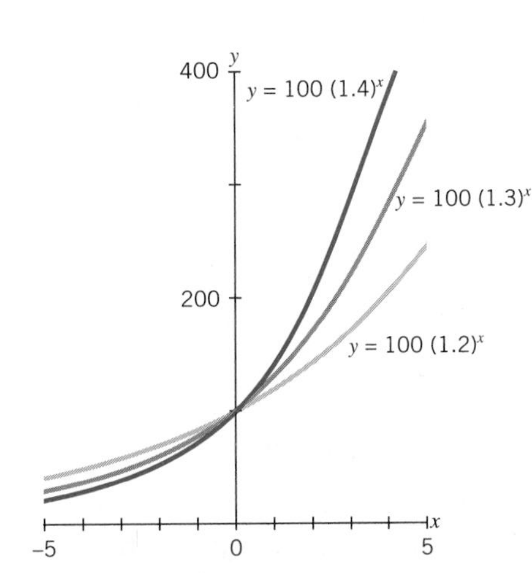

Figure 5.8 Three exponential growth functions.

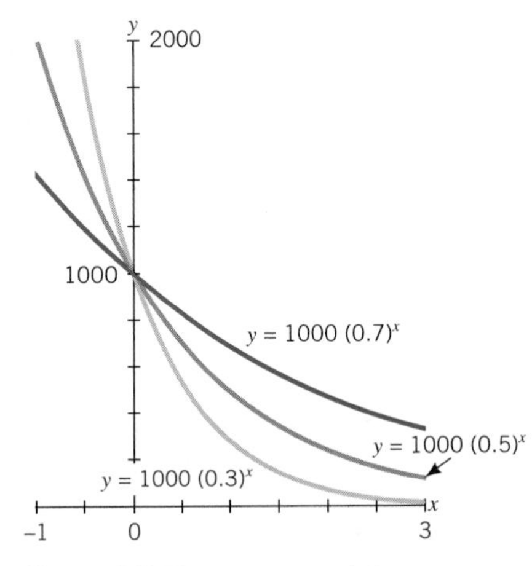

Figure 5.9 Three exponential decay functions.

When $0 < a < 1$: Exponential decay. In an exponential function $y = Ca^x$, if a is positive and less than 1, then when we increase x by 1, multiplying y by a will decrease the value of y. Hence we have decay. The smaller the value of a, the more rapid the decay. For example, if $a = 0.7$, after each unit increase in x, the value of y would be multiplied by 0.7. So x would drop to 70% of its previous size—a loss of

30%. But if $a = 0.5$, then when x increases by 1, y would be multiplied by 0.5, equivalent to dropping to 50% of its previous size—a loss of 50%.

Figure 5.9 compares the graphs of three exponential decay functions. Each has the same initial population of 1000 but decay factors of 0.3, 0.5, and 0.7, respectively.

The Effect of the Coefficient C

In the exponential function $y = Ca^x$, when $x = 0$, we have

$$\begin{aligned} y &= Ca^0 \\ &= C \cdot 1 \\ &= C \end{aligned}$$

What happens if $C = 0$? What happens if $C < 0$?

where C is the value of y when $x = 0$, or the y-intercept. So changing the value of C changes where the graph of the function will cross the vertical axis. In addition, an increase in positive values for C corresponds to larger values for y. Figure 5.10 compares the graphs of three exponential functions with the same growth factor of 1.1 but with C values of 50, 100, and 250, respectively.

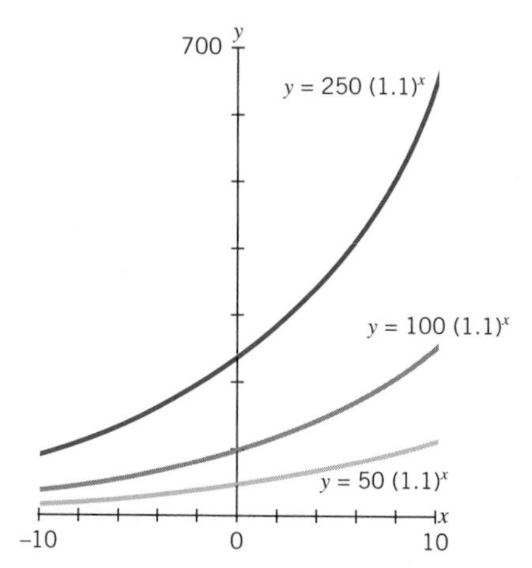

Figure 5.10 Three exponential functions with the same growth factor but different initial populations.

Horizontal Asymptotes

In the exponential decay function $y = 1000 \cdot (\frac{1}{2})^x$ graphed in Figure 5.11, the initial population of 1000 is cut in half each time x increases by 1. So when $x = 0, 1, 2, 3, 4, 5, 6, 7, \ldots$, the corresponding y values are 1000, 500, 250, 125, 62.5, 31.25, 15.625, ... As x gets larger and larger, the y values come closer and closer to, but never reach, zero. We say that as x approaches positive infinity ($x \to +\infty$), y approaches zero ($y \to 0$). The graph of the function $y = 1000 \cdot (\frac{1}{2})^x$ is said to be *asymptotic* to the x-axis, or we say the x-axis is a *horizontal asymptote* to the graph. In general a horizontal asymptote is a horizontal line that the graph of a function approaches for extreme values of the independent variable.

Exponential decay functions of the form $y = Ca^x$ are asymptotic to the x-axis. Similarly, for exponential growth functions, as x approaches negative infinity ($x \to -\infty$), y approaches zero ($y \to 0$). Examine, for instance, the exponential growth

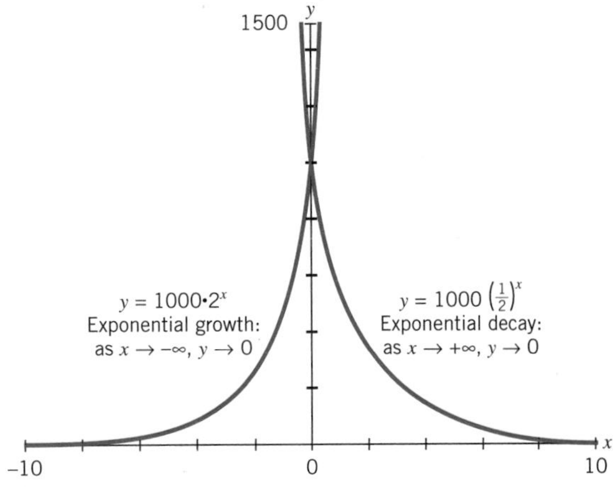

Figure 5.11 The x-axis is a horizontal asymptote for both exponential growth and decay functions of the form $y = Ca^x$.

function $y = 1000 \cdot 2^x$, also graphed in Figure 5.11. At $x = 0, -1, -2, -3, \ldots,$ the corresponding y values are 1000, $1000 \cdot 2^{-1} = 500$, $1000 \cdot 2^{-2} = 250$, $1000 \cdot 2^{-3} = 125, \ldots$. So as $x \to -\infty$, the y values come closer and closer to but never reach zero. So exponential growth functions are also asymptotic to the x-axis.

> Exponential functions of the form $y = Ca^x$ are *asymptotic* to the x-axis.
>
> Exponential growth ($a > 1$): as $x \to -\infty$, $y \to 0$.
> Exponential decay ($0 < a < 1$): as $x \to +\infty$, $y \to 0$.

Algebra Aerobics 5.3

1. a. Without calculating any values, draw a rough sketch of each of the following functions all on the same graph:

$y = 1000(1.5)^x \quad y = 1000(1.1)^x \quad y = 1000(1.8)^x$

b. Do the three curves intersect? If so, where?

c. In the first quadrant which curve should be on the top? Which in the middle? Which on the bottom?

d. In the second quadrant which curve should be on the top? Which in the middle? Which on the bottom?

2. a. Without calculating any values, draw a rough sketch of each of the following functions all on the same graph:

$Q = 250(0.6)^t \quad Q = 250(0.3)^t \quad Q = 250(0.2)^t$

b. Do the three curves intersect? If so, where?

c. In the first quadrant which curve should be on the top? Which in the middle? Which on the bottom?

d. In the second quadrant which curve should be on the top? Which in the middle? Which on the bottom?

3. a. Without calculating any values, draw a rough sketch of each of the following functions all on the same graph:

$P = 50 \cdot 3^t \qquad P = 150 \cdot 3^t$

b. Do the curves intersect anywhere?

c. Describe when and where one curve lies above another.

4. Identify any horizontal asymptotes for the following functions:

a. $Q = 100 \cdot 2^t$

b. $g(r) = (6 \cdot 10^7) \cdot (0.95)^r$

c. $y = 100 - 15x$

5.4 EXPONENTIAL GROWTH OR DECAY EXPRESSED IN PERCENTAGES

Exploration 5.2 offers an alternate strategy using ratios to construct exponential functions.

Exponential functions can be used to model any phenomenon that has a constant growth (or decay) factor, that is, any phenomenon that increases (or decreases) by a fixed *multiple* at regular intervals. All exponential functions can also be expressed in terms of percentages, called *growth* or *decay rates.* Multiplying by 1.05 is equivalent to a 5% increase. For example,

$$1.05 \cdot (\$1000) = (1.00 + 0.05)(\$1000)$$
$$= (1.00 \cdot \$1000) + (0.05 \cdot \$1000)$$

(since 100% is 1 in decimal form)
$$= (100\% \text{ of } \$1000) + (5\% \text{ of } 1000)$$
$$= \$1000 + \$50$$

So $1.05 \cdot$ original amount $=$ original amount $+$ 5% of the original amount

A growth factor of 1.05 corresponds to a growth rate of 5%.

Multiplying by 2 is equivalent to a 100% increase. For example,

$$2 \cdot \$10{,}000 = (1 + 1) \cdot \$10{,}000$$
$$= 1 \cdot \$10{,}000 \quad + 1 \cdot \$10{,}000$$
$$= (100\% \text{ of } \$10{,}000) + (100\% \text{ of } \$10{,}000)$$
$$= \$10{,}000 \quad + \$10{,}000$$

So $\quad 2 \cdot$ original amount $=$ original amount $\quad +$ 100% of the original amount

Hence a growth factor of 2 corresponds to a growth rate of 100%.

Multiplying by 0.8 corresponds to a 20% decrease. For example,

$$0.8 \cdot \$500 = (1 - 0.2) \cdot \$500$$
$$= 1 \cdot \$500 \quad - 0.2 \cdot \$500$$
$$= (100\% \text{ of } \$500) - (20\% \text{ of } \$500)$$
$$= \$500 \quad - \$100$$

So $\quad 0.8 \cdot$ original amount $=$ original amount $-$ 20% of the original amount

Thus a decay factor of 0.8 represents a decay rate of 20%.

In common usage, exponential growth is usually described in terms of its growth rate expressed as a percentage. A typical example is a statement issued by the Congressional Budget Office in 1999 that the cost of Medicaid (the federal health care program for the poor) is expected to grow by 7.8% a year until 2002.

The terms "growth rate" and "growth factor" can be very confusing. Understanding the mathematical relationship between them can help. A growth *rate* of 10% (or 0.10 in decimal form) corresponds to a growth *factor* of $1 + 0.10$ or 1.10. So if r denotes the growth rate (in decimal form), then $1 + r$ is the growth factor.

A decay *rate* of 7% (or 0.07 in decimal form) corresponds to a decay *factor* of $1 - 0.07$, or 0.93. So if r denotes the decay rate (in decimal form), then $1 - r$ is the decay factor.

If r is the *growth rate* then the *growth factor*, a, equals $1 + r$ and

$$y = C \cdot a^x = C(1 + r)^x$$

For example, if r is a growth rate equal to 0.25, then $a = 1.25$.

If r is the *decay rate* then the *decay factor*, a, equals $1 - r$ and

$$y = C \cdot a^x = C(1 - r)^x$$

For example, if r is a decay rate equal to 0.25, then $a = 0.75$.

Example 1

An article in the January 10, 2000, *San Francisco Chronicle* stated that "the 175 million cats, dogs, birds, and other pets who live in America don't know how lucky they are. To an increasing degree, their owners are providing them the best medicine money can buy. . . . Consumer spending [on pet drug sales reached] $1.3 billion in 1998 . . . [With pharmaceutical companies] betting that pet drug sales will grow 15 percent a year, we can expect the industry to concoct more Fido and Kitty pills—in paw-proof containers, of course." Identify the projected annual growth rate and growth factor for pet drug sales. Construct an exponential model and predict sales for the year 2005.

Solution

The projected annual growth rate is 15%, or 0.15 in decimal form. The corresponding growth factor is $1 + 0.15$, or 1.15. If n = number of years since 1998 and S = annual sales in billions of dollars, then $1.3 billion is the initial amount (the sales when $n = 0$). An exponential model would be

$$S = 1.3 \cdot (1.15)^n$$

The year 2005 is 7 years after 1998, so 2005 would correspond to $n = 7$. The projected pet drug sales S for the year 2005 are

$$1.3 \cdot (1.15) \approx 1.3 \cdot 2.66 \approx \$3.46 \text{ billion}$$

Algebra Aerobics 5.4

1. The annual report of the Health Care Financing Administration published in January 2000 stated that overall health care spending in the United States rose from $3,912 per person in 1997 to $4,094 per person in 1998, for a total of $1.1 trillion dollars.
   a. Calculate the percentage increase in health care spending per person from 1997 to 1998.
   b. Assuming this annual percentage increase were to continue for the next few years, construct a function to model the growth in health care spending per person. Identify your variables. What is the growth rate? The growth factor?
   c. Using your model, predict the health care spending per person in 2005.
2. Fill in the following table.

Exponential Function	Initial Value	Growth or Decay?	Growth or Decay Factor	Growth or Decay Rate
$A = 4(1.03)^t$				
$A = 10(0.98)^t$				
	1000		1.005	
	30		0.96	
	$50,000	Growth		7.05%
	200 grams	Decay		49%

5.5 EXAMPLES OF EXPONENTIAL GROWTH AND DECAY

Medicare Costs

The costs for almost every aspect of health care in America have risen dramatically over the last 30 years. One of the central issues in the ongoing debate about containing health care costs is the amount of federal dollars spent on Medicare. Medicare is a federal program that since July 1966 has provided two coordinated

See Excel or graph link file MEDICARE.

Table 5.8
Medicare Costs

Year	Medicare Expenses (billions of dollars)	Year	Medicare Expenses (billions of dollars)
1970	7.7	1984	66.5
1971	8.5	1985	72.2
1972	9.4	1986	76.9
1973	10.8	1987	82.3
1974	13.5	1988	89.4
1975	16.4	1989	102.5
1976	19.8	1990	112.1
1977	23.0	1991	123.0
1978	26.8	1992	138.7
1979	31.0	1993	151.7
1980	37.5	1994	169.2
1981	44.9	1995	184.2
1982	52.5	1996	200.3
1983	59.8	1997	214.6

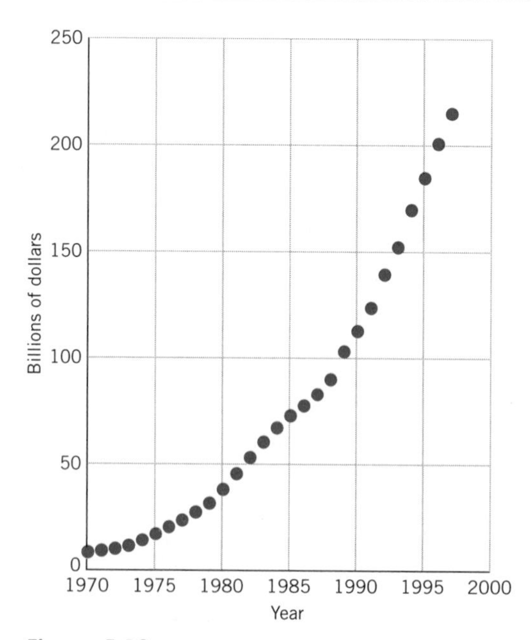

Figure 5.12 Medicare expenditures (in billions of dollars).
Source: U.S. Bureau of the Census, *The Statistical Abstract of the United States,* 1999.

plans for nearly all people age 65 and over: a hospital insurance plan and a voluntary supplementary medical insurance plan. Table 5.8 and the graph in Figure 5.12 show annual Medicare expenses (in billions of dollars) since 1970.

To simplify computations, it is convenient to make the base year, 1970, correspond to $x = 0$. Then 1971 would be represented by $x = 1$, 1972 by $x = 2$, 1980 by $x = 10$, etc. (Table 5.9). A curve-fitting program gives a best fit exponential function as

$$y = 9.1 \cdot (1.13)^x$$

where x = years since 1970 and y = Medicare expenses in billions of dollars. The function is graphed in Figure 5.13. The initial quantity of 9.1 billion is the estimated value for Medicare expenses in 1970 in our *model*. It does not represent the actu-

Table 5.9

Years Since 1970	Medicare Expenses (billions of dollars)	Years Since 1970	Medicare Expenses (billions of dollars)
0	7.7	14	66.5
1	8.5	15	72.2
2	9.4	16	76.9
3	10.8	17	82.3
4	13.5	18	89.4
5	16.4	19	102.5
6	19.8	20	112.1
7	23.0	21	123.0
8	26.8	22	138.7
9	31.0	23	151.7
10	37.5	24	169.2
11	44.9	25	184.2
12	52.5	26	200.3
13	59.8	27	214.6

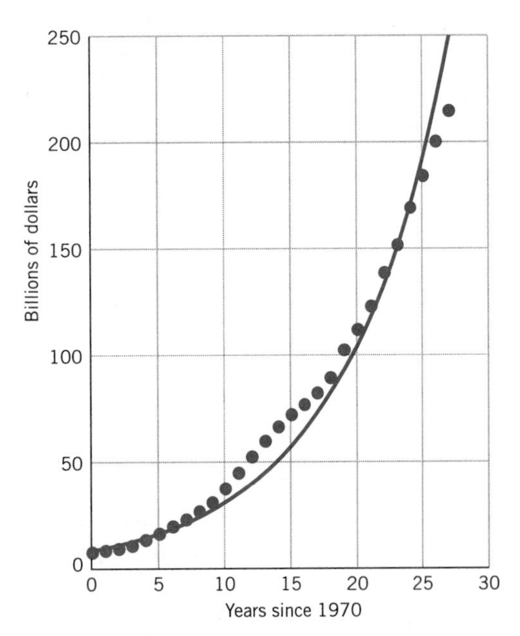

Figure 5.13 Best fit exponential function for Medicare expenses.

al value of 7.7 billion in Table 5.9. The base of 1.13 is the growth factor. According to our model, multiplying each year's Medicare expenses by 1.13 would approximate the next year's expenses. Multiplying by 1.13 is equivalent to calculating 113%. This represents an annual growth rate of 13%.

According to industry sources, the global wireless infrastructure market is growing 22% annually. In the United States every minute 26 more people sign up for wireless phone service. Which of these two statements represents linear and which exponential growth?

A Linear vs. an Exponential Model through Two Points

A town's population increases from 20,000 to 24,000 over a 5-year period. The Town Council expects the population to continue to grow and would like a mathematical description that can be used to predict future population size. The two simplest growth models are linear and exponential.

If we let t = year and P = population size, then when $t = 0$, $P = 20{,}000$ and when $t = 5$, $P = 24{,}000$. This gives us the two points, (0, 20000) and (5, 24000), which we will use to construct our models.

Linear model: The population increases by a fixed **amount** *each year.* Assuming linear growth, we have $P = b + mt$. The starting population, b, is 20,000. The average rate of change or slope, m, through the two points is:

$$m = \frac{\text{change in population}}{\text{change in time}} = \frac{24{,}000 - 20{,}000}{5 - 0} = \frac{4{,}000}{5} = 800 \text{ people/yr}$$

So the equation is:

$$P = 20{,}000 + 800t$$

This linear model says that the original population of 20,000 is increasing by the fixed amount of 800 people each year.

Exponential model: The population increases by a fixed **percentage** *each year.* Assuming exponential growth, we have $P = Ca^t$. The starting population, C, is 20,000. We have enough information to find the 5-year growth factor from which we can then determine the annual growth factor. Since exponential growth is multiplicative, we can calculate the 5-year growth factor by dividing population sizes. So

$$\frac{\text{population in year 5}}{\text{population in year 0}} = \frac{24{,}000}{20{,}000} = 1.2 = \text{5-year growth factor}$$

Since the ratio of the two populations is 1.2, this means that after 5 years the original population is multiplied by 1.2 or, equivalently, increased by 20%. After 5 years we also know that the starting population will be multiplied by a, the *annual* growth factor, 5 times; that is, multiplied by a^5. So we must have

$$a^5 = 1.2$$

Taking the fifth root of both sides gives us

$$(a^5)^{1/5} = (1.2)^{1/5}$$

$$a^1 \approx 1.0371$$

Thus a, the annual growth factor, is 1.0371. The annual growth rate is 0.0371 in decimal form, or 3.71%, expressed as a percentage. The equation is then:

$$P = 20{,}000 \cdot 1.0371^t$$

This exponential model says that the original population of 20,000 is increasing by a fixed 3.71% each year.

We can use the models not only to describe past behavior, but to predict future population sizes. (See Table 5.10 and Figure 5.14.) Note that we assumed the populations were the same in both models in year 0 and year 5. For year 6, the linear and exponential predictions are fairly close: 24,800 vs. approximately 24,890. As we move further beyond year 5, the exponential predictions exceed the linear by a greater and greater amount. By year 25, the linear model predicts a population of 40,000, while the exponential predicts almost 50,000. Both models should be considered as generating only crude future estimates particularly since we constructed the models using only two data points. Clearly the further out we try to predict, the more unreliable the estimates become.

Table 5.10
Town Population

Year	Linear Model $P = 20000 + 800t$	Exponential Model $P = 20000(1.0371)^t$	Year	Linear Model $P = 20000 + 800t$	Exponential Model $P = 20000(1.0371)^t$
0	20,000	20,000	13	30,400	32,110
1	20,800	20,740	14	31,200	33,310
2	21,600	21,510	15	32,000	34,540
3	22,400	22,310	16	32,800	35,820
4	23,200	23,140	17	33,600	37,150
5	24,000	24,000	18	34,400	38,530
6	24,800	24,890	19	35,200	39,960
7	25,600	25,810	20	36,000	41,440
8	26,400	26,770	21	36,800	42,980
9	27,200	27,760	22	37,600	44,570
10	28,000	28,790	23	38,400	46,230
11	28,800	29,860	24	39,200	47,940
12	29,600	30,970	25	40,000	49,720

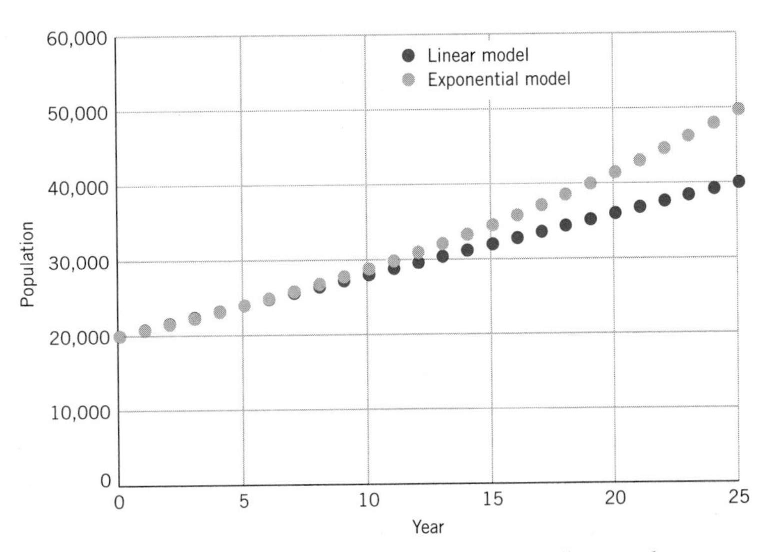

Figure 5.14 Predictions for the town population using linear and exponential models.

Algebra Aerobics 5.5a

1. The United Nations' Department of Economic and Social Affairs stated in 1999 that "the world's population stands at 6 billion and is growing at 1.3% per year, for an annual net addition of 78 million people."
   a. This statement actually contains two contradictory descriptions of predicting world population growth. Why?
   b. Using the information in the statement, construct a linear model for world population growth.
   c. Construct an exponential model for world population growth.
   d. What do both the models predict for the world's population in 2020? In 2050?
2. **a.** Assume that in the accompanying table quantity P is a linear function of time, t.
   i. What is the average rate of change of P with respect to t?
   ii. Fill in the rest of the P values in the table.
   iii. Construct an equation to model the relationship between P and t.
   b. Assume the variable Q is an exponential function of t.
   i. By what is Q multiplied, when t increases by 2 years? You can think of this as a 2 year growth factor.
   ii. Fill in the rest of the Q values in the table.
   iii. What is the annual growth factor?
   iv. Construct an equation to model the relationship between Q and t.

t (years)	P	Q
0	10	10
2	20	20
4		
6		
8		

Radioactive Decay

One of the toxic radioactive byproducts of nuclear fission is strontium-90. A nuclear accident, like the one in Chernobyl, can release clouds of gas containing strontium-90. The clouds deposit the strontium-90 onto vegetation eaten by cows. Humans ingest strontium-90 from the cows' milk. Strontium-90 replaces calcium in bones, causing cancer. Strontium-90 is particularly insidious because it has a *half-life* of approximately 28 years. That means that every 28 years about half (or 50%) of the existing strontium-90 has decayed into nontoxic, stable zirconium-90, but the other half is still in your bones. The following function, f, gives the amount of remaining strontium-90, after T time periods, where a time period represents 28 years. The number 100 represents the initial amount of strontium-90 in milligrams:

$$f(T) = 100(0.5)^T$$

The number 0.5 represents the decay factor for a 28 year period. Since we usually measure time just in years, not periods of years, how can we translate this into an annual decay factor, a? We know that after 28 years any quantity of strontium-90 will be multiplied by a 28 times; that is, multiplied by a^{28}. Hence a^{28} represents the decay factor for 28 years, so we must have

$$a^{28} = 0.5$$

Taking the 28th root of both sides gives us

$$(a^{28})^{1/28} = (0.5)^{1/28}$$
$$a^1 \approx 0.9755$$

Table 5.11
Decay of Strontium-90

Time, t (years)	Strontium-90 (mg)
0	100.000
28	50.000
56	25.000
84	12.500
112	6.250
140	3.125

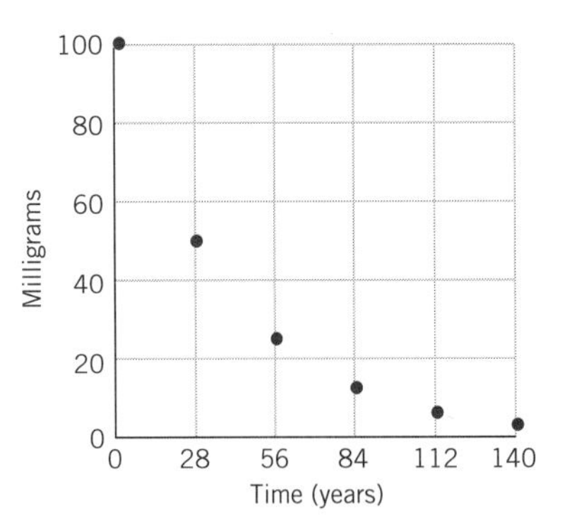

Figure 5.15 Amount of strontium-90.

So a, the annual decay factor, is 0.9755. Hence we can rewrite our original function as

$$f(t) = 100(0.9755)^t$$

where t = the time in years. The annual decay rate would be $1 - 0.9755 = 0.0245$. That means that each year the amount of strontium-90 would be reduced by 2.45%.

Table 5.11 and Figure 5.15 show the decay of 100 mg of strontium-90 over time.

It takes about 28 years for half of the strontium-90 to decay, but half still remains. It takes an additional 28 years for the remaining amount to halve again, still leaving 25 grams out of the original 100.

Every exponential function has a fixed half-life or doubling time. For example, strontium-90 has a fixed half-life of 28 years. The *E. coli* in our initial model have a fixed doubling time of 1 time period, or 20 minutes.

> The *doubling time* of an exponentially growing quantity is the time required for the quantity to double in size.
>
> The *half-life* of an exponentially decaying quantity is the time required for one-half of the quantity to decay.
>
> Exponential functions are the only functions with constant doubling or halving times.

Using a graphing calculator or a spreadsheet, test the rule of 70. Can you find a value for R for which this rule does not work so well?

The "Rule of 70": A Rule of Thumb for Calculating Doubling or Halving Times

A simple way to understand the significance of percentage growth rates is to compute the doubling time. It has been observed that if a quantity is growing at R percent per year, then its doubling time is approximately $70/R$ years. If the quantity is growing at R percent per *month,* then $70/R$ gives its doubling time in *months.* The same reasoning holds for any unit of time.

The rule of 70 is easy to apply. We just need to remember to use values for R in percentage form, such as 5% or 10.3%, not their equivalents in decimal form, such as 0.05 or 0.103.

For now we'll take the rule of 70 on faith. It provides good approximations, especially for smaller values of R (those under 10%). When we return to logarithms in Chapter 6, we'll find out why.

Example 1

Suppose that at age 25, you invest \$1000 in a retirement account, which grows at 5% per year. Since $70/5 = 14$, your investment will double approximately every 14 years. If you retire 42 years later at age 67, then three doubling periods will have elapsed ($3 \cdot 14 = 42$). So your \$1000 investment (disregarding inflation) will have increased in value by a factor of 2^3 and be worth \$8000.

Example 2

According to Mexico's National Institute of Geography, Information and Statistics (at *www.inegi.gob.mx/difusion/ingles*), in 1997 Mexico's population reached 93.7 million. The annual growth rate is listed as approximately 1.4%. The web site states that "if this rate persists, the Mexican population will double in 49.9 years." Does that time period seem about right?

Using the rule of 70, the approximate doubling time for a 1.4% annual growth rate would be $70/1.4 = 50$ years. So 49.9 is a reasonable value for the doubling time.

The rule of 70 also applies when R represents the percentage at which some quantity is decaying. In these cases, $70/R$ equals the half-life.

> The *Rule of 70* states that if a quantity is growing (or decaying) at $R\%$ per year, then the doubling time (or half-life) is approximately
>
> $$\frac{70}{R}$$
>
> provided R is not much bigger than 10.

Example 3

Plutonium, the fuel for atomic weapons, has an extraordinarily long half-life of about 24,400 years. Once the radioactive element plutonium is created from uranium, 25,000 years later almost half the original amount will remain. You can see why there is concern over stored caches of atomic weapons. Using the rule of 70, we can estimate R, plutonium's annual decay rate:

The rule of 70 gives	$70/R = 24{,}400$
multiply by R	$70 = 24{,}400R$
divide and simplify	$R = 70/24{,}400$
	≈ 0.003

Note that R is already in percentage, not decimal form. So the annual decay rate is a tiny 0.003%, or 3 thousandths of one percent.

An atomic weapon is usually designed with a 1% mass margin. That is, it will remain functional until the original fuel has decayed by more than 1%, leaving less than 99% of the original amount. Estimate how many years a plutonium bomb would remain functional.

Algebra Aerobics 5.5b

1. Estimate the doubling time using the rule of 70 when:
   a. $g(x) = 100(1.02)^x$, where x is in years
   b. $M = 10{,}000(1.005)^t$, where t is in months
   c. The annual growth rate is 8.1%
   d. The annual growth factor is 1.065
2. Use the rule of 70 to approximate the growth rate when the doubling time is:
   a. 10 years b. 5 minutes c. 25 seconds
3. Estimate the time it will take an initial quantity to drop to half its value when:
   a. $h(x) = 10(0.95)^x$, where x is in months
   b. $K = 1000(0.75)^t$, where t is in seconds
   c. The annual decay rate is 35%

See Excel or graph link file CELCOUNT

White Blood Cell Counts

On September 27, 1996, a patient was admitted to Brigham and Women's Hospital in Boston for a bone marrow transplant. The transplant was needed to cure myelodysplastic syndrome, in which the patient's own marrow fails to produce enough white blood cells to fight infection; both Carl Sagan and Paul Tsongas had this disease and underwent this treatment.

The patient's own bone marrow was intentionally destroyed using chemotherapy and radiation, and on October 3 the donated marrow was injected. Each day, the hospital carefully monitored the patient's white blood cell count to detect when the transplant took hold and the new marrow became active. The patient's counts are listed in Table 5.12 and plotted in Figure 5.16. Normal counts for a healthy individual are between 4,000 and 10,000 cells per milliliter.

Table 5.12
White Blood Cell Counts

Date	Cell Count (per ml)	Date	Cell Count (per ml)
27 Sept	1,390	18 Oct	290
28 Sept	2,570	19 Oct	310
29 Sept	2,290	20 Oct	450
30 Sept	2,660	21 Oct	580
1 Oct	1,720	22 Oct	740
2 Oct	1,290	23 Oct	1,070
3 Oct	500	24 Oct	1,210
4 Oct	160	25 Oct	1,870
5 Oct	110	26 Oct	2,030
6 Oct	120	27 Oct	2,540
7 Oct	130	28 Oct	3,350
8 Oct	120	29 Oct	5,460
9 Oct	60	30 Oct	6,940
10 Oct	60	31 Oct	8,640
11 Oct	40	1 Nov	7,650
12 Oct	60	2 Nov	7,430
13 Oct	70	3 Nov	8,790
14 Oct	60	4 Nov	10,100
15 Oct	70	5 Nov	9,620
16 Oct	170	6 Nov	9,420
17 Oct	180		

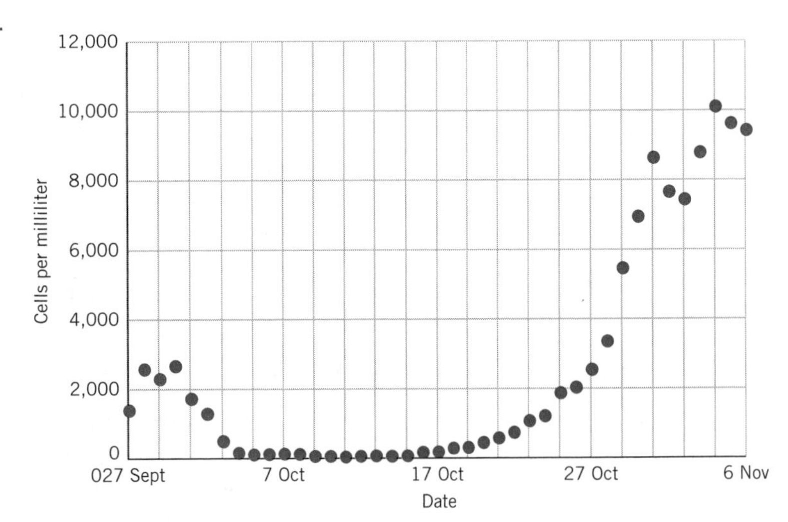

Figure 5.16 White blood cell count.

Table 5.13
Counts Between October 15 and 31

Days *after* October 15	Count (per ml)
0	70
1	170
2	180
3	290
4	310
5	450
6	580
7	740
8	1,070
9	1,210
10	1,870
11	2,030
12	2,540
13	3,350
14	5,460
15	6,940
16	8,640

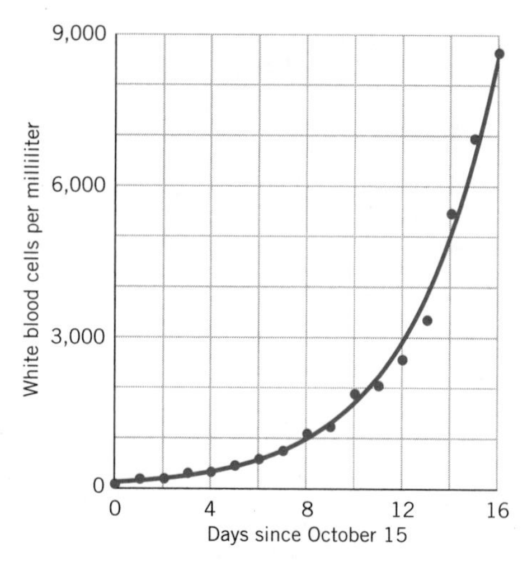

Figure 5.17 White blood cell counts between October 15 and 31.

The white blood cell count dropped steadily until about October 5. Between October 9 and 15, the count fluctuated below 100. October 15 was the beginning of a dramatic increase. The values rose steadily until October 31. After that, the counts fluctuated as the patient's body began to regulate itself. Several months after the transplant, the count stabilized at about 6200.

Clearly the whole data set does not represent exponential growth, but we can reasonably model the data between say October 15 and 31 with an exponential function. Table 5.13 records the number of days after October 15 and the corresponding counts. So October 15 corresponds to 0, October 16 corresponds to 1, etc. This subset of the original data is plotted in Figure 5.17 along with a computer-generated best fit exponential function,

$$y = 105(1.32)^x$$

where x = number of days after October 15 and y = white blood cell count.

The fit looks quite good. The initial quantity of 105 is the white blood cell count (per milliliter) for the model, not the actual white blood cell count of 70 measured by the hospital. This discrepancy is not unusual—remember that a best fit function may not necessarily pass through any of the specific data points. The growth factor is 1.32, so the growth rate is 0.32. So between October 15 and 31, the number of white blood cells was increasing at a rate of 32% a day. According to the rule of 70, the number of white blood cells was doubling roughly every $70/32 \approx 2.2$ days!

Compound Interest

Short-term returns. One of the most common examples of exponential growth is compound interest. Suppose you have \$100 that you could put either in a checking account that earns no interest, a NOW account that earns 3% compounded annually, a savings account that earns 5% compounded annually, or a certificate of deposit (CD) that earns 7% compounded annually. How would your results compare after 10 years? After 25 years?

Assuming you make no withdrawals, the money in the checking account will remain constant at \$100. The money in the interest-bearing accounts will obviously increase. But by how much?

An interest rate of 3% compounded annually means that at the end of each year, you earn 3% on the current value of your account. The interest is automatically deposited in your account. From then on you earn 3% not only on your *principal* (the initial amount you invested), but also on the interest you have already earned. The growth rate is 3%, or 0.03 in decimal form, and the growth factor is 1.03. So each year the current value of the account will be multiplied by 1.03. The functions representing P_r, the value of your account earning interest r, as a function of t, the number of years, would be

$$P_{0.03} = 100 \cdot (1.03)^t$$

$$P_{0.05} = 100 \cdot (1.05)^t$$

$$P_{0.07} = 100 \cdot (1.07)^t$$

In general,

> If
>
> P_0 = original investment
>
> r = interest rate (in decimal form)
>
> t = time periods at which the interest rate is compounded
>
> the resulting value, P_r, of the investment after t time periods is given by the formula
>
> $$P_r = P_0 \cdot (1 + r)^t$$

The interest rate, r, in decimal form is the growth rate and $1 + r$ is the growth factor.

Table 5.14 and Figure 5.18 compare the values of your account over 10 years. At 10 years, the \$100 in the 3% account has risen to \$134.39, while the \$100 in the 7% account has almost doubled to \$196.72. The constant value of the \$100 in the checking account is not listed in the table but appears on the graph as a horizontal line.

Table 5.14
Compound Interest over 10 Years

Years	Value of \$100 at 3%	Value of \$100 at 5%	Value of \$100 at 7%
0	\$100.00	\$100.00	\$100.00
1	\$103.00	\$105.00	\$107.00
2	\$106.09	\$110.25	\$114.49
3	\$109.27	\$115.76	\$122.50
4	\$112.55	\$121.55	\$131.08
5	\$115.93	\$127.63	\$140.26
6	\$119.41	\$134.01	\$150.07
7	\$122.99	\$140.71	\$160.58
8	\$126.68	\$147.75	\$171.82
9	\$130.48	\$155.13	\$183.85
10	\$134.39	\$162.89	\$196.72

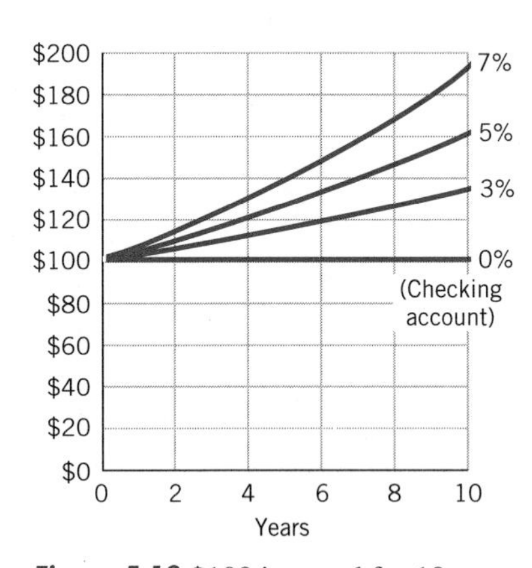

Figure 5.18 \$100 invested for 10 years at 3, 5, and 7% compounded annually.

Long-term returns. The longer the time period over which you invest, the more dramatic the results. Over time small differences in interest rates can produce enormous differences in returns. Table 5.15 and Figure 5.19 show the value of \$100 invested at 3%, 5%, and 7% over 40 years. (Note the difference in scales in Figures 5.18 and 5.19.)

Table 5.15
Compound Interest over 40 Years

Years	Value of $100 at 3%	Value of $100 at 5%	Value of $100 at 7%
0	$100	$100	$100
10	$134	$163	$197
20	$181	$265	$387
30	$243	$432	$761
40	$326	$704	$1497

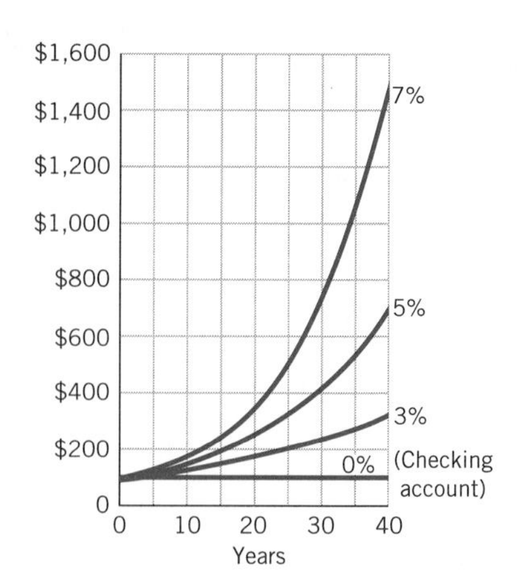

Figure 5.19 $100 invested for 40 years.

If you invested $100 at age 25 for your retirement 40 years later at 3% compounded annually, you'd end up with $326; at 5% you'd have $704; and at 7% you'd have almost $1500, or nearly 15 times as much as you started with. So if you were able to invest $10,000 at 7% for 40 years, you'd have about $150,000; if you won the lottery and could invest $100,000, you'd have almost $1.5 million dollars!

On March 13, 1997, a resident of Melrose, Massachusetts, "scratched" a $5 state lottery ticket and won $1,000,000. The Massachusetts State Lottery Commission notified her that after deducting taxes, they would send her a check for $33,500 every March 13th for 20 years. Is this fair?

Too good to be true? At decent interest rates, if we could all put aside a few thousand dollars each year, it seems we could retire in comfort. Is the picture as rosy as it seems? Not quite. We haven't taken inflation into account.

Compound interest calculations are the same whether you are dealing with inflation or investments. For example, a 5% annual inflation rate would mean that what cost $1 today would cost $1.05 one year from today.

If we think of the three percentages in Table 5.14 as representing inflation rates, then how much would something that costs $100 today cost in 10 years? It would cost $134.39, $162.89, or $196.72, if the annual inflation rate were 3%, 5%, or 7%, respectively. So if you *invest* $100 at 5% for 10 years, it will be worth $162.89. But inflation will drive up costs. If during those 10 years inflation is also 5% a year, then what originally cost $100 will now cost $162.89. In terms of purchasing power you will come out even, with no net gain or loss. So if the inflation rate equals the interest rate, you are not any better off. If the inflation rate were higher than the interest rate, you would actually lose money on your investment. Because of the erosive nature of inflation, most economists usually use *real* or *constant dollars,* which are dollars adjusted for inflation.

Inflation and the Diminishing Dollar

Inflation erodes the purchasing power of the dollar. If you have $1.00 today and annual inflation is 5%, how much will the dollar be worth in a year? In a year, what costs $1.00 today will cost $1.05. Since $1.00/1.05 \approx 0.952$, then $\$1.00 \approx 0.952 \cdot (\$1.05)$. So in a year $1.00 will be worth only $0.952 in today's dollars. The decay factor is 0.952. The exponential function

$$D = 1 \cdot (0.952)^t$$

gives D, the value (or purchasing power) of a future dollar in terms of today's dollars, at year t if there is a steady inflation rate of 5%. In economists' terms, it gives the real purchasing power in today's dollars. Table 5.16 and Figure 5.20 give an indication of

Table 5.16
Inflation's Erosion of the Dollar

Year	Value of Dollar (with 5% inflation)
0	$1.00
1	$0.95
2	$0.91
3	$0.86
4	$0.82
5	$0.78
6	$0.74
7	$0.71
8	$0.67
9	$0.64
10	$0.61
11	$0.58
12	$0.55
13	$0.53
14	$0.50
15	$0.48
16	$0.46
17	$0.43
18	$0.41
19	$0.39
20	$0.37

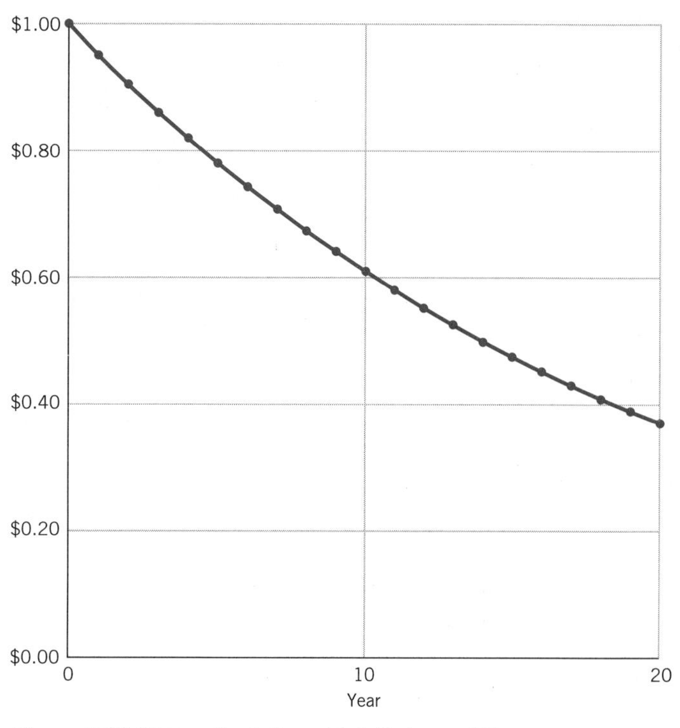

Figure 5.20 Value of a dollar with inflation at 5%.

how rapidly the value of the dollar declines. (Dollar values are rounded to the nearest penny.)

Every 14 years, the purchasing power of your money is cut in half. We can think of 14 years as the half-life of the dollar. So, in 14 years, your dollar is worth 50¢ in today's cents. In 28 years, your dollar is worth only 25¢ in today's cents. This is obviously a serious problem faced by those who retire on a fixed-income pension.

Algebra Aerobics 5.5c

1. Approximately how long would it take for your money to double if the interest rate, compounded annually, were:

   a. 3% **b.** 5% **c.** 7%

2. Suppose you are planning to invest a sum of money. Determine the rate that you need so that your investment doubles in:

   a. 5 years **b.** 10 years **c.** 7 years

3. Construct a function that would represent the resulting value if you invested $1000 for n years at an annually compounded interest rate of:

   a. 4% **b.** 11% **c.** 110%

4. In the early 1980s Brazil's inflation was running rampant at about 10% per *month*.[2] Assuming this inflation rate continued unchecked, construct a function to describe the purchasing power of 100 cruzeiros after n months. (A cruzeiro is a Brazilian monetary unit.) What would 100 cruzeiros be worth after 3 months? After 6 months? After a year?

[2]In such cases sooner or later the government usually intervenes. In 1985 the Brazilian government imposed an anti-inflationary wage and price freeze. When the controls were dropped, inflation soared again, reaching a high in March 1990 of 80% per month! At this level, it begins to matter whether you buy groceries in the morning or wait until that night. On August 2, 1993, the government devalued the currency by defining a new monetary unit, the cruzeiro real, equal to 1000 of the old cruzeiros. Inflation still continued, and on July 1, 1994, yet another unit, the real, was defined equal to 2740 cruzeiros reales. By 1997, inflation had slowed considerably to about 0.1% per month.

Musical Pitch

Exponential functions can also describe the relationship between musical octaves and vibration frequency. The vibration frequency of the note A above middle C is 440 cycles per second (or 440 hertz). The vibration frequency doubles at each octave. So we can define an exponential growth function that gives the vibration frequency, F, in hertz (Hz) as a function of N, the number of octaves above or below A. The growth factor is 2 and the initial frequency is 440. So we have

$$F = 440 \cdot 2^N$$

Table 5.17 and the graph in Figure 5.21 show a few of the values for the vibration frequency.

"E6: Musical Keyboard Frequencies" in *Exponential & Log Functions* offers a multimedia demonstration of this function.

Table 5.17
Musical Octaves

Number of Octaves above or below A (at 440 Hz), N	Vibration Frequency (Hz), F
−3	55
−2	110
−1	220
0	440
1	880
2	1,760
3	3,520

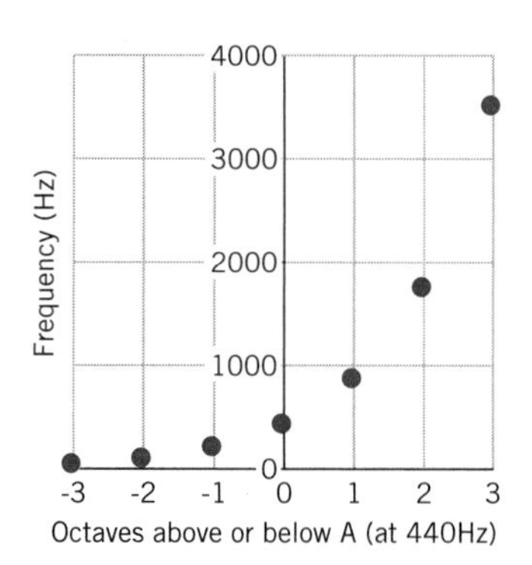

Figure 5.21 Octaves versus frequency.

There are 12 notes in each octave. If we let n = number of notes above or below A, then we have $n = 12N$. Solving for N, we have $N = n/12$. By substituting this expression for N, we can define F as a function of n.

$$\begin{aligned} F &= 440 \cdot 2^{n/12} \\ &= 440 \cdot 2^{(1/12) \cdot n} \\ &= 440 \cdot (2^{1/12})^n \\ &\approx 440 \cdot (1.059)^n \end{aligned}$$

So each note on the "even-tempered" scale has a frequency 1.059 times the frequency of the preceding note.

The Malthusian Dilemma

The most famous attempt to predict growth mathematically was made by a British economist and clergyman, Thomas Robert Malthus, in an essay published in 1798. He argued that the growth of the human population would overtake the growth of food supplies, because the population size was *multiplied* by a fixed amount each year, whereas food production only increased by *adding* a fixed amount each year. In other words, he assumed populations grew exponentially and food supplies grew linearly as functions of time. He concluded that humans were condemned always to breed to the point of misery and starvation, unless the population were reduced by other means, including war or disease.

We can frame his arguments algebraically by letting P_0 represent the original population size, t the time in years, and a the annual growth factor. Then P, the population at time t, is given by

$$P = P_0 \cdot a^t$$

If F_0 represents the amount of food at time $t = 0$ and q the constant quantity added to the food supply each year, then F, the total amount of food, is

$$F = F_0 + q \cdot t$$

Malthus believed the population of Britain, then about 7,000,000, was growing by 2.8% per year ($P_0 = 7{,}000{,}000$ and $a = 1.028$). He counted food supply in units that he defined to be enough food for one person for a year. At that time food supply was adequate, so he assumed that Britons were producing 7,000,000 food units. He thought they could increase food production by about 280,000 units a year ($F_0 = 7{,}000{,}000$ and $q = 280{,}000$). So the two functions are

$$P = 7{,}000{,}000 \cdot (1.028t) \qquad F = 7{,}000{,}000 + 280{,}000 \cdot t$$

Table 5.18 and Figure 5.22 reveal that if the formulas were good models, then after about 25 years population would start to exceed food supply, and some people would starve.

The two centuries since Malthus published his famous essay have not been kind to his theory. The population of the British Isles in 1995 was about 58 million whereas Malthus's model predicted over 100 million people before the year 1900. Improved food production techniques and the opening of new lands to agriculture have kept food production in general growing faster than the population. The distribution of food is a problem and famines still occur with unfortunate regularity in parts of the world, but the mass starvation Malthus predicted has not come to pass.

Table 5.18
Growth in Population vs. Food

Year	Population (millions)	Food Units (millions)
0	7.00	7.00
5	8.04	8.40
10	9.23	9.80
15	10.59	11.20
20	12.16	12.60
25	13.96	14.00
30	16.03	15.40
35	18.40	16.80
40	21.13	18.20
45	24.25	19.60
50	27.85	21.00
100	110.77	35.00

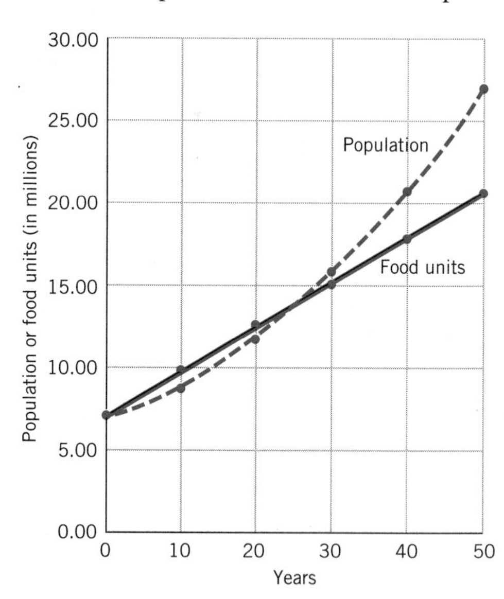

Figure 5.22 Malthus's predictions for population and food.

Trees

Tree structures offer another useful way of visualizing exponential growth. A computer program can generate a tree by drawing two branches at the end of a trunk, then two smaller branches at the ends of each of those branches, and two smaller branches at the ends of the previous branches, and so on until it reaches twig size. This kind of structure produced from self-similar repeating scaled graphic operations is called a *fractal* (Figure 5.23). There are many examples of fractal structures in nature, such as ferns, coastlines, and human lungs.

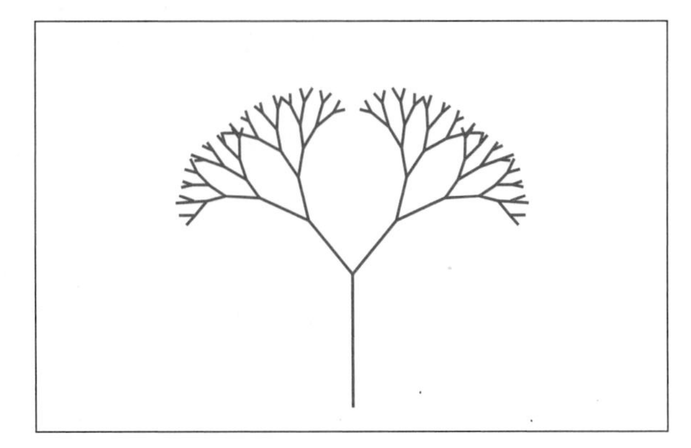

Figure 5.23 A fractal tree.

The fractal tree shown is drawn in successive levels (Figure 5.24):

At level 0 it draws the trunk, 1 line.
At level 1 it draws 2 branches on the previous 1, for a total of 2 new lines.
At level 2 it draws 2 branches on each of the previous 2, for a total of 4 new lines.
At level 3 it draws 2 branches on each of the previous 4, for a total of 8 new lines.

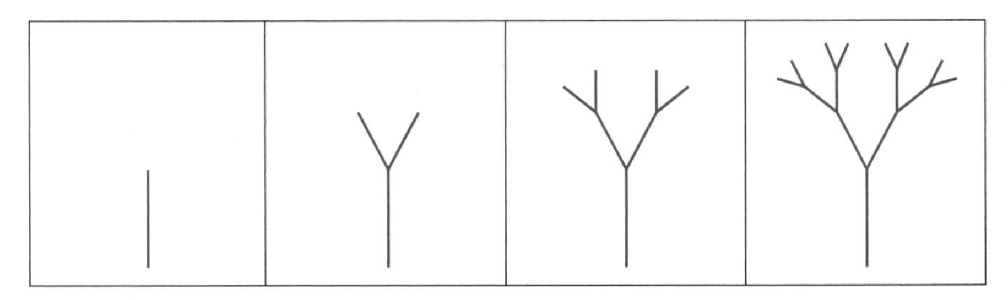

Figure 5.24 Fractal tree: levels 0, 1, 2, and 3.

Table 5.19 shows the relationship between the level, L, and the number of new lines, N, at each level. The formula

$$N = 2^L$$

describes the relationship between level, L, and the number of new lines, N. For example, at the fifth level, there would be $2^5 = 32$ new lines.

Table 5.19

Level, L	New Lines, N	N as Power of 2
0	1	$1 = 2^0$
1	$2 \cdot 1 = 2$	$2 = 2^1$
2	$2 \cdot 2 = 4$	$4 = 2 \cdot 2 = 2^2$
3	$2 \cdot 4 = 8$	$8 = 2 \cdot 2 \cdot 2 = 2^3$
4	$2 \cdot 8 = 16$	$16 = 2 \cdot 2 \cdot 2 \cdot 2 = 2^4$

Example 3

You can think of Figure 5.23 as depicting your family tree. Each level represents a generation. The trunk is you (at level 0). The first two branches are your parents (at level 1). The next four branches are your grandparents (at level 2), and so on. How many ancestors do you have 10 generations back? The answer is $2^{10} = 1024$; that is, you have 1024 great-great-great-great-great-great-great-great-grandparents.

Example 4

An information system that is similar to this process is an emergency phone tree, in which 1 person calls 2 others, each of whom calls 2 others, until everyone in the organization has been called. How many levels of phone calls would be needed to reach an organization with 8000 people?

If you look back at your family tree forty generations ago (roughly 800 years) you had 2^{40} ancestors. This number is larger than all the people that ever lived on the surface of the earth. How can that be?

Solution

If we think of Figure 5.23 and Table 5.19 as representing this phone tree, then each new line (or branch) represents a person. We need to count not just the number of new people N at each level L, but also all the previous people called. At level 0, there is the one person who originates the phone calls. At level 1, there is the original person plus the 2 he or she called, for a total of $1 + 2 = 3$ people. At level 2, there are $1 + 2 + 4 = 7$ people, etc. At level 11, there are $1 + 2 + 4 + 8 + 16 + 32 + 64 + 128 + 256 + 512 + 1024 + 2048 = 4095$ people who have been called. At level 12, there would be $2^{12} = 4096$ new people called, for a total of $4095 + 4096 = 8191$ people called. So it would take 12 levels of the phone tree to reach 8000 people.

Algebra Aerobics 5.5d

1. Assume the tree-drawing process was changed to draw three branches at each level.
   a. Draw a trunk and at least two levels of the tree.
   b. What would the general formula be for N, the number of new lines, as a function of L, the level?
2. In an emergency phone tree, in which 1 person calls 3 others, each of whom calls 3 others, and so on, until everyone in the organization has been called, how many levels of phone calls are required for this phone tree to reach an organization of 8,000 people?

5.6 SEMI-LOG PLOTS OF EXPONENTIAL FUNCTIONS

With exponential growth functions, we very often have the same problem that we did in Chapter 4 when we compared the size of atoms to the size of human beings and to the size of the solar system: The numbers go from very small to very large. For example, in our *E. coli* experiment, how can we determine from a graph whether the growth from 100 to 200 cells follows the same rule as the growth from 100 million to 200 million? It is virtually impossible to display the entire data set on a standard graph.

One solution is to use a logarithmic, or order of magnitude, scale on the vertical axis. Recall that in Chapter 4 we used a logarithmic scale on a single horizontal axis. Moving a fixed distance up or down on a vertical logarithmic scale corresponds to *multiplying* the variable by a constant factor, rather than to *adding* a con-

Table 5.20
Growth of *E. coli*.

t (20-minute Time Periods)	$N = 100 \cdot 2^t$ (No. of *E. coli* Bacteria)
0	100
1	200
2	400
3	800
4	1,600
5	3,200
6	6,400
7	12,800
8	25,600
9	51,200
10	102,400
11	204,800
12	409,600
13	819,200
14	1,638,400
15	3,276,800
16	6,553,600
17	13,107,200
18	26,214,400
19	52,428,800
20	104,857,600
21	209,715,200
22	419,430,400
23	838,860,800
24	1,677,721,600

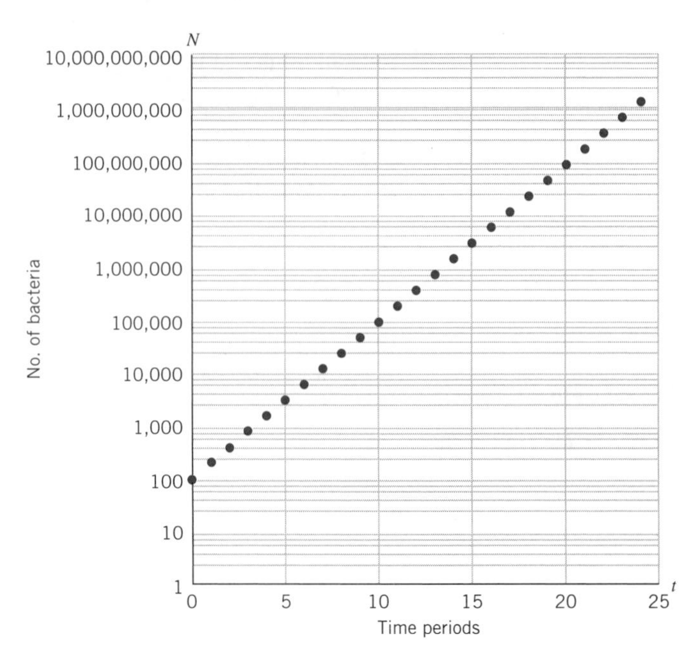

Figure 5.25 Semi-log plot of $N = 100 \cdot 2^t$ number of *E. coli* bacteria.

stant as on a linear scale. In exponential growth the dependent variable is multiplied by a constant factor each time a fixed constant is added to the independent variable. Exponential growth always appears as a line of constant slope when a logarithmic scale is used on the vertical axis and a standard "linear" scale is used on the horizontal axis. We'll take a closer look at why this is true in Chapter 6. Since most graphing software easily does such *log-linear* or *semi-log* plots, this is one of the easiest and most reliable ways to recognize exponential growth in a data set.

Table 5.20 repeats the *E. coli* data from Table 5.1. Figure 5.25 shows a graph of the data plotted on a semi-log graph. Recall that the data fit the exponential function $N = 100 \cdot 2^t$. On the vertical axis, successive powers of 10 appear at equally spaced intervals: 100 million is just as "far" from 10 million as 100 is from 10. We can display the entire data set on a single graph, and its straight-line shape immediately tells us that it represents exponential growth.

> When an exponential function is plotted using a standard linear scale on the horizontal axis and a logarithmic scale on the vertical axis, its graph is a straight line. This type of graph is called a *log-linear* or *semi-log* plot.

Algebra Aerobics 5.6

1. **a.** Using the graph in Figure 5.25, estimate the time interval it takes for the population to increase by a factor of 10.

   b. From the original expression for the population, $N = 100 \cdot 2^t$, over what time interval does the population increase by a factor of 8?

   c. By a factor of 16?

   d. Are these three answers consistent with each other?

Chapter Summary

Exponential Functions. Exponential functions can be used to model any phenomenon that has a constant growth (or decay) factor, that is, any phenomenon that increases (or decreases) by a fixed multiple at regular intervals.

An exponential function has the form

$$y = C \cdot a^x \qquad (a > 0, \qquad a \neq 1)$$

where C is the *y-intercept* or *initial value* (when $x = 0$) and a is the factor by which the value of y is multiplied when x increases by 1. If $a > 1$, we have *exponential growth,* if $0 < a < 1$, we have *exponential decay.*

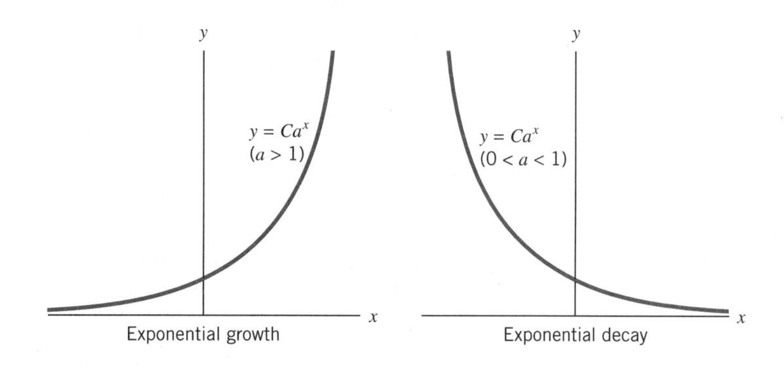

For example, the function $Q = 100 \cdot 1.5^t$ represents exponential growth with an initial population of 100 and a growth factor of 1.5. The function $Q = (6 \cdot 10^7)\,(0.95)^t$ represents exponential decay with an initial population of $6 \cdot 10^7$ and a decay factor of 0.95.

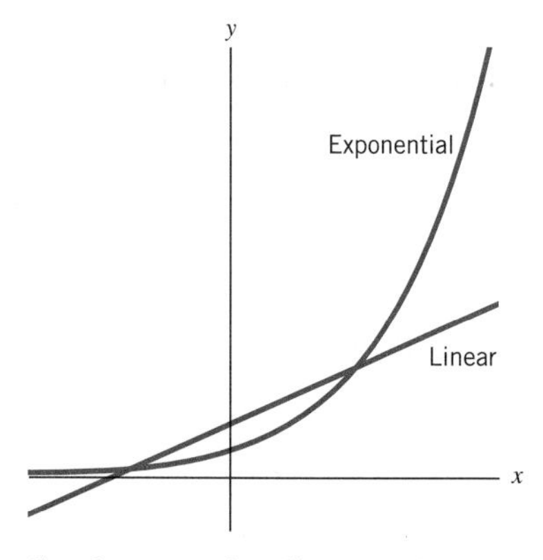

Graph comparing linear and exponential growth.

Linear vs. Exponential Growth. Exponential growth is *multiplicative,* while linear growth is *additive.* In exponential growth for each unit increase in the independent variable we *multiply* the value of the dependent variable by the growth factor. In linear growth we must *add* a fixed amount to the value of the dependent variable. For example, in the exponential function $N = 100 \cdot 2^t$, each time we increase t by 1, we *multiply* the value of N by 2. In the linear function $N = 100 + 2t$, each time we increase t by 1, we *add* 2 to the value of N.

If we compare any linear growth function (whose graph will be a straight line with a positive slope) to any exponential growth function (whose graph will curve upward), sooner or later the exponential curve will permanently lie above the linear graph and continue to grow faster and faster. We say that the exponential function eventually dominates the linear function.

The Graphs of Exponential Functions. For an exponential growth function, the larger the growth factor, a, the more rapid the growth, and the steeper the upward slope of the graph. For an exponential decay function, the smaller the decay factor, a, the more rapid the decay, and the steeper the downward slope of the graph.

Increasing the initial value C in the exponential function $y = Ca^x$, increases the y-intercept and the size of the y value corresponding to each x value.

A *horizontal asymptote* is a horizontal line which the graph of a function approaches for extreme values of the independent variable. The x-axis is a horizontal asympote for both exponential growth and decay functions of the form $y = Ca^x$.

Expressing Exponential Growth in Percentages. An exponential function can also be expressed in terms of percentages. The fixed percentage increase or decrease is called the growth or decay rate. If r is the *growth rate* (the percentage increase in decimal form), then the *growth factor*, a, equals $1 + r$ and

$$y = C \cdot a^x = C(1 + r)^x$$

For example, if r is a *growth rate* equal to 0.25, then $a = 1.25$. If r is the *decay rate* (the percentage decrease in decimal form), then the *decay factor*, a, equals $1 - r$ and

$$y = C \cdot a^x = C(1 - r)^x$$

For example, if r is a decay rate equal to 0.25, then $a = 0.75$.

Special Properties and Examples of Exponential Functions. The *doubling time* of an exponentially growing quantity is the time required for the quantity to double in size. The *half-life* of an exponentially decaying quantity is the time required for one-half of the quantity to decay.

The "rule of 70" offers a simple way to estimate the doubling time or the half-life. If a quantity is growing at R percent per year, then its doubling time is approximately $70/R$ years. If a quantity is growing at R percent per month, then $70/R$ gives its doubling time in months. For example, if a quantity is growing at 5% a year, then the quantity would double in approximately $70/5 = 14$ years.

One of the most common examples of exponential growth is compound interest. If

P_0 = original investment
r = interest rate (in decimal form)
t = time periods at which interest rate is compounded

the resulting value, P_r, of the investment after t time periods is given by the formula

$$P_r = P_0 \cdot (1 + r)^t$$

If you invested \$1000 at 6.5% interest rate compounded yearly, the value, P, of your investment after t years would be given by

$$P = 1000(1.065)^t$$

Tree structures offer another useful way of visualizing exponential growth.

When an exponential function is plotted using a standard scale on the horizontal axis and a logarithmic scale on the vertical axis, its graph is a straight line. This is called a *log-linear* or *semi-log* plot.

EXERCISES

Exercises for Section 5.1

1. Identify the initial population and the growth factor in each of the following exponential functions:.

a. $Q = 275 \cdot 3^T$ **b.** $P = 15{,}000 \cdot 1.04^t$ **c.** $y = (6 \cdot 10^8) \cdot 5^x$

2. Write an equation for an exponential growth function where:

a. The initial population is 350 and the growth factor is $\frac{4}{3}$.

b. The initial population is $5 \cdot 10^9$ and the growth factor is 1.25.

c. The initial population is 150 and the population triples during each time period.

3. A tuberculosis culture increases by a factor of 1.185 each hour.

a. If the initial concentration is $5 \cdot 10^3$ cells/ml, construct an exponential function to describe its growth over time.

b. What would the concentration be after 8 hours?

4. An ancient king of Persia was said to have been so grateful to one of his subjects that he allowed the subject to select his own reward. The clever subject asked for a grain of rice on the first square of a chess board, 2 grains on the second square, 4 on the next, and so on.

a. Construct a function that describes the number of grains of rice, G, as a function of the square, n, on the chess board. (*Note:* There are 64 squares.)

b. Construct a table recording the numbers of grains of rice on the first 10 squares.

c. Sketch your function.

d. How many grains of rice would the king have to provide for the 64th (and last) square?

5. a. Fill in the following table for the linear function $Q = 5 + 1.5t$

t	$Q = 5 + 1.5t$	Average Rate of Change (between $t - 1$ and t)
0		n.a.
1		
2		
3		
4		

b. Fill in the following table for the exponential function $Q = 5 \cdot 1.5^t$

t	$Q = 5 \cdot 1.5^t$	Average Rate of Change (between $t - 1$ and t)
0		n.a.
1		
2		
3		
4		

c. On the same graph, plot the linear function $Q = 5 + 1.5t$ and the exponential function $Q = 5 \cdot 1.5^t$ for $0 \leq t \leq 4$. Do the graphs intersect? Indicate when (if ever) the linear function would lie above the exponential or the exponential would lie above the linear. How would you expect the graphs to behave for values of $t > 4$?

d. Now on a new graph, plot the average rates of change of both the linear and exponential formulas as functions of t. What do these graphs tell you about the linear and exponential functions?

Exercises for Section 5.2

6. Which of the following exponential functions represent growth and which decay?

a. $N = 50 \cdot 2.5^T$

b. $y = 264(5/2)^x$

c. $R = 745(1.001)^t$

d. $g(z) = (3 \cdot 10^5) \cdot (0.8)^z$

e. $f(T) = (1.5 \cdot 10^{11}) \cdot (0.35)^T$

f. $h(x) = 2000(\frac{2}{3})^x$

7. Write an equation for an exponential decay function where:

a. The initial population is 10,000 and the decay factor is $\frac{2}{5}$.

b. The initial population is $2.7 \cdot 10^{13}$ and the decay factor is 0.27.

c. The initial population is 219 and the population drops to one-tenth its previous size during each time period.

8. The accompanying tables show approximate values for the four exponential functions: $f(x) = 5(2^x)$, $g(x) = 5(0.7^x)$, $h(x) = 6(1.7^x)$, and $j(x) = 6(0.6^x)$. Which table is associated with each function?

Function *A*	
x	y
−2	16.67
−1	10.00
0	6.00
1	3.60
2	2.16

Function *B*	
x	y
−2	10.0
−1	7.1
0	5.0
1	3.5
2	2.5

Function *C*	
x	y
−2	1.25
−1	2.50
0	5.00
1	10.00
2	20.00

Function *D*	
x	y
−2	2.1
−1	3.5
0	6.0
1	10.0
2	17.0

9. Determine which of the following functions are exponential. For each exponential function, identify the growth or decay factor and the vertical intercept.

a. $y = 5(x^2)$ **b.** $y = 100 \cdot 2^{-x}$ **c.** $P = 1000(0.999)^t$

10. Determine which of the following functions are exponential. Identify each exponential function as representing growth or decay and find the vertical intercept.

a. $A = 100(1.02^t)$ **c.** $g(x) = 0.3(10^x)$ **e.** $M = 2^P$

b. $f(x) = 4(3^x)$ **d.** $y = 100x + 3$ **f.** $y = x^2$

11. Plutonium-238 is used in bombs and power plants but is dangerously radioactive. It decays very slowly into nonradioactive materials. If you started with 100 grams today, a year from now you would still have 99.2 grams.

a. Construct an exponential function to describe the decay of plutonium-238 over time.

b. How much of the original 100 grams of plutonium-238 would be left after 50 years? After 500 years?

12. Potassium-40 radioactively disintegrates into two stable isotopes: calcium-40 and argon-40. It takes 1.31 billion years for the amount of potassium-40 to drop to half its original size.

a. Construct a function to describe the decay of potassium-40.

b. Approximately what amount of the original potassium-40 would be left after 4 billion years? Justify your answer.

Exercises for Section 5.3

13. Each of the following three exponential functions is in the standard form $y = C \cdot a^x$:

$$y = 2^x \qquad y = 5^x \qquad y = 10^x$$

a. In each case identify C and a.

b. Specify whether each function represents growth or decay. In particular, for each unit increase in x, what happens to y?

c. Do all three curves intersect? If so, where?

d. In the first quadrant, which curve should be on top? Which in the middle? Which on the bottom?

e. Describe any horizontal asymptotes.

f. For each function generate a small table of values.

g. Graph the three functions on the same grid and verify that your predictions in part (d) are correct.

14. Repeat Exercise 13 for the functions $y = (0.5)2^x$, $y = 2 \cdot 2^x$, and $y = 5 \cdot 2^x$.

15. Repeat Exercise 13 for the functions $y = 3^x$, $y = (\frac{1}{3})^x$, and $y = 3 \cdot (\frac{1}{3})^x$.

16. Match each graph with its correct equation:

$$f(x) = 30 \cdot 2^x \qquad h(x) = 100 \cdot 2^x$$

$$g(x) = 30 \cdot (0.5)^x \qquad j(x) = 50 \cdot (0.5)^x$$

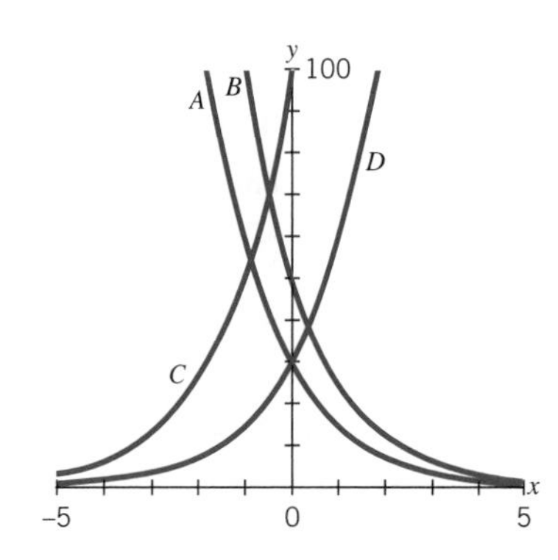

17. Below are sketches of the graphs of four exponential functions. Match each sketch with the function that best describes that graph:

$$P = 5 \cdot (0.7)^x \qquad R = 10 \cdot (1.8)^x$$

$$Q = 5 \cdot (0.4)^x \qquad S = 5 \cdot (3)^x$$

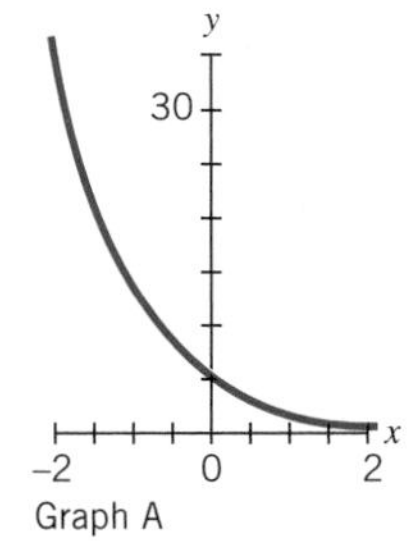

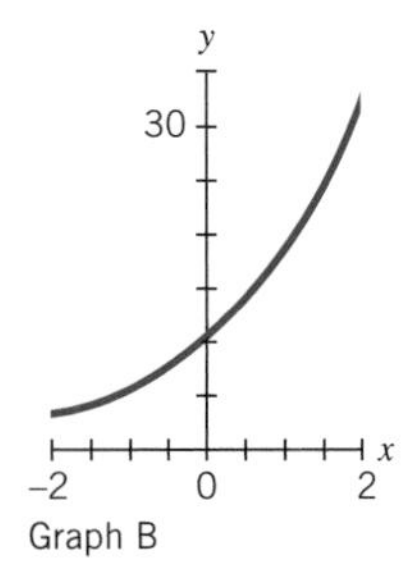

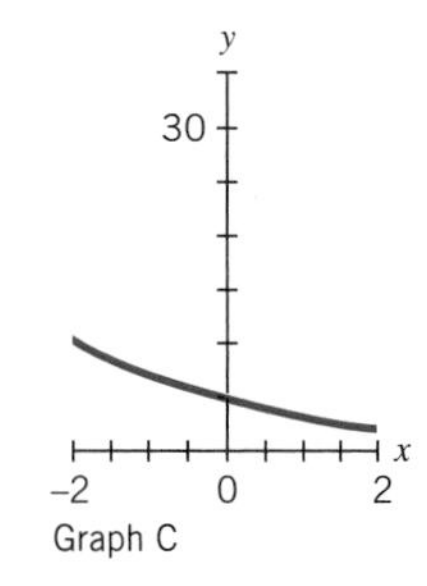

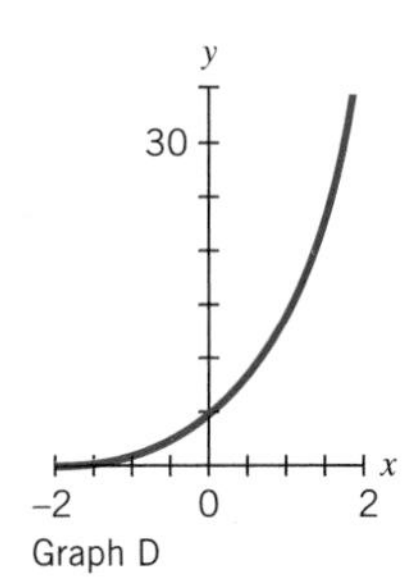

18. Make a table of values and plot each pair of functions on the same coordinate system:

a. $y = 2^x$ and $y = 2x$ for $-3 \leq x \leq 3$

b. $y = (0.5)^x$ and $y = 0.5x$ for $-3 \leq x \leq 3$

c. Which of the four functions that you drew in parts (a) and (b) represent growth?

d. How many times did the graphs that you drew for part (a) intersect? Find the coordinates of any points of intersection.

e. How many times did the graphs that you drew for part (b) intersect? Find the coordinates of any points of intersection.

Exercises for Section 5.4

19. Assume that you start with 1000 units of some quantity Q. Construct an exponential function that will describe the increase in Q over time T if, for each unit increase in T, Q increases by:

a. 300% **b.** 30% **c.** 3% **d.** 0.3%

20. Given each of the following exponential growth functions, identify the growth rate in percentage form:

a. $Q = 10{,}000 \cdot 1.5^T$ **c.** $Q = 10{,}000 \cdot 1.005^T$ **e.** $Q = 10{,}000 \cdot 2.5^T$
b. $Q = 10{,}000 \cdot 1.05^T$ **d.** $Q = 10{,}000 \cdot 2^T$ **f.** $Q = 10{,}000 \cdot 3^T$

21. Generate equations that represent the pollution levels, P, as a function of time, t (in years), such that when $t = 0$, $P = 150$ and:

a. P triples each year
b. P decreases by 12 units each year
c. P decreases by 7% each year
d. the annual average rate of change of P with respect to t is constant at 1.

22. Given an initial value of 50 units for a quantity Q for parts (a)–(d) below, in each case write a function that represents Q as a function of time t. Assume that when t increases by 1,

a. Q doubles **c.** Q increases by 10 units
b. Q increases by 5% **d.** Q is multiplied by 2.5

23. Between 1960 and 1990, the United States grew from about 180 million to 250 million, an increase of approximately 40% in this 30-year period. If the population continues to expand by 40% every 30 years, what would the U.S. population be in the year 2020? In 2050? Use your calculator to estimate when the population of the United States would reach 1 billion.

24. (Optional use of graphing calculator or computer.) According to Mexico's National Institute of Geography, Information and Statistics, in 1997 in Mexico's Quintana Roo state, where the tourist industry in Cancun has created a boom economy, the population was about 770,000 (or 0.77 million) and growing at a rate of 6% per year. In 1997 in Mexico's state Baja California, where many labor-intensive industries are located next to the California border, the population was about 2.2 million and growing at a rate of 3.3% per year. The nation's capital, Mexico City (which with more than 21 million inhabitants in 1997 is considered the world's largest metropolitan area) is growing at a rate of 0.5% per year.

a. Construct three exponential functions to model the growth of Quintana Roo, Baja California, and Mexico City. Identify your variables and their units.

b. Use your functions to predict the populations of Quintana Roo, Baja California, and Mexico City in 2010.

c. Use technology to help you estimate when the population of Mexico City might reach 25 million people. Justify your answer.

25. A pollutant was dumped into a lake, and each year its amount in the lake is reduced by 25%.

a. Construct a general formula to describe the amount of pollutant after n years if the original amount is A_0.

b. How long will it take before the pollution is reduced below 1% of its original level? Justify your answer.

26. (Requires graphing calculator or computer.) A swimming pool is initially shocked with chlorine to bring the chlorine concentration to 3 ppm (parts per million). Chlorine dissipates in reaction to bacteria and sun at a rate of about 15% per day. Above a chlorine concentration of 2 ppm, swimmers experience burning eyes, and below a concentration of 1 ppm, bacteria and algae start to proliferate in the pool environment.

a. Construct an exponential decay function that describes the chlorine concentration (in parts per million) over time.

b. Use technology to construct a table of values that corresponds to monitoring chlorine concentration for at least a 2-week period.

c. How many days will it take for the chlorine to reach a level tolerable for swimmers? How many days before bacteria and algae will start to grow and you will need to add more chlorine? Justify your answers.

27. a. If the inflation rate is 0.7% a month, what is it per year?

b. If the inflation rate is 5% a year, what is it per month?

28. (Parts require use of the Internet and a function graphing program.) A "rule of thumb" used by car dealers is that the trade-in value of a car decreases by 30% each year.

a. Is this decline linear or exponential?

b. Construct a function that would express the value of the car as a function of years owned.

c. Suppose you purchase a car for \$15,000. What would its value be after 2 years?

d. Explain how many years it would take for the car in part (c) to be worth less than \$1000. Explain how you arrived at your answer.

Internet search:

e. Go to the Internet site for the Kelley Blue Book *(www.kbb.com)*.

i. Put in the information about your current car or a car you would like to own. Specify the actual age and mileage of the car. What is the blue book value?

ii. Then, keeping everything else the same, assume the car is 1 year older and increase the mileage by 10,000. What is the new value?

iii. Use technology to find a best fit exponential function to model the value of your car as a function of years owned. What is the annual decay rate?

iv. According to this function, what would the value of your car be 5 years from now?

Exercises for Section 5.5

See Excel or graph link file USPOP.

29. (Requires graphing calculator or computer.) We have seen the accompanying table and graph of the U.S. population before in the text.

Population of the United States, 1790–2000

Year	Population (millions)
1790	3.9
1800	5.3
1810	7.2
1820	9.6
1830	12.9
1840	17.1
1850	23.2
1860	31.4
1870	39.8
1880	50.2
1890	62.9
1900	76.0
1910	92.0
1920	105.7
1930	122.8
1940	131.7
1950	151.9
1960	180.0
1970	204.0
1980	227.2
1990	249.4
2000	274.6 (est.)

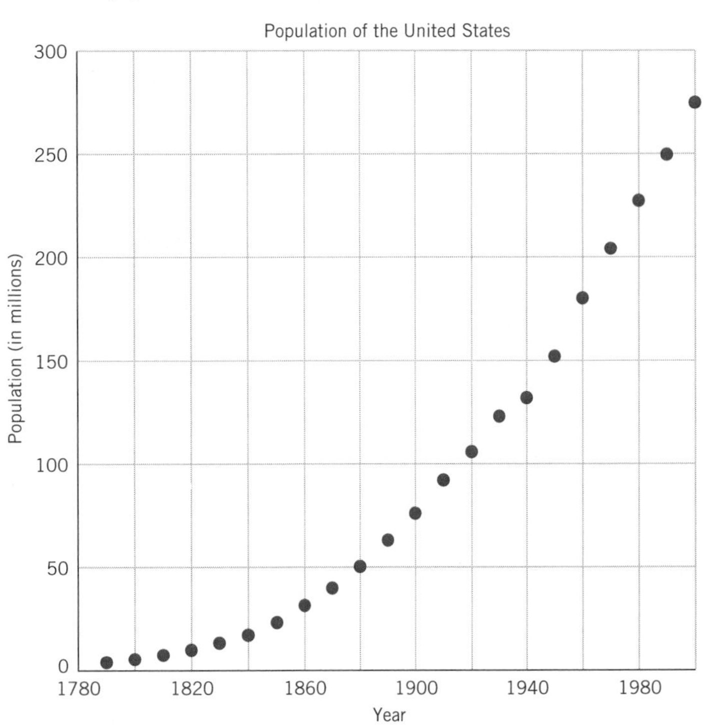

a. Find a best fit exponential function. (You may want to set 1790 as year 0.) Be sure to clearly identify the variables and their units for time and population. What is the annual growth factor? The growth rate? The estimated initial population?

b. Graph your function and the actual U.S. population data on the same grid. Describe how the estimated population size differs from the actual population size. In what ways is this exponential function a good model for the data? In what ways is it flawed?

c. What would your model predict the population would be in the year 2010? In 2025?

30. (Requires results from Exercise 29.) According to a letter published in the Ann Landers column in *The Boston Globe* on Friday, December 10, 1999: "When Elvis Presley died in 1977, there were 48 professional Elvis impersonators. In 1996, there were 7,328. If this rate of growth continues, by the year 2012, one person in every four will be an Elvis impersonator."

a. What was the growth factor in the number of Elvis impersonators for the 19 years between 1977 and 1996?

b. What then would be the *annual* growth factor in Elvis impersonators between 1977 and 1996?

c. Construct an exponential function that describes the growth in Elvis impersonators since 1977.

d. Use your function to estimate the number of Elvis impersonators in 2012.

e. Use your model for the U.S. population from Exercise 29 to determine if in 2012 one out of every four will be an Elvis impersonator? Explain your reasoning.

31. (Requires graphing calculator or computer and results from Exercise 29.) Reliable data about Internet use are hard to come by. But Nua Internet Surveys (at *www.nua.ie/surveys/how_many_online/*) cites estimates of 18 million Internet users in the U.S. in 1995, 76 million users in 1998, and 119.2 million in 1999.

a. Enter the data into a graphing calculator or spreadsheet program. (You may want to initialize 1995 as year 0.) Plot the data. Does it appear that a linear or an exponential function would be a better model?

b. Use the data for just 1995 and 1999 to estimate an annual growth factor, and then an annual growth rate.

c. Use technology to generate a best fit exponential function. What is the annual growth factor? The annual growth rate? How do these compare with your answers in part (b)?

d. If this rate of growth continues, how many Internet users will there be in the United States in the year 2010? In 2025?

e. Using the model from Exercise 29, what percentage of the U.S. population were Internet users in 1995? In 1999?

f. Using your model of U.S. population growth from Exercise 29, what is the estimated percentage of the U.S. population that will be Internet users in 2010? In 2025?

32. Consider the accompanying table.

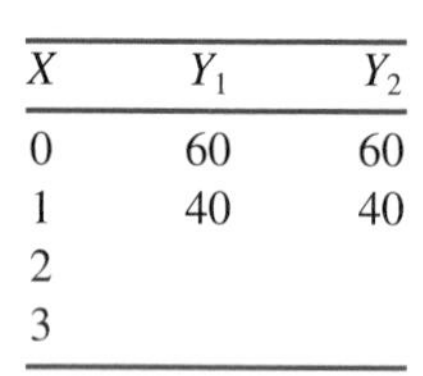

X	Y_1	Y_2
0	60	60
1	40	40
2		
3		

a. Assume that Y_1 is a linear function of X. Fill in the two missing values for Y_1 in the table.

b. Construct an equation for Y_1 as a linear function of X.

c. Assume that Y_2 is an exponential function of X. Fill in the two missing values for Y_2 in the table.

d. Construct the equation for Y_2 as an exponential function of X.

33. Suppose you are offered two similar jobs. The Aerospace Engineering Group offers you a starting salary of \$50,000 per year and raises of \$1000 every 6 months. The Bennington Corporation offers you an initial salary of \$35,000 and a 10% raise every year.

a. Make a table that shows your salary with each company during the first 8 years.

b. For each corporation, write a function that gives your salary after t years.

c. On the same coordinate system, plot the graphs of the functions for years 0 to 20.

d. In what year does the salary at Bennington exceed the salary at Aerospace?

e. Using the models from part (b), determine your salary after 10 years and after 20 years at each corporation.

34. China is the most populous country in the world. In 1998 it had about 1.237 billion people. By the end of 1999 the population had grown to 1.267 billion. One could describe this change as either an increase of 0.030 billion (=30 million) people or as an increase of 2.4%. Use this information to construct models predicting the size of China's population in the future.

a. Identify your variables and units.

b. Construct a linear model.

c. Construct an exponential model.

d. What will China's population be in 2050 according to:

i. your linear model?

ii. your exponential model?

35. If you have a heart attack and your heart stops beating, the amount of time it takes paramedics to restart your heart with a defibrillator is critical. According to a medical report on the evening news, each minute that passes decreases your chance of survival by 10%. From this wording it is not clear whether the decrease is linear or exponential. Assume that the survival rate is 100% if the defibrillator is used immediately.

a. Construct and graph a linear function that describes your chances of survival. After how many minutes would your chance of survival be 50% or less?

b. Construct and graph an exponential function that describes your chances of survival. Now after how many minutes would your chance of survival be 50% or less?

36. The infant mortality rate (the number of deaths per 1000 live births) fell in the United States from 7.6 in 1995 to 7.2 in 1996.

a. Assume that the infant mortality rate is declining linearly over time. Construct an equation modeling the relationship between infant mortality rate and time. Make sure you have clearly identified your variables.

b. Assuming that the infant mortality rate is declining exponentially over time, construct an equation modeling the relationship.

c. Graph both of your models on the same grid.

d. What would each of your models predict for the infant mortality rate in 2000? In 2005?

37. Tritium, the heaviest form of hydrogen, is a critical element in a hydrogen bomb. It decays exponentially with a half-life of about 12.3 years. Any nation wishing to maintain a viable hydrogen bomb has to replenish its tritium supply roughly every 3 years, so world tritium supplies are closely watched. Construct an exponential function that shows the remaining amount of tritium as a function of time as 100 grams of tritium decays (about the amount needed for an average size bomb). Be sure to identify the units for your variables.

38. Cosmic ray bombardment of the atmosphere produces neutrons, which in turn react with nitrogen to produce radioactive carbon-14. Radioactive carbon-14 enters all living tissue through carbon dioxide (via plants). As long as a plant or animal is alive, carbon-14 is maintained in the organism at a constant level. Once the organism dies, however, carbon-14 decays exponentially into carbon-12. By comparing the amount of carbon-14 to the amount of carbon-12, one can determine approximately how long ago the organism died. Willard Libby won a Nobel Prize for developing this technique for use in dating archaeological specimens. The half-life of carbon-14 is about 5730 years. Assume that the initial quantity of carbon-14 is 500 milligrams.

a. Construct an exponential function that describes the relationship between C, the amount of carbon-14 in milligrams, and t, the number of 5730-year time periods.

b. Generate a table of values and plot the function. Choose a reasonable set of values for the domain. Remember that the objects we are dating may be up to 50,000 years old.

c. From your graph or table estimate how many milligrams are left after 15,000 years; after 45,000 years.

d. Now construct an exponential function that describes the relationship between C and T measured in years. What is the annual decay factor? The annual decay rate?

e. Use your function in part (d) to calculate the number of milligrams that would be left after 15,000 years; after 45,000 years.

39. The body eliminates drugs by metabolism and excretion. To predict how frequently a patient should receive a drug dosage, the physician must determine how long the drug will remain in the body. This is usually done by measuring the half-life of the drug, the time required for the total amount of drug to diminish by one-half.

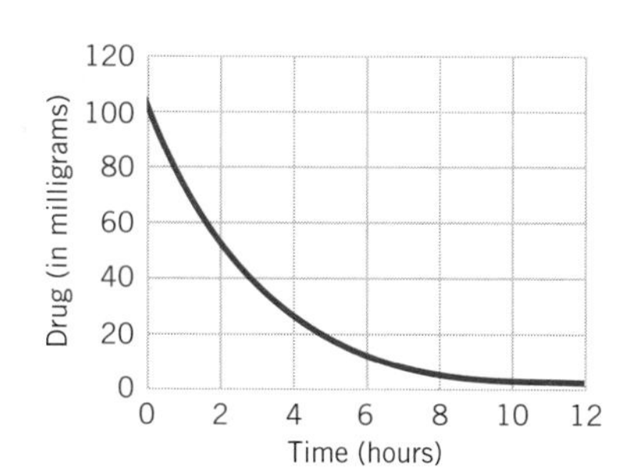

Remaining drug dose in milligrams.

a. Most drugs are considered eliminated after five half-lives, because the amount remaining is probably too low to cause any beneficial or harmful effects. After five half-lives, what percentage of the original dose is left in the body?

b. The accompanying graph shows a drug's concentration in the body over time starting with 100 milligrams.

i. Estimate the half-life of the drug.

ii. Construct an equation that approximates the curve. Specify the units of your variables.

iii. How long would it take for five half-lives to occur? Approximately how many milligrams of the original dose would be left then?

iv. Write a paragraph describing your results to a prospective buyer of the drug.

40. Estimate the doubling time using the rule of 70 when:

a. $P = 2.1(1.0475)^t$, where t is in years

b. $Q = 2.1(1.00475)^T$, where T is in years

41. Use the rule of 70 to approximate the growth rate when the doubling time is:

a. 5730 years **b.** 11,460 years **c.** 5 seconds **d.** 10 seconds

42. Estimate the time it will take an initial quantity to drop to half its value when:

a. $P = 3.02(0.998)^t$ with t in years **b.** $Q = 12(0.75)^T$ with T in decades

World Population

Year	Total Population (millions)
1800	980
1850	1260
1900	1650
1950	2520
1970	3700
1980	4440
1990	5270
2000	6060 (est.)

Source: United Nations Population Division, *www.undp.org/popin*

See Excel or graph link file WORLDPOP.

43. (Requires graphing calculator or computer.) Estimates for world population vary, but the data in the accompanying table are reasonable estimates of the world population from 1800 to 2000.

a. Either enter the data table into the calculator or computer (you may wish to enter 1800 as 0, 1850 as 50, etc.) or use the data file in Excel or graph link form.

b. Generate a best fit exponential function. Record the equation and print out the graph if you can.

c. Interpret each term in the function, and specify the domain and range of the function.

d. What does your model give for the growth rate?

e. Using the graph of your function, estimate the following:

i. The world population in 1750, 1920, 2025, and 2050

ii. The approximate years in which world population attained or will attain 1 billion (i.e., 1000 million), 3.2 billion, 4 billion, and 8 billion

iii. The length of time your model predicts it takes for the population to double in size

See Excel or graph link file REINDEER.

44. (Requires graphing calculator or function graphing program.) In 1911 reindeer were introduced to St. Paul Island, one of the Pribilof Islands off the coast of Alaska in the Bering Sea. There was plenty of food and no hunting or reindeer predators. The size of the reindeer herd grew rapidly for a number of years as given in the accompanying table.

a. Use the reindeer data file (in Excel or graph link form) to plot the data.

b. Find a best fit exponential function and use it to predict the size of the population in each year given in the table. (You may want to set 1911 = 0.)

c. How does the predicted population from part (b) differ from the observed?

Population of Reindeer Herd

Year	Population Size	Year	Population Size	Year	Population Size
1911	17	1921	280	1931	466
1912	20	1922	229	1932	525
1913	42	1923	161	1933	670
1914	76	1924	212	1934	831
1915	93	1925	246	1935	1186
1916	110	1926	254	1936	1415
1917	136	1927	254	1937	1737
1918	153	1928	314	1938	2034
1919	170	1929	339		
1920	203	1930	415		

Source: V.B. Scheffer, The rise and fall of a reindeer herd, *Scientific Monthly,* 73:356–362, 1951.

d. Does your answer in part (c) give you any insights into why the model does not fit the observed data perfectly?

e. Estimate the doubling time of this population.

45. In medicine and biological research, radioactive substances are often used for treatment and tests. In the laboratories of a large east coast university and medical center any waste containing radioactive material with a half-life under 65 days must be stored for 10 half-lives before it can be disposed of with the nonradioactive trash.

a. By how much does this policy reduce the radioactivity of the waste?

b. Fill out the accompanying chart and develop a general formula for pollution amount at any period, given an initial amount, A_0.

Half-Life Period	Pollution Amount
0	A_0, original amount
1	$A_1 = 0.5A_0$
2	$A_2 =$
3	
4	
period n	

46. Belgrade, Yugoslavia (from *USA Today,* September 21, 1993)

A 10 billion dinar note hit the streets today. . . . With inflation at 20% per day, the note will soon be as worthless as the 1 billion dinar note issued last month. A year ago the biggest note was 5,000 dinars. . . . In addition to soaring inflation, which doubles prices every 5 days, unemployment is at 50%.

In the excerpt, inflation is described in two very different ways. Identify these two descriptions in the text and determine whether they are equivalent. Justify your answer.

47. The time it takes for a malignant lung tumor to double in size is 3 months. At the time a lung tumor was detected in a patient, its mass was 10 grams.

a. If untreated, determine the size in grams of the tumor at the listed times in the table. Find a formula to express the tumor mass M (in grams) at any time t (in months).

b. Lung cancer is fatal when it reaches a mass of 2000 grams. If the patient diagnosed with lung cancer went untreated, use your formula to estimate to the nearest quarter year how long he would survive after the diagnosis. How many years is that?

c. By what percentage of its original size has the 10-gram tumor grown when it reaches 2000 grams?

t Time (months)	M Mass (g)
0	10
3	
6	
9	
12	

48. It is now recognized that prolonged exposure to very loud noise can damage hearing. The accompanying table gives the permissible daily exposure hours to very loud noises as recommended by OSHA, the Occupational Safety and Health Administration.

a. Examine the data for patterns. How is D progressing? How is H progressing? Does the data represent a growth or decay phenomenon? Explain your answer.

b. Find a formula to fit the data as closely as possible using D for decibels and H for safe duration hours. Graph your formula and the data on the same axes.

Sound Level, D (decibels)	Maximum Duration, H (hours)
120	0
115	0.25
110	0.5
105	1
100	2
95	4
90	8

49. a. Construct a function that would represent the resulting value if you invested \$5000 for n years at an annually compounded interest rate of:

i. 3.5% **ii.** 6.75% **iii.** 12.5%

b. If you make three different \$5000 investments today at the three different interest rates listed in part (a), how much will each investment be worth in 40 years?

50. A bank compounds interest annually at 4%.

a. Write an equation for the value V of \$100 in t years.

b. Write an equation for the value V of \$1000 in t years.

c. After 20 years will the total interest earned on \$1000 be 10 times the total interest earned on \$100? Why or why not?

51. You have a chance to invest money in a risky investment at 6% interest compounded annually. Or you can invest your money in a safe investment at 3% interest compounded annually.

a. Write an equation that describes the value of your investment after n years if you invest \$100 at 6% compounded annually. Plot the function. Estimate how long it would take to double your money.

b. Write an equation that describes the value of your investment after n years if you invest \$200 at 3% compounded annually. Plot the function on the same graph as in part (a). Estimate the time needed to double your investment.

c. Looking at your graph, will the amount in the first investment in part (a) ever exceed the amount in the second account in part (b)? If so, approximately when?

52. According to the *Arkansas Democrat Gazette* (February 27, 1994):

Jonathan Holdeen thought up a way to end taxes forever. It was disarmingly simple. He would merely set aside some money in trust for the government and leave it there for 500 or 1000 years. Just a penny, Holdeen calculated, could grow to trillions of dollars in that time. But the stash he had in mind would grow much bigger—to quadrillions or quintillions—so big that the government, one day, could pay for all its operations simply from the income. Then taxes could be abolished. And everyone would be better off.

a. Holdeen died in 1967 leaving a trust of \$2.8 million that is being managed by his daughter, Janet Adams. In 1994, the trust was worth \$21.6 million. The trust is being debated in Philadelphia Orphans' Court. Some lawyers who are trying to break the trust have said that it is dangerous to let it go on, because "it would sponge up all the money in the world." Is this possible?

b. In 500 years, how much would the trust be worth? Would this be enough to pay off the current national debt (about 6 trillion dollars in 2000)? What about after 1000 years? Describe the model you have used to make your prediction.

53. Describe how a 6% inflation rate will erode the value of a dollar over time. Approximately when would a dollar be worth only 50¢? This might be called the half-life of the dollar's buying power under 6% inflation.

54. The future value V of a savings plan, where regular payments P are made n times to an account in which interest is compounded each payment period, can be calculated using the formula

$$V = P \cdot \frac{(1 + i)^n - 1}{i}$$

The total number of payments, n, equals the number of payments per year, m, times the number of years, t, so

$$n = m \cdot t$$

The interest rate per compounding period, i, equals the annual interest rate, r, divided by the number of compounding periods a year, m, so

$$i = r/m$$

a. Substitute $n = m \cdot t$ and $i = r/m$ in the formula for V, getting an expression for V in terms of m, t, and r.

b. If a parent plans to build a college fund by putting \$50 a month into an account compounded monthly with a 4% annual interest rate, what will be the value of the account in 17 years?

c. Solve the original formula for P as a function of V, i, and n.

d. Now you are able to find how much must be paid in every month to meet a particular final goal. If you estimate the child will need \$100,000 for college, what monthly payment must the parent make if the interest rate is the same as in part (b)?

55. The following data show the total government debt for the United States:

Year	1970	1975	1980	1985	1990	1995
National debt (\$ billions)	361	541	909	1827	3266	4961

Source: Data from U.S. Department of the Treasury.

a. Plot these data by hand and sketch a curve that roughly approximates the data.

b. Is this a growth or a decay phenomenon? What generic formula might approximate these data?

c. For 1975 to 1995 calculate the ratio of each debt to the one 5 years earlier, such as (1975 debt)/(1970 debt). This is the growth factor for each particular 5-year period. Is the growth factor staying relatively constant? Average the ratios to get an average growth factor G for 5-year time periods.

d. Using your value for G in part (c), plot $y = 361G^x$, for $0 \leq x \leq 5$. Here x represents the number of half-decades since 1970. So $x = 0$ corresponds to 1970; $x = 1$ corresponds to 1975, etc.

e. Find an estimate for G such that $4961 = 361G^5$. Use this new value of G to plot $y = 361G^x$, where again $0 \leq x \leq 5$.

f. Which formula is the better fit to the data? Use this formula to estimate national debts for 2000 and 2005.

g. Do you think your exponential model is an appropriate model for describing the growth of the national debt? Why?

56. The data given show the declining purchasing power of the U.S. dollar since 1983 using the 1983 dollar as a standard. For instance, if a dollar could buy 100 candies in 1983, by 1996 it could only buy 64 of the same candies.

Year	Purchasing Power, P (\$)	P/P_{previous}	Years Since 1983, t	$P_{\text{est}} = G^t$
1983	1.00		0	
1984	0.96	$0.96/1.00 \approx 0.960$	1	
1985	0.93	$0.93/0.96 \approx 0.969$	2	
1986	0.91	$0.91/0.93 \approx 0.978$	3	
1987	0.88	$0.88/0.91 \approx 0.967$	4	
1988	0.85		5	
1989	0.81		6	
1990	0.77		7	
1991	0.73		8	
1992	0.71		9	
1993	0.69		10	
1994	0.67		11	
1995	0.66		12	
1996	0.64		13	

Source: U.S. Bureau of Labor Statistics, derived from the Consumer Price Index (rounded to 2 decimal places).

a. Compute the ratio of the purchasing power each year to the preceding year, P/P_{previous} as shown in the table. What is the average P/P_{previous} for the years shown?

b. If the purchasing power data were truly exponential, the ratio $P/P_{previous}$ would remain constant. In this case, though not constant, it is close to constant. We can use the average $P/P_{previous}$ as the decay factor G in the following formula to estimate purchasing power, P_{est}, as a function of t:

$$P_{est} = G^t$$

Compute P_{est} and plot P and P_{est} vs. t on the same axes to compare how well the exponential formula models the data.

c. What does your formula for P_{est} predict will be the purchasing power of a dollar in 2010?

57. MCI, a phone company that provides long-distance service, introduced a marketing strategy called "Friends and Family." Each person who signed up received a discounted calling rate to 10 specified individuals. The catch was that the 10 people also had to join the "Friends and Family" program.

a. Assume that one individual agrees to join the "Friends and Family" program and that this individual recruits 10 new members, who in turn each recruit 10 new members, and so on. Write a function to describe the number of new people who have signed up for "Friends and Family" at the nth round of recruiting.

b. Now write a function that would describe the total number of people signed up after n rounds of recruiting.

c. How many "Friends and Family" members, stemming from this one person, will there be after 5 rounds of recruiting? After 10 rounds?

d. Write a 60 second summary of the pros and cons of this recruiting strategy. Why will this strategy eventually collapse?

58. In a chain letter one person writes a letter to a number N of other people who are each requested to send the letter to N other people, and so on. In a simple case with $N = 2$, person A1 starts the process.

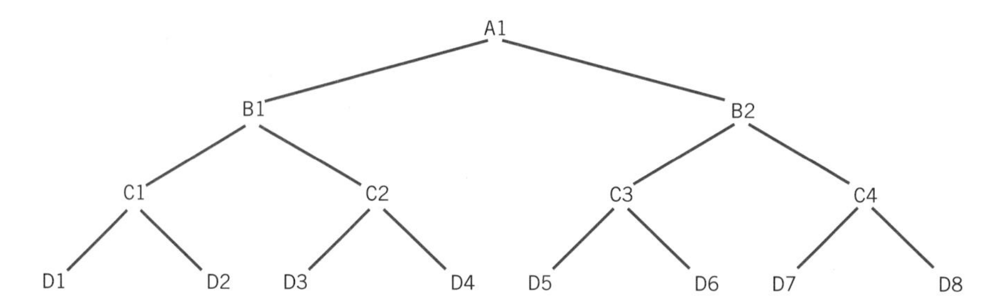

A1 sends to B1 and B2; B1 sends to C1 and C2; B2 sends to C3 and C4, etc. A typical letter has listed in order the chain of senders who sent the letters. So D7 receives a letter that has

A1

B2

C4

If these letters request money they are illegal. A typical request looks like this:

- When you receive this letter send $10 to the person on the top of the list.
- Copy this letter, but add your name to the bottom of the list and leave off the name at the top of the list.
- Send a copy to two friends within 3 days.

a. Construct a mathematical model for the number of new people receiving letters at each level L.

b. If the chain is not broken, how much money should an individual receive?

c. Suppose A1 sent out letters with 2 phony names (say A1a and A1b) with P.O. box addresses she owns. So both B1 and B2 would receive a letter with the list A1, A1a, A1b. If the chain isn't broken, how much money would A1 receive?

d. If the chain continued, how many new people would receive letters at the 25th level?

Internet search:
Chain letters are an example of a "pyramid growth" scheme. A similar business strategy is multi-level marketing. This marketing method uses the customers to sell the product by giving them a financial incentive to promote the product to potential customers or potential salespeople for the product. (See Exercise 57.) Sometimes the distinction between multilevel marketing and chain letters gets blurred. Search the U.S. Postal Service web site *(www.usps.gov)* for "pyramid schemes" for information about what is legal and what is not. Report what you find.

59. The National Council on Pet Population Study and Policy estimates that 7 million unwanted cats and dogs are destroyed at pet shelters every year; other sources put the number killed at anywhere between 5 and 15 million annually. To prevent overpopulation of unwanted domestic pets, and the resulting massive euthanasia, shelters strongly encourage spaying of female pets and neutering of males. To understand how this overpopulation occurs, consider these facts: A female cat starting at 6 months can have 3 to 4 litters of 1 to 9 kittens (with an average of 5 kittens per litter) each year, so that is 15 to 20 new kittens each year. She continues to reproduce, and her female kittens after 6 months are themselves producing kittens, and so on.

To make a conservative estimate of cat population growth we make some assumptions:

- Start with 100 females and 100 males who live to be 5 years old
- Each female has two litters of 4 per year. So every 6 months each female has 4 kittens, 2 of them female, 2 male.

a. Fill out the following table to track the population growth.

	Females		Males	
Time Period, t, 6 months	Female Births	Total Females, F	Male Births	Total Males, M
0		100		100
1	$100 \cdot 2 = 200$	$100 + 100 \cdot 2 = 100 \cdot 3 = 300$	$100 \cdot 2 = 200$	$100 + 200 = 300$
2 (1 yr)	$300 \cdot 2 = 600$	$300 + 300 \cdot 2 = 300 \cdot 3 = 900$	$300 \cdot 2 = 600$	$300 + 600 = 900$
3	$900 \cdot 2 = 1800$	$900 + 900 \cdot 2 = 900 \cdot 3 = 2700$		
4				
5				
6				
7				
8 (4 yr)				

b. Examine the growth pattern revealed by the data, then find a formula to express total number of female cats, F, as a function of the time period, t, and another formula to express the total number of male cats, M, as a function of t.

c. For a town starting out with 100 males and 100 females how many cats in all do your formulas predict after 5 years? By what order of magnitude has the total cat population grown over 5 years?

d. If the town ran a strong campaign to spay female cats, and only 25 of the original 100 females remained fertile, calculate how many total cats there would be after 5 years.

Exercises for Section 5.6

60. According to Rubin and Farber's *Pathology,* "Smoking tobacco is the single largest preventable cause of death in the United States, with direct health costs to the economy of tens of billions of dollars a year. Over 400,000 deaths a year—about one sixth of the total mortality in the United States—occur prematurely because of smoking." The accompanying graph compares the risk of dying for smokers, ex-smokers, and nonsmokers. It shows that individuals who have smoked for 1 year are twice as likely to die as a nonsmoker. Someone who has smoked for 15 years has a more than 3 times greater annual probability of dying than a nonsmoker.

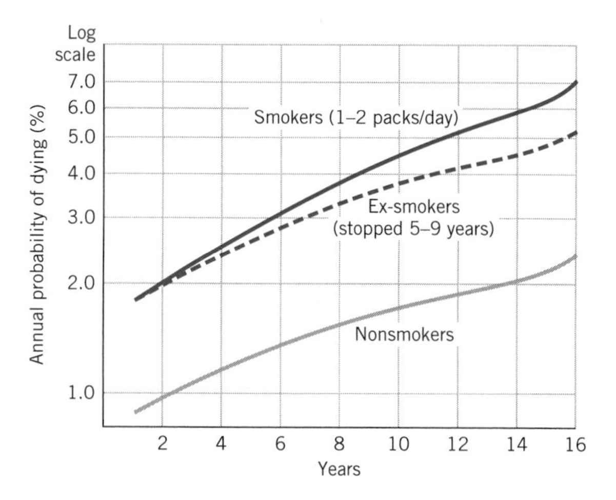

Source: E. Rubin and J. L. Farber, *Pathology,* 3rd ed. (Philadelphia: Lippincott-Raven, 1998), p. 310. Copyright © 1998 by Lippincott-Raven. Reprinted by permission.

a. The graphs for the smokers (1 to 2 packs per day), ex-smokers, and nonsmokers all appear roughly as straight lines on this semi-log plot. What then would be appropriate functions to use to model the increased probability of dying over time for all three groups?

b. The plots for smokers and ex-smokers appear roughly as two straight lines that start at the same point, but the graph for smokers has a steeper slope. How would their two function models be the same and how would they be different?

c. The plots of ex-smokers and nonsmokers appear as two straight lines that are roughly equidistant. How would their two function models be the same and how would they be different?

See Excel or graph link files CELCOUNT and ECOLI.

61. (Requires function graphing program with semi-log plot capability.)

a. Load the file CELCOUNT, which contains all of the white blood cell counts for the bone marrow transplant patient. Now graph the data on a semi-log plot. Which sections(s) of the curve represent exponential growth or decay? Explain how you can tell.

b. Load the file ECOLI, which contains the *E. coli* counts for 24 time periods. Graph the *E. coli* data on a semi-log plot. Which section of this curve represents exponential growth or decay? Explain your answer.

62. (Requires graphing calculator or computer.) The accompanying table shows the U.S. international trade in goods and services.

See Excel or graph link file USTRADE.

U.S. International Trade (Billions of Dollars)

Year	Total Exports	Total Imports	Year	Total Exports	Total Imports
1960	25.9	22.4	1980	271.8	291.2
1961	26.4	22.2	1981	294.4	310.6
1962	27.7	24.4	1982	275.2	299.4
1963	29.6	25.4	1983	266.0	323.8
1964	33.3	27.3	1984	290.9	400.1
1965	35.3	30.6	1985	288.8	410.9
1966	38.9	36.0	1986	309.7	450.3
1967	41.3	38.7	1987	348.8	502.1
1968	45.5	45.3	1988	431.3	547.2
1969	49.2	49.1	1989	489.4	581.6
1970	56.6	54.4	1990	537.2	618.4
1971	59.7	61.0	1991	581.3	611.9
1972	67.2	72.7	1992	615.9	652.9
1973	91.2	89.3	1993	641.8	711.7
1974	120.9	125.2	1994	702.1	800.5
1975	132.6	120.2	1995	793.5	891.0
1976	142.7	148.8	1996	849.8	954.1
1977	152.3	179.5	1997	938.5	1,043.3
1978	178.4	208.2	1998	933.9	1,098.2
1979	224.1	248.7			

Source: U.S. Department of Commerce, Bureau of Economic Analysis.

a. U.S. imports and exports have both been expanding rapidly between 1960 and 1998. Use technology to plot the total U.S. exports and total U.S. imports over time on the same graph.

b. Now change the vertical axis to a logarithmic scale and generate a semi-log plot of the same data as in part (a). What is the shape of the data now and what does this suggest would be an appropriate function type to model U.S. exports and imports?

c. Construct appropriate function models for total U.S. imports and for total exports.

d. The difference between the values of exports and imports measures the *trade balance*. If the balance is negative, it is called a *trade deficit*. The balance of trade has been an object of much concern lately. Calculate the trade balance for each year and plot it over time. Describe the overall pattern.

We have a trade deficit that has been increasing rapidly in recent years. But for quantities that are growing exponentially, the "relative difference" is much more meaningful than the simple difference. In this case the relative difference is

$$\frac{\text{exports} - \text{imports}}{\text{exports}}$$

This gives the trade balance as a fraction (or if you multiply by 100, as a percentage) of exports.

e. Calculate the relative difference. Graph it as a function of time. Does this present a more or less worrisome picture? That is, in particular over the last decade has the relative difference remained stable or is it also rapidly increasing in magnitude?

63. The accompanying graph of the Dow Jones industrial average appeared in the Business section of *The New York Times* on Sunday, February 16, 1997, p. F3. The Dow Jones reflects the performance of 100 mainstream industrial corporations whose stock is traded on the New York Stock Exchange. It had been experiencing phenomenal growth for a number of years and had just reached 7000 points for the first time ever. Since then it has climbed higher still.

A Different Perspective

This chart, plotted on a logarithmic scale, gives the same visual weight to comparable percentage changes. A 100-point rise when the Dow Jones Industrial Average is at 1,000 looks the same here as a 700-point rise when the Dow is at 7,000. Through this lens, the Dow's recent run-up looks less extraordinary.

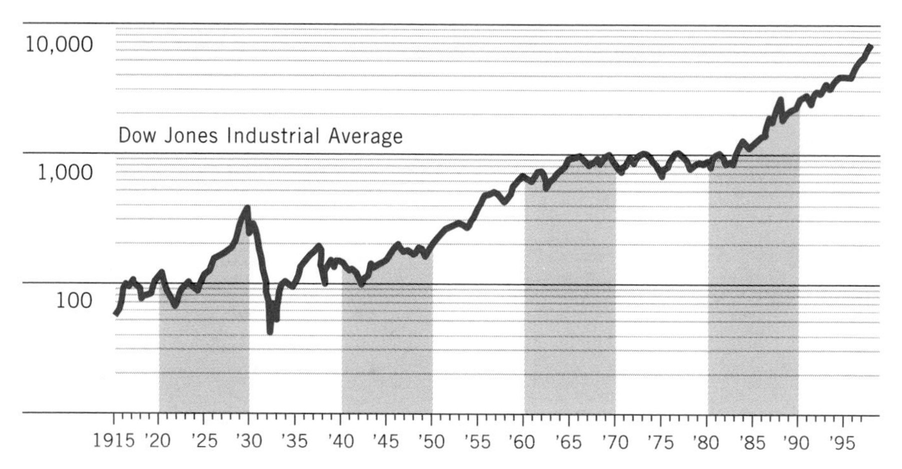

a. What kind of plot is this?

b. Approximately what was the Dow Jones industrial average in 1929? In 1933? What was happening in American society during this time?

c. In what year did the Dow Jones first reach 1000? 3000? 6000?

d. This plot makes the climb in values appear relatively modest by compressing the vertical scale in order to plot numbers of widely different sizes. On this type of graph, when the plot appears to lie roughly along a straight line, what does this tell you about the original data during that time period?

e. The data between 1980 and 1997 lie roughly along a straight line in this graph. Estimate the annual percentage increase in the Dow Jones during those years.

f. The data between 1933 and 1997 appear to be roughly linear on this graph. Estimate the annual percentage increase in the Dow Jones during this 64-year period.

Exploration 5.1

Properties of Exponential Functions

Objectives

- explore the effects of a and C on the graph of the exponential function in the form

$$y = Ca^x \quad \text{where } a > 0 \text{ and } a \neq 1$$

Materials/Equipment

- computer and software "E3: $y = Ca^x$ Sliders" in *Exponential & Log Functions,* or graphing calculator
- graph paper

Procedure

We start by choosing values for a and C and graphing the resulting equations by hand. From these graphs we make predictions about the effects of a and C on the graphs of other equations. Take notes on your predictions and observations so you can share them with the class. Work in pairs and discuss your predictions with your partner.

Making Predictions

1. Start with the simplest case, where $C = 1$. The equation will now have the form

$$y = a^x$$

Make a data table and by hand sketch on the same grid the graphs for $y = 2^x$ (here $a = 2$) and $y = 3^x$ (here $a = 3$). Use both positive and negative values for x. Predict where the graphs of $y = 2.7^x$ and $y = 5^x$ would be located on your graph. Check your work and predictions with your partner.

x	$y = 2^x$	$y = 3^x$
−2		
−1		
0		
1		
2		
3		

How would you describe your graphs? Do they have a maximum or a minimum value? What happens to y as x *increases?* What happens to y as x *decreases?* Which graph shows y changing the fastest compared to x?

2. Now create two functions in the form $y = a^x$, where $0 < a < 1$. Create a data table and graph your functions on the same grid. Make predictions for other functions where $C = 1$ and $0 < a < 1$.

3. Now consider the case where C has a value other than 1 for the general exponential function

$$y = Ca^x$$

Create a table of values and sketch the graphs of $y = 0.5(2^x)$ (in this case $C = 0.5$ and $a = 2$) and $y = 3(2^x)$ (in this case $C = 3$ and $a = 2$). What do all these graphs have in common? What do you think will happen when $a = 2$ and $C = 10$? What do you think will happen to the graph if $a = 2$ and $C = -3$? Check your predictions with your partner.

x	$y = 0.5(2^x)$	$y = 3(2^x)$	$y = -3(2^x)$
−2			
−1			
0			
1			
2			
3			

How would you describe your graphs? Do they have a maximum or a minimum value? What happens to y as x *increases?* What happens to y as x *decreases?* What is the y-intercept for each graph?

Testing Your Predictions

Now test your predictions by using a program called "E3: $y = Ca^x$ Sliders" in the *Exponential & Log Functions* software package or by creating graphs using technology.

1. What effect does a have?

Make predictions when $a > 1$ and when $0 < a < 1$, based on the graphs you constructed by hand. Explore what happens when $C = 1$ and you choose different values for a. Check to see whether your observations about the effect of a hold true when $C \neq 1$.

How does changing a change the graph? When does $y = a^x$ describe growth? When does it describe decay? When is it flat? Write a rule that describes what happens when you change the value for a. You only have to deal with cases when $a > 0$.

2. What effect does C have?

Make a prediction based on the graphs you constructed by hand. Now choose a value for a and create a set of functions with different C values. Graph these functions on the same grid.

How does changing C change the graph? What does the value of C tell you about the graphs of functions in the form $y = Ca^x$? Describe your graphs when $C > 0$ and when $C < 0$. Use technology to test your generalizations.

Exploration-Linked Homework

Write a 60 second summary of your results, and present it to the class.

Exploration 5.2

Recognizing Exponential Patterns in Data Tables

Objectives

- use ratios to identify exponential growth or decay and to construct exponential functions

Materials/Equipment

- calculator or computer with spreadsheet program (optional)

Procedure

Finding Patterns

1. The following tables contain data about the exponential growth of *E. coli* (from Table 5.1) and the exponential decay of the dollar through inflation (from Table 5.16).

Growth of *E. coli* Bacteria

t (Time Periods)	$f(t) = 100 \cdot 2^t$ (No. of Bacteria)	$f(t)/f(t-1)$	$f(t+1)/f(t)$
0	100	n.a.	
1	200		
2	400		
3	800		
4	1,600		
5	3,200		
6	6,400		
7	12,800		
8	25,600		
9	51,200		
10	102,400		n.a.

Erosion of the Dollar through Inflation

t (Year)	$f(t) = 1 \cdot (0.952)^t$ (Value of Dollar)	$f(t)/f(t-1)$	$f(t+1)/f(t)$
0	\$1.00	n.a.	
1	\$0.95		
2	\$0.91		
3	\$0.86		
4	\$0.82		
5	\$0.78		
6	\$0.74		
7	\$0.71		
8	\$0.67		
9	\$0.64		
10	\$.061		n.a.

a. In the third columns of the *E. coli* data and the inflation data, fill in the values of the ratios $f(t) / f(t - 1)$.

b. What do you notice about the ratios in each case? What do the ratios tell you about the functional values as the independent variable t, increases by 1 unit?

c. Fill in the fourth column with values for the ratios $f(t + 1)/f(t)$. Compare the values in columns 3 and 4 in each table. Explain their relationship.

2. Generalizing your results:

a. Using the general form for an exponential function, $f(x) = C \cdot a^x$, write an expression for:

$f(k)$, the value of the function when $x = k$

$f(k + 1)$, the value of the function when $x = k + 1$

$f(k - 1)$, the value of the function when $x = k - 1$

b. Calculate the ratios $f(k) / f(k - 1)$ and $f(k + 1) / f(k)$ For exponential functions, these ratios are called *growth or decay factors.* Is this definition consistent with the definitions of growth and decay factors given in Sections 5.1 and 5.2?

c. How do these ratios relate to the exponential function $f(x) = C \cdot a^x$? State your results as a general rule.

Using Your Results

Use the generalizations for exponential functions that you have found to analyze each of the following data tables.

a. By calculating ratios, determine whether the data in each of the following tables exhibit exponential growth or decay. If the data are exponential, determine the growth or decay factor.

i.

x	$f(x)$
0	1
1	10
2	100
3	1,000
4	10,000

ii.

x	$f(x)$
0	−3
1	−6
2	−12
3	−24
4	−48

iii.

x	$f(x)$
0	25
1	100
2	400
3	1600
4	6400

iv.

x	$f(x)$
0	1.0
1	0.1
2	0.01
3	0.001
4	0.0001

v.

x	$f(x)$
0	15
1	12
2	9
3	6
4	3

vi.

x	$f(x)$
0	0
1	1
2	4
3	9
4	16

b. For each table in part (a) explain in your own words how to find $f(x)$ in terms of x.

c. For each table that exhibits exponential growth or decay, construct an equation to describe the data. Does your equation predict the values in your data table?

d. Using the equations you generated in part (c), extend each of the exponential data tables to include negative values for x.

Chapter 6

Logarithmic Links: Logarithmic and Exponential Functions

Overview

How can we determine the doubling time for a population that is growing exponentially? To answer this question, we need logarithmic functions, the close relatives of exponential functions. Logarithmic functions are also useful in modeling acidity levels and sound intensity. In the sciences logarithmic and exponential functions are usually written using a special number called *e* as the base.

After reading this chapter you should be able to:

- use logarithms to solve exponential equations
- construct both logarithmic and exponential functions using base *e*
- understand the basic properties of common and natural logarithms
- construct an exponential model for continuous compounding
- form the composition of two functions
- determine if a function inverse exists and, if so, find it

6.1 USING LOGARITHMS TO SOLVE EXPONENTIAL EQUATIONS

Estimating Solutions to Exponential Equations

In Chapter 5 we used a simple model for the growth of *E. coli* bacteria given by the exponential equation

$$N = 100 \cdot 2^t$$

where N = number of bacteria, t = time (in 20-minute periods), and the initial number of bacteria is 100. We then looked at different values of t, the independent variable or input, and found the corresponding value for N, the dependent variable or output. For example, when $t = 5$, then the number of *E. coli, N,* equals $100 \cdot 2^5 = 100 \cdot 32 = 3200$.

Starting with a value for N, the output, we could try to find a corresponding value for t, the input. For example, we could ask at what time t will the value for N be 1000? This turns out to be a harder question to answer than might first appear.

We can estimate a value for t that makes $N = 1000$ by using a data table or graph. In Table 6.1, when $t = 3, N = 800$. When $t = 4, N = 1600$. Since N is steadily increasing, $N = 1000$ for a value of t that is somewhere between 3 and 4.

We can also estimate the value for t when $N = 1000$ by looking at a graph of the function $N = 100 \cdot 2^t$ (Figure 6.1). By locating the position on the vertical axis where $N = 1000$, we can move over horizontally to find the corresponding point on the function graph. By moving from this point vertically down to the t-axis, we can estimate the t value for this point. The value for t appears to be approximately 3.3, so the coordinates of the point are roughly (3.3, 1000).

Table 6.1
Values for $N = 100 \cdot 2^t$

t	N
0	100
1	200
2	400
3	800
4	1,600
5	3,200
6	6,400
7	12,800
8	25,600
9	51,200
10	102,400

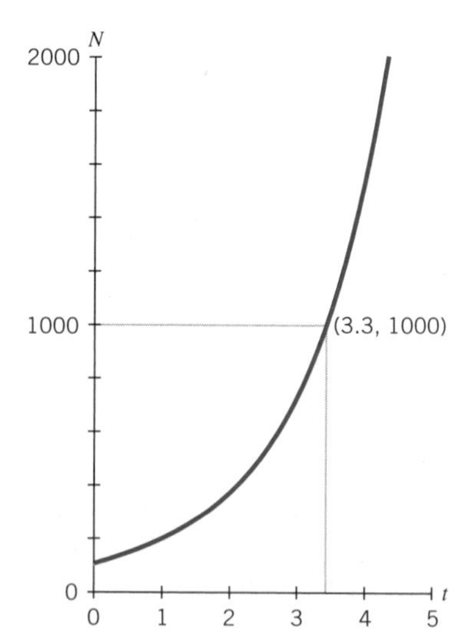

Figure 6.1 Estimating a value for t when $N = 1000$ on a graph of $N = 100 \cdot 2^t$.

An alternate strategy is to set $N = 1000$ and solve the equation for the corresponding value for t.

Start with the equation $\qquad N = 100 \cdot 2^t$

set $N = 1000$ $\qquad 1000 = 100 \cdot 2^t$

divide both sides by 100 $\qquad 10 = 2^t$

Then we are left with the problem of finding a solution to the equation

$$10 = 2^t$$

This equation is unlike any other we have solved. We can estimate the value for t that satisfies the equation. Since $2^3 = 8$ and $2^4 = 16$, then $2^3 < 10 < 2^4$. So $2^3 < 2^t < 2^4$, and therefore t is between 3 and 4, which agrees with our previous estimates.

Using these strategies, we can find approximate solutions to the equation $10 = 2^t$. Strategies for finding exact solutions to such equations require the use of logarithms. Hence we turn for a while to the properties of logarithms after which we will return to the problem of solving exponential equations for the value of the exponent.

Algebra Aerobics 6.1a

1. Given the equation $M = 250 \cdot 3^t$, find values for M when:

   a. $t = 0$ **c.** $t = 2$

   b. $t = 1$ **d.** $t = 3$

2. Use Table 6.1 to determine between which integers the value of t lies when N is:

   a. 2000 **b.** 50,000

3. Use the accompanying graph of $y = 3^x$ to estimate the solution to each of the following equations:

   a. $3^x = 7$

   b. $3^x = 0.5$

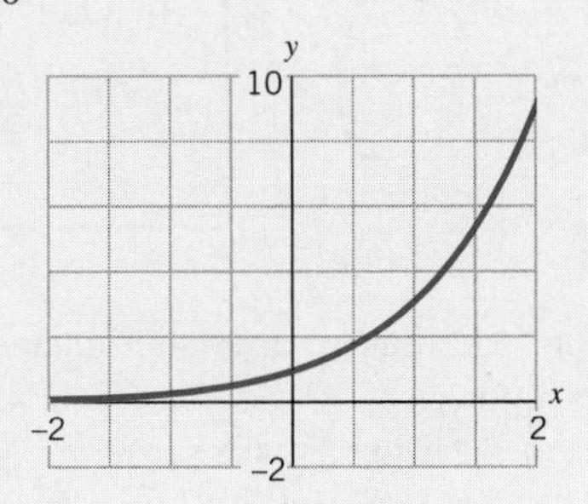

4. Use the accompanying table to determine between which integers the solution to each of the following equations lies:

   a. $5^x = 73$ **b.** $5^x = 0.36$

x	5^x
−2	0.04
−1	0.2
0	1
1	5
2	25
3	125
4	625

Properties of Logarithms

Recall the following:

> The *logarithm base 10 of* x is the power to which we raise 10 in order to produce x.
>
> $$\log_{10} x = c \quad \text{means that} \quad 10^c = x$$
>
> Logarithms base 10 are called *common logarithms* and $\log_{10} x$ is written as $\log x$.

So,

$10^5 = 100{,}000$ is equivalent to saying that $\log 100{,}000 = 5$

$10^{-3} = 0.001$ is equivalent to saying that $\log 0.001 = -3$

Using a calculator we can solve equations such as

$$10^x = 80$$

Rewrite it in equivalent form	$\log 80 = x$
use a calculator to compute the log	$1.903 \approx x$

But to solve equations such as $10 = 2^t$ that involve exponential expressions where the base is not 10, we need to know more about logarithms. As the following tables suggest, the properties of logarithms follow directly from the definition of logarithms and from the properties of exponents:

Properties of Exponents

If a is any positive real number and p and q are any real numbers, then

1. $a^p \cdot a^q = a^{p+q}$
2. $a^p / a^q = a^{p-q}$
3. $(a^p)^q = a^{p \cdot q}$
4. $a^0 = 1$

Corresponding Properties of Logarithms

If A and B are positive real numbers and p is any real number, then

1. $\log (A \cdot B) = \log A + \log B$
2. $\log (A/B) = \log A - \log B$
3. $\log A^p = p \log A$
4. $\log 1 = 0$ (since $10^0 = 1$)

In addition

5. $\log 10^x = x$ (for any real x)
6. $10^{\log x} = x$ $(x > 0)$

Finding the common logarithm of a number means finding its exponent when the number is written as a power of 10. So when you see "logarithm," think "exponent," and the properties of logarithms make sense.

As we learned in Chapter 4, the log of 0 or a negative number is not defined. But when we take the log of a number, we can get 0 or a negative value. For example, $\log 1 = 0$ and $\log 0.1 = -1$.

We will list a rationale for each of the properties of logarithms and prove Properties 1, 5, and 6. We leave the other proofs as exercises.

PROPERTY 1 $\log (A \cdot B) = \log A + \log B$

Try expressing in words all the other properties of logarithms in terms of exponents.

RATIONALE

Property 1 of exponents says that when we multiply two terms with the same base, we keep the base and *add* the exponents; that is, $a^p \cdot a^q = a^{p+q}$.

Property 1 of logs says that if you write $A \cdot B$, A, and B each as powers of 10, then the exponent of $A \cdot B$ is the sum of the exponents of A and B.

PROOF

If we let $\log A = x$, then	$10^x = A$
and if $\log B = y$, then	$10^y = B$
We have two equal products	$A \cdot B = 10^x \cdot 10^y$
by laws of exponents	$= 10^{x+y}$
Now rewrite using logs	$\log(A \cdot B) = x + y$
substitute $\log A$ for x and $\log B$ for y	$\log(A \cdot B) = \log A + \log B$

and we arrive at our desired result.

Example 1

By Property 1 of logarithms $\log(10^2 \cdot 10^3) = \log 10^2 + \log 10^3$.
Verify that this statement is true using properties of exponents and Property 5 of logarithms.

Solution

On the left side $\log(10^2 \cdot 10^3) = \log(10^{2+3}) = \log 10^5 = 5$

On the right side $\log 10^2 + \log 10^3 = 2 + 3 = 5$

Since both sides equal 5, then $\log(10^2 \cdot 10^3) = \log 10^2 + \log 10^3$.

Example 2

Verify that $\log(10^5 \cdot 10^{-7}) = \log 10^5 + \log 10^{-7}$.

Solution

On the left side $\log(10^5 \cdot 10^{-7}) = \log(10^{5-7}) = \log 10^{-2} = -2$

On the right side $\log 10^5 + \log 10^{-7} = 5 - 7 = -2$

Since both sides equal -2, then $\log(10^5 \cdot 10^{-7}) = \log 10^5 + \log 10^{-7}$.

PROPERTY 2 $\log(A/B) = \log A - \log B$

RATIONALE

Property 2 of exponents says that when we divide terms with the same base, we keep the base and *subtract* the exponents, that is, $a^p/a^q = a^{p-q}$. Property 2 of logs says that if we write A and B each as powers of 10, then the exponent of A/B equals the exponent of A *minus* the exponent of B.

Example 3

Verify that $\log(10^2/10^3) = \log 10^2 - \log 10^3$.

Solution

On the left side $\log(10^2/10^3) = \log(10^{2-3}) = \log 10^{-1} = -1$

On the right side $\log 10^2 - \log 10^3 = 2 - 3 = -1$

So both sides are equal.

Example 4

Verify that $\log(10^5/10^{-7}) = \log 10^5 - \log 10^{-7}$.

Solution

$$\log(10^5/10^{-7}) = \log(10^{5-(-7)}) = \log(10^{12}) = 12$$

and $$\log 10^5 - \log 10^{-7} = 5 - (-7) = 12$$

So both sides are equal.

Why does $\log \sqrt{AB}$ lie halfway between $\log A$ and $\log B$?

PROPERTY 3 $\log A^p = p \log A$.

RATIONALE

Since

$$\log A^2 = \log(A \cdot A) = \log A + \log A = 2 \log A$$

$$\log A^3 = \log(A \cdot A \cdot A) = \log A + \log A + \log A = 3 \log A$$

$$\log A^4 = \log(A \cdot A \cdot A \cdot A) = 4 \log A$$

it seems reasonable to expect that in general

$$\log A^p = p \log A$$

Example 5

Simplify each expression, and if possible evaluate with a calculator.

$$\log 10^3 = 3 \log 10 = 3 \cdot 1 = 3$$

$$\log 2^3 = 3 \log 2 \approx 3 \cdot 0.301 \approx 0.903$$

$$\log\sqrt{3} = \log 3^{1/2} = \tfrac{1}{2} \log 3 \approx \tfrac{1}{2} \cdot 0.477 \approx 0.239$$

$$\log x^{-1} = (-1) \cdot \log x = -\log x$$

$$\log 0.01^a = a \log 0.01 = a \cdot (-2) = -2a$$

PROPERTY 4 $\log 1 = 0$

RATIONALE

Since $10^0 = 1$ by definition, the equivalent statement using logarithms is $\log 1 = 0$.

PROPERTIES 5 AND 6

$$\log 10^x = x \text{ and } 10^{\log x} = x$$

RATIONALE FOR $\log 10^x = x$

Finding the logarithm of a number base 10 involves writing the number as 10 to a power and then identifying the exponent. Since 10^x is already written as 10 to the power x, then $\log 10^x = x$.

RATIONALE FOR $10^{\log x} = x$.

By definition, $\log x$ is the number such that when 10 is raised to that power the result is x.

PROOF

Let $\quad y = \log x$

rewrite using definition of logarithm $\quad 10^y = x$

substitute $\log x$ for y $\quad 10^{\log x} = x$

Example 6

Use the properties of logarithms to write the following expression as the sum or difference of several logs:

$$\log\left(\frac{x(y-1)^2}{\sqrt{z}}\right)$$

Solution

By Property 2 $\quad \log\left(\frac{x(y-1)^2}{\sqrt{z}}\right) = \log x(y-1)^2 - \log z^{1/2}$

by Property 1 $\quad = \log x + \log(y-1)^2 - \log z^{1/2}$

by Property 3 $\quad = \log x + 2\log(y-1) - \frac{1}{2}\log z$

We call this process *expanding the expression.*

Example 7

Use the properties of logarithms to write the following expression as a single logarithm.

$$2\log x - \log(x-1)$$

Solution

By Property 3 $\quad 2\log x - \log(x-1) = \log x^2 - \log(x-1)$

by Property 2 $\quad = \log\left(\frac{x^2}{x-1}\right)$

We call this process *contracting the expression.*

Common Error

Probably the most common error in using logarithms stems from confusion over the division property. For example,

$$\log 10 - \log 2 = \log\left(\frac{10}{2}\right)$$

but

$$\log 10 - \log 2 \neq \frac{\log 10}{\log 2}$$

Example 8

Solve for x:

$$1 = \log x + \log(x + 3)$$

Solution

Use Property 1	$1 = \log[x(x + 3)]$
rewrite using definition of log	$10^1 = x(x + 3)$
multiply and subtract 10	$0 = x^2 + 3x - 10$
factor	$0 = (x + 5)(x - 2)$
solve	$x = -5$ or $x = 2$

Since logarithms are not defined for negative numbers, $\log(-5)$ is not defined. So we must discard $x = -5$ as a solution, and $x = 2$ is the only solution.

Algebra Aerobics 6.1b

1. Using only the properties of exponents and the definition of logarithm, verify that:
   a. $\log(10^5/10^7) = \log 10^5 - \log 10^7$
   b. $\log[10^5 \cdot (10^7)^3] = \log(10^5) + 3\log(10^7)$
2. Expand, using the properties of logarithms:
$$\log\sqrt{\frac{2x - 1}{x + 1}}$$
3. Contract, expressing your answer as a single logarithm:
$$\tfrac{1}{3}[\log x - \log(x + 1)]$$
4. Solve for x:
$$\log(x + 12) - \log(2x - 5) = 2$$
5. Show that $\log 10^3 - \log 10^2 \neq \dfrac{\log 10^3}{\log 10^2}$

Answering Our Original Question: Using Logarithms To Solve Exponential Equations

Remember the question that started this chapter? We wanted to find out how many time periods it would take 100 *E. coli* bacteria to become 1000. To find an exact solution, we needed to solve the equation $1000 = 100 \cdot 2^t$, or by dividing by 100, the equivalent equation, $10 = 2^t$. We now have the necessary tools.

Given	$10 = 2^t$
take the logarithm of each side	$\log 10 = \log 2^t$
use Property 3 of logs	$\log 10 = t \log 2$
divide both sides by log 2	$\dfrac{\log 10}{\log 2} = t$
or	$t = \dfrac{\log 10}{\log 2}$ time periods

Thus when $N = 1000$ in the equation $N = 100 \cdot 2^t$, we have $t = \log 10/\log 2$. We have found a strategy using logarithms that produces an *exact* mathematical answer.

It's hard to judge the size of the number (log 10)/(log 2), but we can obtain as accurate an estimate as needed by using a calculator. Since $\log 10 = 1$ and $\log 2 \approx 0.301$,

$$\begin{aligned} t &= \frac{\log 10}{\log 2} \\ &\approx \frac{1}{0.301} \\ &\approx 3.32 \text{ time periods} \end{aligned}$$

which is consistent with our previous estimates of a value between 3 and 4, approximately equal to 3.3.

Since each time period represents 20 minutes, then 3.32 time periods represents $3.32 \cdot (20 \text{ minutes}) = 66.4$ minutes. So in the model, the bacteria would increase from the initial number of 100 to 1000 in a little over 66 minutes.

Example 9

How long would it take for 100 bacteria to increase to 2000 bacteria?

Solution

Given that

$$N = 100 \cdot 2^t$$

where N = number of bacteria and t = time (in 20-minute periods), we must set $N = 2000$ and solve the equation

$$2000 = 100 \cdot 2^t$$

divide both sides by 100	$20 = 2^t$
take the log of both sides	$\log 20 = \log 2^t$
use Property 3 of logs	$\log 20 = t \log 2$
divide by log 2	$\dfrac{\log 20}{\log 2} = t$
evaluate using a calculator	$\dfrac{1.301}{0.301} \approx t$
simplify and switch sides to get	$t \approx 4.32$ time periods

So when $N = 2000$, $t = 4.32$ time periods. Since each time period is 20 minutes, it takes $4.32 \cdot (20 \text{ minutes}) = 86.4$ minutes, or a little over 86 minutes, for the bacteria to increase from 100 to 2000.

We could also arrive at the same answer if we noted that since 2 is the base (or the growth factor) of the exponential function $N = 100 \cdot 2^t$, then the number of bacteria doubles during each time period t. Each time period is 20 minutes. From Example 8 we know it takes about 66 minutes, or 3.3 time periods, for 100 bacteria to grow to 1000. It will take one additional time period of 20 minutes for the bacteria to double from 1000 to 2000. Hence the total time to grow from 100 to 2000 bacteria will be $3.3 + 1 = 4.3$ time periods, or approximately $66 + 20 = 86$ minutes.

Example 10

As we saw in Chapter 5, the equation $P = 250(1.05)^n$ gives the value of \$250 invested at 5% interest (compounded annually) for n years. How many years does it take for the initial \$250 investment to double to \$500?

Solution

a. Estimating the answer.

If $R = 5\%$ per year, then the rule of 70 (discussed in Section 5.5) estimates the doubling time as

$$70/R = 70/5 = 14 \text{ years}$$

b. Calculating a more precise answer.

We can set $P = 500$ and solve the equation

	$500 = 250(1.05)^n$
Divide both sides by 250	$2 = (1.05)^n$
take the log of both sides	$\log 2 = \log(1.05)^n$
use Property 3 of logs	$\log 2 = n \log 1.05$
divide by log 1.05	$\dfrac{\log 2}{\log 1.05} = n$
evaluate with a calculator	$\dfrac{0.3010}{0.0212} \approx n$
divide and switch sides	$n \approx 14.2$ years

So the estimate of 14 years using the rule of 70 was pretty close.

Example 11

In Chapter 5 we used the function $f(t) = 100(0.9755)^t$ to measure the remaining amount of radioactive material as 100 milligrams (mg) of strontium-90 decayed over time t (in years). How many years would it take for there to be only 10 mg of strontium-90 left?

Solution

We set $f(t) = 10$ and solve the equation

	$10 = 100(0.9755)^t$
Divide both sides by 100	$0.1 = (0.9755)^t$

take the log of both sides	$\log 0.1 = \log(0.9755)^t$
use Property 3 of logs	$\log 0.1 = t \log 0.9755$
divide by log 0.9755	$\frac{\log 0.1}{\log 0.9755} = t$
evaluate logs	$\frac{-1}{-0.01077} = t$
divide and switch sides	$t \approx 93$ years

So it takes almost a century for 100 mg of strontium-90 to decay to 10 mg.

Algebra Aerobics 6.1c

1. Solve for t in the following equations:
   a. $60 = 10 \cdot 2^t$
   b. $500(1.06)^t = 2000$
   c. $80(0.95)^t = 10$
2. Using the model $N = 100 \cdot 2^t$ for bacteria growth, where t is measured in 20-minute time periods, how long will it take for the bacteria count
   a. to reach 7000?
   b. to reach 12,000?
3. First use the rule of 70 to estimate how long it would take \$1000 invested at 6% compounded annually to double to \$2000. Then use logs to find a more precise answer.
4. Use the function in Example 11 to determine how long it will take for 100 milligrams of strontium-90 to decay to 1 milligram.

6.2 BASE *e* AND CONTINUOUS COMPOUNDING

A Brief Introduction to *e*

We saw in Chapter 4 that we can construct expressions using any positive number a as a base and any real number as an exponent. Similarly, we can construct logarithms using any positive number a as a base. There are two standard bases used with logarithms. Base 10 may seem most logical when you are first learning about logarithms. But in most scientific applications, another base, a special number called e (named after Euler, a Swiss mathematician), turns out to be a much more natural choice. The value of e is approximately 2.71828. (You can use 2.72 as an estimate for e in most calculations. Your calculator probably has an e^x key for more accurate computations.) The number e is irrational; it cannot be written as the quotient of two integers or as a repeating decimal. Like π, e is a fundamental mathematical constant.

> The irrational number e is a fundamental mathematical constant whose value is approximately 2.71828.

First we'll discuss why e is important, and then we'll see how any exponential function can be written using e.

Continuous Compounding

The number e arises naturally in cases of continuous growth at a specified rate. For example, suppose we invest \$100 in a bank account that pays interest of 6% per year. To compute the amount of money we have at the end of 1 year, we must also know how often the interest is credited to our account, that is, how often it is *compounded.*

The simplest case is the type we have examined in Chapter 5, where the interest is compounded once per year. We earn interest of 6% on our principal, so at the end of 1 year we have

$$\begin{aligned} \$100 + 0.06 \cdot (\$100) &= \$100 \cdot (1 + 0.06) \\ &= \$100 \cdot (1.06) \\ &= \$106 \end{aligned}$$

Now suppose that the interest is compounded twice a year. Since the rate is 6% *per year,* the interest computed at the end of each 6-month period is 6%/2, or 3%. So at the end of the first half year, we have $\$100 \cdot (1 + 0.03) = \$100 \cdot (1.03) = \$103$. At the end of the second half year, we earn 3% interest on our new balance of \$103. So at the end of 1 year our \$100 has become

$$\begin{aligned} \$103 \cdot (1.03) &= (\$100 \cdot 1.03) \cdot (1.03) \\ &= \$100 \cdot (1.03)^2 \\ &= \$106.09 \end{aligned}$$

We earn 9 cents more when interest is credited twice per year than when it is credited once per year; the difference is a result of interest earned during the second half year on the \$3 in interest credited at the end of the first half year. In other words, we're starting to earn interest on interest. To earn the same amount with only annual compounding, we would need an interest rate of 6.09%. We call 6% the *nominal interest rate* and 6.09% the *effective interest rate* or *effective annual yield.*[1] This distinction was not useful before, because when interest is compounded only once a year the nominal and effective rates are the same.

Next, suppose that interest is compounded quarterly, or four times per year. In each quarter, we receive one-quarter of 6%, or 1.5% interest. Each quarter, our investment is multiplied by $1 + 0.015 = 1.015$ and, after the first quarter, we earn interest on the interest we have already received. At the end of 1 year we have received interest four times, so our initial \$100 investment has become

$$\$100 \cdot (1.015)^4 \approx \$106.14$$

In this case, the effective interest rate (or effective annual yield) is about 6.14%.

We may imagine dividing the year into smaller and smaller time intervals and computing the interest earned at the end of 1 year. The effective interest rate will be slightly more each time (Table 6.2).

In general, if we calculate the interest on \$100 n times a year when the nominal interest rate is 6%, we get

$$\$100\left(1 + \frac{0.06}{n}\right)^n$$

[1]Banks call the nominal interest rate the annual percentage rate, or APR. When describing accounts, banks are required by law to list both the nominal (the APR) and the effective interest rates. The effective interest rate is the one that tells you how much interest you will actually earn each year.

Table 6.2
Investing \$100 for One Year at a Nominal Interest Rate of 6%

Number of Times Interest Computed During the Year	Value of \$100 at End of One Year (\$)	Effective Annual Interest Rate (%)
1	$100(1+0.06) = 100(1.06) = 106.00$	6.00
2	$100(1+0.06/2)^2 = 100(1.03)^2 = 100(1.0609) = 106.09$	6.09
4	$100(1+0.06/4)^4 = 100(1.015)^4 \approx 100(1.0614) \approx 106.14$	6.14
6	$100(1+0.06/6)^6 = 100(1.010)^6 \approx 100(1.0615) \approx 106.15$	6.15
12	$100(1+0.06/12)^{12} = 100(1.005)^{12} \approx 100(1.0617) \approx 106.17$	6.17
24	$100(1+0.06/24)^{24} = 100(1.0025)^{24} \approx 100(1.0618) \approx 106.18$	6.18
⋮		
n	$100(1+0.06/n)^n$	

We may imagine increasing the number of periods, n, without limits, so that interest is computed every week, every day, every hour, every second, and so on. The surprising thing is that the term by which 100 gets multiplied, namely

$$\left(1+\frac{0.06}{n}\right)^n$$

does not get arbitrarily large. Examine Table 6.3.

Table 6.3
Value of $(1 + 0.06/n)^n$ as n Increases

Compounding Period	n (Number of Compoundings per Year)	Approximate Value of $(1+0.06/n)^n$
Once a day	365	1.0618313
Once an hour	$365 \cdot 24 =$ 8,760	1.0618363
Once a minute	$365 \cdot 24 \cdot 60 =$ 525,600	1.0618365
Once a second	$365 \cdot 24 \cdot 60 \cdot 60 =$ 31,536,000	1.0618365

Where does e fit in? As n, the number of compounding periods per year, increases, the value of $(1 + 0.06/n)^n$ approaches $1.0618365 \approx e^{0.06}$. You can confirm this on your calculator. As n gets arbitrarily large, we can think of the compounding occurring at each instant. We call this *continuous compounding.*

So if we invest \$100 at 6% continuously compounded, at the end of 1 year we will have

$$\$100 \cdot e^{0.06} \approx \$100 \cdot 1.0618365 = \$106.18365$$

An annual interest rate of 6% *compounded continuously* is equivalent to an annual interest rate of 6.18365% *compounded once a year.* Six percent is the nominal interest rate, and 6.18365% is the effective interest rate, the one that tells you exactly how much money you will make after 1 year.

What if we invest \$100 for x years at a nominal interest rate of 6%? If the interest is compounded n times a year, the annual growth factor is $(1 + 0.06/n)^n$; that is, every year the \$100 is multiplied by $(1 + 0.06/n)^n$. After x years the \$100 is multiplied by $(1 + 0.06/n)^n$ a total of x times, or equivalently, multiplied by $[(1 + 0.06/n)^n]^x = (1 + 0.06/n)^{nx}$. So \$100 will be worth

$$\$100(1+0.06/n)^{nx}$$

If the interest is compounded continuously, the annual growth factor is $e^{0.06}$; that is, every year the \$100 is multiplied by $e^{0.06}$. After x years the \$100 is multiplied by $e^{0.06}$ a total of x times, or equivalently multiplied by $(e^{0.06})^x = e^{0.06x}$. So \$100 will be worth

$$\$100e^{0.06x}$$

In summary, if we wish to invest our \$100 over x years at a nominal interest rate of 6%, then we will have

$\$100(1 + 0.06/n)^{nx}$ if the interest is compounded n times a year

$\$100e^{0.06x}$ if the interest is compounded continuously

Generalizing Our Results

If we invest P dollars at an annual interest rate of r (in decimal form) compounded n times a year, then at the end of 1 year we will have

$$P\left(1 + \frac{r}{n}\right)^n$$

At the end of x years we have

$$P\left(1 + \frac{r}{n}\right)^{nx}$$

where r is the nominal interest rate.

> The value of P dollars invested at an annual interest rate r (expressed in decimal form) compounded n times a year for x years equals
>
> $$y = P\left(1 + \frac{r}{n}\right)^{nx}$$
>
> We call r the *nominal interest rate.*

Just as $(1 + 0.06/n)^n$ approaches $e^{0.06}$ as n gets very large, $(1 + r/n)^n$ approaches e^r.

> $(1 + r/n)^n$ approaches e^r as n increases toward infinity.

Note that in the special case where $r = 1$, as n gets arbitrarily large, $(1 + 1/n)^n$ approaches $e^1 = e$.

> $(1 + 1/n)^n$ approaches e as n increases toward infinity.

So if P dollars are invested at an annual interest rate r (in decimal form) compounded continuously, then after x years we have

$$P \cdot (e^r)^x = P \cdot e^{rx}$$

In summary:

> The value of P dollars invested at an annual interest rate r (expressed in decimal form) compounded continuously for x years equals
>
> $$y = P \cdot e^{rx}$$
>
> We call r the *nominal interest rate.*

Example 1

If you have \$250 to invest, and you are quoted a nominal interest rate of 4%, construct the equations that will tell you how much money you will have if the interest is compounded once a year, quarterly, once a month, or continuously. In each case calculate the value after 10 years.

Solution
See Table 6.4.

Table 6.4
Investing \$250 at a Nominal Interest Rate of 4% for Different Compounding Intervals

Number of Compoundings per Year	\$ Value After x Years	Approximate \$ Value when $x = 10$ Years
1	$250 \cdot (1 + 0.04)^x = 250 \cdot (1.04)^x$	370.06
4	$250 \cdot (1 + 0.04/4)^{4x} = 250 \cdot (1.01)^{4x}$	372.22
12	$250 \cdot (1 + 0.04/12)^{12x} \approx 250 \cdot (1.00333)^{12x}$	372.56
Continuous	$250 \cdot e^{0.04x}$	372.96

Example 2

If the nominal interest rate is 7% compounded continuously, what is the effective interest rate?

Solution
The nominal interest rate of 7% is compounded continuously, so the equation $y = Pe^{0.07x}$ describes the amount y that an initial investment P is worth after x years. Using a calculator, we find that $e^{0.07} \approx 1.073$. The equation could be rewritten as $y = P(1.073)^x$. So the effective interest rate is about 7.3%.

Example 3

You have a choice between two bank accounts. One is a passbook account in which you receive simple interest of 5% per year, compounded once per year. The other is a 1-year certificate of deposit, which pays interest at the rate of 4.8% per year, compounded continuously. Which account is the better deal?

Solution

Since the interest on the passbook account is compounded once a year, the nominal and effective interest rates are both 5%. The equation $y = P(1.05)^x$ can be used to describe the amount y that the initial investment P is worth after x years.

The 1-year certificate of deposit has a nominal interest rate of 4.8%. Since this rate is compounded continuously, the equation $y = Pe^{0.048x}$ describes the amount y that the initial investment P is worth after x years. Since $e^{0.048} \approx 1.049$, the equation can also be written as $y = P(1.049)^x$, and the effective interest rate is 4.9%. So the passbook account is a better deal.[2]

Algebra Aerobics 6.2

1. Find the amount accumulated after 1 year on an investment of \$1000 at 8.5% compounded

   a. Annually **b.** Quarterly **c.** Continuously

2. Find the effective interest rate for each nominal interest rate below with interest compounded continuously.

   a. 4% **b.** 12.5% **c.** 18%

3. The value for e is often defined as the number that $(1 + 1/n)^n$ approaches as n gets arbitrarily large. Use your calculator to fill in the accompanying table. Use your exponent key (x^y or y^x) to evaluate the last column. Is your value consistent with the approximate value for e of 2.71828 given in the text?

n	$1/n$	$1 + (1/n)$	$[1 + (1/n)]^n$
1			
100	0.01	$1 + 0.01 = 1.01$	$(1.01)^{100} \approx 2.7048138$
1,000			
1,000,000			
1,000,000,000			

6.3 THE NATURAL LOGARITHM

The *common logarithm* uses 10 as a base. The *natural logarithm* uses e as a base and is written $\ln x$ rather than $\log_e x$. Scientific calculators have a key that computes $\ln x$.

> The *logarithm base e* of x is the power of e you need to produce x. Logarithms base e are called *natural logarithms* and are written as $\ln x$.
>
> $$\ln x = c \quad \text{means that} \quad e^c = x$$

The properties for natural logarithms are similar to the properties for common logarithms.

[2]When opening an account on which the interest is compounded a finite number of times during the year, you need to be aware of whether or not you lose interest if you withdraw the money early. For example, frequently in accounts that are compounded quarterly, if you withdraw your money before the quarter is up, you will not receive any interest for that quarter.

Properties of Common Logarithms	**Properties of Natural Logarithms**
If A and B are positive real numbers and p is any real number, then:	If A and B are positive real numbers and p is any real number, then:
1. $\log(A \cdot B) = \log A + \log B$	**1.** $\ln(A \cdot B) = \ln A + \ln B$
2. $\log(A/B) = \log A - \log B$	**2.** $\ln(A/B) = \ln A - \ln B$
3. $\log A^p = p \log A$	**3.** $\ln A^p = p \ln A$
4. $\log 1 = 0$ (since $10^0 = 1$)	**4.** $\ln 1 = 0$ (since $e^0 = 1$)
5. $\log 10^x = x$ (for any real x)	**5.** $\ln e^x = x$ (for any real x)
6. $10^{\log x} = x$ $(x > 0)$	**6.** $e^{\ln x} = x$ $(x > 0)$

Like the common logarithm, $\ln A$ is not defined when $A \leq 0$.

Example 1

Property 5 of natural logarithms says that $\ln(e^x) = x$, since x is the power of e needed to produce e^x. Applying this property, we have

$$\ln e = \ln e^1 = 1$$

$$\ln e^4 = 4$$

Property 6 says that $e^{\ln x} = x$, since by definition $\ln x$ is the power of e needed to produce x. As a result we have

$$e^{\ln x^2} = x^2$$

$$e^{3 \ln x} = e^{\ln x^3} = x^3$$

Example 2

If the effective annual interest rate on an account compounded continuously is 0.0521, estimate the nominal interest rate.

Solution

If 0.0521 is the effective interest rate, then the equation $y = P(1.0521)^x$ represents the value, y, of P dollars after x years. To find the nominal interest rate that is continuously compounded, we must write y in the form Pe^{rx}, or equivalently $P(e^r)^x$. So the expression e^r must equal 1.0521. We need to solve for r in the equation

$$1.0521 = e^r$$

Take ln of both sides $\quad \ln 1.0521 = \ln e^r$

use a calculator $\quad 0.0508 \approx \ln e^r$

use Property 5 $\quad 0.0508 \approx r$

So the nominal interest rate is approximately 5.08%. In other words, compounding annually at 5.21% is equivalent to compounding continuously at 5.08%.

Example 3

Expand, using the laws of logarithms:

$$\ln\sqrt{\frac{x+3}{x-2}}$$

Solution

$$\ln\sqrt{\frac{x+3}{x-2}} = \ln\left(\frac{x+3}{x-2}\right)^{1/2} = \frac{1}{2}\ln\left(\frac{x+3}{x-2}\right) = \frac{1}{2}[\ln(x+3) - \ln(x+2)]$$

Example 4

Contract, expressing the answer as a single logarithm:

$$\tfrac{1}{3}\ln(x-1) + \tfrac{1}{3}\ln(x+1)$$

Solution

$$\begin{aligned}\tfrac{1}{3}\ln(x-1) + \tfrac{1}{3}\ln(x+1) &= \tfrac{1}{3}[\ln(x-1) + \ln(x+1)]\\ &= \tfrac{1}{3}\ln[(x-1)(x+1)]\\ &= \ln[(x-1)(x+1)]^{1/3}\\ &= \ln(x^2-1)^{1/3} \quad \text{or} \quad \ln\sqrt[3]{x^2-1}\end{aligned}$$

Example 5

Solve for t:

$$10 = e^t$$

Solution

Given	$10 = e^t$
take ln of both sides	$\ln 10 = \ln e^t$
use Property 5	$\ln 10 = t$
evaluate and switch sides	$t \approx 2.303$

Algebra Aerobics 6.3

1. Evaluate without a calculator:
   - **a.** $\ln e^2$
   - **b.** $\ln 1$
   - **c.** $\ln \frac{1}{e}$
   - **d.** $\ln \frac{1}{e^2}$
   - **e.** $\ln \sqrt{e}$
2. Expand: $\ln \dfrac{\sqrt{x+2}}{x(x-1)}$.
3. Contract, expressing your answer as a single logarithm:
   $$\ln x - 2\ln(2x-1)$$
4. Find the nominal rate on an investment compounded continuously if the effective rate is 6.4%.
5. Determine how long it takes for \$10,000 to grow to \$50,000 at 7.8% compounded continuously.
6. Solve the following equations for x:
   - **a.** $e^{x+1} = 10$
   - **b.** $e^{x-2} = 0.5$

6.4 LOGARITHMIC FUNCTIONS

EXPLORE

Exploration 6.1 and course software "E8: Logarithmic Sliders" in *Exponential & Log Functions* will help you understand the properties of logarithmic functions.

For any $x > 0$, both the common and natural logarithms define unique numbers $\log x$ and $\ln x$, respectively, so we can define two functions:

$$y = \log x \quad \text{and} \quad y = \ln x$$

What will the graphs look like? We know something about the graphs since:

If $x > 1$,	$\log x$ and $\ln x$ are both positive.
If $x = 1$,	$\log 1 = 0$ and $\ln 1 = 0$.
If $0 < x < 1$,	$\log x$ and $\ln x$ are both negative.
If $x \leq 0$,	neither logarithm is defined.

Both graphs lie to the right of the y-axis, since they are only defined for $x > 0$. They both cross the x-axis at $(1, 0)$ and lie above the x-axis to the right of the x-intercept. Between the origin and $(1, 0)$, the graphs lie below the x-axis.

We can use a calculator to construct a table of values and then sketch the graphs of both functions. See Table 6.5 and Figure 6.2.

The graphs of common and natural logarithmic functions share a distinctive shape. Both are asymptotic to the y-axis; that is, as x gets closer and closer to zero, the graphs of $\log x$ and $\ln x$ approach negative infinity. Both graphs come very close to, but never touch, the y-axis. The graphs meet once at the common x-intercept $(1, 0)$. Both functions grow large very slowly when $x > 1$. As $x \to +\infty$, both $\log x$ and $\ln x \to +\infty$. The domain for both functions is all positive real numbers, since the logarithms of zero and negative numbers are not defined. The range for both is all real numbers.

Table 6.5
Evaluating $\log x$ and $\ln x$

x	$y = \log x$	$y = \ln x$
0.001	−3.000	−6.908
0.01	−2.000	−4.605
0.1	−1.000	−2.303
1	0.000	0.000
2	0.301	0.693
3	0.477	1.099
4	0.602	1.386
5	0.699	1.609
6	0.778	1.792
7	0.845	1.946
8	0.903	2.079
9	0.954	2.197
10	1.000	2.303

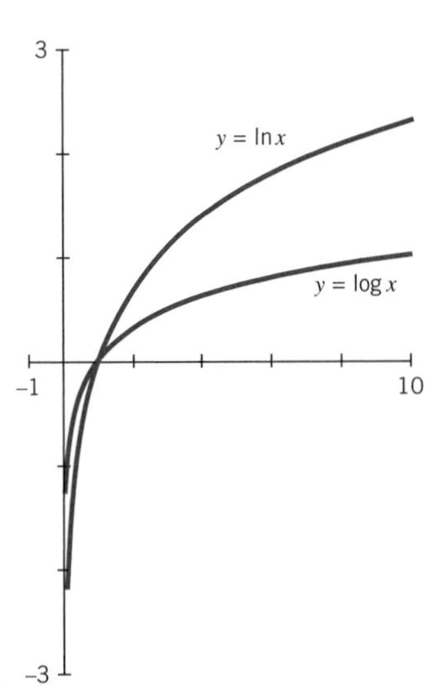

Figure 6.2 Graphs of $y = \log x$ and $y = \ln x$.

Measuring Acidity: The pH Scale

Chemists use the pH scale to measure acidity.[3] The pH is given by the equation

$$pH = -\log[H^+]$$

where $[H^+]$ designates the concentration of hydrogen ions. Chemists use the symbol H^+ for hydrogen ions and the brackets [ ] mean "the concentration of."

The concentration of hydrogen ions is measured in molar units, M. Typical concentrations range from 10^{-15} M to 10 M. Table 6.6 and Figure 6.3 show a set of values and the graph for pH. The graph is the standard logarithmic graph flipped over the horizontal axis, because of the negative sign in front of the log.

Table 6.6
Calculating pH Values

$[H^+]$ (in molar units M)	pH
10^{-15}	15.000
10^{-10}	10.000
1	0.000
5	−0.699
10	−1.000

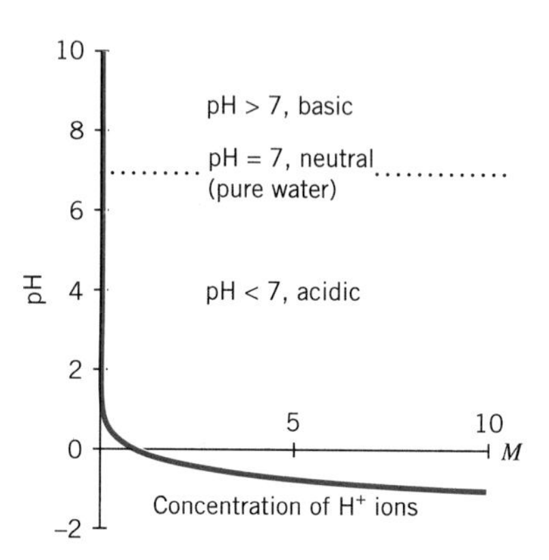

Figure 6.3 Graph of the pH function.

Note that multiplying $[H^+]$ by 10 (increasing $[H^+]$ by one order of magnitude) *decreases* the pH by 1. For example, when $[H^+] = 1$, then $pH = -\log 1 = 0$. When $[H^+] = 10$, then $pH = -\log 10 = -1$.

Pure water has a pH of 7 and is considered neutral. A substance with a $pH > 7$ is called *basic* (or *alkaline*). One with a $pH < 7$ is called *acidic*. The lower the pH, the higher the acidity and the higher the concentration of hydrogen ions.

Example 1

Vinegar has a pH of 3. Compare its H^+ concentration to that of pure water.

Solution

Since vinegar's pH is <7, it is acidic and has a higher concentration of H^+ than pure water. Its pH is 4 *less* than that of pure water, so its H^+ concentration is 10^4, or 10,000 times, that of pure water.

Example 2

Ammonia has a pH of 10. Compare its H^+ concentration to that of pure water.

Solution

Since ammonia's pH is > 7, it is basic (or alkaline). Its pH is 3 *more* than pure water's, so its H^+ concentration (or acidity) is less, only $10^{-3} = 1/10^3$ or 1/1000 that of water.

[3] For more details about the pH scale see Exercises 70 and 89 in Chapter 4.

The rain in many parts of the world is becoming increasingly acidic. The burning of fossil fuels (such as coal and oil) by power plants and automobile exhausts releases gaseous impurities into the air. The impurities contain oxides of sulfur and nitrogen that combine with moisture in the air to form drops of dilute sulfuric and nitric acids. An acid, by definition, releases hydrogen ions in water. High concentrations of hydrogen ions damage plants and water resources (such as the lakes of New England and Sweden) and erode structures (such as the outdoor fountain statues in Rome) by removing oxygen molecules. Some experts feel that the acid rain dilemma may be one of the greatest environmental problems facing the world in the near future.

Example 3

It is critical that health care professionals understand the nonlinearity of the pH function. An arterial blood pH of 7.35 to 7.45 for a patient is quite normal, while a pH of 7.1 means that the patient is severely acidotic and near death.

a. Determine the hydrogen ion concentration, $[H^+]$, first for a patient with an arterial blood pH of 7.4, then for one with a blood pH of 7.1.

b. How many more hydrogen ions are there in blood with a pH of 7.1 than with a pH of 7.4?

Solution

To find $[H^+]$ if the pH is 7.4, we must solve the equation

$$7.4 = -\log[H^+]$$

Multiply both sides by -1 $\qquad -7.4 = \log[H^+]$

then these powers of 10 are equal $\qquad 10^{-7.4} = 10^{\log[H+]}$

use Property 6 of logs $\qquad 10^{-7.4} = [H^+]$

evaluate and switch sides $\qquad [H^+] \approx 0.000000040$

$$\approx 4.0 \cdot 10^{-8}\, M$$

Similarly, if the pH is 7.1, we can solve the equation

$$7.1 = -\log[H^+]$$

to get $\qquad [H^+] \approx 7.9 \cdot 10^{-8}\, M$

Comparing the two concentrations gives us

$$\frac{[H^+] \text{ in a blood with a pH of 7.1}}{[H^+] \text{ in a blood with a pH of 7.4}} \approx \frac{7.9 \cdot 10^{-8}\, M}{4.0 \cdot 10^{-8}\, M} = \frac{7.9}{4.0} \approx 2$$

So there are approximately twice as many hydrogen ions in blood with a pH of 7.1 than in blood with a pH of 7.4.

Measuring Noise: The Decibel Scale

The decibel scale was designed to reflect human perception of sounds.[4] When it is very quiet, it is easy to notice a small increase in sound intensity. The same increase

[4]Two scientists, Weber and Fechner, studied the psychological response to intensity changes in stimuli. Their discovery that the perceived change was proportional to the logarithm of the intensity change of the stimulus is called the Weber–Fechner stimulus law.

in intensity in a noisy environment would not be noticed; it would take a much bigger change to be detected by humans. The same is true for light. If a 50-watt light bulb is replaced with a 100-watt light bulb, it is easy to notice the difference in brightness. But if you replace 500 watts with 550 watts, it would be very hard to distinguish the 50-watt difference. So the decibel scale, like the pH scale for acidity or the Richter scale for earthquakes, is logarithmic; that is, it measures order of magnitude changes.

Noise levels are measured in units called *decibels,* abbreviated dB. The name is in honor of the inventor of the telephone, Alexander Graham Bell. If we designate by I_0 the intensity of a sound at the threshold of human hearing (10^{-16} watts/cm^2), and we let I represent the intensity of an arbitrary sound (measured in watts/cm^2), then the noise level N of that sound measured in decibels is defined to be

$$N = 10 \log\left(\frac{I}{I_0}\right)$$

A noise emission statute enforced by the Massachusetts Registry of Motor Vehicles requires that the noise level of a motorcycle not exceed 82–86 decibels. Jay McMahon, chairman of the Modified Motorcycle Association, says, "It's just another way for the Registry to stick it to motorcycle operators. No one complains when a Maserati or a tractor trailer drives by." Does the statute seem fair to you? Why?

The unitless expression I/I_0 gives the *relative intensity* of a sound compared to the reference value of I_0. For example, if $I/I_0 = 100$, then the noise level, N, is equal to

$$N = 10 \log(100) = 10 \log(10^2) = 10(2) = 20 \text{ dB}$$

Table 6.7 shows relative intensities, the corresponding noise levels (in decibels), and how people perceive these noise levels. Note how much the relative intensity (the ratio I/I_0) of a sound source must increase for people to discern differences. Each time we *add* 10 units on the decibel scale, we *multiply* the relative intensity by 10, increasing it by one order of magnitude.

Table 6.7
How Decibel Levels Are Perceived

Relative Intensity I/I_0	Decibels (dB)	Average Perception
1	0	Threshold of hearing
10	10	Sound-proof room, very faint
100	20	Whisper, rustle of leaves
1,000	30	Quiet conversation, faint
10,000	40	Quiet home, private office
100,000	50	Average conversation, moderate
1,000,000	60	Noisy home, average office
10,000,000	70	Average radio, average factory, loud
100,000,000	80	Noisy office, average street noise
1,000,000,000	90	Loud truck, police whistle, very loud
10,000,000,000	100	Loud street noise, noisy factory
100,000,000,000	110	Elevated train, deafening
1,000,000,000,000	120	Thunder of artillery, nearby jackhammer
10,000,000,000,000	130	Threshold of pain, ears hurt

Example 4

What is the decibel level of a typical rock band playing with an intensity of 10^{-5} watts/cm^2? How much more intense is the sound of the band than an average conversation?

Solution

Given $I_0 = 10^{-16}$ watts/cm^2 and letting $I = 10^{-5}$ watts/cm^2 and N represent the decibel level,

by definition	$N = 10 \log\left(\frac{I}{I_0}\right)$
substitute for I and I_0	$= 10 \log(10^{-5}/10^{-16})$
Property 2 of exponents	$= 10 \log(10^{11})$
Property 5 of logs	$= 10 \cdot 11$
	$= 110$ decibels

So the noise level of a typical rock band is about 110 decibels.

According to Table 6.7, an average conversation measures about 50 decibels. So the noise level of the rock band is 60 decibels higher. Each increment of 10 decibels corresponds to a one order of magnitude increase in intensity. So the sound of a rock band is about six orders of magnitude, or $10^6 = 1{,}000{,}000$ times, more intense than an average conversation.

Example 5

What's wrong with the following statement? "A jet airplane landing at the local airport makes 120 decibels of noise. If we allow three jets to land at the same time, there will be 360 decibels of noise pollution."

Solution

There will certainly be three times as much sound intensity, but would we perceive it that way? According to Table 6.7, 120 decibels corresponds to a relative intensity of 10^{12}. Three times that relative intensity would equal $3 \cdot 10^{12}$. So the corresponding decibel level would be

	$N = 10 \log(3 \cdot 10^{12})$
	$= 10(\log 3 + \log(10^{12}))$
use a calculator	$\approx 10(0.477 + 12)$
	$\approx 10(12.477)$
multiply and round off	≈ 125 decibels

So three jets landing will produce a decibel level of 125, not 360. We would only perceive a slight increase in the noise level.

Algebra Aerobics 6.4

1. How would the graphs of $y = \log x^2$ and $y = 2 \log x$ compare?
2. Draw a rough hand sketch of the graph $y = -\ln x$. Compare it to the graph of $y = \ln x$.
3. A typical pH value for rain or snow in the northeastern United States is about 4. Is this basic or acidic? What is the corresponding hydrogen concentration? How does this compare to the hydrogen concentration of pure water?
4. What is the decibel level of a sound whose intensity is $1.5 \cdot 10^{-12}$ watts/cm^2?
5. If the intensity of a sound increases by a factor of 100, what is the increase in the decibel level? What if the intensity is increased by a factor of 10,000,000?

6.5 WRITING EXPONENTIAL FUNCTIONS USING BASE *e*

Translating from Base *a* to Base *e*

We can use logarithms to show that any exponential function can be rewritten using e as a base.

Consider the bacterial growth we described with the equation $N = 100 \cdot 2^t$. The bacteria don't all double at the same time, precisely at the beginning of each time period t. A continuous growth pattern is much more likely. If we are to rewrite this equation to reflect continuous compounding, we want to use base e. To do that, we need to write 2 as e^r for some r:

If	$2 = e^r$
take ln of both sides	$\ln 2 = \ln e^r$
use Property 5 of ln	$\ln 2 = r$

By substituting $e^r = e^{\ln 2}$ for 2, we can rewrite

$$N = 100 \cdot 2^t \quad \text{as} \quad N = 100(e^{\ln 2})^t = 100e^{(\ln 2)t}$$

Since $\ln 2 \approx 0.693$, we also have $N \approx 100e^{0.693t}$. In general, if

$$y = Ca^x$$

is an exponential function (where $a > 0$), we can always find an r such that $a = e^r$. By substituting for a, we can rewrite y as

$$y = C(e^r)^x = Ce^{rx}$$

We can then write e^r in terms of $\ln a$:

Given	$e^r = a$
take ln of both sides of the equation	$\ln e^r = \ln a$
use Property 5 of ln	$r = \ln a$

See "E4: $y = Ce^{rx}$ Sliders" and "E5: Comparing $y = Ca^x$ to $y = Ce^{rx}$" in *Exponential & Log Functions*.

So $y = Ca^x$ can be rewritten as $y = Ce^{rx} = Ce^{(\ln a)x}$. If $y = Ca^x$ and $a > 1$, the function represents exponential growth. For exponential growth, $\ln a = r$ is greater than zero. When $0 < a < 1$, the function represents decay. For exponential decay, $\ln a = r$ is less than zero.

Given a function $y = Ce^{rx}$, in a financial setting r is called the *nominal interest rate*. In general applications, r is called the *instantaneous growth rate*.

> For any number $a > 0$, the function $y = C \cdot a^x$ may be rewritten $y = C \cdot e^{rx}$ where $r = \ln a$. We call r the *instantaneous growth rate*.
>
> If $a > 1$, then $r > 0$ and the function represents growth.
> If $0 < a < 1$, then $r < 0$ and the function represents decay.
> If $a = 1$, then $r = 0$ and the function is constant.

Writing an exponential function in the form $y = Ca^x$ or $y = Ce^{rx}$ (where $r = \ln a$) is a matter of emphasis, since the graphs and functional values are identical. When we use $y = Ca^x$, we may think of the growth taking place at discrete points in time, whereas the form $y = Ce^{rx}$ emphasizes the notion of continuous growth. For example, $y = 100(1.05)^x$ could be interpreted as giving the value of \$100 invested for x years at 5% compounded annually, whereas its equivalent form

$$y = 100(e^{\ln 1.05})^x \approx 100e^{0.049x}$$

suggests that the money is invested at 4.9% compounded continuously.

Example 1

In Chapter 5 we saw that the function $f(t) = 100(0.9755)^t$ measures the amount of radioactive strontium-90 remaining as 100 milligrams (mg) decay over time t (in years). Rewrite the function using base e.

Solution

We must rewrite 0.9755 as a power of e; that is, we must solve the equation

$$0.9755 = e^k$$

Take ln of both sides	$\ln 0.9755 = \ln (e^k)$
apply Property 5 of ln	$\ln 0.9755 = k$
evaluate ln 0.9755	$-0.0248 \approx k$
we get	$0.9755 \approx e^{-0.0248}$

Substituting into the original function, we have

$$\begin{aligned} f(t) &= 100(0.9755)^t \\ &\approx 100(e^{-0.0248})^t \\ &\approx 100e^{-0.0248t} \end{aligned}$$

where -0.0248 is the instantaneous growth rate. Note that since the original base 0.9755 is less than 1, the function represents decay. So, as we would expect, when the function is rewritten using base e, the value of -0.0248 for the instantaneous growth rate is negative.

Example 2

Use a continuous compounding model to describe the growth of Medicare expenditures.

Solution

In Chapter 5 we found a best fit function for Medicare expenditures to be

$$y = 9.1(1.13)^x$$

where x = years since 1970 and y = Medicare expenditures in billions of dollars. This implies a growth rate of 13% compounded annually. To describe the same growth in terms of continuous compounding, we need to rewrite 1.13 as a power of e. Hence we need to solve

$$1.13 = e^r$$

Take ln of both sides	$\ln 1.13 = \ln e^r$
use Property 5 of ln	$\ln 1.13 = r$
evaluate ln 1.13	$0.122 \approx r$
substitute for r	$1.13 \approx e^{0.122}$

Substituting into the original function gives us

$$y = 9.1e^{0.122x}$$

So the instantaneous growth rate is 12.2%. Note that since the original base, 1.13, is greater than 1 and represents exponential growth, the instantaneous growth rate is positive.

Example 3

We now have the tools to prove the rule of 70 introduced in Chapter 5. Recall the rule said that if a quantity is growing (or decaying) at R percent per time period, then the time it takes the quantity to double (or halve) is approximately $70/R$ time periods. For example, if a quantity increases by a rate, R, of 7% each month, the doubling time is about $70/7 = 10$ months.

Proof

Let f be an exponential function of the form

$$f(t) = Ce^{rt}$$

Then C is the initial quantity, r is the instantaneous growth rate, and t represents time. Recall that r is in decimal form, and R is the equivalent amount expressed as a percentage. Assume $f(t)$ represents exponential growth, so $r > 0$. Since the doubling time for an exponential function is constant, we need only calculate the time for any given quantity to double. In particular, we can determine the time it takes for the initial amount C (at time $t = 0$) to become twice as large; that is, we can calculate the value for t such that

$$f(t) = 2C$$

Set	$Ce^{rt} = 2C$	
divide by C	$e^{rt} = 2$	
take ln of both sides	$\ln e^{rt} = \ln 2$	
evaluate and use ln Property 5	$rt \approx 0.693$	(1)

Here, $R = 100r$, so $r = R/100$. We round 0.693 up to 0.70.

Substitute in Equation (1)	$(R/100) \cdot t \approx 0.70$
multiply both sides by 100	$R \cdot t \approx 70$
divide by R	$t \approx 70/R$

So the time t it takes for the initial amount to double is approximately $70/R$, which is what the rule of 70 claims. We leave the similar proof about half-lives, where $r < 0$ represents decay, to the exercises.

Determining the Equation of an Exponential Function through Two Points

Question: The amount of carbon dioxide (CO_2) in the atmosphere has been increasing steadily since at least 1860, primarily because of the increased burning of fossil fuels such as coal, oil, and gasoline. In 1890 the CO_2 concentration (in parts per million by volume) was roughly 290. This means that out of, say, one million cubic feet of air, 290 cubic feet were CO_2. By 1970 the concentration had risen to about 320 ppm. Many scientists describe the increase as exponential and believe that the CO_2 concentration will continue to rise at the same pace for the next 100 to

200 years. Construct a model of CO_2 concentration, starting from 1860, that describes the increase in CO_2 concentration over time.

Solution. If we let our independent variable x = number of years since 1860, then 1890 corresponds to $x = 30$ and 1970 corresponds to $x = 110$. Let Q denote our dependent variable, the CO_2 concentration in ppm. Then when $x = 30$, $Q = 290$, and when $x = 110$, $Q = 320$. Our goal then is to construct an exponential function through the two points (30, 290) and (110, 320). The function will be of the basic form

$$Q = Ce^{rx}$$

where Q is the CO_2 concentration x years after 1860 and C is the initial CO_2 concentration in 1860.

Since the points (30, 290) and (110, 320) must satisfy our equation, substituting for x and Q, we have

$$290 = Ce^{30r} \quad \text{and} \quad 320 = Ce^{110r}$$

In order to solve for r, we start by dividing Ce^{110r} by Ce^{30r}:

$$\frac{Ce^{110r}}{Ce^{30r}} = \frac{320}{290}$$

We can simplify the left-hand side:

cancel the C's $\qquad \dfrac{Ce^{110r}}{Ce^{30r}} = \dfrac{e^{110r}}{e^{30r}}$

use Property 2 of exponents $\qquad = e^{110r-30r}$

$\qquad = e^{80r}$

Simplifying the right-hand side gives us

$$\frac{320}{290} \approx 1.1034$$

Setting the simplified expressions for the left-hand side and the right-hand side equal, we have

$$e^{80r} \approx 1.1034$$

Take ln of both sides $\qquad \ln e^{80r} \approx \ln 1.1034$

use Property 5 of ln $\qquad 80r \approx 0.09844$

divide by 80 $\qquad r \approx 0.00123$

With that value for the exponent r, our function model will be of the form $Q = Ce^{0.00123x}$. We need to find a value for the coefficient C, the initial amount of CO_2 concentration in 1860.

Substitute one of the two points in our equation for Q to solve for C:

Substitute the point (30, 290) $\qquad 290 = Ce^{(0.00123)(30)}$

multiply $\qquad = Ce^{0.0369}$

use a calculator to evaluate $e^{0.0369}$ $\qquad 290 \approx C(1.038)$

divide by 1.038 and switch sides $\qquad C \approx 280$

So the concentration of CO_2 in 1860 is about 280 parts per million by volume.

The desired function is then

$$Q = 280e^{0.00123x}$$

Search the Internet to find the current carbon dioxide levels in the atmosphere. How close are they to what your function would predict?

To double check that this is a reasonable answer, verify that the other point (110, 320) satisfies the equation; that is, using your calculator, evaluate $280e^{0.00123x}$ when $x = 110$ to verify that you get approximately 320.

Algebra Aerobics 6.5

1. Identify each of the following exponential functions as representing growth or decay:
   a. $M = Ne^{-0.029t}$
   b. $K = 100(0.87)^r$
   c. $Q = 375e^{0.055t}$
2. Rewrite each of the following as a continuous-growth model using base e:
   a. $y = 1000(1.062)^t$
   b. $y = 50(0.985)^t$
3. Iodine-131 is a radioactive substance that decays exponentially. If the amount of iodine-131 left after 8 days is 2.40 micrograms and after 20 days is 0.88 microgram, construct a function using e to describe its decay. How many micrograms of iodine-131 were there initially, and what is the instantaneous decay rate?

6.6 AN INTRODUCTION TO COMPOSITION AND INVERSE FUNCTIONS

The two functions $y = \log x$ and $y = 10^x$ are intimately linked. What one function "does" to a number x, the other function "undoes." For example, $\log 10^x = x$ (by Property 5 of logs): that is, if you take the log of 10^x, you get back x. Similarly, $10^{\log x} = x$ (Property 6): If you raise 10 to the power $\log x$, you also get x. When two functions behave this way, we say that they are *inverses* of each other. We can examine this inverse relationship using graphs, tables, and functional notation.

Comparing $y = \log x$ and $y = 10^x$

Imagine plotting the functions $y = \log x$ and $y = 10^x$ on graph paper and then folding the graph paper along the dotted line $y = x$ (see Figure 6.4). The two graphs would coincide perfectly. We say that the graphs are mirror images across the line $y = x$.

We can verify this symmetry numerically by comparing pairs of points on the two graphs. The point (0, 1) satisfies the equation $y = 10^x$ (since $10^0 = 1$) and

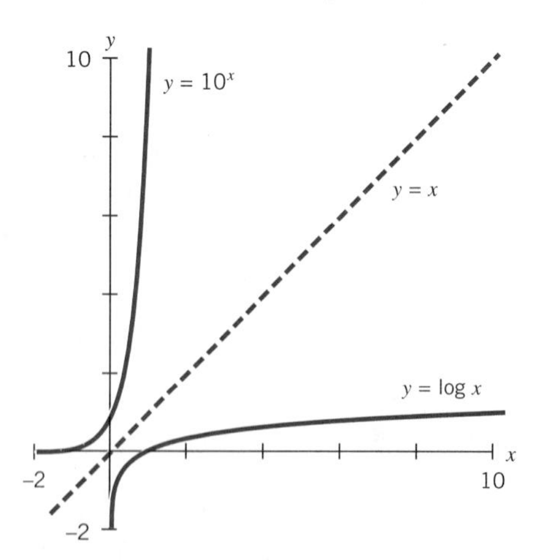

Figure 6.4 The graphs of $y = \log x$ and $y = 10^x$ are mirror images across the dotted line $y = x$.

Table 6.8
Points Satisfying
$y = 10^x$ and $y = \log x$

$y = 10^x$	$y = \log x$
(0, 1)	(1, 0)
(1, 10)	(10, 1)
(2, 100)	(100, 2)
(−1, 0.1)	(0.1, −1)
(−2, 0.01)	(0.01, −2)

hence lies on the equation's graph. What would be the mirror image of (0, 1) across the line $y = x$? Think again about folding the graph on the dotted line, and you'll see that the mirror image is (1, 0). Since $\log 1 = \log 10^0 = 0$, the point (1, 0) lies on the graph of $y = \log x$.

Similarly the point (1, 10), lies on the graph of $y = 10^x$ since $10^1 = 10$. Its mirror image, the point (10, 1), lies on the graph of $y = \log x$, since $\log 10 = \log 10^1 = 1$. Table 6.8 shows a list of similar pairs of points that satisfy $y = 10^x$ and $y = \log x$.

Suppose we take an arbitrary point (a, b). If you plot both the points (a, b) and (b, a) on graph paper (Figure 6.5) and fold the paper at the line $y = x$, the points would coincide. So the points (a, b) and (b, a) are mirror images of each other across the line $y = x$.

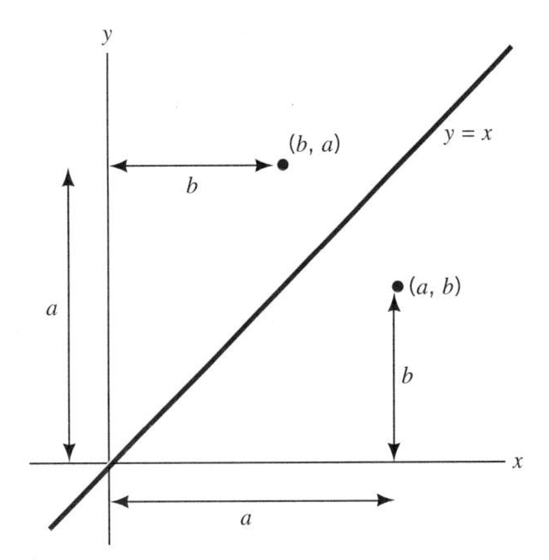

Figure 6.5 A plot of the points (a, b) and (b, a) and the line $y = x$.

Now suppose that our point (a, b) satisfies the equation $y = \log x$. That means that $b = \log a$, so we could rewrite (a, b) as $(a, \log a)$. Does its mirror image, the point $(b, a) = (\log a, a)$, satisfy $y = 10^x$? We can check by substituting $\log a$ for x and a for y. We get $a = 10^{\log a}$, which is a true statement. Similarly, we can show that if (a, b) satisfies $y = 10^x$, then (b, a) satisfies $y = \log x$. The graphs of $y = 10^x$ and $y = \log x$ then are, as we have claimed, mirror images across the line $y = x$.

We can use similar arguments to show that the graphs of $y = e^x$ and $y = \ln x$ are also mirror images of each other.

The points (a, b) and (b, a) are mirror images of each other across the line $y = x$.

When (a, b) satisfies $y = 10^x$, then (b, a) satisfies $y = \log x$, and vice versa. Similarly when (a, b) satisfies $y = e^x$, then (b, a) satisfies $y = \ln x$, and vice versa.

So in each pair of functions

$$y = 10^x \quad \text{and} \quad y = \log x$$

$$y = e^x \quad \text{and} \quad y = \ln x$$

the graphs are mirror images across the line $y = x$.

The Composition of Two Functions

It's easier to see what's happening algebraically to inverse functions if we use functional notation and introduce the idea of composing two functions.

Let $M(p)$ be a "mother" function. Ignoring complexities like adoption or, possibly in the future, cloning, assume that if p is a person, $M(p)$ is that person's mother, a unique individual. So the input for M is a person, and the output is that person's mother. We can define a similar "father" function, $F(p)$, where $F(p)$ is p's father, again assumed to be unique. The input for F is a person, and the output is that person's father. Since $M(p)$ is a person, let's call her q, then we can apply F to q to get $F(q) = F(M(p))$. For example, if you are the person, then

$$\begin{aligned} F(M(\text{you})) &= F\ (\text{your mother}) \\ &= \text{the father of your mother} \end{aligned}$$

We can think of this combination of the two functions F and M as a single new function, called a composition of F and M, denoted by $F \circ M$. The input to $F \circ M$ would be a person, p, and the output would be the person's mother's father, $F(M(p))$. Formally we define

$$(F \circ M)(p) = F(M(p))$$

What do these symbols mean? They say that in order to evaluate $F \circ M$ at p, first determine p's mother, $M(p)$, and then determine the father of p's mother, $F(M(p))$.

What would the composite functions $M \circ F$, $M \circ M$, and $F \circ F$ mean?

$$(M \circ F)(p) = M(F(p)) = M(p\text{'s father}) = \text{mother of } p\text{'s father}$$

$$(M \circ M)(p) = M(M(p)) = M(p\text{'s mother}) = \text{mother of } p\text{'s mother}$$

$$(F \circ F)(p) = F(F(p)) = F(p\text{'s father}) = \text{father of } p\text{'s father}$$

So $(F \circ M)$(you) and $(M \circ M))$(you) are your grandparents on your mother's side, and $(M \circ F)$(you) and $(F \circ F)$(you) are your grandparents on your father's side.

In general:

> If $f(x)$ and $g(x)$ are two functions, then the function $f \circ g$, called a *composition* of f and g, is defined by
>
> $$(f \circ g)(x) = f(g(x))$$

Example 1

a. Let $f(x) = x^2$ and $g(x) = 1/(x + 1)$.

b. Evaluate $(f \circ g)(3)$.

c. Determine a general expression for $(f \circ g)(x)$.

d. Does $(f \circ g)(x) = (g \circ f)(x)$?

Solution

a. We have $\qquad (f \circ g)(3) = f(g(3))$

evaluate $g(x)$ when $x = 3$ $\qquad = f\left(\dfrac{1}{3+1}\right)$

simplify $\qquad = f(\frac{1}{4})$

evaluate $f(x)$ when $x = \frac{1}{4}$ $\quad = (\frac{1}{4})^2$

$= \frac{1}{16}$

b. By definition $\quad (f \circ g)(x) = f(g(x))$

evaluate g $\quad = f\left(\frac{1}{x+1}\right)$

evaluate f $\quad = \left(\frac{1}{x+1}\right)^2$

c. By definition $\quad (g \circ f)(x) = g(f(x))$

evaluate f $\quad = g(x^2)$

evaluate g $\quad = \frac{1}{x^2+1}$

Since $\quad \left(\frac{1}{x+1}\right)^2 \neq \frac{1}{x^2+1}$ (try evaluating both sides when $x = 1$)

then $\quad (f \circ g)(x) \neq (g \circ f)(x)$

Algebra Aerobics 6.6a

1. Given $f(x) = 2x + 3$ and $g(x) = x^2 - 4$, find:
   - **a.** $f(g(2))$
   - **b.** $g(f(2))$
   - **c.** $f(g(3))$
   - **d.** $f(f(3))$
   - **e.** $(f \circ g)(x)$
   - **f.** $(g \circ f)(x)$

2. Given $F(x) = \frac{2}{x-1}$ and $G(x) = 3x - 5$, find:
   - **a.** $(F \circ G)(x)$
   - **b.** $(G \circ F)(x)$
   - **c.** Are the two composite functions in parts (a) and (b) equal?

Inverse Functions

Whenever two functions are inverses of each other, whatever one function does, the other undoes. So the composition of two inverse functions evaluated at x should equal x. It is this intuitive idea that is the basis of the formal definition of inverse functions.

Converting between dollars and yen. On Friday January 14, 2000, the conversion rate was one U.S. dollar to 105.70 Japanese yen. So if d = number of dollars, then the function $F(d) = 105.70d$ would convert dollars to yen. Similarly, if y = number of yen, then the function $G(y) = y/105.70$ would convert yen to dollars. Ideally, if we converted dollars to yen and yen back to dollars (assuming there are no transaction fees), we would end up with the number of dollars we started with. We could represent this symbolically by

$(G \circ F)(d) = G(F(d))$ starting with d dollars

$= G(105.70d)$ convert dollars to yen

$= \frac{105.70d}{105.70}$ convert yen back to dollars

$= d$ to get the original amount back

For example, if we converted \$150 to yen and back to dollars, we'd have

$$\begin{aligned}
(G \circ F)(150) &= G(F(150)) && \text{starting with \$150} \\
&= G(105.70 \cdot 150) && \text{convert to yen} \\
&= G(15855) && \text{to get 15,855 yen} \\
&= \frac{15855}{105.70} && \text{convert yen back to dollars} \\
&= 150 && \text{to get \$150}
\end{aligned}$$

We could start with yen, convert to dollars, and then convert dollars back to the same number of yen: that is,

$$\begin{aligned}
(F \circ G)(y) &= F(G(y)) && \text{starting with } y \text{ yen} \\
&= F\left(\frac{y}{105.70}\right) && \text{convert yen to dollars} \\
&= 105.70 \cdot \frac{y}{105.70} && \text{convert dollars back to yen} \\
&= y
\end{aligned}$$

So F and G are inverse functions of each other.

In general, we define any two functions $f(x)$ and $g(x)$ as inverse functions of each other if

$$(f \circ g)(x) = x \quad \text{and} \quad (g \circ f)(x) = x$$

We have f is the inverse of g and g is the inverse of f. We sometimes write the inverse of a function f as f^{-1}. So $f = g^{-1}$ and $g = f^{-1}$. Be cautious using this notation, since -1 is *not* an exponent. It's merely a symbol to indicate inverses. It's a good idea to read f^{-1} as "f inverse" instead of "f to the minus 1" to avoid confusion.

> Any two functions $f(x)$ and $g(x)$ are inverse functions of each other if
>
> $$(f \circ g)(x) = x \quad \text{and} \quad (g \circ f)(x) = x$$
>
> We write $g^{-1} = f$ and $f^{-1} = g$.

Example 2

Verify that $F(x) = 32 + \frac{9}{5}x$, which converts degrees from Celsius to Fahrenheit, and $C(x) = \frac{5}{9}(x - 32)$, which converts Fahrenheit to Celsius, are inverse functions of one another.

Solution

$$\begin{aligned}
(C \circ F)(x) &= C(F(x)) \\
&= C(32 + \tfrac{9}{5}x) \\
&= \tfrac{5}{9}((32 + \tfrac{9}{5}x) - 32) \\
&= \tfrac{5}{9}(\tfrac{9}{5}x) \\
&= x
\end{aligned}$$

$$\begin{aligned}(F \circ C)(x) &= F(C(x)) \\ &= F(\tfrac{5}{9}(x - 32)) \\ &= 32 + \tfrac{9}{5}(\tfrac{5}{9}(x - 32)) \\ &= 32 + (x - 32) \\ &= x\end{aligned}$$

So $C = F^{-1}$, the inverse of F, and $F = C^{-1}$, the inverse of C.

We can now use our definition to formally verify that $f(x) = 10^x$ and $g(x) = \log x$ are inverse functions of each other. We have

$$\begin{aligned}(g \circ f)(x) &= g(f(x)) && \text{definition of composition} \\ &= g(10^x) && \text{substitute } 10^x \text{ for } f(x) \\ &= \log(10^x) && \text{definition of } g \\ &= x && \text{Property 5 of logarithms}\end{aligned}$$

and when $x > 0$,

$$\begin{aligned}(f \circ g)(x) &= f(g(x)) && \text{definition of composition} \\ &= f(\log x) && \text{substitute } \log x \text{ for } g(x) \\ &= 10^{\log x} && \text{definition of } f \\ &= x && \text{Property 6 of logarithms}\end{aligned}$$

For example,

$$\begin{aligned}(g \circ f)(2) &= g(f(2)) = g(10^2) = \log 10^2 = 2 \\ (g \circ f)(0.1) &= g(f(0.1)) = g(10^{0.1}) = \log(10^{0.1}) = 0.1 \\ (f \circ g)(2) &= f(g(2)) = f(\log 2) = 10^{\log 2} = 2 \\ (f \circ g)(0.1) &= f(g(0.1)) = f(\log 0.1) = 10^{\log 0.1} = 0.1\end{aligned}$$

By the same method, it's easy to verify that $f(x) = e^x$ and $g(x) = \ln x$ are also inverse functions of one another. In summary:

> Any two functions $f(x)$ and $g(x)$ are inverse functions of each other if
>
> $$(f \circ g)(x) = x \quad \text{and} \quad (g \circ f)(x) = x$$
>
> We write $f^{-1} = g$ and $g^{-1} = f$.
>
> The graphs of $f(x)$ and $f^{-1}(x)$ are mirror images across the line $y = x$. If the point (a, b) lies on the graph of one function, then the point (b, a) lies on the graph of the other. Equivalently $f(a) = b$ if and only if $f^{-1}(b) = a$.
>
> In particular in each pair
>
> $$f(x) = 10^x \quad \text{and} \quad g(x) = \log x$$
>
> $$f(x) = e^x \quad \text{and} \quad g(x) = \ln x$$
>
> the functions are inverses of each other.

Determining if a Function Has an Inverse

Not every function has an inverse. For the mother function, M(p), the input is a person and the unique output is the person's mother. If the inverse of M existed, it would be a child function, say $C(p)$, whose input would be a mother and whose output would be a child of the mother. But $C(p)$ is not a function, since a mother may have more than one child. If you have siblings, then $(C \circ M)$ (you) might equal your brother or sister. Hence M as defined has no inverse.

Example 3

How could you look at a graph of f and decide if f^{-1} exists? *Hint:* If f^{-1} passes the vertical line test, what sort of line test would the graph of f have to pass?

Does the function $y = x^2$ have an inverse?

Solution

If the function $y = x^2$ has an inverse, its graph would be the mirror image of the graph of $y = x^2$ across the line $y = x$. Think of graphing $y = x^2$ with wet ink, folding the paper on the line $y = x$, and imagine the imprint the ink would leave (Figure 6.6). The graph of the potential inverse must be a function in its own right and hence pass the vertical line test. It clearly fails, since there are any number of vertical lines that would cross the graph of the potential inverse more than once. So $y = x^2$ does not have an inverse function.

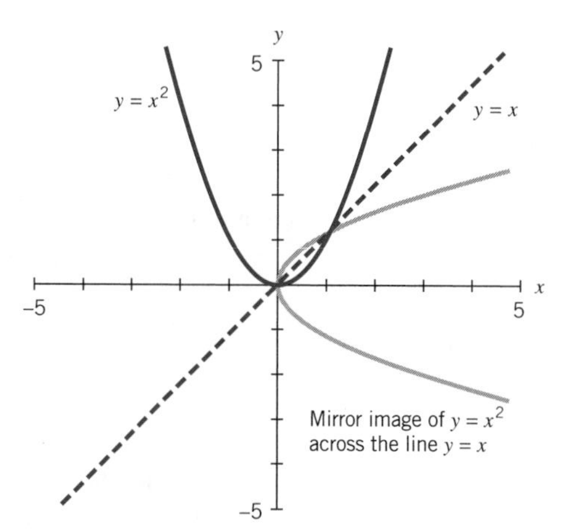

Figure 6.6 The mirror image of the graph of $y = x^2$ is not a function.

Finding the Formula for the Inverse of a Function

Does the function $f(x) = \dfrac{1}{x-1}$ have an inverse f^{-1}? If so, construct a formula for f^{-1}.

Figure 6.7 shows a graph of $f(x) = \dfrac{1}{x-1}$. It's a little harder now to imagine what the mirror image of $f(x)$ across the line $y = x$ would look like, so it's useful to employ an algebraic strategy to look for an inverse. If $f(x) = y$, then $f^{-1}(y) = x$. So substituting y for $f(x)$ we need to solve the equation

$$y = \frac{1}{x-1}$$

for x in terms of y.

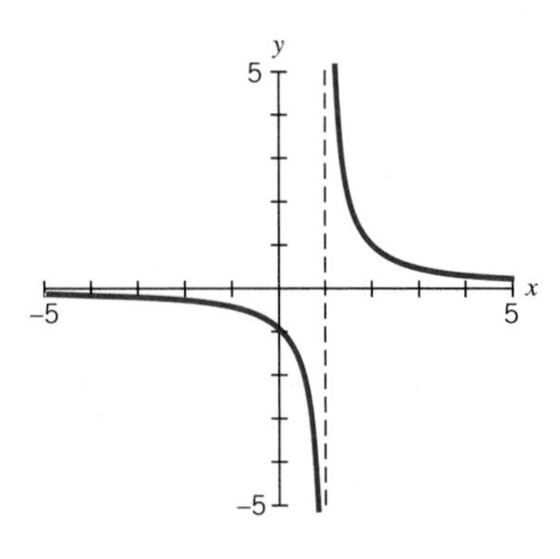

Figure 6.7 The graph of $f(x) = \dfrac{1}{x - 1}$.

Given	$y = \dfrac{1}{x - 1}$
multiply both sides by $(x - 1)$	$y(x - 1) = 1$
multiply through	$yx - y = 1$
add y to both sides	$yx = 1 + y$
divide both sides by y	$x = \dfrac{1 + y}{y}$

The equation associates each value of y with a unique value x, so we can define the function $f^{-1}(y)$ where

$$f^{-1}(y) = \frac{1 + y}{y}$$

We can verify that $(f \circ f^{-1})(y) = y$ (for $y \neq 0$) and $(f^{-1} \circ f)(x) = x$ (for $x \neq 1$). So f^{-1} and f are inverse functions. (See Algebra Aerobics 6.6b, Problem 2.)

Example 4

The temperature T of a cup of hot coffee (at 150°Fahrenheit) left to stand in a 70°F room for t minutes can be modeled by the equation

$$T = 70 + 80e^{-0.0924t}$$

where 80 is the difference between the temperature of the coffee and the temperature of the room.

We can think of this as a function $f(t) = T$ that describes temperature T as a function of time t.

(a) Construct an inverse function $f^{-1}(T) = t$ that describes time t as a function of temperature T.

(b) How long will it take for the coffee to reach 75°F?

Solution

(a) To find $f^{-1}(T)$ we would need to solve our original equation for t.

	$T = 70 + 80e^{-0.0924t}$
Subtract 70 from both sides	$T - 70 = 80e^{-0.0924t}$

divide both sides by 80	$(T - 70)/80 = e^{-0.0924t}$
take ln of both sides	$\ln[(T - 70)/80] = \ln(e^{-0.0924t})$
apply Property 5 of ln	$\ln[(T - 70)/80] = -0.0924t$
divide by -0.0924	$\dfrac{\ln[(T - 70)/80]}{-0.0924} = t$

For each value of $T > 70$, there is a unique value of t. So we can define $f^{-1}(T) = t$ where

$$t = -\frac{\ln[(T - 70)/80]}{0.0924}$$

(b) When T has cooled from 150°F to 75°F, then the corresponding value for time, t, is given by

$$\begin{aligned} f^{-1}(75) &= -\frac{\ln[(75 - 70)/80]}{0.0924} \\ &= -\frac{\ln(5/80)}{0.0924} \\ &= -\frac{\ln 0.0625}{0.0924} \\ &\approx -\frac{(-2.773)}{0.0924} \\ &\approx 30 \text{ minutes} \end{aligned}$$

So it takes about 30 minutes for the coffee to cool from 150°F to 75°F in a 70°F room.

Algebra Aerobics 6.6b

1. In each part, verify that f and g are inverse functions by showing that $(f \circ g)(x) = x$ and $(g \circ f)(x) = x$.
   a. $f(x) = 2x + 1$ and $g(x) = \dfrac{x - 1}{2}$
   b. $f(x) = \sqrt[3]{x + 1}$ and $g(x) = x^3 - 1$
2. Verify that $f(x) = \dfrac{1}{x - 1}$ and $f^{-1}(y) = \dfrac{1 + y}{y}$ are inverse functions of one another.
3. Given the accompanying graph of function f, sketch the graph of its inverse.

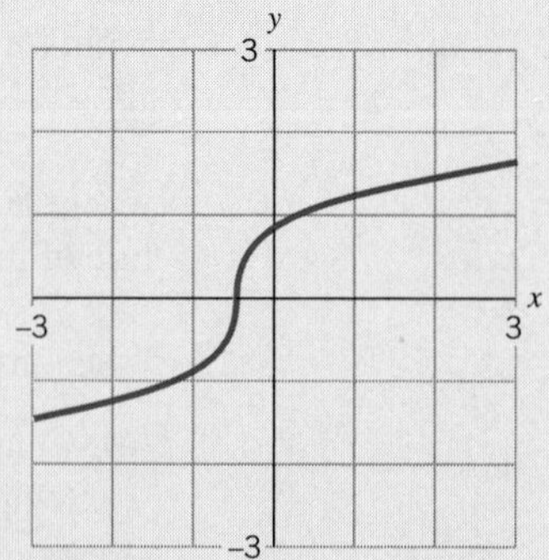

4. Find the inverse, if it exists, for each of the following functions. *Hint:* Start by letting $y = f(x)$.
   a. $f(x) = x^3$
   b. $f(x) = \dfrac{1}{x + 1}$
   c. $f(x) = 150(0.097)^x$

Chapter Summary

Common Logarithms. The *logarithm base 10 of* x is the power of 10 you need to produce x; that is,

$$\log_{10}x = c \quad \text{means that} \quad 10^c = x$$

Logarithms base 10 are called *common logarithms* and $\log_{10}x$ is written as $\log x$. The properties of logarithms follow directly from the definition of logarithms and from the properties of exponents.

Properties of Exponents

If a is any positive real number and p and q are any real numbers, then:

1. $a^p \cdot a^q = a^{p+q}$
2. $a^p/a^q = a^{p-q}$
3. $(a^p)^q = a^{p \cdot q}$
4. $a^0 = 1$

Corresponding Properties of Logarithms

If A and B are positive real numbers and p is any real number, then:

1. $\log (A \cdot B) = \log A + \log B$
2. $\log (A/B) = \log A - \log B$
3. $\log A^p = p \log A$
4. $\log 1 = 0$ (since $10^0 = 1$)

In addition

5. $\log 10^x = x$ (for any real x)
6. $10^{\log x} = x$ $(x > 0)$

The log of 0 or a negative number is not defined. But when we take the log of a number, we can get 0 or a negative value.

While we can estimate solutions to equations of the form $10 = 2^t$, where a variable occurs in the exponent, we can use logarithms to find an exact answer. For example:

Given	$10 = 2^t$
take logarithm of both sides	$\log 10 = \log(2^t)$
apply Property 3 of logs	$\log 10 = t \log 2$
divide by log 2	$t = \dfrac{\log 10}{\log 2}$

Using a calculator, we can obtain as accurate an estimate as needed. We have

$$t \approx \frac{1}{0.301} \approx 3.32$$

Base e: Natural Logarithms. The irrational number e is a fundamental mathematical constant whose value is approximately 2.71828.

The value of P dollars invested at an annual interest rate r (in decimal form) compounded n times a year for x years equals

$$y = P(1 + r/n)^{nx}$$

where r is the *nominal interest rate.* We can solve for x given a particular value of y by taking the log of both sides. The actual amount of interest earned per year is called the *effective interest rate.*

The value of $(1 + 1/n)^n$ approaches e as n approaches infinity.

The number e is used to describe *continuous compounding.* For example, the value of P dollars invested at an annual interest rate r (expressed in decimal form) compounded continuously for x years can be calculated using the formula

$$y = P \cdot e^{rx}$$

where r is called the nominal interest rate.

For any number $a > 0$, the function $y = C \cdot a^x$ may be rewritten $y = C \cdot e^{rx}$, where $r = \ln a$. We call r the *instantaneous growth rate.*

If $a > 1$, then $r > 0$ and the function represents growth.
If $0 < a < 1$, then $r < 0$ and the function represents decay.
If $a = 1$, then $r = 0$ and the function is constant, $y = C$.

We define the *logarithm base e* or the *natural logarithm* of x as the power of e you need to produce x. The natural log of x is written as $\ln x$. The properties of natural logarithms correspond directly to those of common logarithms.

Properties of Natural Logarithms

If A and B are positive real numbers and p is any real number, then:

1. $\ln(A \cdot B) = \ln A + \ln B$
2. $\ln(A/B) = \ln A - \ln B$
3. $\ln A^p = p \ln A$
4. $\ln 1 = 0$ (since $e^0 = 1$)
5. $\ln e^x = x$
6. $e^{\ln x} = x$

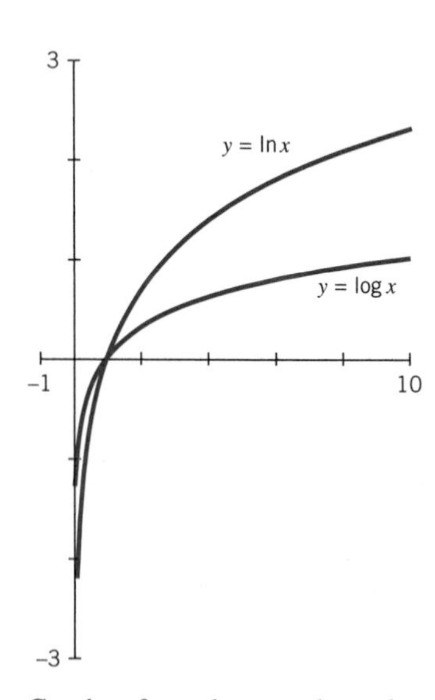

Graphs of $y = \log x$ and $y = \ln x$

As with common logarithms, $\ln x$ is not defined if $x \leq 0$.

Logarithmic Functions. For any $x > 0$, both the common and natural logarithms define unique numbers $\log x$ and $\ln x$, respectively. So we can define two functions

$$y = \log x \quad \text{and} \quad y = \ln x \quad (\text{where } x > 0 \text{ for both functions})$$

The graphs of these functions have similar shapes. Both are defined only when $x > 0$, are increasing everywhere, and are asymptotic to the y-axis.

If $x > 1$, $\log x$ and $\ln x$ are both positive.
If $x = 1$, $\log 1 = 0$ and $\ln 1 = 0$.
If $0 < x < 1$, $\log x$ and $\ln x$ are both negative.
If $x \leq 0$, neither logarithm is defined.

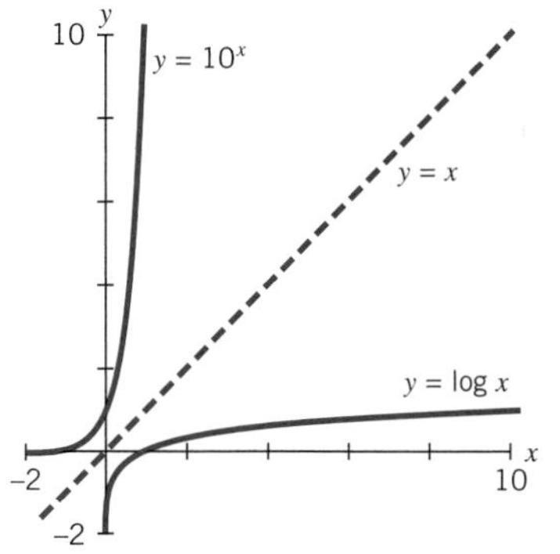

The graphs of $y = \log x$ and $y = 10^x$ are mirror images across the dotted line $y = x$.

Composition and Inverse Functions. Two functions are inverses of each other if what one function "does," the other "undoes." More formally, if $f(x)$ and $g(x)$ are two functions, the function $f \circ g$ is called a *composition* of f and g and is defined by

$$(f \circ g)(x) = f(g(x))$$

Then two functions $f(x)$ and $g(x)$ are *inverse functions* of each other if

$$(f \circ g)(x) = x \quad \text{and} \quad (g \circ f)(x) = x$$

We write $f^{-1} = g$ and $g^{-1} = f$.

The graphs of $f(x)$ and $f^{-1}(x)$ are mirror images across the line $y = x$. If the point (a, b) lies on the graph of one function, then the point (b, a) lies on the graph of the other. Equivalently $f(a) = b$ if and only if $f^{-1}(b) = a$. In particular in each pair

$$f(x) = 10^x \quad \text{and} \quad g(x) = \log x$$

$$f(x) = e^x \quad \text{and} \quad g(x) = \ln x$$

the functions are inverses of each other.

EXERCISES

Exercises for Section 6.1

1. Use the accompanying table to estimate the number of years it would take \$100 to become \$300 at an interest rate compounded annually at:

a. 3% **b.** 5% **c.** 7%

Compound Interest over 40 Years

Years	Value of \$100 at 3% (\$)	Value of \$100 at 5% (\$)	Value of \$100 at 7% (\$)
0	100	100	100
10	134	163	197
20	181	265	387
30	243	432	761
40	326	704	1,497

2. Generate a table of values to estimate the half-life of a substance that decays according to the function $y = 100(0.8)^x$, where x is the number of time periods and each time period is 12 hours. How long will it be before there is less than 1 gram of the substance remaining?

3. The accompanying graph shows the concentration of a drug in the human body as the initial amount of 100 mg dissipates over time. Estimate when the concentration becomes:

a. 60 mg **b.** 40 mg **c.** 20 mg

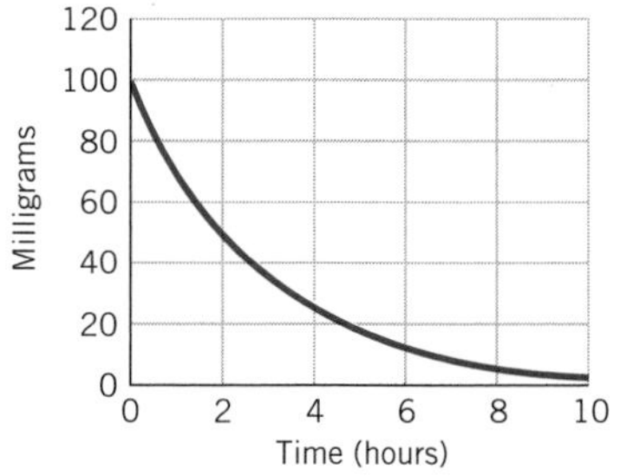

Remaining drug dosage in milligrams.

4. a. Plot the graph of $y = 6(1.3)^x$ for $0 \le x \le 4$. If x represents time in seconds, estimate the doubling time from the graph.

b. Now plot $y = 100(1.3)^x$ and estimate the doubling time from the graph.

c. Compare your answers to parts (a) and (b).

5. Determine x if we know that $\log x$ equals:

a. -3 **b.** 6 **c.** $\frac{1}{3}$ **d.** 0 **e.** 1 **f.** -1

6. Given that $\log 5 \approx 0.699$, without using a calculator determine the value of:

a. $\log 25$ **b.** $\log \frac{1}{25}$ **c.** $\log 10^{25}$ **d.** $\log 0.0025$

Check your answer with a calculator.

7. Prove Property 2 of logarithms:

$$\log(A/B) = \log A - \log B \qquad (A \text{ and } B \text{ both} > 0)$$

8. Expand using the properties of logarithms:

a. $\log\left(x^2y^3\sqrt{z-1}\right)$ **b.** $\log\frac{A}{\sqrt[3]{BC}}$

9. Contract using the properties of logarithms and express your answer as a single logarithm:

a. $3\log K - 2\log(K+3)$ **b.** $-\log m + 5\log(3+n)$

10. For each of the following equations either prove that it is correct (by using the properties of logarithms and exponents) or else show that it is not correct (by finding specific numerical values for the variables that make the values of the two sides of the equation different):

a. $\log\left(\frac{x}{y}\right) = \frac{\log x}{\log y}$

b. $\log x - \log y = \log\left(\frac{x}{y}\right)$

c. $\log(2x) = 2\log x$

d. $2\log x = \log(x^2)$

e. $\log\frac{x+1}{x+3} = \log(x+1) - \log(x+3)$

f. $\log\left(x\sqrt{x^2+1}\right) = \log x + \frac{1}{2}(x^2+1)$

g. $\log(x^2+1) = 2\log x + \log 1$

11. Prove Property 3 of common logarithms: $\log A^p = p\log A$ (where $A > 0$).

12. Solve for t:

a. $10^t = 4$

b. $3(2^t) = 21$

c. $1 + 5^t = 3$

d. $10^{-t} = 5$

e. $5^t = 7^{t+1}$

f. $6 \cdot 2^t = 3^{t-1}$

13. Solve for x:

a. $\log x = 3$

b. $\log(x+1) = 3$

c. $3\log x = 5$

d. $\log x + \log(2x+1) = 0$

e. $\log(x+1) - \log x = 1$

f. $\log x - \log(x+1) = 1$

14. Solve for x:

a. $2^x = 7$

b. $\left(\sqrt{3}\right)^{x+1} = 9^{2x-1}$

c. $12(1.5)^{x+1} = 13$

d. $\log(x+3) + \log 5 = 2$

e. $\log x + \log(x-6) = 1$

f. $\log(x-1) = 2$

15. Returning to Exercise 1, now *calculate* the number of years it would take \$100 to become \$300 at an interest rate compounded annually at:

a. 3% **b.** 5% **c.** 7%

16. If the amount of drug remaining in the body after t hours is given by $f(t) = 100\left(1/\sqrt{2}\right)^t$ (graphed in Exercise 3), then calculate:

a. the number of hours it would take for the initial 100 mg to become

i. 60 mg **ii.** 40 mg **iii.** 20 mg

b. the half-life of the drug

17. In Chapter 5 we saw that the function $N = N_0 \cdot 1.5^t$ described the actual number N of *E. coli* bacteria in an experiment after t time periods (of 20 minutes each) starting with an initial bacteria count of N_0.

a. What is the doubling time?

b. How long would it take for there to be 10 times the original amount of bacteria?

18. (Requires graphing calculator or computer.) A woman starts a training program for a marathon. She starts in the first week by doing 10-mile runs. Each week she increases her run length by 20% of the distance for the previous week.

a. Write a formula for her run distance, D, as a function of week, W.

b. Use technology to graph your function and use the graph to estimate the week in which she will reach a marathon length of approximately 26 miles.

c. Now use your formula to calculate the week in which she will start running 26 miles.

19. The half-life of bismuth-214 is about 20 minutes.

a. Construct a function to model the decay of bismuth-214 over time. Be sure to specify your variables and their units.

b. For any given sample of bismuth-214, how much is left after 1 hour?

c. How long will it take to reduce the sample to 25% of its original size?

d. How long will it take to reduce the sample to 10% of its original size?

20. The atmospheric pressure at sea level is approximately 14.7 lb/in.^2 and the pressure is reduced by half for each 3.6 miles above sea level.

a. Construct a model that describes the atmospheric pressure as a function of miles above sea level.

b. At how many miles above sea level would the atmospheric pressure have dropped in half, to 7.35 lb/in.^2?

21. A department store has a discount basement where the policy is to reduce the selling price, S, of an item by 10% of its current price each week. If the item has not sold after the 10th reduction, the store gives the item to charity.

a. For a \$300 suit construct a function for the selling price, S, as a function of week, W.

b. After how many weeks might the suit first be sold for less than \$150? What is the selling price at which the suit might be given to charity?

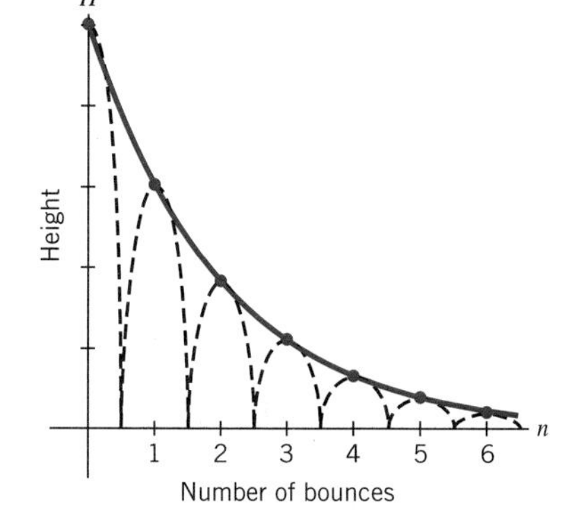

22. If you drop a rubber ball on a hard, level surface, it will usually bounce repeatedly. Each time it bounces, it rebounds to a height that is a percentage of the previous height. This percentage is called the rebound height.

a. Assume you drop the ball from a height of 5 feet and that the rebound height is 60%. Construct a table of values that shows you the rebound height for the first four bounces.

b. Construct a function to model the ball's rebound height, H, on the nth bounce.

c. How many bounces would it take for the ball's rebound height to be 1 foot or less?

d. Construct a general function that would model a ball's rebound height, H, on the nth bounce, where H_0 is the initial height of the ball and r is the ball's rebound height.

Exercises for Section 6.2

23. Assume \$10,000 is invested at a nominal interest rate of 8.5%. Write the equations that give the value of the money after n years and determine the effective interest rate if the interest is compounded:

a. annually **c.** quarterly

b. semiannually **d.** continuously

24. Assume you invest $2000 at 3.5% compounded continuously.

a. Construct an equation that describes the value of your investment at year t.

b. How much will $2000 be worth after 1 year? 5 years? 10 years?

c. How long will it take $2000 to double to $4000?

d. How long will it take P dollars to double to $2P$ dollars?

25. The half-life of uranium-238 is about 5 billion years. Assume you start with 10 grams of U-238 that decays continuously.

a. Construct an equation to describe the amount of U-238 remaining after x billion years.

b. How long would it take for 10 grams of U-238 to become 1 gram?

26. You want to invest money for your newborn child so that she will have $50,000 for college on her 18th birthday. Determine how much you should invest if the best annual rate that you can get on a secure investment is:

a. 6.5% compounded annually

b. 9% compounded quarterly

c. 7.9% compounded continuously

27. Determine the doubling time for money invested at the rate of 12% compounded:

a. annually **b.** quarterly **c.** continuously

28. a. Phosphorus-32 is used to mark cells in biological experiments. If phosphorus-32 has a continuous daily decay rate of 0.0485, what is its half-life?

b. Phosphorus-32 can be quite dangerous to work with if the experimenter fails to use the proper shields, since its high-energy radiation extends out to 610 cm. Because disposal of radioactive wastes is increasingly difficult and expensive, laboratories often store the waste until it is within acceptable radioactive levels for disposal with nonradioactive trash. For instance, the rule of thumb for the laboratories of a large east coast university and medical center is that any waste containing radioactive material with a half-life under 65 days must be stored for 10 half-lives before disposal with the nonradioactive trash.

i. For how many days would phosphorus-32 have to be stored?

ii. What percentage of the original phosphorus-32 would be left at that time?

29. Find the nominal interest rate if a bank advertises that the effective interest rate on an account compounded continuously is:

a. 3.43% on a checking account **b.** 4.6% on a savings account

30. An investment pays 6% compounded four times a year.

a. What is the annual growth rate?

b. What is the annual growth factor?

c. Develop a formula to represent the total principal after each compounding period.

d. If you invest $2000 for a child's college fund, how much will it total after 15 years?

e. For how many years would you have to invest to increase the total to $5000?

f. If the interest is compounded continuously, what is the formula for total principal? How long would it take to grow the total to $5000?

Exercises for Section 6.3

31. Write an equivalent equation in exponential form.

a. $n = \log 35$ **b.** $\ln(N/N_0) = -kt$

32. Write an equivalent equation in logarithmic form.

a. $N = 10^{-t/c}$ **b.** $I = I_0 \cdot e^{-k/x}$

33. Prove that $\ln(A \cdot B) = \ln A + \ln B$, where A and B are positive real numbers.

34. Expand:

a. $\ln\sqrt[3]{\dfrac{x^2-1}{x+2}}$ **b.** $\ln\left(\dfrac{x}{y\sqrt{2}}\right)^2$ **c.** $\ln\left(\dfrac{K^2L}{M+1}\right)$

35. Contract, expressing your answer as a single logarithm.

a. $\frac{1}{4}\ln(x+1) + \frac{1}{4}\ln(x-3)$ **b.** $3\ln R - \frac{1}{2}\ln P$ **c.** $\ln N - 2\ln N_0$

36. For each of the following equations either prove that it is correct (by using the properties of logarithms and exponents) or else show that it is not correct (by finding specific numerical values for x that make the values of the two sides of the equation different):

a. $e^{x+\ln x} = x \cdot e^x$

b. $\ln\left(\dfrac{(x+1)^2}{x}\right) = 2\ln(x+1) - \ln x$

c. $\ln\left(\dfrac{x}{x+1}\right) = \dfrac{\ln x}{\ln(x+1)}$

d. $\ln(x + x^2) = \ln x + \ln x^2$

e. $\ln(x + x^2) = \ln x + \ln(x+1)$

37. Solve for x:

a. $e^x = 10$

b. $10^x = 3$

c. $2 + 4^x = 7$

d. $\ln x = 5$

e. $\ln(x+1) = 3$

f. $\ln x - \ln(x+1) = 4$

38. Solve for t:

a. $5^{t+1} = 6^t$

b. $e^{t^2} = 4$

c. $5 \cdot 2^{-t} = 4$

d. $ln\left(\dfrac{t}{t-2}\right) = 1$

e. $\ln t - \ln(t-2) = 1$

Exercises for Section 6.4

39. The stellar magnitude M of a star is approximately $-2.5 \log (B/B_0)$, where B is the brightness of the star and B_0 is a constant.

a. If you plotted B on the horizontal and M on the vertical axis, where would the graph cross the B axis?

b. Without calculating any other coordinates, draw a rough sketch of the graph of M. What is the domain?

c. As the brightness B increases, does the magnitude M increase or decrease? Is a sixth magnitude star brighter or dimmer than a first magnitude star?

d. If the brightness of a star is increased by a factor of 5, by how much does the magnitude increase or decrease?

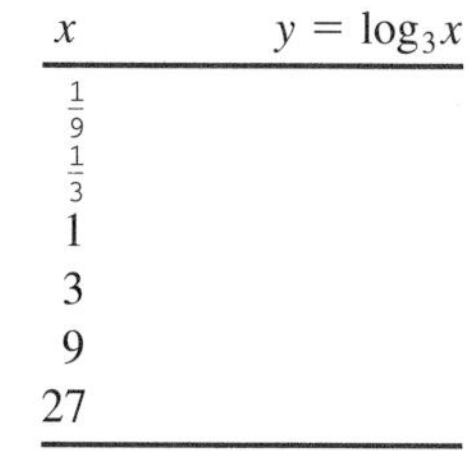

x	$y = \log_3 x$
$\frac{1}{9}$	
$\frac{1}{3}$	
1	
3	
9	
27	

40. Logarithms can be constructed using any positive number except 1 as a base:

$$\log_a x = y \quad \text{means that} \quad a^y = x$$

a. Complete the table in the margin and sketch the graph of $y = \log_3 x$

b. Now make a small table and sketch the graph of $y = \log_4 x$. (*Hint:* To simplify computations, try using powers of 4 for values of x.)

41. (Adapted from H. D. Young, *University Physics,* Vol. 1. Reading, MA: Addison-Wesley, 1992, p. 591.) If you listen to a 120-decibel sound for about 10 minutes, your threshold of hearing will typically shift from 0 dB up to 28 dB for a while. If you are exposed to a 92-dB sound for 10 years, your threshold of hearing will be permanently shifted to 28 dB. What intensities correspond to 28 and 92 dB? (See Section 6.4)

42. In all the sound problems so far, we did not take into account the distance between the sound source and the listener. Sound intensity is inversely proportional to the square of the distance from the sound source; that is, $I = k/r^2$, where I is intensity, r is the distance from the sound source, and k is a constant.

Suppose that you are sitting a distance R from the TV, where its sound intensity is I_1. Now you move to a seat twice as far from the TV, a distance $2R$ away, where the sound intensity is I_2.

a. What is the relationship between I_1 and I_2?

b. What is the relationship between the decibel levels associated with I_1 and I_2?

43. If there are a number of different sounds being produced simultaneously, the resulting intensity is the sum of the individual intensities. How many decibels louder is the sound of quintuplets crying than the sound of one baby crying?

44. An ulcer patient has been told to avoid acidic foods. If he drinks coffee, with a pH of 5.0, it bothers him, but he can tolerate both tap water, with a pH of 5.8, and milk, with a pH of 6.9.

a. Will a mixture of half coffee and half milk be at least as tolerable as tap water?

b. What pH will the half coffee/half milk mixture have?

c. In order to make 10 oz of a milk/coffee drink with a pH of 5.8, how many ounces of each are required?

d. Lemon juice has a pH of 2.1. If you make diet lemonade by mixing $\frac{1}{4}$ cup of lemon juice with 2 cups of tap water, with a pH of 5.8, will the resulting acidity be more or less than orange juice, with a pH of 3?

Exercises for Section 6.5

45. Rewrite each of the following functions using base e:

a. $N = 10(1.045)^t$ **b.** $Q = 5 \cdot 10^{-7} \cdot (0.072)^A$

46. Sketch a graph of each the following. Find, as appropriate, the doubling time or half-life.

a. $A = 50e^{0.025t}$ **b.** $A = 100e^{-0.046t}$

47. Identify each of the following functions as representing growth or decay:

a. $Q = Ne^{-0.029t}$ **b.** $h(r) = 100(0.87)^r$ **c.** $f(t) = 375e^{0.055t}$

48. Identify each function as representing growth or decay. Then determine the growth or decay factor.

a. $A = A_0(1.0025)^{20t}$ **c.** $A = A_0(0.992)^{t/2}$ **e.** $A = A_0e^{0.015t}$

b. $A = A_0(1.0006)^{t/360}$ **d.** $A = A_0e^{-0.063t}$

49. The barometric pressure, p, in millimeters of mercury at height h in kilometers above sea level is given by the equation $p = 760e^{-0.128h}$. At what height is the barometric pressure 200 mm?

50. After t days, the amount of thorium-234 in a sample is $A(t) = 35e^{-0.029t}$ micrograms.

a. How much is there initially?

b. How much is there after a week?

c. When is there just 1 microgram left?

d. What is the half-life of thorium-234?

51. Assume $f(t) = Ce^{rt}$ is an exponential decay function (so $r < 0$). Prove the rule of 70 for halving times; that is, if a quantity is decreasing at $R\%$ per time period t, then the number of time periods it takes for the quantity to halve is approximately $70/R$. (*Hint:* $R = 100r$.)

52. In 1859, the Victorian landowner Thomas Austin imported 12 wild rabbits into Australia and let them loose to breed. Since they had no natural enemies, the population increased very rapidly. By 1949 there were approximately 600 million rabbits.

a. Find an exponential function to model this situation.

b. If the growth had gone unchecked, what would the rabbit population be in 2000?

Internet search: Find out what was done to curb the population of rabbits in Australia.

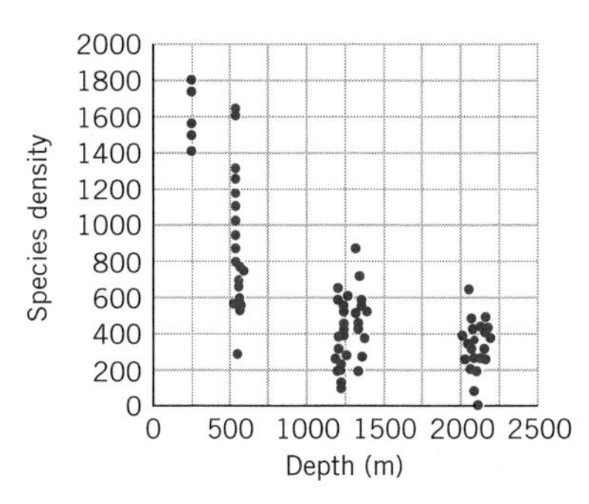

Source: Data collected by Ron Etter, University of Massachusetts Boston, Boston Biology Department.

53. Biologists believe that, in the deep sea, species density decreases exponentially with the depth. The accompanying graph shows data collected in the North Atlantic. Sketch an exponential decay function through the data. Then identify two points on your curve, and generate an equation for your function.

54. (Requires use of graphing calculator or computer.) According to Moore's Law, the computing power built into chips doubles every 18 months. The accompanying table shows the computing power of chips (measured in calculations per second) throughout the 1990s.

Year	Chip Type	Chip Computing Power (millions of calculations per second)
1993	Intel Pentium	66
1995	Pentium Pro	233
1997	Pentium II	525
1999	Pentium III	1700

Source: Intel Corporation.

a. Graph the data points. Justify why an exponential function would be an appropriate model.

b. Construct an exponential function to model the data in three different ways. In each case let t = number of years since 1993.

i. Using the doubling time given by Moore's Law, construct an exponential function, $P_1(t)$, for chip computing power.

ii. Letting 1993 be the base year, identify the coordinates of two points, and use those points to construct a function model, $P_2(t)$, for chip computing power. Does this model verify Moore's Law; that is, does the computing power approximately double every year and a half?

iii. Use technology to find a best fit exponential function, $P_3(t)$, to the data. Does this model come closer to verifying Moore's Law than $P_2(t)$?

55. The number of neutrons in a nuclear reactor can be predicted from the equation $n = n_0 e^{(\ln 2)t/T}$ where n = number of neutrons at time t (in seconds), n_0 = the number of neutrons at time $t = 0$, and T = the reactor period, the doubling time of the neutrons (in seconds). When $t = 2$ seconds, $n = 11$, and when $t = 22$ seconds, $n = 30$. Find the initial number of neutrons, n_0, and the reactor period, T, both rounded to the nearest whole number.

56. According to Rubin and Farber's *Pathology,* "death from cancer of the lung, more than 85% of which is attributed to cigarette smoking, is today the single most common cancer death in both men and women in the United States." The accompanying graph shows the annual death rate (per thousand) from lung cancer for smokers and nonsmokers.

a. The death rate for nonsmokers is roughly a linear function of age. After replacing each range of ages with a reasonable middle value (e.g., you could use 60 to approximate 55 to 64), estimate the coordinates of two points on the graph of nonsmokers and construct a linear model. Interpret your results.

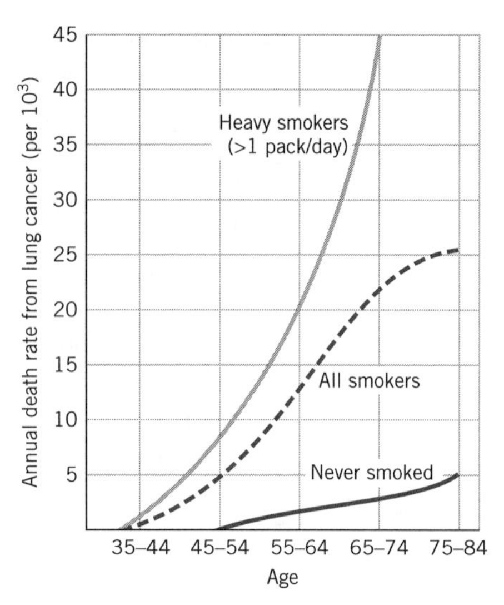

Source: E. Rubin and J. L. Farber, *Pathology,* 3rd ed. (Philadelphia: Lippincott-Raven, 1998), p. 312. Copyright © 1998 by Lippincott-Raven. Reprinted by permission.

b. By contrast, those who smoke more than one pack per day show an exponential rise in the annual death rate from lung cancer. Estimate the coordinates for two points on the graph for heavy smokers, and use the points to construct an exponential model. Interpret your results.

57. The following data from the National Center for Health Statistics, U.S. Department of Health and Human Services, shows the rise of total medical expenditures in the United States since 1965 in billions of dollars.

Year	1965	1970	1975	1980	1985	1990	1995
Billions of dollars	41.1	73.2	130.7	247.3	428.7	699.5	991.4

a. It is clear that the United States is spending more on health care as time goes on. Does this mean that we as individuals are paying more? How would you find out?

b. Is it fair to say that the expenses shown are growing exponentially? Use graphing techniques to find out; measure time in years since 1960.

c. Around 1985 the high cost of health care became increasingly a political issue, and insurance companies began to introduce "managed care" in an attempt to cut costs. Find a mathematical model assuming a continuous growth for health costs from 1960 to 1995. What does your model predict for health costs in 1995? How closely does this reflect the actual cost in 1995?

Exercises for Section 6.6

58. Given $f(x) = 3x - 2$ and $g(x) = (x + 1)^2$, find:

a. $f(g(1))$ **d.** $f(f(2))$

b. $g(f(1))$ **e.** $(f \circ g)(x)$

c. $f(g(2))$ **f.** $(g \circ f)(x)$

59. Given $F(x) = 2x + 1$ and $G(x) = \dfrac{x - 1}{x + 2}$ find:

a. $F(G(1))$ **d.** $F(F(0))$

b. $G(F(-2))$ **e.** $(F \circ G)(x)$

c. $F(G(2))$ **f.** $(G \circ F)(x)$

x	$f(x)$	$g(x)$
−1	−2	1
0	1	2
1	2	−1
2	0	−1

60. From the accompanying table, find:

a. $f(g(1))$
b. $g(f(1))$
c. $f(g(0))$
d. $g(f(0))$
e. $f(f(2))$

x	$f(x)$	$g(x)$
0	2	1
1	1	0
2	3	3
3	0	2

61. From the accompanying table, find:

a. $f(g(1))$
b. $g(f(1))$
c. $f(g(0))$
d. $g(f(0))$
e. $f(f(2))$

62. Using the accompanying graphs, find:

a. $g(f(-2))$ **b.** $f(g(1))$ **c.** $g(f(0))$ **d.** $g(f(1))$

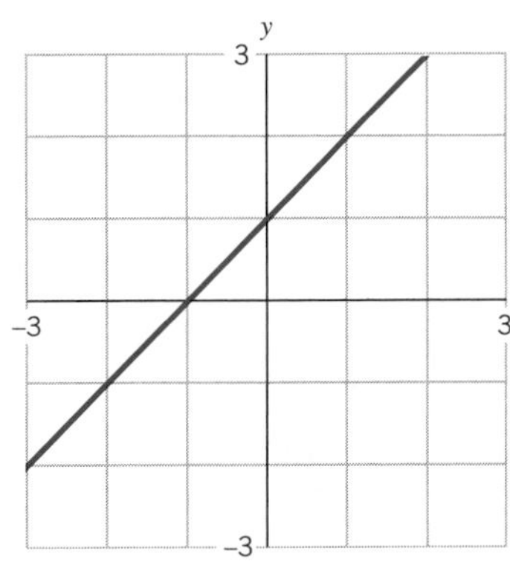

Graph of $f(x)$

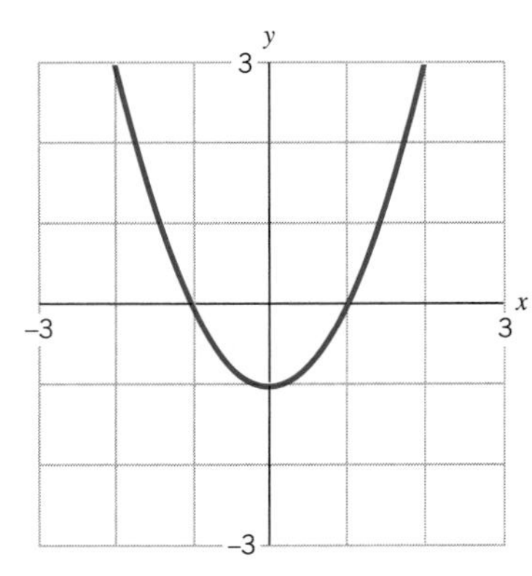

Graph of $g(x)$

63. Using the accompanying graphs, find:

a. $g(f(2))$ **b.** $f(g(-1))$ **c.** $g(f(0))$ **d.** $g(f(1))$

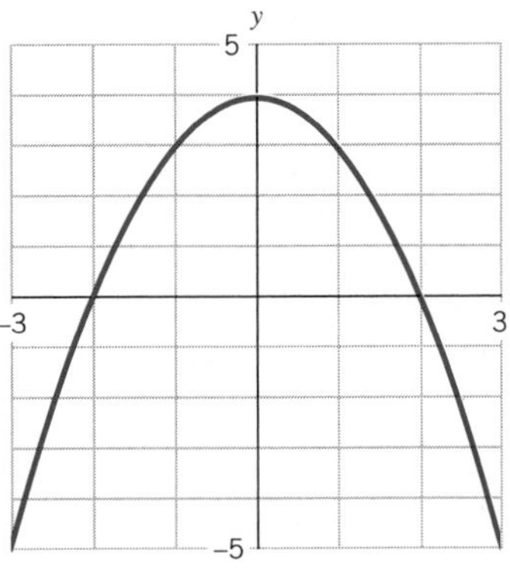

Graph of $f(x)$

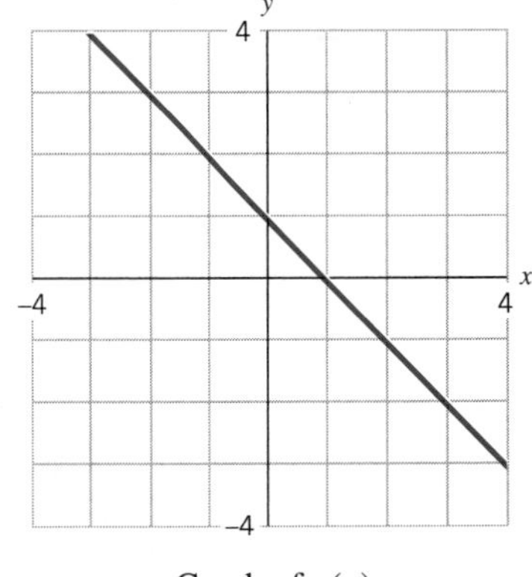

Graph of $g(x)$

64. The exchange rate a bank gave for the Canadian dollar in January 2000 was 1.4525 Canadian dollars for 1 U.S. dollar. The bank also charges a constant fee of 3 U.S. dollars per transaction.

a. Construct a function F that converts U.S. dollars, d, into Canadian dollars.

b. Construct a function G that converts Canadian dollars, c, into U.S. dollars.

c. What would the function $F \circ G$ do? Would its input be U.S. or Canadian dollars (i.e., d or c)? Construct a formula for $F \circ G$.

d. What would the function $G \circ F$ do? Would its input be U.S. or Canadian dollars (i.e., d or c)? Construct a formula for $G \circ F$.

65. If you are blowing up a balloon that has an uninflated radius of 5 inches, then the inflated radius $r = 5 + p$, where p is the pressure you exert with your lungs. The volume, V, of the balloon can be described in terms of its radius as $V = \frac{4}{3}\pi r^3$.

a. Construct a formula for the volume, V, of the balloon in terms of the pressure, p, with which you blow.

b. If we think in terms of functions, we can let $r = f(p) = 5 + p$ and $V = g(r) = \frac{4}{3}\pi r^3$. What composition of f and g would give you the volume, V?

66. The windchill temperature is the apparent temperature caused by the extra cooling from the wind. A rule of thumb for estimating the windchill temperature for a temperature, t, that is above 0° Fahrenheit, is $W(t) = t - 1.5S_0$, where S_0 is any given windspeed in miles per hour.

a. If the windspeed is 25 mph, and the temperature is 10°F, what is the windchill temperature?

We know how to convert Celsius to Fahrenheit; that is, $t = F(x)$, where $F(x) = 32 + \frac{9}{5}x$, where x is the number of degrees Celsius and $F(x)$ is the equivalent in degrees Fahrenheit.

b. Construct a function that would give the windchill temperature as a function of degrees Celsius.

c. If the windspeed is 40 mph and the temperature is -10°C, what is the windchill temperature?

67. Salt is applied to roads to decrease the temperature at which icing occurs. Assume that with no salt icing occurs at 32°F, and each unit increase in the density of salt applied decreases the icing temperature by 5°F.

a. Construct a formula for icing temperature, T, as a function of salt density, s.

Trucks spread salt on the road, but they do not necessarily spread it uniformly across the whole road surface. If the edges of the road get half as much salt as in the middle, we can describe salt density $s = S(x)$ as a function of the distance, x, from the center of the road by $S(x) = [1 - \frac{1}{2}(x/k)^2]\, S_d$, where k is the distance from the centerline to the road edges and S_d is the salt density applied in the middle of the road.

b. What would the expression for $S(x)$ be if the road were 40 feet wide?

c. What will the value for x be at the middle of the 40-foot-wide road? At the edge of the road? Verify that at the middle of the road the value of the salt density $S(x)$ is S_d and that at the edge the value of $S(x)$ is $\frac{1}{2}S_d$.

d. Construct a function that describes the icing temperature, T, as a function of x, the distance from the center of the 40-foot-wide road.

e. What is the icing temperature at the middle of the 40-foot-wide road? At the edge?

68. Show that f and g are inverses of each other.

a. $f(x) = 3x + 2$ and $g(x) = \dfrac{x - 2}{3}$

b. $f(x) = \sqrt{x - 1}$ (where $x > 1$) and $g(x) = x^2 + 1$ (where $x > 0$)

69. Show that f and g are inverses of each other.

a. $f(x) = 2x - 1$ and $g(x) = \dfrac{x + 1}{2}$

b. $f(x) = \sqrt[3]{4x + 5}$ and $g(x) = \dfrac{x^3 - 5}{4}$

70. Show that in each of the following pairs the functions are inverses of each other:

a. $f(x) = 10^{x/2}$ and $g(x) = \log(x^2)$

b. $F(t) = e^{3t}$ and $G(t) = \ln(t^{1/3})$

c. $H(r) = \frac{1}{2}\ln r$ and $J(r) = e^{2r}$

71. Verify that $f(x) = e^x$ and $g(x) = \ln x$ are inverse functions.

72. Show that $F(x) = a^x$ and $G(x) = \log_a x$ are inverse functions. (*Note:* See Exercise 40 for a definition of $\log_a x$.) The software "E10: Inverse Functions $y = a^x$ and $y = \log_a x$" in *Exponential and Log Functions* can help you visualize the relationship between the two functions.

73. In each part create a table of values for the inverse of function $f(x)$:

a.

x	$f(x)$
−2	5
−1	1
0	2
1	4

b.

x	$f(x)$
0	5
1	3
−2	2
4	−7

74. Given the accompanying graph of function $f(x)$, sketch the graph of its inverse.

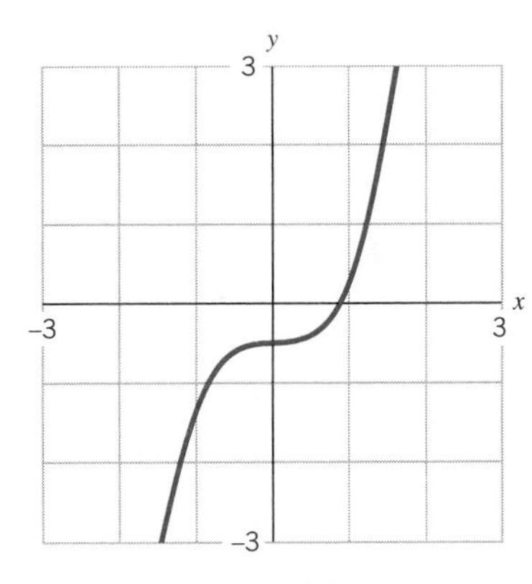

Graph of $f(x)$

75. Given the accompanying graph of function $g(x)$, sketch the graph of its inverse.

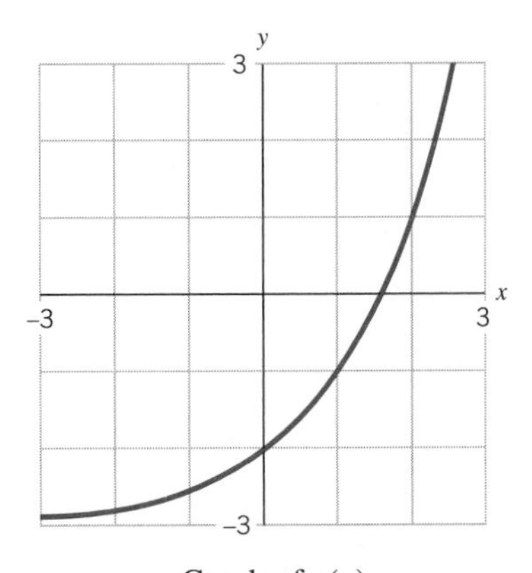

Graph of $g(x)$

76. Find the inverse of each of the following functions:

a. $f(x) = 3x - 1$ **b.** $g(x) = x^3 - 4$ **c.** $h(x) = \dfrac{2}{x-1}$

77. Find the inverse of each of the following functions:

a. $f(x) = 4x - 7$ **b.** $g(x) = 8x^3 - 5$ **c.** $h(x) = \dfrac{x+1}{2x}$

78. If an object is put in an environment at a fixed temperature A (the "ambient temperature"), then its temperature, T, at time t is modeled by Newton's Law of Cooling:

$$T = A + Ce^{-kt}$$

where k is a positive constant. Note that T is a function of t and that as $t \to +\infty$, then $e^{-kt} \to 0$, so the temperature T gets closer and closer to the ambient temperature, A.

a. Assume that a hot cup of tea (at 160°F) is left to cool in a 75°F room. If it takes 10 minutes for it to reach 100°F, determine the constants A, C, and k in the equation for Newton's Law of Cooling. What is Newton's Law of Cooling in this situation?

b. Sketch the graph of your function.

c. What is the temperature of the tea after 20 minutes?

d. Determine the inverse of the function you derived in part (a).

e. Use the inverse function from part (d) to determine the time at which the tea would reach 80°F.

79. Newton's Law of Cooling also works for objects being heated. (See Exercise 78.) At time $t = 0$, a potato at 70°F (room temperature) is put in an oven at 375°F. Thirty minutes later, the potato is at 220°.

a. Determine the constants A, C, and k in Newton's Law. Write down Newton's Law for this case.

b. When is the potato at 370°F?

c. When is the potato at 374°F?

d. According to your model, when (if ever) is the potato at 375°F?

e. Sketch a graph of your function in part (a).

f. Sketch a graph of the inverse function.

Exploration 6.1

Properties of Logarithmic Functions

Objective

- explore the effects of a and c on the graphs of the following logarithmic functions:

$$y = c\log(ax) \quad \text{and} \quad y = c\ln(ax) \quad \text{where } a > 0 \quad \text{and} \quad x > 0$$

Materials/Equipment

- graphing calculator or computer with "E8: Logarithmic Sliders" in *Exponential & Log Functions* in course software or a function graphing program
- graph paper

Procedure

Making Predictions

1. *The effect of* a *on the equation* y = *log*(ax), *where* a > 0 *and* x > 0.

a. Why do we need to restrict a and x to positive values?

b. Using the properties of logarithms, write the expression $\log(ax)$ as the sum of two logs. Discuss with your partner what effect you expect a to have on the graph. Now let $y = \log(2x)$ and $y = \log(3x)$ (where a is 2 and 3, respectively). Using $\log 2 \approx 0.301$ and $\log 3 \approx 0.477$, complete the accompanying data table and, by hand, sketch the three graphs on the same grid. Do your results confirm your predictions? What do you expect to happen to the graph of $y = \log(ax)$ if larger and larger positive values are substituted for a?

Evaluating $y = \log(ax)$ when $a = 1, 2,$ and 3

x	$y = \log x$	$y = \log(2x)$	$y = \log(3x)$
0.001	−3.000		
0.010	−2.000		
0.100	−1.000		
1.000	0.000		
5.000	0.699		
10.000	1.000		

c. Discuss with your partner how you think the graphs of $y = \log x$ and $y = \log(ax)$ will compare if $0 < a < 1$. Complete the following small data table and sketch the graphs of $y = \log x$ and $y = \log(x/10)$ on the same grid. Do your predictions and your graph agree? Predict what would happen to the graph of $y = \log(ax)$ if smaller and smaller positive values were substituted for a.

Evaluating $y = \log(ax)$ when $a = 1$ and $\frac{1}{10}$

x	$y = \log x$	$y = \log(x/10)$
0.001	−3.000	
0.010	−2.000	
0.100	−1.000	
1.000	0.000	
5.000	0.699	
10.000	1.000	

d. How do you think your findings for $y = \log(ax)$ relate to the function $y = \ln(ax)$?

2. *The effect of c on the equation $y = c \log x$*

a. When $c > 0$:

Talk over with your partner your predictions for what will happen as c increases. Fill in the following table and on the same grid do a quick sketch of the functions $y = \log x$, $y = 2 \log x$, and $y = 3 \log x$ (where $c = 1, 2, 3$, respectively). Were your predictions correct? What do you expect to happen to the graph of $y = c \log x$ as you substitute larger and larger positive values for c?

Evaluating $y = c \log x$ when $c = 1, 2,$ and 3

x	$y = \log x$	$y = 2 \log x$	$y = 3 \log x$
0.001	−3.000		
0.010	−2.000		
0.100	−1.000		
1.000	0.000		
5.000	0.699		
10.000	1.000		

b. When $c < 0$:

Why can c be negative when a had to remain positive? How do you think the graphs of $y = \log x$ and $y = c \log x$ will compare if $c < 0$? Complete the following small data table and sketch the graphs of $y = -\log x$ and $y = -2 \log x$ on the same grid. Check your results and your predictions with your partner. What will happen to the graph of $y = c \log x$ if you substitute a negative value for c? What happens if c remains negative but $|c|$ gets larger and larger (for example, $c = -10, -150, -5000$, etc.)?

Evaluating $y = c \log x$ when $c = -1$ and -2

x	$y = -\log x$	$y = -2 \log x$
0.001		
0.010		
0.100		
1.000		
5.000		
10.000		

c. How do your findings on $y = c \log x$ relate to the function $y = c \ln x$?

3. *Generalizing your results.*

   Talk over with your partner the effect of varying both a and c on the general functions $y = c\ \log(ax)$ and $y = c\ \ln(ax)$. Try predicting the shapes of the graphs of such functions as $y = 3\ \log(2x)$ or $y = -2\ \log(x/10)$. Have each partner construct a small table, and graph the results of one such function. Compare your findings.

Testing Your Predictions

Test your predictions by either using "E8: Logarithmic Sliders" in *Exponential & Log Functions* or creating your own graphs with a graphing calculator or a function graphing program.

1. Try changing the value for a in different functions of the form $y = \log(ax)$ or $y = \ln(ax)$. Be sure to try values of $a > 1$ and values such that $0 < a < 1$.
2. Try changing the value for c in functions of the form $y = c \log x$ and $y = c \ln x$. Let c assume both positive and negative values.

Summarizing Your Results

Write a 60 second summary describing the effect of varying a and c in functions of the form $y = c\ \log(ax)$ and $y = c\ \ln(ax)$.

Exploration-Linked Homework

1. Use your knowledge of logarithmic functions to predict the shapes of the graphs of the following:

   a. *The decibel scale:* given by the function $N = 10 \log\dfrac{I}{I_0}$, where I is the intensity of a sound and I_0 is the intensity of sound at the threshold of human hearing.

   b. *Stellar magnitude:* approximated by the function $M = -2.5 \log\dfrac{B}{B_0}$, where B is the brightness of a star and B_0 is a constant.

   Use technology to check your predictions.

2. Explore the effect of changing the base, that is, the effect of changing b in $y = c\ \log_b(ax)$.

Chapter 7

Power Functions

Overview

Power functions are those in which the dependent variable equals a constant times the independent variable raised to a power. They help us answer questions such as: "Why is the most important rule of scuba diving 'never hold your breath'?" and "Why do small animals have faster heartbeats and higher metabolic rates than large ones?"

After reading this chapter you should be able to:

- recognize the properties of power functions
- construct and interpret graphs of power functions
- understand direct and inverse proportionality
- use logarithmic scales to determine whether a linear, power, or exponential function would be the best function model for a set of data
- develop a sense about the relationship between size and shape

7.1 THE TENSION BETWEEN SURFACE AREA AND VOLUME

Why do small animals have faster heartbeats and higher metabolic rates than large ones? Why are the shapes of the bodies and organs of large animals often quite different from those of small ones? Why are wood logs hard to set on fire, whereas small sticks and chips burn easily and sawdust can explode? To find answers to these questions, we examine how the relationship between surface area and volume changes as objects increase in size.

Scaling Up a Cube

Let's look at what happens to the surface area and volume of a simple geometric figure, the cube, as we increase its size. In Figure 7.1 we have drawn a series of cubes where the lengths of the edges are 1, 2, 3, and 4 units.[1]

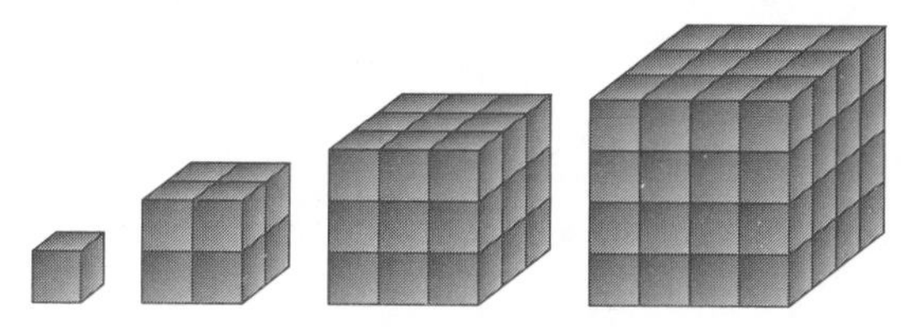

Figure 7.1 Four cubes for which the lengths of the edges are 1, 2, 3, and 4 units, respectively.

Surface area of a cube. If we were painting a cube, the surface area would tell us how much area we would have to cover. Each cube has six identical faces, so

$$\text{surface area} = 6 \cdot (\text{area of one face})$$

If the edge length of one face of the cube is x, then the surface area of that face is x^2. So the total surface area $s(x)$ of the cube can be defined as

$$s(x) = 6x^2$$

The second column of Table 7.1 contains the surface areas of cubes for which the lengths of the edges are 1, 2, 3, 4, 6, 8, and 10 units, and Figure 7.2 shows a graph of the function $s(x)$.

Since the surface area of a cube with edge length x can be represented as

$$s(x) = \text{constant} \cdot x^2$$

we say that the surface area is *directly proportional* to the *square* (or second power) of the length of its edge. In Table 7.1 observe what happens to the surface area when we double the length of an edge. If we double the length from 1 to 2 units, the surface area becomes four times larger, increasing from 6 to 24 square units. If we double the length from 2 to 4 units, the surface area is again four times larger, increasing this time from 24 to 96 square units. In general, if we double the length of the edge from x to $2x$, the surface area will increase by a factor of 2^2, or 4:

Surface area of a cube with edge length x is: $\quad s(x) = 6x^2$

Surface area of a cube with edge length $2x$ is: $\quad s(2x) = 6(2x)^2$

[1]It doesn't matter which unit we use, but if you prefer, you may think of "unit" as being "centimeter" or "foot."

Apply exponent	$= 6 \cdot 2^2 \cdot x^2$
simplify and rearrange terms	$= 4(6x^2)$
substitute $s(x)$ for $6x^2$	$s(2x) = 4 \cdot s(x)$
	$= 4 \cdot$ (surface area of a cube with edge length x)

Table 7.1
Surface Area and Volume of a Cube with Edge Length *x*

Edge Length x (unit)	Surface Area $s(x) = 6x^2$ (sq. unit)	Volume $v(x) = x^3$ (cubic unit)
1	6	1
2	24	8
3	54	27
4	96	64
6	216	216
8	384	512
10	600	1000

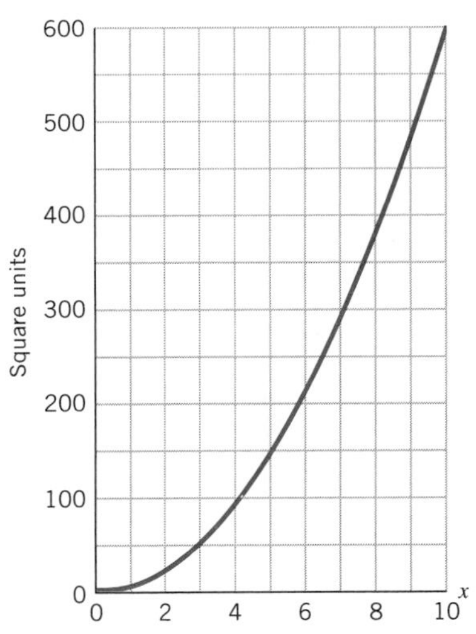

Figure 7.2 Graph of $s(x) = 6x^2$.

Volume of a cube. If the edge length of one face of the cube is x, then the volume $v(x)$ of the cube is given by

$$v(x) = x^3$$

The third column of Table 7.1 contains the volumes for cubes of certain sizes, and Figure 7.3 is the graph of the volume function, $v(x)$.

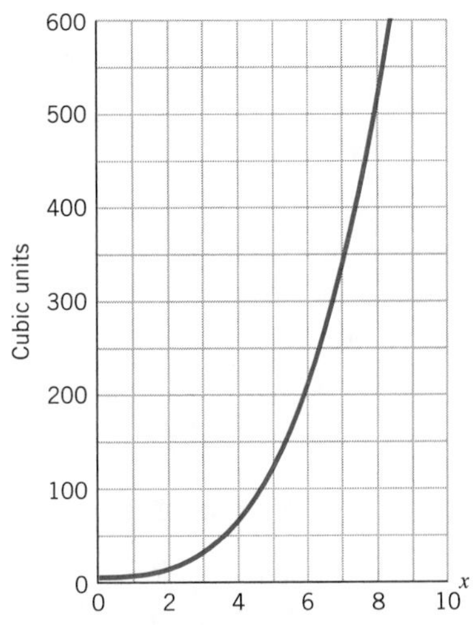

Figure 7.3 Graph of $v(x) = x^3$.

Since

$$\text{volume} = \text{constant} \cdot x^3 \quad \text{(the constant in this case is 1)}$$

we say that the volume is *directly proportional* to the *cube* (or the third power) of the length of its edge.

What happens to the volume when we double the length? In Table 7.1, if we double the length from 1 to 2 units, the volume increases by a factor of 2^3 or 8, increasing from 1 to 8 cubic units. If we double the length from 2 to 4 units, the volume again becomes 8 times larger, increasing from 8 to 64 cubic units. In general, if we double the edge length from x to $2x$, the volume will increase by a factor of 8:

Volume of a cube with edge length x is:	$v(x) = x^3$
Volume of a cube with edge length $2x$ is:	$v(2x) = (2x)^3$
Apply exponent	$= 2^3 \cdot x^3$
simplify	$= 8x^3$
substitute $v(x)$ for x^3	$v(2x) = 8 \cdot v(x)$
	$= 8 \cdot$(volume of a cube with edge length x)

Surface area/volume. When we double the length of the edge, the surface area increases by a factor of 4, but the volume increases by a factor of 8. As we increase the length of the edge, the volume eventually grows faster than the surface area. Hence the ratio

$$\frac{\text{surface area}}{\text{volume}}$$

decreases as the side length increases. See the fourth column of Table 7.2.

In Exploration 7.1 you can study further the effects of scaling up an object.

Table 7.2

Ratio of Surface Area to Volume of Cube with Length of Edge x

Edge Length x	Surface Area $s(x) = 6x^2$	Volume $v(x) = x^3$	$\frac{\text{Surface Area}}{\text{Volume}}$ $\frac{s(x)}{v(x)} = \frac{6x^2}{x^3} = \frac{6}{x}$
1	6	1	6.00
2	24	8	3.00
3	54	27	2.00
4	96	64	1.50
6	216	216	1.00
8	384	512	0.75
10	600	1000	0.60

Size and Shape

What we learned about the cube is true for any three-dimensional object, no matter what the shape. In general:

For any shape, as an object becomes larger while keeping the same shape, the ratio of its surface area to its volume decreases.

Thus a larger object has relatively less surface area than a smaller one. This fact allows us to understand some basic principles of biology and to answer the questions we asked at the beginning of this section.

Biological functions such as respiration and digestion depend upon surface area but must service the body's entire volume.[2] The biologist J. B. S. Haldane wrote that "comparative anatomy is largely the story of the struggle to increase surface in proportion to volume." This is why the shapes of the bodies and organs of large animals are often quite different from those of small ones. Many large species have adapted by developing complex organs with convoluted exteriors, thus greatly increasing the organs' surface areas. Human lungs, for instance, are heavily convoluted to increase the amount of surface area, thereby increasing the rate of exchange of gases. Stephen Jay Gould writes, "the villi of our small intestine increase the surface area available for absorption of food (small animals neither have nor need them)."

Stephen Jay Gould's essay "Size and Shape" in *Ever Since Darwin: Reflections in Natural History* offers an interesting perspective on the relationship between the size and shape of objects.

Body temperature depends upon the ratio of surface area to volume. Animals generate the heat needed for their volume by metabolic activity and lose heat through their skin surface. Small animals have more surface area in proportion to their volume than do large animals. Since heat is exchanged through the skin, small animals lose heat proportionately faster than large animals and have to work harder to stay warm. Hence their heartbeats and metabolic rates are faster. As a result, smaller animals burn more energy per unit mass than larger animals.

The surface area/volume tension relates to other physical phenomena as well. To set an object on fire, some part of the volume must be raised to the ignition point. Heat is absorbed through the surface, so large pieces, which have relatively less surface area than smaller ones, are more difficult to ignite. That is why small pieces of wood called tinder are used to light log fires. Tiny dust particles floating in the air can be quite flammable and are responsible for a number of dramatic explosions of grain elevators in the Midwest.

Algebra Aerobics 7.1

1. The surface area and volume of a sphere are both functions of the radius. They can be described by the functions

$$v(r) = \tfrac{4}{3}\pi r^3$$

where $v(r)$ represents the volume of a sphere with radius r, and

$$s(r) = 4\pi r^2$$

where $s(r)$ represents the surface area of a sphere with radius r.

a. When you double the radius of a sphere, what happens

i. to the surface area?

ii. to the volume?

b. **i.** Which eventually grows faster, the surface area or the volume?

ii. As a result, what happens to the surface area/volume ratio as the radius increases?

[2]For those who want to investigate how species have adapted and evolved over time, see D. W. Thompson, *On Growth and Form* (New York: Dover Publications, 1992) and T. McMahon and J. Bonner, *On Size and Life* (New York: Scientific American Books, 1983).

7.2 POWER FUNCTIONS WITH POSITIVE POWERS

In the last section, we saw that the surface area function, $s(x)$, and the volume function, $v(x)$, of a cube can be represented by the functions

$$s(x) = 6x^2$$

and

$$v(x) = x^3$$

where x is the edge length of the cube. These functions are examples of *power functions.*

The general form of an equation for a power function is

$$\text{dependent variable} = \text{constant} \cdot (\text{independent variable})^{\text{power}}$$

How would you describe a power function with an exponent of 0?

> A *power function* $y = f(x)$ can be represented by an equation in the form
>
> $$y = kx^p$$
>
> where k and p are any constants.

For example, the functions

$$v = \tfrac{4}{3}\pi r^3 \qquad F = 0.2d^{-2} \qquad A = 25M^{1/2} \qquad y = 5x$$

are all power functions with powers 3, -2, $\frac{1}{2}$, and 1, respectively.

The next few sections focus on power functions with positive integer exponents. In Section 7.5 we study power functions with negative integer exponents.

Example 1

Assume a person weighing 200 pounds is standing at the center of a fir plank with a $2'' \times 12''$ cross section that spans a length of L feet:

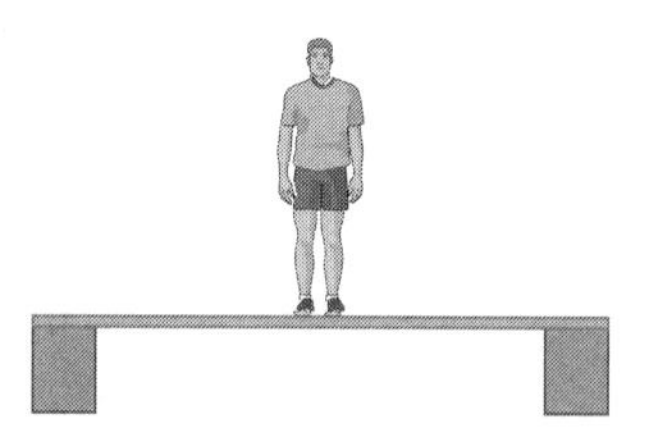

Architects use the following power function to predict the number of inches of downwards deflection (or bending) of the plank:

$$D_2 = 0.00132\, L^3$$

a. Graph the D_2 deflection formula.

b. A rule used by architects for estimating acceptable deflection in inches of a beam L feet long bent downward as a result of carrying a load is

$$D_{\text{safe}} = 0.05L$$

Add a plot of D_{safe} to your graph from part (a).

c. Estimate the maximum span that is safe for a 200-pound person on a $2'' \times 12''$ fir plank.

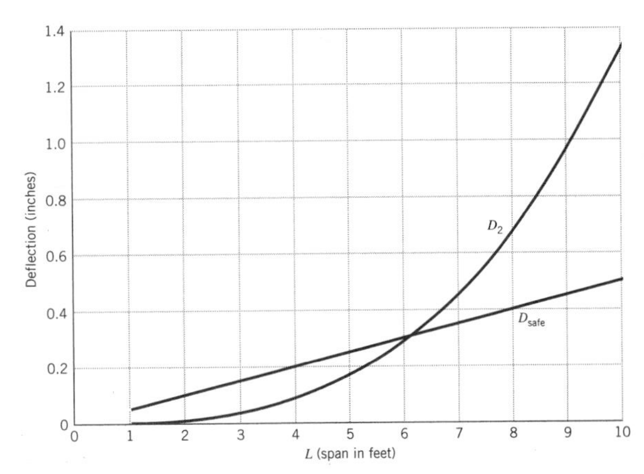

Figure 7.4 Graph of the deflection, D_2, for a 2″ thick fir plank and the safe level of deflection (D_{safe}).

Weight is directly proportional to length cubed, but the ability to support the weight, as measured by the cross-sectional area of the bones, is proportional to length squared. Why does this mean that Godzilla or King Kong could only exist in the movies?

Solution

a., b. Figure 7.4 shows the graphs of $D_2 = 0.00132L^3$ and $D_{safe} = 0.05L$ for values of L between 1 and 10 feet.

c. From the graph we can see that above a span of about 6 feet the deflection for the 2″ × 12″ fir plank, D_2, starts to exceed the safe deflection levels, D_{safe}.

Direct Proportionality

In Chapter 2, for linear functions, we said that y is *directly proportional to* x if y equals a constant times x. For example, if $y = 4x$, then y is directly proportional to x. We can extend the same concept to any power function with positive exponents. If $y = kx^p$ and p is positive, we say that y is *directly proportional to* x^p.

In Section 7.1, we saw that the surface area of a cube is directly proportional to the square of its edge length and that the volume is directly proportional to the cube of its edge length. The symbol $\propto$ is used to indicate direct proportionality.

> If
>
> $$y = kx^p$$
>
> where k is a constant and $p > 0$, we say that y is *directly proportional to* x^p. We write this as
>
> $$y \propto x^p$$
>
> The term k is called the constant of proportionality.

Example 2

Write formulas to represent the following relationships:

a. The circumference, C, of a circle is directly proportional to its radius, r.

b. The area, A, of a circle is directly proportional to the radius, r, squared.

c. The volume, V, of a liquid flowing through a tube is directly proportional to the fourth power of the radius, r, of the tube.

Solution

a. $C = kr$, where $k = 2\pi$.

b. $A = kr^2$, where $k = \pi$.

c. $V = kr^4$ for some constant k.

Example 3

When you pedal a bicycle, it's much easier to pedal with the wind than into the wind. That's because as the wind blows against an object, it exerts a force upon it. This force, F, is directly proportional to the wind velocity, V, squared.

a. Construct an equation to describe the relationship between F and V.

b. If the wind velocity doubles, by how much does the force go up?

Solution

a. $F = kV^2$ for some constant k.

b. If the velocity doubles, say from v to $2v$, then the force goes up by a factor of 2^2, or 4. If $V = v$, then $F = kv^2$. If $V = 2v$, then $F = k(2v)^2 = k \cdot 2^2 \cdot v^2 = 4\,kv^2$. This means that pedaling into a 20-mph wind requires four times as much effort as pedaling into a 10-mph wind.

Example 4

The radius of the core of Earth is over half the radius of Earth as a whole, yet the core is only about 16% of the total volume of Earth. How is this possible?

Solution

We can think of both Earth and its core as approximately spherical in shape. The function

$$v = \tfrac{4}{3}\pi r^3$$

describes the volume, v, of a sphere as directly proportional to the cube of its radius, r. So if the core of Earth has radius R, then the core's volume is approximately

$$v_{\text{core}} \approx \tfrac{4}{3}\pi R^3$$

If the radius of Earth were $2R$ or exactly twice that of its core, Earth's volume would be

$$v_{\text{earth}} \approx \tfrac{4}{3}\pi(2R)^3$$

rewrite $(2R)^3$
$$= \tfrac{4}{3}\pi(2)^3R^3$$

rearrange terms
$$= (2^3)(\tfrac{4}{3})\,\pi R^3$$

substitute v_{core} for $\frac{4}{3}\pi R^3$ and 8 for 2^3
$$v_{\text{earth}} \approx 8v_{\text{core}}$$

So if the radius of Earth were twice the radius of the core, the volume of Earth would be approximately 8 times the volume of the core. Equivalently the volume of the core would be 1/8th or 12.5%, of the volume of Earth. Since the radius of the core is a little more than half the radius of Earth, 16% of the Earth's volume is a reasonable estimate for the volume of the core.

Example 5

In Chapter 4 we encountered the following formula used by police. It estimates the speed, S, at which a car must have been traveling given the distance, d, the car skids on a dry tar road after the brakes have been applied:

$$S = \sqrt{30d} \approx 5.48d^{1/2}$$

Speed, S, is in miles per hour and distance, d, is in feet.

a. Use the language of proportionality to describe the relationship between S and d.

b. If your skid marks were 50 feet long, approximately how fast were you going when you applied the brakes?

c. What happens to S if d doubles? Quadruples?

d. Is d directly proportional to S?

Solution

a. The speed, S, is directly proportional to $d^{1/2}$ and the constant of proportionality is 5.48.

b. If the length of the skid marks, d, is 50 feet, then the estimated speed $S \approx 5.48 \cdot (50)^{1/2} \approx 5.48 \cdot 7.07 \approx 39$ mph.

c. If the skid marks double in length from d to $2d$, then the estimate for the speed the car was going increases from $\sqrt{30d}$ to $\sqrt{30 \cdot (2d)} = \sqrt{2} \cdot \sqrt{30d}$. So the estimated speed goes up by a factor of $\sqrt{2} \approx 1.414$ (or equivalently by about 41.4%).

If the skid marks quadruple in length from d to $4d$, then the estimated speed of the car goes from $\sqrt{30d}$ to $\sqrt{30 \cdot (4d)} = \sqrt{4} \cdot \sqrt{30d} = 2\sqrt{30d}$. So the speed estimate goes up by a factor of 2, or by 100%.

So if the skid marks doubled from 50 to 100 feet, the speed estimate would go up from 39 mph to $1.414 \cdot 39 \approx 55$ mph. If the skid marks quadrupled from 50 to 200 feet, then the speed estimate would double from about 39 to almost 78 mph.

d. Given $\quad S = \sqrt{30d}$

square both sides of the equation $\quad S^2 = 30d$

divide both sides by 30 $\quad \dfrac{S^2}{30} = d$

or $\quad d = \dfrac{S^2}{30}$

So d is directly proportional to S^2 but not to S.

Direct Proportionality with More Than One Variable

When a quantity depends on more than one other quantity, we no longer have a simple power function. For example, the volume, V, of a cylindrical can depends on both the radius, r, of the base and the height, h. The equation describing this relationship is

$$V = \text{area of base} \cdot \text{height}$$

$$V = \pi r^2 h$$

We say V is directly proportional to both r^2 and h.

Example 6

In Example 1 in this section we looked at the deflection caused by a 200-pound person standing at the center of a 2″ × 12″ fir plank.[3] If the person weighs P pounds, then the downward deflection, D_2, is given by the formula

$$D_2 = (6.6 \cdot 10^{-6}) \cdot P \cdot L^3$$

where D_2 is in inches, P in pounds, and L in feet.

[3]The deflection formula for the plank is a quite accurate model for deflections up to about an inch. After that it starts to produce unrealistic D values very quickly since L is cubed.

a. Express this relationship in terms of direct proportionality.

b. Verify that if $P = 200$ pounds, then $D_2 = 0.00132L^3$, the special case we saw in Example 1.

c. What happens to the value of D_2 if we increase P by 50%? If we increase L by 50%?

Solution

a. The deflection, D_2, is directly proportional to P and to the cube of L.

b. If $P = 200$, we have

$$\begin{aligned} D_2 &= (6.6 \cdot 10^{-6}) \cdot 200 \cdot L^3 \\ &= 6.6 \cdot 200 \cdot 10^{-6} \cdot L^3 \\ &= 1320 \cdot 10^{-6} \cdot L^3 \\ &= 0.00132L^3 \end{aligned}$$

c. Increasing the person's weight, P, by 50% means multiplying the value of P by 1.50. Then the amount of deflection, D_2, will increase from $(6.6 \cdot 10^{-6}) \cdot P \cdot L^3$ to

$$(6.6 \cdot 10^{-6}) \cdot (1.5\ \text{P}) \cdot L^3 = 1.5[(6.6 \cdot 10^{-6}) \cdot P \cdot L^3].$$

Hence D_2 increases also by a factor of 1.5, or by 50%.

Similarly, increasing the span, L, by 50% means multiplying the value of L by 1.50. But now the amount of deflection, D_2, will increase from $(6.6 \cdot 10^{-6}) \cdot P \cdot L^3$ to

$$\begin{aligned} (6.6 \cdot 10^{-6}) \cdot P \cdot (1.5L)^3 &= (6.6 \cdot 10^{-6}) \cdot P \cdot (1.5)^3 (L)^3 \\ &\approx (6.6 \cdot 10^{-6}) \cdot P \cdot 3.4L^3 \\ &\approx 3.4[(6.6 \cdot 10^{-6}) \cdot P \cdot L^3] \end{aligned}$$

Hence increasing the span, L, by 50% increases the deflection, D_2, by about a factor of 3.4, or by 240%.

Algebra Aerobics 7.2

1. In each case, indicate whether or not the function is a power function. If it is, identify the independent and dependent variables, the constant of proportionality, and the power.

   a. $A = \pi r^2$ **c.** $z = w^5 + 10$ **e.** $y = 3x^5$

   b. $y = z^5$ **d.** $y = 5^x$

2. Identify which (if any) of the following equations represent direct proportionality:

   a. $y = 5.3x^2$ **b.** $y = 5.3x^2 + 10$ **c.** $y = 5.3x^{-2}$

3. Given $g(x) = 5x^3$:

   a. Calculate $g(2)$ and compare this value to $g(4)$.

   b. Calculate $g(5)$ and compare this value to $g(10)$.

   c. What happens to the value of $g(x)$ if x doubles in value?

   d. What happens to $g(x)$ if x is divided by 2?

4. Express in your own words the relationship between y and x in the following functions:

   a. $y = 3x^5$ **b.** $y = 2.5x^3$ **c.** $y = \frac{x^5}{4}$

5. Now express each of the relationships in Problem 4 in terms of direct proportionality.

6. **a.** Solve $P = aR^2$ for R.

   b. Solve $V = (1/3)\pi r^2 h$ for h.

   c. Solve $V = (1/3)\pi r^2 h$ for r.

   d. Solve $Y = Z(a^2 + b^2)$ for Z.

   e. Solve $Y = Z(a^2 + b^2)$ for a.

7.3 VISUALIZING POSITIVE INTEGER POWERS

Moving to the Abstract

The power functions we graphed in Section 7.1 represented physical quantities (volume and surface area) and hence their domains were constrained. What will the graphs of abstract power functions look like where x can be infinitely large or assume negative values? Let's examine two basic power functions, $f(x) = x^2$ and $g(x) = x^3$.

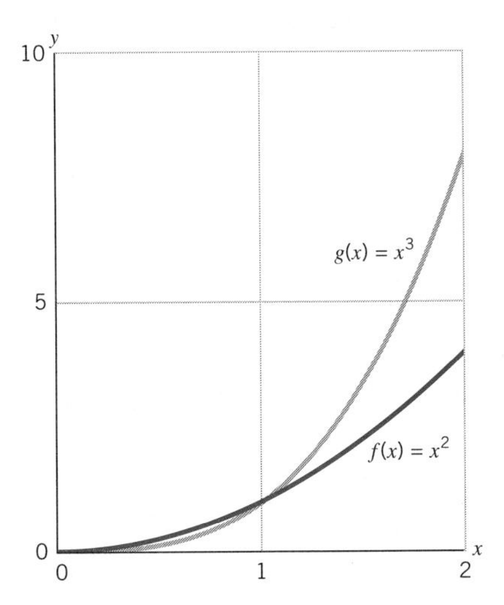

Figure 7.5 Comparing the graphs of $f(x) = x^2$ and $g(x) = x^3$ when $x \geq 0$.

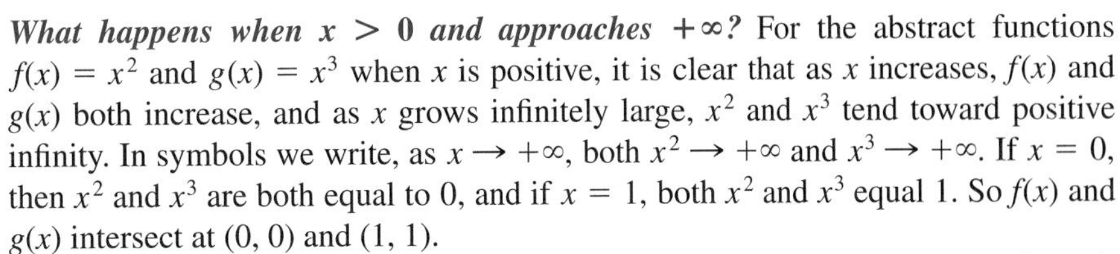

What happens when x > 0 and approaches +∞? For the abstract functions $f(x) = x^2$ and $g(x) = x^3$ when x is positive, it is clear that as x increases, $f(x)$ and $g(x)$ both increase, and as x grows infinitely large, x^2 and x^3 tend toward positive infinity. In symbols we write, as $x \to +\infty$, both $x^2 \to +\infty$ and $x^3 \to +\infty$. If $x = 0$, then x^2 and x^3 are both equal to 0, and if $x = 1$, both x^2 and x^3 equal 1. So $f(x)$ and $g(x)$ intersect at (0, 0) and (1, 1).

We can see from the graph in Figure 7.5 that when $0 < x < 1$, then $x^3 < x^2$, but when $x > 1$, then $x^3 > x^2$. We say that the graph of $g(x) = x^3$ eventually dominates the graph of $f(x) = x^2$.

What happens when x < 0 and approaches −∞? When x is negative, the functions $f(x) = x^2$ and $g(x) = x^3$ exhibit quite different behaviors. (See Table 7.3 and Figure 7.6 below.) When x is negative, x^2 is positive. As $x \to -\infty$, $x^2 \to +\infty$. When x is negative, x^3 is also negative. As $x \to -\infty$, $x^3 \to -\infty$. The range for $f(x) = x^2$ is the set of nonnegative numbers. The range for $g(x) = x^3$ is all real numbers.

Table 7.3

x	$f(x) = x^2$	$g(x) = x^3$
−4	16	−64
−3	9	−27
−2	4	−8
−1	1	−1
0	0	0
1	1	1
2	4	8
3	9	27
4	16	64

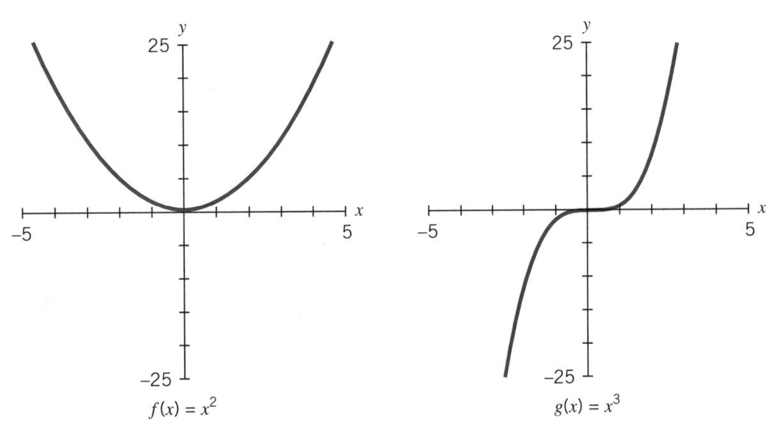

Figure 7.6 The graphs of $f(x) = x^2$ and $g(x) = x^3$.

Algebra Aerobics 7.3a

1. If $f(x) = 4x^3$, evaluate the following:
   a. $f(2)$ **b.** $f(-2)$ **c.** $f(s)$ **d.** $f(3s)$
2. If $g(t) = -4t^3$, evaluate the following:
   a. $g(2)$ **b.** $g(-2)$ **c.** $g(\frac{1}{2}t)$ **d.** $g(5t)$
3. **a.** For both of the following functions, generate small tables, including positive and negative values for x, and graph the functions on the same grid:

   $$f(x) = 4x^2 \quad \text{and} \quad g(x) = 4x^3$$

   b. What happens to $f(x)$ and to $g(x)$ as $x \to +\infty$?
   c. What happens to $f(x)$ and to $g(x)$ as $x \to -\infty$?
   d. Specify the domain and range of each function.
   e. Where do the graphs of these functions intersect?
   f. For what values of x is $g(x) > f(x)$?

Odd vs. Even Powers

The graphs of power functions with positive integer exponents fall into basic categories: those with odd and those with even powers. Why?

Exploration 7.2 can help reinforce your understanding of power functions with positive integer exponents.

If n is a positive integer and if x is positive, then x^n is positive for all values of n. But if x is negative, we have to consider whether n is even or odd. If n is even, then x^n is positive, since

$$(\text{negative number})^{\text{even power}} = \text{positive number}$$

If n is odd, then x^n is negative, since

$$(\text{negative number})^{\text{odd power}} = \text{negative number}$$

So whether the exponent of a power function is odd or even will affect the shape of the graph.

If we graph the simplest power functions, $y = x, y = x^2, y = x^3, y = x^4, y = x^5, y = x^6$, and so on, we see quickly that the graphs fall into two groups: the odd powers and the even powers (Figure 7.7).

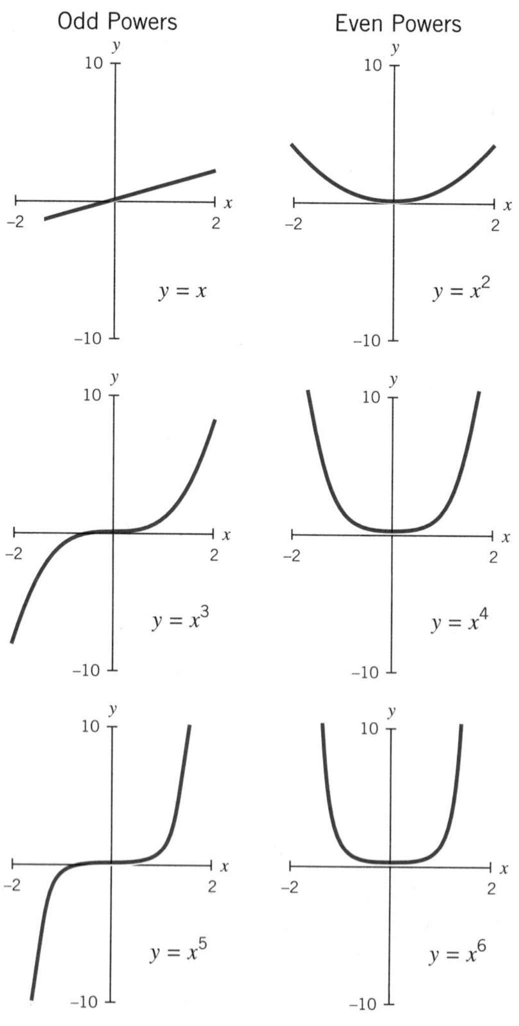

Figure 7.7 Graphs of odd and even power functions.

The graphs of all power functions (with either odd or even positive integer exponents) go through the origin.

Symmetry of the graphs. For the odd positive powers, x^1 (usually written as just x), $x^3, x^5, x^7, \ldots$, as x increases, y increases. The graphs are rotationally symmetrical about the origin; that is, if you hold the graph fixed at the origin and then rotate it 180°, you end up with the same graph. All the graphs of odd powers with exponents greater than 1 have a similar overall appearance.

The graphs of even powers, $x^2, x^4, x^6, x^8, \ldots$, are U-shaped. For all positive power functions of even degree, if we start with negative values of x and increase x, y first decreases and then, as x becomes positive, y increases. The graphs are symmetric about the y-axis. If you think of the y-axis as a dividing line, the "left" side of the graph is a mirror image, or reflection, of the "right" side.

The Effect of the Coefficient *k*

The program "P1: *k* & *p* Sliders" in *Exponential & Log Functions* can help you visualize the graphs of $y = kx^p$ for different values of k and p.

The power functions $y = x, y = x^2, y = x^3, y = x^4, y = x^5$, and $y = x^6$ all have a coefficient of 1 since we can think of them as $y = 1 \cdot x, y = 1 \cdot x^2, y = 1 \cdot x^3$, and so on. We now consider the effect of different values for the coefficient k on graphs of power functions in the form

$$y = kx^p \quad \text{where } p \text{ is a positive integer}$$

Case 1, $k > 0$: Comparing $y = kx^p$ to $y = x^p$ when k is positive. We know from our work with power functions of degree 1, that is, linear functions of the form $y = kx$, that k affects the steepness of the line. For values of $k > 1$, the larger the value for k, the more vertical the graph of $y = kx^p$ becomes compared to $y = x^p$. As k increases, the steepness of the graph increases. We say the graph is *stretched vertically.*

When $0 < k < 1$, the graph of $y = kx^p$ is flatter than the graph of $y = x^p$ and lies closer to the x-axis. We say the graph is *compressed vertically.*

In general, for any positive k, the larger the value for k, the steeper the graph of the power function $y = kx^p$. The graphs in Figure 7.8 illustrate this effect for power functions of degrees 3 and 4.

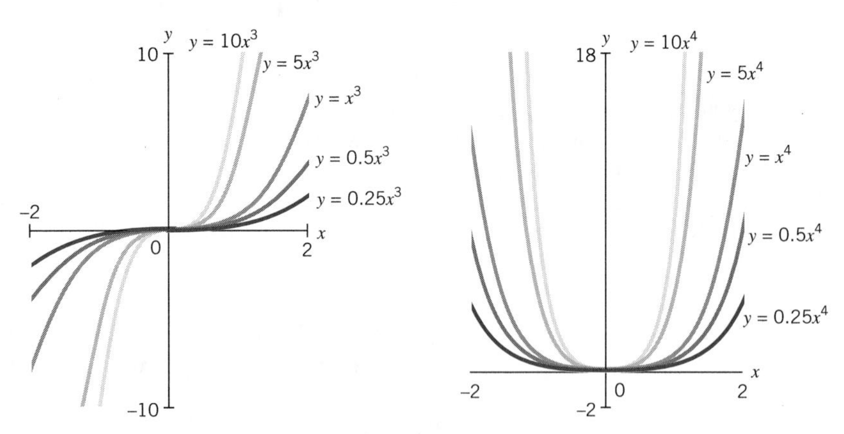

Figure 7.8 When $k > 0$, the larger the value of k, the more closely the power function $y = kx^p$ "hugs" the y-axis for both odd and even powers.

Case 2, $k < 0$: Comparing $y = kx^p$ to $y = x^p$ when k is negative. We know from our work with linear power functions of the form $y = kx$ that when k is negative, multiplying x by k not only changes the steepness of the line but also "flips," or "reflects," the line across the horizontal axis. The graphs of $y = kx$ and $y = -kx$ are mirror images across the x-axis. Similarly, the graphs of $y = kx^p$ and $y = -kx^p$ are mirror images of each other across the x-axis. For example, $y = -3x$ is the mirror image of $y = 3x$, and $y = -7x^3$ is the mirror image of $y = 7x^3$. Figure 7.9 shows various pairs of power functions of the type $y = kx^p$ and $y = -kx^p$.

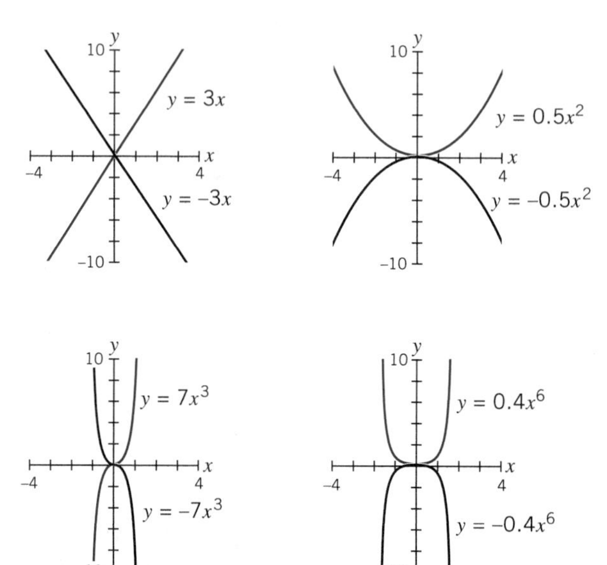

Figure 7.9 In each case, the graphs of $y = kx^p$ and $y = -kx^p$ (shown in blue and black respectively) are mirror images of each other across the x-axis.

Algebra Aerobics 7.3b

1. Draw a quick sketch by hand of the power functions:
   a. $y = x^9$ **b.** $y = x^{10}$.
   Check your work using a graphing calculator or computer. Try switching among various window sizes to compare different sections of the graph.

2. Draw a rough hand sketch of:
   a. $y = x$, $y = 4x$, and $y = -4x$ (all on the same grid).
   b. $y = x^4$, $y = 0.5x^4$, and $y = -0.5x^4$ (all on the same grid).
   Check the graphs in parts (a) and (b) using a graphing calculator or function graphing program.

7.4 COMPARING POWER AND EXPONENTIAL FUNCTIONS

Question

Which eventually grows faster, a power function or an exponential function?

Discussion

Although power and exponential functions may appear to be similar in construction, in each function the independent variable assumes a very different role. For power functions the independent variable, x, is the *base* which is raised to a fixed power. Power functions have the form

$$\text{dependent variable} = k \cdot (\text{independent variable})^{\text{power}}$$

$$y = kx^p$$

where k and p are any constants. For power functions that describe growth, k and p are both positive.

For exponential functions the independent variable, x, is the *exponent* applied to a fixed base. Exponential functions have the form

$$\text{dependent variable} = C \cdot a^{\text{independent variable}}$$

$$y = C \cdot a^x$$

where a and C are constants with $a > 0$ and $a \neq 1$. For exponential growth, $C > 0$ and $a > 1$.

Consider the functions $y = x^3$, a power function; $y = 3^x$, an exponential function; and $y = 3x$, a linear function. Table 7.4 compares the role of the independent variable, x, in the three functions.

Table 7.4

x	Linear Function $y = 3x$ (x is multiplied by 3)	Power Function $y = x^3$ (x is the *base* raised to third power	Exponential Function $y = 3^x$ (x is the *exponent* for base 3)
0	$3 \cdot 0 = 0$	$0 \cdot 0 \cdot 0 = 0$	$3^0 = 1$
1	$3 \cdot 1 = 3$	$1 \cdot 1 \cdot 1 = 1$	$3^1 = 3$
2	$3 \cdot 2 = 6$	$2 \cdot 2 \cdot 2 = 8$	$3^2 = 9$
3	$3 \cdot 3 = 9$	$3 \cdot 3 \cdot 3 = 27$	$3^3 = 27$
4	$3 \cdot 4 = 12$	$4 \cdot 4 \cdot 4 = 64$	$3^4 = 81$
5	$3 \cdot 5 = 15$	$5 \cdot 5 \cdot 5 = 125$	$3^5 = 243$

Visualizing the difference. Table 7.4 and Figure 7.10 show that the power function $y = x^3$ and the exponential function $y = 3^x$ both grow very quickly relative to the linear function $y = 3x$.

Yet there is a vast difference between the growth of an exponential function and the growth of a power function. In Figure 7.11 we zoom out on the graphs.

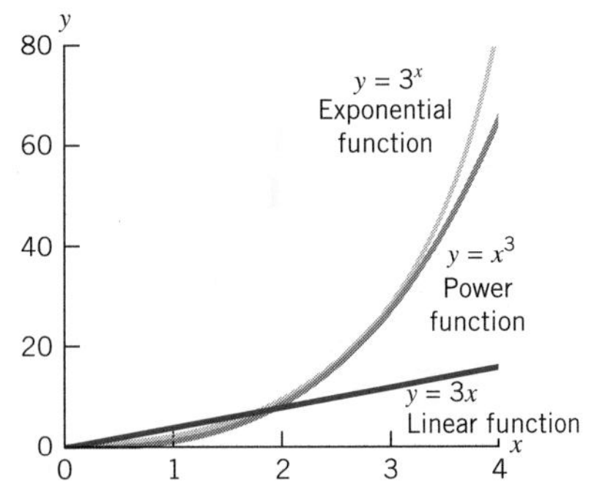

Figure 7.10 A comparison of $y = 3x$, $y = x^3$, and $y = 3^x$.

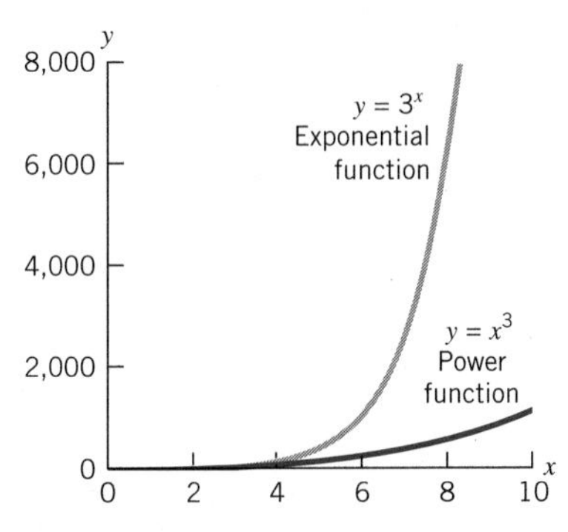

Figure 7.11 "Zooming out" on the graph.

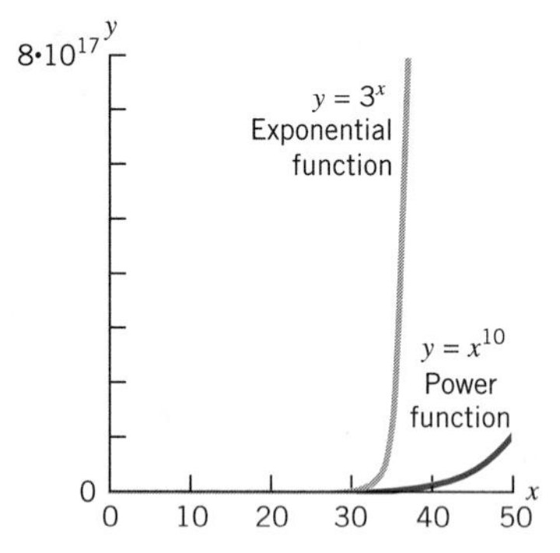

Figure 7.12 Graph of $y = x^{10}$ and $y = 3^x$.

Notice that now the scale on the x-axis goes from 0 to 10 (rather than just 0 to 4 as in Figure 7.10) and the y-axis extends to 8,000, which is still not large enough to show the value of 3^x once x is slightly greater than 8.

The exponential function $y = 3^x$ clearly dominates the power function $y = x^3$. The exponential function continues to grow so rapidly that its graph appears almost vertical relative to the graph of the power function.

What if we had picked a larger exponent for the power function? Would the exponential function still overtake the power function? The answer is yes. Let's compare, for instance, the graphs of $y = 3^x$ and $y = x^{10}$. If we zoomed in on the graph in Figure 7.12, we could see that for a while the graph of $y = 3^x$ lies below the graph of $y = x^{10}$. For example, when $x = 2$, then $3^2 < 2^{10}$. But eventually $3^x > x^{10}$. We see on the graph that somewhere after $x = 30$, the values for 3^x become substantially larger than the values for x^{10}.

In summary:

> Any exponential growth function will eventually dominate any power function.

Example 1

Construct a table using values of $x \geq 0$ for each of the following functions:

$$y = 2^x \quad \text{and} \quad y = x^2$$

Then plot the functions on the same grid and answer the following questions:

a. If $x > 0$, for what value(s) of x is $2^x = x^2$?

b. Does one function eventually dominate? If so, which one and after what value of x?

Solution

a. For positive values of x, if $x = 2$ or $x = 4$, then $2^x = x^2$.

b. In Table 7.5 and Figure 7.13, we can see that if $x > 4$, then $2^x > x^2$, so the function $y = 2^x$ will dominate the function $y = x^2$ after $x = 4$.

Table 7.5

x	$y = x^2$	$y = 2^x$
0	0	1
1	1	2
2	4	4
3	9	8
4	16	16
5	25	32
6	36	64

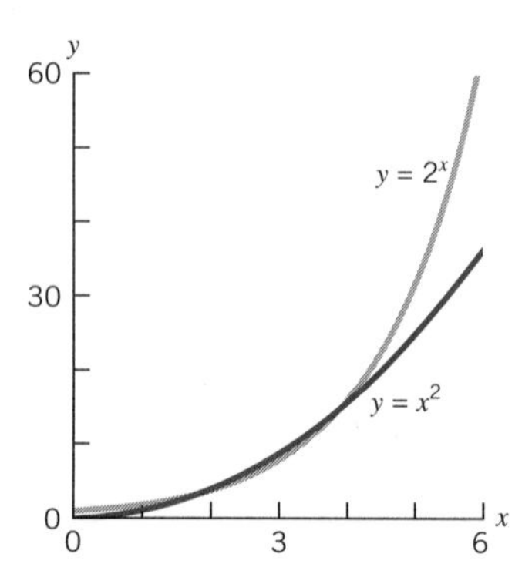

Figure 7.13 Comparison of $y = x^2$ and $y = 2^x$.

Algebra Aerobics 7.4

1. **a.** Construct a table using nonnegative values of x for each of the following functions. Then plot the functions on the same grid.

$$y = 4^x \quad \text{and} \quad y = x^3$$

b. Does one function eventually dominate? If so, which one and after approximately what value of x?

2. Which function eventually dominates?

a. $y = x^{10}$ or $y = 2^x$?

b. $y = (1.000\ 005)^x$ or $y = x^{1{,}000{,}000}$?

7.5 POWER FUNCTIONS WITH NEGATIVE INTEGER POWERS

Recall that the general form of an equation for a power function is

$$\text{dependent variable} = \text{constant} \cdot (\text{independent variable})^{\text{power}}$$

In Sections 7.1 through 7.4 we focused on functions where the power was a positive integer. We now consider power functions where the power is a negative integer.

Using the rules for negative exponents, we can rewrite power functions in the form

$$y = kx^{\text{negative power}}$$

where k is a constant, as

$$y = \frac{k}{x^{\text{positive power}}}$$

For example, $y = 3x^{-2}$ can be rewritten $y = 3/x^2$. In this form it is easier to make calculations and to see what happens to y as x increases or decreases in value.

In Section 7.1 we constructed functions $s(x) = 6x^2$ and $v(x) = x^3$ to describe the surface area and volume of a cube with length of edge x. We can extend our discussion about the ratio of surface area to volume by constructing a new function $r(x)$, where

$$\begin{aligned} r(x) &= \frac{\text{surface area}}{\text{volume}} \\ &= \frac{s(x)}{v(x)} \\ &= \frac{6x^2}{x^3} \\ &= \frac{6}{x} \text{ or } 6x^{-1} \end{aligned}$$

Then $r(x) = 6x^{-1}$ is an example of a power function where the power, -1, is a negative integer.

Table 7.6 and Figure 7.14 help us see that as x, the edge length, increases, $r(x)$, the ratio of surface area to volume, decreases.

Table 7.6

Edge Length x	$\frac{\text{Surface Area}}{\text{Volume}}$ $r(x) = 6/x$
1	6.0
2	3.0
3	2.0
4	1.5
5	1.2
6	1.0
10	0.6

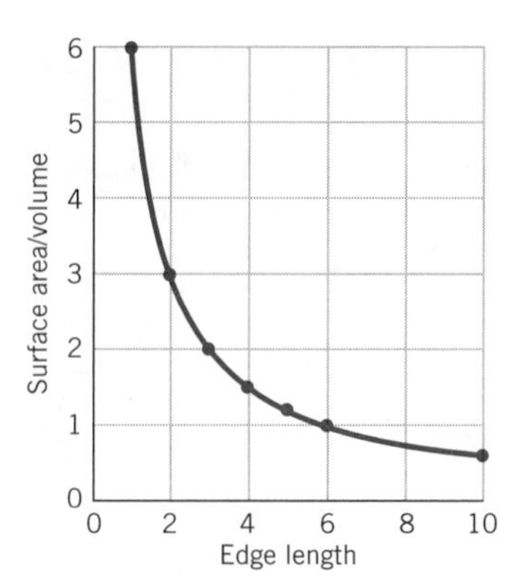

Figure 7.14 The graph of $r(x) = 6/x$ shows that the ratio of surface area/volume decreases as the length of edge, x, increases.

In Section 7.2 we noted that for any power function in the form

$$y = \text{constant} \cdot x^p$$

where p is positive, we say that y is *directly proportional* to x^p. For power functions in the form

$$y = \frac{\text{constant}}{x^p}$$

where p is positive, we say that y is *inversely proportional* to x^p. For example, if $y = 8/x^3$, we say that y is *inversely proportional to* x^3.

> Let p be a positive number. If
>
> $$y = kx^p$$
>
> we say that y is *directly proportional to* x^p. If
>
> $$y = \frac{k}{x^p} = kx^{-p}$$
>
> we say that y is *inversely proportional to* x^p.

Example 1

Write formulas to represent the following relationships:

a. Boyle's Law says that the volume, V, of a fixed quantity of gas is inversely proportional to the pressure, P, applied to it.

b. The force, F, keeping an electron in orbit is inversely proportional to the square of the distance, d, between the electron and the nucleus.

c. The acceleration, a, of an object is directly proportional to the force, F, applied upon the object and inversely proportional to the object's mass, m.

Solution

a. $V = \dfrac{k}{P}$ for some constant k

b. $F = \dfrac{k}{d^2}$ for some constant k

c. $a = \dfrac{kF}{m}$ for some constant k

Example 2

The Cardinal Rule of Scuba Diving

The most important rule in scuba diving is, "Never, ever, hold your breath." Why?

Solution

To answer this question, we need to examine the behavior of gases under pressure. Imagine filling up a balloon with a cubic foot of air and then pulling the balloon deeper and deeper under water. What would you expect to happen? As the balloon moves deeper, the pressure of the surrounding water increases, compressing the air in the balloon. The increase in depth, and the resulting increase in pressure, decreases the volume of air.

Pressure can be measured in units called *atmospheres.* One atmosphere (abbreviated 1 atm) equals 15 lb/in^2, which is the atmospheric pressure at Earth's surface at sea level. Starting with 1 atm at the surface, each additional 33 feet of water depth increases the pressure by 1 atm. So at 33 feet under water, the pressure is 2 atm, at 66 feet under water, 3 atm, and so on. Table 7.7 describes the relationships among depth, pressure, and volume. Figure 7.15 graphs volume vs. pressure.

The relationship between volume and pressure can be modeled by the function

$$V = \frac{1}{P} \quad \text{or equivalently} \quad V = P^{-1}$$

where V is the volume in cubic feet and P is the atmospheric pressure measured in atmospheres. This is a special case of Boyle's Law, where k, the constant of proportionality, is 1. The volume, V, is *inversely proportional* to the pressure, P. As the pressure increases, the volume decreases. Examine Table 7.7 and Figure 7.15. If we start with 1 ft^3 of air at 1 atm and double the pressure to 2 atm, the volume of air drops in half to $\frac{1}{2}$ ft^3. At 4 atm, the volume will drop to one-fourth of the original size (or to $\frac{1}{4}$ ft^3). Why does this matter to divers?

Suppose you are swimming in a pool, take a lung full of air at the surface, and then dive down to the bottom. As you descend, the buildup of pressure will decrease the volume of air in your lungs. When you ascend back to the surface, the volume of air in your lungs will expand back to its original size and everything is fine.

But when you are scuba diving, you are constantly breathing air that has been pressurized at the surrounding water pressure. If you are scuba diving 33 feet below the surface of the water, the surrounding water pressure is at 2 atm, twice that at the surface. What will happen then if you fill your lungs from your tank, hold your

Table 7.7
Pressure Versus Volume

Depth (ft)	Pressure (atm)	Volume (ft^3)
0	1	1
33	2	$\frac{1}{2}$
66	3	$\frac{1}{3}$
99	4	$\frac{1}{4}$

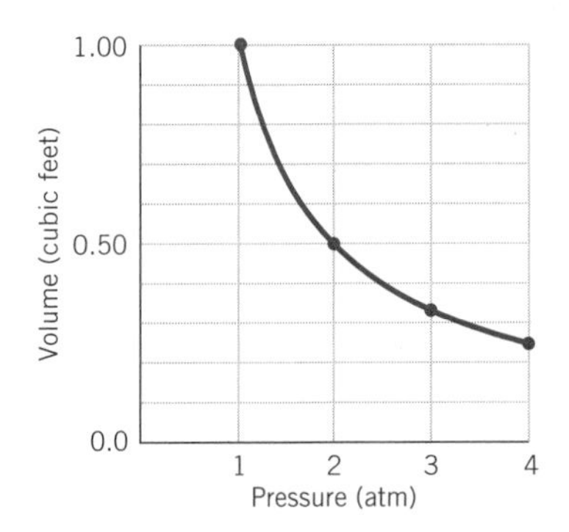

Figure 7.15 Graph of volume vs. pressure for a balloon descending under water.

breath, and ascend to the surface? When you reach the surface, the pressure will drop in half, from 2 atm down to 1 atm, so the volume of air in your lungs will double, rupturing your lungs! Hence the first rule of scuba diving: "Never, ever, hold your breath."

Algebra Aerobics 7.5a

1. Assume you are scuba diving at 99 feet below the surface of the water where the pressure is equal to 4 atm. If you use your tank to inflate a balloon with 1 ft^3 of compressed air, by how much will the volume have increased by the time it reaches the surface (assuming it doesn't burst)?
2. Fill in the following data table:

a.

x	$r(x) = 6/x$
0.01	
0.25	
0.50	
1.00	
2.00	
5.00	
10.00	

b. When $x > 0$, what happens to $r(x)$ as x increases in value?

c. When $x > 0$, what happens to $r(x)$ as x gets closer and closer to zero?

d. Is the function $r(x)$ defined for $x = 0$?

Inverse square laws. When the dependent variable is inversely proportional to the square of the independent variable, the functional relationship is called an *inverse square law.* Inverse square laws are quite common in the sciences. Some examples are shown below.

Example 3

Seeing the Light

The intensity of light, I, is inversely proportional to the square of the distance, d, between the light source and the observer.

a. Write an equation to describe the relationship between I and d.

b. What happens to the intensity of light if the distance between the observer and the source is doubled? Tripled?

Solution

a. $I = k/d^2$

b. If you are d feet away from a light source and the distance doubles to $2d$ feet, the intensity decreases from k/d^2 to

$$\frac{k}{(2d)^2} = \frac{k}{4d^2} = \frac{1}{4} \cdot \left(\frac{k}{d^2}\right)$$

So the intensity drops to one-fourth of the original intensity as the distance doubles.

If the distance triples from d to $3d$, the intensity decreases from k/d^2 to

$$\frac{k}{(3d)^2} = \frac{k}{9d^2} = \frac{1}{9} \cdot \left(\frac{k}{d^2}\right)$$

So the intensity drops to one-ninth of the original intensity as the distance triples.

For example, if you are reading a book that is 3 feet away from a lamp, and you move the book to 6 feet away (doubling the distance between the book and the light), the light will be one-fourth as intense. If you move the book from 3 to 9 feet away (tripling the distance), the light will be only one-ninth as intense. The reverse is also true; for example, if the book is 6 feet away and the light seems too dim for reading, by cutting the distance in half (to 3 feet), the illumination will be four times as intense.

Example 4

Gravitational Force between Objects

The gravitational force between you and Earth is inversely proportional to the square of the distance between you and the center of Earth.

a. Express this relationship as a power function.

b. What happens to the gravitational force as the distance between you and the center of Earth increases by a factor of 10?

Solution

a. The power function

$$F(d) = \frac{k}{d^2} = kd^{-2}$$

describes the gravitational force $F(d)$ between you and Earth in terms of a constant k times d^{-2}, where d is the distance between you and the center of Earth.

b. Given $\qquad F(d) = \dfrac{k}{d^2}$

if we increase d by a factor of 10 $\qquad F(10d) = \dfrac{k}{(10d)^2}$

apply exponent $\qquad = \dfrac{k}{100d^2}$

rearrange terms $\qquad = \dfrac{1}{100} \cdot \dfrac{k}{d^2}$

substitute $F(d)$ for k/d^2 $\qquad F(10d) = \dfrac{1}{100} \cdot F(d)$

The gravitational force $F(d)$ decreases to $\frac{1}{100}$ of its original value if d increases by a factor of 10.

Suppose you were an astronaut who started at the surface of Earth (roughly 4000 miles from Earth's center) and traveled to 40,000 miles from Earth's center (about $\frac{1}{7}$th of the way to the moon). The pull of Earth's gravity there would be $\frac{1}{100}$th of that on Earth's surface.

Example 5

Why Many Inverse Square Laws Work

Inverse square laws in physics often depend upon a power source and simple geometry. Imagine a single point as a source of power, emitting perhaps heat, sound, or light. We can think of the power radiating out from the point as passing through an

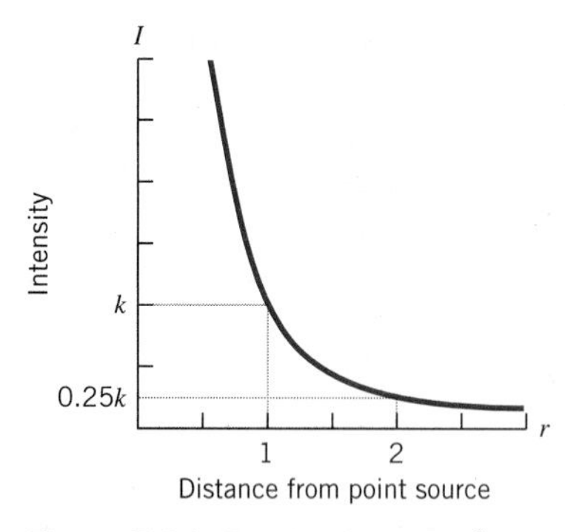

Figure 7.16 Graph of $I = k/r^2$, the relationship between intensity and distance from a point source.

infinite number of concentric spheres. The farther away you are from the point source, the lower the intensity of the power, since it is spread out over the surface area of a sphere that increases in size as you move away from the point source. Therefore the intensity, I, of the power you receive at any point on the sphere will be a function of the distance you are from the point source. The distance can be thought of as the radius, r, of a sphere with the point source at its center:

$$\text{intensity} = \frac{\text{power from a source}}{\text{surface area of sphere}}$$

$$I = \frac{\text{power}}{4\pi r^2}$$

$$= \frac{(\text{power}/4\pi)}{r^2}$$

If the power from the source is constant, we can simplify the expression by substituting a constant $k = \text{power}/(4\pi)$ and rewrite our equation as

$$I = \frac{k}{r^2}$$

The intensity you receive, I, is inversely proportional to the square of your distance from the source if the power from the point source is constant. If you double the distance, the intensity you receive is one-fourth of the original intensity. If you triple the distance, the intensity you receive is one-ninth of the original intensity. In particular, if $r = 1$, then $I = k$; if r doubles to become 2, then $I = k/4 = 0.25k$. The graph of $I = k/r^2$ is sketched in Figure 7.16.

Example 6

Direct and Inverse Proportionality in the Same Equation: The Relative Effect of the Moon and the Sun on Tides

The force that creates ocean tides on the surface of Earth varies inversely with the cube of the distance from Earth to any other large body in space and varies directly with the mass of the other body.

a. Construct an equation that describes this relationship.

b. The sun has a mass $2.7 \cdot 10^7$ times larger than the moon's, but the sun is about 390 times farther away from Earth than the moon. Would you expect the sun or the moon to have a greater effect on Earth's tides?

Solution

a. Using T for the tide-generating force, d for the distance between Earth and another body, and m for the mass of the other body, we have

$$T = \frac{km}{d^3} \quad \text{or} \quad T = kmd^{-3}$$

for some constant of proportionality k.

b. Now compare the tide generating force of the moon, T_{moon}, to the tide generating force of the sun, T_{sun}. If we let d = distance from Earth to the moon, then

for the moon $$T_{moon} = \frac{k \cdot \text{mass of the moon}}{d^3}$$

for the sun	$T_{\text{sun}} = \dfrac{k \cdot \text{mass of the sun}}{(390d)^3}$
Substitute for mass of sun	$T_{\text{sun}} = \dfrac{k \cdot \text{mass of the moon} \cdot 2.7 \cdot 10^7}{(390d)^3}$
apply exponents	$= \dfrac{k \cdot \text{mass of the moon} \cdot 2.7 \cdot 10^7}{390^3 d^3}$
rearrange terms	$= \dfrac{2.7 \cdot 10^7}{390^3} \cdot \dfrac{k \cdot \text{mass of the moon}}{d^3}$
substitute T_{moon}	$T_{\text{sun}} = \dfrac{2.7 \cdot 10^7}{390^3} \cdot T_{\text{moon}}$

We can further simplify the coefficient.

use scientific notation	$T_{\text{sun}} = \dfrac{2.7 \cdot 10^7}{(3.9 \cdot 10^2)^3} \cdot T_{\text{moon}}$
apply exponent and round	$\approx \dfrac{2.7 \cdot 10^7}{60 \cdot 10^6} \cdot T_{\text{moon}}$
use law of exponents	$\approx \dfrac{27}{60} \cdot T_{\text{moon}}$
so	$T_{\text{sun}} \approx \dfrac{9}{20} T_{\text{moon}}$

Thus, despite the fact that the sun is more massive than the moon, because it is much farther away from Earth than the moon, its effect on Earth's tides is about one-half that of the moon's.

Algebra Aerobics 7.5b

1. a. Rewrite each of the following expressions using positive exponents:

i. $15x^{-3}$ **ii.** $-10x^{-4}$ **iii.** $3.6x^{-1}$

b. Rewrite each of the following expressions using negative exponents:

i. $\dfrac{1.5}{x^2}$ **ii.** $-\dfrac{6}{x^3}$ **iii.** $-\dfrac{2}{3x^2}$

2. The time in seconds, t, needed to fill a tank with water is inversely proportional to the square of the diameter, d, of the pipe delivering the water. Write an equation describing this relationship.

3. A light is 4 feet above the book you are reading. The light seems too dim, so you move the light 2 feet closer to the book. What is the change in light intensity?

4. If $g(x) = 3/x^4$, what happens to $g(x)$ when:

a. x doubles? **b.** x is divided by 2?

5. a. Fill in the following data table:

x	$f(x) = 1/x^2$	$g(x) = 1/x^3$
0.01		
0.25		
0.50		
1.00		
2.00		
10.00		

b. Are the abstract functions $f(x) = 1/x^2$ and $g(x) = 1/x^3$ defined when $x = 0$? Are they defined for negative values of x?

c. Describe what happens to the functions $f(x)$ and $g(x)$ as x approaches positive infinity. What happens to $f(x)$ and $g(x)$ when $0 < x < 1$?

7.6 VISUALIZING NEGATIVE INTEGER POWER FUNCTIONS

Moving to the Abstract

All the graphs of power functions with negative exponents we've seen so far were restricted to the first quadrant since they represented physical phenomena.

Let's take a close look at two abstract power functions with negative exponents, $f(x) = x^{-1}$ and $g(x) = x^{-2}$, where x can now assume negative values. If we rewrite these functions as $f(x) = 1/x$ and $g(x) = 1/x^2$, it is clear that the domain cannot include zero since these functions are undefined when $x = 0$. Two important questions are: "What happens to these functions when x is close to 0?" and "What happens as x approaches $+\infty$ or $-\infty$?"

What happens when $x > 0$ and approaches $+\infty$? From Table 7.8 and Figure 7.17, when x is positive and increasing, both $f(x)$ and $g(x)$ *decrease.* For both functions, as $x \to +\infty$, $1/x$ and $1/x^2$ grow smaller and smaller, approaching, but never reaching, zero. Both graphs are asymptotic to the x-axis.

Table 7.8
$x \geq 0$

x	$f(x) = 1/x$	$g(x) = 1/x^2$
0	Undefined	Undefined
$\frac{1}{100}$	100	10,000
$\frac{1}{4}$	4	16
$\frac{1}{3}$	3	9
$\frac{1}{2}$	2	4
1	1	1
2	$\frac{1}{2}$	$\frac{1}{4}$
3	$\frac{1}{3}$	$\frac{1}{9}$
4	$\frac{1}{4}$	$\frac{1}{16}$
100	$\frac{1}{100}$	$\frac{1}{10,000}$

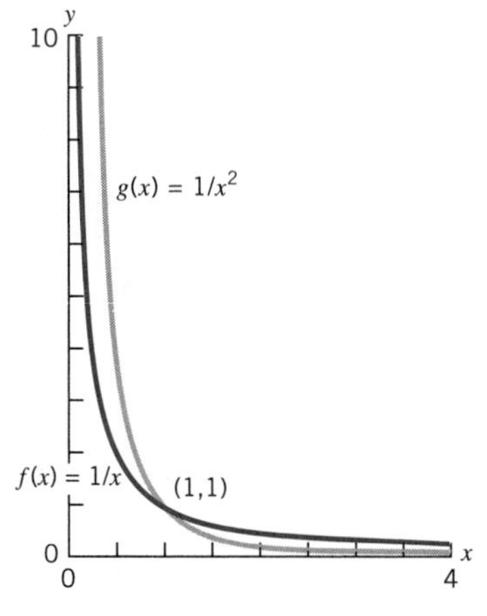

Figure 7.17 Graphs of $f(x) = 1/x$ and $g(x) = 1/x^2$ when $x > 0$.

What happens when $0 < x < 1$? We can see in Table 7.8 and Figure 7.17 that when x is positive but grows smaller and smaller, approaching zero, the values for $f(x) = 1/x$ and $g(x) = 1/x^2$ get larger and larger. Let's examine what happens when we use fractions between 0 and 1 to calculate values for $f(x) = 1/x$ and $g(x) = 1/x^2$:

$$f\left(\frac{1}{4}\right) = \frac{1}{1/4} = 1 \div \frac{1}{4} = 1 \cdot \frac{4}{1} = 4$$

$$f\left(\frac{1}{100}\right) = \frac{1}{1/100} = 1 \div \frac{1}{100} = 1 \cdot \frac{100}{1} = 100$$

$$g\left(\frac{1}{4}\right) = \frac{1}{(1/4)^2} = 1 \div \frac{1}{16} = 1 \cdot \frac{16}{1} = 16$$

$$g\left(\frac{1}{100}\right) = \frac{1}{(1/100)^2} = 1 \div \frac{1}{10,000} = 1 \cdot \frac{10,000}{1} = 10,000$$

If x is positive and approaches zero, then $f(x) = 1/x$ and $g(x) = 1/x^2$ approach positive infinity. So both graphs are asymptotic to the y-axis.

What happens when $x < 0$? When x is negative, the values of $f(x) = 1/x$ will also be negative. So when $x < 0$ (Table 7.9), the graph of $f(x)$ will lie below the x-axis. As $x \to 0$ through negative values, $1/x \to -\infty$. The function $g(x) = 1/x^2$ is positive for all values of x, so the graph of $g(x)$ will always lie above the x-axis. Whether x is positive or negative, as $x \to 0$, $1/x^2 \to +\infty$. As $x \to -\infty$, both $f(x)$ and $g(x)$ approach zero.

Table 7.9
$x < 0$

x	$f(x) = 1/x$	$g(x) = 1/x^2$
-100	$-\frac{1}{100}$	$\frac{1}{10{,}000}$
-4	$-\frac{1}{4}$	$\frac{1}{16}$
-3	$-\frac{1}{3}$	$\frac{1}{9}$
-2	$-\frac{1}{2}$	$\frac{1}{4}$
-1	-1	1
$-\frac{1}{2}$	-2	4
$-\frac{1}{3}$	-3	9
$-\frac{1}{4}$	-4	16
$-\frac{1}{100}$	-100	$10{,}000$

The graphs of $f(x)$ and $g(x)$ in Figure 7.18 exhibit very different shapes for negative values of x. The graphs of these functions approach the x-axis and the y-axis but never reach them. Thus, both curves are asymptotic to both axes.

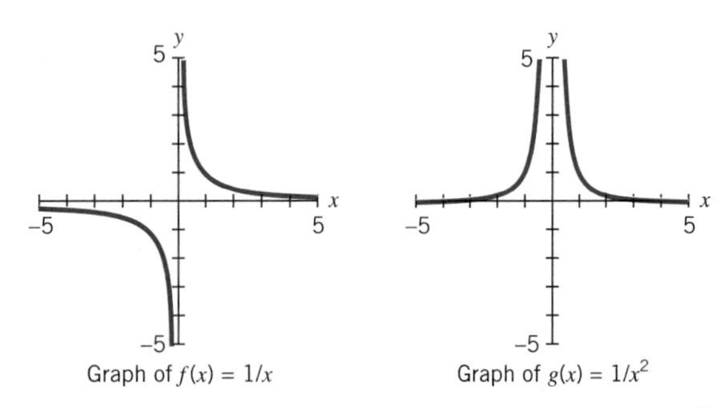

Figure 7.18 The graphs of $f(x) = 1/x$ and $g(x) = 1/x^2$.

Odd vs. Even Powers

If we examine the graphs of the simplest negative power functions $y = x^{-1}$, $y = x^{-2}$, $y = x^{-3}$, $y = x^{-4}$, $y = x^{-5}$, and $y = x^{-6}$ (Figure 7.19), we see that the negative power functions, like the positive power functions, fall into two groups: the odd powers and the even powers.

For both the even and odd negative powers, the functions are not defined when $x = 0$; hence, the domains never include 0. As $x \to 0$, either $y \to +\infty$ or $y \to -\infty$.

In Exploration 7.3 you examine the effect of k on the graphs of negative power functions.

The graphs of these functions approach the y-axis but never reach it. Thus, each curve has the y-axis as a *vertical asymptote.*

For both even and odd negative powers, y is never equal to zero. As $x \to +\infty$ or $x \to -\infty$, the values of y come closer and closer to zero but never reach zero. The graphs of these functions approach the x-axis but never touch it. Thus each curve has the x-axis as a *horizontal asymptote.*

All of the odd negative powers have rotational symmetry about the origin. If you hold the graph fixed at the origin and rotate it 180°, you end up with the same image.

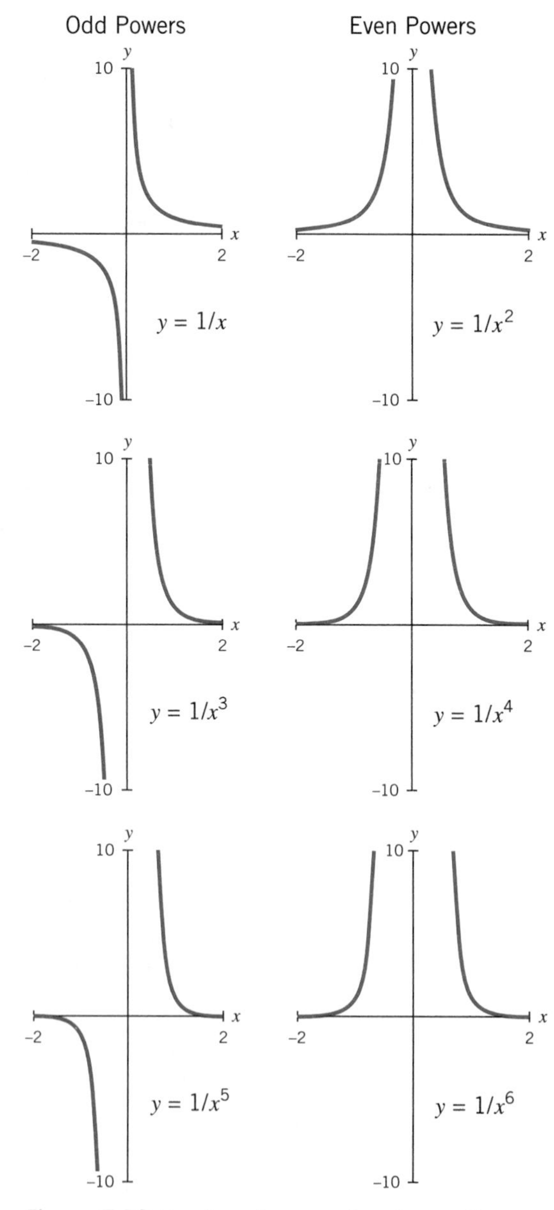

Figure 7.19 Graphs of power functions with even and odd powers.

All of the even negative power functions are symmetrical about the y-axis. If you think of the y-axis as the dividing line, the "left" side of the graph is a mirror image, or reflection, of the "right" side.

Algebra Aerobics 7.6

1. Rewrite each of the following expressions with positive exponents and then evaluate each expression for $x = -2$:

   a. x^{-2} **b.** x^{-3} **c.** $4x^{-3}$ **d.** $-4x^{-3}$

2. Sketch the graph for each equation:

   a. $y = x^{-10}$ **b.** $y = x^{-11}$

3. On the same grid, graph the following equations:

   a. $y = x^{-2}$ and $y = x^{-3}$

   Where do these graphs intersect? How are the graphs similar and how are they different?

   b. $y = 4x^{-2}$ and $y = 4x^{-3}$

   Where do these graphs intersect? How are the graphs similar and how are they different?

7.7 USING LOGARITHMIC SCALES TO FIND THE BEST FUNCTION MODEL

Throughout this course, we have examined several different families of functions, including linear, exponential, logarithmic, and power. In this section we address the question: "How can we determine a reasonable functional model for a given set of data?" The simple answer is: "Look for a way to plot the data so it appears as a straight line."

We look for straight-line representations not because the world is intrinsically linear, but because straight lines are easy to recognize and manipulate. In Chapter 2, we considered linear functions, whose graphs are straight lines when plotted on a "standard" plot, with linear scales on both axes. In Chapter 5, we saw how the graph of an exponential function appeared as a straight line when plotted on a *semi-log plot* using a logarithmic scale on the vertical axis and a standard linear scale on the horizontal axis. Here we shall see that the graph of a power function appears as a straight line when plotted using a logarithmic scale on both axes. We call such a graph a *log-log plot.*

The course software "E11: Semi-log Plots of $y = Ca^x$" in *Exponential & Log Functions* and "P2: Log-log Plots of Power Functions" in *Power Functions* can help you visualize the ideas in this section.

Let's examine the following functions:

$$y = 3 + 2x \quad \text{linear function}$$

$$y = 3 \cdot 2^x \quad \text{exponential function}$$

$$y = 3x^2 \quad \text{power function}$$

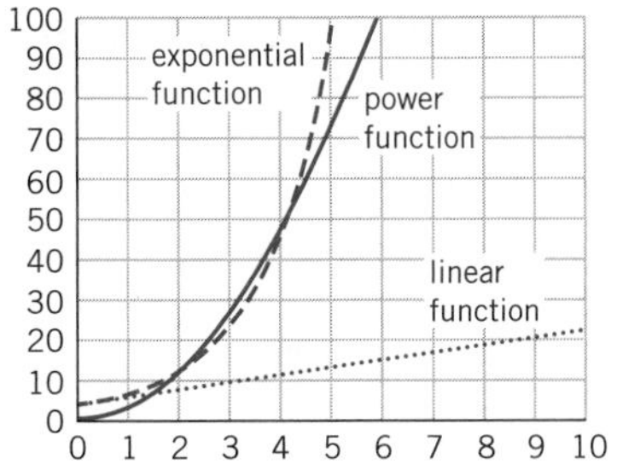

In Figure 7.20, we plot the three functions using a linear scale on both the x- and y-axes. Only the linear function appears as a straight line; the power and exponential functions curve steeply upward.

Figure 7.20 Graph of a linear, a power, and an exponential function on a "standard plot," with linear scales on both axes.

Semi-log Plots

On the semi-log plot in Figure 7.21, we see that the exponential function now is a straight line, and the power and linear functions curve downward. From Chapter 4, we know that moving a fixed distance on a logarithmic scale corresponds to *multiplying* by a constant factor. In exponential growth, the dependent variable is multiplied by the growth factor each time 1 unit is added to the independent variable.

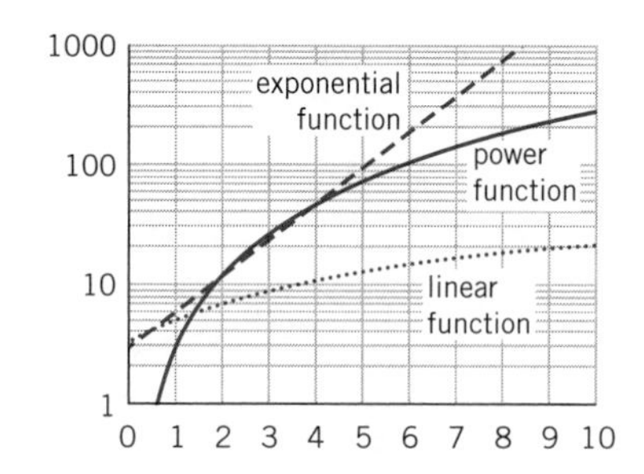

Figure 7.21 Graph of a linear, a power, and an exponential function on a semi-log plot, with a logarithmic scale on the vertical and linear scale on the horizontal axis.

Thus, as we saw in Chapter 5, exponential growth will always appear as a straight line on a *semi-log* plot, where the independent variable is plotted on a linear axis and the dependent variable on a logarithmic axis. This is one of the easiest and most reliable ways to recognize exponential growth in a data set.

A logarithmic scale has the added advantage of allowing us to display clearly a wide range of values. In Figure 7.20 the vertical axis only goes to 100 units, whereas in Figure 7.21, the vertical axis extends to 1000 units.

The properties of logarithms help us understand why exponential functions appear as straight lines on a semi-log plot. Let's examine our exponential function:

$$y = 3 \cdot 2^x \quad (1)$$

Take the logarithm of both sides $\quad \log y = \log(3 \cdot 2^x)$

use Property 1 of logs $\quad = \log 3 + \log 2^x$

use Property 3 of logs $\quad \log y = \log 3 + x(\log 2) \quad (2)$

Equation (2) is an equivalent form of Equation (1). If we use a calculator to evaluate log 3 and log 2, we can rewrite Equation (2) as

$$\log y \approx 0.48 + 0.30x$$

Representing $\log y$ as Y, we can rewrite this equation as

$$Y \approx 0.48 + 0.30x \quad (3)$$

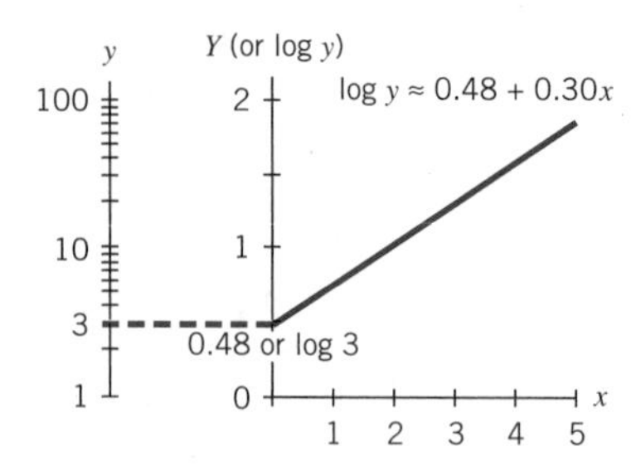

Figure 7.22 Comparing y on a logarithmic scale to $\log y$ on a linear scale.

It's easier then to see that Y (or $\log y$) depends linearly on x. On a standard linear plot, the graph of the points (x, Y) [or equivalently $(x, \log y)$] that satisfy Equation (3) is a straight line. Its slope is 0.30, or log 2, the logarithm of the growth factor of the original exponential function, Equation (1). (See Figure 7.22.)

Figure 7.22 also shows the direct translation between plotting y on a logarithmic scale vs. plotting $\log y$ on a linear scale. Many graphing calculators do not have the ability to switch between using linear and logarithmic scales, but one can in effect change the scale on the vertical axis from linear to logarithmic by plotting $\log y$ instead of y. Notice that on the $\log y$ axis, the units are now evenly spaced, and we can use the standard strategies for finding the slope and vertical intercept.

Using Figure 7.22, we can find the slope and the vertical intercept. When $x = 0$, the vertical intercept on the Y (or $\log y$) axis is log 3, which is about 0.48. This is the value of Y in the equation $Y \approx 0.48 + 0.30x$ when $x = 0$. On the y-axis the vertical intercept is 3, which is the value of y in the original equation $y = 3 \cdot 2^x$ when $x = 0$.

The slope of the line $Y \approx 0.48 + 0.30x$ is log 2, which is about 0.30. Thinking in terms of the x and Y axes, each time we add 1 unit to x, we must add log 2 (≈ 0.30) units to Y in order to stay on the line.

If we think in terms of the x- and y-axes, adding 1 unit to x means multiplying y by 2 in order to return to the line. So adding 1 to x translates to *adding* log 2 to $\log y$ in Equation (2) or *multiplying* y by 2 in Equation (1), where $y = 3 \cdot 2^x$. These translations make sense if we remember that logarithms are exponents.

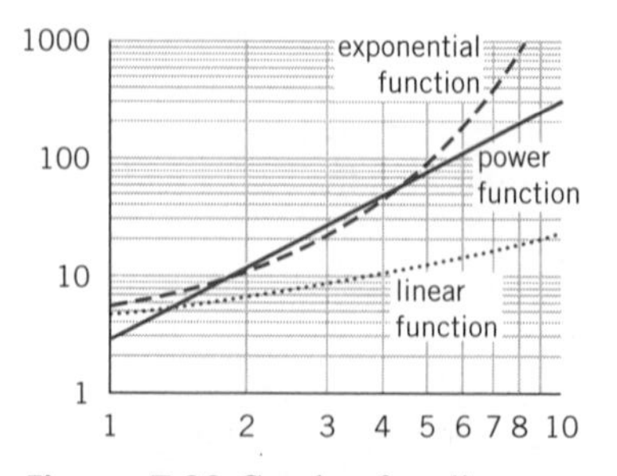

Figure 7.23 Graph of a linear, a power, and an exponential function on a log-log plot.

Log-log Plots

On the log-log plot in Figure 7.23, the power function now appears as a straight line.

Again we can use the properties of logarithms to understand why power functions appear as straight lines on a log-log plot. Let's analyze our power function:

$$y = 3x^2 \tag{4}$$

Take the logarithm of both sides	$\log y = \log(3x^2)$
use Property 1 of logs	$= \log 3 + \log x^2$
use Property 3 of logs	$\log y = \log 3 + 2 \log x \quad (5)$

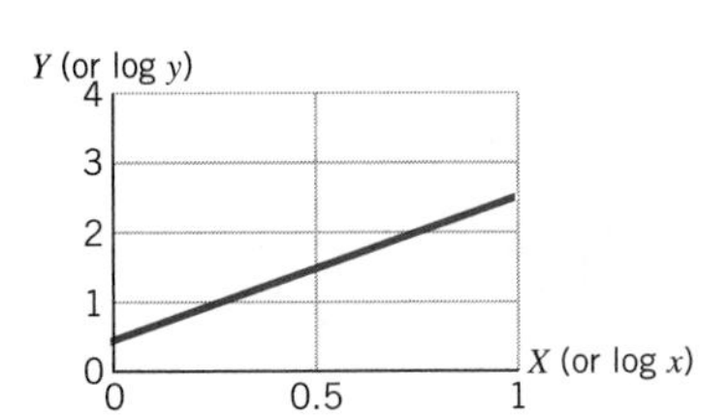

Figure 7.24 The graph of $Y \approx 0.48 + 2X$.

Equations (4) and (5) are equivalent ways of saying the same thing. If we evaluate log 3, we get $\log y \approx 0.48 + 2 \log x$. By letting $X = \log x$ and $Y = \log y$, we can rewrite the equation as

$$Y \approx 0.48 + 2X \tag{6}$$

This helps us see that Y depends linearly on X, or equivalently, that $\log y$ depends linearly on $\log x$. If we plot Y vs. X (or equivalently $\log y$ vs. $\log x$) the slope, 2, of the straight line described by Equation (6) and graphed in Figure 7.24 is the exponent of the original power function (4).

Since the slope is 2, each time X increases by 1 unit, Y increases by 2 units. Equivalently, when $\log x$ increases by 1 unit, $\log y$ increases by 2 units. Each time we increase $\log x$ by 1, we increase x by a factor of 10. In Equation (4) each time x increases by a factor of 10, y increases by a factor of 10^2.

The linear function looks almost like a straight line on our log-log plot in Figure 7.23. This is perhaps not surprising, since if the constant term were 0, the linear function would be a power function $y = mx$, where the power of x is 1.

Conclusion

We can tell the difference among the graphs of these three types of functions by plotting them on different types of axes. To summarize:

On what type of plot would a logarithmic function appear as a straight line?

> The graph of a linear function appears as a straight line on a standard plot.
>
> The graph of an exponential function appears as a straight line on a semi-log plot. The slope of the line is the logarithm of the growth factor.
>
> The graph of a power function appears as a straight line on a log-log plot. The slope of the line is the exponent of the power function.

Algebra Aerobics 7.7a

1. Identify the type of function for the following:
   a. $y = 4x^3$ **b.** $y = 3x + 4$ **c.** $y = 4 \cdot 3^x$
2. Which of the functions in Problem 1 would have a straight line graph on a standard linear plot? On a semi-log plot? On a log-log plot?

   If possible, use technology to check your previous answers. Remember that if you cannot switch from a linear to a logarithmic scale on an axis, then plot the log of the number instead. For example, plotting $(x, \log y)$ will produce a graph equivalent to a semi-log plot. Plotting $(\log x, \log y)$ will produce a graph equivalent to a log-log plot.
3. For each straight line graph in Problem 2, use the original equation in Problem 1 to predict the slope of the line.

Using Semi-log and Log-log Plots to Investigate Data

Analyzing Weight and Height Data

In 1938, Katherine Simmons and T. Wingate Todd measured the average weight and height of children in Ohio between the ages of 3 months and 13 years. Table 7.10 shows their data.

DATA SETS DATA SETS

See Excel or graph link file CHILDSTA.

Table 7.10
Measuring Children

	Weight (kg)		Height (cm)	
Age (yr)	Boys	Girls	Boys	Girls
$\frac{1}{4}$	6.5	5.9	61.3	59.3
1	10.8	9.9	76.1	74.2
2	13.2	12.5	87.4	86.2
3	15.2	14.7	96.2	95.5
4	17.4	16.8	103.9	103.2
5	19.6	19.2	110.9	110.3
6	22.0	22.0	117.2	117.4
7	24.8	24.5	123.9	123.2
8	28.2	27.9	130.1	129.3
9	31.5	32.1	136.0	135.7
10	35.6	35.2	141.4	140.8
11	39.2	39.5	146.5	147.8
12	42.0	46.6	151.1	155.3
13	46.6	52.0	156.7	159.9

Source: Data adapted from D. W. Thompson, *On Growth and Form* (New York: Dover Paperback, 1992), p. 105.

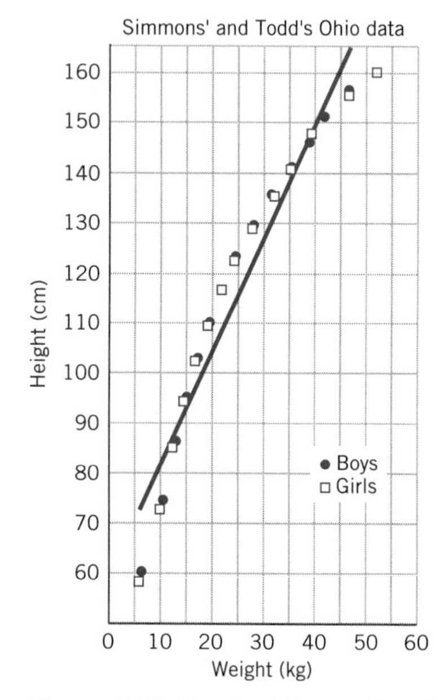

Figure 7.25 Standard linear plot of Simmons' and Todd's height vs. weight data for boys and girls.

Let's examine the relationship between height and weight. Figure 7.25 shows height vs. weight for boys and for girls together on a linear scale. The filled circles are data for boys, the open squares for girls. Table 7.10 shows that, in general, baby girls are smaller than baby boys and teenage girls are larger than teenage boys. But by superimposing the height/weight data for boys and girls in Figure 7.25, we see that at the same weight the heights of boys and girls are roughly the same.

The line in Figure 7.25 shows the linear model that best approximates the combined data. The line does not describe the data very well, and it is not reasonable to consider that height is a linear function of weight for growing children. Other models may describe the data better.

Figure 7.26 shows the same data, this time using a linear scale for weight and a logarithmic scale for height. The line shows the best model of the form

$$\log H = b + mW \tag{1}$$

where H is the height (in centimeters) and W is the weight (in kilograms). To solve for H, we use each side of Equation (1) as the exponent for the base 10. We get:

$$10^{\log H} = 10^{b+mW}$$

Use Property 6 of logs $\quad H = 10^{b+mW}$

Properties 1 and 3 of exponents $\quad = 10^b \cdot (10^m)^W$

So Equation (1) is equivalent to the exponential model

$$H = Ca^W \quad \text{where} \quad C = 10^b \quad \text{and} \quad a = 10^m$$

On this semi-log plot, the exponential model appears as a straight line. Clearly, it does not fit the data well, and it is not reasonable to suppose that height is an exponential function of weight.

Why is there no zero on an axis with a logarithmic scale?

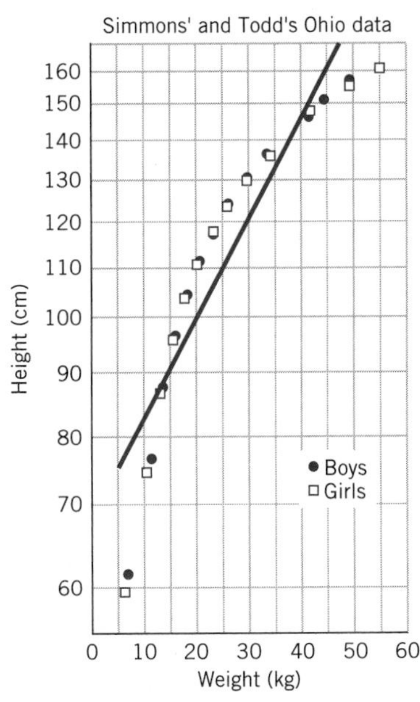

Figure 7.26 Semi-log plot of Simmons' and Todd's height vs. weight data for boys and girls.

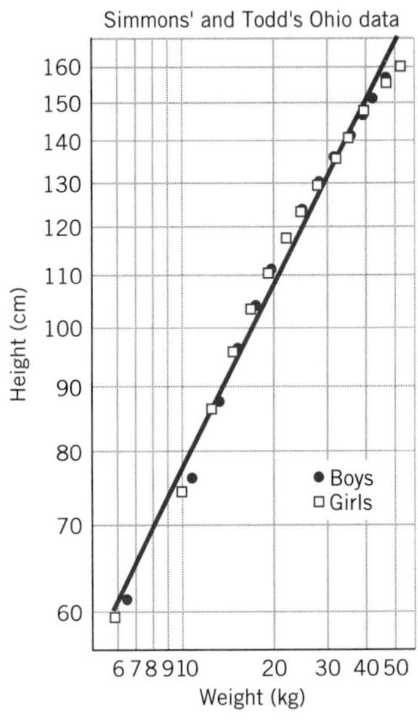

Figure 7.27 Log-log plot of Simmons' and Todd's height vs. weight data.

Figure 7.27 again shows the same data, but this time using logarithmic scales on both axes. The line shows the best model of the form

$$\log H = b + m \log W \tag{2}$$

If we use each side of the equation as an exponent for the base 10, we get

$$10^{\log H} = 10^{b+m \log W}$$

Use Property 6 of logs $\quad H = 10^{b+m\log W}$

Property 1 of exponents $\quad = 10^b \cdot 10^{m \log W}$

Property 3 of logs $\quad = 10^b \cdot 10^{\log W^m}$

Property 6 of logs $\quad H = 10^b \cdot W^m$

If we let $C = 10^b$, we get the following power model, which is equivalent to Equation (2):

$$H = C \cdot W^m \tag{3}$$

We can find the slope of the regression line in Figure 7.27 by translating the units on the axes into log(weight) and log(height). We would then have evenly spaced units and could use the standard strategies (see Figure 7.22). For example, measuring the axes in units of weight and height, we can estimate the coordinates for two points on the regression line as (8, 70) and (40, 150). If we instead measure the axes in units of log(weight) and log(height), the corresponding coordinates are (log 8, log 70) and (log 40, log 150). So the slope of the regression line, the value for m in the equivalent Equations (2) and (3), is approximately:

$$\frac{\log 150 - \log 70}{\log 40 - \log 8} = \frac{\log (150/70)}{\log (40/8)} \approx \frac{\log 2.143}{\log 5} \approx \frac{0.331}{0.699} \approx 0.47$$

Thus

$$H \approx C \cdot W^{0.47}$$

In this case, the line is a reasonably good approximation to the data and certainly the best fit of the three alternatives. It is also plausible to argue that height is a power law function of weight for growing children. Approximating 0.47 by 0.5 since the data are fairly rough, we may argue that

$$H \propto W^{1/2} \quad \text{or} \quad W \propto H^2$$

where the symbol $\propto$ means "is directly proportional to."

Let's think about whether this is a reasonable exponent. Suppose children grew *self-similarly;* that is, they kept the same shape as they grew from the age of 3 months to 13 years. Then, as we have discussed in Section 7.1, their volume, and hence their weight, would be proportional to the cube of their height:

$$W = kH^3 \quad \text{(self-similar growth)}$$

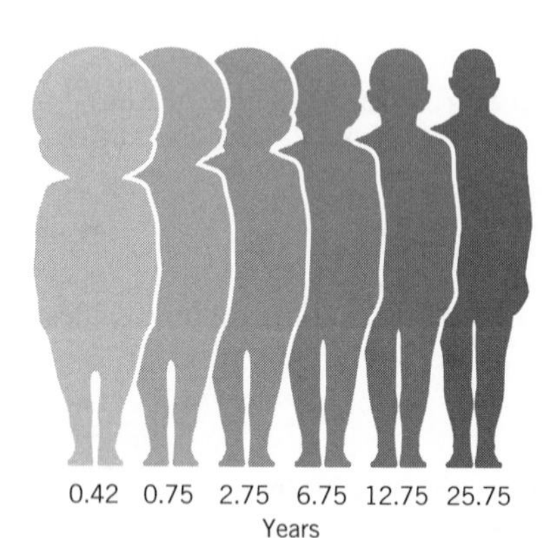

Figure 7.28 The change in human body shape with increasing age. *Source:* Medawar, P. B., 1945. "Size and Shape in Age" in *Essays on Growth and Form Presented to D'Arcy Thompson* (W. E. LeGros Clark and P. B. Medawar, eds.) pp. 157–187. Oxford: Clarendon Press.

But of course children do not grow self-similarly; they become proportionately more slender as they grow from babies to young adults. (See Figure 7.28.) Their weight therefore grows less rapidly than would be predicted by self-similar growth, and we expect an exponent less than 3 for weight as a function of height.

On the other hand, suppose that children's bodies grew no wider as their height increased. Since weight is proportional to height times cross-sectional area, with constant cross-sectional area we would expect weight to increase linearly with height. So $W \propto H^1$. But certainly, children do become wider as they grow taller, so we expect the exponent of H to be greater than 1. Therefore the experimentally determined exponent of close to 2 seems reasonable.

There is also a fourth possibility for plotting our height–weight data: We may use a linear scale for the height axis and a logarithmic scale for the weight axis. On such a plot, a straight line will represent a model in which weight is an exponential function of height.

If we do this plot, the data fall very close to a straight line. Based only on the data, we might argue that an exponential function is a reasonable model. However, it is very difficult to construct a plausible physical explanation of such a model, and it would not help us understand how children grow. It may be, of course, that this relationship reveals some unsuspected physical or biological law. Perhaps you, the reader, will be the one to find some previously unsuspected explanation.

Algebra Aerobics 7.7b

1. Interpret the slopes of 1.2 and 1.0 on the accompanying graph in terms of arm length and body height. Note: Slopes refer to the slopes of the lines where the units of the axes are translated to log (body height) and log (arm length).

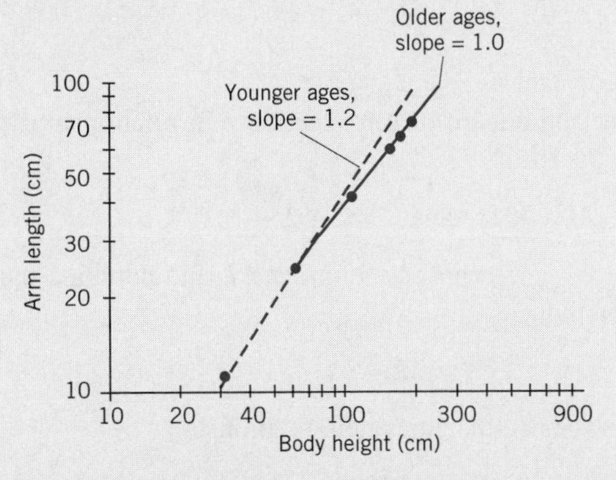

Source: T. A. McMahon and J. T. Bonner, *On Size and Life* (New York: Scientific American Books, 1983), p. 32.

2. For each of the accompanying graphs examine the scales on the axes. Then decide whether a linear, exponential, or power function would be the most appropriate model for the data.

a.

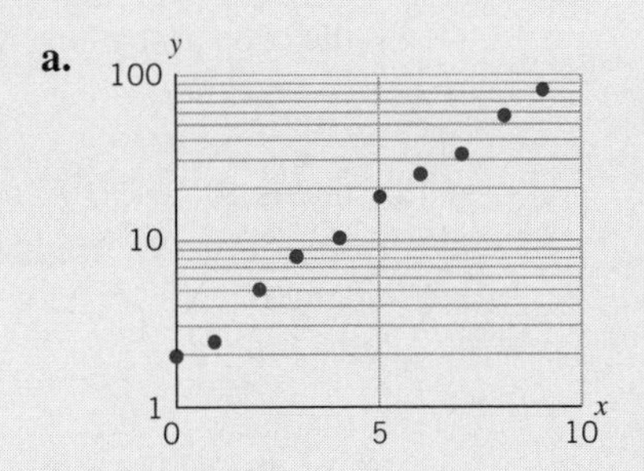

b.

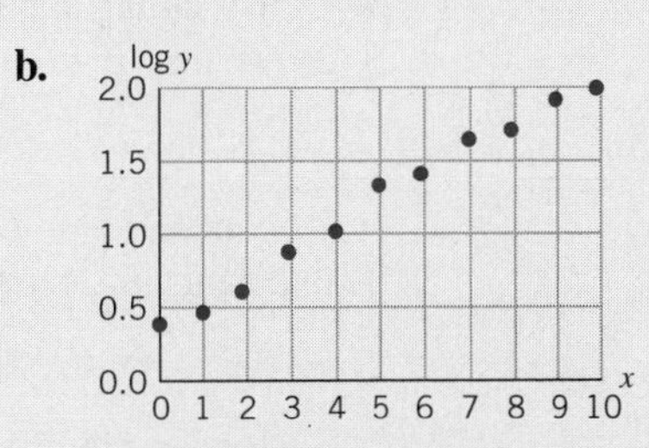

c.

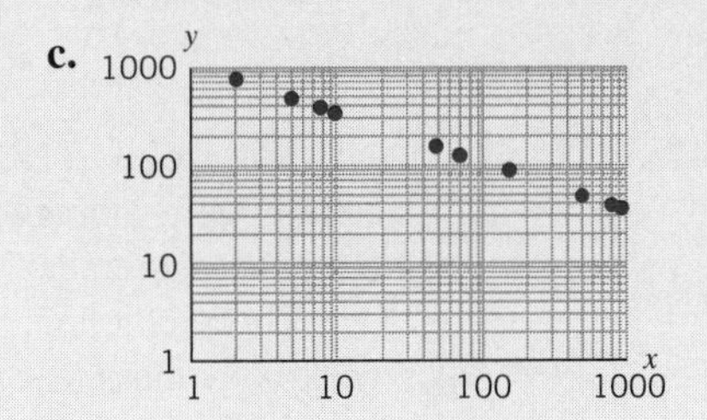

7.8 ALLOMETRY: THE EFFECT OF SCALE

Allometry is the study of how the relationships among different physical attributes of an object change with scale. In biology, allometric studies naturally focus on relationships in living organisms. Biologists look for general "laws" that can describe the relationship, for instance, between height and weight (as in Section 7.7) or surface area and volume (as in Section 7.1) as the organism size increases. The relationships may be within one species or across species. These laws are characteristically power functions. Sometimes we can predict the exponent by simple reasoning about physical properties and verify our prediction by examining the data; other times we can measure the exponent from data but cannot yet give a simple explanation for the observed value.

Surface Area vs. Body Mass

In Section 7.1 we made the argument that larger animals have relatively less surface area than smaller ones. Let's see if we can describe this relationship as a power law and then look at some real data to see if our conclusion was reasonable.

Since all animals have roughly the same mass per unit volume, their mass should be directly proportional to their volume. We will substitute mass for volume in our discussion since we can determine mass easily by weighing an animal. Since volume is measured in cubical units of some length, if animals of different sizes have roughly the same shape, then we would expect mass, M, to be proportional to the cube of the length, L, of the animal,

$$M \propto L^3 \tag{1}$$

that is, $M = k_1 \cdot L^3$ for some constant k_1.

Also, we would expect surface area, S, to be proportional to the square of the length,

$$S \propto L^2 \tag{2}$$

that is, $M = k_2 L^2$ for some constant k_2.

By taking the cube root and square root respectively in relationships (1) and (2), we may rewrite them in the form

$$M^{1/3} \propto L \quad \text{and} \quad S^{1/2} \propto L$$

Since $M^{1/3} \propto L$ and $L \propto S^{1/2}$, we can now eliminate L and combine these two statements into the prediction that

$$S^{1/2} \propto M^{1/3}$$

By squaring both sides, we have the equivalent relationship

$$S \propto M^{2/3}$$

It is useful to eliminate length as a measure of the size of an animal, since it is a little more ambiguous than mass. Should a tail, for instance, be included in the length measurement?

This prediction, that surface area is directly proportional to the two-thirds power of body mass, can be tested experimentally. Figure 7.29 shows data collected for a wide range of mammals, from mice with a body mass on the order of 1 gram to elephants, whose body mass is more than a million grams (1 metric ton). Here surface area, S, measured in square centimeters, is plotted on the vertical axis and body mass, M, in grams, is on the horizontal. The scales on both axes are logarithmic.

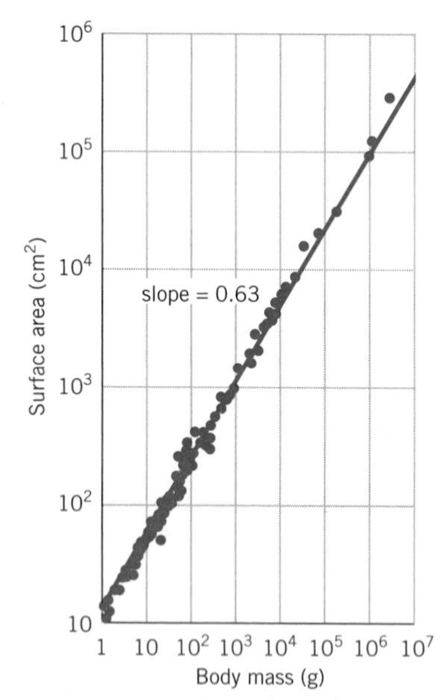

Figure 7.29 A log-log plot of body surface vs. body mass for a wide range of mammals.
Source: McMahon, T. A., 1973. "Size and Shape in Biology," *Science*, 179: pp. 1201–1204.

The relationship looks quite linear. Since this is a log-log plot, that implies that S is directly proportional to M^p for some power p. The best fit line has a slope of approximately $\frac{2}{3}$, which implies that

$$S \propto M^{2/3}$$

confirming our simple prediction of a two-thirds power law. We have approximate verification of the fact that mammals of different body masses have roughly the same proportion of surface area to the two-thirds power of body mass. This power function is just another way of describing our finding in Section 7.1 on the relationship between surface area and volume.

To see that the slope of the line corresponds to the exponent $\frac{2}{3}$, notice that at the lower left corner it passes through the point whose coordinates are (1, 10). In the middle of the graph, it passes through the point whose coordinates are $(10^3, 10^3)$, and near the upper right corner it passes through the point with coordinates $(10^6, 10^5)$. That is, every time the horizontal coordinate increases by a factor of 10^3, the vertical coordinate increases by a factor of 10^2. The line is described by the equation

$$S = 10M^{2/3} \tag{3}$$

where area S is measured in square centimeters and mass M in grams.

Metabolic Rate vs. Body Mass

Another example of scaling among animals of widely different sizes is metabolic rate as a function of body mass. This relationship is very important in understanding the mechanisms of energy production in biology. Figure 7.30 shows a log-log plot of metabolic heat production, H, in kilocalories per day, versus body mass, M, in kilograms for a range of land mammals.

The graph again looks quite linear, so we would expect H to be directly proportional to a power of M. The slope of this line, and hence the exponent of M, is $\frac{3}{4} = 0.75$. As body mass increases by four factors of 10, say from 10^{-1} to 10^3 kg, metabolic rate increases by three factors of 10 from 10 to 10^4 kcal/day. Thus $H \propto M^{3/4}$ where M is mass in kilograms and H is metabolic heat production in kilocalories (thousands of calories) per day.

This scaling relationship is called "Kleiber's law," after the American veterinary scientist who first observed it in 1932. It has been verified by many series of subsequent measurements, though its cause is not fully understood.

Animals have evolved biological modifications partially to avoid the consequences of scaling laws. For example, our argument for the two-thirds power law of surface area vs. body mass was based on the assumption that the shapes of animals stay roughly the same as their size increases. But elephants have relatively thicker legs than gazelles in order to provide the extra strength needed to support their much larger weight. Such changes in shape are very important biologically but are too subtle to be seen in the overall trend of the data we have shown. Despite these adaptations, inexorable scaling laws give an upper limit to the size of land animals, since eventually the animal's weight would become too heavy to be supported by its body. That's why large animals such as whales must live in the ocean and why giant creatures in science fiction movies couldn't exist in real life.

Allometric laws provide an overview of the effects of scale, valid over several orders of magnitude. They help us compare important traits of elephants and mice, or of children and adults, without getting lost in the details. They help us to understand the limitations imposed by living in three dimensions.

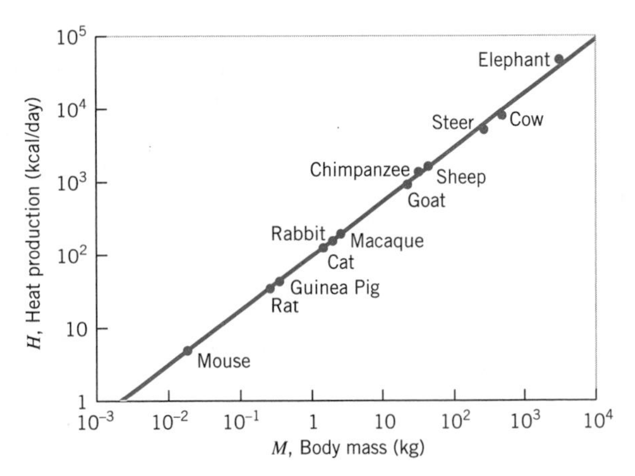

Figure 7.30 A log-log plot of heat production vs. body mass.

Source: Kleiber, M., 1951. "Body Size and Metabolism," *Hilgardia*, 6: pp. 315–353.

Algebra Aerobics 7.8

1. What is the approximate surface area for a human being whose mass is 70 kg (7×10^4 g)? First estimate the answer by locating the proper point on the graph in Figure 7.29; then compute it using Equation (3), $S = 10M^{2/3}$. How do your answers compare? Translate your answers into pounds and square inches.
2. The accompanying graph shows the relationship between heart rate (in beats per minute) and body mass (in kilograms) for mammals ranging over many orders of magnitude in size.

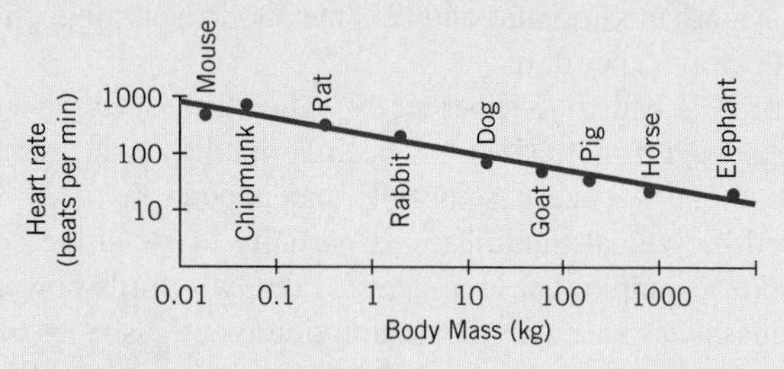

Source: Robert E. Ricklefs, *Ecology* (San Francisco: W. H. Freeman, 1990), p.66.

   a. Would a linear, power, or exponential function be the best model for the relationship between heart rate and body mass?
   b. The slope of the best fit line shown on the graph is -0.23. Construct the basic form of the function model.
   c. Interpret the slope of -0.23 in terms of heart rate and body mass.
3. Using the graph in Figure 7.30, what rate of heat production does this model predict for a 70-kg human being? Considering that each kilocalorie of heat production requires consumption of one food calorie, does this seem about right?

Chapter Summary

Power Functions. A power function has the form $y = kx^p$, where k and p are any constants and $p > 0$.

- If $y = kx^p$, we say that y is *directly proportional to* x^p.
- If $y = k/x^p = kx^{-p}$, we say that y is *inversely proportional to* x^p.

Here, k is called the *constant of proportionality.*

Any exponential growth function will eventually dominate any power function.

Positive Integer Powers: $y = kx$, $y = kx^2$, $y = kx^3$, . . . Let $y = kx^p$, where p is a positive integer and $k > 0$. Then for odd powers:

- as x increases, y increases.
- the graph has rotational symmetry about the origin.

For even powers:

- as x increases from negative values to positive values, y decreases and then increases as x becomes positive.
- the graph is symmetric about the y-axis and is U-shaped.

If $k > 0$, then the larger the value for k, the steeper the graph of $y = kx^p$.

Changing the sign of k "flips" the graph over the x-axis; that is, the graphs of $y = kx^p$ and $y = -kx^p$ are mirror images of each other across the x-axis.

Negative Integer Powers: $y = kx^{-1}$, $y = kx^{-2}$, $y = kx^{-3}$, . . . Power functions in the form

$$y = kx^{\text{negative power}}$$

where k is a constant, can be rewritten as

$$y = \frac{k}{x^{\text{positive power}}}$$

Power functions in this form are not defined when $x = 0$; hence, their domains do not include zero.

Graphs of power functions of the form

$$y = \frac{k}{x^p}$$

where p is a positive integer, are *asymptotic* to both the x- and y-axes.

For odd powers, the graphs have rotational symmetry about the origin.

For even powers, the graphs are symmetrical about the y-axis.

Semi-log and Log-log Graphs. To find the best function to model a relationship between two variables, we look for a way to plot the data so they appear as a straight line:

- The graph of a linear function appears as a straight line on a standard plot that has linear scales on both axes.
- The graph of an exponential function appears as a straight line on a semi-log plot that has a linear scale on the horizontal axis and a logarithmic scale on the vertical axis. The slope is the logarithm of the growth factor.
- The graph of a power function appears as a straight line on a log-log plot that has logarithmic scales on both axes. The slope is the exponent of the power function.

Allometry is the study of how the relationships among different physical attributes of an object change with scale. These relationships can often be described by power functions.

EXERCISES

Exercises for Section 7.1

1. The radius of a certain protein molecule is 0.000 000 000 1 = 10^{-10} meter. Assuming the molecule is roughly spherical, answer the following questions. Express your answers in scientific notation.

a. Find its surface area, S, in square meters, where $S = 4\pi r^2$.

b. Find its volume, V, in cubic meters, where $V = \frac{4}{3}\pi r^3$.

c. Find the ratio of the surface area to the volume.

2. The radius of Earth is about 6400 kilometers. Assume that Earth is spherical. Express your answers to the questions below in scientific notation.

a. Find the surface area of Earth in square meters.

b. Find the volume of Earth in cubic meters.

c. Find the ratio of the surface area to the volume.

3. If the radius of a sphere is x meters, what happens to the surface area and volume of a sphere when you:

a. quadruple the radius?

b. multiply the radius by n?

c. divide the radius by 3?

d. divide the radius by n?

4. Assume a box has a square base and the length of a side of the base is equal to twice the height of the box.

a. Write functions for the surface area and the volume that are dependent on the height, h.

b. Construct a small table and graph for each of the two functions.

c. If the height triples, what happens to the surface area? What happens to the volume?

d. As the height increases, what will happen to the ratio of surface area/volume?

5. Consider a cylinder with volume $V = \pi r^2 h$? What happens to its volume when you double its height, h? When you double its radius, r?

6. The volume, V, of a cylindrical can is

$$V = \pi r^2 h$$

and the total surface area, S, of the can is

$$\begin{aligned} S &= \text{area of curved surface} + 2 \cdot (\text{area of base}) \\ &= 2\pi rh + 2\pi r^2 \end{aligned}$$

where r is the radius of the base and h is the height.

a. Assume the height is three times the radius. Write the volume and the surface area as functions of the radius.

b. Does the volume or the surface area eventually grow faster as r increases? Show why.

7. Consider a circle with radius r. Its circumference, C, is $2\pi r$ and the area, A, of the region within it is πr^2. Fill in the following table and then describe what happens when you double, triple, or halve the radius.

Radius	Circumference, C	Area, A	Ratio of Circumference C to Circumference of a Circle with Radius R	Ratio of Area A to Area of a Circle with Radius R
R	$2\pi R$	πR^2	1	1
$2R$				
$3R$				
$0.5R$				

8. To celebrate Groundhog Day, a gourmet candy company makes a solid, quarter-pound chocolate groundhog that is 2 inches tall.

a. If the company wants to introduce a solid chocolate groundhog that is twice as tall, 4 inches high, how much chocolate is required?

b. How tall a chocolate groundhog could be made with twice the original quarter-pound amount (for a total of 8 ounces)?

Exercises for Section 7.2

9. Evaluate the following expressions for $x = 2$ and $x = -2$:

a. $5x^2$ **b.** $5x^3$ **c.** $-5x^2$ **d.** $-5x^3$

10. Identify which of the following are power functions. For each power function, identify the value for *k*, the constant of proportionality, and the value for *p*, the power.

a. $y = -3x^2$ **b.** $y = 3x^{10}$ **c.** $y = x^2 + 3$ **d.** $y = 3^x$

11. Assume *Y* is directly proportional to X^3.

a. Express this relationship as a function where *Y* is the dependent variable.

b. If $Y = 10$ when $X = 2$, then find the value of the constant of proportionality in part (a).

c. If *X* is increased by a factor of 5, what happens to the value of *Y*?

d. If *X* is divided by 2, what happens to the value of *Y*?

e. Rewrite your equation from part (a) solving for *X*.

12. The distance, *d*, a ball travels down an inclined plane is proportional to the square of the total time, *t*, of the motion.

a. Express this relationship as a function where *d* is the dependent variable.

b. If the ball starting at rest travels a total of 4 feet in 0.5 second, find the value of the constant of proportionality in part (a).

c. Using your equation in part (a), solve for *t*.

13. a. Assume *L* is directly proportional to x^5. What is the effect of doubling *x*?

b. Assume *M* is directly proportional to x^p, where *p* is a positive integer. What is the effect of doubling *x*?

14. In "Love That Dirty Water," the *Chicago Reader,* April 5, 1996, Scott Berinato interviewed Ernie Vanier, Captain of the towboat *Debris Control.* The Captain said, "We've found a lot of bowling balls. You wouldn't think they'd float, but they do."

When will a bowling ball float in water? The bowling rule book specifies that a regulation ball must have a circumference of exactly 27 inches. Recall that the circumference, *C*, of a circle with radius *r* is $C = 2\pi r$. The volume of a sphere with radius *r* is $\frac{4}{3}\pi r^3$.

a. What is a regulation bowling ball's radius in inches?

b. What is the volume of a regulation bowling ball in cubic inches? (Retain at least two decimal places in your answer.)

c. What is the weight in pounds of a volume of water equivalent in size to a regulation bowling ball? (Water weighs 0.03612 lb/in.^{3})

d. A bowling ball will float when its weight is less than or equal to the weight of an equivalent volume of water. What is the heaviest weight of a regulation bowling ball that will float in water?

e. Typical men's bowling balls are 15 or 16 pounds. Women commonly use 12-pound bowling balls. What will happen to the men's and to the women's bowling balls? Will they sink or float?

15. The intensity of light from a point source is inversely proportional to the square of the distance from the light source. If the intensity is 4 watts per square meter at a distance of 6 m from the source, find the intensity at a distance of 8 m from the source. Find the intensity at a distance of 100 m from the source.

16. Suppose you are traveling in your car at speed *S* and you suddenly brake hard, leaving skid marks on the road. A "rule of thumb" for the distance, *D*, that the car would skid is given by

$$D = \frac{S^2}{30f}$$

where D = distance the car skidded in feet, S = speed of the car in miles per hour, and f is a number called the coefficient of friction that depends upon the road surface and condition. For a dry tar road, $f \approx 1.0$. For a wet tar road, $f \approx 0.5$. (We saw a variation of this problem in Example 5, Section 7.2.)

a. What is the equation giving distance skidded as a function of speed for a dry tar road? For a wet tar road?

b. Generate a small table of values for both functions, including speeds between 0 and 100 miles per hour.

c. Plot both functions on the same grid.

d. Why do you think the coefficient of friction is less for a wet road than for a dry road? What effect does this have on the graph in part (c)?

e. In the accompanying table, estimate the speed given the distances skidded on dry and on wet tar roads. Describe the method you used to find these numbers.

Distance Skidded (ft)	Estimated Speed (mph)	
	Dry Tar	Wet Tar
25		
50		
100		
200		
300		

f. If one car is going twice as fast as another when they both jam on the brakes, how much farther will the faster car skid? Explain. Does your answer depend on whether the road is dry or wet?

17. Assume a person weighing P pounds is standing at the center of a 4″ × 12″ fir plank that spans a distance of L feet. The downward deflection (in inches) of the plank can be described by

$$D_4 = (5.25 \cdot 10^{-7}) \cdot P \cdot L^3$$

a. Graph the D_4 deflection formula, assuming $P = 200$ pounds. Put values of L from 0 to 20 feet on the horizontal axis. Graph the deflection (in inches) on the vertical axis.

b. We have seen that a rule used by architects for estimating acceptable deflection, D, in inches of a beam L feet long bent downward as a result of carrying a load is

$$D_{\text{safe}} = 0.05L$$

Add to your graph from part (a) a plot of D_{safe}.

c. Is it safe for a 200-pound person to sit in the middle of a 4″ × 12″ fir plank that spans 20 feet? What maximum span is safe for a 200-pound person on a 4″ × 12″ plank?

18. In Problem 17 we looked at how much a single load (a person's weight) could bend a plank downward by standing at its midpoint. Now we look at what a continuous load, such as a solid row of books spread evenly along a shelf, can do. A long row of paperback fiction weighs about 10 pounds for each foot of the row. Typical hardbound books weigh about 20 lb/ft, and oversize hardbounds such as atlases, encyclopedias, or dictionaries weigh around 36 lb/ft. The following function is used to model the deflection in inches of a 1″ × 12″ common pine board spanning a length of L feet carrying a continuous row of books:

$$D = (4.87 \cdot 10^{-4}) \cdot W \cdot L^4$$

where W is the weight per foot of the type of books along the shelf. This deflection model is quite good for deflections up to 1 inch; beyond that the fourth power causes the deflection value to increase very rapidly into unrealistic numbers.

a. How much deflection does the formula predict for a shelf span of 30 inches with oversize books? Would you recommend a stronger, thicker shelf?

b. Plot $D_{\text{hardbound}}$, $D_{\text{paperback}}$, and D_{oversize} on the same graph. Put L on the horizontal axis with values up to 4 feet.

c. At what length, L, will each of the kinds of books above cause a deflection of 0.5 inch in a 1″ × 12″ pine shelf?

19. Construct formulas to represent the following relationships:

a. The distance, d, traveled by a falling object is directly proportional to the square of the time, t, traveled.

b. The energy, E, released is directly proportional to the mass, m, of the object and the speed of light, c, squared.

c. The area, A, of a triangle is directly proportional to its base, b, and height, h.

d. The reaction rate, R, is directly proportional to the concentration of oxygen, $[O_2]$, and the square of the concentration of nitric oxide, [NO].

e. When you drop a small sphere in a dense fluid like oil, it eventually acquires a constant velocity, v, that is directly proportional to the square of its radius, r.

20. When a variable is directly proportional to the product of two or more variables, we say that the variable is *jointly* proportional to those variables. Express each of the following as an equation.

a. x is jointly proportional to y and the square of z.

b. V is jointly proportional to l, w, and h.

c. w is jointly proportional to the square of x and the cube root of y.

d. the volume of a cylinder is jointly proportional to its height and the square of its radius.

Exercises for Section 7.3

21. Graph and compare the following:

$$y_1 = x^2 \qquad y_2 = -x^2 \qquad y_3 = 2x^2 \qquad y_4 = \tfrac{1}{2}x^2 \qquad y_5 = -\tfrac{1}{2}x^2$$

22. In each part sketch the three graphs on the same grid and clearly label each function. Describe how the three graphs are alike and not alike.

a. $y = x^1 \qquad y = x^3 \qquad y = x^5$

b. $y = x^2 \qquad y = x^4 \qquad y = x^6$

c. $y = x^3 \qquad y = 2x^3 \qquad y = -2x^3$

d. $y = x^2 \qquad y = 4x^2 \qquad y = -4x^2$

If possible, check your results using a graphing calculator or function graphing program.

23. Evaluate each of the following functions at 0, 0.5, and 1. Then, on the same grid, graph each over the interval [0, 1]. Compare the graphs.

$$y_1 = x \qquad y_2 = x^{1/2} \qquad y_3 = x^{1/3} \qquad y_4 = x^{1/4}$$

$$y_5 = x^2 \qquad y_6 = x^3 \qquad y_7 = x^4$$

24. Now graph the functions in Exercise 23 over the interval [1, 4] and compare the graphs.

Exercises for Section 7.4

25. Graph the following functions on the same grid. Use a graphing calculator or function graphing program if available.

$$y = x^4 \quad \text{and} \quad y = 4^x$$

a. For positive values of x, where do your graphs intersect? Do they intersect more than once?

b. For positive values of x, describe what happens to the right and left of any intersection points. You may need to change the scales on the axes or change the windows on the graphing calculator in order to see what is happening.

c. Which eventually dominates, $y = x^4$ or $y = 4^x$?

26. **a.** Which eventually dominates, $y = (1.001)^x$ or $y = x^{1000}$?

b. As the independent variable approaches $+\infty$, which function eventually approaches zero faster, an exponential decay function or a power function with negative integer exponent?

27. Use a table or graphing utility to determine where $2^x > x^2$ and where $3^x > x^3$ for nonnegative values of x.

28. Match the following data tables with the appropriate function.

Table 1

x	y
1	2
2	1
3	$\frac{2}{3}$
4	$\frac{1}{2}$

Table 2

x	y
1	2
2	4
3	8
4	16

Table 3

x	y
1	$\frac{1}{2}$
2	1
3	$\frac{3}{2}$
4	2

Table 4

x	y
1	2
2	4
3	6
4	8

Table 5

x	y
1	1
2	4
3	9
4	16

a. $f(x) = 2x$

b. $f(x) = \frac{x}{2}$

c. $f(x) = \frac{2}{x}$

d. $f(x) = x^2$

e. $f(x) = 2^x$

29. If x is positive, for what values of x is $3 \cdot 2^x < 3 \cdot x^2$? For what values of x is $3 \cdot 2^x > 3 \cdot x^2$?

Exercises for Section 7.5

30. **a.** B is inversely proportional to x^4. What is the effect on B of doubling x?

b. Z is inversely proportional to x^p, where p is a positive integer. What is the effect on Z of doubling x?

31. A light fixture is mounted on a 10-foot-high ceiling over a 3-foot-high counter. How much will the illumination (the light intensity) increase if the light fixture is lowered to 4 feet above the counter?

32. The frequency, F (the number of oscillations per unit of time), of an object of mass m attached to a spring is inversely proportional to the square root of m.

a. Write an equation describing the relationship.

b. If a mass of 0.25 kilogram attached to a spring makes three oscillations per second, find the constant of proportionality.

c. Find the number of oscillations per second made by a mass of 0.01 kilogram that is attached to the spring discussed in part (b).

33. Boyle's Law says that if the temperature is held constant, then the volume, V, of a fixed quantity of gas is inversely proportional to the pressure, P. That is, $V = k/P$ for some constant k. What happens to the volume if:

a. the pressure triples?

b. the pressure is multiplied by n?

c. the pressure is halved?

d. the pressure is divided by n?

34. The pressure of the atmosphere around us is relatively constant at 15 lb/in.^{2} at sea level, or 1 atmosphere of pressure (1 atm). In other words, the column of air above 1 square inch of Earth's surface is exerting 15 pounds of force on that square inch of Earth. Water

Water Depth (ft)	Pressure (atm)
0	1
33	2
66	3
99	4

is considerably more dense. As we saw in Section 7.5, pressure increases at a rate of 1 atm for each additional 33 feet of water. The adjacent table shows a few corresponding values for water depth and pressure.

a. What type of relationship does the table describe?

b. Construct an equation that describes pressure, P, as a function of depth, D.

c. In Section 7.5 we looked at a special case of Boyle's Law for the behavior of gases, $P = 1/V$ (where if $V = 1$ cubic foot, then $P = 1$ atm).

i. Use Boyle's Law and the equation you found in part (b) to construct an equation for volume, V, as a function of depth, D.

ii. When $D = 0$ feet, what is V?

iii. When $D = 66$ feet, what is V?

iv. If a snorkeler takes a lung full of air at the surface, dives down to 10 feet, and returns to the surface, describe what happens to the volume of air in her lungs.

v. A large balloon is filled with a cubic foot of compressed air from a scuba tank at 132 feet below water level, sealed tight, and allowed to ascend to the surface. Use your equation to predict the change in the volume of its air.

35. Waves on the open ocean travel with a velocity that is directly proportional to the square root of their wavelength, the distance from one wave crest to the next. (See D. W. Thompson, *On Growth and Form,* New York: Dover Publications, 1992.)

a. Since the time between successive waves equals wavelength/velocity, show that time is also directly proportional to the square root of the wavelength.

b. On one day waves crash on the beach every 3 seconds. On the next day, the waves crash every 6 seconds. On the open ocean, how much farther apart do you expect the wave crests to be on the second day than on the first?

36. Construct a formula to represent each of the following statements:

a. Rate, r, is directly proportional to distance, d, and inversely proportional to time, t.

b. The gravitational force, F, between two planets is directly proportional to both of their masses (m_1 and m_2) and is inversely proportional to the square of the distance, d, between them.

c. The intensity, I, of electromagnetic radiation is directly proportional to energy, E, and inversely proportional to both cross-sectional area, A, and length, L.

37. a. Construct an equation to represent a relationship where x is directly proportional to y and inversely proportional to z.

b. Assume that $x = 4$ when $y = 16$ and $z = 32$. Find k, the constant of proportionality.

c. Using your equation from part (b), find x when $y = 25$ and $z = 5$.

38. a. Construct an equation to represent a relationship where w is jointly proportional to y and z and inversely proportional to the square of x.

b. Assume that $w = 10$ when $y = 12$, $z = 15$, and $x = 6$. Find k, the constant of proportionality.

c. Using your equation from part (b), find x when $w = 2$, $y = 5$, and $z = 6$.

39. (Requires a graphing calculator or computer.)
Oil is forced into a closed tube containing air. The height, H, of the air column is inversely proportional to the pressure, P.

a. Construct a general equation to describe the relationship between H and P.

b. Using technology, construct a best fit function for the accompanying data collected on height and pressure.

Height, H (in.)	Pressure, P (lb/in./in.)
13.3	6.7
10.7	8.1
9.2	10.2
7.1	12.9
6.3	14.4
5.1	17.6
4.1	21.4
3.5	25.1
3.0	29.4

40. (Requires graphing calculator or computer.)
Recall that Boyle's Law states that for a fixed mass of gas at a constant temperature, the volume, V, of the gas is inversely proportional to the pressure, P, exerted on the gas, that is, that $V = k/P$ for some constant k. Use technology to determine a best fit function model for the accompanying measurements collected on the volume of air as the pressure on it was increased.

Pressure, P (atm)	Volume of Air, V (cm^3)
1.0098	20
1.1610	18
1.3776	16
1.6350	14
1.9660	12
2.3828	10
2.9834	8
3.9396	6
5.0428	4
6.2687	3

Exercises for Section 7.6

41. A cube of edge length x has a surface area $s(x) = 6x^2$ and a volume $v(x) = x^3$. We constructed the function $r(x) = s(x)/v(x) = 6x^2/x^3 = 6/x$. Consider $r(x)$ as an abstract function. What is the domain? Construct a small table of values including negative values of x, and plot the graph. Describe what happens to $r(x)$ when x is positive and $x \to 0$. What happens to $r(x)$ when x is negative and $x \to 0$?

42. **a.** Make a table and sketch a graph for each of the following functions. Be sure to include negative and positive values for x, as well as values for x that lie close to zero.

$$y_1 = \frac{1}{x} \qquad y_2 = \frac{1}{x^2} \qquad y_3 = \frac{1}{x^3} \qquad y_4 = \frac{1}{x^4}$$

b. Describe the domain and range of each function.

c. Describe the behavior of each function as x approaches positive infinity and as x approaches negative infinity.

d. Describe the behavior of each function when x is near 0.

43. **a.** Generate tables and graphs for each of the following functions:

$$g(x) = 5x \qquad h(x) = x/5 \qquad t(x) = 1/x \qquad f(x) = 5/x$$

b. Describe the ways in which the graphs in part (a) are alike and the ways in which they are not alike.

44. In each part, sketch the three graphs on the same grid and label each function. Describe how the three graphs are similar and how they are different.

a. $y = x^{-1}$ $\quad y = x^{-3}$ $\quad y = x^{-5}$

b. $y = x^0$ $\quad y = x^{-2}$ $\quad y = x^{-4}$

c. $y = 2x^{-1}$ $\quad y = 4x^{-1}$ $\quad y = -2x^{-1}$

45. On the same grid, graph $y = x^2$ and $y = x^{-2}$.

a. Over what interval does each function increase? Decrease?

b. Where do the graphs intersect?

c. What happens to each function as x approaches positive infinity? Negative infinity?

46. Match each of the following functions with its graph:

a. $y = 3(2^x)$ **c.** $y = 2x - 3$ **e.** $y = x^{-3}$

b. $y = 2 - x$ **d.** $y = x^3$ **f.** $y = x^{-2}$

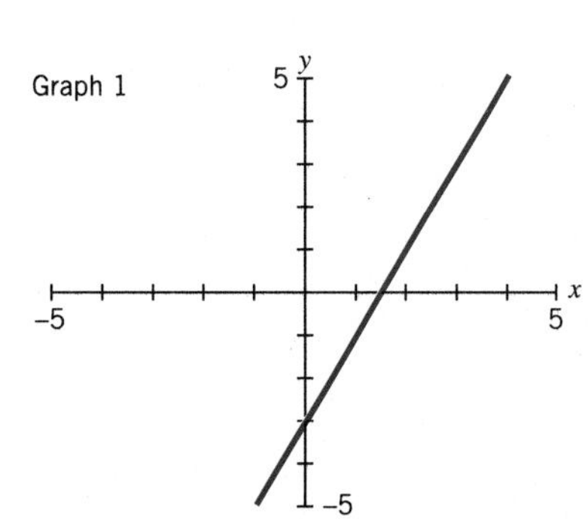

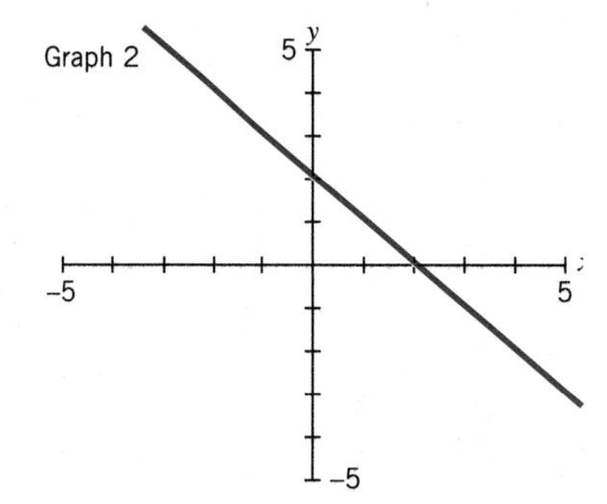

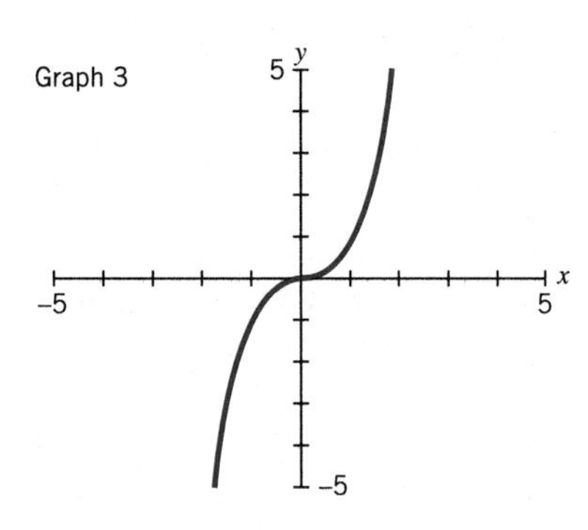

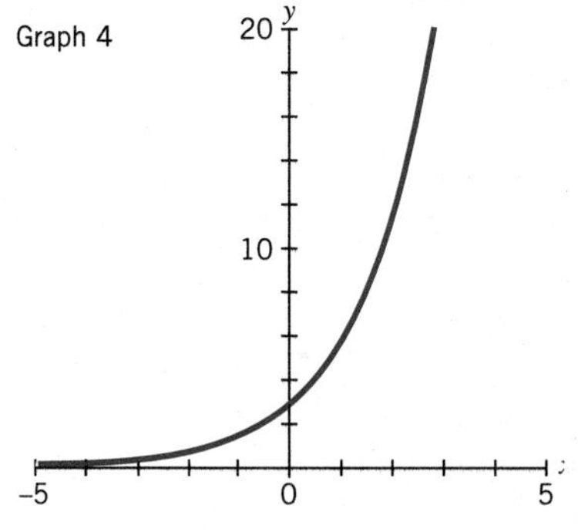

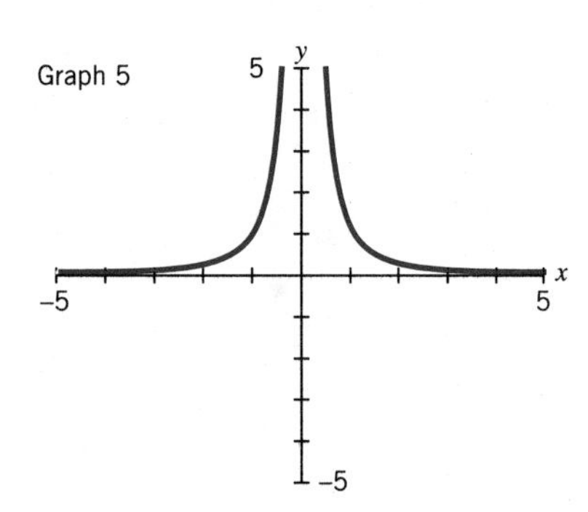

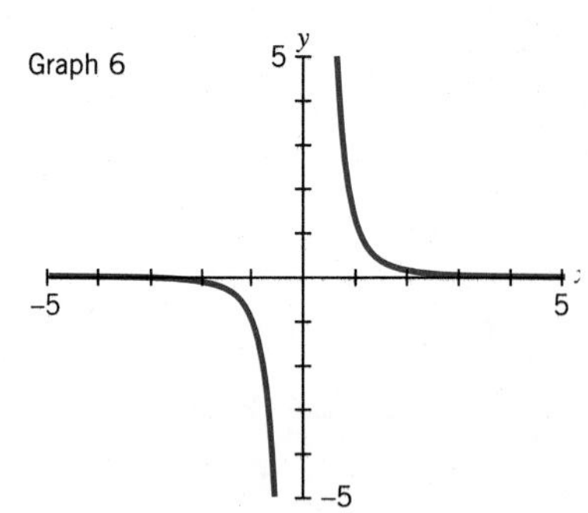

47. Which of the graphs of the following pairs of equations intersect? If the graphs intersect, find the point or points of intersection.

a. $y = 2x$ $y = 4x^2$ **d.** $y = x^{-1}$ $y = x^{-2}$

b. $y = 4x^2$ $y = 4x^3$ **e.** $y = 4x^{-2}$ $y = 4x^{-3}$

c. $y = x^{-2}$ $y = 4x^2$

Exercises for Sections 7.7 and 7.8

48. The accompanying two graphs show the same data on U.S. annual death rates from cancer of the large intestine as a function of age. Would an exponential or power function be a better model for the data? Justify your answer.

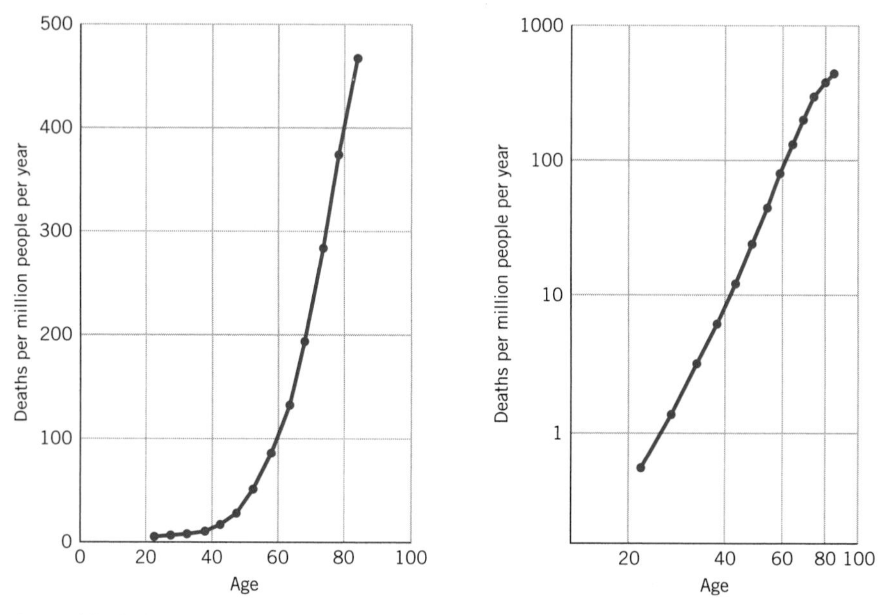

Annual U.S. death rate from cancer of the large intestine in relation to age, 1968.
Source: J. Cairns, *Cancer and Society* (San Francisco: W. H. Freeman and Company, 1978), p. 37.

49. Does the accompanying graph suggest that an exponential or a power function would be a better choice to model infant mortality rates in the United States from 1915 to 1977? Explain your answer.

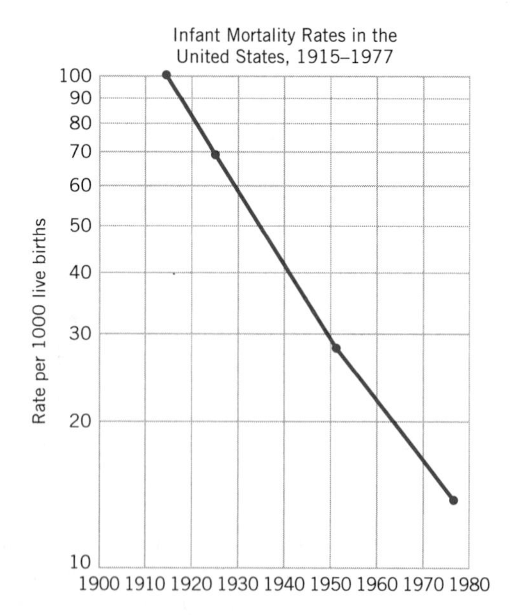

Source: Healthy People: The Surgeon General's Report on Health Promotion and Disease Prevention, DHEW (PHS) Publication 79-55071, Department of Health, Education, and Welfare, Washington, DC, 1979.

50. (Requires a computer or graphing calculator.)
Assume that $f(1) = 5$ and $f(3) = 45$.

a. Find an equation for f assuming f is:

i. a linear function

ii. an exponential function

iii. a power function

b. Then use technology to verify that you get a straight line when you plot

i. your linear function a on a standard plot

ii. your exponential function on a semi-log plot

iii. your power function on a log-log plot

51. (Optional use of technology.) The following data give the typical masses for some birds and their eggs:

a. Calculate the third column in the data table to see if the ratio of egg mass to adult bird mass is the same for all of the birds. Write a sentence that describes what you discover.

Species	Adult Bird Mass (g)	Egg Mass (g)	Egg/Adult Ratio
Ostrich	113,380.0	1,700.0	0.015
Goose	4,536.0	165.4	
Duck	3,629.0	94.5	
Pheasant	1,020.0	34.0	
Pigeon	283.0	14.0	
Hummingbird	3.6	0.6	

Source: W. A. Calder III, *Size, Function and Life History* (Boston: Harvard University Press, 1984).

b. Graph egg mass (vertical axis) and adult bird mass (horizontal axis) using each of the following types of plots:

i. standard linear

ii. semi-log

iii. log-log

c. Examine the three graphs you made and determine which looks closest to linear. Find a linear equation in the form $Y = b + mX$ to model the line. Remember that any value, such as the vertical intercept, that you read off a log scale has been converted to the log of the number by the scale. So, for example, in Figure 7.24 on a log-log plot, $Y = b + mX$ is actually of the form

$$\log y = b + m(\log x)$$

where $Y = \log y$ and $X = \log x$.

d. Once you have a linear model, transform it to a form that gives egg mass as a function of adult bird mass.

e. What egg mass does your formula predict for a 12.7 kilogram turkey?

f. A giant hummingbird (Patagona gigas) lays an egg of mass 2 grams. What size does your formula predict for the mass of an adult bird?

52. Find the equation of the line in each of the accompanying graphs. Rewrite each equation, expressing y in terms of x.

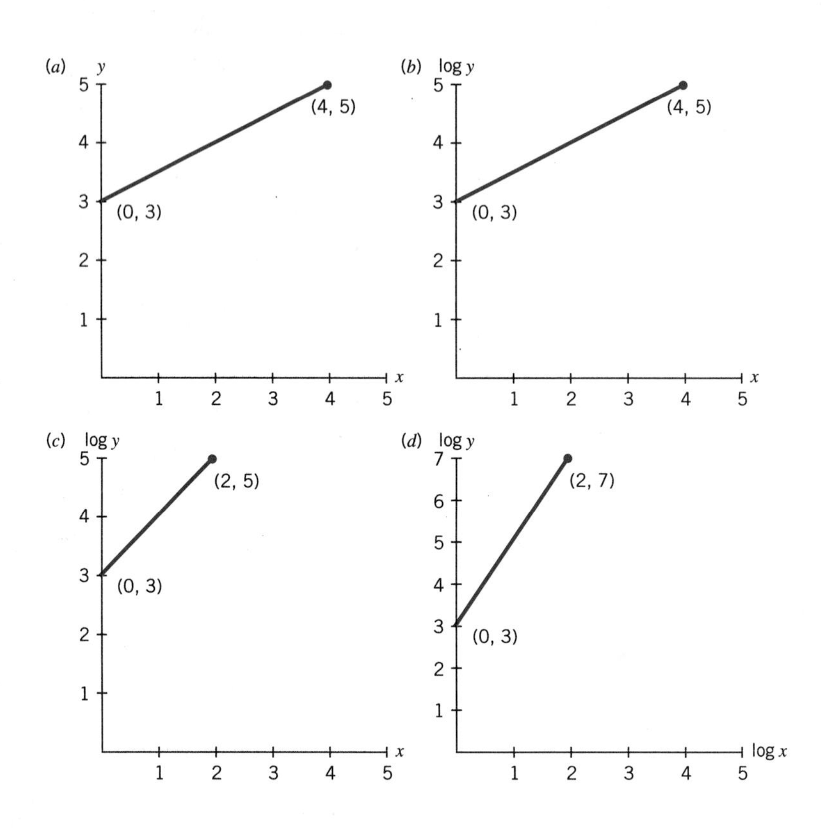

53. Use the accompanying graphs to answer the following questions:

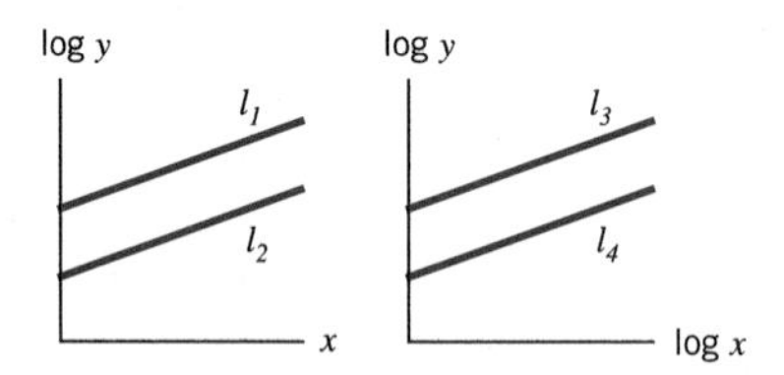

a. Assume l_1 and l_2 are straight lines that are parallel. In each case what type of equation would describe y in terms of x? How are the equations corresponding to l_1 and l_2 similar? How are they different?

b. Assume l_3 and l_4 are also parallel straight lines. For each case, what type of equation would describe y in terms of x? How are the equations corresponding to l_3 and l_4 similar? How are they different?

54. The accompanying graph shows oxygen consumption vs. body mass in mammals.

a. Would a power or exponential function be the best model of the relationship between oxygen consumption and body mass?

b. The slope of the best fit line shown on the graph is approximately $\frac{3}{4}$. Construct the basic form of the functional model that you chose in part (a).

c. Interpret the slope in terms of oxygen consumption and body mass. In particular, by how much does oxygen consumption increase when body mass increases by a factor of 10? By a factor of 10^4?

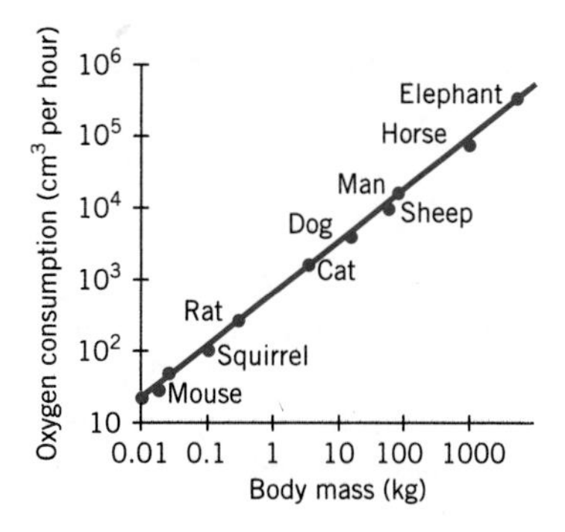

Source: R. E. Rickles, *Ecology* (San Francisco: W.H. Freeman, 1990), p. 67.

55. a. The figure below on the left shows the relationship between population density and length of an organism. The slope of the line is -2.25. Express the relationship between population density and length in terms of direct proportionality.

b. The figure below on the right shows the relationship between population density and body mass of mammals. The slope of the line is -0.75. Express the relationship between population density and body mass in terms of direct proportionality.

c. Are your statements in (a) and (b) consistent with the fact that body mass is directly proportional to length^3? *Hint:* Calculate the cube of the length to the -0.75 power.

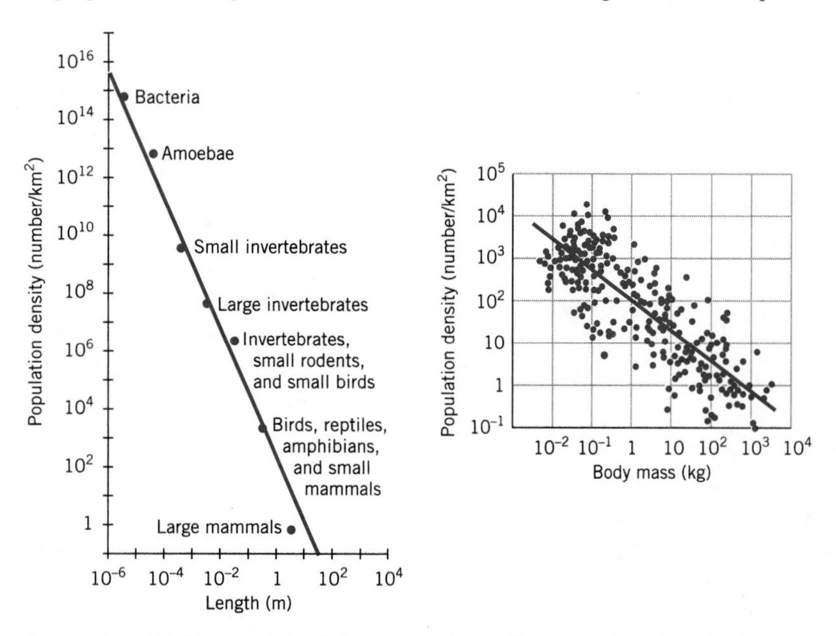

Source: T. A. McMahon and John T. Bonner, *On Size and Life* (New York: Scientific American Books, 1983), p. 228.

Exploration 7.1

Scaling Objects

Objectives

- find and use general formulas for scaling different types of objects

Procedure

1. *Scaling factors*

   When a picture or three-dimensional object is enlarged or shrunk, each linear dimension is multiplied by a constant called the *scaling factor.* For the two squares below, the scaling, F, is 3. That means that any linear measurement (for example, the length of the side or of the diagonal) is 3 times larger in the bigger square.

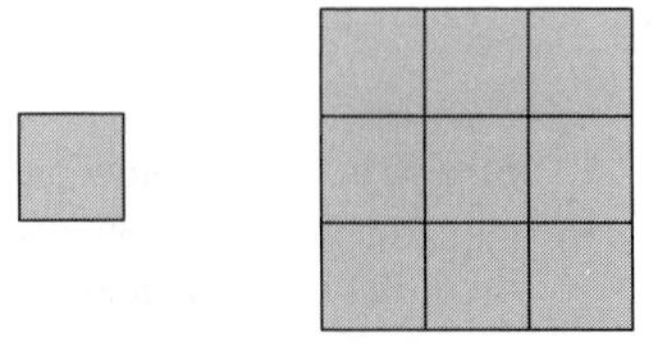

Scaling up a square by a factor of 3

   a. What is the relationship between the areas of the two squares above? Show that for all squares

   $$\text{area scaled} = (\text{original area}) \cdot F^2 \quad \text{where } F \text{ is the scaling factor}$$

   b. Given a scaling factor, F, find an equation to represent the surface area of a scaled-up cube, S_1, in terms of the surface area of the original object, S_0.

   c. Given a scaling factor, F, find an equation to represent the volume of a scaled-up cube, V_1, in terms of the volume of the original object, V_0.

2. *Representing volumes*

   We can describe the volume, V, of any three-dimensional object as $V = kL^3$, where V depends on any length, L, which describes the size of the object. The coefficient, k, depends upon the shape of the object and the particular length L and the measurement units we choose. For a statue of a deer, for example, L could represent the deer's width or the length of an antler or a tail. If we let L_0 represent the overall height of the deer in Figure 1, then L_0^3 is the volume of a cube

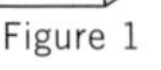

Figure 1

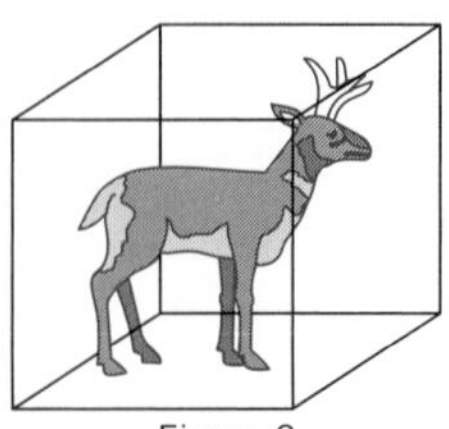

Figure 2

with edge length L_0 that contains the entire deer. The actual volume, V_0, of the deer is some fraction k of L_0^3; that is, $V = kL_0^3$ for some constant k.

3. *Finding the scaling factor*

If we scale up the deer to a height L_1 (see Figure 2), then the volume of the scaled-up deer equals kL_1^3. The scaled-up object inside the cube will still occupy the same fraction, k, of the volume of the scaled-up cube.

a. If an original object has volume V_0 and some measure of its length is L_0, then we know $V_0 = kL_0^3$. Rewrite the equation, solving for the coefficient, k, in terms of L_0 and V_0.

b. Write an equation for the volume, V_1, of a replica scaled to have length L_1. Substitute for k the expression you found in part (a). Simplify the expression such that V_1 is expressed as some term times V_0.

c. The ratio L_1/L_0 is the *scaling factor.* A scaling factor is the number by which each linear dimension of an original object is *multiplied* when the size is changed. Write an equation for the volume of a scaled object, V_1, in terms of V_0, the volume of the original object and F, the scaling factor.

4. *Using the scaling factor*

a. Draw a circle with a 1-inch radius. Then draw a circle with an area four times as large. What is the change in the radius?

b. A photographer wants to blow up a 3-cm by 5-cm photograph. If he wants the final print to be double the area of the original photograph, what scaling factor should he use?

c. A sculptor is commissioned to make a bronze statue of George Washington sitting on a horse. To fit into its intended location, the final statue must be 15 feet long from the tip of the horse's nose to the end of its tail. In her studio, the sculptor experiments with smaller statues that are only 1 foot long. Suppose that the final version of the small statue requires 0.15 cubic feet of molten bronze. When the sculptor is ready to plan construction of the larger statue, how can she figure out how much metal she needs for the full-size one?

d. (Adapted from COMAP, *For All Practical Purposes,* W. H. Freeman, New York, 1988, p. 370.) One of the famous problems of Greek antiquity was the *duplication of the cube.* Our knowledge of the history of the problem comes down to us from Eratosthenes (circa 284 to 192 B.C.), who is famous for his estimate of the circumference of Earth. According to him, the citizens of Delos were suffering from a plague. They consulted the oracle, who told them that to rid themselves of the plague, they must construct an altar to a particular god. That altar must be the same shape as the existing altar but double the volume. What should the scaling factor be for the new altar?

e. Pyramids have been built by cultures all over the world, but the largest and most famous were built in Egypt and Mexico.

i. The Great Pyramid of Khufu at Giza in Egypt has a square base 755 feet on a side. It originally rose about 481 feet high (the top 31 feet have been destroyed over time). Find the original volume of the Great Pyramid. The volume V of a pyramid is given by $V = (B \cdot H)/3$, where B is the area of the base and H is its height.

ii. The third largest pyramid of Giza, the Pyramid of Menkaure, occupies approximately one-quarter of the land area covered by the Great Pyramid. Menkaure's pyramid is the same shape as the original shape of the Great Pyramid but it is scaled down. Find the volume of the Pyramid of Menkaure.

Exploration 7.2

Predicting Properties of Power Functions

Objectives

- construct power functions with positive integer exponents
- find patterns in the graphs of power functions

Materials

- graphing calculator or function graphing program or "P1: k & p Sliders" in *Power Functions*
- graph paper

Procedure

Working in Pairs

We will be constructing power functions in the form $y = kx^p$ where k and p are constants and p is a positive integer. In each case, if possible, check your work using technology.

1. a. For each of the following, write the equation of a power function such that:

i. the graph of your function is symmetric across the y-axis.

ii. the graph of your function is symmetric around the origin.

b. For each of the following, construct the equations for two power functions such that:

i. the graphs of your two functions are mirror images of each other across the x-axis.

ii. the graphs of your two functions are mirror images of each other across the y-axis.

2. Using the same power for each function, construct two different power functions such that:

a. both functions have even powers and the graph of one function "hugs" the y-axis more closely than the other function.

b. both functions have odd powers and the graph of one function "hugs" the y-axis more closely than the other function.

3. Using the same value for the coefficient, k, construct two different power functions such that:

a. both functions have even powers and the graph of one function "hugs" the y-axis more closely than the other function.

b. both functions have odd powers and the graph of one function "hugs" the y-axis more closely than the other function.

4. a. Choose a value for k that is greater than 1 and construct the following functions:

i. $y = kx^0 = k$ $\quad y = kx^2$ $\quad y = kx^4$

ii. $y = kx^1$ $\quad y = kx^3$ $\quad y = kx^5$

b. Graph the functions in (i) on the same grid. Choose a scale for your axes so you can examine what happens when $0 < x < 1$. Now regraph, choosing a scale for your axes so you can examine what happens when $x > 1$.

c. Graph the functions in (ii) on the same grid. Choose a scale for your axes so you can examine what happens when $0 < x < 1$. Now regraph, choosing a scale for your axes so you can examine what happens when $x > 1$.

d. Describe your findings.

Class Discussion

Compare your findings. As a class develop a 60 second summary describing the results.

Exploration-Linked Homework

A Challenge Problem

In Chapter 7 we primarily studied power functions with integer exponents. Try extending your analyses to power functions with positive fractional exponents. Start with fractions that in reduced form have 1 in the numerator. What would the graphs of $y = x^{1/2}$ and $y = x^{1/4}$ look like? What are the domains? The ranges? What would the graphs of $y = x^{1/3}$ or $y = x^{1/5}$ look like? What about their domains and ranges? Generalize your results for fractions of the form $1/n$.

Exploration 7.3

Visualizing Power Functions with Negative Integer Powers

Objectives

- examine the effect of k on negative integer power functions

Materials

- graphing calculator or function graphing program
- graph paper

Related Software

- "P1: k & p Sliders" in *Power Functions*

Procedure

Class Demonstration: Constructing Negative Integer Power Functions

1. A power function has the form $y = kx^p$, where k and p are constants. Consider the following power functions with negative integer exponents where k is 4 and 6, respectively:

$$y = 4x^{-2} \quad \text{and} \quad y = 6x^{-2}$$

   We can also write these as

$$y = \frac{4}{x^2} \quad \text{and} \quad y = \frac{6}{x^2}$$

   What are the constraints on the domain and the range for each of these functions?

2. Construct a table for these functions using positive and negative values for x. How do different values for k lead to different values for y for these two functions? Sketch a graph for each function. Check your graphs with a function graphing program or graphing calculator.
   - **a.** Describe the overall behavior of these graphs. In each case, when is y increasing? When is y decreasing? How do the graphs behave for values of x near 0?
   - **b.** Describe how these graphs are similar and how they are different.

Working in Small Groups

In the following exploration you will predict the effect of the coefficient k on the graphs of power functions with negative integer exponents. In each part, compare your findings, and then write down your observations.

1. **a.** Choose a value for p in the function $y = k/x^p$, where p is a positive integer. Construct several functions where p has the same value but k assumes different *positive* values (as in the example above where $k = 4$ and $k = 6$). What effect do you think k has on the graphs of these equations? Graph the functions. As you choose larger and larger values for k, what happens to the graphs? Try choosing values for k between 0 and 1. What happens to the graphs?

b. Using the same value of p as in part (a), construct several functions with *negative* values for the constant k. What effect do you think k has on the graphs of these equations? Graph the functions. Describe the effect of k on the graphs of your equations. Do you think your observations about k will hold for any value of p?

c. Choose a new value for p and repeat your experiment. Are your observations still valid? Compare your observations with those of your partners. Have you examined both odd and even negative integer powers?

In your own words, describe the effect of k on the graphs of functions of the form $y = k/x^p$, where p is a positive integer. What is the effect of the sign of the coefficient k?

2. a. Choose a value for k where $k > 1$ and rewrite each function in the form $y = k/x^p$.

i. $y = kx^0 = k$ $\quad$ $y = kx^{-2}$ $\quad$ $y = kx^{-4}$

ii. $y = kx^{-1}$ $\quad$ $y = kx^{-3}$ $\quad$ $y = kx^{-5}$

b. Graph the functions in part (i) on the same grid. Choose a scale for your graphs such that you can examine what happens when $0 < x < 1$. Now choose scales for your axes so you can examine what happens when $x > 1$.

c. Graph the functions in part (ii) on the same grid. Choose scales for your graphs such that you can examine what happens when $0 < x < 1$. Now choose scales for your axes so you can examine what happens when $x > 1$.

d. Describe your findings.

Analysis

Compare the findings of each of the small groups. As a class, develop a 60 second summary on the effect of k on power functions in the form $y = k/x^p$, where p is a positive integer.

Quadratic and Other Polynomial Functions

Overview

By summing together power functions, we create polynomial functions. In this chapter we focus in particular on the properties of quadratic functions, polynomial functions of degree 2. Their graphs have a distinctive U shape called a parabola, a form visible in the arc of a basketball foul shot.

After reading this chapter, you should be able to:

- describe the basic properties of polynomial functions
- identify the zeros of a polynomial function
- recognize, evaluate, and graph quadratic functions
- determine the vertex and the intercepts of a quadratic function
- convert quadratic functions from one form into another

8.1 POLYNOMIAL FUNCTIONS

Suppose you make deposits of \$2000 a year in a savings account, starting today. What annual interest must you earn if you want to have \$10,000 in the account after 4 years?[1] Assuming the interest rate, r, is compounded annually and remains the same throughout the 4 years, the problem is to find the value of r that will give you \$10,000 after 4 years. Let $x = 1 + r$ be the annual multiplier or growth factor. For example, if the account pays 8% interest, then $x = 1 + 0.08$. We can write the balance after 1 year in terms of x as

$$\begin{aligned}\text{amount after 1 year} &= (\text{initial amount})(\text{growth factor}) + \text{new deposit}\\ &= 2000x + 2000\end{aligned}$$

During the second year you will earn interest on the amount you had after 1 year. At the end of the second year you will also deposit another \$2000. So the total amount after 2 years will be

$$\begin{aligned}\text{amount after 2 years} &= (\text{amount after 1 year})(\text{growth factor}) + \text{new deposit}\\ &= (2000x + 2000)x + 2000\\ &= 2000x^2 + 2000x + 2000\end{aligned}$$

After 3 years, interest will again be earned and a fourth deposit made:

$$\begin{aligned}\text{amount after 3 years} &= (\text{amount after 2 years})(\text{growth factor}) + \text{new deposit}\\ &= (2000x^2 + 2000x + 2000)x + 2000\\ &= 2000x^3 + 2000x^2 + 2000x + 2000\end{aligned}$$

A pattern is emerging. At the end of 4 years, assuming you do not make a final deposit, you will have

$$\text{amount after 4 years} = 2000x^4 + 2000x^3 + 2000x^2 + 2000x$$

If you reached your goal of \$10,000 at the end of 4 years, then you will have

$$10{,}000 = 2000x^4 + 2000x^3 + 2000x^2 + 2000x$$

Subtracting 10,000 from both sides you get:

$$0 = 2000x^4 + 2000x^3 + 2000x^2 + 2000x - 10{,}000 \qquad (1)$$

If you can solve this equation for x, you can find out what interest rate r will give you \$10,000 at the end of 4 years, since $x = 1 + r$. Dividing both sides of Equation (1) by 2000, you get

$$0 = x^4 + x^3 + x^2 + x - 5 \qquad (2)$$

A solution to Equation (2) will be a value x for the function

$$f(x) = x^4 + x^3 + x^2 + x - 5$$

such that $f(x) = 0$. So x-intercepts for $f(x)$ correspond to solutions to Equation (2). We can estimate the x-intercepts of $f(x)$ by graphing the function (Figure 8.1) and then zooming in (Figure 8.2).

[1]Example modeled after a problem in E. Connally, D. Hughes-Hallett, A. M. Gleason, et al., *Functions Modeling Change: A Preparation for Calculus,* preliminary ed. (John Wiley & Sons, New York, 1998), pp. 412–413.

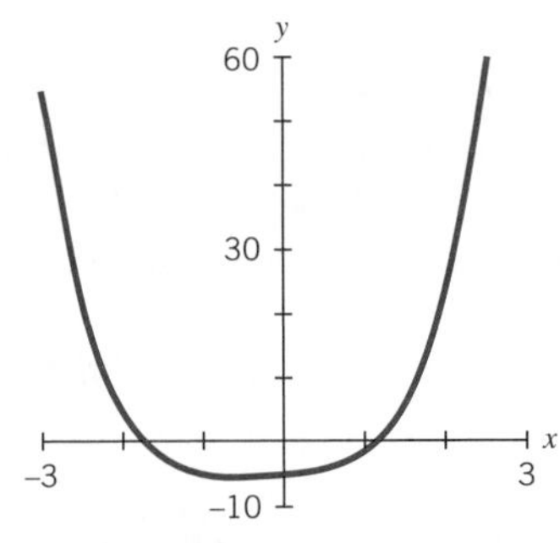

Figure 8.1 Graph of $f(x) = x^4 + x^3 + x^2 + x - 5$.

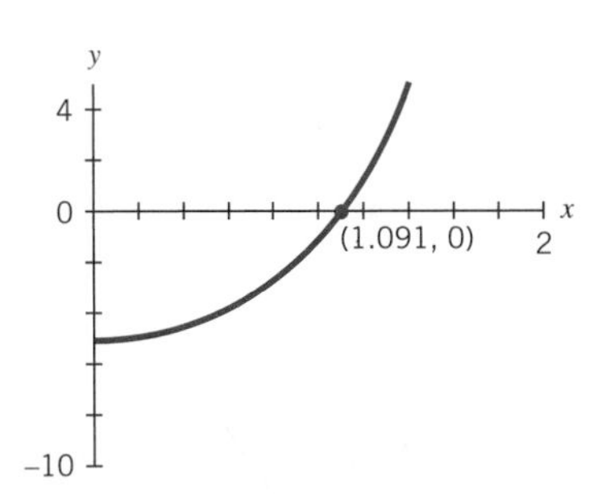

Figure 8.2 Zooming in on the x-intercept between 1 and 2.

Only positive solutions have meaning in this model, and there is only one positive x-intercept, at approximately (1.091, 0). We have $x = 1 + r$. Since $x = 1.091$, $r = x - 1 = 0.091$ in decimal form, or 9.1% when written as a percent. So you would need an annual interest rate of approximately 9.1% to end up with $10,000 by investing $2000 each year for 4 years.

Adding Power Functions: Polynomials

The previous function example $f(x) = x^4 + x^3 + x^2 + x - 5$ can be thought of as the sum of power functions

$$f_1(x) + f_2(x) + f_3(x) + f_4(x) + f_5(x)$$

where $f_1(x) = x^4$, $f_2(x) = x^3$, $f_3(x) = x^2$, $f_4(x) = x$, $f_5(x) = -5x^0 = -5$. We call $f(x)$ a *polynomial function* of degree 4 since the highest power of the independent variable is 4. In general, we can create a polynomial function by adding together power functions with non-negative integer exponents.

> A *polynomial function* $y = f(x)$ can be represented by an equation in the form
>
> $$y = a_n x^n + a_{n-1}x^{n-1} + \cdots + a_1 x^1 + a_0$$
>
> where each coefficient $a_n, a_{n-1}, \ldots, a_0$ is a constant and n is a non-negative integer, called the *degree* of the polynomial (provided $a_n \neq 0$).

The constant a_0 is often called "the constant term."

Polynomials of certain degrees have special names:

Polynomials of Degree	Are Called	Example
0	Constant	$y = 3$
1	Linear	$y = -4x - 8$
2	Quadratic	$y = 3x^2 + 5x - 10$
3	Cubic	$y = 5x^3 - 4x - 14$
4	Quartic	$y = -2x^4 - x^3 + 4x^2 + 4x - 14$
5	Quintic	$y = 8x^5 - 3x^4 + 4x^2 - 4$

Notice that the example for the cubic function, $y = 5x^3 - 4x - 14$, does not include an x^2 term. In this case the coefficient for the x^2 term is zero, since this function could be rewritten as $y = 5x^3 + 0x^2 - 4x - 14$.

Early algebraists believed that higher degree polynomials were not relevant to the physical world and hence were useless: "Going beyond the cube just as if there were more than three dimensions . . . is against nature."[2] But as we saw in the previous example, there are real applications for polynomials of degree greater than 3.

Algebra Aerobics 8.1a

1. For the following polynomials, specify the degree of the polynomial, and evaluate each function when $x = -1$:
   a. $f(x) = 11x^5 + 4x^3 - 11$
   b. $y = -5x^3 + 7x^4 + 1$
   c. $g(x) = -2x^4 - 20$
   d. $z = -2x^2 + 3x - 4$
2. Let $f(x) = x^3 - 3x^2$ and $g(x) = 2x^2 + 4$.
   a. Write out the function $h(x)$ if $h(x) = f(x) + g(x)$ and simplify the expression.
   b. What is $h(0)$? $h(2)$? $h(-2)$?

Visualizing Polynomial Functions

What can we predict about the graph of a polynomial function from its equation? Examine the graphs of polynomials of different degrees in Figure 8.3. What can we observe from each of these pairs?

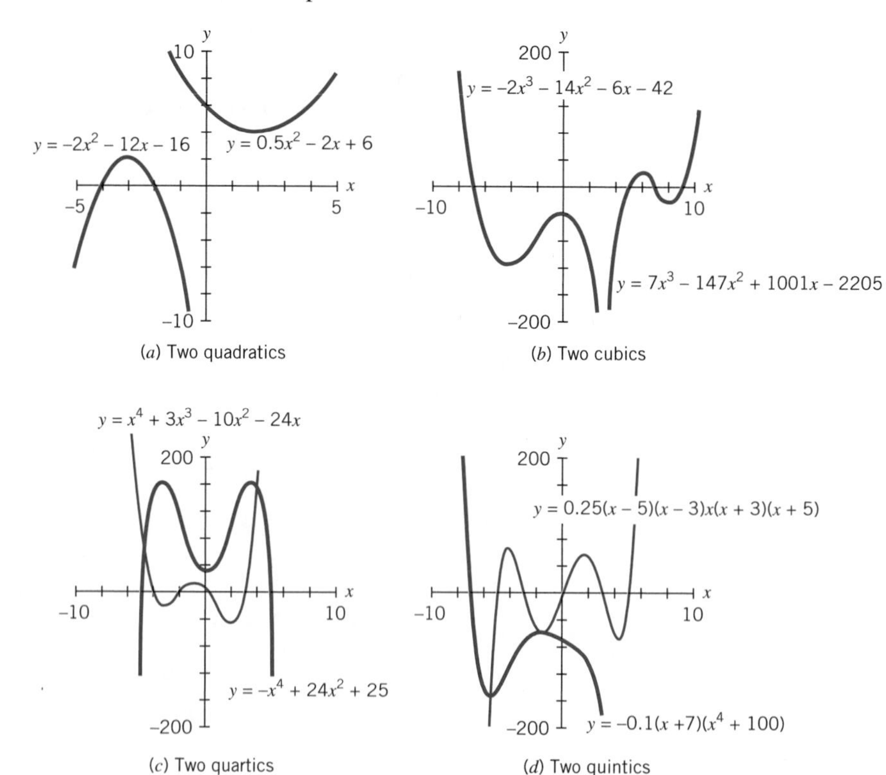

Figure 8.3 Graphs of pairs of polynomial functions: (*a*) of degree 2 (quadratics); (*b*) of degree 3 (cubics); (*c*) of degree 4 (quartics); (*d*) of degree 5 (quintics).

[2]Stifel, as cited by M. Kline, *Mathematical Thought from Ancient to Modern Times* (Oxford: Oxford University Press, 1972).

1. The first thing we might notice is the number of "wiggles" or bumps on each graph. The quadratics bend once, the cubics seem to bend twice, the quartics three times, one quintic seems to bend four times, and the other quintic appears to bend twice. In general, a polynomial function of degree n will have at most $n - 1$ "wiggles" or bumps.

2. Second, we might notice the number of times each graph crosses the x-axis. Each quadratic crosses at most two times; the cubics each cross at most three times; the quartics cross at most four times; and the quintics cross at most five times. In general, a polynomial function of degree n will cross the x-axis at most n times.

3. Finally, imagine zooming way out on the graph, to look at it on a global scale. After all, the sections of the x-axis displayed in the four graphs of Figure 8.3 are really quite small, the largest containing values of x only between -10 and $+10$. Suppose we consider values of x between -1000 and $+1000$ or between $-1{,}000{,}000$ and $+1{,}000{,}000$. What will the graphs look like?

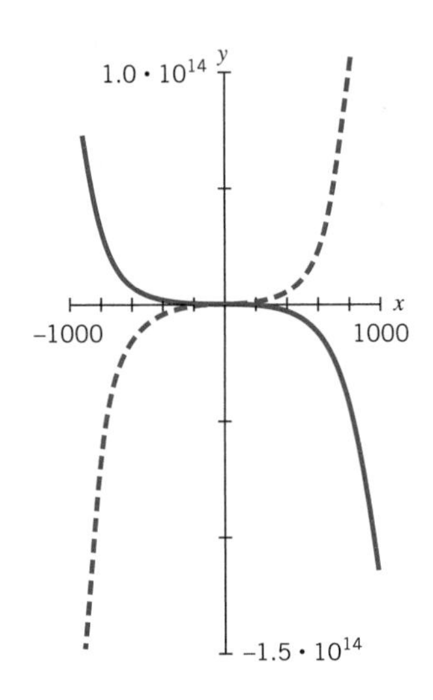

Figure 8.4 Zooming out on the graphs of the two quintics in Figure 8.3.

Figure 8.4 displays a graph of the two quintics in Figure 8.3d. Here the displayed values of x extend between -1000 and $+1000$. The wiggles are no longer noticeable. The dominant feature to notice now is whether the two "arms" of the function extend indefinitely up or indefinitely down. For polynomial functions of odd degree, such as the quintics in Figure 8.4, the two arms extend in opposite directions, one up and one down. For polynomial functions of even degree, both arms extend in the same direction, either both up or both down. This is because given a polynomial

$$y = a_n x^n + a_{n-1} x^{n-1} + \cdot \cdot \cdot + a_1 x^1 + a_0$$

of degree n, for large values of x the values of the leading term $a_n x^n$ will dominate the values of the other terms of smaller degree. In other words, for large values of x the function behaves like the simple power function $y = a_n x^n$. The degree of a polynomial determines its global shape.

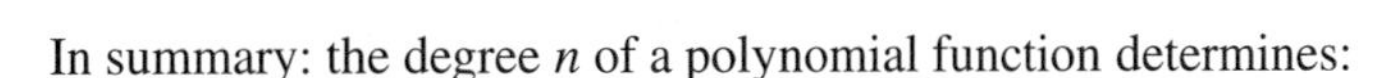

In summary: the degree n of a polynomial function determines:

- that the graph will have at most $n - 1$ "wiggles" or bumps
- that the graph will cross the x-axis at most n times
- the global shape of the graph of the polynomial

Intercepts of Polynomial Functions

Finding the vertical (or y) intercept: When $x = 0$. The y-intercept of the function $y = f(x)$ is the point at which the graph of $f(x)$ crosses the y-axis. At any point on the y-axis, $x = 0$. So to find the y-intercept, we must find $f(0)$; that is, we must determine the output of f when the input is 0. If $f(x)$ is a general polynomial function of the form

$$y = f(x) = a_n x^n + a_{n-1} x^{n-1} + \cdots + a_1 x^1 + a_0$$

the y-intercept will occur at $f(0) = a_n(0)^n + a_{n-1}(0)^{n-1} + \cdots + a_1(0)^1 + a_0 = a_0$, where a_0 is the constant term. Hence the coordinates of the y-intercept are $(0, a_0)$. We often shorten this to say that the y-intercept is a_0.

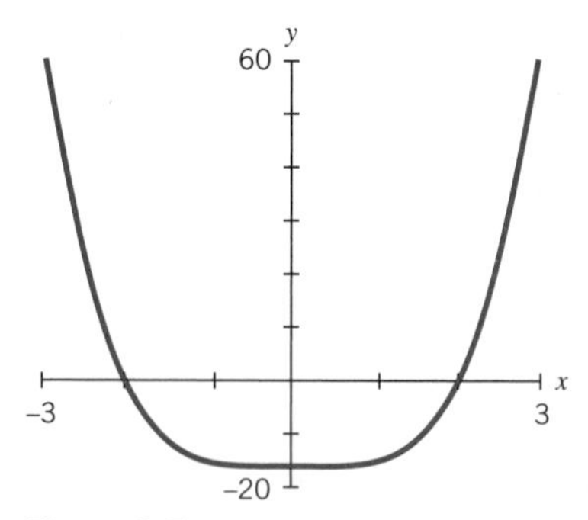

Figure 8.5 The graph of $y = x^4 - 16$ has a y-intercept at -16 and x-intercepts at $+2$ and -2.

For example, the function $y = x^4 - 16$ has a y-intercept at $(0, -16)$, or we say the y-intercept is -16 (Figure 8.5).

> Given a polynomial function of the form $y = f(x)$, where
>
> $$y = a_n x^n + a_{n-1} x^{n-1} + \cdots + a_1 x^1 + a_0$$
>
> (and n is a non-negative integer)
>
> the y-intercept is the point $(0, a_0)$.

Finding the horizontal (or x) intercepts: When f(x) = 0. The x-intercepts of a function $y = f(x)$ are the points at which the graph of $f(x)$ crosses the x-axis. At any point on the x-axis, $y = 0$. So the x-intercepts occur when $y = f(x) = 0$. To find the x-intercepts, we must find the values of x that make f equal to 0; that is, we must determine what input values for x, give an output value of zero.

The values of x for which $f(x) = 0$ are called the *zeros* of the function. Not all the zeros may be real numbers. However, each real zero determines an x-intercept.

> Given a polynomial function
>
> $$f(x) = a_n x^n + a_{n-1} x^{n-1} + \cdots + a_1 x^1 + a_0$$
>
> (where n is a non-negative integer)
>
> the *zeros* of the function are the values of x that make $f(x) = 0$. Each real zero corresponds to an x-intercept.

For example, for the relatively simple function $f(x) = x^4 - 16$, we can set $f(x) = 0$ and solve the corresponding equation:

Given	$f(x) = x^4 - 16$
set $f(x) = 0$	$0 = x^4 - 16$
add 16 to both sides	$16 = x^4$
take fourth root	$x = \pm 2$

Both $+2$ and -2 are zeros of the function. Since they are both real numbers, they represent x-intercepts. So the graph of $f(x)$ crosses the x-axis at $(2, 0)$ and $(-2, 0)$. Since every point on the x-axis has a y-coordinate of 0, we often refer to an x-intercept only by its x-coordinate. We say that the function has x-intercepts at $+2$ and -2. (See Figure 8.5.)

The zeros of the function $f(x) = a_n x^n + a_{n-1} x^{n-1} + \cdots + a_1 x^1 + a_0$ correspond to solutions of the equation

$$0 = a_n x^n + a_{n-1} x^{n-1} + \cdots + a_1 x^1 + a_0$$

These solutions are called the *roots* of the equation. For example, the numbers 2 and -2 are the roots of the equation $0 = x^4 - 16$.

The solutions of a polynomial equation

$$0 = a_n x^n + a_{n-1} x^{n-1} + \cdots + a_1 x^1 + a_0$$

(where n is a non-negative integer)

are called the *roots* of the equation.

Estimating x-intercepts from graphs. As we saw in the opening discussion on the compound interest rate problem, we can estimate horizontal intercepts from function graphs. A graphing calculator or computer (preferably with a zoom function) is very useful in helping to identify the number of x-intercepts and their approximate values. For example, if we graph the function $y = x^4 - 2x^2 - 3$, we see that it crosses the x-axis twice, which means that it has two x-intercepts (Figure 8.6).

If we zoom in on an x-intercept, we can obtain increasingly accurate, though still approximate, values for that intercept (Figure 8.7 and 8.8).

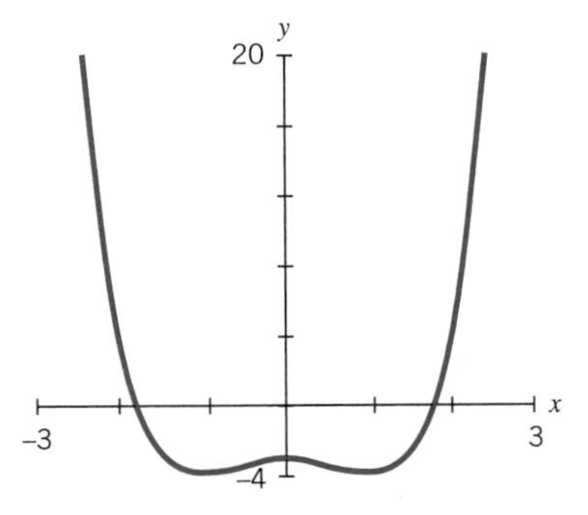

Figure 8.6 The graph of $y = x^4 - 2x^2 - 3$ shows two x-intercepts.

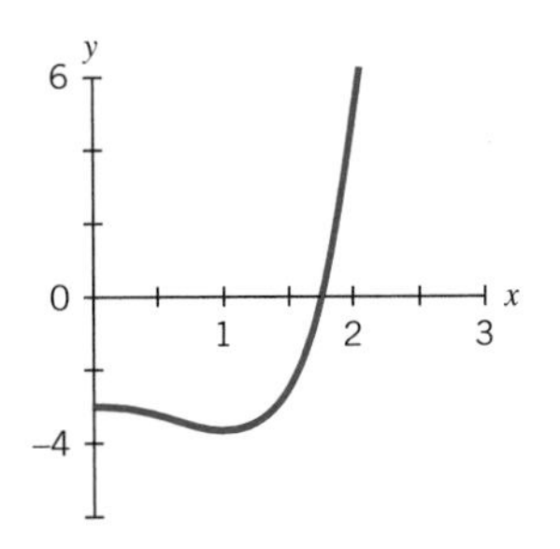

Figure 8.7 Zooming in on the x-intercept to the right tells us that its value is somewhere between 1.5 and 2.0

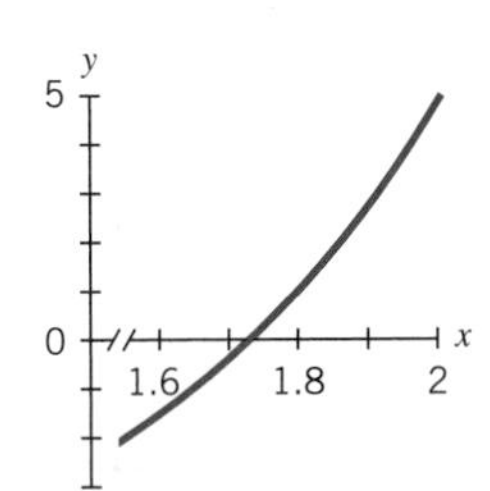

Figure 8.8 Zooming in again indicates that the x-intercept occurs between 1.7 and 1.8.

Determining x-intercepts from polynomials in factored form. We calculated x-intercepts for lines in Chapter 2, and later in this chapter we will learn strategies for calculating x-intercepts for polynomials of degree 2. Life isn't as easy when dealing with polynomial functions of higher degrees. The formulas for the zeros of third- and fourth-degree polynomials are extremely complicated. It has been proved that there are no general algebraic formulas for the zeros of polynomials of degree 5 or higher. There are however "algebraic approximation methods that allow us to calculate values for the x-intercepts or real zeros of functions accurate to as many decimal places as we wish. In some cases factoring allows us to find exact x-intercepts for polynomials. In particular if a function can be written as a product of linear factors, then each factor corresponds to a real zero or x-intercept of the function. Why is this the case? As the following example illustrates, the reason depends upon the *zero product rule.*

> *Zero Product Rule*
>
> For any two real numbers r and s, if the product $rs = 0$, then r or s or both must equal 0.

Example 1

Find the x-intercepts of the cubic function $g(x) = x(x + 2)(2x - 3)$.

Solution

To find the zeros of $g(x) = x(x + 2)(2x - 3)$, we must set $g(x) = 0$ and solve for x:

Let $g(x) = 0$	$0 = x(x + 2)(2x - 3)$				
Apply the zero product rule twice	$x = 0$	or	$x + 2 = 0$	or	$2x - 3 = 0$
Solve each equation	$x = 0$	or	$x = -2$	or	$x = \frac{3}{2}$

The three factors x, $x + 2$, and $2x - 3$ give us the corresponding real zeros, 0, -2, and $\frac{3}{2}$, which we can see as x-intercepts in Figure 8.9.

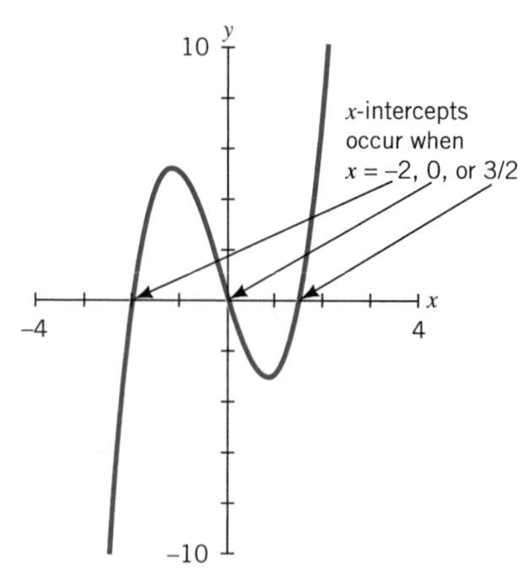

Figure 8.9 The x-intercepts of the function $g(x) = x(x + 2)(2x - 3)$.

Creating a polynomial function given its x-intercepts. Given any finite set of points on the horizontal axis, we can construct a polynomial function with those points as its horizontal intercepts.

Example 2

Create a polynomial function with x-intercepts at 0, -2, and 1.

Solution

If $x = 0$ is an x-intercept, then x is a factor of the polynomial function.
If $x = -2$ is an x-intercept, then $(x + 2)$ is also a factor.
If $x = 1$ is an x-intercept, then $(x - 1)$ is a factor as well.
So the function $f(x) = x(x + 2)(x - 1)$ has x-intercepts at 0, -2, and 1.

There are many different polynomial functions with the same horizontal intercepts.

Example 3

Find other polynomial functions with x-intercepts at 0, -2 and 1.

Solution

The functions $g(x) = 3x(x + 2)(x - 1)$ and $h(x) = -5x(x + 2)(x - 1)$ are also polynomial functions with x-intercepts at 0, -2, and 1. Note that they are both multiples of the function $f(x) = x(x + 2)(x - 1)$ from the previous example: $g(x) = 3 \cdot f(x)$ and $h(x) = -5 \cdot f(x)$. In fact, any of the infinitely many functions of the form $k \cdot f(x)$ will have x-intercepts at 0, -2, and 1. Figure 8.10 shows a graph of three such functions, namely $f(x)$, $g(x)$, and $h(x)$.

Are the only functions with x-intercepts at 0, -2, and 1 of the form $k \cdot f(x)$ where k is a constant and $f(x) = x(x + 2)(x - 1)$?

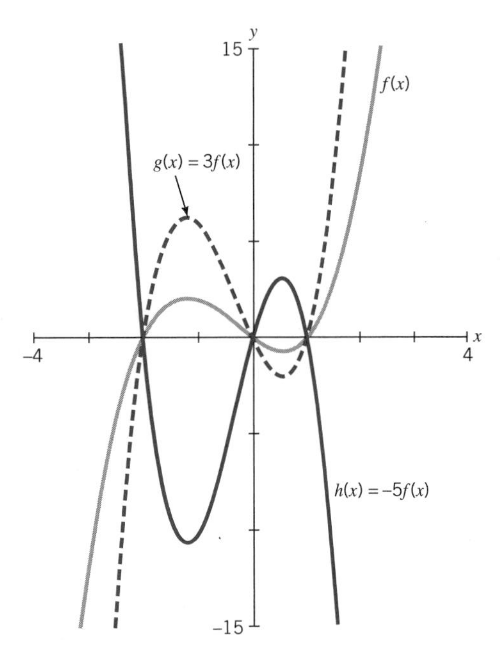

Figure 8.10 The graphs of three polynomial functions all with x-intercepts at 0, -2, and 1.

From Examples 2 and 3, we see that we can create a polynomial function with any finite set of real numbers as its zeros. Looking at it another way, we can also create new polynomials by multiplying together existing polynomials.

Algebra Aerobics 8.1b

1. For each of the following functions, identify the y-intercept. Then use a function graphing program (and its zoom feature) to estimate the number of x-intercepts and their approximate values.
   a. $y = 3x^3 - 2x^2 - 3$ **b.** $f(x) = x^2 + x + 3$
2. Identify the degree and the x-intercepts of each of the following polynomial functions. Graph each function to verify your work.
   a. $y = 3x + 6$
   b. $f(x) = (x + 4)(x - 1)$
   c. $g(x) = (x + 5)(x - 3)(2x + 5)$
   d. $h(x) = 0.25(x - 5)(x - 3)x(x + 3)(x + 5)$
3. Construct three polynomial functions, all with x-intercepts at -3, 0, 5, and 7. Use technology to plot them.

8.2 PROPERTIES OF QUADRATIC FUNCTIONS

Polynomial functions of degree 2 are called quadratics. In the last section we saw polynomials written in two different forms: in the "standard form" as the sum of power functions and in the "factored form" as the product of two or more polynomial expressions. A quadratic function in standard or *a-b-c* form is written as $f(x) = ax^2 + bx + c$ (for some constants a, b, and c). A quadratic in factored form can be written as the product of two linear expressions. Each form has advantages. From the standard form, it's easy to find the vertical intercept; from the factored form it's easy to find any horizontal intercepts.

In this section we first examine the graphs of quadratic functions and then look at examples of the two most common types of quadratic applications: area and motion.

EXPLORE

You may wish to use "Q1: *a, b, c* Sliders" in *Quadratic Functions* and Exploration 8.1, Part 1, in parallel with this section.

Visualizing Quadratic Functions

The graphs of quadratic functions have a distinctive U shape called a *parabola.* Some parabolas are wide, some are narrow, some open upward *(concave up)*, and some open downward *(concave down).*

The vertex and axis of symmetry. Each parabola has a *vertex* that is either the lowest or highest point on the curve. When a parabola opens upward, the vertex is the lowest point on the parabola. We say the function has a *minimum* at the vertex. When the parabola opens downward, the vertex is the highest point on the parabola. Hence the function has a *maximum* at the vertex.

Now imagine drawing a vertical line through the vertex and folding the parabola along that line. The right half of the curve will fall exactly on the left half. The right half of the curve is a mirror image of the left half. The vertical line of the fold is called the *axis of symmetry* (Figure 8.11). The graph is said to be *symmetrical* about its axis of symmetry.

Horizontal and vertical intercepts. While each parabola always has a vertical or y-intercept, it may have 0, 1, or 2 horizontal or x-intercepts (see Figure 8.19). If the parabola has two x-intercepts, then the intercepts will lie an equal distance, d, to the left and right of the axis of symmetry (see Figure 8.12).

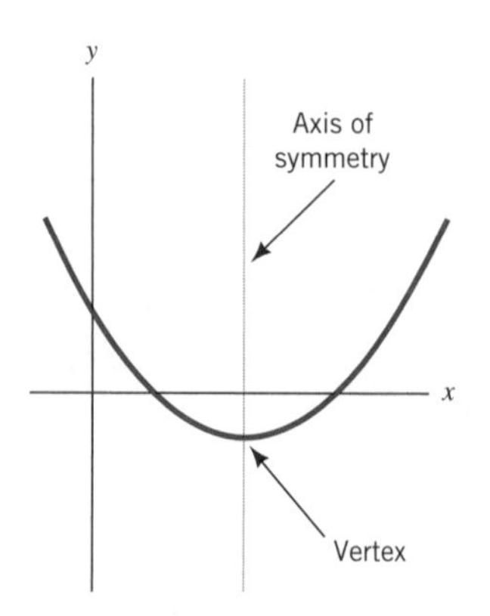

Figure 8.11 Each parabola has a vertex that lies on an axis of symmetry.

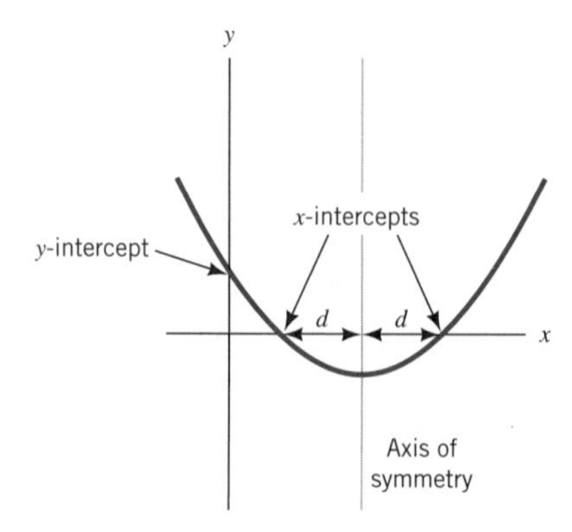

Figure 8.12 Each parabola has a vertical intercept and possibly horizontal intercepts.

The Effect of the Coefficient a.

When $a > 0$: Given a quadratic function, $y = ax^2 + bx + c$, when a is positive, the parabola opens upward. What happens as we vary the size of a? Figure 8.13 shows the graphs of four parabolas where $a > 0$, namely, $y = 0.25x^2$, $y = x^2$, $y = 2x^2$, and $y = 4x^2$. Each parabola opens upward, and the larger the value of a, the narrower the graph. Conversely, the smaller the value of a, the flatter the graph. Even if we set b and c to values other than zero, varying a still has the same effect. The parabola will still open upward and get narrower as the value of a increases.

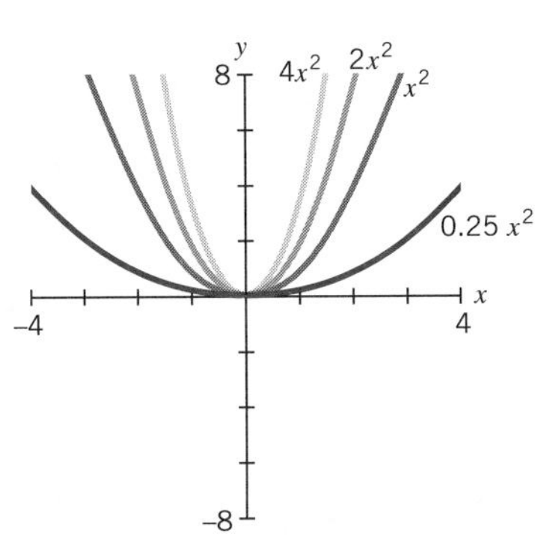

Figure 8.13 Parabolas with $a > 0$.

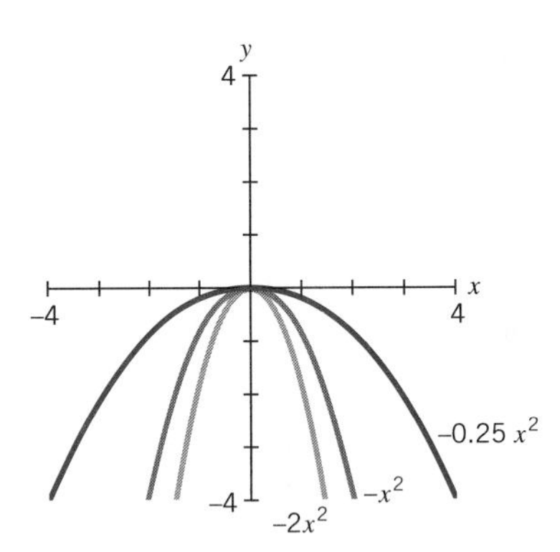

Figure 8.14 Parabolas with $a < 0$.

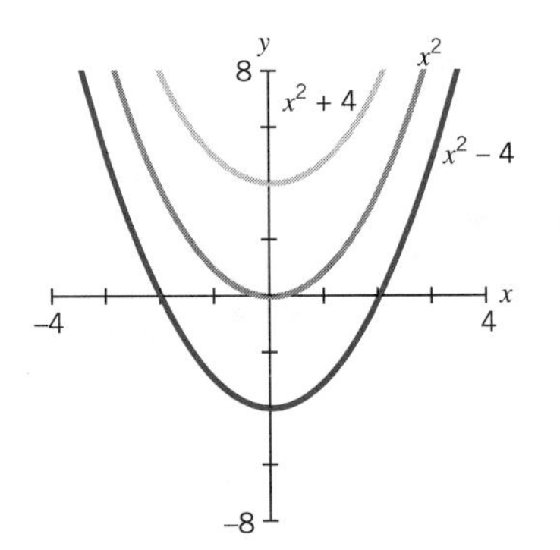

Figure 8.15 Parabolas with different values for c.

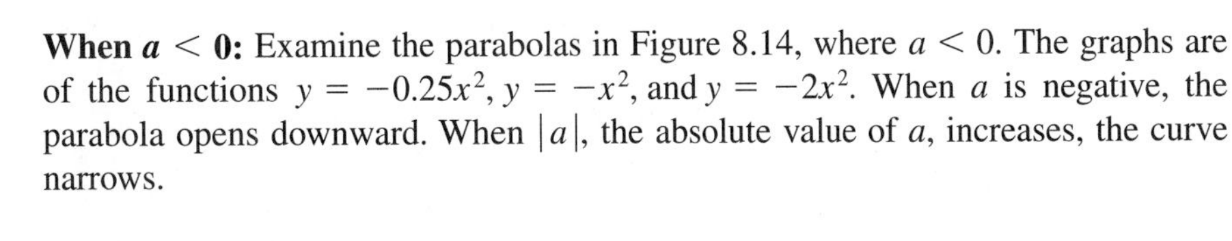

When $a < 0$: Examine the parabolas in Figure 8.14, where $a < 0$. The graphs are of the functions $y = -0.25x^2$, $y = -x^2$, and $y = -2x^2$. When a is negative, the parabola opens downward. When $|a|$, the absolute value of a, increases, the curve narrows.

The Effect of the Constant Term c.

Changing c changes only the vertical position of the graph, not its shape (Figure 8.15). The parabola $y = x^2 + 4$ is the graph of $y = x^2$ raised up four units. Similarly, the graph of $y = x^2 - 4$ is the graph of $y = x^2$ lowered down four units. The constant term c has the same effect for any value of a and b. Think of c as an "elevator" term. Without changing the shape of the curve, increasing the value of c shifts the parabola up and decreasing the value of c shifts the parabola down.

SOMETHING TO THINK ABOUT

Using "Q1: a, b, c Sliders" in *Quadratic Functions* in the course software, can you describe the effect on the parabola of changing the value for b while you hold a and c fixed?

The graph of a quadratic function $y = ax^2 + bx + c$:

- is called a *parabola*
- has a U-like shape
- has a lowest or highest point called the *vertex*
- is symmetric about a vertical line, called the *axis of symmetry,* that runs through the vertex
- opens up if $a > 0$ and down if $a < 0$
- becomes narrower as $|a|$, the absolute value of a, is increased
- maintains its shape but is shifted up if c is increased and down if c is decreased

Estimating the Vertex and the Intercepts

The earliest problems we know of that lead to quadratic equations are on Babylonian tablets dating from 1700 B.C. The writings suggest a problem similar to the following example.

Maximizing Area. Suppose you wish to enclose a rectangular region and are constrained by a fixed perimeter of 24 meters. How would you find the dimensions of the rectangle that will contain the greatest area?

If the rectangular region has length L and width W, then

	$2L + 2W = \text{perimeter}$
Substitute	$2L + 2W = 24$
divide by 2	$L + W = 12$
Subtract L from both sides	$W = 12 - L$
Since area, A, is given by	$A = L \cdot W$
substitute for W	$= L \cdot (12 - L)$
multiply through	$= 12L - L^2$

So area, A, is a quadratic function of L.

Table 8.1 shows values of L from 0 to 12 meters and corresponding values for W and A.

Figure 8.16 shows a graph of area, A, versus length, L. Since in the equation $A = 12L - L^2$ the coefficient of L^2 is negative (-1), the parabola opens downward. From the table and the graph, it appears that the vertex of the parabola is at (6, 36); that is, at a length of 6 meters the area reaches a maximum of 36 square meters. We have $W = 12 - L$, so when $L = 6$, $W = 6$. Hence the rectangle has maximum area when the length equals the width, or in other words when the rectangle is a square.

Table 8.1

Length, L (m)	Width, W (m)	Area, A (m^2)
0	12	0
1	11	11
2	10	20
3	9	27
4	8	32
5	7	35
6	6	36
7	5	35
8	4	32
9	3	27
10	2	20
11	1	11
12	0	0

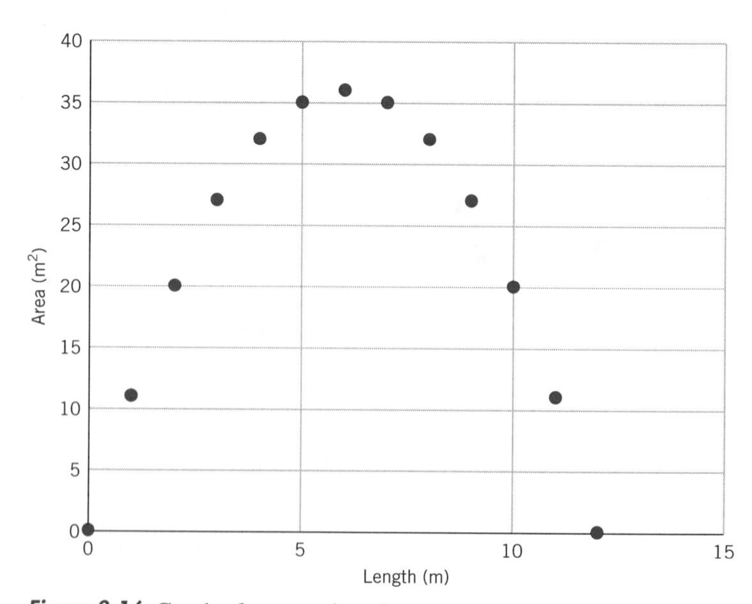

Figure 8.16 Graph of area vs. length.

"The Mathematics of Motion" following this chapter offers an extended exploration into constructing mathematical models to describe the motion of freely falling bodies.

Motion of a projectile. Figure 8.17 shows a plot of the height above the ground (in feet) of a projectile for the first 5 seconds of its trajectory. How could we

a. estimate the height from which the projectile was launched and

b. estimate when the projectile will hit the ground?

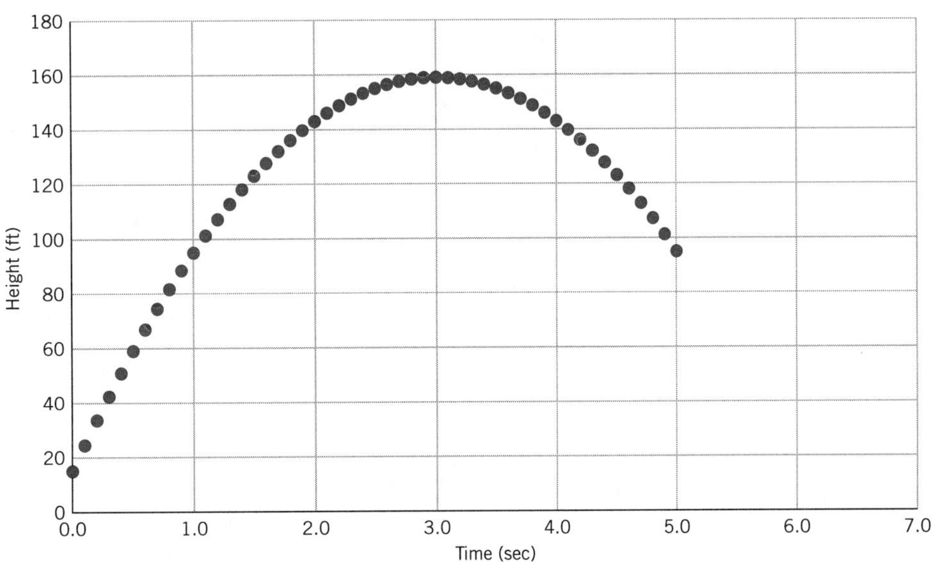

Figure 8.17 The height of a projectile.

a. The initial height of the projectile corresponds to the vertical intercept of the parabola (where $t = 0$). A quick look at Figure 8.17 gives us an estimate of somewhere between 10 and 20 feet.

We can also model the path of a projectile with a quadratic function. If we use technology to generate a best fit function, we get $H(t) = -16t^2 + 96t + 15$, where t = time (in seconds) and $H(t)$ = height (in feet). Since $H(t)$ is in the standard form, the constant term of 15 gives the vertical intercept. In other words, when $t = 0$ seconds, the initial height is $H(0) = 15$ feet.

b. Figure 8.18 shows the graph of $H(t) = -16t^2 + 96t + 15$, our function model for the projectile's path. The projectile will hit the ground when the height above the ground $H(t) = 0$. This occurs at a horizontal intercept of the parabola. One horizontal intercept (not shown) would occur at a negative value of time, t, which would be meaningless here. From the graph we can estimate that the other intercept occurs at $t \approx 6.2$ seconds.

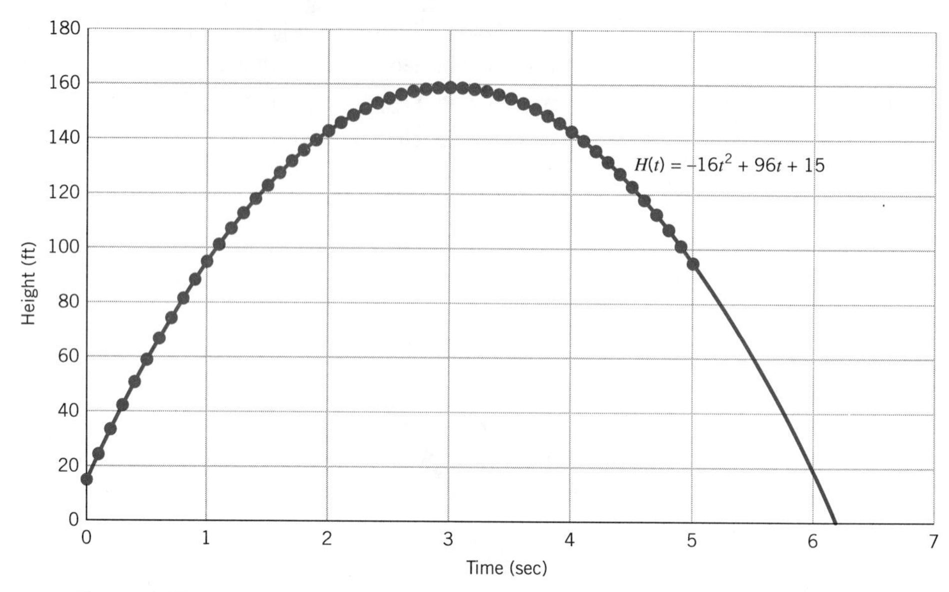

Figure 8.18 A graph of the function model for the projectile's path.

Algebra Aerobics 8.2

1. Using a graphing tool, plot each of the following quadratic functions. Estimate the coordinates of the vertex for each parabola and give the equation of the axis of symmetry. Is the parabola concave up or concave down? Specify the number of x-intercepts and identify the y-intercept.

   a. $y = -x^2 + 2$ **c.** $y = 0.5x^2 - 2x + 3$

   b. $y = x^2 + 2x + 1$

2. Without drawing the graph, describe whether the graph of each function in parts (a) to (d):

   i. has a maximum or minimum at the vertex

   ii. is narrower or broader than $y = x^2$

   a. $y = 2x^2 - 5$

   b. $y = 0.5x^2 + 2x - 10$

   c. $y = 3 + x - 4x^2$

   d. $y = -0.2x^2 + 11x + 8$

Without drawing the graphs, in Problems 3 to 7, compare the graph of part (b) to the graph of part (a):

3. **a.** $y = x^2 + 2$ **b.** $y = 2x^2 + 2$
4. **a.** $f(x) = x^2 + 3x + 2$ **b.** $g(x) = x^2 + 3x + 8$
5. **a.** $d = t^2 + 5$ **b.** $d = -t^2 + 5$
6. **a.** $f(z) = -5z^2$ **b.** $g(z) = -0.5z^2$
7. **a.** $h = -3t^2 + t - 5$ **b.** $h = -3t^2 + t - 2$

8.3 FINDING THE HORIZONTAL INTERCEPTS OF A QUADRATIC FUNCTION

A quadratic function may have one, two, or no horizontal (or x) intercepts as shown in Figure 8.19.

We saw in Section 8.1 that when we look for x-intercepts, we ask, "What value(s) of x make the function $f(x)$ equal to zero?" Or we could ask, "What are the input value(s) that make the output of a function zero?" When the output value of a function is equal to zero, the corresponding input values are called the zeros of the function.

In the previous section, we estimated the x-intercepts from the graph of a quadratic function. In this section, we find the x-intercepts by using the quadratic formula and by factoring.

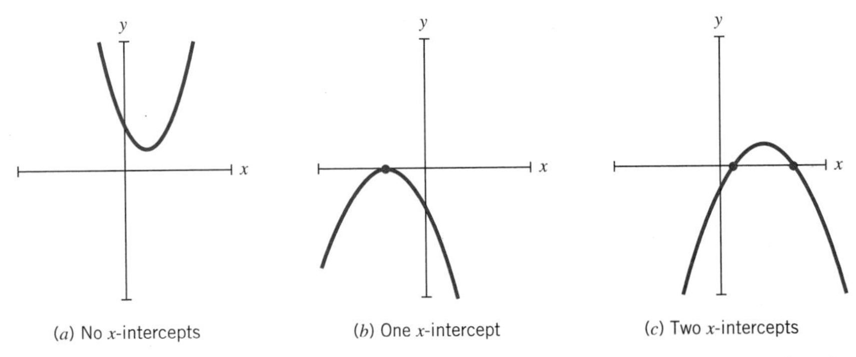

Figure 8.19 Graphs of quadratic functions showing the three possible cases for the number of horizontal or x-intercepts.

Using the Quadratic Formula

To find the x-intercepts of $f(x) = ax^2 + bx + c$, we set $f(x) = 0$ and solve for x. The solutions to the equation $0 = ax^2 + bx + c$ are called the *roots* of the equation. We can use the quadratic formula to find the roots.

See "Why the Formula for the Vertex and the Quadratic Formula Work."

The Quadratic Formula

For any quadratic equation of the form $0 = ax^2 + bx + c$ $(a \neq 0)$, the solutions, or *roots*, of the equation are given by

$$x = \frac{-b \pm \sqrt{b^2 - 4ac}}{2a}$$

The term under the radical sign, $b^2 - 4ac$, is called the *discriminant.*

The symbol $\pm$ lets us use one formula to write the two roots as

$$x = \frac{-b + \sqrt{b^2 - 4ac}}{2a} \quad \text{and} \quad x = \frac{-b - \sqrt{b^2 - 4ac}}{2a}$$

Note on terminology. The numbers 3 and -3 are called the *roots* or *solutions* of the *equation* $x^2 - 9 = 0$. The numbers 3 and -3 are called the *zeros* of the *function* $f(x) = x^2 - 9$. The zeros of the function f are the roots of the equation $f(x) = 0$.

Example 1

a. What are the horizontal intercepts (if there are any) for the function $h(t) = 34 + 32t - 16t$ where t is the number of seconds and $h(t)$ is the height above ground (in feet) of a projectile?

b. What significance do the intercepts have?

Solution

a. To find the horizontal intercepts, we set $h(t) = 0$ and solve the resulting equation $0 = 34 + 32t - 16t^2$. If we rearrange the terms, we get:

$$0 = -16t^2 + 32t + 34$$

In this form it is easier to see how to apply the quadratic formula.

We have $a = -16$, $b = 32$, and $c = 34$. Since a is negative, the graph opens downward.

Using the quadratic formula $\quad t = \dfrac{-b \pm \sqrt{b^2 - 4ac}}{2a}$

substitute for a, b, and c $\quad = \dfrac{-32 \pm \sqrt{32^2 - 4(-16)(34)}}{(2)(-16)}$

$$= \frac{-32 \pm \sqrt{1024 + 2176}}{-32}$$

(note that the discriminant = 3200, a positive real number) $\quad = \dfrac{-32 \pm \sqrt{3200}}{-32}$

evaluate with a calculator $\quad t \approx \dfrac{-32 \pm 56.6}{-32}$

So there are two roots,

one at $\quad \dfrac{-32 + 56.6}{-32} = \dfrac{24.6}{-32} \approx -0.77$

and the other at $\quad \dfrac{-32 - 56.6}{-32} = \dfrac{-88.6}{-32} \approx 2.77$

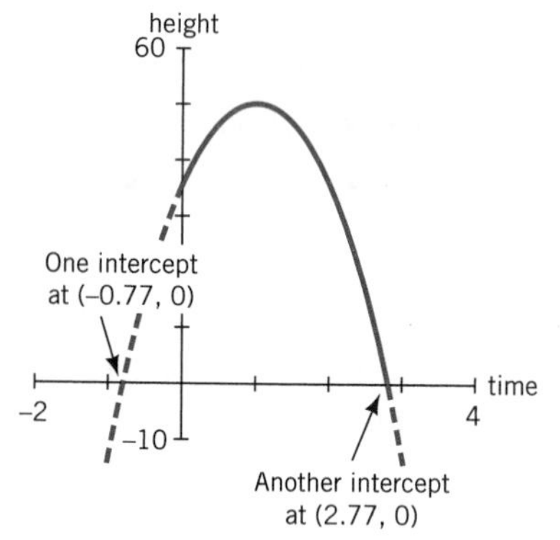

Figure 8.20 Graph of the equation $h(t) = 34 + 32t - 16t^2$, with two horizontal intercepts.

Therefore the parabola crosses the horizontal axis at approximately $(-0.77, 0)$ and $(2.77, 0)$, as shown in Figure 8.20.

b. The t-intercept at $(-0.77, 0)$ lies outside the model, since it represents a negative value for time, t. The positive t-intercept says that when $t = 2.77$ seconds, $h = 0$ feet. In other words, at 2.77 seconds the object hits the ground.

Example 2

Find the x-intercepts of $f(x) = x^2 + 3x + 2.25$.

Solution

We can find the x-intercepts by solving the equation $0 = x^2 + 3x + 2.25$, where $a = 1$, $b = 3$, and $c = 2.25$. The quadratic formula says that the solution occurs at

$$x = \frac{-b \pm \sqrt{b^2 - 4ac}}{2a}$$

$$= \frac{-3 \pm \sqrt{(3)^2 - 4(1)(2.25)}}{(2)(1)}$$

$$= \frac{-3 \pm \sqrt{9 - 9}}{2}$$

$$= \frac{-3 \pm \sqrt{0}}{2} \quad \text{(note that the discriminant = 0)}$$

$$= \frac{-3}{2}$$

$$= -1.5$$

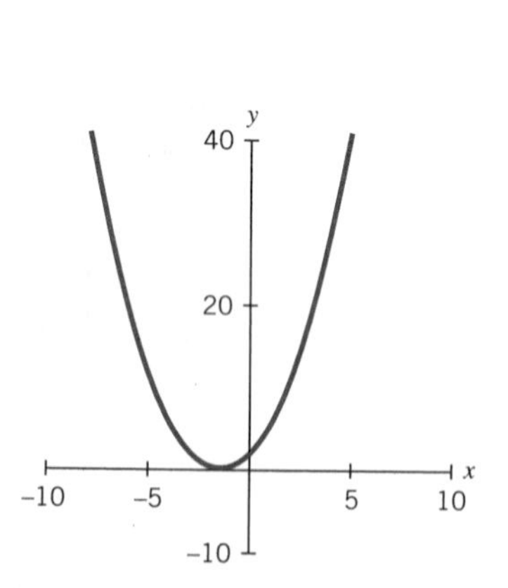

Figure 8.21 Graph of $f(x) = x^2 + 3x + 2.25$, with one x-intercept at the vertex.

In this case, the quadratic formula produces only one value, -1.5, so there is only one x-intercept, at $(-1.5, 0)$, which is also the vertex of the parabola. We know that the parabola opens upward since $a > 0$. (see Figure 8.21).

Example 3 Find the x-intercepts of the function $f(x) = -x^2 - 6x - 10$.

Solution

We can find the intercepts of the function by using the quadratic formula to solve the equation $0 = -x^2 - 6x - 10$. In this case $a = -1, b = -6$, and $c = -10$. Thus, we have

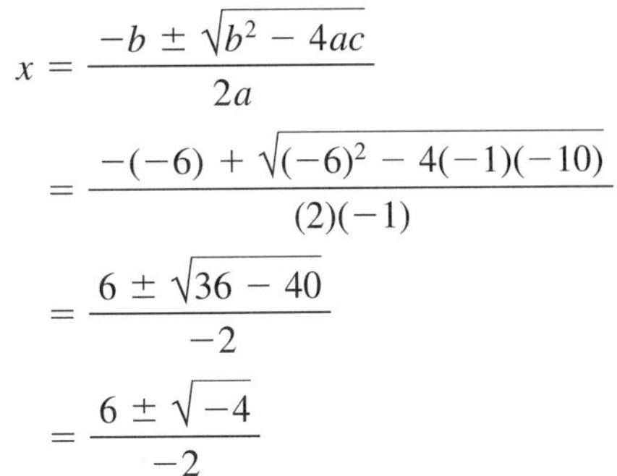

$$\begin{aligned} x &= \frac{-b \pm \sqrt{b^2 - 4ac}}{2a} \\ &= \frac{-(-6) + \sqrt{(-6)^2 - 4(-1)(-10)}}{(2)(-1)} \\ &= \frac{6 \pm \sqrt{36 - 40}}{-2} \\ &= \frac{6 \pm \sqrt{-4}}{-2} \end{aligned}$$

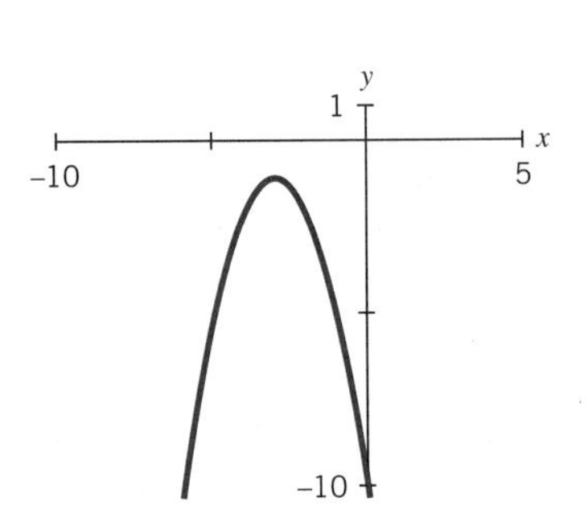

Figure 8.22 Graph of $y = -x^2 - 6x - 10$, with no x-intercepts.

Here the discriminant, -4, is negative, so taking its square root presents a problem, as $\sqrt{-4}$ is not a real number since there is no real number, r, such that $r^2 = -4$. Therefore the solutions or roots $(6 \pm \sqrt{-4})/(-2)$ are not real. Since there are no real values for x such that $f(x) = 0$, there are no x-intercepts, as we can see in Figure 8.22.

Imaginary and complex numbers. Mathematicians were uncomfortable with the notion that certain quadratic equations did not have solutions. So they literally invented a number system in which such equations would be solvable. In the process, they created new numbers, called *imaginary numbers*. The imaginary number i is defined as a number such that

$$i^2 = -1$$

or equivalently

$$i = \sqrt{-1}$$

A number such as $\sqrt{-4}$ is also an imaginary number. We can write $\sqrt{-4}$ as

$$\sqrt{(4)(-1)} = \sqrt{4}\,\sqrt{-1} = 2\sqrt{-1} = 2i$$

When a number is called imaginary it sounds as if it does not exist. But imaginary numbers are just as legitimate as real numbers. Imaginary numbers are used to extend the real number system to a larger system called the complex numbers.

A complex number is defined as any number that can be written in the form

$$z = a + bi$$

where a and b are real numbers and $i = \sqrt{-1}$.

The real part of z is the number a, and the imaginary part is the number b.

For example,

$$-2 + 7i$$

$$4 + \sqrt{-9} = 4 + 3\sqrt{-1} = 4 + 3i$$

$$\sqrt{-25} = 5\sqrt{-1} = 5i = 0 + 5i$$

are all complex numbers.

Solutions to quadratic equations may not be real numbers. In Example 3, the solutions of the quadratic are both complex numbers. Their values are

$$x = \frac{6 + \sqrt{-4}}{-2} = \frac{6}{-2} + \frac{\sqrt{-4}}{-2} = -3 - \frac{2i}{2} = -3 - i$$

and

$$x = \frac{6 - \sqrt{-4}}{-2} = \frac{6}{-2} - \frac{\sqrt{-4}}{-2} = -3 + \frac{2i}{2} = -3 + i$$

So for $f(x) = -x^2 - 6x - 10$, there is no real number x such that $f(x) = 0$, and hence its graph has no x-intercepts. But there are two complex numbers, $-3 - i$ and $-3 + i$, such that $f(-3 - i) = 0$ and $f(-3 + i) = 0$. In this example, $f(x)$ has no real zeros but does have two non-real zeros. The number of distinct real zeros determines the number of x-intercepts.

To find the zeros of the function $f(x) = ax^2 + bx + c$, we use the quadratic formula to solve the equation $0 = ax^2 + bx + c$. The discriminant $b^2 - 4ac$, the term under the radical, can be used to determine the number of real roots and thus the number of x-intercepts.

If the discriminant $b^2 - 4ac = 0$, there is only one real root, at

$$x = \frac{-b}{2a}$$

and the graph has one x-intercept. This only happens when the vertex is on the x-axis. (See Example 2.)

If the discriminant $b^2 - 4ac > 0$, then $\sqrt{b^2 - 4ac}$ is a real non-zero number, which means that there are two real roots and thus two x-intercepts. (See Example 1.)

If the discriminant $b^2 - 4ac < 0$, then $\sqrt{b^2 - 4ac}$ is not a real number, which means that there are no real roots, and the graph has no x-intercepts. (See Example 3.)

In summary:

> To find the x-intercepts of a quadratic function
>
> $$f(x) = ax^2 + bx + c$$
>
> we use the quadratic formula to find the solutions, or roots, of the associated equation
>
> $$0 = ax^2 + bx + c$$
>
> If $b^2 - 4ac > 0$, there are two real roots and the graph has two x-intercepts.
> If $b^2 - 4ac = 0$, there is one real root and the graph has one x-intercept.
> If $b^2 - 4ac < 0$, there are no real roots and the graph has no x-intercepts.

Algebra Aerobics 8.3a

1. Return to Figure 8.19 and for each function specify the number of real zeros.
2. Evaluate the discriminant and then predict the number of x-intercepts for each function. Use the quadratic formula to find all the zeros of each function and identify the coordinates of any x-intercept(s).
   a. $y = 4 - x - 5x^2$
   b. $y = 4x^2 - 28x + 49$
   c. $y = 2x^2 + 5x + 4$
3. Find and interpret the horizontal and vertical intercepts for the following height equations.
   a. $h = -4.9t^2 + 50t + 80$ (h is in meters and t is in seconds)
   b. $h = 150 - 80t - 490t^2$ (h is in centimeters and t is in seconds)

The Factored Form

When a quadratic equation is in factored form or can be put into factored form, we can easily find the zeros of the function. A quadratic function written as the product of two linear terms, both involving the independent variable, is said to be in factored form. For example, the function

$$f(x) = (x - 4)(x + 5)$$

is in factored form. To find the zeros, set $f(x) = 0$ and solve for x:

$$0 = (x - 4)(x + 5) \tag{1}$$

We can apply the zero product rule (see Section 8.1) to Equation (1) to solve for x:

$$x - 4 = 0 \quad \text{or} \quad x + 5 = 0$$
$$x = 4 \qquad\qquad x = -5$$

The x-intercepts are at 4 and -5, or, equivalently, the function crosses the x-axis at (4, 0) and (−5, 0).

To change between the a–b–c form of the quadratic function and the factored form, some knowledge of factoring and multiplying binomials is necessary. We offer a short review of these skills.

Factoring Review

Multiplying Binomials. When we multiply binomials, we use the distributive law. For example, to multiply $(x + 2)(x + 5)$, we need to multiply each term in the first expression by each term in the second expression:

Apply distributive law	$(x + 2)(x + 5) = x(x + 5) + 2(x + 5)$
apply distributive law again	$= x^2 + 5x + 2x + 10$
simplify	$= x^2 + 7x + 10$

We say that $x^2 + 7x + 10$ is the *product* of $x + 2$ and $x + 5$ or that $x + 2$ and $x + 5$ are *factors* of $x^2 + 7x + 10$.

We can generalize to any two binomials:

$$(a + b)(c + d) = a(c + d) + b(c + d)$$
$$= ac + ad + bc + bd$$

One strategy for remembering all four terms in the product is to consider those four terms as products of the first (F), outside (O), inside (I), and last (L) two terms of the factors:

$$\overset{\text{F}\quad\text{L}}{(a+b)(c+d)} = \underset{\text{F}}{ac} + \underset{\text{O}}{ad} + \underset{\text{I}}{bc} + \underset{\text{L}}{bd}$$

(I = inside terms b and c; O = outside terms a and d)

Factoring quadratics. To convert $ax^2 + bx + c$ into factored form requires thinking, practice, and a few hints. It is often a trial-and-error process. We usually restrict ourselves to finding factors with integer coefficients.

First, look for common factors in all of the terms.

For example, $10x^2 + 2x$ can be factored as $2x(5x + 1)$.

Second, look for two linear factors.

This is easiest to do when the coefficient of x^2 is 1. For example, to factor $x^2 + 7x + 12$, we want to rewrite it as

$$(x + m)(x + n)$$

for some m and n. Note that the coefficients of both x's in the factors equal 1, since x times x is equal to the x^2 in the original expression. Now we need to determine values for the constants m and n. We know that when we multiply $m \cdot n$ we need to get 12. Since 12 is positive, m and n must have the same sign. Since the coefficient of the sum of the outside and inside terms is positive 7, m and n must both be positive. So we consider pairs of positive integers whose product is 12, namely, 1 and 12, 2 and 6, or 3 and 4. We can then narrow our list of factors of 12 to those whose sum equals 7, the coefficient of the x term. Only the factors 3 and 4 fit this criterion. We can factor our polynomial as:

$$x^2 + 7x + 12 = (x + 3)(x + 4)$$

We can check that these factors work by multiplying them out.

Third, when factoring a binomial, look for the special case of the difference of two squares.

In this case the middle terms cancel out when multiplying:

$$\begin{aligned} x^2 - 25 &= (x - 5)(x + 5) \\ &= x^2 - 5x + 5x - 25 \\ &= x^2 - 25 \end{aligned}$$

In general,

$$x^2 - n^2 = (x - n)(x + n)$$

We can use these strategies to see if a quadratic function can be easily factored. If a quadratic function is in factored form, we can then easily identify the horizontal intercepts.

Example 4

Write the function $f(x) = 5700x^2 + 3705x$ in factored form. Identify the x- and y-intercepts. Sketch a graph of the function.

Solution

Given $\qquad f(x) = 5700x^2 + 3705x$

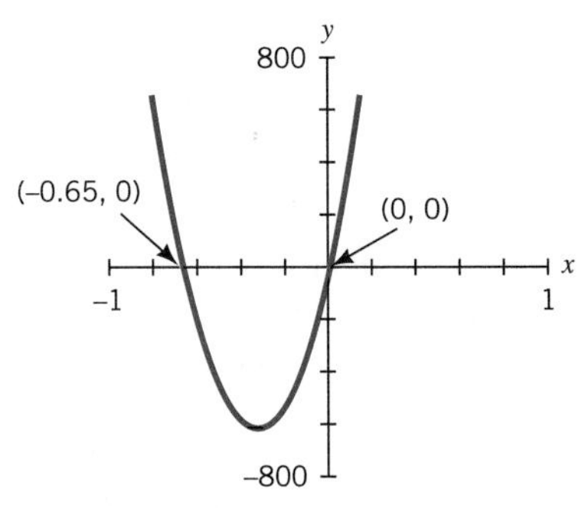

Figure 8.23 Graph of $f(x) = 5700x^2 + 3705x$.

factor out $15x$	$= 15x(380x + 247)$
factor out 19	$= 15x \cdot 19\,(20x + 13)$
	$= 285x\,(20x + 13)$

To find the x-intercepts of the function, set $f(x)$ equal to zero and solve the resulting equation:

Given	$f(x) = 285x\,(20x + 13)$
set $f(x) = 0$	$0 = 285x\,(20x + 13)$
apply zero product rule to get	$285x = 0$ or $20x + 13 = 0$
solve for x	$x = 0$ or $20x = -13$
	$x = \dfrac{-13}{20}$
	$= -0.65$

The function crosses the x-axis when $x = -0.65$ and when $x = 0$, or equivalently at the points $(-0.65, 0)$ and the origin (0, 0). The origin is also the y-intercept. The parabola opens up (since the coefficient of x^2 is positive) and crosses the x-axis twice, so we know that the vertex must be below the x-axis (see Figure 8.23).

Example 5

Write the function $y = -3x^2 + 12x - 12$ in factored form and identify the x- and y-intercepts and the vertex. Sketch the graph.

Solution

Given	$y = -3x^2 + 12x - 12$
factor out -3	$= -3(x^2 - 4x + 4)$
factor the remaining trinomial	$= -3(x - 2)(x - 2)$

and we have y in factored form.

To find the x-intercepts, set $y = 0$ and solve the equation for x:

Given	$y = -3(x - 2)(x - 2)$
set $y = 0$	$0 = -3(x - 2)(x - 2)$
apply the zero product rule	$x - 2 = 0$
solve for x	$x = 2$

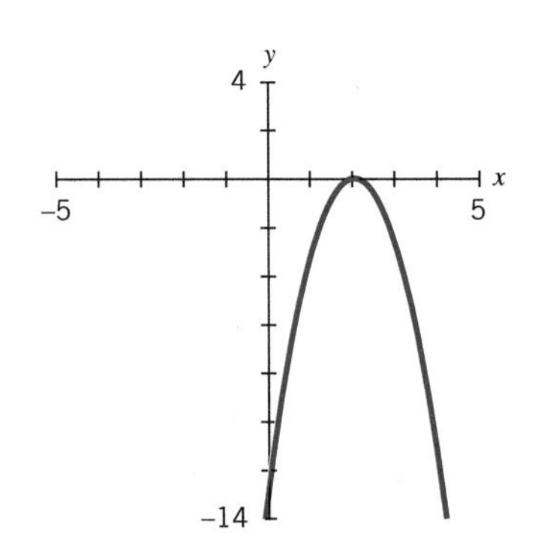

Figure 8.24 Graph of $y = -3x^2 + 12x - 12$.

The function $y = -3x^2 + 12x - 12$ intersects the x-axis once when $x = 2$. In this case there is only one x-intercept at (2, 0), which is also the vertex of the parabola. (See Figure 8.24.) The parabola opens down, since $a = -3$, and it intersects the y-axis at $(0, -12)$.

Algebra Aerobics 8.3b

1. When possible, put the function in factored form with integer coefficients and find the horizontal intercepts.
   a. $y = -16t^2 + 50t$
   b. $y = x^2 + x - 6$
   c. $y = 2x^2 + x - 5$
   d. $h(t) = 69 - 9t^2$
   e. $f(x) = -2x^2 + 12x + 54$
   f. $g(x) = 64x^2 + 16x + 4$
2. Find the y-intercept of each parabola.
   a. $y = 2x - 3x^2$
   b. $y = 5 - x - 4x^2$
   c. $y = \frac{2}{3}x^2 + 6x - \frac{11}{3}$

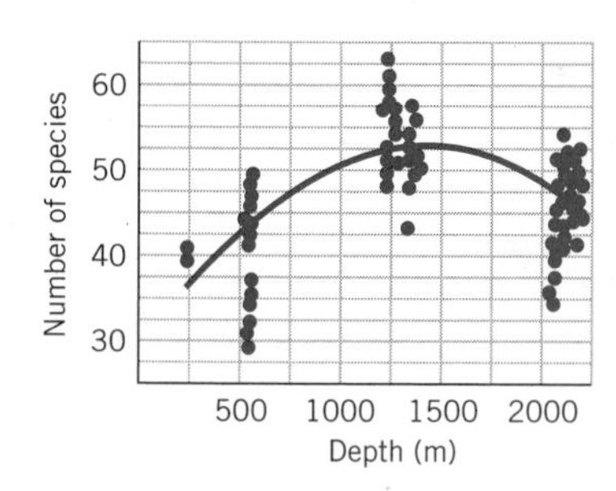

Figure 8.25 A best fit quadratic function for species diversity as a function of depth in northern regions.

Source: R. J. Etter and F. Grassle, "Patterns of Species Diversity in the Deep Sea as a Function of Sediment Particle Size Diversity," *Nature,* Vol. 360, pp. 576–578, Dec. 10, 1992. Reprinted with permission from Macmillan Magazines Limited.

8.4 FINDING THE VERTEX

Why the Vertex Is Important

Maximum and Minimum Values. Since the vertex is the point at which a parabola reaches a maximum or minimum value, the vertex often assumes particular significance in a model. For example, biologists have discovered a quadratic relationship between species diversity and ocean depth. The graph in Figure 8.25 shows the number of different species as a function of the depth of the water in cold northern oceans. As the depth increases, the number of species initially increases, reaches a maximum at the vertex (at a depth of approximately 1250 meters), and then decreases.[3]

Using a Formula to Find the Vertex

The following formula can be used to find the coordinates of the vertex of a parabola.

See "Why the Formula for the Vertex and the Quadratic Formula Work."

> The vertex of the quadratic function
> $$y = ax^2 + bx + c$$
> has coordinates $\left(-\dfrac{b}{2a}, -\dfrac{b^2}{4a} + c\right)$

Example 1

The function $h(t) = 34 + 32t - 16t^2$ describes the height of a projectile (in feet) after t seconds. In Example 1 in the previous section, we found the horizontal intercepts for this function to be $(-0.77, 0)$ and $(2.77, 0)$. Now find the vertex and the vertical intercept and describe their significance.

Solution

Remember that a is the coefficient of the squared term, which in this case happens to be the third term. So $h(t) = c + bt + at^2$ and $a = -16, b = 32$, and $c = 34$. Since a is negative, we know that the graph opens downward, so the vertex represents a maximum value.

Using the formula for the horizontal coordinate of the vertex, we have

$$-\frac{b}{2a} = -\frac{32}{2(-16)} = \frac{-32}{-32} = 1$$

Using the formula for the vertical coordinate of the vertex, we get

$$-\frac{b^2}{4a} + c = -\frac{32^2}{4(-16)} + 34 = \frac{-1024}{-64} + 34 = 16 + 34 = 50$$

So the vertex is at $(1, 50)$.

Since the formula for the vertical coordinate of the vertex is pretty complicated to remember, you may prefer to only use the formula $-\frac{b}{2a}$ to find the horizontal coordinate. To find the vertical coordinate of the vertex, you can evaluate the

[3]According to the authors of the article, each point represents data from a single core sample with a cross section of 0.25 m^2. Each sample costs more than $100,000 to collect and analyze!

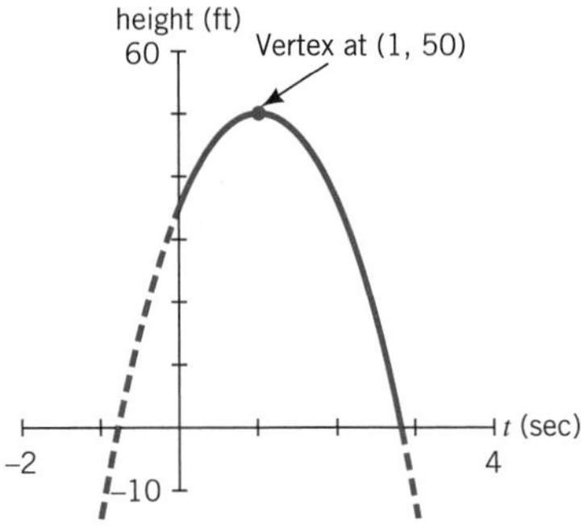

function at the appropriate value of the horizontal coordinate, which in this case is $t = 1$. Setting $t = 1$,

$$h(1) = 34 + 32(1) - 16(1)^2$$
$$= 34 + 32 - 16$$
$$= 50$$

Both methods give the same result.

The vertex represents the point at which the object reaches a maximum height. So the coordinates of the vertex, (1, 50), tell us that at 1 second, the object reaches a maximum height of 50 feet. See Figure 8.26.

To find the vertical intercept, we evaluate the function at $h(0)$.

$$h(0) = 34 + 32(0) - 16(0)^2$$
$$= 34$$

In this problem, the vertical intercept represents the initial height (in feet) of the projectile. Thus, 34 feet is the initial height of the projectile or the height at 0 seconds.

Note that our height model consists only of the solid part of the curve. Values of negative time or negative height do not have meaning in this problem.

Figure 8.26 Graph of the function $h(t) = 34 + 32t - 16t^2$.

Example 2

Find the vertex and sketch the graph of $f(x) = x^2 - 10x + 100$.

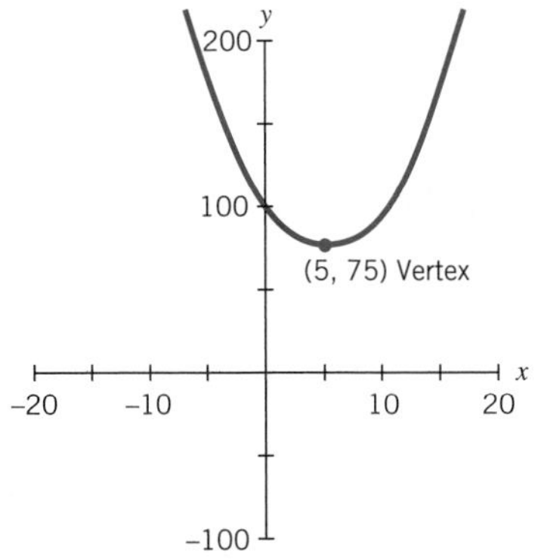

Solution

Here $a = 1$, $b = -10$, and $c = 100$. Since the coefficient a is positive, we know that the graph opens upward and the vertex represents a minimum value.

Using the formula for the horizontal coordinate of the vertex, we have

$$x = -\frac{b}{2a} = -\frac{-10}{2(1)} = \frac{10}{2} = 5$$

To find the vertical coordinate, we can find the value of $f(x)$ when $x = 5$.

Given	$f(x) = x^2 - 10x + 100$
let $x = 5$	$f(5) = 5^2 - 10(5) + 100$
simplify	$= 75$

The coordinates of the vertex are (5, 75). Figure 8.27 shows a sketch of the graph.

Figure 8.27 Graph of $y = x^2 - 10x + 100$.

Algebra Aerobics 8.4a

1. Find the vertex of the graph of each of the following quadratic functions:
   a. $f(x) = 2x^2 - 4$
   b. $g(z) = -z^2 + 6$
   c. $w = 4t^2 + 1$
2. Find the vertex of the graph of each of the following functions and then sketch the graphs on the same grid:
   a. $y = x^2 + 3$
   b. $y = -x^2 + 3$
3. In parts (a) to (d) determine the vertex and whether the graph opens upward or downward. Then predict the number of x-intercepts. Graph the function to confirm your answer. Estimate the values of the x-intercepts.
   a. $f(x) = -3x^2$
   b. $f(x) = -2x^2 - 5$
   c. $f(x) = x^2 + 4x - 7$
   d. $f(x) = 4 - x - 2x^2$
4. Find the vertex and horizontal intercepts and draw a rough sketch of the graph of each of the following functions:
   a. $y = x^2 + 3x + 2$
   b. $f(x) = 2x^2 - 4x + 5$
   c. $g(t) = -t^2 - 4t - 7$

The Vertex Form: The a–h–k Form

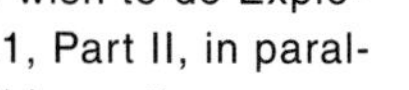

You may wish to do Exploration 8.1, Part II, in parallel with this section.

There is another convenient form for writing quadratic functions that allows us easily to identify not only whether a quadratic function has a maximum or minimum at its vertex but also the specific coordinates of the vertex. A quadratic function written

$$f(x) = a(x - h)^2 + k$$

is said to be in a–h–k form. In this section we'll see that the coordinates of the vertex are (h, k).

Getting to the a–h–k Form: Parabolic Shifts

How can we transform a function in a–b–c form into the a–h–k form? Why does a function written in a–h–k form have a vertex at (h, k)? Do the vertex coordinates (h, k) coincide with the vertex coordinates given earlier in terms of a, b, and c? The answers to these questions lie in vertical and horizontal shifts of a simple quadratic function of the form $f(x) = ax^2$.

"Q3: *a, h, k* Sliders" in *Quadratic Functions* can help you visualize quadratic functions in the *a–h–k* form.

Shifting a parabola horizontally h units. The simplest quadratic functions are power functions, such as

$$f(x) = 2x^2$$

where $a = 2$ and b and c are both zero. The graph of this function is a parabola that opens upward (since the coefficient of x^2 is positive) and therefore has a minimum at its vertex. The vertex is at (0, 0), the origin.

Suppose we compare $f(x) = 2x^2$ to a new function $g(x)$, where

$$g(x) = 2(x - 3)^2$$

The function $g(x)$ tells us to subtract 3 from x, square the result, and then multiply by 2. The graph of this function is the same as the graph of the original function $f(x) = 2x^2$, except that it is shifted *to the right* by 3 units.

To remember the direction of the shift, compare the positions of the vertices of $f(x)$ and $g(x)$. Since $g(x) \geq 0$, its vertex represents a minimum that occurs when $g(x) = 0$. Since $g(3) = 2(3 - 3)^2 = 2 \cdot 0 = 0$, the vertex for $g(x)$ is at (3, 0). So the vertex has been shifted 3 units to the right, from (0, 0) for $f(x)$ to (3, 0) for $g(x)$. (See Figure 8.28.)

What if we compare the function $f(x) = 2x^2$ to a new function $j(x)$, where

$$j(x) = 2(x + 5)^2$$

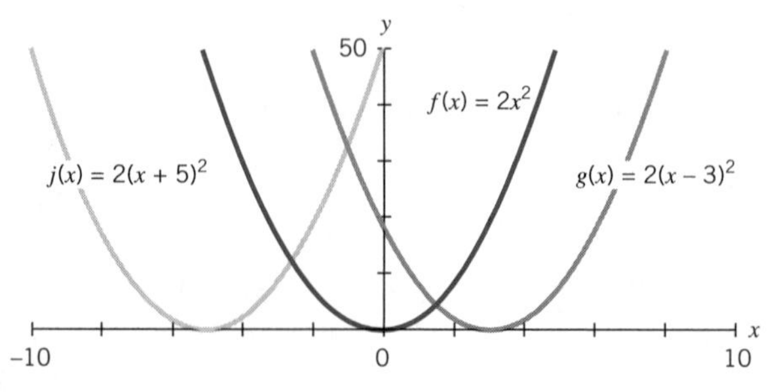

Figure 8.28 Graphs of $f(x) = 2x^2$, $g(x) = 2(x - 3)^2$, and $j(x) = 2(x + 5)^2$.

the function $j(x)$ tells us to add 5 to x, square the result, and multiply by 2. The graph of this function is the same as the graph of the original function $f(x) = 2x^2$, except that it is shifted *to the left* by 5 units. The vertex of $j(x)$ is at $(-5, 0)$, or 5 units to the left of $(0, 0)$, the vertex of $f(x)$ (see Figure 8.28).

Figure 8.28 shows, on the same grid, the graphs of the functions $f(x)$, $g(x)$, and $j(x)$, where

$$f(x) = 2x^2$$

$$g(x) = 2(x - 3)^2$$

$$= 2[x - (+3)]^2 \quad \text{graph of } f(x) \text{ shifted 3 units to the right}$$

$$j(x) = 2(x + 5)^2$$

$$= 2[x - (-5)]^2 \quad \text{graph of } f(x) \text{ shifted 5 units to the left}$$

How would the graph of $y = 2(x - 4)^3$ compare to the graph of $y = 2x^3$? What about the graph of $y = 2(x + 4)^3$?

In general, if we compare the function $f(x) = ax^2$ to a new function

$$g(x) = a(x - h)^2$$

the graph of $g(x)$ is the graph of $f(x)$ shifted horizontally h units. The shift will be to the right if h is positive and to the left if h is negative. In particular, the vertex of $g(x)$ at $(h, 0)$ is the vertex of $f(x)$ at $(0, 0)$ shifted horizontally h units.

Shifting a parabola vertically k units. Suppose we wanted to shift the graph vertically, in the y direction. Recall from Section 8.2 that the constant term acts as an "elevator"; that is, in order to raise or lower a function, we simply add a constant term. For example, the graph of the function $y = 2(x - 3)^2 + 10$ is the graph of $y = 2(x - 3)^2$ shifted up 10 units (Figure 8.29). The graph of $y = 2(x - 3)^2 - 5$ is the graph of $y = 2(x - 3)^2$ shifted down 5 units.

In general, to shift the function $f(x) = a(x - h)^2$ vertically k units, we add a constant term k. So the graph of the function

$$g(x) = a(x - h)^2 + k$$

is the same as the graph of $f(x)$ shifted vertically k units. The graph is shifted up if k is positive and down if k is negative. In particular, the vertex $(h, 0)$ for $f(x)$ is vertically shifted k units to become (h, k), the vertex for $g(x)$.

How do you think the graph of $y = 2(x - 4)^3 + 7$ compares to the graph of $y = 2x^3$? What about the graph of $y = 2(x - 4)^3 - 5$? Check your predictions by plotting the functions.

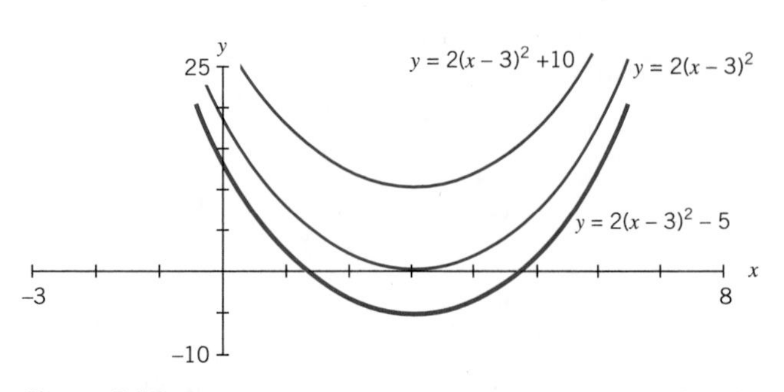

Figure 8.29 Three identically shaped parabolas, one 10 units above $y = 2(x - 3)^2$ and one 5 units below.

Putting all this together we have:

For a graphic illustration of the shift from $a–b–c$ to $a–h–k$ form, see "Q7: From $a–b–c$ to $a–h–k$ Form" and "Q8: $y = ax^2$ vs. $y = a(x - h)^2 + k$" in *Quadratic Functions*.

> A quadratic function in the $a–h–k$ form,
>
> $$y = a(x - h)^2 + k$$
>
> has a vertex at (h, k) and the graph is symmetrical about the vertical line $x = h$.
>
> Its graph is the same as the graph of $y = ax^2$ that has been:
>
> shifted h units horizontally (to the right if $h > 0$, to the left if $h < 0$)
> shifted k units vertically (up if $k > 0$, down if $k < 0$)

Example 3

Identify the number of x-intercepts for the function $y = (x - 3)^2 - 2$.

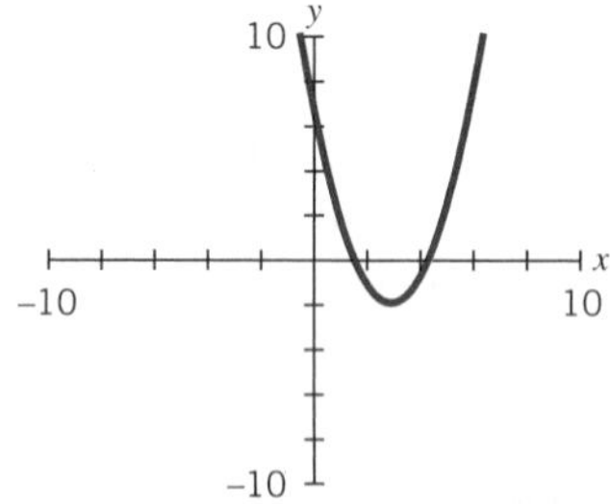

Figure 8.30 Graph of $y = (x - 3)^2 - 2$

Solution

The quadratic function $y = (x - 3)^2 - 2$ is in the $a–h–k$ form, where $a = 1$, $h = 3$, and $k = -2$. Its vertex is at $(3, -2)$. Since the value of the coefficient a is positive, the graph of the function opens upward and the vertex represents a minimum.

Since the vertex is below the x-axis and the graph opens upward, the function has two x-intercepts. A sketch of the graph is shown in Figure 8.30.

Example 4

Identify the vertex of the function $y = -3(x + 5)^2 + 10$.

Solution

We have $a = -3$ and $k = 10$. We have to be careful in identifying the value for h. If we think of $x + 5$ as $x - (-5)$, then we have written the expression exactly in the $x - h$ format, and it is clear that $h = -5$. The function then has a vertex at $(-5, 10)$. Since $a < 0$, the parabola opens downward and the vertex represents a maximum.

Algebra Aerobics 8.4b

For problems 1 to 3, graph parts (a), (b), and (c) on the same grid. Compare the positions of the vertices of parts (b) and (c) to that of part (a).

1. **a.** $y = x^2$
   b. $y = (x + 3)^2$
   c. $y = (x - 2)^2$
2. **a.** $f(x) = 0.5x^2$
   b. $f(x) = 0.5(x - 1)^2$
   c. $f(x) = 0.5(x + 4)^2$
3. **a.** $r = -2t^2$
   b. $r = -2(t + 1.2)^2$
   c. $r = -2(t - 0.9)^2$

For problems 4 and 5, graph the four parabolas on the same coordinate plane, and then compare the graphs of parts (b), (c), and (d) to that of part (a).

4. **a.** $y = x^2$
   b. $y = (x - 2)^2$
   c. $y = (x - 2)^2 + 4$
   d. $y = (x - 2)^2 - 3$
5. **a.** $y = -x^2$
   b. $y = -(x + 3)^2$
   c. $y = -(x + 3)^2 - 1$
   d. $y = -(x + 3)^2 + 4$
6. Identify the number of x-intercepts for the following functions:
   a. $y = 3(x - 1)^2 + 5$
   b. $y = -2(x + 4)^2 - 1$
   c. $y = -5(x + 3)^2$
   d. $y = 3(x - 1)^2 - 2$

Converting from the $a-h-k$ to the $a-b-c$ Form of a Quadratic

Every function written in $a-h-k$ form can be rewritten as a function in $a-b-c$ form if we multiply out and group terms with the same power of x.

Example 5

Rewrite the quadratic $f(x) = 3(x + 7)^2 - 9$ in the $a-b-c$ form.

Solution

The function $f(x)$ is in the $a-h-k$ form where $a = 3$, $h = -7$, and $k = -9$:

Given	$f(x) = 3(x + 7)^2 - 9$
write out the factors	$= 3(x + 7)(x + 7) - 9$
multiply the factors	$= 3(x^2 + 14x + 49) - 9$
distribute the 3	$= 3x^2 + 42x + 147 - 9$
group the constant terms	$= 3x^2 + 42x + 138$

This function is in the $a-b-c$ format with $a = 3$, $b = 42$, and $c = 138$.

Two Strategies for Getting from the $a-b-c$ to the $a-h-k$ Form of a Quadratic

Now suppose we want to go the other way. Suppose we are given a quadratic function in the $a-b-c$ form:

$$g(x) = ax^2 + bx + c$$

and we want to write it in the $a-h-k$ form. We want to write

$$g(x) = a(x - h)^2 + k$$

for some values of a, h, and k.

Strategy 1: "Completing the Square". We can convert the function $f(x) = x^2 + 14x + 9$ into $a-h-k$ form using a method called completing the square.

When a function is in $a-h-k$ form, the term $(x - h)^2$ is a perfect square; that is, $(x - h)^2$ is the product of the expression $x - h$ times itself. So we examine separately the expression $x^2 + 14x$ and ask what constant term we would need to add to it in order to make it a perfect square.

A perfect square is in the form:

$$(x + r)^2 = x^2 + 2rx + r^2$$

for some number r. Notice that the coefficient of x is two times the number r in our expression for a perfect square.

To complete the square in the expression $x^2 + 14x$, we need to find r. Since $2r = 14$, $r = 7$. To complete the square, we square 7 and get 49.

$$(x + 7)^2 = x^2 + 14x + 49$$

We can now translate $f(x) = x^2 + 14x + 9$ into the $a-k-h$ form by adding 49 in order to make a perfect square, and then subtracting 49, in order to preserve equality. So we have:

Given	$f(x) = x^2 + 14x + 9$
add and subtract 49	$= x^2 + 14x + (49 - 49) + 9$
regroup terms	$= (x^2 + 14x + 49) - 49 + 9$
factor and simplify	$= (x + 7)^2 - 40$

We now have $f(x)$ in a–h–k form. The vertex is at $(-7, -40)$.

Strategy 2: Using the Formula for the Vertex. We can convert $g(x) = 3x^2 - 12x + 5$ to the a–h–k form by using the formula for the coordinates of the vertex. Since the coefficient a is the same in both the a–b–c and the a–h–k forms, we have $a = 3$. The coordinates of the vertex of a quadratic in the a–b–c form are given by $(-\frac{b}{2a}, -\frac{b^2}{4a} + c)$. For $g(x)$ we have $a = 3$, $b = -12$, and $c = 5$. So

$$-\frac{b}{2a} = -\frac{-12}{2 \cdot 3} = \frac{12}{6} = 2$$

and

$$-\frac{b^2}{4a} + c = -\frac{(-12)^2}{4 \cdot 3} + 5 = -\frac{144}{12} + 5 = -12 + 5 = -7$$

By multiplying out and combining like terms, verify that

$$y = a\left(x + \frac{b}{2a}\right)^2 - \frac{b^2}{4a} + c$$

and

$$y = ax^2 + bx + c$$

are two forms of the same function.

The vertex is at $(2, -7)$. In the a–h–k form the vertex is at (h, k) so we must have $h = 2$ and $k = -7$. Substituting for a, h, and k in $g(x) = a(x - h)^2 + k$, we get

$$g(x) = 3(x - 2)^2 - 7$$

and the transformation is complete.

In general, to convert any quadratic function $f(x) = ax^2 + bx + c$ in the a–b–c form into an equivalent a–h–k form, the value of the coefficient a remains the same, $h = -\frac{b}{2a}$ and $k = -\frac{b^2}{4a} + c$, and therefore

$$f(x) = a\left(x + \frac{b}{2a}\right)^2 - \frac{b^2}{4a} + c$$

Example 6

Convert the function $g(t) = -2t^2 + 12t - 23$ into a–h–k form.

Solution

This function is more difficult to convert by completing the square, since a is not 1. We need first to factor out -2 *from the t terms only,* getting

$$g(t) = -2(t^2 - 6t) - 23 \qquad (1)$$

It is the expression $t^2 - 6t$ for which we must complete the square. Since

$$t^2 - 6t + 9 = (t - 3)^2$$

we must add the constant term 9 *inside the parentheses* in Equation (1) in order to make the binomial a perfect square. Since everything inside the parentheses is multiplied by -2, we have essentially subtracted 18 from our original function, and so we need to add 18 *outside the parentheses* in order to preserve equality. So we have:

Given	$g(t) = -2t^2 + 12t - 23$
factor out -2 from t terms	$= -2(t^2 - 6t) - 23$

add 9 inside parentheses and
18 outside parentheses $= -2(t^2 - 6t + 9) + 18 - 23$

factor and simplify $= -2(t - 3)^2 - 5$

We now have $g(t)$ in a–h–k form. The vertex is at $(3, -5)$.

Example 7

Find the equation of the parabola in Figure 8.31. Write it in a–h–k and a–b–c form.

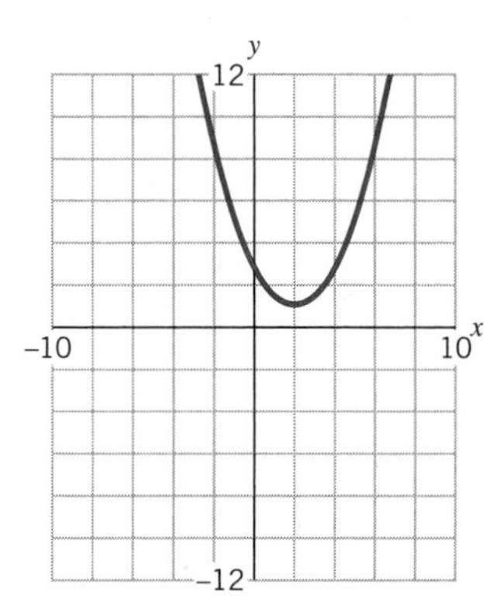

Figure 8.31 A mystery parabola.

Solution

The vertex of the graph appears to be at (2, 1) and the graph opens upward. Using this estimate of the vertex, we can substitute the coordinates of the vertex in the a–h–k form of the quadratic equation to get

$$y = a(x - 2)^2 + 1 \qquad (1)$$

How can we find the value for a? If we can identify values for any other point (x, y) that lies on the parabola, we can substitute these values into Equation (1) and solve it for a. The y-intercept, (0, 3), is a convenient point to pick. Setting $x = 0$ and $y = 3$, we get

$$3 = a(0 - 2)^2 + 1$$

$$3 = 4a + 1$$

$$2 = 4a$$

so $a = 0.5$

The equation in the a–h–k form is

$$y = 0.5(x - 2)^2 + 1$$

If we wanted it in the equivalent a–b–c form, we could square, multiply, and collect like terms to get

$$y = 0.5x^2 - 2x + 3$$

Algebra Aerobics 8.4c

1. Convert the following functions into a–h–k form by completing the square:
   a. $f(x) = x^2 + 2x - 1$
   b. $j(z) = 4z^2 - 8z - 6$
   c. $h(x) = -3x^2 - 12x$
2. Express each of the following functions in the form $y = ax^2 + bx + c$.
   a. $y = 2(x - \frac{1}{2})^2 + 5$
   b. $y = -\frac{1}{3}(x + 2)^2 + 4$
3. Express each of the following functions in the form $y = a(x - h)^2 + k$.
   a. $y = x^2 + 6x + 7$
   b. $y = 2x^2 + 4x - 11$
4. Find the coordinates of the vertex and the x- and y-intercepts, and graph the following functions:
   a. $y = x^2 + 8x + 11$
   b. $y = 3x^2 + 4x - 2$
5. Find the coordinates of the vertex and the x- and y-intercepts, and graph the following functions:
   a. $y = 0.1(x + 5)^2 - 11$
   b. $y = -2(x - 1)^2 + 4$

8.5 AVERAGE RATE OF CHANGE OF A QUADRATIC FUNCTION

In previous chapters, we saw that the average rate of change of a linear function is constant and of an exponential function it is exponential. In this section we examine the average rate of change of a quadratic function.

Let's begin with the simplest quadratic, $y = f(x) = x^2$. Table 8.2 shows the average rate of change of f over each interval in the table from x to $x + 1$. It is clearly not constant.

Now let's examine the average rate of change of the average rate of change. We find that it is constant, that is, as x increases by 1, the average rate of change increases at a constant rate of 2 (see column 4 of Table 8.2). The average rate of change is a linear expression in x.

Table 8.2

x	$y = x^2$	Average Rate of change from x to $x + 1$	Average Rate of Change of the Average Rate of Change
-3	9	$\dfrac{4-9}{-2-(-3)} = -5$	
-2	4	$(1-4)/1 = -3$	$\dfrac{-3-(-5)}{-2-(-3)} = 2$
-1	1	$(0-1)/1 = -1$	$(-1-(-3))/1 = 2$
0	0	$(1-0)/1 = 1$	$(1-(-1))/1 = 2$
1	1	$(4-1)/1 = 3$	$(3-1)/1 = 2$
2	4	$(9-4)/1 = 5$	$(5-3)/1 = 2$
3	9		

Let's examine another quadratic, $y = f(x) = 3x^2 - 8x - 23$. Column 3 of Table 8.3 contains the average rate of change of f over each interval from x to $x + 1$. As the value of x increases by 1, the value for the average rate of change increases by 6. So the average rate of change, while not constant, increases at a constant rate of 6. The average rate of change is a linear expression in x.

Table 8.3

x	$y = 3x^2 - 8x - 23$	Average Rate of Change	Average Rate of Change of the Average Rate of Change
-3	28	-23	
-2	5	-17	6
-1	-12	-11	6
0	-23	-5	6
1	-28	1	6
2	-27	7	6
3	-20		

We have seen numerically in two examples that the average rate of change of a quadratic function is a linear function. We now demonstrate algebraically that this is true for *every* quadratic function.

Suppose that y is a quadratic function of x, written using functional notation as $y = f(x) = ax^2 + bx + c$. In the previous examples we fixed an interval size of 1 over which to calculate the average rate of change, since it is easy to make comparisons. Now we pick a constant interval size r and for each position x compute the average rate of change of f over the interval from x to $x + r$.

First we must compute $f(x + r)$:

$$f(x + r) = a(x + r)^2 + b(x + r) + c$$

Apply exponent $\quad = a(x^2 + 2rx + r^2) + b(x + r) + c$

multiply through $\quad = ax^2 + 2arx + ar^2 + bx + br + c$

regroup terms $\quad = ax^2 + bx + c + (2ax + b)r + ar^2$

Then the average rate of change of $f(x)$ between x and $x + r$ is

$$\frac{\text{change in } f(x)}{\text{change in } x} = \frac{f(x + r) - f(x)}{(x + r) - x}$$

$$= \frac{[ax^2 + bx + c + (2ax + b)r + ar^2] - (ax^2 + bx + c)}{r}$$

$$= \frac{(2ax + b)r + ar^2}{r}$$

$$= 2ax + (b + ar) \tag{1}$$

Figure 8.32 shows the relationship between $f(x)$ and the slope of the line segment connecting $(x, f(x))$ and $(x + r, f(x + r))$. This is a linear function of x, with

$$\text{slope} = 2a \qquad y\text{-intercept} = b + ar$$

Note that the slope depends only on the original equation (in particular only on the value of a). The y-intercept depends not only on a and b in the original equation, but on r, the interval size over which we calculate the average rate of change.

If we took smaller and smaller values for r, then the term ar would get closer and closer to zero, and hence the y-intercept $b + ar$ would get closer and closer to b. For very small r's the average rate of change would get closer and closer to the linear expression $2ax + b$.

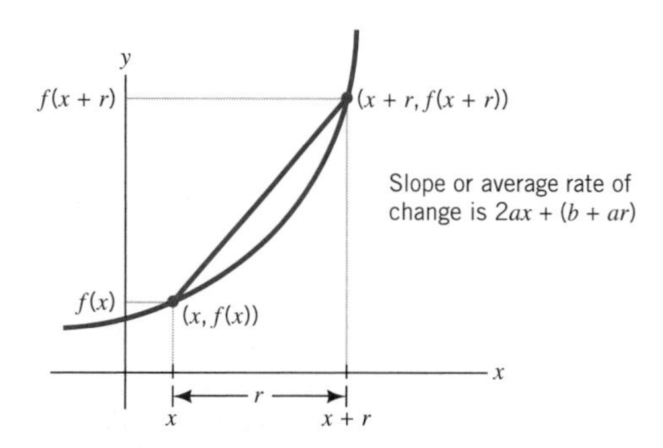

Figure 8.32 The slope of the line segment connecting two points on the parabola separated by a horizontal distance of r is $2ax + (b + ar)$.

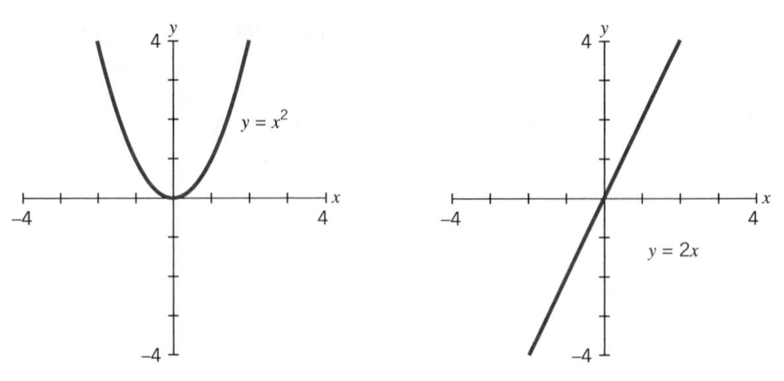

Figure 8.33 Graph of $y = x^2$ and graph of its average rate of change.

For the functions $y = x^2$ and $y = 3x^2 - 8x - 23$, the expressions for their respective average rates of change as functions of x are $y = 2x$ and $y = 6x - 8$. (See Figures 8.33 and 8.34.)

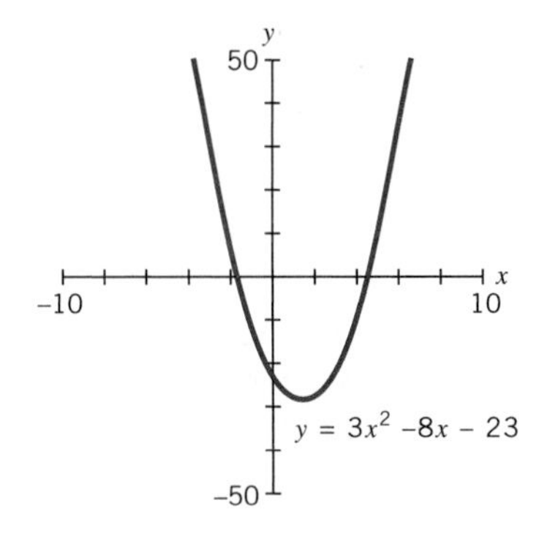

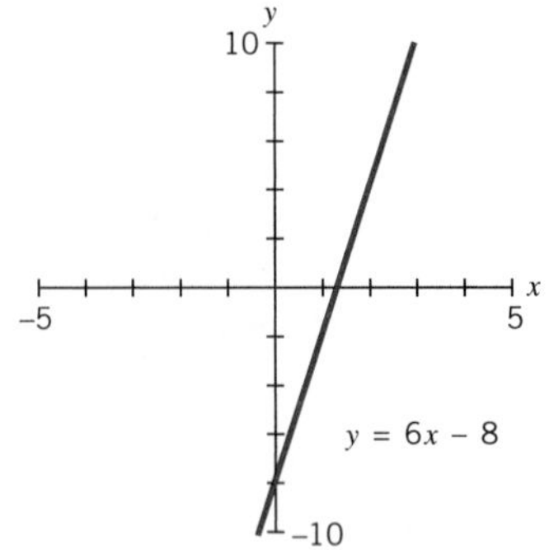

Figure 8.34 Graph of $y = 3x^2 - 8x - 23$ and graph of its average rate of change.

> Given a quadratic function $y = ax^2 + bx + c$ (where a, b, and c are constants and $a \neq 0$), the average rate of change is linear, and over very small intervals approaches $2ax + b$.

This is a central idea that you meet again in calculus and that you use in the Extended Exploration that follows this chapter.

Chapter Summary

Polynomial functions. A *polynomial function* $y = f(x)$ can be represented by an equation in the form

$$y = a_n x^n + a_{n-1}x^{n-1} + \cdots + a_1 x^1 + a_0$$

where each coefficient $a_n, a_{n-1}, \ldots, a_0$ is a constant and n is a non-negative integer, called the *degree* of the polynomial (provided $a_n \neq 0$).

The degree n of a polynomial function determines:

- that the graph will have at most $n - 1$ "wiggles" or bumps
- that the graph will cross the x-axis at most n times
- the global shape of the polynomial

Any value of x for which $f(x) = 0$ is called a *zero* of the function, f. Each x-intercept corresponds to a zero of a function.

Quadratic functions. The standard form of a *quadratic function,* or polynomial function of degree 2, is,

$$y = ax^2 + bx + c$$

where a, b, and c are constants and $a \neq 0$.

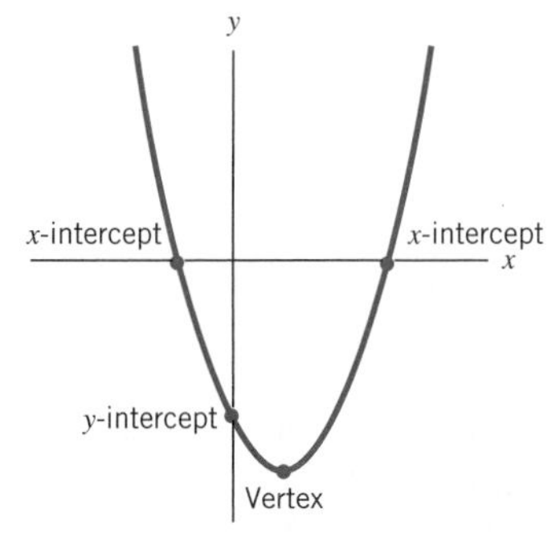

Graph of a quadratic with two x-intercepts.

The graph of a quadratic function is called a *parabola* and

- has a highest or lowest point called the vertex
- is U-shaped and symmetric about the axis of symmetry that runs through the vertex
- opens upward if $a > 0$ and downward if $a < 0$
- becomes narrower as $|a|$ (the absolute value or magnitude of a) increases
- is shifted up if c is increased and down if c is decreased.

Finding the intercepts and the vertex of a quadratic function. The y-intercept is the point with coordinates $(0, c)$ and is often abbreviated as just c.

A quadratic function may have one, two, or no horizontal or x-intercepts

To find the x-intercepts of a quadratic function $f(x) = ax^2 + bx + c$, we must solve or find the *roots* of the related equation $0 = ax^2 + bx + c$. The *quadratic formula* says that the roots of the equation are given by

$$x = \frac{-b \pm \sqrt{b^2 - 4ac}}{2a}$$

The roots may or may not be real. The non-real roots will be complex numbers in the form $a + bi$, where a and b are real numbers, and $i = \sqrt{-1}$. The real part is a and the imaginary part is b.

The term $b^2 - 4ac$ in the quadratic formula is called the *discriminant.*

If $b^2 - 4ac > 0$, there are two real roots and the graph has two x-intercepts.

If $b^2 - 4ac = 0$, there is one real root and the graph has one x-intercept.

If $b^2 - 4ac < 0$, there are no real roots and the graph has no x-intercepts.

The coordinates of the vertex are

$$\left(-\frac{b}{2a}, -\frac{b^2}{4a} + c\right)$$

The vertex form of a quadratic function is

$$y = a(x - h)^2 + k \qquad (a \neq 0)$$

where (h, k) is the vertex.

EXERCISES

Exercises for Section 8.1

1. Evaluate the following expressions for $x = 2$ and $x = -2$:

a. x^{-3} **b.** $4x^{-3}$ **c.** $-4x^{-3}$ **d.** $-4x^3$

2. Evaluate the following polynomials for $x = 2$ and $x = -2$ and specify the degree of each polynomial:

a. $y = 3x^2 - 4x + 10$ **b.** $y = x^3 - 5x^2 + x - 6$ **c.** $y = -2x^4 - x^2 + 3$

3. Let $f(x) = 3x^5 + x$ and $g(x) = x^2 - 1$.

a. Construct the following functions:

$$j(x) = f(x) + g(x) \qquad k(x) = f(x) - g(x) \qquad l(x) = f(x) \cdot g(x)$$

b. Evaluate $j(2)$, $k(3)$, and $l(-1)$.

4. A patient with an acute poison ivy rash is given a 5-day "prednisone taper." The daily dosage of the drug prednisone is respectively 20 mg, 15 mg, 10 mg, 5 mg, and 5 mg. If a percent is the daily absorption rate, then $x = (1 - a/100)$ is the percentage (in decimal form) of the drug left in the body after each day. For example, if the absorption rate was 40%, then $x = 1 - 0.40$. Then on day 2 the patient would have $20x$ remaining of the original 20 mg of prednisone, plus the additional 15 mg he took that day. On day 3 he'd have $20x^2$ left of his 20-mg dosage from day 1, plus $15x$ left from his 15-mg dosage from day 2, plus 10 mg he took that day. If we let $\text{Day}_i(x)$ = total amount in the body for day i, then

$$\text{Day}_1(x) = 20$$

$$\text{Day}_2(x) = 20x + 15$$

$$\text{Day}_3(x) = 20x^2 + 15x + 10$$

$$\text{Day}_4(x) = ?$$

$$\text{Day}_5(x) = ?$$

a. Construct the functions $\text{Day}_4(x)$ and $\text{Day}_5(x)$.

b. Use technology to graph $\text{Day}_5(x)$. Then assuming the patient's daily absorption rate is 60%:

i. Use the graph to estimate the total dosage of prednisone remaining on day 5.

ii. Use the function to calculate the total dosage of prednisone remaining on day 5.

c. Construct the functions $\text{Day}_6(x)$ and $\text{Day}_n(x)$ where $n > 5$. How does the polynomial structure of these functions differ from the earlier functions?

5. Match each of the following with its graph.

a. $y = 2x - 3$ **c.** $y = 3(2^x)$

b. $y = 2 - x$ **d.** $y = (x^2 + 1)(x^2 - 4)$

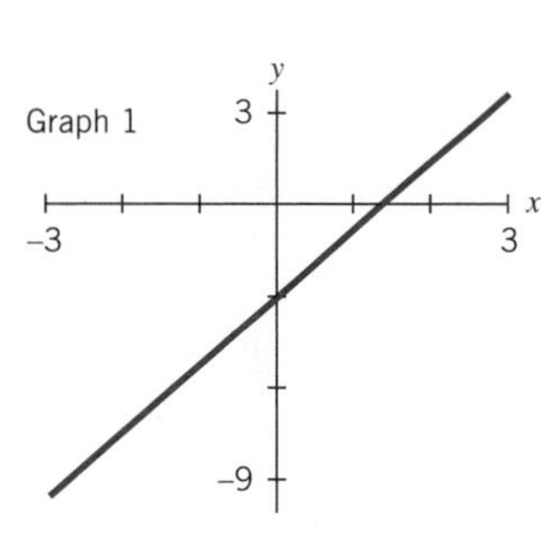

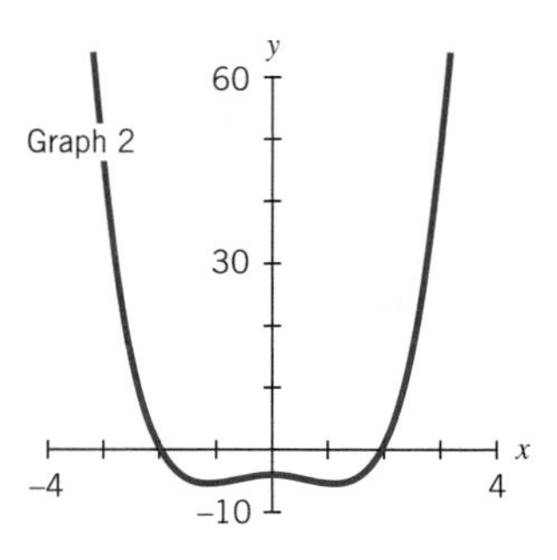

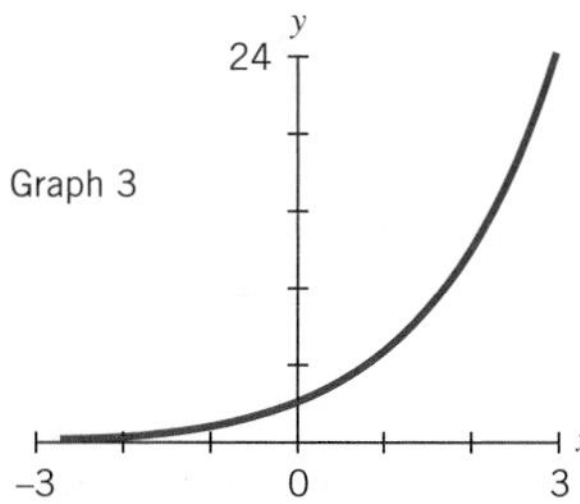

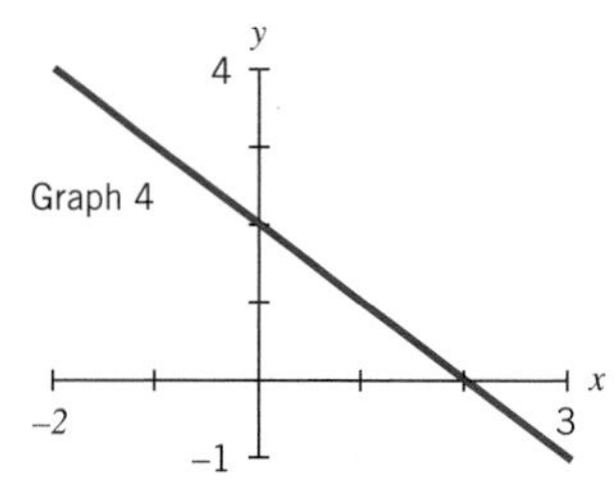

6. Match each of the following with its graph:

a. $f(x) = x^2 + 3x + 1$

b. $f(x) = -2x^2$

c. $f(x) = \frac{1}{3}x^3 + x - 3$

d. $f(x) = -\frac{1}{2}x^3 + x - 3$

e. $f(x) = (x^2 + 1)(x^2 - 4)$

f. $f(x) = 3(2^x)$

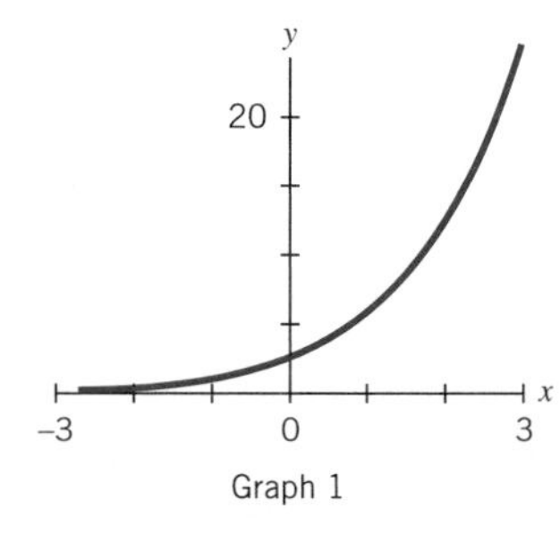

Graph 1

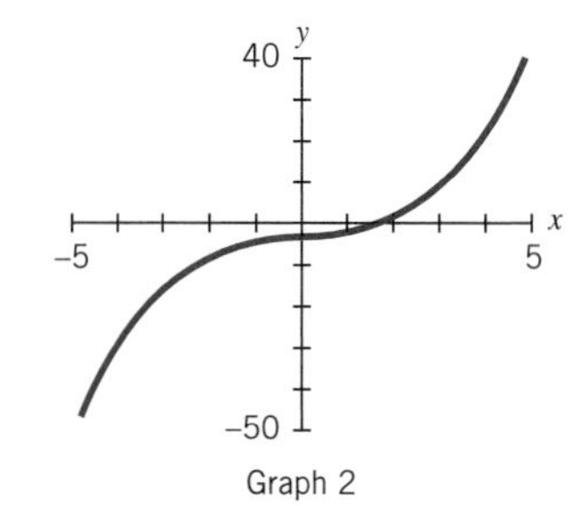

Graph 2

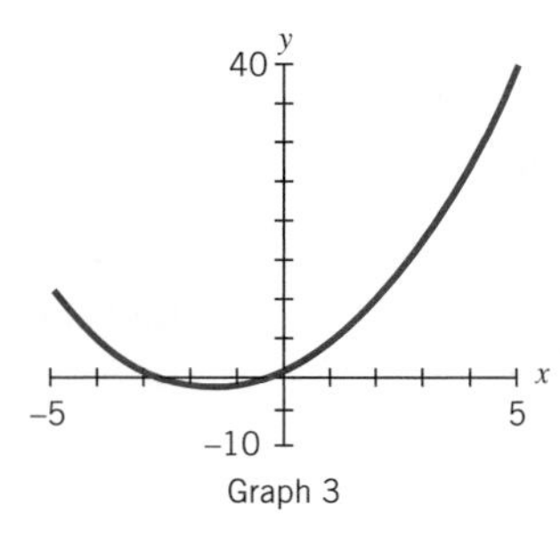

Graph 3

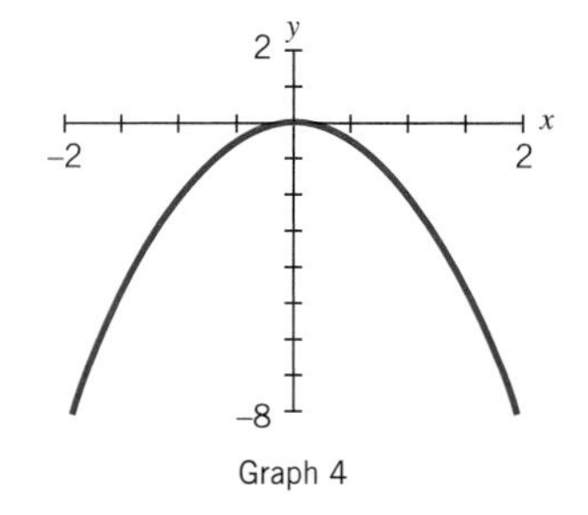

Graph 4

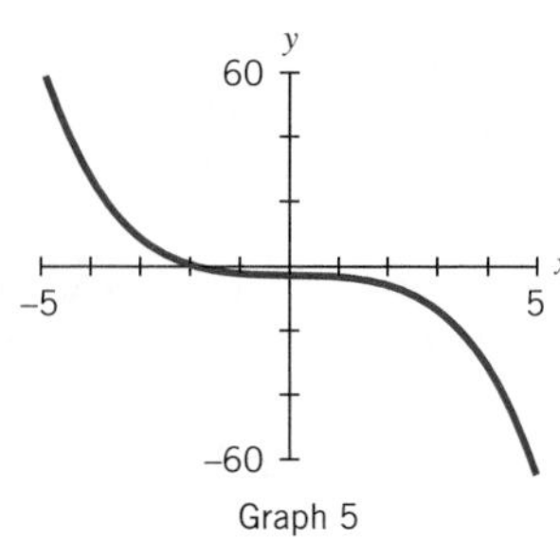

Graph 5

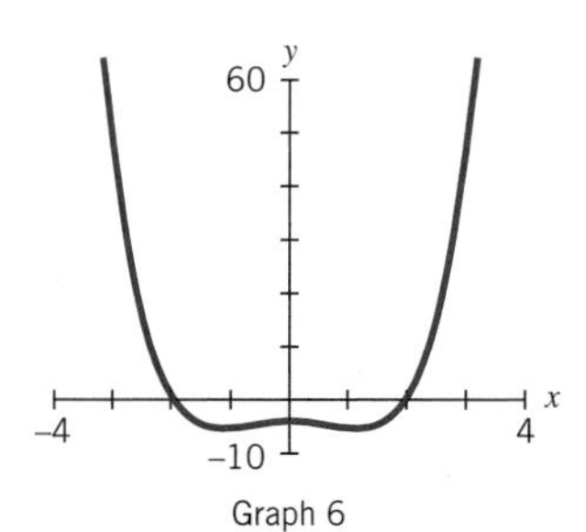

Graph 6

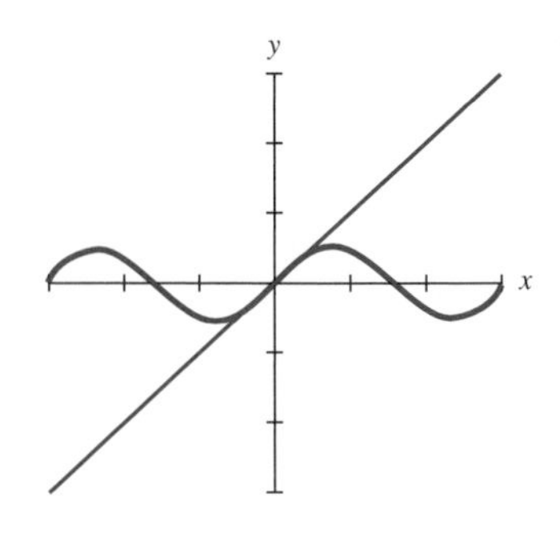

7. One method of graphing functions is called "addition of ordinates." For example, to graph $y = x + 1/x$ using this method, we would first graph $y_1 = x$. On the same coordinate plane, we would then graph $y_2 = 1/x$. Then we would estimate the y coordinates (called ordinates) for several selected x coordinates by adding geometrically on the graph itself the values of y_1 and y_2 rather than by substituting numerically. This technique is often used when graphing the sum or difference of two different types of functions by hand, without the use of a calculator.

a. Use this technique to sketch the graph of the sum of the two functions graphed in the accompanying figure.

b. Use this technique to sketch the graph of $y = -x^2 + x^3$ for $-2 \leq x \leq 2$. Then use a graphing tool and compare.

c. Use this technique to graph $y = 2^x - x^2$ for $-2 \leq x \leq 2$.

8. The stopping distance of a car is composed of two parts: first the distance traveled during the reaction time, after seeing the reason to stop and before applying the brakes, then the distance it takes the car to stop once the brakes have been applied.

a. An average reaction time is 0.75 second. Find a formula to express D_{reaction}, the distance (in feet) traveled during the reaction time as a function of car speed, S (in miles per hour). To do this, it is helpful to find the conversion factor from miles per hour to feet per second. Also recall that distance traveled = speed · time traveled.

b. Average data from the Automotive Encyclopedia on braking distance in feet, D_{brake}, for a car traveling at speeds up to 70 mph can be approximated with the formula

$$D_{\text{brake}} = 0.78S + 0.0043S^2 + 0.00067S^3$$

Generate graphs of D_{brake} and D_{reaction}.

c. Find the complete formula for stopping distance in feet as a function of speed in miles per hour:

$$D_{\text{stopping}} = D_{\text{reaction}} + D_{\text{brake}}$$

Use technology to graph D_{stopping} for speeds up to 100 mph.

d. What stopping distance does the formula predict for a car traveling at 100 mph?

9. Polynomial expressions of the form $an^3 + bn^2 + cn + d$ can be used to express positive integers, such as 4573, using different powers of 10:

$$4573 = 4 \cdot 1000 + 5 \cdot 100 + 7 \cdot 10 + 3$$

or
$$4573 = 4 \cdot 10^3 + 5 \cdot 10^2 + 7 \cdot 10^1 + 3 \cdot 10^0$$

or if $n = 10$
$$4573 = 4 \cdot n^3 + 5 \cdot n^2 + 7 \cdot n^1 + 3 \cdot n^0$$

Notice that in order to represent any positive number the coefficient multiplying each power of 10 can be any integer from 0 to 9. Decimals such as 407.32 can be expressed by including negative powers of 10:

$$407.32 = 4 \cdot 10^2 + 0 \cdot 10^1 + 7 \cdot 10^0 + 3 \cdot 10^{-1} + 2 \cdot 10^{-2}$$

or if $n = 10$
$$407.32 = 4 \cdot n^2 + 0 \cdot n^1 + 7 \cdot n^0 + 3 \cdot n^{-1} + 2 \cdot n^{-2}$$

a. Express 8701.5 as a polynomial in n assuming $n = 10$.

b. Express 239 as a polynomial in n assuming $n = 10$. If $n = 2$, what would be the value of the polynomial?

Computers use a similar polynomial system, called binary numbers, to represent numbers as sums of powers of 2. The number 2 is used because each minuscule switch in a computer can have two states, on or off; the symbol 0 signifies off, and the symbol 1 signifies on. Each binary number is built up from a row of switch positions set at 0 or 1 as multipliers for different powers of 2. For instance, in the binary number system 13 is represented as 1 1 0 1, which stands for

$$1 \cdot 2^3 + 1 \cdot 2^2 + 0 \cdot 2^1 + 1 \cdot 2^0 = 1 \cdot 8 + 1 \cdot 4 + 0 \cdot 2 + 1 \cdot 1 = 13.$$

c. What number does the binary notation 1 1 0 0 1 represent?

d. Find a way to write 35 as the sum of powers of 2; then give the binary notation.

10. A typical retirement scheme for state employees is based on three things: age at retirement, highest salary attained, and total years on the job. The annual retirement allowance is computed by the rule

$$\text{total years worked} \cdot \text{retirement age factor} \cdot \text{highest salary}$$

but is limited to a maximum of 80% of highest salary. We also have:

$$\text{total years worked} = \text{retirement age} - \text{starting age}$$

$$\text{retirement age factor} = 0.001 \cdot (\text{retirement age} - 40)$$

$$\text{salary at retirement} = \text{starting salary} + \text{annual raises for total years worked.}$$

a. For employees who started at age 30 in 1973 with a salary of \$12,000 and worked steadily receiving a \$2000 raise every year, find a formula to express retirement allowance, R, as a function of employee retirement age, A.

b. Graph R vs. A. At what age do these employees hit the limit of 80% of highest salary?

c. If the rule changes so that instead of highest salary, you use the average of the three highest years of salary, how would your formula for R as a function of A change?

11. (Requires computer or graphing calculator.) Use a function graphing program to estimate the x-intercepts for each of the following. Make a table showing the degree of the polynomial and the number of x-intercepts. What can you conclude?

$$y = 2x + 1$$

$$y = x^2 - 3x - 4$$

$$y = x^3 - 5x^2 + 3x + 5$$

$$y = 0.5x^4 + x^3 - 6x^2 + x + 3$$

12. (Requires computer or graphing calculator.) Repeat Exercise 11 for the following functions. How do your results compare to those for Exercise 11? What modification must be made to the conclusion?

$$y = 3x + 5$$

$$y = x^2 + 2x + 3$$

$$y = x^3 - 2x^2 - 4x + 8$$

$$y = (x - 2)^2(x + 1)^2$$

13. Use a function graphing program (and its zoom feature) to estimate the number of x-intercepts and their approximate values for:

a. $y = 3x^3 - 2x^2 - 3$ **b.** $f(x) = x^2 + 5x + 3$

14. Identify the x-intercepts of the following functions; then graph the functions to check your work.

a. $y = 3x + 6$ **b.** $y = (x + 4)(x - 1)$ **c.** $y = (x + 5)(x - 3)(2x + 5)$

15. The real zeros of a polynomial correspond to x-intercepts on the graph of a function.

a. If the degree of a polynomial is odd, then at least one of its zeros must be real. Explain why this is true.

b. Sketch a polynomial function that has no real zeros and whose degree is:

i. 2 **ii.** 4

c. Sketch a polynomial function of degree 3 that has exactly:

i. One real zero **ii.** Three real zeros

d. Sketch a polynomial function of degree 4 that has exactly two real zeros.

16. In each part, construct a polynomial function with the indicated characteristics.

a. Crosses the x-axis at least three times

b. Crosses the x-axis at -1, 3, and 10

c. Has a y-intercept of 4 and degree of 3

d. Has a y-intercept of -4 and degree of 5

17. A function is said to be *even* if $f(-x) = f(x)$ and *odd* if $f(-x) = -f(x)$. Use these definitions to answer the questions below.

a. Show that the even power functions are even.

b. Show that the odd power functions are odd.

c. What do you predict about the symmetries in the graphs of even power functions? What do you predict about the symmetries in the graphs of odd power functions? Check your predictions with a function graphing program or graphing calculator.

d. Show whether each of the following functions is even, odd, or neither:

i. $f(x) = x^4 + x^2$ **iii.** $h(x) = x^4 + x^3$

ii. $u(x) = x^5 + x^3$ **iv.** $g(x) = 10 \cdot 3^x$

e. For each function that you have identified as even or odd, what do you predict about the symmetry of its graph? Check your predictions with a function graphing program or graphing calculator.

Exercises for Section 8.2

The programs "Q2: Finding *a,b,c*" and "Q4: Finding 3 Points: *a-b-c* Form," both in *Quadratic Functions,* can give you skills practice in the relationship between the graph and the standard form of the quadratic equation.

18. a. For each of the following functions, evaluate $f(2)$ and $f(-2)$:

i. $f(x) = x^2 - 5x - 2$ **ii.** $f(x) = 3x^2 - x$ **iii.** $f(x) = -x^2 + 4x - 2$

b. For each of the following functions, evaluate $g(1)$ and $g(-1)$:

i. $g(x) = x^4 - 2x^3 + x$ **iii.** $g(x) = -2x^3 + 5x^2 - 4x + 3$

ii. $g(x) = x^3 + 2x^2 - 10$

19. For each set of functions, describe how the graphs are similar and how they are different.

a. $f(x) = x^2$ $g(x) = 3x^2$ $h(x) = 0.5x^2$

b. $f(x) = 3x^2 + 4$ $g(x) = -3x^2 + 4$ $h(x) = -3x^2 - 4$

c. $f(x) = x^2$ $g(x) = 2x$ $h(x) = 2^x$

20. Match each of the following graphs with an equation. Explain your reasoning for each of your choices. (Note that the grid lines are two units apart.)

$f(x) = 2x^2 - 8x - 2$ $j(x) = -0.5x^2 - 2x + 3$

$g(x) = 2x^2 - 8x + 3$ $k(x) = 2x - 5$

$h(x) = 0.5x^2 - 2x + 3$ $l(x) = 0.5x^2 - 2x - 3$

$i(x) = -2x^2 - x + 2$

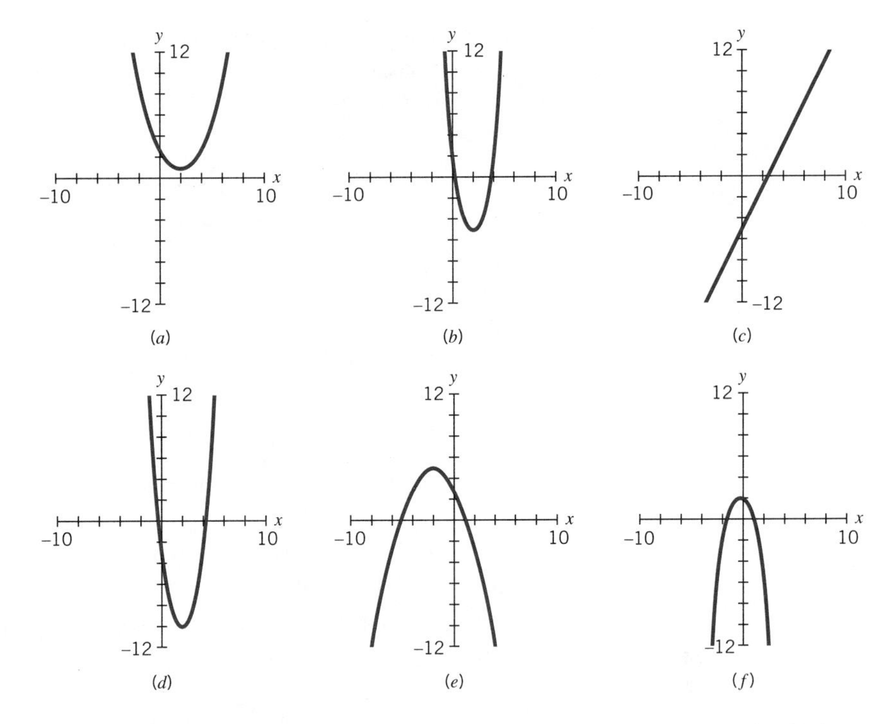

21. On the same graph, sketch the plots of the following functions and label each with its equation.

$$y = 2x^2 \qquad y = -2x^2 + 3$$
$$y = -2x^2 \qquad y = -2x^2 - 3$$

22. Without drawing the graph, list the following parabolas in order, from the narrowest to the broadest. Verify your results with technology.

a. $y = x^2 + 20$

b. $y = 0.5x^2 - 1$

c. $y = \frac{1}{3}x^2 + x + 1$

d. $y = 4x^2$

e. $y = 0.1x^2 + 2$

f. $y = -2x^2 - 5x + 4$

23. Construct the equations of two quadratic functions each with a y-intercept of 10, one with a graph that turns up and the other with a graph that turns down. Draw a rough sketch of each function and identify how many horizontal intercepts each function has.

24. For each part draw a rough sketch of a graph of any function of the type

$$f(x) = ax^2 + bx + c$$

a. Where $a > 0$, $c > 0$, and the function has no real zeros

b. Where $a < 0$, $c > 0$, and the function has two real zeros

c. Where $a > 0$ and the function has one real zero

25. A piece of wire 20 inches long is cut into two pieces and bent to make two squares.

a. If the length of one of the pieces is x inches, express the total area, A, of the two squares in the accompanying figure as a function of x.

b. What is the domain of your function?

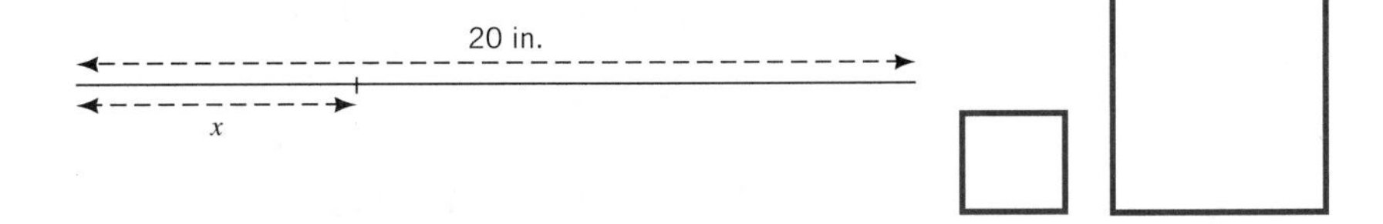

26. A dog breeder has 200 feet of fencing. He wants to use the whole 200 feet to construct a rectangle and three interior separators that together form four rectangular pens. See the accompanying figure.

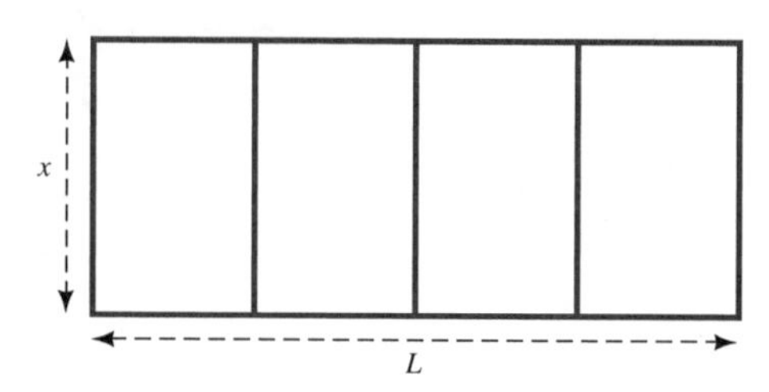

a. If x is the width of the larger rectangle, express the length, L, of the larger rectangle in terms of x.

b. Express the total area, $A(x)$, of the four pens as a polynomial in terms of x.

c. What is the domain of the function $A(x)$?

27. In Section 8.2 we found the dimensions for enclosing the maximum possible area when the perimeter of a rectangle was fixed as 24 meters. What would the dimensions be for the maximum area if:

a. The perimeter was held constant at 200 meters?

b. The perimeter was held constant at P meters?

To solve Exercises 28 to 30, construct and then graph the appropriate equations.

28. A gardener wants to grow carrots along the side of her house. To protect the carrots from wild rabbits, the plot must be enclosed by a wire fence. The gardener wants to use 16 feet of fence material left over from a previous project. Assuming that she constructs a rectangular plot, using the side of her house as one edge, estimate the area of the largest plot she can construct.

29. A box has a base of 10 feet by 20 feet and a height of h feet. If we decrease each side of the base by h feet, estimate the value of h that will maximize the volume of the box. Recall that the volume of a rectangular solid equals the area of the base times the height.

30. Estimate the dimensions of a cylinder with the maximum possible volume if the total surface area (including the two bases) is held constant at 80 square meters. Recall that

$$\begin{aligned} \text{volume of a cylinder} &= \text{area of base} \cdot \text{height} \\ &= \pi r^2 \cdot h \\ \text{surface area of a cylinder} &= \text{circumference of cylinder} \cdot \text{height} + 2 \cdot \text{area of base} \\ &= 2\pi r h + 2\pi r^2 \end{aligned}$$

where r = radius and h = height.

31. The amount of weight a wooden beam can carry evenly distributed along its length depends on the strength of the wood, the beam length, L, the cross section breadth, b, and depth, d. For commonly available wood that can take about 1200 lb/in.^{2} in bending, the weight, W, a beam can carry is

$$W = \frac{1600 b d^2}{L}$$

where b, d, and L are all in inches and W is in pounds.

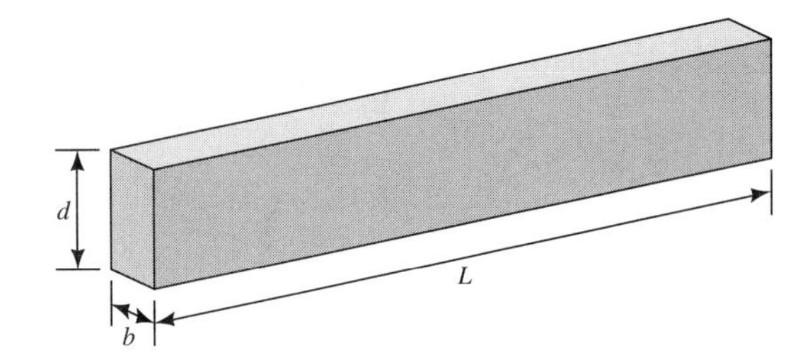

a. If you have a 10-foot-long beam, with breadth of 4 inches and depth of 10 inches, how much weight can it carry?

b. If the contractor installs the same beam lying on its side, so that now $b = 10$ inches and $d = 4$ inches, how much can it carry? By what percentage has the weight capacity been reduced by installing the beam incorrectly?

c. If you have a 20-foot-long beam, with $b = 4$ inches and $d = 10$ inches, how much can it carry? This beam is twice the length of the beam in part (a). Describe how increasing the length of the beam has affected its weight-bearing capacity.

d. If you want a 16-foot-long beam, with $b = 4$ inches, to carry the same load as the beam in part (a), what depth, d, is needed?

32. (Requires graphing calculator or computer.) The following data show the average growth of the human embryo prior to birth:

Embryo Age (weeks)	Weight (g)	Length (cm)
8	3	2.5
12	36	9
20	330	25
28	1000	35
36	2400	45
40	3200	50

Source: Reprinted with permission from Kimber et al., *Textbook of Anatomy and Physiology* (Upper Saddle River, NJ: Prentice Hall, 1955), "Embryo Age, Weight and Height," p. 785.

a. Plot weight vs. age and find the closest quadratic model that you can to approximate the data.

b. According to your model, what would an average 32-week embryo weigh?

c. Comment on the domain for which your formula is reliable.

d. Plot length vs. age; then construct a mathematical model for the length vs. age of an embryo from 20 to 40 weeks.

e. Using your model, compute the age at which an embryo would be 42.5 centimeters long.

Exercises for Section 8.3

33. Solve the following equations using the quadratic formula. (*Hint:* Rewrite each equation so that one side of the equation is zero.)

a. $6t^2 - 7t = 5$

b. $3x(3x - 4) = -4$

c. $(z + 1)(3z - 2) = 2z + 7$

d. $(x + 2)(x + 4) = 1$

34. Solve using the quadratic formula:

a. $x^2 - 3x = 12$

b. $3x^2 = 4x + 2$

c. $3(x^2 + 1) = x + 2$

d. $(3x - 1)(x + 2) = 4$

e. $\dfrac{1}{x-2} = \dfrac{x+1}{x-1}$

f. $\dfrac{x^2}{3} + \dfrac{x}{2} - \dfrac{1}{6} = 0$

g. $\dfrac{1}{x^2} - \dfrac{3}{x} = \dfrac{1}{6}$

35. Estimate all real solutions to the following equations using a graphical approach:

a. $x^2 - 5x + 6 = 0$ **b.** $3x^2 - 2x + 5 = 0$ **c.** $3x^2 - 12x + 12 = 0$

36. Use the discriminant to predict the number of x-intercepts for each function. Then use the quadratic formula to find all the zeros. Identify the coordinates of any x- and y-intercepts.

a. $y = 2x^2 + 3x - 5$ **b.** $f(x) = -16 + 8x - x^2$ **c.** $f(x) = x^2 + 2x + 2$

37. Determine analytically the coordinates of the x- and y-intercepts of the following parabolas:

a. $y = 3x^2 + 2x - 1$ **b.** $y = 3(x - 2)^2 - 1$

38. In parts (a) to (e), graph a parabola with the given characteristics. Then write an equation of the form $y = ax^2 + bx + c$ for a parabola with the given characteristics.

a. $a > 0, b^2 - 4ac > 0, c > 0$

b. $a > 0, b^2 - 4ac > 0, c < 0$

c. $a > 0, b^2 - 4ac < 0, b \neq 0$

d. $a < 0, b^2 - 4ac = 0, c \neq 0$

e. $b \neq 0, c < 0, b^2 - 4ac > 0$

39. Solve the following quadratic equations by factoring:

a. $x^2 - 9 = 0$

b. $x^2 - 4x = 0$

c. $3x^2 = 25x$

d. $x^2 + x = 20$

e. $4x^2 + 9 = 12x$

f. $3x^2 = 13x + 10$

g. $(x + 1)(x + 3) = -1$

h. $x(x + 2) = 3x(x - 1) - 3$

40. These quadratics are given in factored form. Find the x-intercepts directly, without using the quadratic formula. Will the vertex lie above or below the x-axis? Find the vertex and sketch the graph, labeling the x-intercepts.

a. $y = (x + 2)(x + 1)$ **b.** $y = 3(1 - 2x)(x + 3)$

41. Factor the quadratic expression in each of the following and then sketch the graph, labeling the axes and horizontal intercepts:

a. $y = x^2 + 6x + 8$

b. $z = 3x^2 - 6x - 9$

c. $f(x) = x^2 - 3x - 10$

d. $w = t^2 - 25$

e. $r = 4s^2 - 100$

f. $g(x) = 3x^2 - x - 4$

42. a. Construct a quadratic function with zeros at $x = 1$ and $x = 2$.

b. Is there more than one possible quadratic function for part (a)? Why or why not?

43. We dealt previously with systems of lines and ways to determine the coordinates of points where lines intersect. Once you know the quadratic formula, it's possible to determine where a line and a parabola, or two parabolas, intersect. As with two straight lines, at the point where the graphs of two functions intersect (*if* they intersect), the functions share the same x value and the same y value.

a. Find the intersection of the parabola $y = 2.0x^2 - 3.0x + 5.1$ and the line $y = -4.3x + 10.0$.

b. Plot both functions, labeling any intersection point(s).

44. Open up the *FAM1000 Census Graphs* course software file and in "F3: Regression with Multiple Subsets" examine mean personal wages vs. age for all men, and then for all women. Note the linear regression formulas and the corresponding correlation coefficients for these two cases.

a. Given the shape of the data, it is also reasonable to consider quadratic models for wages (in dollars) vs. age (in years) for women and men:

$$\text{women's wages} = -23.15(\text{age})^2 + 2012(\text{age}) - 19{,}470$$

$$\text{men's wages} = -42.49(\text{age})^2 + 3811(\text{age}) - 41{,}120$$

At what age does the quadratic model for women predict the same wages as the linear regression model? Calculate your answer and sketch a graph showing the answer.

b. Is there an age at which both quadratic models predict the same income for men and women? If so, what is it?

c. Comment on what domains might produce reliable values for linear and quadratic models in this case.

Exercises for Section 8.4

The program "Q9: Finding 3 Points: *a-h-k* Form" in *Quadratic Functions* will give you practice in finding points that satisfy a quadratic written in the $a-h-k$ form.

45. Given $f(x) = -x^2 + 8x - 15$, first estimate (by graphing), then determine analytically:

a. the x-intercepts

b. the y-intercept

c. the coordinates of the vertex

46. Write each of the following quadratic equations in function form (i.e., solve for y in terms of x). Find the vertex and the y-intercept and x-intercepts using any method. Finally, using these points, draw a rough sketch of the quadratic function.

a. $y + 12 = x(x + 1)$

b. $2x^2 + 6x + 14.4 - 2y = 0$

c. $y + x^2 - 5x = -6.25$

d. $y - 8x = x^2 + 15$

e. $y + 1 = (x - 2)(x + 5)$

47. Marketing research by a company has shown that the profit, $P(x)$ (in thousands of dollars), made by the company is related to the amount spent on advertising, x (in thousands of dollars), by the equation $P(x) = 230 + 20x - 0.5x^2$. What expenditure (in thousands of dollars) for advertising gives the maximum profit? What is the maximum profit?

48. Tom has a taste for adventure. He decides that he wants to bungee jump off the Missouri River bridge. If at any time t (in seconds) from the moment he jumps his height $h(t)$ (in feet) above the water level is given by the function $h(t) = 20.5t^2 - 123t + 190.5$, how close to the water will Tom get?

49. A manager wishes to determine the revenue made on the sale of supercomputers when the revenue, R (in millions of dollars), is given by $R(x) = 48x - 3x^2$, where x represents the number of supercomputers sold. How many supercomputers must be sold to maximize revenue? According to this model, what is the maximum revenue (in millions of dollars)?

50. The accompanying graph of the data file INCLINE shows the motion of a cart after it was given a push toward a motion detector positioned at the top of an inclined plane. The distance (in feet) of the cart from the top of the plane is plotted vs. time (in seconds). The motion can be modeled with a quadratic function.

a. Estimate the coordinates of the vertex. Describe what is happening to the cart at the vertex.

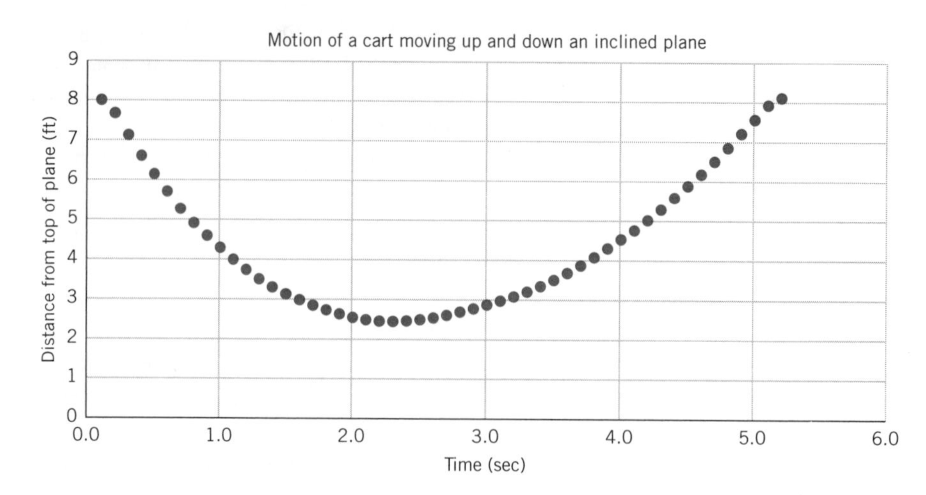

b. Use technology to generate a best fit quadratic to the data.

c. Calculate the coordinates of the vertex.

51. A baseball hit straight up in the air is at height

$$h = 4 + 50t - 16t^2$$

feet above the ground level at time t seconds after being hit. This formula is valid for $t \geq 0$ until the ball hits the ground.

a. What is the value for h when $t = 0$? What does this value represent in this context?

b. When does the baseball hit the ground?

c. When, if ever, is it 30 feet high? 90 feet high?

d. What is the maximum height that the baseball reaches? When does it reach that height?

52. The following function represents the relationship between time t (in sec) and height h (in feet) for objects thrown upward on the planet Pluto. For an initial velocity of 20 ft/sec and an initial height above ground of 25 feet, we get

$$h = -t^2 + 20t + 25$$

a. Find the coordinates of the point where the graph intersects the h-axis.

b. Find the coordinates of the vertex of the parabola.

c. At how many points will the graph cross the t-axis? If the graph crosses the t-axis, find the value of the horizontal intercept(s).

d. Sketch the graph. Label the axes.

e. Interpret the vertex in terms of time and height.

f. For what values of t does the mathematical model make sense?

53. In ancient times, after a bloody defeat that made her flee her city, the queen of Carthage, Dido, found refuge on the shores of Northern Africa. Sympathetic to her plight, the local inhabitants offered to build her a new city on their land, along the shore of the Mediterranean Sea, or inland, as long as the perimeter of her city (which had to be rectangular in shape) was no larger than the length of a ball of string that she could make using fine strips from one and only one cow hide. After lots of hard labor, queen Dido made the thinnest string possible out of the cow hide, and the string length turned out to be about a mile long. She was also told that she could use all the sea shore she wanted free. For this reason, Dido chose the length of her city, which she named Carthage, to run along the sea shore, and thus used the whole string to enclose only the three other sides of a rectangular piece of land. Having the length along the sea shore, she decided to make the width exactly half the length. This way, she claimed, she would have the maximum possible area the ball of string would allow her to enclose. Was Dido right?

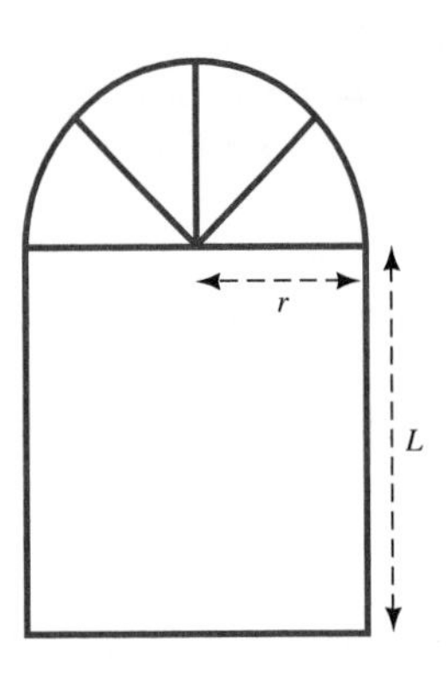

54. A Norman window has the shape of a rectangle surmounted by a semicircle of diameter equal to the width of the rectangle (see the accompanying figure). If the perimeter of the window is 20 feet (including the arch), what dimensions will admit the most light (maximize the area)? *Hint:* Express L in terms of r. Recall that the circumference of a circle $= 2\pi r$, and the area of a circle $= \pi r^2$, where r is the radius of the circle.

55. A pilot has crashed in the Sahara Desert. She still has her maps, and she knows her position, but her radio is destroyed. Her only hope for rescue is to hike out to a highway that passes near her position. She needs to determine the closest point on the highway, and how far away it is.

a. The highway is a straight line passing through a point 15 miles due north of her and another point 20 miles due east. Draw a sketch of the situation on graph paper, placing the pilot at the origin and labeling two points on the highway.

b. Construct an equation that represents the highway (using x for miles east and y for miles north).

c. Now use the Pythagorean Theorem to represent the square of the distance, d, of the pilot to any point (x, y) on the highway.

d. Substitute the expression for y from part (b) into the equation from part (c) in order to write d^2 as a quadratic in x.

e. If we minimize d^2, we minimize the distance d. So let $D = d^2$ and write D as a quadratic function in x. Now find the minimum value for D.

f. What are the coordinates of the closest point on the highway, and what is the distance, d, to that point?

56. (Requires graphing calculator or computer.) The accompanying figure is a plot of the data in the file BOUNCE, which shows the height of a bouncing racquetball (in feet) over time (in seconds). The path of the ball between each pair of bounces can be modeled using a quadratic function.

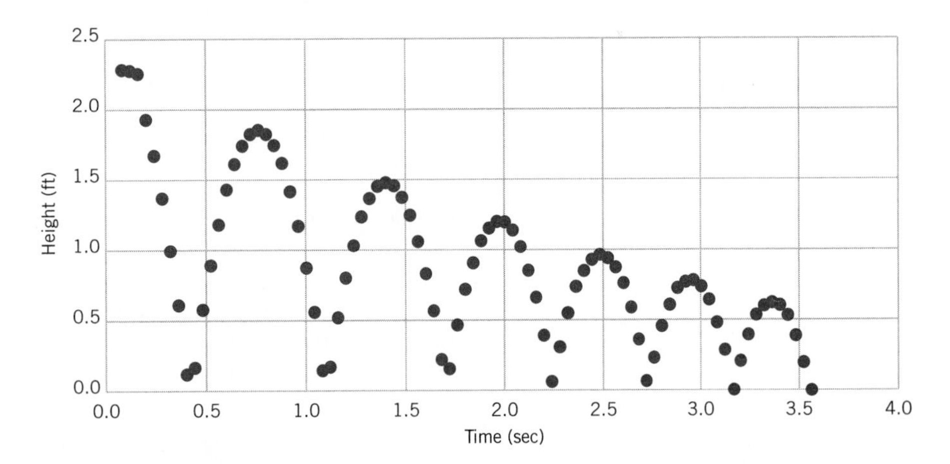

a. Select from the file BOUNCE the subset of the data that represents the motion of the ball between the first and second bounces (that is, between the first and second times the ball hits the floor). Use technology to generate a best fit quadratic function to this subset.

b. From the graph, estimate the maximum height the ball reaches between the first and second bounces.

c. Use the best fit function to calculate the maximum height the ball reaches between the first and second bounces.

57. At low speeds an automobile engine is not at its peak efficiency; efficiency initially rises with speed and then declines at the higher speeds. When efficiency is at its maximum, consumption rate of gas measured in gallons per hour is at a minimum. The gas consumption rate of a particular car can be modeled by the following equation, where G is gas consumption rate in gallons per hour and M is speed in miles per hour:

$$G = 0.00020M^2 - 0.013M + 1.07$$

a. Construct a graph of gas consumption rate vs. speed. Estimate the minimum gas consumption rate from your graph and the speed at which it occurs.

b. Using the equation for G, calculate the speed at which the gas consumption rate is at its minimum. What is the minimum gas consumption rate?

c. If you travel for 2 hours at peak efficiency, how much gas will you use and how far will you go?

d. If you travel at 60 mph, what is your gas consumption rate? How long does it take to go the same distance that you calculated in part (c)? (Recall that travel distance = speed × time traveled.) How much gas is required for the trip?

e. Compare the answers for parts (c) and (d), which tell you how much gas is used for the same length trip at two different speeds. Is gas actually saved for the trip by traveling at the speed that gives minimum gas consumption rate?

f. Using the function G, generate data for gas consumption rate measured in gallons per mile by computing the following table:

Speed of Car	Measures of the Rate of Gas Consumption	
(mph)	(gal/hr)	(gal/hr)/mph = gal/mile
0		
10		
20		
30		
40		
50		
60		
70		
80		

Plot gallons per mile vs. miles per hour. At what speed is gallons per mile at a minimum?

g. Add a fourth column to the data table above. This time compute miles/gal = mph/(gal/hr). Plot miles per gallon vs. miles per hour. At what speed is miles per gallon at a maximum? This is the inverse of the preceding question; we are normally used to maximizing miles per gallon instead of minimizing gallons per mile. Does your answer make sense in terms of what you found for parts (b) and (f)?

58. Identify the vertex and specify whether it represents a maximum or minimum and then sketch the graph of each of the following functions:

a. $y = (x - 2)^2$

b. $y = \frac{1}{2}(x - 2)^2 + 3$

c. $y = -2(x + 1)^2 + 5$

d. $y = -0.4(x - 3)^2 - 1$

59. a. Write each of the following functions in both the a–b–c and a–h–k forms. Is one form easier than another for finding the vertex? The x- and y-intercepts?

$$y_1 = 2x^2 - 3x - 20 \qquad y_3 = 3x^2 + 6x + 3$$

$$y_2 = -2(x - 1)^2 - 3 \qquad y_4 = -(2x + 4)(x - 3)$$

b. Find the vertex and x- and y-intercepts and construct a graph for each function in part (a). If you have access to a graphing calculator or function graphing program, check your work.

60. Convert the following functions from the $a-b-c$ to the $a-h-k$ or vertex form:

a. $f(x) = x^2 + 6x + 5$ **c.** $y = 3x^2 - 12x + 12$ **e.** $h(x) = 3x^2 + 6x + 5$

b. $g(x) = x^2 - 3x + 7$ **d.** $y = 2x^2 + 3x - 5$

61. **a.** Find the equation of the parabola with a vertex of (2, 4) that passes through the point (1, 7).

b. Construct two different quadratic functions both with a vertex at (2, − 3) such that the graph of one function is concave up and the graph of the other function is concave down.

62. For each part construct a function that satisfies the given conditions.

a. Has a constant rate of increase of $15,000/year

b. Is a quadratic that opens up and has a vertex at (1, − 4)

c. Is a quadratic that opens down and has a single root

d. Is a quadratic with no real roots

e. Is a quadratic with zeros at 1 and − 3

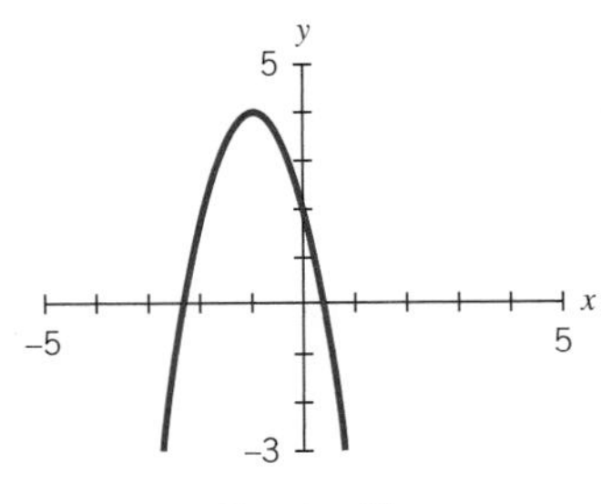

Exercise 64

63. Determine the equation of a parabola whose vertex is at (2, 3) and that passes through the point (4, − 1). Show your work, including a sketch of the parabola.

64. Construct an equation for the accompanying parabola.

65. Find the equation $y = ax^2 + bx + c$ of a parabola with $a = 4$ and x-intercepts (2, 0) and (− 3, 0). Explain your reasoning. Sketch the parabola.

66. Students noticed that the path of water from a water fountain seemed to form a parabolic arc. They set a flat surface at the level of the water spout and measured the maximum height of the water from the flat surface as 8 inches and the distance from the spout to where the water hit the flat surface as 10 inches. Construct a function model for the stream of water.

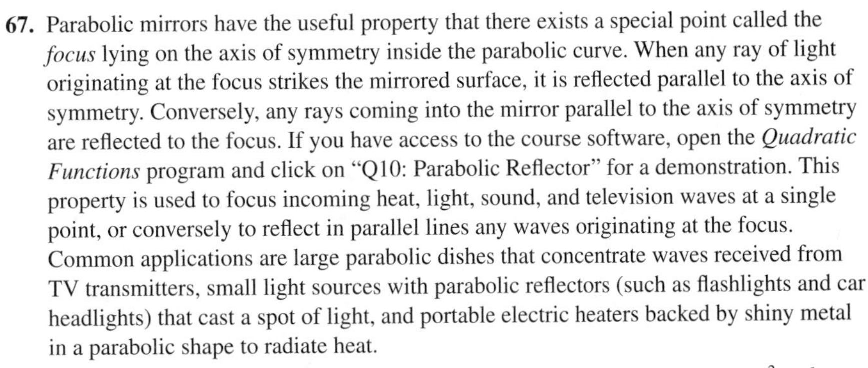

67. Parabolic mirrors have the useful property that there exists a special point called the *focus* lying on the axis of symmetry inside the parabolic curve. When any ray of light originating at the focus strikes the mirrored surface, it is reflected parallel to the axis of symmetry. Conversely, any rays coming into the mirror parallel to the axis of symmetry are reflected to the focus. If you have access to the course software, open the *Quadratic Functions* program and click on "Q10: Parabolic Reflector" for a demonstration. This property is used to focus incoming heat, light, sound, and television waves at a single point, or conversely to reflect in parallel lines any waves originating at the focus. Common applications are large parabolic dishes that concentrate waves received from TV transmitters, small light sources with parabolic reflectors (such as flashlights and car headlights) that cast a spot of light, and portable electric heaters backed by shiny metal in a parabolic shape to radiate heat.

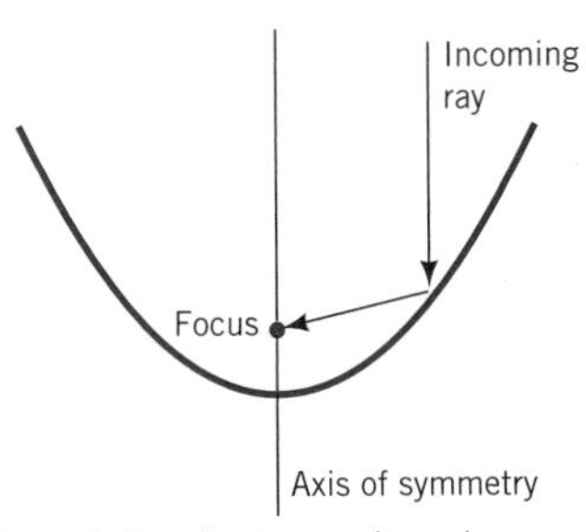

A parabolic reflector: any incoming ray parallel to the axis of symmetry will be reflected to the focus.

a. We know that the coordinates of the vertex of a quadratic function $y = ax^2 + bx + c$ are $(-\frac{b}{2a}, -\frac{b^2}{4a} + c)$. The focus is $|\frac{1}{4a}|$ units above the vertex (if the graph opens up) or below the vertex (if the graph opens down). The distance between the focus and the vertex is called the *focal length*. Find an algebraic expression in terms of a, b, and c for the coordinates of the focus.

b. Calculate the focal length and the coordinates of the focus for the following parabolas:

i. $y = x^2 + 3x + 2$ **ii.** $y = 2x^2 - 4x + 5$

c. Sketch a graph of each parabola in part (b) indicating the position of the focus.

d. Describe in words how the focal length varies depending on how narrow or wide the parabolic curve is.

68. (Assumes Exercise 67.) Suppose you are given a lightbulb 4 inches long with a spherical end having a 1-inch radius. Think of the filament as being located at the center of this sphere. You want to mount the lightbulb in a parabolic reflector so that the vertex is at the base of the bulb and the filament is at the focus. The reflector is to extend 1 inch above the glass end of the bulb.

a. Sketch a picture of the lightbulb and the surrounding parabola.

b. What focal length is needed to get the filament at the focus of the parabola?

c. What is the equation of a parabola that gives this focal distance? (*Hint:* You may want to put the vertex at the origin.)

d. What is the diameter of the wide end of the reflector?

e. How much clearance to the side of the reflector will the bulb have at the widest part of the bulb?

69. A designer proposes a parabolic satellite dish 5 feet wide across the top and 1 foot 3 inches deep.

a. What is the equation for the dish? Center it at the origin for convenience.

b. What is the focal length of the dish?

c. What is the diameter of the dish at the focus?

70. An industrial designer is trying to design a reflector for an electrical heater with a single, long, straight heating element. The heater is to be ceiling mounted in factory environments and is to throw parallel heat rays out to the work area below. The client wants the 4-foot-long heating element to be mounted in a reflector that sticks out no more than 9 inches from the ceiling, including a 1-inch-thick insulation between the reflector and the ceiling (see the accompanying figures). The heater is to be no more than 4 feet long and 2 feet wide. Find a formula from which to manufacture the reflector, and diagram where the heating element would be placed.

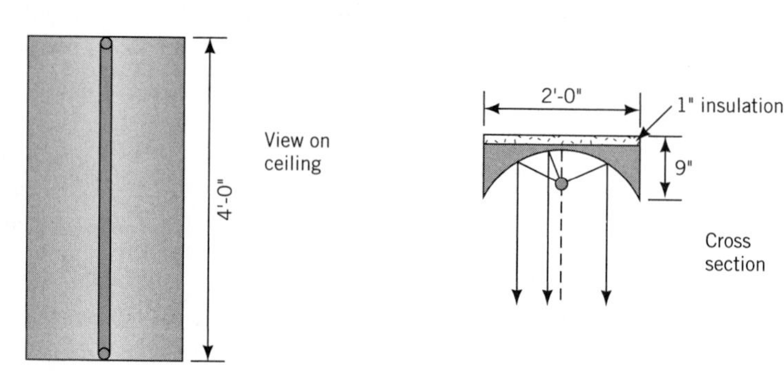

Exercises for Section 8.5

71. Complete the table below for the function $y = 3 - x - x^2$.

x	y	Average Rate of Change	Average Rate of Change of Average Rate of Change
-3	-3		
-2	1	$\frac{1-(-3)}{-2-(-3)} = \frac{4}{1} = 4$	$\frac{2-4}{-1-(-2)} = \frac{-2}{1} = -2$
-1	3	$\frac{3-1}{-1-(-2)} = \frac{2}{1} = 2$	$\frac{0-2}{0-(-1)} =$
0	3	$\frac{3-3}{0-(-1)} = \frac{0}{1} = 0$	
1	1		
2			
3			

Plot the average rate of change on the vertical axis and x on the horizontal axis. What type of function does the graph represent? What is its slope (the rate of change of the average rate of change)?

72. Complete the table below for the function $Q = 2t^2 + t + 1$.

t	Q	Average Rate of Change, R
-3		
-2		
-1		
0		
1		
2		
3		

a. Plot $Q = 2t^2 + t + 1$. What type of function is this?

b. Plot the average rate of change, R, on the vertical axis and t on the horizontal axis. What type of function is this? Construct a formula for R as a function of t. What does the formula tell you about the average rate of change of the average rate of change?

Exploration 8.1

Properties of Quadratic Functions

Objectives

- Part I: Explore the effects of a, b, and c on the graphs of quadratic equations in the $a-b-c$ form: $y = ax^2 + bx + c$.
- Part II: Explore the effects of a, h, and k on the graphs of quadratic equations in the $a-h-k$ form: $y = a(x - h)^2 + k$.

Materials/Equipment

- graphing calculator or computer and several programs in *Quadratic Functions*
- graph paper

Procedure

Part I: Exploring Quadratics in the Form $y = ax^2 + bx + c$

1. Making predictions.

a. Make a data table and hand sketch on the same grid $y = x^2$ and $y = 4x^2$. Use both positive and negative values for x. Make a prediction of what the graph of $y = 2x^2$ will look like.

b. How would you describe your graphs? What is their basic shape? When are the values for y increasing? Decreasing? Do the functions have a maximum or a minimum value? Which graph shows y changing the fastest compared to x?

c. Make a data table and sketch the graph of $y = -x^2$. Make predictions for what the graphs of $y = -4x^2$ and $y = -2x^2$ will look like.

d. Compare the graphs in part (c) to those in part (a).

e. Summarize how changes in a affect the graph of the equation.

2. Testing your predictions.

Use a graphing calculator or a computer and "Q1: a, b, c Sliders" in the course software package *Quadratic Functions* to check your predictions.

a. What effect does a have? Using technology graph several equations each of which has the same value for b and c but different values for a. Summarize how changing a changes the graph. What happens to the graph if you change the sign of a? What happens if you keep the sign of a but increase the absolute value of a? Is there any point on the graph that doesn't change as you increase or decrease the size of a?

b. What effect does c have? Use technology to graph several equations with the same values for a and b, but different values for c. How does changing c change the graph?

c. What effect does b have?

3. Class discussion.

Summarize your conclusions in your own words. Does the rest of the class agree?

Part I: Exploration-Linked Homework

1. Write a 60 second summary of your conclusions.
2. Open "Q2: Finding a, b, c" in *Quadratic Functions.* The software will pick a random quadratic function and display its graph. You predict the corresponding values of a, b, and c using the sliders, graph your predicted quadratic, and then see whether your guess is correct. Repeat, using integer and real values, until you are consistently right.

Part II: Exploring Quadratics in the Form $y = a(x - h)^2 + k$

Work with a partner, recording your results as you go. Again you'll start graphing by hand and then use technology to confirm your predictions.

1. **Making predictions.**
   - **a.** Make a data table and hand sketch on the same grid $y = 2x^2$, $y = 2(x - 1)^2$, and $y = 2(x + 1)^2$. Use both positive and negative values for x. Make a prediction about the graphs of $y = 2(x - 3)^2$ and $y = 2(x + 3)^2$.
   - **b.** What effect does replacing x with $(x - 1)$ have? What has happened to the vertex? The shape of the curve? What if you replace x with $(x + 1)$? With $(x - 3)$? With $(x + 3)$? In general, how does the graph of $y = 2x^2$ compare with the graph of $y = 2(x - h)^2$ if h is positive? If h is negative?
   - **c.** Using what you know from Part I about the effect of adding a constant term to a quadratic, on a new graph sketch $y = 2x^2$, $y = 2x^2 + 4$, and $y = 2x^2 - 5$. In general, how does the graph of $y = 2x^2$ compare with the graph of $y = 2x^2 + k$ if k is positive? If k is negative?
   - **d.** Now (without generating a data table) draw a rough sketch of the graphs of the quadratic functions $y = 2(x - 1)^2 + 3$ and $y = 2(x + 1)^2 - 5$.

2. **Testing your predictions.**

   Use a graphing calculator or a computer and "Q3: a, h, k Sliders" in the course software package *Quadratic Functions* to test your predictions.
   - **a.** What effect does a have? Using a graphing calculator or course software, plot several equations with the same values for h and k but different values for a.
   - **b.** What effect does h have? Use technology to plot several equations with the same values for a and k, but different values for h.
   - **c.** What effect does k have?
   - **d.** What effect do h and k have together? What do you think will be the combined effect of changing both of them?

3. **Class discussion.**

   Summarize your conclusions in your own words. Does the rest of the class agree?

Part II: Exploration-Linked Homework

Write a 60 second summary of your conclusions.

An Extended Exploration: The Mathematics of Motion

Overview

In this extended exploration we use the laboratory methods of modern physicists to collect and analyze data about freely falling bodies and then examine the questions asked by Galileo about bodies in motion.

After conducting this exploration, you should be able to:

- understand the importance of the scientific method
- describe the relationship between distance and time for freely falling bodies
- derive equations describing the velocity and acceleration of a freely falling body

THE SCIENTIFIC METHOD

Today we take for granted that scientists study physical phenomena in laboratories using sophisticated equipment. But in the early 1600s, when Galileo did his experiments on motion, the concept of laboratory experiments was unknown. In his attempts to understand nature, Galileo asked questions that could be tested directly in experiments. His use of observation and direct experimentation and his discovery that aspects of nature were subject to quantitative laws were of decisive importance not only in science but in the broad history of human ideas.

The Greeks and medieval thinkers believed that basic truths existed within the human mind and that these truths could be uncovered through reasoning, not empirical experimentation. Their scientific method has been described as a "qualitative study of nature." Greek and medieval scientists were interested in *why* objects fall. They believed that a heavier object fell faster than a lighter one because "it has weight and it falls to the Earth because it, like every object, seeks its natural place, and the natural place of heavy bodies is the center of the Earth. The natural place of a light body, such as fire, is in the heavens, hence fire rises." [1]

Galileo changed the question from *why* things fall to *how* things fall. This question suggested other questions that could be tested directly by experiment: "By alternating questions and experiments, Galileo was able to identify details in motion no one had previously noticed or tried to observe." [2] His quantitative descriptions of objects in motion led not only to new ways of thinking about motion but also to new ways of thinking about science. His process of careful observation and testing began the critical transformation of science from a qualitative to a quantitative study of nature.[3] Galileo's decision to search for quantitative descriptions "was the most profound and the most fruitful thought that anyone has had about scientific methodology." [4] This approach became known as the scientific method.

THE FREE FALL EXPERIMENT

Instructions for conducting the free fall experiment are in the last section.

The software "Q11: Freely Falling Objects" in *Quadratic Functions* provides a simulation of the free fall experiment.

In this extended exploration, you will conduct a modern version of Galileo's free fall experiment. This classic experiment records the distance that a freely falling object falls during each fraction of a second. The experiment can be performed either with a graphing calculator connected to a motion sensor or in a physics laboratory with an apparatus that drops a heavy weight and records its position on a tape.

In this experiment, Galileo sought to answer the following questions:

> How can we describe mathematically the distance an object falls over time?
> Do freely falling objects fall at a constant speed? If the speed of freely falling objects is not constant, is it increasing at a constant rate?

You can try to find answers to these questions by collecting and analyzing your own data or by using the data provided on Excel files for the computer or on graph link files for the graphing calculator. Instructions for using technology to collect and analyze data are provided in the last section. The following discussion will help you analyze your results and help you understand the answers to Galileo's questions.

[1]M. Kline, *Mathematics for the Nonmathematician* (Dover Publications, New York, 1967), p. 287.

[2]E. Cavicchi, "Watching Galileo's Learning," in the Anthology of Readings.

[3]Galileo's scientific work was revolutionary not only in science but also in the politics of the time; his work was condemned by the ruling authorities and he was arrested.

[4]M. Kline, ibid., p. 288.

Interpreting Data from a Free Fall Experiment

The sketch of a tape given in Figure 1 gives data collected by a group of students from a falling object experiment. Each dot represents how far the object fell in each succeeding 1/60th of a second.

Since the first few dots are too close together to get accurate measurements, we start measurements instead at the sixth dot, which we call dot_0. At this point, the object is already in motion. This dot is considered to be the starting point, and the time, t, at dot_0, is set at 0 seconds. The next dot represents the position of the object 1/60th of a second later. Time increases by 1/60th of a second for each successive dot. In addition to assigning a time to each point, we also measure the total distance fallen, d, from the point designated dot_0. For every dot we have two values: the time, t, and the distance fallen, d. At dot_0, we have $t = 0$ and $d = 0$.

The time and distance measurements from the tape are recorded in Table 1 and plotted on the graph in Figure 2. Time, t, is the independent variable, and distance, d, is the dependent variable. The graph gives a representation of the data collected on distance fallen over time, not a picture of the physical motion of the object. The graph of the data looks more like a curve than a straight line, so we expect the average rates of change between different pairs of points to be different. We know how

t = 0, d = 0
t = 1/60, d = 1.72
t = 2/60, d = 3.75
t = 3/60, d = 6.10
t = 4/60, d = 8.67
t = 5/60, d = 11.58

Figure 1 Tape from a free fall experiment.

See Excel or graph link file FREEFALL, which contains the data in Table 1.

Table 1

Time (sec)	Total Distance Fallen (cm)
0.0000	0.00
0.0167	1.72
0.0333	3.75
0.0500	6.10
0.0667	8.67
0.0833	11.58
0.1000	14.71
0.1167	18.10
0.1333	21.77
0.1500	25.71
0.1667	29.90
0.1833	34.45
0.2000	39.22
0.2167	44.22
0.2333	49.58
0.2500	55.15
0.2667	60.99
0.2833	67.11
0.3000	73.48
0.3167	80.10
0.3333	87.05
0.3500	94.23

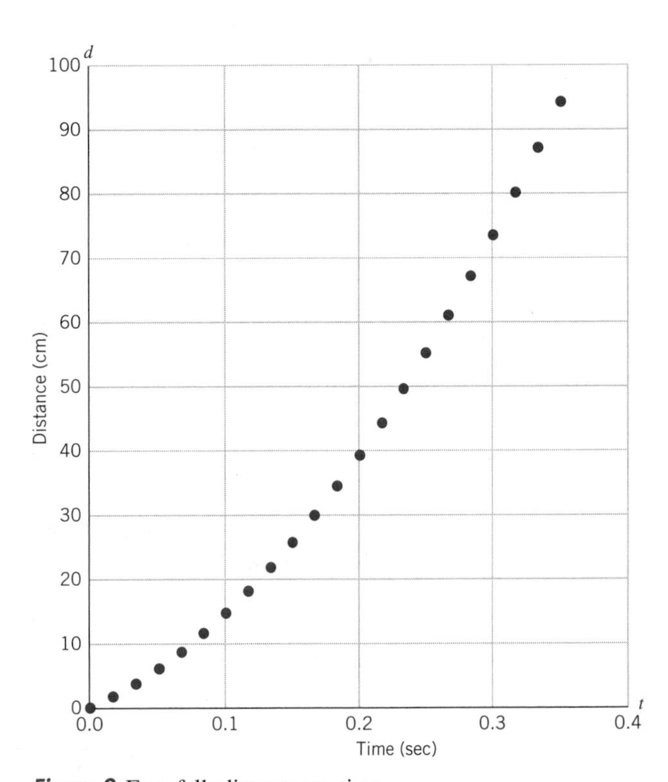

Figure 2 Free fall: distance vs. time.

Table 2

t	d	Average Rate of Change
0.0500 0.1333	6.10 21.77	$\frac{21.77 - 6.10}{0.1333 - 0.0500} \approx 188$ cm/sec
0.0833 0.2333	11.58 49.58	$\frac{49.58 - 11.58}{0.2333 - 0.0833} \approx 253$ cm/sec
0.2167 0.3333	44.22 87.05	$\frac{87.05 - 44.22}{0.3333 - 0.2167} \approx 367$ cm/sec

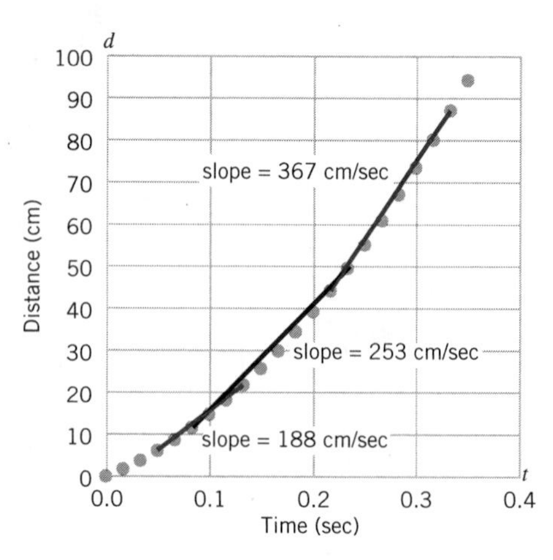

Figure 3 Slopes (or average velocities) between three pairs of end points.

to calculate the average rate of change between two points and that it represents the slope of a line segment connecting the two points:

$$\text{average rate of change} = \frac{\text{change in distance}}{\text{change in time}} = \text{slope of line segment}$$

Table 2 and Figure 3 show the increase in the average rate of change over time for three different pairs of points. The time interval nearest the start of the fall shows a relatively small change in the distance per time step and therefore a relatively gentle slope of 188 cm/sec. The time interval farthest from the start of the fall shows a greater change of distance per time step and a much steeper slope of 367 cm/sec.

In this experiment the average rate of change has an additional important meaning. For objects in motion, the change in distance divided by the change in time is also called the *average velocity* over that time period. For example, in the calculations in Table 2, the average rate of change of 188 cm/sec represents the average velocity of the falling object between 0.0500 and 0.1333 seconds.

$$\textit{average velocity} = \frac{\text{change in distance}}{\text{change in time}}$$

Important Questions

Do objects fall at a constant speed?[5] The rate-of-change calculations and the graph in Figure 3 indicate that the average rate of change of position with respect to time, the velocity, of the falling object is not constant. Moreover, the average velocity appears to be increasing over time. In other words, as the object falls, it is moving

[5]In everyday usage, "speed" and "velocity" are used interchangeably. In physics, "velocity" gives the direction of motion by the sign of the number, positive for forward, negative for backward. "Speed" means the absolute value, or magnitude, of the velocity. So speed is never negative, whereas velocity can be positive or negative.

faster and faster. Our calculations agree with Galileo's observations. He was the first person to show that the velocity of a freely falling object is not constant.

This finding prompted Galileo to ask more questions. One of these questions was: If the velocity of freely falling bodies is *not constant,* is it increasing at a *constant rate?* Galileo discovered that the velocity of freely falling objects does increase at a constant rate. If the rate of change of velocity with respect to time is constant, then the graph of velocity vs. time is a straight line. The slope of that line is constant and equals the rate of change of velocity with respect to time. A theory of gravity has been built around Galileo's discovery of a constant rate of change for the velocity of a freely falling body. This constant of nature, the gravitational constant of Earth, is denoted by g and is approximately 980 cm/sec^2.

Deriving an Equation Relating Distance and Time

If you are interested in learning more about how Galileo made his discoveries, read Elizabeth Cavicchi's "Watching Galileo's Learning."

Galileo wanted to describe mathematically the distance an object falls over time. Using mathematical and technological tools not available in Galileo's time, we can describe the distance fallen over time in the free fall experiment using a "best fit" function for our data. Galileo had to describe his finding in words. Galileo described the free fall motion first by direct measurement and then abstractly with a time-squared rule. "This discovery was revolutionary, the first evidence that motion on Earth was subject to mathematical laws."[6]

Using Galileo's finding that distance is related to time by a time-squared rule, we find a best fit polynomial function of degree 2 (a quadratic) for the free fall data. A function graphing program gives the following best fit quadratic function for the free fall data in Table 1:

$$d = 487.8t^2 + 98.72t - 0.0528$$

Figure 4 shows a plot of the data and the function. If your curve fitting software program or graphing calculator does not provide a measure of closeness of fit, such as the correlation coefficient for regression lines, you may have to rely on a visual judgment. Rounding the coefficients to the nearest unit, we obtain the equation

$$\begin{aligned} d &= 488t^2 + 99t - 0 \\ &= 488t^2 + 99t \end{aligned} \qquad (1)$$

We now have a mathematical model for our free fall data.

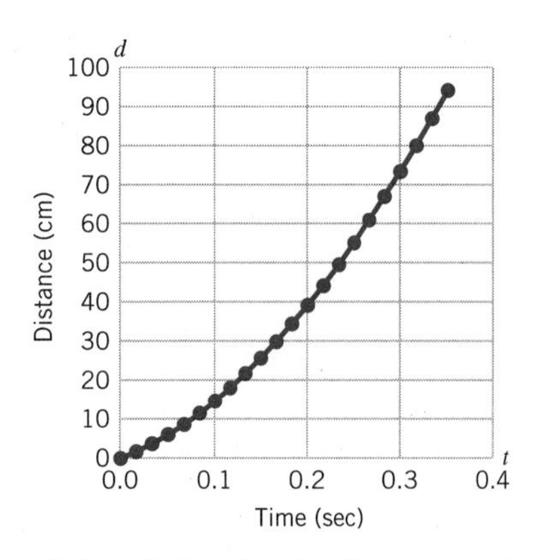

Figure 4 Best fit function for distance vs. time.

[6]E. Cavicchi, "Watching Galileo's Learning," in Anthology of Readings.

What are the units for each term of the equation? Since d is in centimeters, each term on the right-hand side of Equation (1) must also be in centimeters. Since t is in seconds, the coefficient 488 of t^2, must be in centimeters per square second:

$$\frac{\text{cm}}{\cancel{\text{sec}^2}} \cdot \frac{\cancel{\text{sec}^2}}{1} = \text{cm}$$

the coefficient 99 of t, must be in centimeters per second, and the constant term, 0, in centimeters.

What if we ran the experiment again? How would the results compare? In one class, four small groups did the free fall experiment, plotted the data, and found a corresponding best fit second-degree polynomial. The functions are listed below, along with Equation (1). In each case we have rounded the coefficients to the nearest unit. All of the constant terms rounded to 0.

$$d = 488t^2 + 99t \tag{1}$$

$$d = 486t^2 + 72t \tag{2}$$

$$d = 484t^2 + 173t \tag{3}$$

$$d = 486t^2 + 73t \tag{4}$$

$$d = 495t^2 + 97t \tag{5}$$

Examine the coefficients of each of the terms in these equations. All the functions have similar coefficients for the t^2 term, very different coefficients for the t term, and zero for the constant term. Why is this the case? Using concepts from physics, we can describe what each of the coefficients represents.

The coefficients of the t^2 term found in Equations (1) to (5) are all close to one-half of 980 cm/sec^2, or half of g, Earth's gravitational constant. The data from this simple experiment give very good estimates for half of g.

The coefficient of the t term represents the initial velocity, v_0, of the object when $t = 0$. In Equation (1), $v_0 = 99$ cm/sec. Recall that we didn't start to take measurements until the sixth dot, the dot we called dot_0. So at dot_0, where we set $t = 0$, the object was already in motion with a velocity of approximately 99 cm/sec. The initial velocities, or v_0 values, in Equations (2) to (5) range from 72 to 173 cm/sec. Each v_0 represents approximately how fast the object was moving when $t = 0$, the point chosen to begin recording data in each of the various experiments.

The constant term rounded to zero in each of Equations (1) to (5). On the tape when we set $t = 0$, we set $d = 0$. So we expect that in all our equations the constant terms, which represent the distance at time zero, are approximately zero. If we substitute zero for t in Equations (1) to (5), the value for d is indeed zero. If we looked at additional experimental results, we might encounter some variation in the constant term but all should have values of approximately zero.

Galileo's discoveries are the basis for the following equations relating distance and time:

> The general equation of motion of freely falling bodies that relates distance fallen, d, to time, t, is
>
> $$d = \tfrac{1}{2}gt^2 + v_0 t$$
>
> where v_0 is the initial velocity and g is the acceleration due to gravity on Earth.

For example, in our particular model the equation is $d = 488t^2 + 99t$, where 488 approximates half of g in centimeters per square second, and 99 approximates the initial velocity in centimeters per second.

Returning to Galileo's Question

If the velocity for freely falling bodies is not constant, is it increasing at a constant rate? Galileo discovered that the rate of change of the velocity of a freely falling object is constant. In this section we confirm his finding with data from the free fall experiment.

Velocity: Change in Distance over Time

See C3: "Average Velocity and Distance" in *Rates of Change.*

If the rate of change of velocity is constant, then the graph of velocity vs. time should be a straight line. Previously we calculated the average rates of change of distance with respect to time (or average velocities) for three arbitrarily chosen pairs of points. Now, in Table 3 we calculate the average rates of change for all the pairs of adjacent points in our free fall data. The results are in column 4. Since each computed velocity is the average over an interval, for increased precision we associate each velocity with the midpoint time of the interval instead of one of the end points. In Figure 5, we plot velocity from the fourth column against the midpoint times from the third column. The graph is strikingly linear.

Generating a best fit linear function and rounding to the nearest unit, we obtain the equation

$$\text{average velocity} = 977t + 98$$

Table 3

Time, t (sec)	Distance Fallen, d (cm)	Midpoint Time, t (sec)	Velocity, v (cm/sec)
0.0000	0.00		
		0.0083	103.2
0.0167	1.72		
		0.0250	121.8
0.0333	3.75		
		0.0417	141.0
0.0500	6.10		
		0.0583	154.2
0.0667	8.67		
		0.0750	174.6
0.0833	11.58		
		0.0917	187.8
0.1000	14.71		
		0.1083	203.4
0.1167	18.10		
		0.1250	220.2
0.1333	21.77		
		0.1417	236.4
0.1500	25.71		
		0.1583	251.4
0.1667	29.90		
		0.1750	273.0
0.1833	34.45		
		0.1917	286.2
0.2000	39.22		
		0.2083	300.0
0.2167	44.22		
		0.2250	321.6
0.2333	49.58		
		0.2417	334.2
0.2500	55.15		
		0.2583	350.4
0.2667	60.99		
		0.2750	367.2
0.2833	67.11		
		0.2917	382.2
0.3000	73.48		
		0.3083	397.2
0.3167	80.10		
		0.3250	417.0
0.3333	87.05		
		0.3417	430.8
0.3500	94.23		

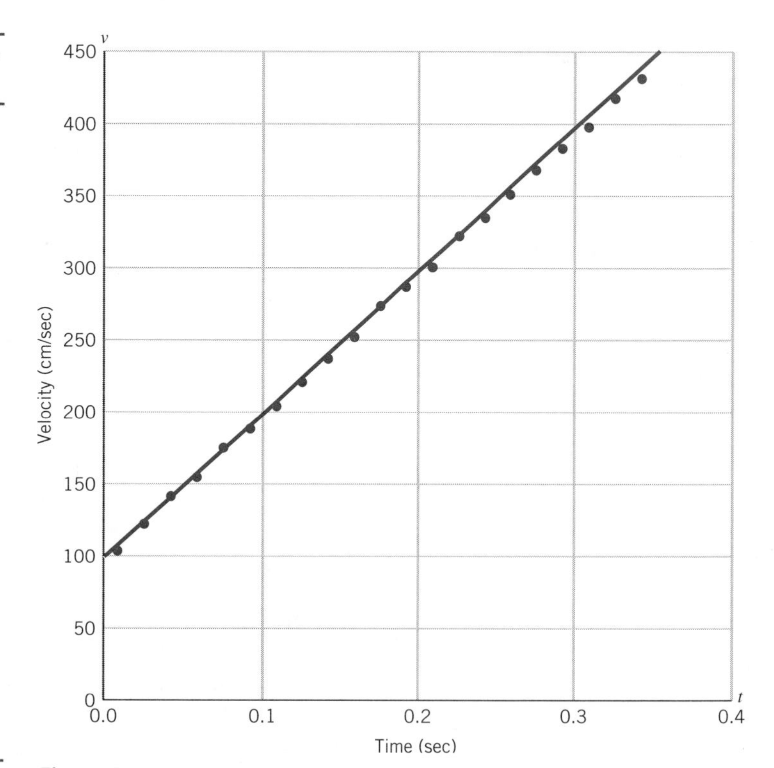

Figure 5 Best fit linear function for average velocity vs. time.

for the average velocity (in centimeters per second) at time t where t is in seconds. The graph of this function appears in Figure 5. The slope of the line is constant and equals the rate of change of velocity with respect to time. So while the velocity is not constant, its rate of change with respect to time *is* constant.

The coefficient of t, 977, is the slope of the line and in physical terms represents g, the acceleration due to gravity. The conventional value for g is 980 cm/sec^2. So the velocity of the freely falling object increases by about 980 cm/sec during each second of free fall.

With this equation we can estimate the velocity at any given time t. When $t = 0$, then $v = 98$ cm/sec. This means that the object was already moving at about 98 cm/sec when we set $t = 0$. In our experiment, the velocity when $t = 0$ depends on where we choose to start measuring our dots. If we had chosen a dot closer to the beginning of the free fall, we would have had an initial velocity smaller than 98 cm/sec. If we had chosen a dot farther away from the start, we would have had an initial velocity larger than 98 cm/sec. Note that 98 cm/sec closely matches the value of 99 cm/sec that we found for the initial velocity for these data.

The general equation that relates v, the velocity of a freely falling body, to t, time, is

$$v = gt + v_0$$

where v_0 = initial velocity (velocity at time $t = 0$) and g is the acceleration due to gravity.

Acceleration: Change in Velocity over Time

Acceleration means a change in velocity or speed. If you push the accelerator pedal in a car down just a bit, the speed of the car increases slowly. If you floor the pedal, the speed increases rapidly. The rate of change of velocity with respect to time is called *acceleration.* Calculating the average rate of change of velocity with respect to time gives an estimate of acceleration. For example, if a car is traveling at 20 mph and 1 hour later the car has accelerated to 60 mph, then

$$\frac{\text{change in velocity}}{\text{change in time}} = \frac{(60 - 20)\text{ mph}}{1\text{ hr}} = (40\text{ mph})/\text{hr} = 40\text{ mi/hr}^2$$

In 1 hour, the velocity of the car changed from 20 to 60 mph, so its average acceleration was 40 mph/hr, or 40 mi/hr^2.

$$\textit{average acceleration} = \frac{\text{change in velocity}}{\text{change in time}}$$

Table 4 uses the average velocity data and midpoint time from Table 3 to calculate average accelerations. Figure 6 shows the plot of average acceleration in centimeters per square second (the third column) vs. time in seconds (the first column).

The data lie along a roughly horizontal line. The average acceleration values vary between a low of 756 cm/sec^2 and a high of 1296 cm/sec^2 with a mean of

Table 4

Midpoint Time, t (sec)	Average Velocity, v (cm/sec)	Average Acceleration (cm/sec^2)
0.0083	103.2	n.a.
0.0250	121.8	1116
0.0417	141.0	1152
0.0583	154.2	792
0.0750	174.6	1224
0.0917	187.8	792
0.1083	203.4	936
0.1250	220.2	1008
0.1417	236.4	972
0.1583	251.4	900
0.1750	273.0	1296
0.1917	286.2	792
0.2083	300.0	828
0.2250	321.6	1296
0.2417	334.2	756
0.2583	350.4	972
0.2750	367.2	1008
0.2917	382.2	900
0.3083	397.2	900
0.3250	417.0	1188
0.3417	430.8	828

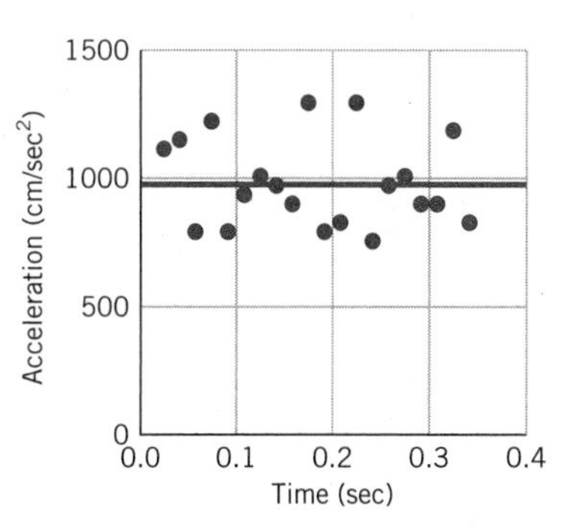

Figure 6 Average acceleration for free fall data.

982.8. Rounding off, we have

$$\text{acceleration} \approx 980 \text{ cm/sec}^2$$

This expression confirms that for each additional 1 second of fall, the velocity of the falling object increases by approximately 980 cm/sec. The longer it falls, the faster it goes. We have verified a characteristic feature of gravity near the surface of Earth: it causes objects to fall at a velocity that increases every second by about 980 cm/sec. We say that the acceleration due to gravity near Earth's surface is 980 cm/sec^2.

In order to express g in feet per square second, we need to convert 980 centimeters into feet. We start with the fact that 1 ft = 30.48 cm. So the conversion factor for centimeters to feet is 1 ft/30.48 cm = 1. If we multiply 980 cm by 1 ft/30.48 cm to convert centimeters to feet, we get

$$980 \text{ cm} = (980 \cancel{\text{cm}})\left(\frac{1 \text{ ft}}{30.48 \cancel{\text{cm}}}\right) \approx 32.15 \text{ ft} \approx 32 \text{ ft}$$

So a value of 980 cm/sec^2 for g is equivalent to approximately 32 ft/sec^2.

> The conventional values for g, the acceleration due to gravity near the surface of Earth, are
>
> $$g = 32 \text{ ft/sec}^2$$
>
> or equivalently
>
> $$g = 980 \text{ cm/sec}^2 = 9.8 \text{ m/sec}^2$$

The numerical value used for the constant g depends on the units being used for the distance, d, and the time, t. The exact value of g also depends on where it is measured.[7]

Deriving an Equation for the Height of an Object in Free Fall

Assume we have the following motion equation relating distance fallen in centimeters and time in seconds:

$$d = 490t^2 + 45t$$

Also assume that when $t = 0$, the height, h, of the object was 110 cm above the ground. Until now, we have considered the distance from the point the object was dropped, a value that *increases* as the object falls. How can we describe a different distance, the *height above ground* of an object, as a function of time, a value that *decreases* as the object falls?

At time zero, the distance fallen is zero and the height above the ground is 110 centimeters. After 0.05 second, the object has fallen about 3.5 centimeters, so its height would be $110 - 3.5 = 106.5$ cm. For an arbitrary distance d, we have $h = 110 - d$. Table 5 gives associated values for time, t, distance, d, and height, h. The graphs in Figure 7 show distance vs. time and height vs. time.

The equation $d = 490t^2 + 45t$ relates distance fallen and time. How can we find an expression relating height and time?

We know that the relationship between height and distance is $h = 110 - d$. We can substitute the expression for d into the height equation:

$$\begin{aligned} h &= 110 - d \\ &= 110 - (490t^2 + 45t) \\ &= 110 - 490t^2 - 45t \qquad (6) \end{aligned}$$

Since we can switch the order in which we add terms, we could of course write this equation as $h = -490t^2 - 45t + 110$. The constant term, here 110 cm, represents the initial height when $t = 0$. By placing the constant term first as in Equation (6), we emphasize 110 cm as the initial or starting value. Height equations often appear in the form $h = c + bt + at^2$ to emphasize the constant term c as the starting height. This is similar to writing linear equations in the form $y = b + mx$ to emphasize the constant term b as the base, or starting, value. Rewriting Equation (6), we have

$$h = 110 - 45t - 490t^2$$

Note that in our equation for height, the coefficients of both t and t^2 are negative. If we consider what happens to the height of an object in free fall, this makes

[7]Because the Earth is rotating, is not a perfect sphere, and is not uniformly dense, there are variations in g according to latitude and elevation. The following are a few examples of local values for g:

Location	North Latitude (deg)	Elevation (m)	g (cm/sec^2)
Panama Canal	9	0	978.243
Jamaica	18	0	978.591
Denver, CO	40	1638	979.609
Pittsburgh, PA	40.5	235	980.118
Cambridge, MA	42	0	980.398
Greenland	70	0	982.534

Source: H. D. Young, *University Physics,* Vol. I, 8th ed. (Reading, MA: Addison Wesley, 1992), p. 336.

Table 5

Time, t (sec)	Distance Fallen, d (cm) ($d = 490t^2 + 45t$)	Height above Ground, h (cm) ($h = 110 - 45t - 490t^2$)
0.00	0.0	110.0
0.05	3.5	106.5
0.10	9.4	100.6
0.15	17.8	92.2
0.20	28.6	81.4
0.25	41.9	68.1
0.30	57.6	52.4
0.35	75.8	34.2
0.40	96.4	13.6

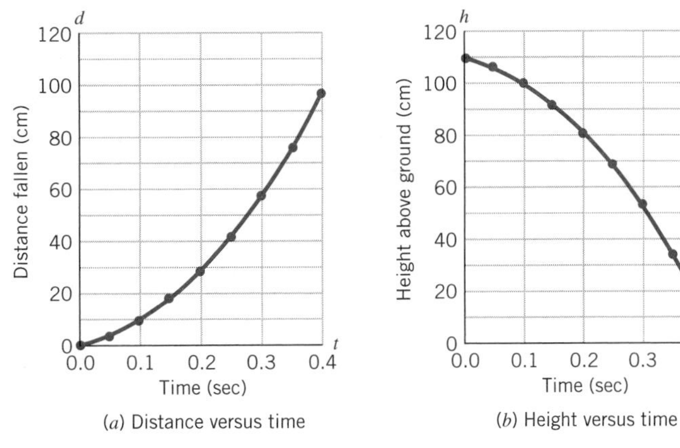

Figure 7 Representations of free fall data.

sense. As time increases, the height decreases. (See Table 5 and Figure 7.) When we were measuring the increasing distance an object fell, we did not take into account the direction in which it was going (up or down). We cared only about the magnitudes (the absolute values) of distance and velocity, which were positive. But when we are measuring a decreasing height or distance, we have to worry about direction. In these cases we define downward motion to be negative and upward motion to be positive. In the height equation $h = 110 - 45t - 490t^2$, the constant term, the initial height is 110 cm; and the change in height resulting from the initial velocity, $-45t$, is negative because the object was moving down when we started to measure it. The change in height caused by acceleration, $-490t^2$, is negative because the effect of gravity is downward motion–it reduces the height of a falling object as time increases.

Once we have introduced the notion that downward motion is negative and upward motion is positive, we can also deal with situations in which the initial velocity is upward and the acceleration is downward. The velocity equation is

$$v = -gt + v_0$$

where v_0 could be either positive or negative, depending on whether the object is thrown upward or downward, and the sign for the g term is negative because gravity accelerates downward in the negative direction.

In summary:

> The general equations of motion of freely falling bodies that relate height, h, and velocity, v, to time, t, are
>
> $$h = h_0 + v_0 t - \tfrac{1}{2}gt^2$$
>
> $$v = -gt + v_0$$
>
> where h_0 = initial height, g = acceleration due to gravity, and v_0 = initial velocity (which can be positive or negative).

Working with an Initial Upward Velocity

If we want to use the general equation to describe the height of a thrown object, we need to understand the meaning of each of the coefficients. Suppose a ball is thrown upward with an initial velocity of 97 cm/sec from a height of 87 cm above the ground. Describe the relationship between the height of the ball and time with an equation.

The initial height of the ball is 87 cm when $t = 0$, so the constant term is 87 cm. The coefficient of t, or the initial velocity term, is $+97$ cm/sec since the initial motion is upward. The coefficient of t^2, the gravity term, is -490 cm/sec^2, since the effect of gravity is downward motion.

Substituting these values into the equation for height, we get

$$h = 87 + 97t - 490t^2$$

Table 6 gives a series of values for heights corresponding to various times.

The graph of the heights at each time in Figure 8 should not be confused with the trajectory of a thrown object. The actual motion we are talking about is a purely vertical motion–straight up and straight down. The graph shows that the object travels up for a while before it starts to fall. This corresponds with what we all know from practical experience throwing balls. The upward (positive) velocity is decreased by the pull of gravity until the object stops moving upward and begins to fall. The downward (negative) velocity is increased in magnitude by the pull of gravity until the object strikes the ground.

Table 6

t (sec)	h (cm)
0.00	87.00
0.05	90.63
0.10	91.80
0.15	90.53
0.20	86.80
0.25	80.63
0.30	72.00
0.35	60.93
0.40	47.40
0.45	31.43
0.50	13.00

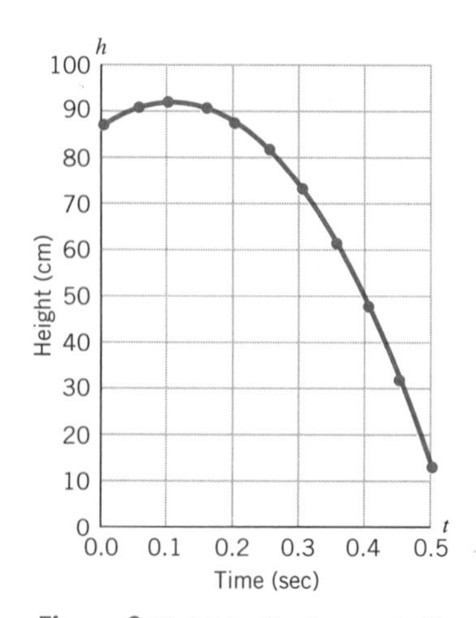

Figure 8 Height of a thrown ball.

COLLECTING AND ANALYZING DATA FROM A FREE FALL EXPERIMENT

Objective

- to describe mathematically how objects fall

Equipment/Materials

- graphing calculator with best fit function capabilities or computer with spreadsheet and function graphing program
- notebook for recording measurements and results

If using precollected data, see Excel or graph link file FREEFALL

Equipment needed for collecting data in physics laboratory:

a. free fall apparatus
b. meter sticks 2 meters long
c. masking tape

Equipment needed for collecting data with CBL® (Calculator-Based Laboratory System®):

a. CBL® unit with AC-9201 power adapter
b. Vernier CBL® ultrasonic motion detector
c. graphing calculator
d. extension cord and some object to drop, such as a pillow or rubber ball

Preparation

If collecting data in a physics laboratory, schedule a time for doing the experiment and have the laboratory assistant available to set up the equipment and assist with the experiment. If collecting data with a CBL® unit with graphing calculators, instructions for using a CBL® unit are in the Instructor's Manual.

Procedure

The following procedures can be used for collecting data in a physics laboratory.[8] If you are collecting data with a CBL®, collect the data and go to the Results section. If you are using the precollected data in the file FREEFALL, go directly to the Results section.

Collecting the data

Since the falling times are too short to record with a stopwatch, we use a free fall apparatus. Every 1/60th of a second a spark jumps between the falling object or "bob" and the vertical metal pole supporting the tape. Each spark burns a small dot on the tape, recording the bob's position. The procedure is to:

1. Position the bob at the top of the column in its holder.
2. Pull the tape down the column so that a fresh tape is ready to receive spark dots.
3. Be sure that the bob is motionless before you turn on the apparatus.

[8]These procedures are adapted from "Laboratory Notes for Experiment 2: The Kinematics of Free Fall," *University of Massachusetts, Boston,* Elementary Physics 181.

4. Turn on the spark switch and bob release switch as demonstrated by the laboratory assistant.
5. Tear off the length of tape recording the fall of the bob.

Obtaining and recording measurements from the tapes

The tape is a record of the distance fallen by the bob between each 1/60th of a second spark dot. Each pair of students should measure and record the distance between the dots on the tape. Let d = the distance fallen in centimeters and t = time in seconds.

1. Fasten the tape to the table using masking tape.
2. Inspect the tape for missing dots. *Caution:* The sparking apparatus sometimes misses a spark. If this happens, take proper account of it in numbering the dots.
3. Position the 2-meter stick on its edge along the dots on the tape. Use masking tape to fasten the meter stick to the table, making sure that the spots line up in front of the bottom edge of the meter stick so you can read their positions off of the stick.
4. Beginning with the sixth visible dot, mark the time for each spot on the tape; that is, write $t = 0/60$ sec by the sixth dot, $t = 1/60$ by the next dot, $t = 2/60$ by the next dot, and so on, until you reach the end of the tape.

   Note: The first five dots are ignored in order to increase accuracy of measurements. One cannot be sure that the object is released exactly at the time of the spark, instead of between sparks, and the first few dots are too close together to get accurate measurements. When the body passes the sixth dot, it already has some velocity, which we call v_0, and this point is arbitrarily taken as the initial time, $t = 0$.

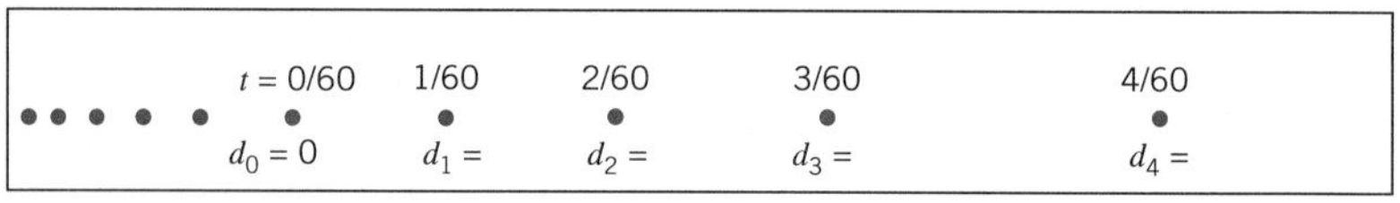

	$t = 0/60$	1/60	2/60	3/60	4/60
• • • • •	•	•	•	•	•
	$d_0 = 0$	$d_1 =$	$d_2 =$	$d_3 =$	$d_4 =$

5. Measure the distances (accurate to a fraction of a millimeter) from the sixth dot to each of the other dots. Record each distance by the appropriate dot on the tape.
6. Recheck your measurements.
7. Clean your work area.

Results

Use your notebook to keep a record of your data, observations, graphs, and analysis of the data.

a. Record the data obtained from your measurements on the tape or from using a CBL® unit. If you are entering your data into a function graphing program or a spreadsheet, you can use a printout of the data and staple it into your laboratory notework. Your data should include time, t, and distance fallen, d, as in the following table:

t (sec)	d (cm)
0/60	0
1/60	—
2/60	—
....	—
To last record	—

This table assumes regular time intervals of 1/60th of a second. Check your equipment to see whether it uses a different interval size.

b. Note at which dot on the tape you started to make your measurements.

Analysis of Data

1. By hand:

a. Graph your data, using the vertical axis for distance fallen, d, in centimeters and the horizontal axis for time, t, in seconds. What does your graph suggest about the average rate of change of distance with respect to time?

b. Calculate the average rate of change for distance, d, with respect to time, t, for three pairs of points from your data table:

$$\text{average rate of change} = \frac{\text{change in distance}}{\text{change in time}} = \frac{\Delta d}{\Delta t}$$

Show your work. This average rate of change is called the *average velocity* of the falling object between these two points. Do your calculations support your answer in part (a)?

c. Jot down your observations from your graph and calculations in your notebook. Staple your graph into your notebook.

2. With graphing calculators or computers:

a. Use technology to graph your data for the free fall experiment. Plot time, t, on the horizontal axis and distance fallen, d, on the vertical axis.

b. Find a best fit function for distance fallen vs. time.

c. Use your spreadsheet or graphing calculator to calculate the average rate of change in distance over each of the small time intervals. This average rate of change is the average velocity over each of these time intervals.

d. Plot average velocity vs. time, with time on the horizontal axis and average velocity on the vertical axis.

e. Jot down your observations from your graphs and calculations in your notebook. Be sure to specify the units for any numbers you recorded.

Conclusions

Summarize your conclusions from the experiment:

- Describe what you found out from your graph of distance vs. time and your calculations for the average rate of change of distance with respect to time. Is the average rate of change of distance with respect to time the same for each small time interval?
- What does your graph of the average velocity vs. time tell you about the average velocity of the freely falling body? Is the average rate of change in velocity from one interval to the next roughly constant?
- In light of the readings and class discussion, interpret your graphs for distance and average velocity and interpret the coefficients in the equation you found for distance.

In his own version of this experiment, Galileo sought to answer the following questions:

How can we describe mathematically the distance an object falls over time?

Do freely falling objects fall at a constant speed?
If the velocity of a freely falling object is not constant, is it increasing at a constant rate?

Use your results to answer these questions.

EXERCISES

1. Complete the accompanying table. What happens to the average velocity of the object as it falls?

Time (sec)	Distance Fallen (cm)	Average Velocity (average rate of change for the previous 1/30th of a second)
0.0000	0.00	n.a.
0.0333	3.75	$(3.75 - 0.00)/(0.0333 - 0.0000) \approx 113$ cm/sec
0.0667	8.67	$(8.67 - 3.75)/(0.0667 - 0.0333) \approx 147$ cm/sec
0.1000	14.71	
0.1333	21.77	
0.1667	29.90	

2. The essay "Watching Galileo's Learning" examines the learning process that Galileo went through to come to some of the most remarkable conclusions in the history of science. Write a summary of one of Galileo's conclusions about motion. Include in your summary the process by which Galileo made this discovery and some aspect of your own learning or understanding of Galileo's discovery.

3. The equation $d = 490t^2 + 50t$ describes the relationship between distance fallen, d, in centimeters, and time, t, in seconds, for a particular freely falling object.

a. Interpret each of the coefficients and specify its units of measurement.

b. Generate a table for a few values of t between 0 and 0.3 second.

c. Graph distance vs. time by hand. Check your graph using a computer or graphing calculator if available.

4. A freely falling body has an initial velocity of 125 cm/sec. Assume that $g = 980$ cm/sec^2.

a. Write an equation that relates d, distance fallen in centimeters, to t, time in seconds.

b. How far has the body fallen after 1 second? After 3 seconds?

c. If the initial velocity were 75 cm/sec, how would your equation in part (a) change?

5. The equation $d = 4.9t^2 + 1.7t$ describes the relationship between distance fallen, d, in meters, and time, t, in seconds, for a particular freely falling object.

a. Interpret each of the coefficients and specify its units of measurement.

b. Generate a table for a few values of t between 0 and 0.3 second.

c. Graph distance vs. time by hand. Check your graph using a computer or graphing calculator if available.

d. Relate your answers to earlier results in this chapter.

6. In the equation of motion, $d = \frac{1}{2}gt^2 + v_0t$, we specified that distance was measured in centimeters, velocity in centimeters per second, and time in seconds. Rewrite this as an equation that shows only units of measure. Verify that you get centimeters = centimeters.

7. The equation $d = \frac{1}{2}gt^2 + v_0t$ could also be written using distances measured in meters. Rewrite the equation showing only units of measure and verify that you get meters = meters.

8. The equation $d = \frac{1}{2}gt^2 + v_0t$ could be written using distance measured in feet. Rewrite the equation showing only units of measure and verify that you get feet = feet.

9. A freely falling object has an initial velocity of 50 cm/sec.

a. Write two motion equations, one relating distance and time and the other relating velocity and time.

b. How far has the object fallen and what is its velocity after 1 second? After 2.5 seconds? Be sure to identify units in your answers.

10. A freely falling object has an initial velocity of 20 ft/sec.

a. Write one equation relating distance fallen (in feet) and time (in seconds) and a second equation relating velocity (in feet per second) and time.

b. How many feet has the object fallen and what is its velocity after 0.5 second? After 2 seconds?

11. A freely falling object has an initial velocity of 12 ft/sec.

a. Construct an equation relating distance fallen and time.

b. Generate a table by hand for a few values of the distance fallen between 0 and 5 seconds.

c. Graph distance vs. time by hand. Check your graph using a computer or graphing calculator if available.

12. Use the information in Exercise 11 to do the following:

a. Construct an equation relating velocity and time.

b. Generate a table by hand for a few values of velocity between 0 and 5 seconds.

c. Graph velocity vs. time by hand. If possible, check your graph using a computer or graphing calculator.

13. If the equation $d = 4.9t^2 + 11t$ represents the relationship between distance and time for a freely falling body, in what units is distance now being measured? How do you know?

14. The distance that a freely falling object with no initial velocity falls can be modeled by the quadratic function $d = 16t^2$, where t is measured in seconds and d in feet. There is a closely related function $v = 32t$ that gives the velocity, v, in feet per second at time t, for the same freely falling body.

a. Fill in the missing values in the following table:

Time, t (sec)	Distance, d (ft)	Velocity, v (ft/sec)
1		
1.5		
2		
		80
	144	

b. When $t = 3$ describe the associated values of d and v and what they tell you about the object at that time.

c. Sketch both functions, distance vs. time and velocity vs. time, on two different graphs. Label the points from part (b) on the curves.

d. You are standing on a bridge looking down at a river below. How could you use a pebble to estimate how far you are above the water?

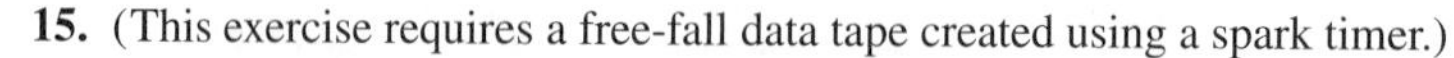

One screen in "Q11: Freely Falling Objects" in *Quadratic Functions* simulates this activity.

15. (This exercise requires a free-fall data tape created using a spark timer.)

a. Make a graph from your tape: cut the tape with scissors crosswise at each spark dot, so you have a set of strips of paper that are the actual lengths of the distances fallen by the object during each time interval. Arrange them evenly spaced in increasing order, with the bottom of each strip on a horizontal line. The end result should look like a series of steps. You could paste or tape them down on a big piece of paper or newspaper.

b. Use a straight edge to draw a line that passes through the center of the top of each strip. Is the line a good fit? Each separate strip represents the distance the object fell during a fixed time interval, so we can think of the strips as representing change in distance over time, or an average velocity. Interpret the graph of the line you have constructed in terms of the free fall experiment.

16. In the Anthology Reading "Watching Galileo's Learning," Cavicchi notes that Galileo generated a sequence of odd integers from his study of falling bodies. Show that in general the odd integers can be constructed from the difference of the squares of successive integers, that is, that the terms $(n + 1)^2 - n^2$ (where $n = 1, 2, 3, \ldots$) generate a sequence of odd integers.

17. The data from a free fall tape generate the following equation relating distance fallen in centimeters and time in seconds:

$$d = 485.7t^2 + 7.6t$$

a. Give a physical interpretation of each of the coefficients along with their appropriate units of measurement.

b. How far has the object fallen after 0.05 second? 0.10 second? 0.30 second?

18. What would the free fall equation $d = 490t^2 + 90t$ become if d were measured in feet instead of centimeters?

19. In the equation $d = 4.9t^2 + 500t$, time is measured in seconds and distance in meters. What does the number 500 represent?

20. In the height equation $h = 300 + 50t - 4.9t^2$, time is measured in seconds and height in meters.

a. What does the number 300 represent?

b. What does the number 50 represent? What does the fact that 50 is positive tell you?

21. The height of an object that was projected vertically from the ground with initial velocity of 200 m/sec is given by the equation $h = 200t - 4.9t^2$, where t is in seconds.

a. Find the height of the object after 0.1, 2, and 10 seconds.

b. Sketch a graph of height vs. time.

c. Use the graph to determine the maximum height of the projectile and the approximate number of seconds that the object traveled before hitting the ground.

22. The height of an object that was shot downward from a 200-meter platform with an initial velocity of 50 m/sec is given by the equation $h = -4.9t^2 - 50t + 200$, where t is in seconds and h is in meters. Sketch the graph of height vs. time. Use the graph to determine the approximate number of seconds that the object traveled before hitting the ground.

23. Let $h = 85 - 490t^2$ be a motion equation describing height, h, in centimeters and time, t, in seconds.

a. Interpret each of the coefficients and specify its units of measurement.

b. What is the initial velocity?

c. Generate a table for a few values of t between 0 and 0.3 second.

d. Graph height vs. time by hand. Check your graph using a computer or graphing calculator if possible.

24. Let $h = 85 + 20t - 490t^2$ be a motion equation describing height, h, in centimeters and time, t, in seconds.

a. Interpret each of the coefficients and specify its units of measurement.

b. Generate a table for a few values of t between 0 and 0.3 second.

c. Graph height vs. time by hand. Check your graph using a computer or graphing calculator if possible.

25. At $t = 0$, a ball is thrown upward at a velocity of 10 ft/sec from the top of a building 50 feet high. The ball's height is measured in feet above the ground.

a. Is the initial velocity positive or negative? Why?

b. Write the motion equation that describes height, h, at time, t.

26. The concepts of velocity and acceleration are useful in the study of human childhood development. The accompanying figure shows (*a*) a standard growth curve of weight over time; (*b*) the rate of change of weight over time (the *growth rate* or *velocity*); and (*c*) the rate of change of the growth rate over time (or *acceleration*). Describe in your own words what each of the graphs shows about a child's growth.

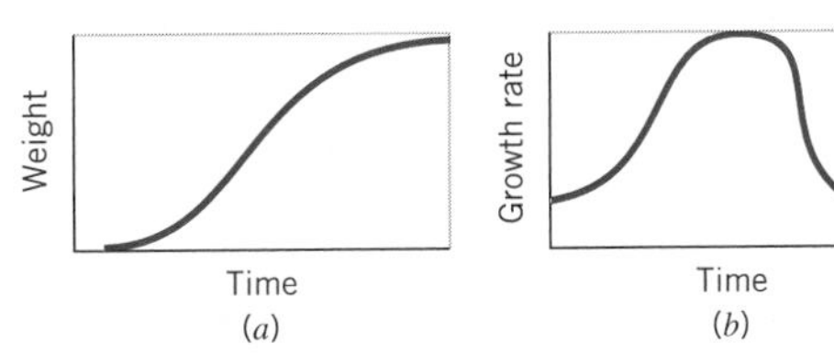

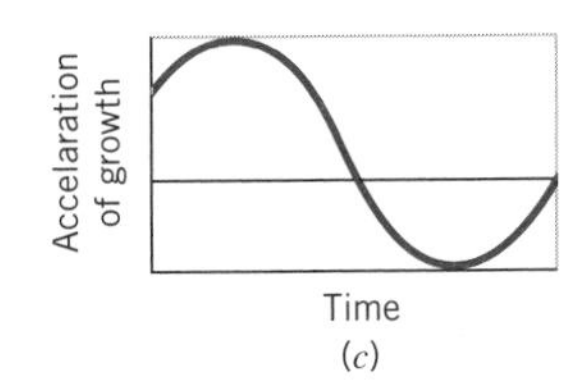

Source: Adapted from B. Bogin, "The Evolution of Human Childhood," *BioScience,* Vol. 40, p. 16.

27. The relationship between the velocity of a freely falling object and time is given by

$$v = -gt - 66$$

where g is the acceleration due to gravity and the units for velocity are centimeters per second.

a. What value for g should be used in the equation?

b. Generate a table of values for t and v, letting t range from 0 to 4 seconds.

c. Graph velocity vs. time and interpret your graph.

d. What was the initial condition? Was the object dropped or thrown? Explain your reasoning.

28. A certain baseball is at height $h = 4 + 64t - 16t^2$ feet at time t in seconds. Compute the average velocity for each of the following time intervals and indicate for which intervals the baseball is rising and for which it is falling. In which interval was the average velocity the greatest?

a. $t = 0$ to $t = 0.5$

b. $t = 0$ to $t = 0.1$

c. $t = 0$ to $t = 1$

d. $t = 1$ to $t = 2$

e. $t = 2$ to $t = 3$

f. $t = 1$ to $t = 3$

g. $t = 4$ to $t = 4.01$

29. At $t = 0$, an object is in free fall 150 cm above the ground, falling at a rate of 25 cm/sec. Its height, h, is measured in centimeters above the ground.

a. Is its velocity positive or negative? Why?

b. Write an equation that describes its height, h, at time t.

c. What is the average velocity from $t = 0$ to $t = \frac{1}{2}$? How does it compare to the initial velocity?

30. In 1974 in Anaheim, California, Nolan Ryan threw a baseball at just over 100 mph. If he had thrown the ball straight upward at this speed, it would have risen to a height of over 335 feet and taken just over 9 seconds to fall back to Earth. Choose another planet and see what would have happened if he had been able to throw a baseball straight up at 100 mph on that planet. In your computations, use the table for the acceleration due to gravity on other planets in Exercise 32.

31. Suppose an object is moving with constant acceleration, a, and its motion is initially observed at a moment when its velocity is v_0. We set time, t, equal to 0, at this point when velocity equals v_0. Then its velocity t seconds after the initial observation is $V(t) = at + v_0$. (Note that the product of acceleration and time is velocity.) Now suppose we want to find its average velocity between time 0 and time t. The average velocity can be measured in two ways. First, we can find the average of the initial and final velocities by calculating a numerical average or mean: that is, we add the two velocities and divide by 2. So, between time 0 and time t,

$$\text{average velocity} = \frac{v_0 + V(t)}{2} \tag{1}$$

We can also find the average velocity by dividing the change in distance by the change in time. Thus, between time 0 and time t,

$$\text{average velocity} = \frac{\Delta \text{ distance}}{\Delta \text{ time}} = \frac{d - 0}{t - 0} = \frac{d}{t} \tag{2}$$

If we substitute the expression for average velocity (from time 0 to time t) given by Equation (1) into Equation (2), we get

$$\frac{d}{t} = \frac{v_0 + V(t)}{2} \tag{3}$$

We know that $V(t) = at + v_0$. Substitute this expression for $V(t)$ in Equation (3) and solve for d. Interpret your results.

32. *The force of acceleration on other planets.* We have seen that the function $d = \frac{1}{2}gt^2 + v_0 t$ (where g is the acceleration due to Earth's gravity and v_0 is the object's initial velocity) is a mathematical model for the relationship between time and distance fallen by freely falling bodies near Earth's surface. This relationship also holds for freely falling bodies near the surfaces of other planets. We would need to replace g, the acceleration of Earth's gravitational field, with the acceleration for the planet under consideration. The following table gives the acceleration due to gravity for planets in our solar system:

Acceleration Due to Gravity

	m/sec^2	ft/sec^2
Mercury	3.7	12.1
Venus	8.9	29.1
Earth	9.8	32.1
Mars	3.7	12.1
Jupiter	24.8	81.3
Saturn	10.4	34.1
Uranus	8.5	27.9
Neptune	11.6	38.1
Pluto	0.6	2.0

Source: The Astronomical Almanac, U.S. Naval Observatory, 1981.

a. Choose units of measurement (meters or feet) and three of the planets (other than Earth). For each of these planets, find an equation for the relationship between the distance an object falls and time. Construct a table as shown below. Assume for the moment that the initial velocity of the freely falling object is 0.

Name of Planet	Function Relating Distance and Time (sec)	Units for Distance

b. Using a function graphing program or a graphing calculator, plot the three functions, with time on the horizontal axis and distance on the vertical axis. What domain makes sense for your models? Why?

c. On which of your planets will an object fall the farthest in a given time? On which will it fall the least distance in a given time?

d. Examine the graphs and think about the *similarities* that they share. Describe their general shape. What happens to d as the value for t increases? As the value for t decreases?

e. Think about the *differences* among the three curves. What effect does the coefficient of the t^2 term have on the shape of the graph, i.e., when the coefficient gets larger (or smaller), how is the shape of the curve affected? Which graph shows d increasing the fastest compared to t?

33. An object that is moving on the ground is observed to have (initial) velocity of 60 cm/sec and to be accelerating at a constant rate of 10 cm/sec^2.

a. Determine its velocity after 5 seconds; after 60 seconds; after t seconds.

b. Find the average velocity for the object between 0 and 5 seconds.

34. Find the distance traveled by the object described in Exercise 33 after 5 seconds by using two different methods.

a. Use the formula: distance equals rate times time. For rate, use the value found in Exercise 33(b). For time, use 5 seconds.

b. Write an equation of motion $d = \frac{1}{2}at^2 + v_0t$ using $a = 10$ cm/sec^2 and $v_0 = 60$ cm/sec and evaluate when $t = 5$. Does your answer agree with part (a)?

35. An object is observed to have an initial velocity of 200 m/sec and to be accelerating at 60 m/sec^2.

a. Write an equation for its velocity after t seconds.

b. Write an equation for the distance traveled after t seconds.

36. You may have noticed that when a basketball player or dancer jumps straight up in the air, in the middle of a blurred impression of vertical movement, the jumper appears to "hang" for an instant at the top of the jump.

a. If a player jumps 3 feet straight up, how long does it take him to fall back down to the ground from the top of the jump? At what downward velocity does he hit the ground?

b. At what initial upward velocity does the player have to leap to achieve a 3-foot-high jump? How long does the total jump take from takeoff to landing?

c. How much vertical distance is traveled in the first third of the total time the jump takes? In the middle third? In the last third?

d. Now explain in words why it is that the jumper appears suspended in space at the top of the jump.

37. Old Faithful, the most famous geyser at Yellowstone National Park, regularly shoots up a jet of water 120 feet high.

a. At what speed must the stream of water be traveling out of the ground to go that high?

b. How long does it take to reach maximum height?

38. A vehicle trip is composed of the following parts:

i. Accelerate from 0 to 30 mph in 1 minute

ii. Travel at 30 mph for 12 minutes

iii. Accelerate from 30 to 50 mph in $\frac{1}{3}$ minute

iv. Travel at 50 mph for 5 minutes

v. Decelerate from 50 to 0 mph in $\frac{1}{2}$ minute

a. Make a graph of speed vs. time for the trip. You may find it helpful to convert the speeds to feet per second, and the minutes to seconds.

b. What are the accelerations for parts (i), (iii), and (v) of the trip?

c. How much distance is covered in each part of the trip and what is the total trip distance?

39. In general, for straight motion of a vehicle with constant acceleration, a, the velocity, v, at any time, t, is the original velocity, v_0, plus acceleration multiplied by time: $v = v_0 + at$. The distance traveled in time t is $d = v_0 t + \frac{1}{2}at^2$.

a. A criminal going at speed v_c passes a police car and immediately accelerates with constant acceleration a_c. If the police car has constant acceleration $a_p > a_c$, starting from 0 mph, how long will it take to pass the criminal? Give t in terms of v_c, a_p, and a_c.

b. At what time are the police and the criminal traveling at the same speed? If they are traveling at the same speed, does it mean the police have caught up with the criminal? Explain.

Anthology of Readings

CONTENTS

Additional readings are on the Web at www.wiley.com/college/kimeclark

1. Excerpt from *Chance News,* an electronic magazine
   Prepared by J. Laurie Snell, written by Travis Linn and Jerry Johnson, University of Nevada, Reno. Reprinted with permission.

2. "The Median Isn't the Message"
   Discover magazine. Copyright © 1985 by Dr. Stephen J. Gould, Cambridge, MA

3. "Fragmented War on Cancer"
   Hamilton Jordan, *The New York Times,* May 29, 1996
   Copyright © 1996 by The New York Times Company. Reprinted by permission.

4. "Promote Cancer Treatment, Not Cancer Phobia"
   William London, *The New York Times,* June 1, 1996
   Reprinted by permission of the author.

5. "U.S. Government Definitions of Census Terms"
   Copyright © 1995 by Anthony Roman. Reprinted by permission.

6. "Slopes"
   Copyright © 1995 by Meg Hickey. Reprinted by permission.

7. Excerpts from the *Statistical Portrait–Fall 1999* University of Massachusetts–Boston
   Reprinted with permission from the Office of Institutional Research, University of Massachusetts–Boston

8. Excerpts from the *University of Southern Mississippi Factbook, 1998–1999*
   Reprinted with permission from the University of Southern Mississippi

9. An Excerpt from *Performing Arts–The Economic Dilemma*
William Baumol and William G. Bowen. Reprinted with permission from the Twentieth Century Fund.

10. "His Stats Can Oust a Senator or Price a Bordeaux"
Peter Aseltine, *The Times,* May 8, 1994

11. "How a Flat Tax Would Work, for You and for Them"
David Cay Johnston, *The New York Times,* January 21, 1996

12. "Flat Tax Goes from 'Snake Oil' to G.O.P. Tonic"
Leslie Wayne, *The New York Times,* November 14, 1999.

13. An Excerpt from the book *Powers of Ten: About the Relative Size of Things in the Universe*
Philip and Phylis Morrison and the Office of Charles and Ray Eames.

14. "The Cosmic Calendar"
Chapter from Carl Sagan's *The Dragons of Eden: Speculations on the Evolution of Human Intelligence*

15. "Watching Galileo's Learning"

CHANCE News*

Prepared by J. Laurie Snell, with help from William Peterson, Fuxing Hou and Ma. Katrina Munoz Dy, as part of the CHANCE Course Project supported by the National Science Foundation.

Please send comments and suggestions for articles to:

jlsnell@dartmouth.edu

Back issues of Chance News and other materials for teaching a CHANCE course are available from the Chance Web Data Base.

http://www.geom.umn.edu/locate/chance

===

" Data, data everywhere, but not a thought to think "

Jesse Shera's paraphrase of Coleridge.

===

We found this quote in John Paulos' new book *A Mathematician Reads the Newspaper.*

FROM OUR READERS

Jerry Johnson sent us the following excerpts from a discussion on a journalism listserve group.

I teach statistics at the University of Texas at Arlington. Two weeks ago I read in the science section of a local paper an article defining the difference between the median, the mean, and the average. Everything was fine until he defined the mean as the average of the largest and smallest numbers in a set of data. I have always used and taught that the (arithmetic) mean is the same as the average of the numbers. When I talked to him about this, he indicated that this was the definition given in an Associated Press list of definitions. Is this the definition used by journalists?

Thanks for everyone who answered my question concerning the mean. I have contacted the Associated Press and hope to change their definition. One of the problems seems to come from dictionaries that define the mean as midway between extremes. I contacted Merriam-Webster and got one editor there to agree that midway between extremes is in a philosophical sense and not a mathematical sense. Webster's New World Dictionary has a more specific definition as "a middle or intermediate position as to place, time, quantity, kind, value,..."

After discussion with Norm Goldstein, Director of APN Special Projects, the Associated Press Style Book will be modified to indicate that the calculation of the mean is identical to that of the average.

* Chance News *is an electronic magazine. For instructions to subscribe to this magazine, send an E-mail to jlsnell@dartmouth.edu*

The Median Isn't the Message

Stephen Jay Gould

In 1982, I learned I was suffering from a rare and serious cancer. After surgery, I asked my doctor what the best technical literature on the cancer was. She told me, with a touch of diplomacy, that there was nothing really worth reading. I soon realized why she had offered that humane advice: my cancer is incurable, with a median mortality of eight months after discovery.

My life has recently intersected, in a most personal way, two of Mark Twain's famous quips. One I shall defer to the end of this essay. The other (sometimes attributed to Disraeli), identifies three species of mendacity, each worse than the one before—lies, damned lies, and statistics.

Consider the standard example of stretching truth with numbers—a case quite relevant to my story. Statistics recognizes different measures of an "average," or central tendency. The *mean* is our usual concept of an overall average—add up the items and divide them by the number of sharers (100 candy bars collected for five kids next Halloween will yield 20 for each in a just world). The *median*, a different measure of central tendency, is the halfway point. If I line up five kids by height, the median child is shorter than two and taller than the other two (who might have trouble getting their mean share of the candy). A politician in power might say with pride, "The mean income of our citizens is $15,000 per year." The leader of the opposition might retort, "But half our citizens make less than $10,000 per year." Both are right, but neither cites a statistic with impassive objectivity. The first invokes a mean, the second a median. (Means are higher than medians in such cases because one millionaire may outweigh hundreds of poor people in setting a mean; but he can balance only one mendicant in calculating a median).

The larger issue that creates a common distrust or contempt for statistics is more troubling. Many people make an unfortunate and invalid separation between heart and mind, or feeling and intellect. In some contemporary traditions, abetted by attitudes stereotypically centered upon Southern California, feelings are exalted as more "real" and the only proper basis for action—if it feels good, do it—while intellect gets short shrift as a hang-up of outmoded elitism. Statistics, in this absurd dichotomy, often become the symbol of the enemy. As Hilaire Belloc wrote, "Statistics are the triumph of the quantitative method, and the quantitative method is the victory of sterility and death."

This is a personal story of statistics, properly interpreted, as profoundly nurturant and life-giving. It declares holy war on the downgrading of intellect by telling a small story about the utility of dry, academic knowledge about science. Heart and head are focal points of one body, one personality.

Stephen Jay Gould teaches biology, geology, and the history of science at Harvard University.

In July 1982, I learned that I was suffering from abdominal mesothelioma, a rare and serious cancer usually associated with exposure to asbestos. When I revived after surgery, I asked my first question of my doctor and chemotherapist: "What is the best technical literature about mesothelioma?" She replied, with a touch of diplomacy (the only departure she has ever made from direct frankness), that the medical literature contained nothing really worth reading.

Of course, trying to keep an intellectual away from literature works about as well as recommending chastity to *Homo sapiens*, the sexiest primate of all. As soon as I could walk, I made a beeline for Harvard's Countway medical library and punched mesothelioma into the computer's bibliographic search program. An hour later, surrounded by the latest literature on abdominal mesothelioma, I realized with a gulp why my doctor had offered that humane advice. The literature couldn't have been more brutally clear: mesothelioma is incurable, with a median mortality of only eight months after discovery. I sat stunned for about fifteen minutes, then smiled and said to myself: so that's why they didn't give me anything to read. Then my mind started to work again, thank goodness.

If a little learning could ever be a dangerous thing, I had encountered a classic example. Attitude clearly matters in fighting cancer. We don't know why (from my old-style materialistic perspective, I suspect that mental states feed back upon the immune system). But match people with the same cancer for age, class, health, socioeconomic status, and, in general, those with positive attitudes, with a strong will and purpose for living, with commitment to struggle, with an active response to aiding their own treatment and not just a passive acceptance of anything doctors say, tend to live longer. A few months later I asked Sir Peter Medawar, my personal scientific guru and a Nobelist in immunology, what the best prescription for success against cancer might be. "A sanguine personality," he replied. Fortunately (since one can't reconstruct oneself at short notice and for a definite purpose), I am, if anything, even-tempered and confident in just this manner.

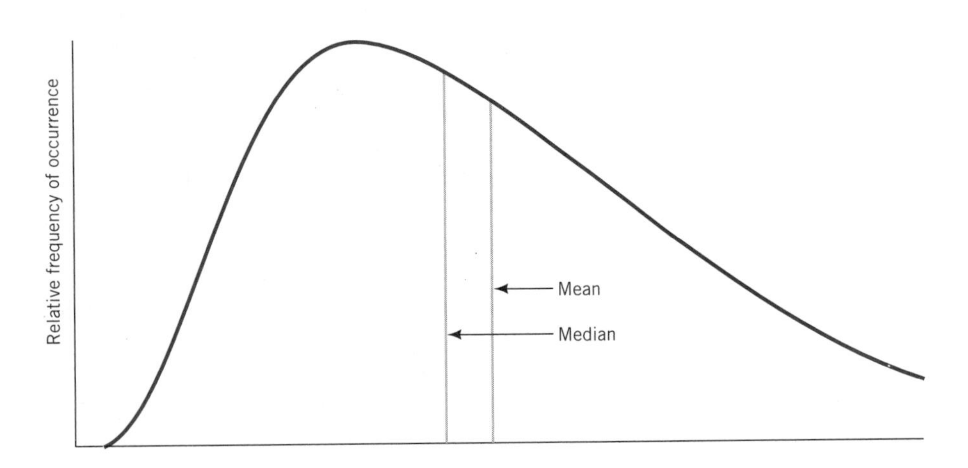

Hence the dilemma for humane doctors: since attitude matters so critically, should such a sombre conclusion be advertised, especially since few people have sufficient understanding of statistics to evaluate what the statements really mean? From years of experience with the small-scale evolution of Bahamian land snails treated quantitatively, I have developed this technical knowledge—and I am convinced that it played a major role in saving my life. Knowledge is indeed power, in Bacon's proverb.

The problem may be briefly stated: What does "median mortality of eight months" signify in our vernacular? I suspect that most people, without training in statistics, would read such a statement as "I will probably be dead in eight months"—the very conclusion that must be avoided, since it isn't so, and since attitude matters so much.

I was not, of course, overjoyed, but I didn't read the statement in this vernacular way either. My technical training enjoined a different perspective on "eight months median mortality." The point is a subtle one, but profound—for it embodies the distinctive way of thinking in my own field of evolutionary biology and natural history.

We still carry the historical baggage of a Platonic heritage that seeks sharp essences and definite boundaries. (Thus we hope to find an unambiguous "beginning of life" or "definition of death," although nature often comes to us as irreducible continua.) This Platonic heritage, with its emphasis on clear distinctions and separated immutable entities, leads us to view statistical measures of central tendency wrongly, indeed opposite to the appropriate interpretation in our actual world of variation, shadings, and continua. In short, we view means and medians as the hard "realities," and the variation that permits their calculation as a set of transient and imperfect measurements of this hidden essence. If the median is the reality and variation around the median just a device for its calculation, the "I will probably be dead in eight months" may pass as a reasonable interpretation.

But all evolutionary biologists know that variation itself is nature's only irreducible essence. Variation is the hard reality, not a set of imperfect measures for a central tendency. Means and medians are the abstractions. Therefore, I looked at the mesothelioma statistics quite differently—and not only because I am an optimist who tends to see the doughnut instead of the hole, but primarily because I know that variation itself is the reality. I had to place myself amidst the variation.

When I learned about the eight-month median, my first intellectual reaction was: fine, half the people will live longer; now what are my chances of being in that half. I read for a furious and nervous hour and concluded, with relief: damned good. I possessed every one of the characteristics conferring a probability of longer life: I was young; my disease had been recognized in a relatively early stage; I would receive the nation's best medical treatment; I had the world to live for; I knew how to read the data properly and not despair.

Another technical point then added even more solace. I immediately recognized that the distribution of variation about the eight-month median would almost surely be what statisticians call "right skewed." (In a symmetrical distribution, the profile of variation to the left of the central tendency is a mirror image of variation to the right. In skewed distributions, variation to one side of the central tendency is more stretched out—left skewed if extended to the left, right skewed if stretched out to the right.) The distribution of variation had to be right skewed, I reasoned. After all, the left of the distribution contains an irrevocable lower boundary of zero (since mesothelioma can only be identified at death or before). Thus there isn't much room for the distribution's lower (or left) half—it must be scrunched up between zero and eight months. But the upper (or right) half can extend out for years and years, even if nobody ultimately survives. The distribution must be right skewed, and I needed to know how long the extended tail ran—for I had already concluded that my favorable profile made me a good candidate for that part of the curve.

The distribution was, indeed, strongly right skewed, with a long tail (however small) that extended for several years above the eight month median. I saw no reason why I shouldn't be in that small tail, and I breathed a very long sigh of relief. My technical knowledge had helped. I had read the graph correctly. I had asked the right question and found the answers. I had obtained, in all probability, that most precious of all possible gifts in the circumstances—substantial time. I didn't have to stop and immediately follow Isaiah's injunction to Hezekiah—set thine house in order: for thou shalt die, and not live. I would have time to think, to plan, and to fight.

One final point about statistical distributions. They apply only to a prescribed set of circumstances—in this case to survival with mesothelioma under conventional modes of treatment. If circumstances change, the distribution may alter. I was placed on an experimental protocol of treatment and, if fortune holds, will be in the first cohort of a new distribution with high median and a right tail extending to death by natural causes at advanced old age.

It has become, in my view, a bit too trendy to regard the acceptance of death as something tantamount to intrinsic dignity. Of course I agree with the preacher of Ecclesiastes that there is a time to love and a time to die—and when my skein runs out I hope to face the end calmly and in my own way. For most situations, however, I prefer the more martial view that death is the ultimate enemy—and I find nothing reproachable in those who rage mightily against the dying of the light.

The swords of battle are numerous, and none more effective than humor. My death was announced at a meeting of my colleagues in Scotland, and I almost experienced the delicious pleasure of reading my obituary penned by one of my best friends (the so-and-so got suspicious and checked; he too is a statistician, and didn't expect to find me so far out on the left tail). Still, the incident provided my first good laugh after the diagnosis. Just think, I almost got to repeat Mark Twain's most famous line of all: the reports of my death are greatly exaggerated.

The New York Times Op Ed Wednesday May 29, 1996

A Fragmented War on Cancer

By Hamilton Jordan

ATLANTA

It has been 25 years since President Richard Nixon declared war on cancer. Having had two different cancers, I am a survivor of that war and a grateful beneficiary. My first, an aggressive lymphoma, was treated with an experimental therapy developed at the National Cancer Institute. Ten years later, my prostate cancer was detected early by the simple P.S.A. blood test, a diagnostic tool supported by Federal grants.

But I am also a symbol of the limited success of that war. The treatments I received were merely updated versions of the methods used 25 years ago. A powerful cocktail of chemicals killed my lymphoma while ravaging my body. A surgeon, using an elegant procedure with no permanent side effects, cut out my prostate.

Scientists are still looking for both the "magic bullet" that kills only cancer cells and the genetic switch that turns off random cancer growth or prevents genetic flaws from causing cancer. While significant progress has been made, twice as many people will be diagnosed with cancer this year as in 1971, and twice as many will die. One in three women and one in two men will have cancer in their lifetimes. The raw data suggest we are on the verge of an epidemic. What happened to the "war"?

Hamilton Jordan, who was President Jimmy Carter's chief of staff, is a board member of Capcure, a nonprofit organization that finances prostate cancer research.

Groups compete for a shrinking pie.

- Our rhetoric exceeded our commitment. Dr. Donald Coffey, a cancer researcher, says we promised a war but financed only a few skirmishes. The Federal budget expresses our national priorities: The Federal Aviation Administration, for example, will spend $8.92 billion to make air travel safe. The chances of dying in an airline accident are one in two million. But the National Cancer Institute will spend only $2.2 billion — one-tenth of a cent of every Federal tax dollar — to find a cure for the disease that kills more Americans in a month than have died in all commercial aviation accidents in our history.
- We created expectations not based on scientific reality. Our political leaders failed to appreciate the simple reality that there is not just one cancer but more than 100 cancers that have all defied a single solution. At the same time, the numerous organizations representing those different cancers have fought among themselves for bigger slices of a shrinking pie instead of forging a consensus on behalf of a larger pie.
- Huge successes with some cancers have been offset by rises in others. In addition, mortality from other diseases has declined, leaving an aging, cancer-prone population.

With adequate financing, breakthroughs in cancer prevention and treatment are likely over the next decade. Yet promising research that would have been automatically financed a decade ago is rejected today because of belt-tightening, discouraging brilliant young investigators from entering cancer research in the first place.

Is one-tenth of one cent enough to find a cure for a disease that will strike 40 percent of Americans? You will not think so when cancer strikes you or your loved ones. □

The New York Times Editorials/ Letters, Saturday June 1, 1996

Promote Cancer Treatment, Not Cancer Phobia

To the Editor:

In "A Fragmented War on Cancer" (Op-Ed, May 29), Hamilton Jordan says, "The raw data suggest we are on the verge of an epidemic."

His assertions that cancer will be diagnosed in twice as many people and that twice as many will die of it this year than in 1971 are misleading.

More people will be diagnosed with and die of cancer this year than in 1971 because the population is now much larger. Moreover, cancer is much more common for older age groups than younger age groups, and the proportion of Americans who are in older age groups is larger this year than in 1971.

After you take into account population size and the number of people in different age groups, recent national cancer data suggest a slight downward trend in overall cancer mortality during the current decade.

A reported increase in the diagnosis of cancer usually reflects changes in diagnostic and reporting practices rather than an actual increase in cancer incidents.

It was also misleading for Mr. Jordan to present as evidence of an emerging cancer epidemic the often quoted figures, "One in three women and one in two men will have cancer in their lifetimes." Such high lifetime risks figures actually mean many people will live long enough lives to be likely to develop cancer. These figures will mask much lower risk of people of any given age developing cancer in the next 10, 20 or 30 years.

While I support Mr. Jordan's continued efforts to promote progress in cancer prevention and treatment, I encourage him to be more careful lest he unwittingly promote cancer phobia.

William M. London Dir. of Public Health, American Council on Science and Health
New York, May 29, 1996

U.S. Government Definitions of Census Terms

by Anthony Roman (1994), Center for Survey Research, University of Massachusetts, Boston.

Housing Unit

The government defines *housing unit* as a "house, an apartment, or a single room occupied as separate living quarters." The exact definition becomes more complex when one considers what constitutes separate living quarters. Perhaps the easiest way to define a housing unit is to describe what it is not. The following are not housing units and therefore the people living in them do not make up households: 1) most units in rooming or boarding houses where people share kitchen facilities, 2) units in transient hotels or motels, 3) college dormitories, 4) bunk houses, 5) group quarters living arrangements such as military housing, convents, prisons, and mental institutions, and 6) units within other housing units which do not have direct access from a hallway or outside. Most other living arrangements should come under the definition of a housing unit and therefore a household.

Household & Family

A *household* includes all persons occupying a housing unit. A *family* consists of all people in a household who are related by blood, marriage, or adoption. By strict definition, a family needs to have a minimum of two people, a householder and at least one other person. A householder is defined as the person in whose name the housing unit is owned or rented. If more than one such person exists, or if none exist, than any adult household member can be designated as a householder.

A household may contain no families. This, in fact, is not uncommon. Although they may consider themselves one, a group of single unrelated people or an unmarried couple living together are not considered a family. A household may also contain more than one family, although this is rare. In most cases in which multiple families share a single housing unit, there is at least one member of one family who is related in some way to a member of another family. By definition then this group becomes one large family. For example, if a husband and wife rent a house and the wife's cousin and her two children come to live with them, this constitutes one large family since all members are related by blood, marriage, or adoption.

Size of Household

The *size of a household* is the number of persons who are residing in the household at the time of interview and who do not usually live elsewhere. A visitor staying temporarily at someone's house is not part of the household. A person who is away on vacation or in a hospital is a member of the household if he or she usually lives there. The status of college students may be the hardest to determine. If they usually live away from the household, they are not part of the household.

Household vs. Family Income

Household income is the sum of all incomes earned by all who live in the household. If a husband and wife each earn $30,000, their child earns $5,000, and an unrelated boarder living in an extra bedroom earns $15,000, then the household income is $80,000. All four members of the household are considered to be living in a household with an income of $80,000, even though all household members may not have access to that entire amount.

For the purposes of income reporting, *family income* is considered to be the sum of all incomes earned by an individual and all other family members. In the previous example, the husband, wife, and child would all be considered as part of a family whose income is $65,000. Although by strict definition, a family of one cannot exist, for the purposes of the distribution of total household income, the boarder would be considered a family of one whose income is $15,000.

Employment Status

A person's *employment status* falls into one of three distinct categories: employed, unemployed, or not in the labor force. An employed person is anyone at least 16 years old who worked last week for pay or for his or her own family's business, regardless of the number of hours worked. Persons with a job but who did not work last week due to illness, vacation, or other reasons, are also considered employed. An unemployed person is one who did not work at all last week, but who was available to accept a job and has looked for work during the last 4 weeks. A person is not in the labor force if he or she did not work at all last week and either hasn't looked for work during the last four weeks or did not want a job. Retired persons, housewives, and students are the most common examples of persons not in the labor force. A special class of persons called "discouraged workers" has recently been added to those who are not in the labor force. Discouraged workers are people who did not work last week, and have been out of work for a long time. They may have looked for work in the past, but have not looked in the last four weeks.

What constitutes "looking for work?" Interviewing for a job, sending resumes to companies, or answering newspaper ads are all considered looking for work. Reading the newspaper "help wanted" ads without following up is not considered looking for work.

Race

The Census Bureau classifies people into one of five distinct *racial categories*: 1) White, 2) Black, 3) American Indian, Eskimo or Aleut, 4) Asian or Pacific Islander, and 5) Other. Included among whites are those who claim their race is Canadian, German, Near Eastern Arab, or Polish among many others. Included among blacks are those who claim their race is Jamaican, Haitian, or Nigerian among many others. Included among Asian and Pacific Islanders are people who claim their race is Chinese, Filipino, Vietnamese, Hawaiian or Samoan among many others. Included among the "other" race category, are people who consider themselves to be multiracial and many people of Hispanic origin.

Ethnicity

An individual's *ethnicity* refers to what the person considers to be his or her origin. A person may consider him or herself Polish, Irish, African, Hispanic, English or one of many other ethnic origins. Origin means different things to different people. For example, origin can be interpreted as a person's ancestry, nationality group, lineage, or country of birth of the individual, his or her parents or ancestors. A person can have dual ethnicity, but may identify more closely with one and therefore consider themselves to be of that ethnicity.

Ethnicity is distinct from race. People of a given ethnicity may be of any race. A person can be black and English, or white and Greek. A person of Hispanic ethnicity may also consider her or himself to be black, white or neither of the two. The Census Bureau asks only one specific question on ethnicity which is, "Do you consider yourself to be of Hispanic origin?" This question is asked of all respondents along with questions of race. This produces a more accurate count of people of Hispanic origin than asking the single question, "Are you white, black, Hispanic, or something else?" Many people of Hispanic origin will choose white or black to this single question.

There is no single list of ethnic categories now in use. Many people use race, Hispanic origin, and a question pertaining to origin to infer ethnicity. This way Poles can either be separated from Czechs or combined as Eastern Europeans.

Wages vs. Total Income

Wages are defined as the amount of money earned from jobs or services performed. Total income includes money received from all sources. Total income includes wages as well as interest on bank accounts, stock dividends, royalties, retirement distributions, inheritances and government subsidies such as welfare payments.

Slopes

by Meg Hickey

As children we learn about slopes as something hard to walk up and fun to slide down. Because humans find it much easier to walk on flat surfaces, we are physically aware of the increase in effort required to climb hilly or sloped ground. Some sloped ground has recreational possibilities, such as skiing or sledding, or dramatic interest in a marathon, such as Heartbreak Hill in the Boston Marathon. Though we experience slopes in different ways, we all know that once you have expended energy to go up them, they offer the potential to roll or slide down, and, the steeper the slope, the faster the movement downwards.

There are a lot of practical uses of slopes that involve their potential for moving people or things downwards, by the force of gravity upon them. In buildings, sloped roofs are designed to shed water and snow to the ground. Streets are sloped from the middle down to the gutters so that rain can run along the gutters into the catch basins. Planted areas and paved parking areas need to be sloped at least 1 foot down for every 100 feet for the surface to drain well. Sloped pipes also use gravity to move drain water and sewage down into the main sanitary sewer in the street. A slope as small as 1/8″ down for every foot of horizontal length is sufficient for plumbing pipes to drain well: note that a slope of 1/8″ per foot is equivalent to 1/8″ per 12″ or 1 unit down for every 96 horizontal units, roughly the same as for surface drainage.

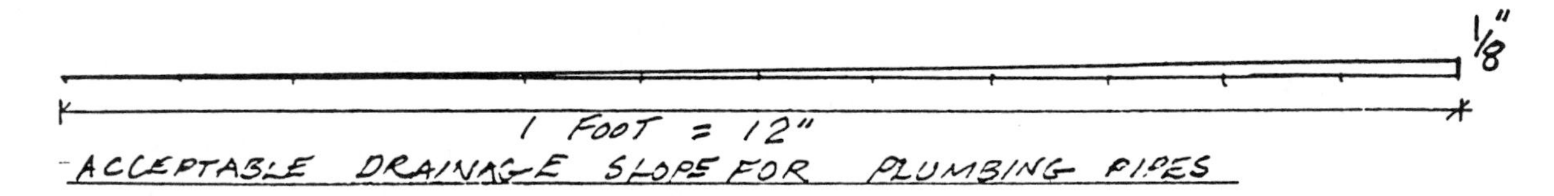

There are a number of ways to specify how steep a slope is:

With roofs slope is often referred to as "pitch" and might be specified as 5:12 which means 5 feet up for every 12 horizontal feet. Sometimes carpenters refer to slopes as "rise over run" meaning the ratio of vertical to horizontal. Slopes can be laid out with a carpenter's framing square.

When a slope is specified as a percentage, such as 40%, then the vertical rise is 40 feet for a horizontal run of 100 feet. You can use any other units like meters or yards and you will get the same slope.

Meg Hickey is a mechanical engineer and architect who teaches Architectural Engineering at The Massachusetts College of Art.

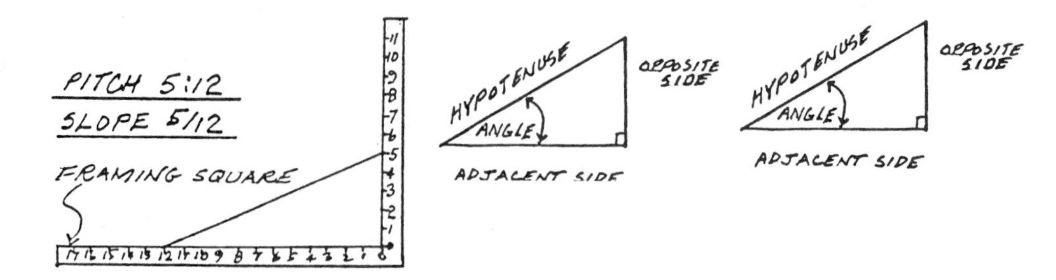

Slopes can also be specified by the angle at which they rise from the horizontal. The angle can be found by trigonometry very easily if you know the rise and run of a slope, since the tangent of an angle is the rise/run. On a triangle showing the rise, run, and angle of a slope, the tangent is sometimes called the "opposite side over the adjacent side" referring to the sides of the triangle relative to the angle.

The angle of a slope is the *arctangent* of the slope ratio. Many calculators have an ATAN function which gives angles from tangents; if your calculator doesn't have it, you can look it up in math books with trigonometric tables.

For a triangle with 40 rise and 100 run, the tangent is:

$$40/100 = 0.40$$
$$\text{ATAN}(0.40) = 20.8^\circ$$

What is the slope, from the horizontal, of this italic type?

Roof plan A is the view looking down on the roof of the Victorian house shown (both on page 507). The arrows show the down direction of drainage, and the dotted path shows the path of a raindrop starting on the top of the tower and rolling down to the ground. Notice the small section of ridged roof behind the chimney: it is called a "cricket"; if it was not there the water and snow would drain into the back of the chimney and cause a leakage problem.

Now look at roof plan B (next page) which shows three proposed additions to an L shaped building. You should be able to identify three potential leakage problems with this plan. Where will the snow build up?

If an earth slope is too steep it may cause erosion. Roots of trees and shrubs can help to prevent erosion of planted banks, but if the slope is greater than 50%, retaining walls are required to stop the earth from sliding. Because slope is critical to ground water drainage and building and planting potential, architects start with a contour map of a site, usually available from the U.S. Geological Survey map service. From these you can draw a sideways cut-through view, called a "section" showing the slope of the land.

One of the obvious problems in placing a building on sloped land is that you need to avoid having all the surface water runoff running into your cellar—as in Section A. What is done is to slope the ground away from the house, all the way around it, as in Section B.

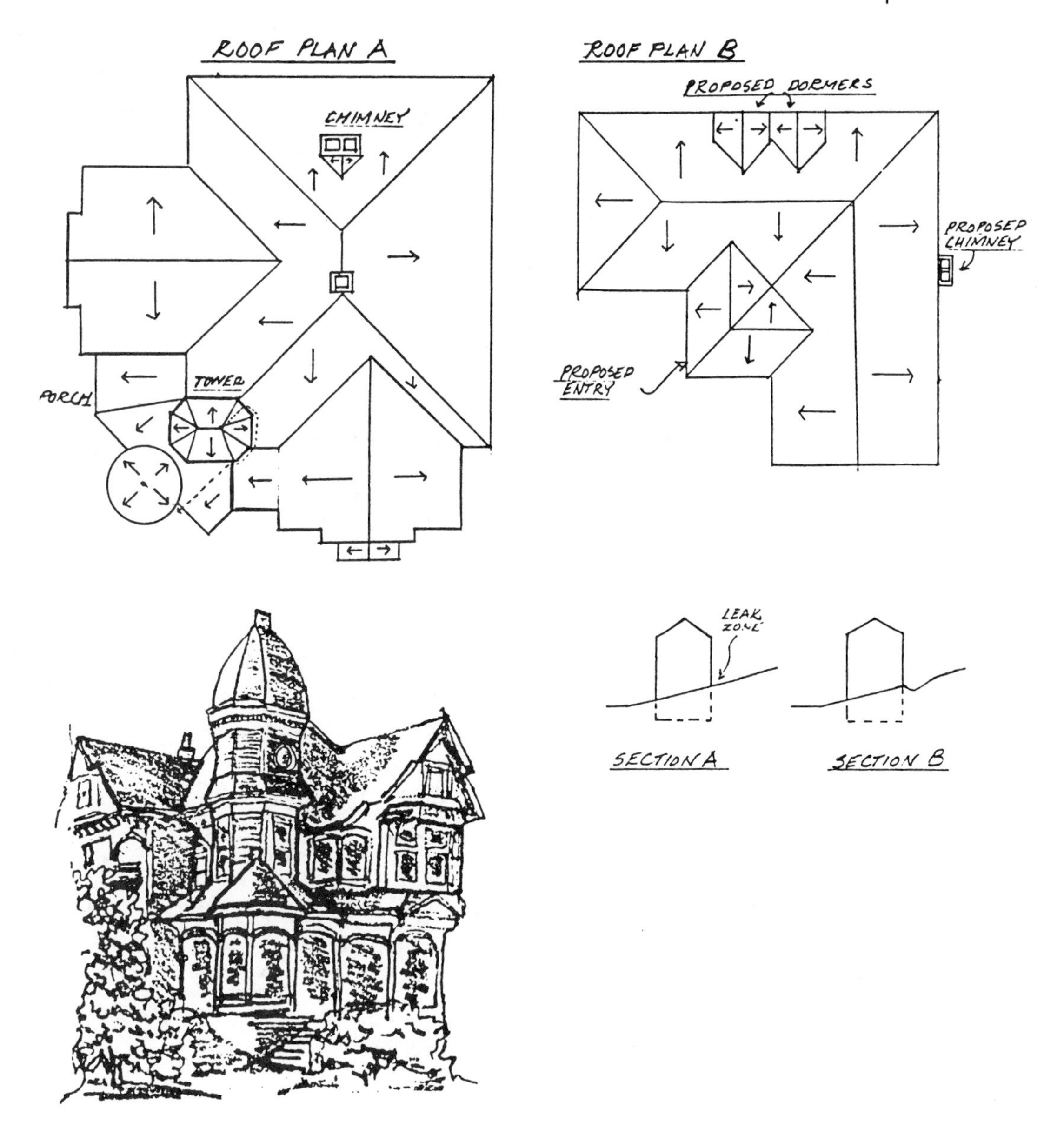

If you look at the contour map of Amesbury on the following page you can compute the slope of the ski runs on the northeast side of Powwow Hill. The scale is one centimeter represents 200 meters, and the contour lines are 3 meter differences in height. The shorter run starts on the 69m. contour line and ends down at the 27m. line for a rise of $69 - 27 = 45$m.; it covers about 1 centimeter on the map, so the horizontal distance is 200m. The average slope then is $45/200 = .225$ or 22.5%.

Disregarding the slight inconvenience of ending up in the lake, calculate the slope % on the opposite, southwest side of the hill, from the 60 meter contour line to the base.

After the Vietnam war many young veterans confined to wheelchairs successfully lobbied for much needed improvements in the architectural standards for wheelchair access to public

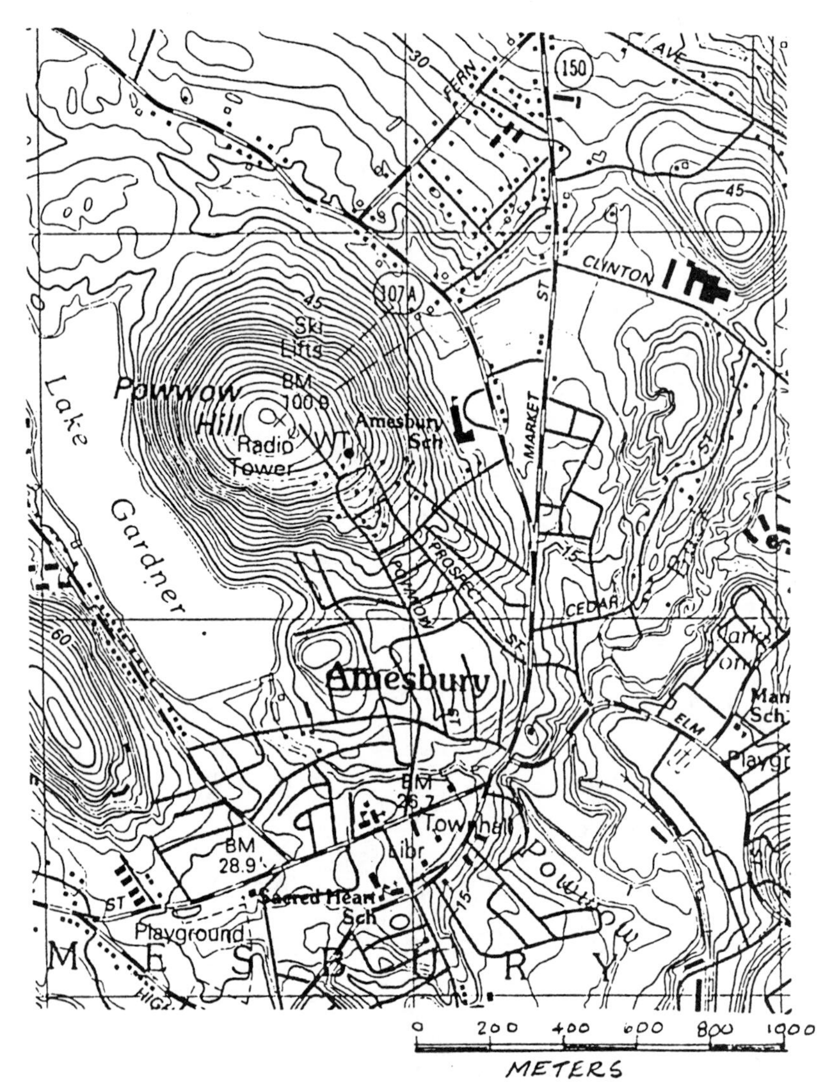

From Massachusetts Highway Department & the U.S. Geological Survey

buildings and housing. As a result many intersections now have curb cuts where a section of curb has been replaced with a sloped ramp from sidewalk to street area, and the steepness of ramps for rising into buildings has been changed from 1 foot of rise for every 10 feet of horizontal length to 1 foot of rise for every 12 feet of ramp length. This gentler slope makes it easier for people with less strength to feel more in control of the downwards roll of the chair, and more able to roll the chair up the ramp. Curb cuts, because they are shorter, are allowed to be steeper but should be kept under 15%. If wheelchair ramps are specified 1:12 and the front door is 3 feet above the ground, then the ramp must be $12 \times 3'$ or 36 feet long.

Ramps that rise around a central space have been used in two famous buildings: the Guggenheim art museum in New York City has a spiralling ramped gallery 5 stories tall, and the Boston Aquarium has a series of ramps winding around a huge fish tank about 3 stories tall. The museum was designed by Frank Lloyd

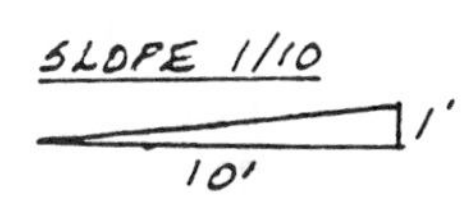

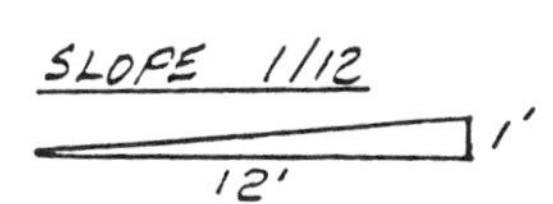

Wright, and the aquarium by the Cambridge Seven Associates. Both of these places are worth a visit.

Another practical use of slopes is in making 3D objects that have angled surfaces rather than just horizontal and vertical planes. People who sew clothing are familiar with producing sloped surfaces in skirts or trousers where the fabric must be sloped in from the larger hip to the smaller waist. For example, if a 40″ hip must be reduced to a 30″ waist, 10″ must be gradually removed over the 7″ distance from hip to waist. If the entire 10″ was removed only at the side seams, the garment would fit badly over the butt and stomach. These parts can be fitted with sloped "darts," which taper the fabric in gently to the waist. Darts are made by folding the fabric and stitching at an angle down to the edge of the fold. The tapering resulting from a dart is twice the width of the top of the dart.

For the skirt shown: if the front and back are each 20″ wide with a 1″ taper on each side from the hip to the top, there remains 18″ at the top of each piece; this is still too big since half of the 30″ waist is 15″. The front and back pieces each have to be reduced from 18″ to 15″ with darts. If the front has 2 darts, one each side, each dart can take up half of the 3″ difference required, or 1.5″. The top of the dart when folded is 1.5″/2 = .75″.

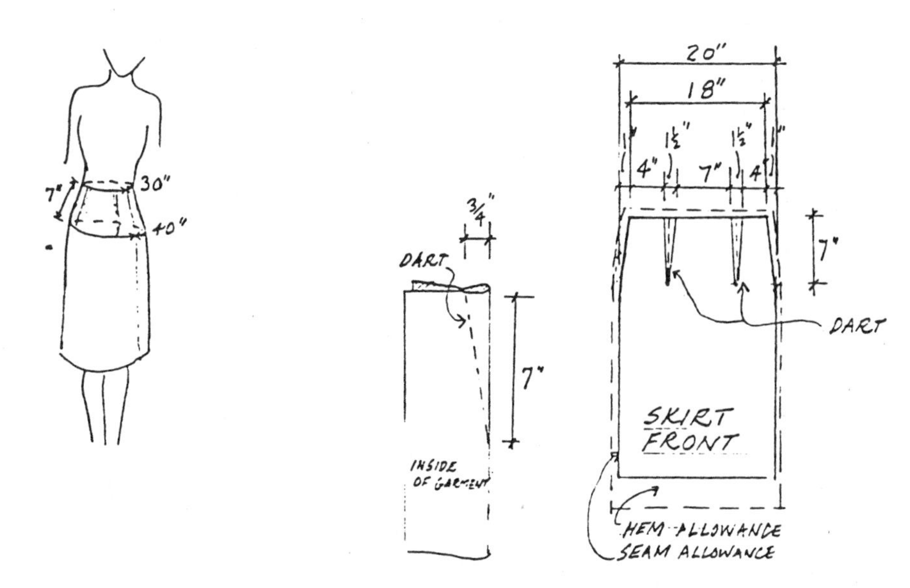

Incidentally, the old-fashioned word for a basic clothing pattern is "sloper," deriving from the old English "slupan"—to slip. This word is not only the root of clothing to slip over your body, but you can see how words like slip and slope are related to the same root.

Another example of slopes in clothing is the high heel. If cowboy boots are sold with a 2″ heel in both men's and women's sizes, ranging from a women's size 5, 7.5″ long, to a men's size 13, 12.5″ long, you can calculate the difference in slope for each case.

The small size has a slope of 2/7.5 = .29 = 29%; the large size has a slope of 2/12.5 = .16 = 16%.

What angles are the 2 slopes?

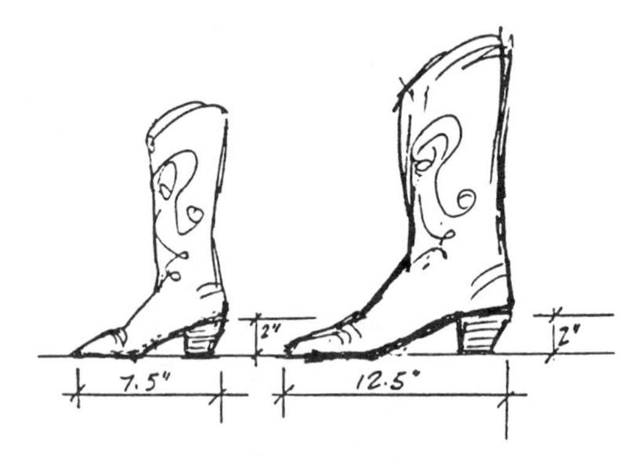

What size heel would a woman with a size 5 foot have to buy in order to have a slope of 16%?

All of the examples of slopes cited so far have been physical, and measurable in space as one dimension divided by another, in effect, a ratio. The word "rate," which comes from the same root as "ratio," is often used for slopes on graphs. We talk about "birth rate" (16.2 live births per 1000 population in 1991 in the U.S.), "tax rate" (5 cents per dollar = 5%), or "rate of travel" (miles/hr or mph).

Any 2 related quantities which can be graphed, such as U.S. population versus time, or position of mercury in a thermometer

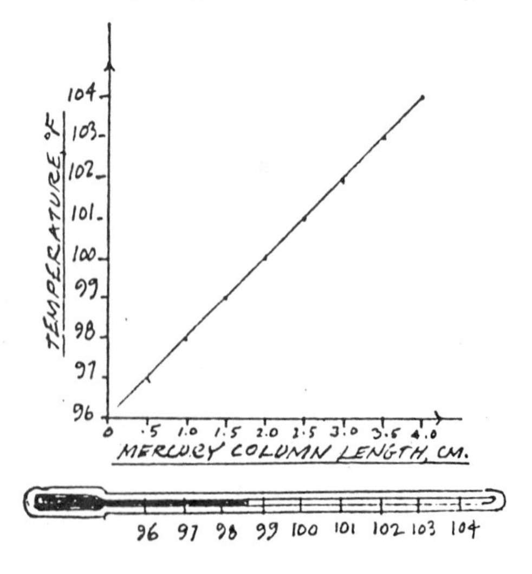

versus temperature, can produce a plotted line which has one or more slopes along it. If the line is not straight you can find the slope for short segments.

For example, if you set your car trip meter to 0, then read the mileage every 5 minutes for a 25 mile trip from the center of one town to the suburbs of another, then plot distance versus time, you can find the average speed during any 5 minute interval, or you can find the overall average speed of the whole trip.

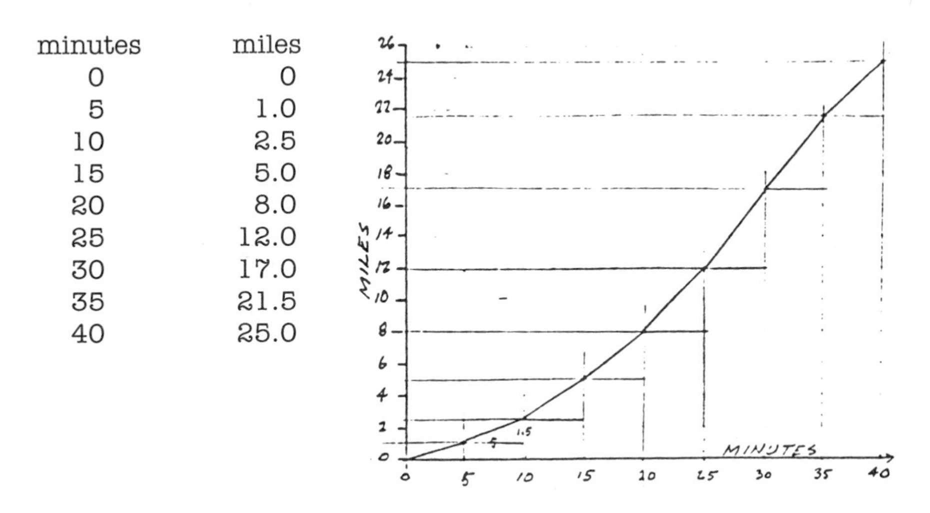

minutes	miles
0	0
5	1.0
10	2.5
15	5.0
20	8.0
25	12.0
30	17.0
35	21.5
40	25.0

The average speed for the whole trip is:

$$25 \text{ miles}/40 \text{ minutes} = 5/8 \text{ mile per minute}$$

or, to get it in mph, multiply by 60 minutes per hour:

$$60 \text{ min/hr} \times 5/8 \text{ mile/min} = 37.5 \text{ mph}$$

You know that at some times you were travelling faster or slower than this average. You can get a more accurate picture of the variations in speed by looking at individual time segments.

During the first 5 minutes 1 mile was covered. The average speed is 1 mile per 5 minutes, or 1/5 mile per minute: this can be read as the slope of the first line segment. At the same rate for an hour, since there are 60 minutes in an hour, you would go $60 \times 1/5$ mile = 12 miles. The speed is then 12 miles per hour (mph).

During the next segment $2.5 - 1 = 1.5$ miles was covered in 5 minutes. The speed was:

$$(1.5/5) \text{ miles/min} \times 60 \text{ min/hr} = 18 \text{ mph.}$$

Calculate the average mph for the segment from 25 to 30 minutes.

Obviously, if you took mileage data at smaller time intervals, you could get a more accurate reading of the speed as it changes. The branch of mathematics called differential calculus teaches how to find slopes for infinitesimal segments of a graphed line or curve.

Once you know the slope of a line you have a ratio you can apply to similar right angled triangles of the same slope but different size as shown on page 512. The law of Pythagoras, $A \times A + B \times B = C \times C$, and the proportional rules of similar triangles, $A/a = B/b = C/c$, can be used to find unknown dimensions. Here is one worked out example, and one for you to try:

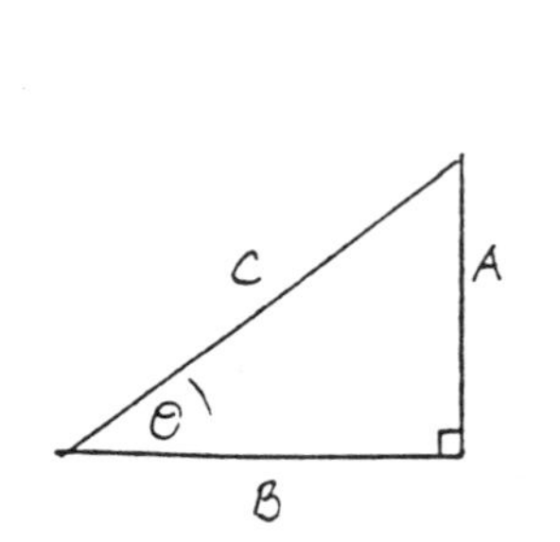

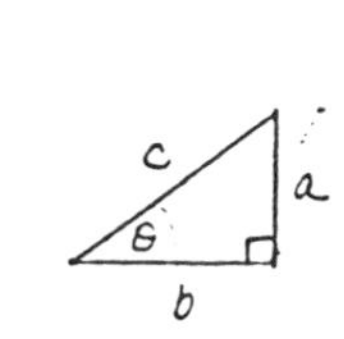

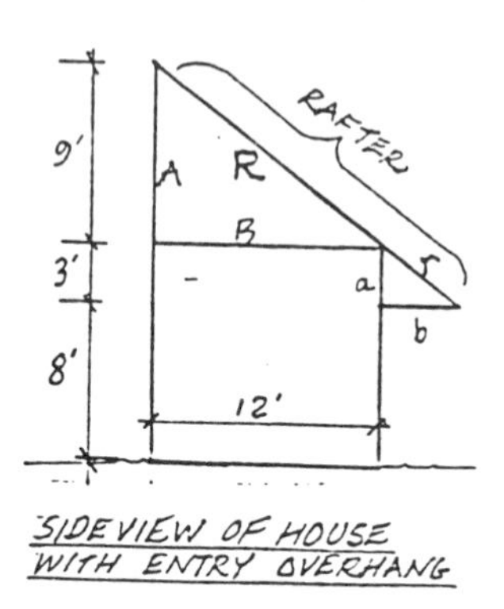

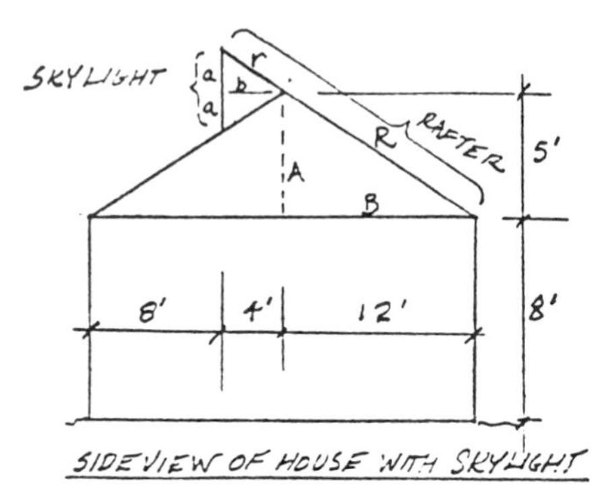

For the house with entry overhang shown above, how long is the total rafter, R + r?

Using Pythagoras: $A \times A + B \times B = R \times R$

$$9 \times 9 + 12 \times 12 = R \times R = 81 + 144 = 225$$

$$R = \sqrt{225} = 15$$

Using similar triangles, because triangles A B R and a b r have the same slope:

$$a/A = r/R$$

$$3/9 = r/15$$

$$15 \cdot (3/9) = r = 5$$

So: Total Rafter $= R + r = 15 + 5 = 20$

For the house section with skylight, how long is the total rafter, R + r, and what is the height of the skylight, a + a?

References

Architectural Graphic Standards, Ramsey & Sleeper, John Wiley & Sons, N.Y.

Vogue Pattern Book, Vogue Pattern Service.

Webster's Unabridged Dictionary, Merriam-Webster Inc.

United States Coastal & Geodetic Survey.

Frank Lloyd Wright, Solomon R. Guggenheim Museum.

Daughters of Painted Ladies, Pomada, Larsen, Keister; E.P. Dutton, N.Y.

Sketches by Meg Hickey, building illustrations by Myrna Kustin and Juanita Jones.

The University of Massachusetts Boston

STATISTICAL PORTRAIT
FALL 1999

Office of Institutional Research and Policy Studies (OIRP)

SAT Scores of New Freshmen by College/Program*

		SAT Scores of New Freshmen by College/Program, Ten-Year Trend (Excluding the DSP Program, Learning Disabled and Foreign Students)																	
		1982	1983	1984	1985	1986	1987	1988	1989	1990	1991	1992	1993	1994	1995	1996	1997*	1998*	1999**
College of Arts & Science	SATVerbal	439	447	450	449	453	453	464	464	449	447	432	431	433	430				
	SATMath	463	460	468	478	468	475	479	488	471	473	477	464	483	474				
	Combined	902	907	918	927	921	928	943	952	920	920	909	895	916	904				
	[N]			[417]	[399]	[429]	[444]	[330]	[299]	[294]	[262]	[240]	[256]	[211]	[251]				
Recentered CAS	SATVerbal											516	511	510	509	518	502	513	521
	SATMath											509	493	510	504	500	504	511	513
	Combined											1,025	1,004	1,020	1,013	1,018	1,006	1024	1034
	[N]											[227]	[243]	[210]	[248]	[288]	[244]	[328]	[418]
College of Management	SATVerbal	456	463	449	449	448	453	449	409	436	428	423	418	412	413				
	SATMath	516	502	508	508	512	529	513	507	520	502	518	500	490	515				
	Combined	972	965	957	957	960	982	962	916	956	930	941	918	902	928				
	[N]			[80]	[105]	[108]	[93]	[68]	[54]	[47]	[36]	[32]	[52]	[31]	[32]				
Recentered CM	SATVerbal											501	496	487	490	510	450	468	481
	SATMath											536	524	513	536	541	521	546	537
	Combined											1,037	1,020	1,000	1,026	1,051	971	1014	1018
	[N]											[30]	[52]	[31]	[32]	[34]	[51]	[54]	[43]
Human Performance & Fitness	SATVerbal		407	407	390	436	425	432	385	453	363	397	399	413	416				
	SATMath		416	423	392	439	467	476	426	442	407	407	453	489	463				
	Combined		823	830	782	875	892	908	811	895	770	804	852	902	879				
	[N]			[9]	[4]	[15]	[11]	[20]	[18]	[6]	[6]	[15]	[12]	[7]	[12]				
Recentered HPF	SATVerbal											497	480	490	495	416	445	440	534
	SATMath											467	492	513	494	454	520	477	514
	Combined											964	972	1003	989	870	965	917	1048
	[N]											[12]	[11]	[7]	[12]	[5]	[2]	[6]	[5]
College of Nursing	SATVerbal		448	493	501	441	419	444	454	439	488	457	420	387	408				
	SATMath		463	482	478	480	459	478	457	443	522	502	466	537	521				
	Combined		911	975	979	921	878	922	911	882	1010	959	886	924	929				
	[N]			[13]	[9]	[13]	[15]	[11]	[9]	[7]	[6]	[17]	[13]	[3]	[8]				
Recentered CN	SATVerbal											536	499	463	486	536	489	478	531
	SATMath											525	497	553	548	521	501	496	496
	Combined											1061	996	1016	1034	1057	990	974	1027
	[N]											[17]	[13]	[3]	[8]	[9]	[10]	[19]	[8]

*1997 and 1998 SAT scores exclude DSP and Category 3 freshmen.

**1998 SAT scores exclude data on DSP students only.

Undergraduate Admissions Summary

	1988	1989	1990	1991	1992	1993	1994	1995	1996	1997	1998	1999
FRESHMEN												
Applied	3,274	3,272	2,775	2,487	2,378	2,494	2,356	2,439	2,740	2,668	2,977	3,461
Decision Ready	2,932	2,804	2,420	2,181	2,035	2,237	1,963	2,110	2,347	2,305	2,466	2,724
Admitted	1,601	1,648	1,516	1,388	1,366	1,496	1,320	1,363	1,541	1,441	1,460	1,694
Admit Rate	54.6%	58.8%	62.6%	63.6%	67.1%	66.9%	67.2%	64.6%	65.7%	62.5%	59.2%	62.2%
Enrolled	818	823	751	745	734	800	662	691	743	637	674	789
Admitted but Deferred	41	53	68	70	53	34	41	23	57	44	61	30
Yield Rate	51.1%	49.9%	49.5%	53.7%	53.7%	53.5%	50.2%	50.7%	48.2%	44.2%	46.2%	46.6%
Not Admitted:												
Denied	1,331	1,156	904	793	669	741	643	747	806	864	1,006	1,030
Incomplete Applications	342	468	355	306	343	257	393	329	393	363	511	737
TRANSFERS												
Applied	3,683	4,118	3,653	3,189	3,375	3,382	2,875	2,761	2,890	2,994	3,515	3,790
Decision Ready	3,012	3,398	3,149	2,808	2,912	3,018	2,480	2,374	2,510	2,576	2,872	2,964
Admitted	2,224	2,623	2,610	2,455	2,555	2,741	2,249	2,081	2,234	2,224	2,479	2,601
Admit Rate	73.8%	77.2%	82.9%	87.4%	87.7%	90.8%	90.7%	87.7%	89.0%	86.3%	86.3%	87.8%
Enrolled	1,319	1,619	1,552	1,442	1,505	1,666	1,397	1,225	1,371	1,347	1,574	1,590
Admitted but Deferred	170	182	218	166	170	110	93	79	121	102	161	89
Yield Rate	59.3%	61.7%	59.5%	58.7%	58.9%	60.8%	62.1%	58.9%	61.4%	60.6%	63.5%	61.1%
Not Admitted:												
Denied	788	775	539	353	357	277	231	293	276	352	393	363
Incomplete Applications	671	720	504	381	463	364	395	387	380	418	643	826
TOTAL UNDERGRADUATES												
Applied	6,957	7,390	6,428	5,676	5,753	5,876	5,231	5,200	5,630	5,662	6,492	7,251
Decision Ready	5,944	6,202	5,569	4,989	4,947	5,255	4,443	4,484	4,857	4,881	5,338	5,688
Admitted	3,825	4,271	4,126	3,843	3,921	4,237	3,569	3,444	3,775	3,665	3,939	4,295
Admit Rate	64.4%	68.9%	74.1%	77.0%	79.3%	80.6%	80.3%	76.8%	77.7%	75.1%	73.8%	75.5%
Enrolled	2,137	2,442	2,303	2,187	2,239	2,466	2,059	1,916	2,114	1,984	2,248	2,379
Admitted but Deferred	211	235	286	236	223	144	134	102	178	146	222	119
Yield Rate	55.9%	57.2%	55.8%	56.9%	57.1%	58.2%	57.7%	55.6%	56.0%	54.1%	57.1%	55.4%
Not Admitted:												
Denied	2,119	1,931	1,443	1,146	1,026	1,018	874	1,040	1,082	1,216	1,399	1,393
Incomplete Applications	1,013	1,188	859	687	806	621	788	716	773	781	1,154	1,563

Enrollment Trends at UMASS Boston (State Funded Enrollment)

	1982	1983	1984	1985	1986	1987	1988	1989	1990	1991	1992	1993	1994	1995	1996	1997	1998	1999
Total Enrollment	**11,135**	**11,496**	**11,711**	**12,547**	**12,919**	**13,574**	**12,451**	**12,584**	**12,478**	**11,606**	**11,775**	**12,136**	**12,142**	**11,602**	**11,736**	**11,843**	**12,499**	**12,923**
Full-Time Enrollment																		
(Headcount)	7,287	6,941	7,064	6,831	6,969	7,448	7,007	6,964	7,002	6,556	6,561	6,657	6,532	6,064	6,105	6,230	6,467	6,808
FTE Enrollment	8,156	8,525	8,654	8,804	8,983	9,526	8,921	8,921	8,863	8,300	8,439	8,607	8,552	8,095	8,186	8,287	8,714	9,045
Matriculated																		
Undergraduate	9,524	9,316	9,108	8,834	9,065	9,615	9,283	9,514	9,216	8,589	8,693	8,972	8,556	8,007	7,990	7,949	8,286	8,715
% Full-Time	67.2%	70.0%	70.6%	68.4%	66.8%	68.0%	67.9%	66.6%	68.0%	67.9%	66.9%	65.1%	66.2%	64.6%	66.0%	66.8%	66.1%	66.0%
Matriculated Graduate	431	455	654	921	1,010	1,474	1,678	1,756	1,802	1,890	1,897	1,958	2,035	2,258	2,395	2,515	2,612	2,683
% Full-Time	63.1%	27.9%	32.0%	30.7%	33.4%	30.9%	27.0%	24.0%	26.4%	27.4%	30.6%	32.9%	32.5%	30.8%	27.8%	28.3%	29.2%	30.5%
Non-degree Students	1,180	1,725	1,949	2,792	2,844	2,485	1,490	1,314	1,460	1,127	1,185	1,206	1,551	1,337	1,351	1,379	1,601	1,525
% Undergraduate	93.2%	90.0%	83.0%	78.0%	82.3%	79.9%	75.0%	77.5%	76.7%	76.5%	68.9%	69.3%	67.0%	74.0%	68.7%	68.6%	71.3%	71.87%
Total Undergraduate Students																		
HCT Enrollment				11,003	11,406	11,601	10,399	10,532	10,336	9,451	9,509	9,808	9,595	8,997	8,918	8,895	9,428	9,811
FTE Enrollment				7,878	8,005	8,252	7,605	7,613	7,489	6,891	6,933	7,035	6,849	6,312	6,325	6,349	6,679	6,961
Full-Time Students											5,966	5,995	5,860	5,353	5,425	5,495	5,687	5,979
% Female				53.9%	54.6%	55.8%	56.9%	56.5%	55.6%	55.0%	53.5%	53.0%	53.3%	53.7%	55.6%	55.1%	55.8%	56.3%
Mean Age				27	27	27	27	27	27	28	28	28	28	29	29	28	28	27
Median Age				24	24	24	24	24	25	25	25	25	25	25	25	25	24	24
Total Graduate Students																		
HCT Enrollment				1,544	1,513	1,973	2,052	2,052	2,142	2,155	2,266	2,328	2,547	2,605	2,818	2,948	3,071	3,112
HCT Doctoral						23	25	37	53	71	93	139	171	213	234	296	327	349
HCT Master's/CAGS/																		
Cert						1,474	1,653	1,719	1,749	1,819	1,804	1,819	1,864	2,045	2,161	2,219	2,284	2,334
FTE Enrollment				926	978	1,274	1,316	1,308	1,374	1,409	1,506	1,572	1,703	1,783	1,861	1,938	2,034	2,084
Full-Time Students											595	662	672	711	680	735	780	829
% Female				58.1%	58.6%	60.4%	62.6%	62.0%	64.0%	64.7%	64.2%	63.6%	63.3%	62.3%	63.6%	64.3%	65.1%	67.2%
Mean Age				35	34	34	34	35	34	35	34	34	34	35	35	35	35	35
Median Age				33	33	33	32	33	32	33	32	32	32	32	32	32	32	31

Distribution of High School Rank and SAT Scores: Entering Freshmen

HIGH SCHOOL RANK

Percentile Rank in Class

	1986	1987	1988	1989	1990	1991	1992	1993	1994	1995	1996	1997*	1998*	1999**
Top 10%	14%	13%	12%	10%	9%	6%	9%	10%	9.3%	5.9%	9.6%	9.6%	6.4%	8.0%
11–20%	13%	16%	18%	17%	14%	22%	12%	12%	12.6%	17.1%	15.5%	17.4%	18.3%	13.9%
21–40%	31%	32%	33%	31%	32%	37%	27%	30%	31.1%	30.2%	25.1%	29.9%	31.7%	33.3%
41–60%	26%	23%	24%	24%	28%	23%	27%	21%	23.5%	25.9%	32.0%	24.0%	28.7%	25.9%
61–80%	11%	12%	9%	14%	12%	10%	18%	18%	16.9%	17.1%	14.2%	16.2%	10.9%	12.9%
Bottom 20%	5%	4%	4%	4%	5%	2%	7%	9%	6.6%	3.9%	3.7%	3.0%	4.0%	6.0%
[N]		[498]	[403]	[337]	[295]	[235]	[254]	[231]	[183]	[205]	[219]	[167]	[202]	[201]

SAT VERBAL: RECENTERED

Percentiles	1986	1987	1988	1989	1990	1991	1992	1993	1994	1995	1996	1997*	1998*	1999**
75%							560	570	580	560	580	570	570	570
50%							520	500	520	510	510	490	510	520
25%							460	450	450	450	460	440	440	470
[N]							[294]	[322]	[254]	[300]	[337]	[308]	[408]	[475]

SAT MATH: RECENTERED

Percentiles	1986	1987	1988	1989	1990	1991	1992	1993	1994	1995	1996	1997*	1998*	1999**
75%							560	540	560	560	560	560	560	560
50%							500	500	500	500	500	500	510	510
25%							460	450	460	450	450	450	450	460
[N]							[294]	[322]	[254]	[300]	[337]	[308]	[408]	[475]

SAT COMBINED: RECENTERED

Percentiles	1986	1987	1988	1989	1990	1991	1992	1993	1994	1995	1996	1997*	1998*	1999**
75%							1110	1090	1120	1110	1110	1110	1100	1110
50%							1020	990	1010	995	1020	980	1020	1020
25%							930	920	930	920	930	910	910	940
[N]							[294]	[322]	[254]	[300]	[337]	[308]	[408]	[475]

*1997 and 1998 SAT scores exclude DSP and Category 3 freshmen.

**1999 scores exclude data on DSP students only.

The University of Southern Mississippi

STATISTICAL PORTRAIT FOR 1998–1999

The University of Southern Mississippi

10-YEAR TREND: ENROLLMENT BY CLASSIFICATION

Hattiesburg Campus
Fall 1989 - Fall 1998

FALL SEMESTER	FRESHMEN	SOPHOMORE	JUNIOR	SENIOR	GRADUATE	TOTAL	% Change
1989	1,950	1,659	2,571	3,597	1,767	11,544	0.48%
1990	1,773	1,722	2,725	3,898	1,794	11,912	3.19%
1991	1,818	1,605	2,886	4,148	1,891	12,348	3.66%
1992	1,550	1,470	2,522	4,299	1,839	11,680	-5.41%
1993	1,580	1,385	2,468	4,172	1,882	11,487	-1.65%
1994	1,685	1,405	2,410	3,956	2,131	11,587	0.87%
1995	1,797	1,472	2,575	4,024	2,245	12,113	4.54%
1996	1,843	1,551	2,611	4,225	2,267	12,497	3.17%
1997	1,895	1,649	2,759	4,307	2,344	12,954	3.66%
1998	**1,940**	**1,638**	**2,825**	**4,412**	**2,081**	**12,896**	**-0.45%**

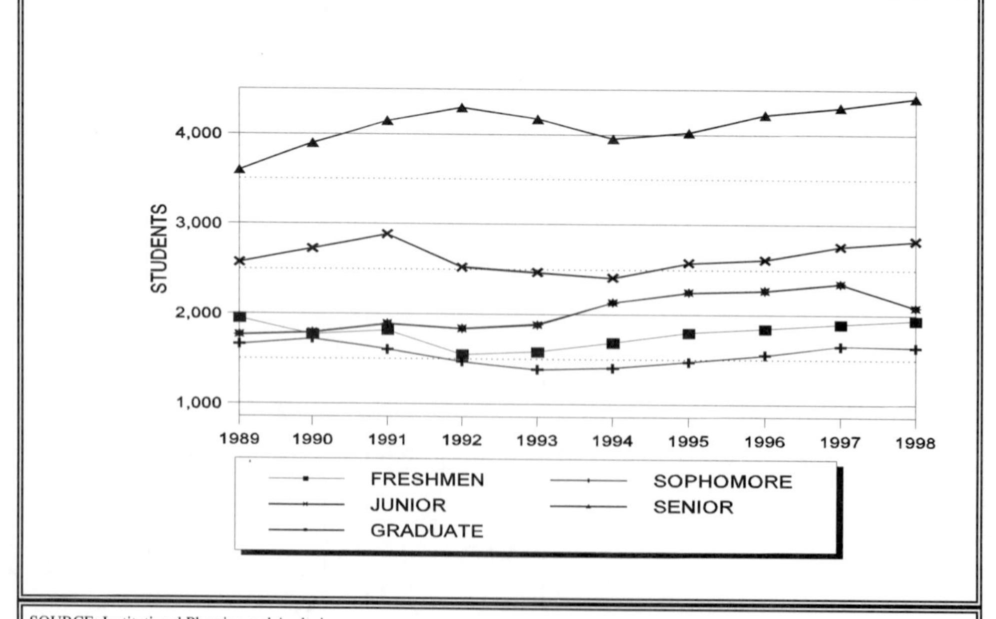

SOURCE: Institutional Planning and Analysis
DATE: Wednesday, November 18, 1998

10YR_CLS

The University of Southern Mississippi

10-YEAR TREND:
ENROLLMENT BY ETHNIC GROUP

Hattiesburg Campus
Fall 1989 - 1998

FALL SEMESTER	WHITE		ASIAN		BLACK		AMERICAN INDIAN		HISPANIC		TOTAL
1989	9,576	83.0%	248	2.1%	1,650	14.3%	21	0.2%	49	0.4%	11,544
1990	9,872	82.9%	260	2.2%	1,702	14.3%	20	0.2%	58	0.5%	11,912
1991	10,146	82.2%	307	2.5%	1,811	14.7%	24	0.2%	60	0.5%	12,348
1992	9,468	81.1%	329	2.8%	1,791	15.3%	26	0.2%	66	0.6%	11,680
1993	9,187	80.0%	311	2.7%	1,874	16.3%	33	0.3%	82	0.7%	11,487
1994	9,202	79.4%	306	2.6%	1,953	16.9%	23	0.2%	103	0.9%	11,587
1995	9,454	78.0%	291	2.4%	2,218	18.3%	38	0.3%	112	0.9%	12,113
1996	9,626	77.0%	325	2.6%	2,385	19.1%	33	0.3%	128	1.0%	12,497
1997	9,953	76.8%	335	2.6%	2,484	19.2%	44	0.3%	138	1.1%	12,954
1998	9,860	76.5%	313	2.4%	2,556	19.8%	39	0.3%	128	1.0%	12,896

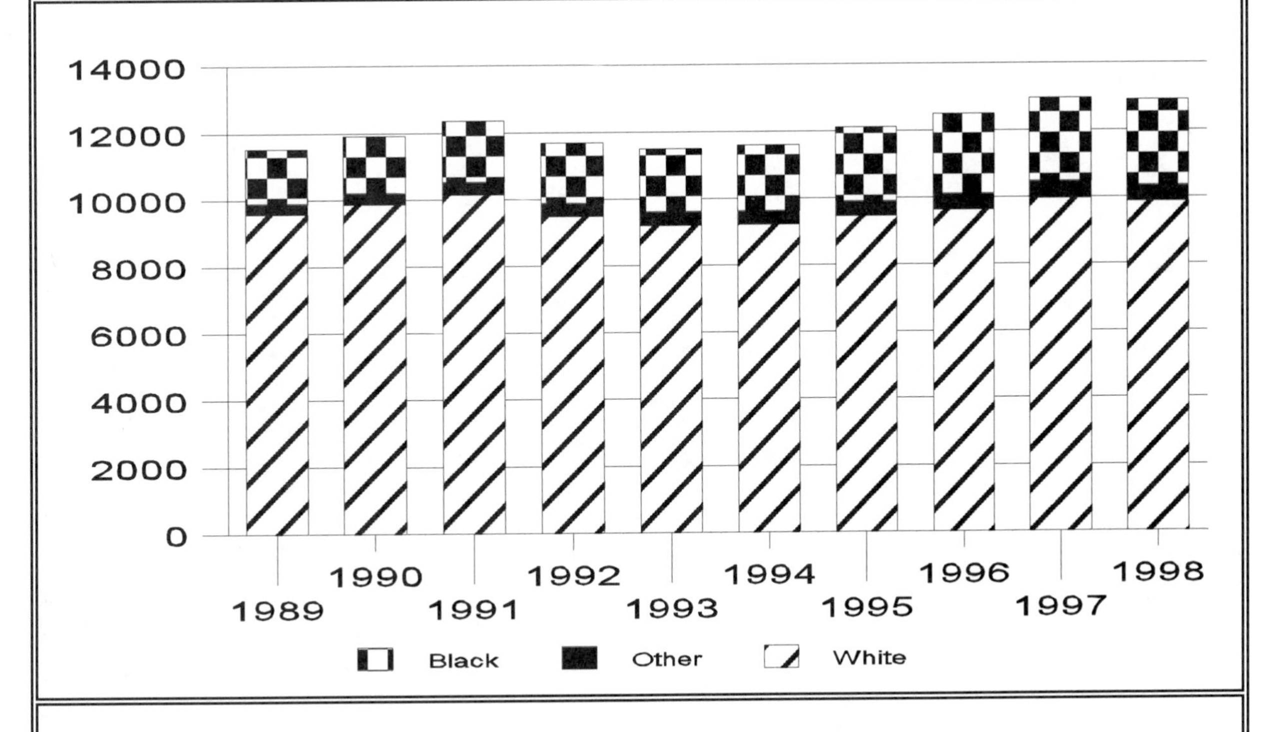

SOURCE: Institutional Planning and Analysis, SD3
DATE: Wednesday, November 18, 1998

10YR_ETH

The University of Southern Mississippi
10-YEAR TREND:
ENROLLMENT BY GENDER

Hattiesburg Campus
Fall 1989 - 1998

FALL SEMESTER	MEN		WOMEN		TOTAL
1989	5,009	43.4%	6,535	56.6%	11,544
1990	5,098	42.8%	6,814	57.2%	11,912
1991	5,309	43.0%	7,039	57.0%	12,348
1992	5,055	43.3%	6,625	56.7%	11,680
1993	5,023	43.7%	6,464	56.3%	11,487
1994	5041	43.5%	6546	56.5%	11,587
1995	5192	42.9%	6921	57.1%	12,113
1996	5215	41.7%	7282	58.3%	12,497
1997	5283	40.8%	7671	59.2%	12,954
1998	5284	41.0%	7612	59.0%	12,896

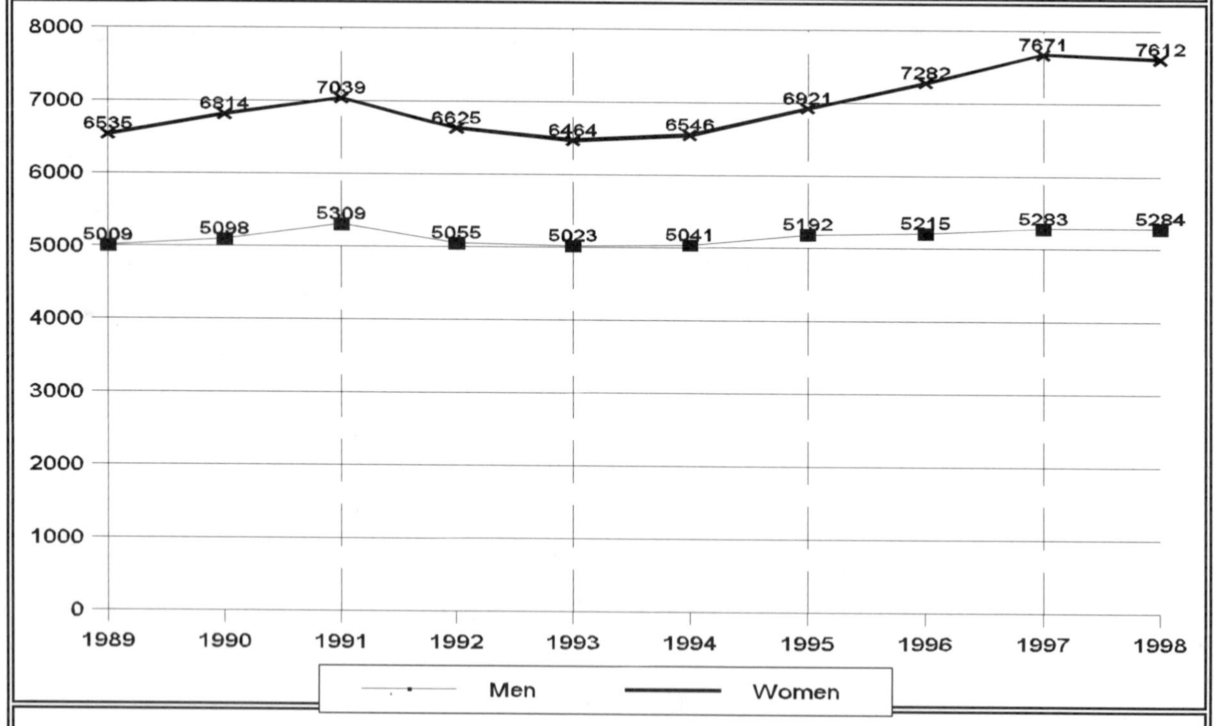

SOURCE: Institutional Planning and Analysis, SD1
DATE: March 31, 1999

10YR_GND

The University of Southern Mississippi

RETENTION OF FIRST-TIME ENTERING FRESHMEN
(All First-Time Full-Time Freshmen)

Hattiesburg Campus
Fall 1983-1998

	ALL FIRST-TIME FULL-TIME FRESHMEN N	AVERAGE ACT	CUMULATIVE GRADUATION AND CONTINUATION RATES							
			AFTER 1ST YEAR	AFTER 2ND YEAR	AFTER 4TH YEAR		AFTER 5TH YEAR		AFTER 6TH YEAR	
			CONTINUED	CONTINUED	GRADUATED	CONTINUED	GRADUATED	CONTINUED	GRADUATED	CONTINUED
1983	1,179	19.7	72.3%	61.7%	21.3%	29.3%	37.9%	9.7%	42.7%	5.0%
1984	1,151	19.8	72.7%	59.3%	18.1%	31.8%	35.6%	10.6%	40.7%	6.0%
1985	1,093	19.7	71.6%	57.9%	20.8%	29.4%	38.0%	10.9%	43.5%	6.6%
1986	1,015	20.1	72.0%	60.6%	18.9%	31.4%	35.0%	13.1%	42.3%	6.3%
1987	1,033	20.3	75.7%	65.0%	19.9%	34.7%	39.9%	12.7%	47.0%	6.5%
1988	1,183	20.0	74.0%	61.5%	19.5%	34.9%	39.5%	10.0%	44.2%	5.2%
1989	1,204	20.4	76.5%	63.0%	22.2%	32.1%	41.2%	11.8%	47.3%	6.3%
1990	1,084	21.6	78.9%	63.1%	21.7%	32.2%	39.5%	13.1%	46.2%	6.9%
1991	1,124	21.8	71.3%	58.4%	16.9%	33.3%	35.4%	12.0%	40.4%	5.5%
1992	1,004	21.8	71.9%	55.6%	17.7%	31.4%	33.2%	12.9%	40.1%	6.6%
1993	1,017	21.7	72.9%	61.5%	17.9%	33.3%	36.7%	12.7%		
1994	1,063	21.7	73.9%	60.5%	19.2%	36.8%				
1995	1,235	21.5	72.7%	61.8%						
1996	1,229	21.7	77.1%	64.3%						
1997	1,314	21.6	72.3%							
1998	1,333	21.7								

The University of Southern Mississippi

DEGREES AWARDED:
5-YEAR PROFILE BY TYPE OF DEGREE

Fiscal Year 1993-94 through 1997-98

	1993-94	1994-95	1995-96	1996-97	1997-98
UNDERGRADUATE DEGREES	**2,383**	**2,274**	**2,068**	**2,176**	**2,236**
Bachelor of Arts	176	193	227	373	406
Bachelor of Fine Arts	33	31	28	20	30
Bachelor of Music	16	17	17	12	11
Bachelor of Music Education	15	19	15	14	18
Bachelor of Science	1,570	1,513	1,266	1,234	1,206
BS in Business Administration	394	326	314	316	369
Bachelor of Science in Nursing	138	145	173	155	165
Bachelor of Social Work	41	30	28	52	31
MASTER'S DEGREES	**744**	**704**	**775**	**941**	**915**
Master of Arts	36	39	45	42	36
Master of Art Education	4	2	0	1	1
MA in the Teaching of Languages	10	17	29	56	63
Master of Business Administration	51	28	35	52	46
Master of Education	191	216	232	346	287
Master of Fine Arts	6	4	6	6	5
Master of Library Science	41	29	38	45	44
Master of Music	8	11	3	7	8
Master of Music Education	12	11	13	12	7
Master of Professional Accountancy	13	20	16	20	24
Master of Public Health	9	13	22	20	20
Master of Science	282	257	244	250	279
Master of Science in Nursing	38	20	41	28	48
Master of Social Work	43	37	51	56	47
SPECIALIST DEGREES	**9**	**14**	**26**	**44**	**29**
DOCTORAL DEGREES	**95**	**103**	**127**	**98**	**113**
Doctor of Education	11	10	11	7	11
Doctor of Music Education	1	1	0	0	0
Doctor of Musical Arts	3	3	4	0	2
Doctor of Philosophy	80	89	112	91	100
TOTAL DEGREES AWARDED	**3,231**	**3,095**	**2,996**	**3,259**	**3,293**

SOURCE: Institutional Planning and Analysis, Degrees Awarded by Major, STU0510
DATE: Wednesday, January 20, 1999

DEG_TYPE

Performing Arts— The Economic Dilemma

by William J. Baumol and William G. Bowen

A Study of Problems Common to Theater, Opera, Music, and Dance

APPENDIX X–1

Number of Performances and Audience Size

While the typical length of season is currently increasing, primarily in response to the demands of the performers, it is by no means obvious that the increased number of performances will bring with it a corresponding increase in the size of the total audience. There is some reason to expect that when more performances are provided, more tickets will be sold. If most of the orchestras that are lengthening their seasons are also developing new types of concerts, new series and different types of orchestral service, these may attract new persons to attend or they may lead others to attend more often. The new performances will probably take place on nights of the week or at times of the year which are convenient for some persons who would otherwise not have been able to come. But, other things being equal, we may well suspect that more frequent performance will serve, in part, just to spread the concert audience more thinly.

Since, as we saw in Chapter VIII, cost per concert does decrease up to a point as the number of concerts goes up, it is important to see whether the added concerts can be expected to bring an additional audience of average size along with them. Obviously, if attendance per concert declines sufficiently as the number of concerts grows, economies of scale will not help finances.

We compared average attendance per concert with the number of concerts per season for two of the major orchestras for which we had obtained statistically significant cost functions (see the appendix to Chapter VIII). Appendix Figure X–A shows what we found for the same orchestra whose unit cost curve is depicted in Figure VIII–4. It is evident from the diagram that average attendance does fall somewhat with the number of concerts.[1] Furthermore, the decline shown in the diagram is probably a considerable understatement of the rate of decrease involved in the underlying relationship. Many of the historical increases in number of con-

[1] Actually the decline is small in terms of the historical relationship shown in Appendix Figure X–A. An increase of 1 per cent in number of concerts yields at 1964 attendance levels a decrease of about 0.3 per cent in attendance per concert, a loss of about 8 persons per concert. With an average of slightly more than 100 concerts per season, this would reduce attendance by a total of about 800 admissions. Since, however, the new concert would on the average be attended by about 2,500 persons, it would bring in a net gain of some 1,700 paid admissions.

APPENDIX FIGURE X–A

RELATION BETWEEN ATTENDANCE PER CONCERT AND NUMBER OF CONCERTS, FOR A MAJOR ORCHESTRA

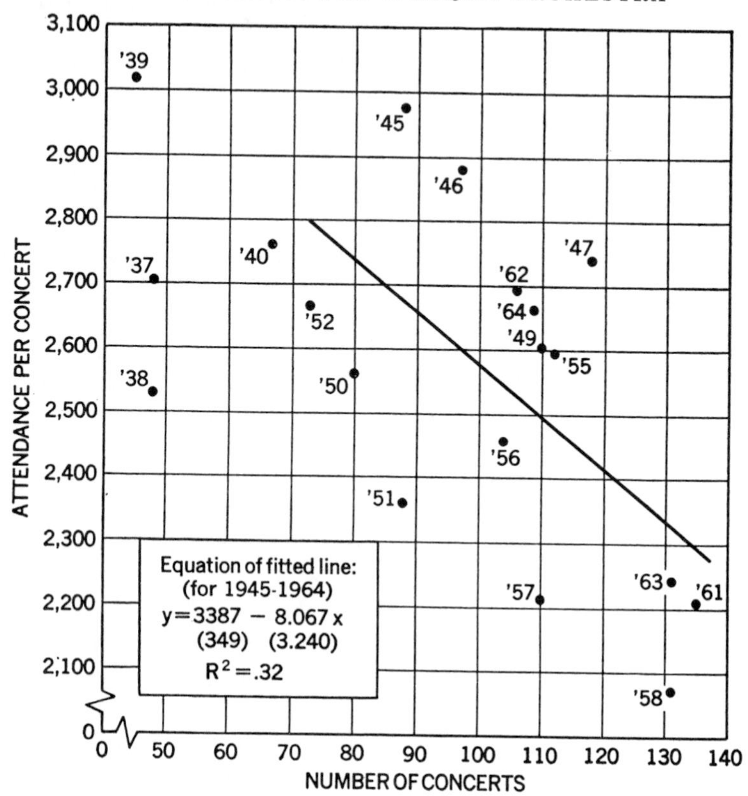

certs occurred only in response to popular demand when the potential audience was already available. If the increased number of performances had been undertaken haphazardly, without regard for autonomous demand changes, and if the new concert had been required, as it were, to hunt up its own audience, one may surmise that the resulting decrease in attendance per concert would have been far more marked. But as more orchestras strive for year-round operation, the number of concerts per season is bound to increase, whether or not they are accompanied by a considerable decline in attendance per concert.

While a larger supply of performances does increase the *total* audience, it undoubtedly yields diminishing returns in terms of audience per concert, and so it is not always an unmixed financial blessing.

His Stats Can Oust a Senator or Price a Bordeaux

By Peter Aseltine
Staff Writer

Princeton-Metro The Times
Sunday May 8, 1994

A Princeton economics professor uses statistics to uncover the mysteries of human behavior—from voting patterns that changed the shape of Pennsylvania's Senate, to the price of fine French wines.

PRINCETON BOROUGH—Orley Ashenfelter's days are numbered.

Behind the tragedy of discrimination and the bumbling of bureaucrats, the Princeton econometrician sees numbers.

Behind the bouquet of a fine wine—more numbers.

Ashenfelter's ability to see the telling patterns of the world and interpret them with numbers has led people to seek the Princeton University professor's expertise for a wide variety of very practical reasons.

Last month a federal judge in Philadelphia relied partly on Ashenfelter's analysis of voting patterns to give a Pennsylvania Senate seat to a losing Republican candidate. The judge found the Democratic winner had used fraud in soliciting absentee votes in the election in Philadelphia's Second District.

The case raised a difficult question, Ashenfelter says, because while the fraud justified unseating the Democrat, William Stinson, it was unclear whether the proper remedy was to seat the Republican, Bruce Marks, or hold a new election.

Marks had outpolled Stinson on the voting machines, 19,691 to 19,127. But in absentee ballots, Stinson received 1,391 votes to Marks' 366, winning the November 1993 special election by 461 votes.

U.S. District Judge Clarence C. Newcomer found that Stinson workers had improperly influenced absentee voters, mostly in Latino and African-American neighborhoods, in some cases marking the voters' ballots or forging their names.

In February, Newcomer threw out the absentee votes and ordered that Marks be seated. The following month, however, a federal appellate court ruled that while the judge was not wrong to unseat Stinson, he should not have seated Marks without first analyzing whether the fraud actually had swayed the election.

"It was a tricky business for the judge, because the issue really was whether it was reasonable to assume that Marks had actually won that election," Ashenfelter said. "The circuit court wanted some evidence that, in the absence of the fraud on the voters, there would have been a presumption that Marks could have been expected to get at least enough of the absentee votes so that he wouldn't be overwhelmed."

Ashenfelter, a 51-year-old economics professor, was appointed as an expert for the new hearing by Newcomer, who had taken one of the statistics seminars that Ashenfelter regularly teaches for federal judges. Two other statistics experts testified, one hired by Marks and one hired by a group of voters opposed to him.

Ashenfelter has testified in court about a dozen times, mostly in discrimination cases, which often rely on statistical analysis.

What Ashenfelter did in this case was quite simple, he says. He used a standard statistical technique called regression analysis to compare machine votes to absentee votes for each of the 22 senatorial elections held in Philadelphia in the last decade. What he found was that the difference between Democratic and Republican votes on the machines generally is a good indicator of the difference in absentee votes.

Ashenfelter used a scatter diagram to show that when the difference between Democratic and Republican machine votes is plotted against the difference in absentee votes, points representing the elections tend to fall around a line representing an ideal correlation between the machine and absentee tallies.

The graph shows that the 1993 election falls well outside the pattern of typical elections. Ashenfelter's analysis would predict a 133-vote advantage for Marks in absentee votes, given his 564-vote margin on the machines. Instead, Stinson had a 1,025-vote advantage.

Ashenfelter calculated that the probability was less than 1 percent that the deviation of 1,158 votes between his predicted result and the actual result was simply due to random changes in voting patterns. He calculated that the probability that Stinson could have received enough absentee votes to win was about 6 percent.

Ashenfelter said surveys of absentee voters conducted by The Philadelphia Inquirer and the Republicans provided information that was extrapolated in court to produce estimates of fraudulent votes that were surprisingly close to his deviation figure of 1,158.

Ashenfelter said there was an amusing moment in court when a lawyer for the Democratic-controlled board of elections tried to suggest that Ashenfelter's analysis would not establish public confidence in a decision to seat Marks. The circuit court had called in its opinion for "evidence and an analysis . . . worthy of the confidence of the electorate."

"He had asked me in my deposition whether I considered myself an expert on establishing the public confidence, and I had said, 'No,'" Ashenfelter said. "So in cross-examination, he got up, and he said, in the usual lawyer way, 'Isn't it true, Dr. Ashenfelter, that you are no more of an expert on what would establish the public confidence than I am?' He had been having his way with me, so at that point I just thought, if a guy asks you a ques-

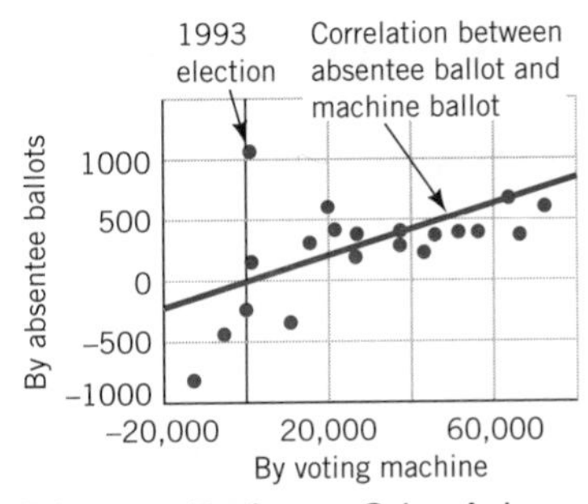

Princeton Professor Orley Ashenfelter used a scaller diagram to show that when the difference between Democratic and Republican machine votes is plotted against the difference in absentee votes for senatorial elections held in Philadelphia in the last decade, points representing the elections tend to fall around a line representing an ideal correlation between the machine and absentee tallies. The graph shows that the 1993 election falls well outside the shaded area where 95% of the results would be expected to fall. The probability that the deviation in the 1993 results was simply due to random changes in voting patterns is less than 1 percent, Ashenfelter calculated.

tion like that, you've got to let him have it. So I said, 'Well, I may be more of an expert than you are.' Well, the courtroom just cracked up."

Newcomer ultimately issued a new order to seat Marks, based on the testimony of the experts. The appeals court upheld the decision and Marks was sworn in on April 28, returning control of the Pennsylvania Senate to the Republicans.

Ashenfelter is modest about his role in the important case. He seems more excited about the prospect of using his court analysis as a teaching tool. He said many other professors have asked for it.

"It's very simple as these things go," he said. "The reason people want to use it for teaching is because it's hard to come up with such a simple example that's so telling for an actual problem."

Ashenfelter spent only a week developing his analysis for the election case. He currently is working on a much bigger project for the state of New York, evaluating the state's Wicks Law, which requires that state construction projects be done without a prime contractor. By forcing the state to act as prime contractor, the law has probably cost New York taxpayers a billion dollars, Ashenfelter said.

"Expecting bureaucrats to manage a construction project is a pretty expensive proposition," he said.

Ashenfelter's greatest claim to fame, however, is a regression analysis that predicts the prices of Bordeaux wine vintages by charting rainfall and temperature during each growing season. His analysis has proved quite accurate and he publishes his predictions twice each year in a newsletter called "Liquid Assets."

The newsletter has nearly a thousand subscribers, including some wine-industry insiders in Bordeaux. But Ashenfelter said the industry still refuses to use the information in setting prices.

"Most people in the industry refuse to believe this," he said. "They will not change prices to reflect quality, and they have never had to do it in the past. So the whole system is built on this bizarre sucker operation. The wine writers—I guess they're the ones who go along with it. It's a mystery to me."

Meanwhile, however, Ashenfelter is storing away some underpriced vintages. His days may be numbered in one sense, but he's looking forward to enjoying the fruits of his numerical labors for years to come.

How a Flat Tax Would Work, For You and for Them

By David Cay Johnston

Ideas and Trends *The New York Times*
Sunday, January 21, 1996

The big issue of the Presidential campaign in these first weeks of 1996 has been how Americans should tax themselves. And most of the talk is about a flat tax on incomes, with a large deduction for individuals so that lower-income people would pay no income tax.

There are about as many flat tax variants floating around as there are would-be occupants of the White House. But all derive from "The Flat Tax" (Hoover Institution Press, 1995), by Robert Hall and Alvin Rabushka, both economists at the Hoover Institution at Stanford University, who devised the concept in 1981 and have been refining it ever since.

The two economists say their plan, with a 19 percent flat tax rate, would spur investment and growth by taxing income only once, eliminating taxes on capital gains and taxing individuals only on wages, salaries and pensions. It would allow no deduction for mortgage interest and charitable contributions. Interest and dividends would not be taxed directly, but would be affected by a business tax, also at 19 percent. (Businesses would no longer be able to deduct the cost of fringe benefits, except pensions, for their employees, however.)

So how would a flat tax affect the typical family? And how would it affect the taxes of the Presidential candidates who have pushed the issue to center stage?

The New York Times recalculated the taxes that some of the major candidates would have owed in 1994 had the Hall-Rabushka flat tax been in effect, consulting with the Republican candidates' advisers and making minor adjustments to eliminate anomalies in the complex tax returns that are irrelevant to a flat tax.

The two candidates most closely identified with the flat tax—Steve Forbes, the magazine owner, and Patrick J. Buchanan, the

Sources: Tax Foundation (average family's return).
Those candidates that provided tax returns were consulted on redefining 1994 income into categories appropriate to the flat tax form.

television commentator—would not make their tax returns available. Mr. Forbes said he would never disclose his returns; Mr. Buchanan said he would make his public if he wins the Republican nomination.

President Clinton and three prominent Republican candidates— Senator Bob Dole and Phil Gramm and Lamar Alexander, the former Education Secretary—made available copies of their 1994 Federal income tax returns. (Mr. Clinton opposes the flat tax, as does Mr. Alexander. Mr. Dole has said he would go along with a flat tax. Mr. Gramm favors a 16 percent flat tax.)

All four candidates would enjoy considerable savings in personal income taxes under the Hall-Rabushka flat tax. A comparison also requires calculating the effect of the Hall-Rabushka business tax on business income, but even with both taxes, the taxes owed by each of the four candidates would decline—although probably none would save as much as Mr. Forbes, a fervent flat-tax advocate who is one of the wealthiest men in the United States.

Senator Dole and his wife, who had $607,000 in income in 1994, would save about $49,000; the Clintons whose income was $267,000, would save about $21,000.

Senator Gramm, who with his wife reported $305,000 in income, would save about $40,000; his aides emphasized that under his own proposal, the Senator's taxes would decline even more sharply.

Mr. Alexander's savings would be about $184,000 on an income of about $1 million. The savings would presumably be significantly greater for Mr. Forbes, who has an income of $1.4 million as chief executive officer of Forbes Inc. and a net worth that Fortune magazine last week estimated at $439 million.

The picture is not so bright for the Smiths, a hypothetical family of four used by the Tax Foundation, a conservative research group, in its analyses of the various flat-tax plans. The Smiths, who have $50,000 in total income, nearly all from wages, would see no change in their individual tax bill. And if interest rates decline as predicted, the Smiths, who collected $1,425 in interest in 1994, would come out a few dollars behind.

Flat Tax Goes From 'Snake Oil' to G.O.P. Tonic

By LESLIE WAYNE

from *The New York Times*, November 14, 1999

Four years ago, when Steve Forbes first promoted the idea of a flat tax, he was ridiculed by his fellow Republicans. "A truly nutty idea," said Lamar Alexander. Others, like Bob Dole and Newt Gingrich, heaped scorn, calling it everything from "snake oil" to "voodoo."

But in last month's New Hampshire forum among five Republican presidential candidates, when the question "Do you favor a flat tax?" was posed, three candidates said they favored throwing out the existing federal income tax code and replacing it with a one-rate plan.

"Sure, I'm for a flat tax," said Senator John McCain, who said he wanted a system so simple that a tax return could fit on a postcard. The conservative Gary Bauer boasted that his 16 percent flat tax was better than anyone else's.

And Mr. Forbes, who made the flat tax the centerpiece of his 1996 presidential bid, said he was happy that other Republicans were "coming on board."

What a difference four years makes. The flat tax—having a single income tax rate for all American taxpayers instead of the current graduated set of rates—is slowly gaining ground as a campaign issue in the Republican presidential race, much the way that school vouchers moved from a back-burner issue popular mainly among conservatives to one that is now embraced by all Republican candidates.

For Mr. Forbes, who has since added other conservative issues to his agenda, this is a bittersweet victory. While other candidates are stealing some of his flat-tax thunder, he can take credit for putting the issue on the political map. "It's almost an unalloyed blessing," Mr. Forbes, who advocates a flat 17 percent income tax, said in an interview. "But people just like it. They feel they are burdened by taxes and are really becoming very enthusiastic about it."

So far, the idea has yet to filter down to Congressional races, and the Republican front-runner, Gov. George W. Bush of Texas, does not support it. But analysts say it has undeniably potent appeal.

"The flat tax is not quite Republican dogma yet," said Ed Gillespie, a Republican strategist and president of Policy Impact Communications, a public relations firm. "But it is something that most Republican candidates want to be for. The flat tax resonates strongly with core Republicans, and it has broad appeal across party and demographic lines. We've seen this before with school vouchers."

Grover Norquist, executive director of Americans for Tax Reform, a Republican anti-tax group, added: "Not everyone will have a completely written-out plan like Forbes. But they are all saying, 'That's the direction I want to go.'"

That direction is to reduce federal income tax rates across the board and provide a huge tax cut on a scale not seen in more than a decade. Tax experts on all points of the political spectrum say the rich will benefit more than the poor. Flat-tax proponents are playing to sentiment against Washington, especially the Internal Revenue Service.

One of Mr. Forbes's biggest applause lines is his vow to abolish the I.R.S.: "Take this monster, kill it, put a stake through its heart and make sure it never rises again to terrorize the American people."

And less money flowing to Washington means less government—another goal of Republican conservatives. Backers of the flat tax say the economy would prosper even more if tax dollars now sent to Washington were left in private hands.

Liberal critics dislike the flat tax for the same reasons that it is gaining popularity among conservatives: because it would deprive the government of money, and top income tax rates would drop. The Forbes plan would reduce tax receipts to the government by $50 billion a year.

"There are two things that the flat tax is fronting for," said William G. Gale, a tax expert at the Brookings Institution, a Washington research group. "One is tax cuts for the rich–though while not talking about it and putting it into the guise of simplicity or fairness. And it is an attack on government. It makes more sense to complain about how high taxes are rather than to attack all these wonderful benefits the government is bestowing on the people. The flat tax is a proxy for both those things."

A Republican who remains cool to a flat tax is Mr. Bush, who advocates flattening marginal tax rates by reducing them and eliminating estate taxes. But he has not endorsed a single-rate tax and has said he will have a more detailed tax proposal before the year's end.

Congress, too, has not embraced the idea, even though major elements of a flat tax—eliminating estate taxes and reducing taxes on wages and capital gains—have been proposed by various Republicans on Capitol Hill.

The flat tax that Mr. Forbes is pushing would change the entire theory of taxation embedded in the United States Tax Code. It would abolish the current progressive tax system—one in which lower-income people are taxed at lower rates and higher-income people at higher rates—and replace it with a single tax rate for everyone, rich or poor.

Current federal income tax rates on wages go from 15 percent to 39.6 percent. For instance, a family earning under $43,000 has a tax rate of 15 percent, one earning over $104,000 has a tax rate of 31 percent and a family earning over $283,000 pays at a rate of 39.6 percent. These are the marginal rates, or what people pay on the last dollars earned.

Of course, taxpayers now have a variety of ways to whittle their taxable income to fall into a lower tax bracket—everything from mortgage interest payments, to donations to charities, to payments for child care, education and health care can be deducted to reduce taxable income.

A flat tax would replace this progressive system with a one-rate plan; under Mr. Forbes's proposal, for instance, both Bill Gates and the factory worker would pay the same 17 percent of income in taxes. In addition, there would be no deductions, like those for home mortgages, although some politicians, including Mr. Forbes, are beginning to soften on the issue.

Earnings from investments—capital gains from stock and bond sales and dividend and interest income—would not be taxed at all. Capital gains are now taxed at a 20 percent rate, and interest and dividends at the same rate as wages.

"The tax system we have now is steeply progressive," said Steve Moore, a flat-tax advocate and director of fiscal policies at the Cato Institute, a Washington nonprofit organization. "The wealthiest 1 percent are paying around one-third of all taxes, and the bottom 50 percent of the people pay only 4 percent of the burden. Anything you do to flatten tax rates will lead to lower taxes on the rich and more to those on the bottom."

Moreover, under a flat tax, it is possible for a wealthy person who lives solely off income from an

POLICY POSITIONS

Tax Proposals From G.O.P. Candidates

Here is what Republican candidates for president are saying about taxes, according to each candidate's campaign Web site:

GARY BAUER "Our current tax code is needlessly complex and deeply unfair. It has been written by lobbyists and special interests who could care less about average Americans."

He proposes: A 16 percent flat tax on individuals and corporations. No corporate deductions allowed. Families would receive a $1,400 per person tax credit. Home mortgage and charitable deductions would be retained. Capital gains would be taxed at a 16 percent rate. Estate taxes would be eliminated.

GEORGE W. BUSH "It is conservative to cut taxes and compassionate to give people more money to spend."

He proposes: A cut in marginal tax rates, but does not specify rates. A phase-out of estate taxes. An increase in tax incentives for charitable giving and expansion of tax breaks for education.

STEVE FORBES "With more personal control you will be free to choose a less burdensome tax system that's simple, honest, fair and a real tax cut."

He proposes: A 17 percent flat tax on individuals and corporations, with existing deductions eliminated. Investment income would not be taxed. Estate taxes would be eliminated. Each taxpayer would get an exemption of $13,000 per adult and $5,000 per child. Corporations could deduct capital investments in one year.

JOHN McCAIN "It is essential that we provide American families with relief from the excessive rate of taxation that saps job growth and robs them of the opportunity to provide for their needs."

He proposes: Expansion of the 15 percent tax bracket, which is the lowest. Elimination of estate taxes and the so-called marriage penalty. Simplification of the income tax code and elimination of "special interest loopholes."

investment portfolio to pay no taxes at all, while someone working in a factory or driving a cab faces a 17 percent tax bill. To many, the prospect of such glaring inequities between rich and working class is the plan's political Achilles' heel.

"I don't think the flat tax will work for a lot of reasons," said Michael J. Graetz, a professor at Yale Law School and a tax policy official in the Bush administration. "As it's been proposed, it involves a massive tax cut for the very best-off people in society. There will be questions as to why a tax cut ought to be concentrated in that group."

He said taxpayers in the $75,000 to $150,000 income range would be hurt the most under a flat tax.

This is an especially sensitive political point for Mr. Forbes, whose $440 million net worth comes from inherited wealth, and for Republicans in general, as the party tries to shake an elitist image and appeal to new and less-affluent constituencies.

For corporations, taxes would fall to 17 percent from 35 percent under the Forbes plan. In addition, the flat tax would allow corporations to deduct the costs of capital improvements in a single year, rather than spreading them out over time. This would favor young and fast-growing companies, which could see their tax bills wiped out entirely, over older businesses, whose years of heavy capital spending are behind them. A company like, say, Intel, a computer chip maker, would benefit under a flat tax, while older manufacturers like the General Motors Corporation could be hurt, tax experts say.

Corporations would see all other deductions wiped out under a flat tax, and many critics worry that once the tax code no longer provides deductions for workers' benefits–especially health insurance–companies will stop providing them.

"Flat-tax advocates are being completely deceptive when they say they are closing loopholes," said Robert McIntyre, director of Citizens for Tax Justice, a Washington nonprofit group. "Au contraire. Calling it a loophole-closing plan is a misnomer. This is more of a loophole-consolidations plan. It's saving the tax breaks that matter–those that favor the high end and some corporations."

Mr. Forbes defends his flat tax by saying that the economic growth it would unleash would be so great that all Americans, even the poorest, would benefit from new-found wealth. And he said $50 billion in lost tax revenues to the government from a flat tax could be covered by the current $200 billion budget surplus–which, Mr. Forbes said, should be dedicated to tax relief.

He also predicted that a flat tax would cause the nation's economy, which is already operating at full employment, to grow by 4 percent to 4.5 percent a year.

Yet those rosy predictions have been greeted with skepticism. "Forbes is proposing an enormous tax cut that dwarfs anything Congressional Republicans were proposing this year," said J. D. Foster, a former Bush tax adviser and head of the Tax Foundation, a Washington nonprofit organization.

"And he says that the economy, which is now cruising at a 3 percent real growth rate, will suddenly jump to 4.5 percent. Most economists have dismissed that out of hand. Short of annexing Mexico and some Asian countries, I don't know where you get that growth rate."

Moreover, even flat-tax proponents say that Mr. Forbes's 17 percent rate is too drastic a cut in government financing. A 1996 Treasury analysis said a flat tax rate of 20.8 percent would be needed to raise the same revenues as the current system. Even a flat-tax supporter like Mr. Moore of the Cato Institute said a 20 percent tax rate is more realistic–"17 percent is probably too low."

And if Republicans succumb to political pressure and begin to give back popular deductions—like the mortgage interest deduction—tax experts estimate that the number might have to climb even higher.

Even Mr. Forbes is softening his plan. He is offering personal exemptions of up to $36,000 for a family of four. This means that the actual taxes paid under his plan could drop to as low as 10.9 percent for a family earning more than $100,000, a 33 percent tax cut, according to Mr. Forbes's campaign literature. His plan would also allow taxpayers to stick with the current tax system if it worked to their financial advantage.

"What you are beginning to see is a kinder and gentler flat tax," Mr. Moore said. "It isn't the same product talked about four years ago; it's a little more of a politically strategic version. The question is how do you get popular appeal for the flat tax, without all the baggage?"

Powers of Ten

A book about the relative size of things in the universe
and the effect of adding another zero

Philip and Phylis Morrison
and
The Office of Charles and Ray Eames
based on the film *Powers of Ten*
by The Office of Charles and Ray Eames

POWERS OF 10: HOW TO WRITE NUMBERS LARGE AND SMALL

This book uses a notation based on counting how many times 10 must be multiplied by itself to reach an intended number: For example, 10×10 equals 10^2, or 100; and $10 \times 10 \times 10$ equals 10^3, or 1000. Multiplying a number by itself produces a *power* of that number: 10^3 is read out loud as "ten to the third power," and is another way to say one thousand. In this case, there is no great advantage, but it is much easier and clearer to write or say 10^{14} than 100,000,000,000,000 or one hundred trillion. After 10^{14}, we even run low on names. The number written above in smaller type—the 14 in the last example—is called an *exponent,* and the powers notation is often called *exponential notation.*

It is not hard to grasp the positive powers of ten—10^4, 10^7, 10^{19}—and how they work; but the negative powers—10^{-2} or 10^{-3}—are another matter. If the exponent tells how many times the 10 is to be self-multiplied, what can an exponent of -5 (negative five) mean? The system requires a negative exponent to signal division by 10 a certain number of times: 10^{-1} equals 1 divided by 10, or 0.1 (one-tenth); 10^{-2} equals 0.1 divided by 10, or 0.01 (one-hundredth). Because *adding* 1 to the exponent easily works out to be the equivalent of multiplying by 10, it is self-consistent that subtracting 1 there works out to a division by 10. It is all a matter of placing zeros. Adding another terminal zero is simply to multiply by 10: 100×10 equals 1,000. Putting another zero next after the decimal point is to *divide* by 10: $0.01 \div 10$ equals 0.001. The powers notation makes these operations even clearer.

But what of 10^0? That seems a strange number. However, notice it is equal to 10^1 (10) divided by 10 (or to 10^{-1} multiplied by 10). Although surprising, it is at least logical that 10^0 should be equal to 1.

Because you can make any power of ten ten times larger by adding 1 to its exponent ($10^4 \times 10 = 10^5$), it follows that to mul-

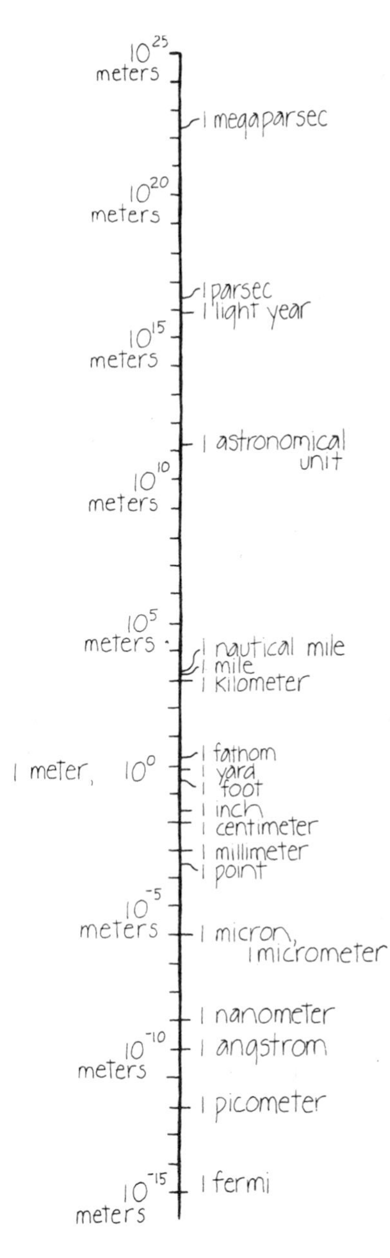

tiply by 100 you add 2 to the exponent: $10^3 \times 100 = 10^5$ or $1000 \times 100 = 100{,}000$. In general, you can multiply one power of ten by another simply by adding their exponents: $10^6 \times 10^3 = 10^9$. Subtracting the exponents is the equivalent of division: $10^7 \div 10^5 = 10^2$.

All numbers, not only numbers that are exact powers of ten, like 100 or 10,000, can be written with the help of exponential notation. The number 4000 is 4×10^3; 186,000 is 1.86×10^5. This convenient scheme is referred to as scientific notation.

All of this can be extended to basic multipliers other than ten: $2^4 = 2 \times 2 \times 2 \times 2$ (the fourth power of two); $12^2 = 12 \times 12$, and 8^{-1} = one-eighth. (But note that $2^0 = 1$, $12^0 = 1$, and $8^0 = 1$.)

Logarithms arise from extensions of this scheme.

The symbol $\sim$ is mathematicians' shorthand for "approximately" or "about."

UNITS OF LENGTH

Grow you own food and build your own house, and no formal units of measurement much interest you; such is the general rule of thumb. But commerce has implied agreement on units of measurement. The legal yard has long been displayed for the use of Londoners, and the meter is still open to public comparison on a wall of a Paris building.

The system we call metric is the work of the savants of Revolutionary Paris in the 1790s. Even their determination to celebrate both novelty and reason met limits: Our modern second, minute, and hour remain resolutely nondecimal. That was no oversight—the metric day of ten hours, each of a hundred minutes with a hundred seconds to the minute, was formally adopted. But the scheme met fierce resistance. About the only costly mechanism every middle-class family then proudly owned was a clock or watch, not to be rendered at once useless by any mere claim of consistency! Practice won out over theory.

In much the same way, people who today frequently use units in a particular context are not always persuaded to sacrifice appropriateness to consistency. We list here a few nonstandard units of linear measurement that retain their utility even in these metric days, some even within the sciences.

Cosmic Distances

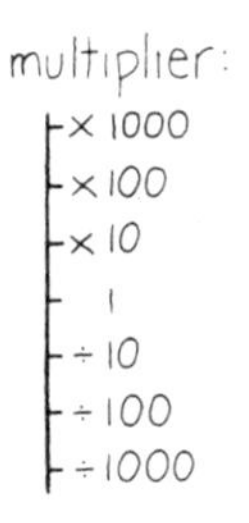

Parsec The word is a coinage from *par*allax of one *sec*ond. The *parsec* is in common use among astronomers because it hints at the surveyor's basic technique of measuring stellar distance by using triangulation. The standard parallax is the apparent shift in direction of a distant object at six-month intervals as the observer moves with the orbiting earth. It is defined so that the radius of the earth's orbit seen from a distance of one parsec spans an angle of one second of arc. The nearest known star to the sun is more than one parsec away.

Light-year This graspable interstellar unit rests on the relationship between cosmic distance and light travel over time.

The speed of light in space is 3.00×10^8 meters per second; in one year light thus moves 9.46×10^{15} meters, which is usually rounded off to 10^{16} meters, especially since only a few cosmic distances are so well known that the roundoff is any real loss of accuracy.

Astronomical unit The mean distance between sun and earth is a fine baseline for surveying the solar system; it is a typical length among orbits. 1 AU = 1.50×10^{11} meters. *Note:* 1 parsec = 3.26 light-years = 206,300 AU. The interstellar and the solar-system scales plainly differ; intergalactic distances run to megaparsecs.

Terrestrial Lengths

Miles, leagues, etc. These are units suited for earthbound travel, for distances at sea, or for road distances between cities. Nobody ever measured cloth by the mile, or train rides by the parsec.

Yards, feet, meters These rest on human scale, in folklore the length of some good king's arm. They suit well the sizes of rooms, people, trucks, boats. Textiles are yard goods. The meter was defined more universally, but clearly it was meant to supplant the yard and the foot. It was related to the size of the earth: One quadrant of the earth's circumference was defined as exactly 10^7 meters, or 10^4 kilometers. In 1981 the meter is defined with great precision in terms of the wavelength of a specific atomic spectral line. It is "1,650,763.73 wavelengths in vacuum of the radiation corresponding to the transitions between the levels $2p_{10}$ and $5d_5$ of the krypton-86 atom."

Inches, centimeters, etc. The same king's thumb? Human-scale units intended for the smaller artifacts of the hand: paper sizes, furniture, hats, or pies.

Line, millimeter, point Small units for fine work are relatively modern. The seventeenth-century French and English line was a couple of millimeters, and the printer's point measure is about 0.35 mm. Film, watches, and the like are commonly sized by millimeters. The pioneer microscopist Antony van Leeuwenhoek used sand grains as his length comparison, coarse and fine: He counted one hundred of the fine grains to the common inch of his place and time. Smaller measurement units are generally part of modern science, and thus usually metric.

Atomic Distances

Angstroms, fermis, etc. Once atoms became the topic of meaningful measurement, new small units of length naturally came into specialized use. The Swedish physicist Anders Ångström a century ago pioneered wavelength measurements of the solar spectrum. He expressed his results in terms of a length unit just 10^{-10} meters long. It has remained in widespread informal use bearing his name: convenient because atoms measure a few angstroms (Å) across. The impulse for such useful jargon words is by no means ended; nuclear particles are often measured in fermis, after the Italian physicist Enrico Fermi. 1 fermi equals 10^{-15} meters.

Angles and Time

Angles are measured, especially in astronomy, by a nonmetric system that goes back to Babylon. A circle is 360 degrees; 1 degree = 60 minutes of arc; 1 minute = 60 arc-seconds. An arc-second is roughly the smallest angle that the image of a star occupies, smeared as it is by atmospheric motion. This page, viewed from about twenty-five miles away, would appear about one arc-second across.

Time measurement shares the cuneiform usage of powers of sixty. Note that a year of 365.25 days of 24 hours, each of 60 minutes with 60 seconds apiece, amounts to about 3.16×10^7 seconds.

What seest thou else
In the dark backward and abysm of time?
—Wm. Shakespeare
The Tempest

The Cosmic Calendar*

Carl Sagan

The world is very old, and human beings are very young. Significant events in our personal lives are measured in years or less; our lifetimes in decades; our family genealogies in centuries; and all of recorded history in millennia. But we have been preceded by an awesome vista of time, extending for prodigious periods into the past, about which we know little—both because there are no written records and because we have real difficulty in grasping the immensity of the intervals involved.

Yet we are able to date events in the remote past. Geological stratification and radioactive dating provide information on archaeological, palenotological and geological events; and astrophysical theory provides data on the ages of planetary surfaces, stars, and the Milky Way Galaxy, as well as an estimate of the time that has elapsed since that extraordinary event called the Big Bang—an explosion that involved all of the matter and energy in the present universe. The Big Bang may be the beginning of the universe, or it may be a discontinuity in which information about the earlier history of the universe was destroyed. But it is certainly the earliest event about which we have any record.

The most instructive way I know to express this cosmic chronology is to imagine the fifteen-billion-year lifetime of the universe (or at least its present incarnation since the Big Bang) compressed into the span of a single year. Then every billion years of Earth history would correspond to about twenty-four days of our cosmic year, and one second of that year to 475 real revolutions of the Earth about the sun. On the next two pages I present the cosmic chronology in three forms: a list of some representative pre-December dates; a calendar for the month of December; and a closer look at the late evening of New Year's Eve. On this scale, the events of our history books—even books that make significant efforts to deprovincialize the present—are

* From: Carl Sagan, *The Dragons of Eden: Speculations on the Evolution of Human Intelligence,* Ballantine Books, NY, 1977.

Pre-December Dates

Big Bang	January 1
Origin of the Milky Way Galaxy	May 1
Origin of the solar system	September 9
Formation of the Earth	September 14
Origin of life on Earth	~September 25
Formation of the oldest rocks known on Earth	October 2
Date of oldest fossils (bacteria and blue-green algae)	October 9
Invention of sex (by microorganisms)	~November 1
Oldest fossil photosynthetic plants	November 12
Eukaryotes (first cells with nuclei) flourish	November 15

~ = *approximately*

Cosmic Calendar
DECEMBER

Sunday	Monday	Tuesday	Wednesday	Thursday	Friday	Saturday
	1 Significant oxygen atmosphere begins to develop on Earth.	2	3	4	5 Extensive vulcanism and channel formation on Mars.	6
7	8	9	10	11	12	13
14	15	16 First worms.	17 Precambrian ends. Paleozoic Era and Cambrian Period begin. Invertebrates flourish.	18 First oceanic plankton. Trilobites flourish.	19 Ordovician Period. First fish, first vertebrates.	20 Silurian Period. First vascular plants. Plants begin colonization of land.
21 Devonian Period begins. First insects. Animals begin colonization of land.	22 First amphibians. First winged insects.	23 Carboniferous Period. First trees. First reptiles.	24 Permian Period begins. First dinosaurs.	25 Paleozoic Era ends. Mesozoic Era begins.	26 Triassic Period. First mammals.	27 Jurassic Period. First birds.
28 Cretaceous Period. First flowers. Dinosaurs become extinct.	29 Mesozoic Era ends. Cenozoic Era and Tertiary Period begin. First cetaceans. First primates.	30 Early evolution of frontal lobes in the brains of primates. First hominids. Giant mammals flourish.	31 End of the Pliocene Period. Quaternary (Pleistocene and Holocene) Period. First humans.			

December 31

Origin of *Proconsul* and *Ramapithecus,* probable ancestors of apes and men	~1:30 P.M.
First humans	~10:30 P.M.
Widespread use of stone tools	11:00 P.M.
Domestication of fire by Peking man	11:46 P.M.
Beginning of most recent glacial period	11:56 P.M.
Seafarers settle Australia	11:58 P.M.
Extensive cave painting in Europe	11:59 P.M.
Invention of agriculture	11:59:20 P.M.
Neolithic civilization; first cities	11:59:35 P.M.
First dynasties in Sumer, Ebla and Egypt; development of astronomy	11:59:50 P.M.
Invention of the alphabet; Akkadian Empire	11:59:51 P.M.
Hammurabic legal codes in Babylon; Middle Kingdom in Egypt	11:59:52 P.M.
Bronze metallurgy; Mycenaean culture; Trojan War; Olmec culture; invention of the compass	11:59:53 P.M.
Iron metallurgy; First Assyrian Empire; Kingdom of Israel; founding of Carthage by Phoenicia	11:59:54 P.M.
Asokan India; Ch'in Dynasty China; Periclean Athens; birth of Buddha	11:59:55 P.M.
Euclidean geometry; Archimedean physics; Ptolemaic astronomy; Roman Empire; birth of Christ	11:59:56 P.M.
Zero and decimals invented in Indian arithmetic; Rome falls; Moslem conquests	11:59:57 P.M.
Mayan civilization; Sung Dynasty China; Byzantine empire; Mongol invasion; Crusades	11:59:58 P.M.
Renaissance in Europe; voyages of discovery from Europe and from Ming Dynasty China; emergence of the experimental method in science	11:59:59 P.M.
Widespread development of science and technology; emergence of a global culture; acquisition of the means for self-destruction of the human species; first steps in spacecraft planetary exploration and the search for extraterrestrial intelligence	Now: The first second of New Year's Day

so compressed that it is necessary to give a second-by-second recounting of the last seconds of the cosmic year. Even then, we find events listed as contemporary that we have been taught to consider as widely separated in time. In the history of life, an equally rich tapestry must have been woven in other periods—for example, between 10:02 and 10:03 on the morning of April 6th or September 16th. But we have detailed records only for the very end of the cosmic year.

The chronology corresponds to the best evidence now available. But some of it is rather shaky. No one would be astounded if, for example, it turns out that plants colonized the land in the Ordovician rather than the Silurian Period; or that segmented worms appeared earlier in the Precambrian Period than indicated. Also, in the chronology of the last ten seconds of the cosmic year, it was obviously impossible for me to include all significant events; I hope I may be excused for not having explicitly mentioned advances in art, music and literature or the historically significant American, French, Russian and Chinese revolutions.

The construction of such tables and calendars is inevitably humbling. It is disconcerting to find that in such a cosmic year the Earth does not condense out of interstellar matter until early September; dinosaurs emerge on Christmas Eve; flowers arise on December 28th; and men and women originate at 10:30 P.M. on New Year's Eve. All of recorded history occupies the last ten seconds of December 31; and the time from the waning of the Middle Ages to the present occupies little more than one second. But because I have arranged it that way, the first cosmic year has just ended. And despite the insignificance of the instant we have so far occupied in cosmic time, it is clear that what happens on and near Earth at the beginning of the second cosmic year will depend very much on the scientific wisdom and the distinctly human sensitivity of mankind.

Watching Galileo's Learning

Elizabeth Cavicchi

1 Introduction

By closely following Stillman Drake's biographies of Galileo, this essay interprets Galileo's free fall studies. Drake's biographies piece together how Galileo came to understand that the distance an object has fallen increases as the square of its descent time. He infers this story from calculations recorded among Galileo's working papers, including some that were not included in the definitive twenty volume set of Galileo's *Opera*, edited by Favaro in 1934.

Prior to Drake's studies, the interpretation of the historian Koyre prevailed among Galilean scholars. Koyre maintained that Galileo never made any observations of motion (except with the telescope); he says Galileo derived laws of motion through thought alone. For Koyre, Galileo's innovation in science lay in this introspective method of reasoning.

Koyre influenced the textbook treatment of Galileo's work. Many physics texts do not mention Galileo's free fall experiments. While some recent texts refer to his experiments, they simplify Galileo's process of learning by leaving out details and context which might assist readers in following how Galileo's experimenting and thinking developed. Texts also omit reference to the mathematics in use at the time, which were tools for Galileo's thought, and to the contemporary politics and thought, which Galileo's work challenged. Such simplifications and omissions are among the many ways textbooks distort their presentations of how people learned, and can learn, about the phenomena of nature. As a result, unable to imagine how Galileo could have come to the law of how things fall just by postulating it correctly, students may doubt their own potential to learn through exploring and questioning how things happen.

By comparison with the methods of analysis in use today, the mathematical tools and physical picture available to Galileo seem very limited. The task of stripping away our twentieth century sophistication, to better approximate the outlook and thinking of Galileo, is formidable. As a biographer tracking Galilean documents Drake acquired this outlook. However, he does not carefully provide clues that would facilitate a reader's understanding of Galileo's world view. My effort to understand Drake's argument seems parallel to Galileo's effort to understand motion. We begin from awareness of what we do not know, which deepens as we notice more confusion in our thinking and more complexity in the subject we are trying to understand. Through this deepening of what we do not know, we come to ways of making thoughts and questions that can take our understanding further. In watching Galileo's learning, I am repeatedly bewildered as my assumptions prevent me from seeing what Galileo did not know. I am also astonished by Galileo's creativity in using the tools he did have to find something new.

Elizabeth Cavicchi

Accelerated motion is subtle; Galileo devoted most of a lifetime to its study. We can re-experience some of the subtlety and complexity Galileo encountered by observing our own students as they try to make sense of it. I did this as a project with one student from the algebra course this reading is intended to supplement. I did not provide the student – Hazel Garland – with explanations of motion. Instead as we experimented and observed together, the experimenting itself became the beginnings of our further thought and experimenting.

Hazel and I released a weight, with a long paper tape attached, so that it fell freely through a gap in a spark-timer. At 1/60 s time intervals, a spark jumped through the gap, marking a dot on the paper tape. When we examined these paper tapes, Hazel found inconsistencies between the dot patterns on different tapes: this intrigued her. As I watched her developing analysis of these dots, I began to realize how ideas which seemed 'obvious' to me as a teacher – such as that the timer's periodicity was independent of the moment when we released the weight – were not at all obvious to Hazel. I saw how understandings of time intervals and motion developed through her careful and animated thought about what was on the paper tapes. As I tried to understand free fall from her perspective, I was repeatedly surprised by her creativity in working through what seemed – to me – to be limitations in what she knew. I came to see these 'limitations' as productive beginnings for what she came to know ever more deeply.

Through watching Hazel's experimenting, I came to appreciate how, by their notice of details in what natural phenomena do, learners can form new questions and work out ideas that develop what they know. While writing this essay, I realized that the engagement of learning through experimenting that I saw in Hazel's work was also evident in what Galileo did. I began to understand that the confusions and difficulties I encountered in interpreting Galileo's experimenting, and Drake's account of it, were integral to the work and process of how learning develops. These confusions cannot be simplified away or replaced by a neat logical sequence. The complexity of what happens in the learners' thoughts and work makes evident the depth of their response to the complexity of the physical motions they explore. It also makes possible their continued efforts at learning.

2 Falling Objects

When younger than we can now remember, we played by dropping a toy and noisily goading an attentive older companion to retrieve it. While the game now seems a test of patience, seeing something fall was a new experience for us, which we observed without asking either why or how things fall. When, as an older child, we asked what makes things fall, the answer was "gravity". When we asked what gravity was, we were told "it makes things fall". Gravity was an empty word, devoid of explanation.

Better (or more complete, or less circular) answers to this question have evaded not only parents, but also people who explicitly study things and try to explain them. This simple question is too grand. Why does it take effort to get things to go where we want– lifting, dragging, carrying, bicycling– while they fall by themselves? No strings or other apparatus visibly pulls them down. Seeking a mechanism for falling is premature if we are not yet sure what is happening when something falls. Falling is so quick that it is not easy to say what is happening as something falls, even if it is watched carefully. Asking the question differently– "How do things fall?"– provides a clearer way to start towards an answer: by careful description of the phenomenon itself.

3 Early Explanations

The seeming simplicity of the "grand" question– why things fall– motivated early historical attempts to talk about falling in a way that made sense to ordinary people. The Greeks said the weight of an object makes it go towards earth's center. Things that have lightness, like fire, move away from the center. Weightier things fall faster than lighter ones. Aristotle (384-322 B.C.) extended this explanation with logic, to make many other statements about motion and change. His mostly incorrect view was considered authoritative for about a millennium.

People who looked carefully found that some of these ideas just did not match with nature. Two sixteenth century Italian instructors dropped objects differing only in weight from a high window or tower. The objects struck the ground almost at once. These observations were augmented when Galileo Galilei (1564-1642, Figure 1, [7]) began investigating what happens when something falls.

4 Galileo and the Inclined Plane

Galileo revised the question; he asked "how" things fall, instead of "why". This question suggested other questions that could be tested directly by experiment. By alternating questions and experiments, Galileo was able to identify details in motion no one had previously noticed or tried to observe. His questions about free fall suggested questions about other motions, such as motions Aristotle had classed as "natural". These motions start spontaneously without a push, such as a ball rolling down a slope, or a pendulum swinging. What he learned about one motion drove his study of the others.

In one experiment, performed near 1602, Galileo rolled balls down a ramp. He mounted lute strings crosswise to the motion, so that a distinct sound was produced each time a ball rolled over a string. He adjusted positions of the strings until time intervals between sounds were roughly equal. Perhaps Galileo

Figure 1: A portrait of Galileo, taken from his book on the sunspots, *Istoria e dimostrazioni intorno alle macchie solari*, published in Rome in 1613. By permission of the Houghton Library, Harvard University.

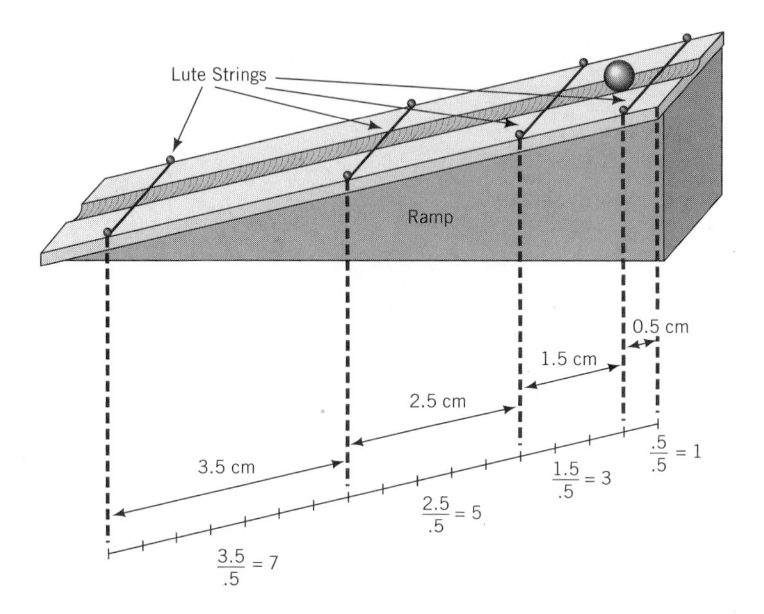

Figure 2: Whenever a ball rolling down a ramp bumps over a string, it makes a sound. The strings are positioned so the sounds mark out equal intervals in time. If we measure the distances between each pair of strings, the ratio of each distance to the first one makes a series of odd integers.

hummed a tune, with beats about a half-second apart, to estimate constant time intervals.

Once he had equalized the rolling time between each pair of strings, he measured the distance between each pair of adjacent strings. The distances between strings were unequal: each consecutive pair of strings was further apart than the preceding pair of strings (Figure 2). When he divided all the distances between pairs of strings by the first distance, the quotients were the odd integers 1, 3, 5, 7, ...

This discovery was revolutionary, as the first evidence that motion on Earth was subject to mathematical laws. Until this time, people assumed mathematics governed only motions in the heavens; terrestrial motions were considered disorderly. Galileo did not simply accept this belief, but tried to test it with direct observation.

Galileo interpreted the separation between successive pairs of lute strings as a measure of the ball's speed while rolling between the two strings. The numerical pattern showed Galileo that the ball's speed increased, but it did not tell him how it increased. To understand this subtlety he again had to contradict physical theory of his contemporaries.

People already realized that one time or position measurement could change to another time or position by imperceptibly small, continuous increments. But speed was depicted as a motion that occurred in a perceptible time interval. While an object's speed could change between time intervals, they did not think speed could change instantaneously as we know it does today. Instantaneous change in speed was ill-defined for them, because they could not conceive of a speed without being able to see the motion directly. Galileo eventually understood and resolved this apparent contradiction, and concluded that speed changes continuously during free fall and natural motions. He wrote:

> I suppose (and perhaps I shall be able to demonstrate this) that the naturally falling body goes continually increasing its *velocità* according as the distance increases from the point which it parted[6].

5 Mathematical Tools

Galileo's ability to construct patterns from his measurements was limited by the mathematics available to him at the time. Although European mathematicians were then developing algebra beyond its Arabic and Hindu origins, and Galileo was aware of their work, he never used algebra in his physical studies. He never expressed any results using equations. His analysis of data and publications relied on a more classical training– popular at the time– in the logic, geometry, and numerical proportions of Euclid's *Elements* (fl. 300 B.C.).

One result of these mathematical limitations was that Galileo could not define speed the way we do today, as a distance divided by a time (for example miles *per* hour). In manipulating numbers according to the Euclidean theory of proportions, it was not legitimate to construct ratios of measurements with different units, such as dividing a distance by a time. Ratios could only be constructed by dividing one distance measurement by another like measurement (for example (20 meters)/(10 meters)), or one time by another time (for example (6 seconds)/(2 seconds)). The units of distance or time then canceled out, yielding a pure, dimensionless number.

Galileo's interpretation of speed as the separation between lute strings prevented him from seeing it as the distance traveled while the object rolled between the strings. If he had been able to identify that this one measurement conveyed data pertaining to two different physical quantities (speed and distance) at once, he might have worked out the relation inherring between distance and time two years earlier than he did. But, persisting with questions about what his observations revealed about motion, he continued experimenting by rolling balls down ramps of different lengths and of different vertical heights. He incorporated results from all these experiments into his mature, formal understanding of motion, but at the time he performed them, they did not lead him to the relation he sought.

6 Pendulum Swings

After failing to work out a lawful expression for motion along the incline, in 1604 Galileo began looking for it in the motion of pendulums. He had already experimented extensively with pendulums. He had already been the first person to demonstrate that a pendulum swings through a small arc in very nearly the same time that it swings through a wide arc.

Galileo timed pendulum swings with a water clock he designed. A water pipe was fed from a large elevated reservoir and plugged with a finger. When it was turned on by removing his finger from the pipe, water spurted from the pipe into a bucket. He unplugged the pipe for the duration of a pendulum swing and weighed the water that flowed into the bucket. As long as the flow rate was steady, weights of water were a measure of time; the longer the run, the more water, and hence the more weight (Figure 3). Galileo checked consistency of water flow through the pipe with a built-in timer: his own pulse. He used this clock to measure times for a wide variety of physical phenomena, such as free fall, balls rolling down ramps, and pendulum swings.

He measured one-quarter of a pendulum swing: from its moment of release to the moment when it was vertical. He mounted the pendulum so that when it hung vertically (like a plumb bob), the bob just touched a wall (Figure 4). He started his water clock when he released the pendulum and stopped it when he heard the bob bang against the wall. The sound helped him make measurements repeatably and reliably.

Since Galileo used the same clock for measurement of the pendulum and other natural motions, the times for all his measurements were expressed in the same 'units', which suggested that he could compare times for *different kinds* of natural motion, for example, pendulum swings and falling objects. Using a gear mechanism he designed, he tried to adjust the length of a pendulum string so that the time of its quarter swing exactly matched the time it took for an object to fall a given distance, by matching the weights of water that flowed during both motions. But his free fall measurement required too long a pendulum to be practical, so he tried matching the pendulum swing with half the free fall time he had measured, which corresponds to a fall of one meter. The length of the resulting pendulum string was a little less than one meter.

Galileo then made a pendulum twice as long as this one, and timed its quarter swing. He decided to double the pendulum length, rather than to try shorter lengths, because the mathematics of the time did not condone the construction of fractional lengths such as 1/2 or 1/4 of a given length. He found that swing

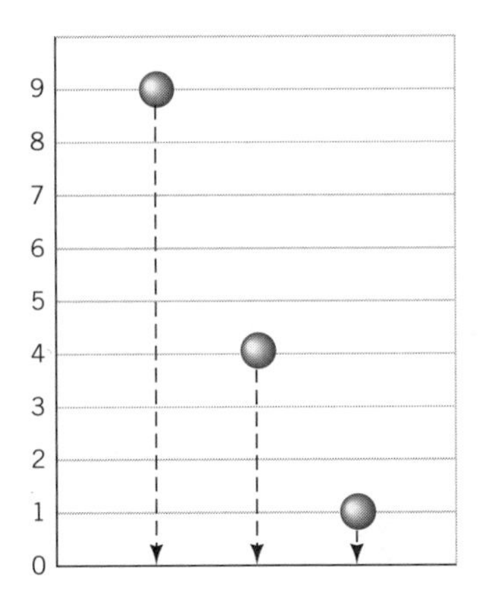

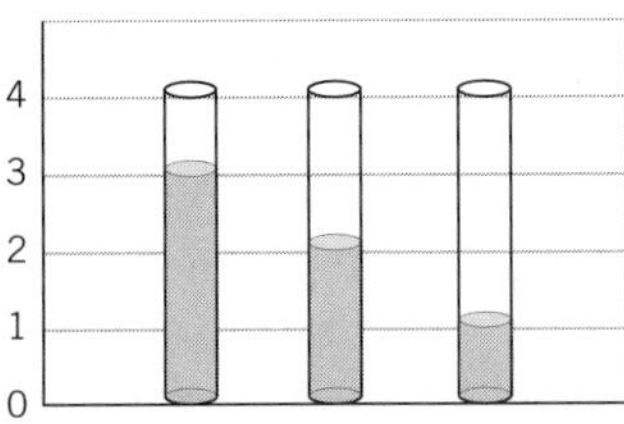

Figure 3: The water clock was started when a weight was released, and stopped when it hit the ground. More water was collected during long falls than during short falls. The amount of water is a measure of the elapsed time.

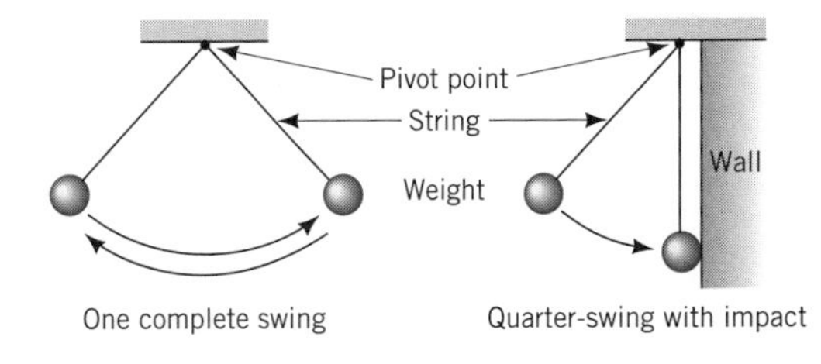

Figure 4: A complete pendulum swing and one quarter pendulum swing. When the bob completes one quarter swing, it hits a wall, making a sound.

time was proportional to a 'geometrical mean' involving pendulum length, which he computed using a straightedge and compass on a geometric drawing. This geometrical computation is equivalent to taking the square root of pendulum length. Today we say that a pendulum's period is proportional to the square root of its length. Galileo used his version of this rule to correctly predict the swing time of a thirty foot pendulum which he may have hung in the courtyard of the University of Padua where he worked.

Galileo linked the pendulum swing to the free fall by performing the experiment backward. He used the water clock to time a weight as it fell through the same height as the length of one of the pendulums he had already timed. He expressed the ratio of the two times (the free fall time over the pendulum time) as a ratio of the whole number weights 942/850. While he did not reduce this ratio to the decimal form 1.108, he was quite close to the true ratio $\pi/2\sqrt{2} = 1.1107$.

For Stillman Drake, Galileo's biographer, this ratio is the closest analogy in Galileo's work to something like a physical constant. Physical constants are an artifact of algebra. They drop out of ratios. Thus, although we now commonly say that Galileo discovered the constant acceleration due to gravity, g, he did not and could not. The mathematics, which he so carefully applied, and which revealed so much to him, masked numerical constancy.

Galileo combined his method of relating a pendulum's length to its swing time with the ratio he found between fall and pendulum times. He thus calculated the height a weight would have to fall from in order to match the swing time of one of his pendulums. When he analyzed the ratio calculation, he found that the falling time appeared twice. Time was multiplied by time; time was *squared*. In following the mathematical rules for combining ratios, he had constructed a new physical law: falling distance is proportional to the square of falling time. He had also constructed the new physical quantity 'time squared'. Until Galileo discovered this pattern in his calculation, no one had ever squared a physical quantity other than length, whose square is usually area.

7 Extending the Law

Galileo wondered if the same rule applied to the data collected in the inclined plane/lute string experiments. There he had computed the sequence of odd integers 1, 3, 5, 7, ... from ratios of the first distance to each successive distance covered in equal times. This pattern suggests to us that the link between distance and time involves a square: sums of the series of odd integers produce the series of squared integers (1, $1+3=4$, $1+3+5=9$, $1+3+5+7=16$). Galileo did not identify this pattern when he originally performed the experiment, because he interpreted the lute string separations as speeds rather than distances.

Galileo recorded the integer squares (1, 4, 9, 16, ...) beside his original distance measurements on the notebook page containing data from the lute string experiment. He found that the product of the first distance and each successive integer square (1, 4, 9, 16, ...) yielded the same number that he had originally measured as the ball's cumulative distance traveled from the start of motion. These squared integers behaved like the squares of times in free fall (Figure 5). Balls rolling down an incline and falling weights exhibited the same relation: distance traveled is proportional to the square of time. We can also say that the pendulum motion is somehow similar; the square root of its length is proportional to its swing time. All these 'natural' motions were somehow similar.

Galileo commented, though not in these modern words, that if you plot the total distance traveled for several consecutive and equal time intervals, the dots fall along a parabola (Figure 6). But to Galileo this comment was purely geometric; it did not provide him with a definition of speed or a characterization of projectile motion. For some time he questioned whether a object's speed in free fall could be represented by the total distance it had fallen, or by the distance it had fallen in the most recent interval of time. If the first choice were correct, the speeds of successive time intervals would also fall along a parabola, just like the distance measurements, while the second choice would place the speeds on a straight line of increasing slope (Figure 7).

Galileo tried to measure speed of a falling object by observing the effects of

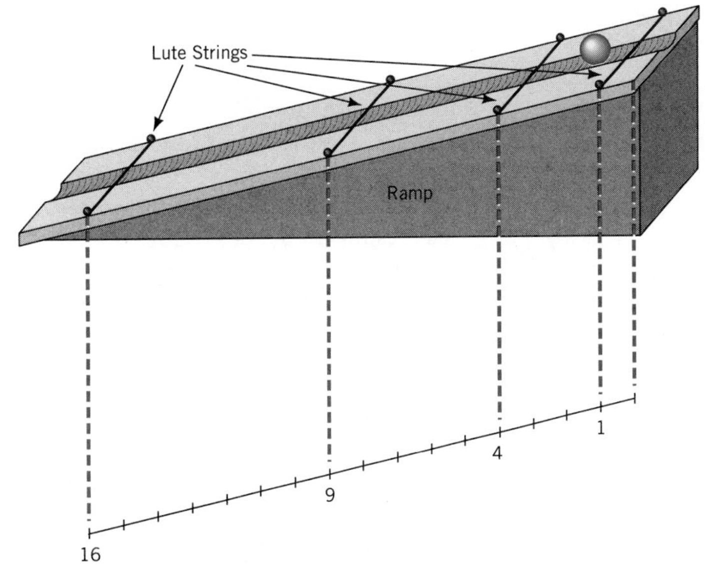

Figure 5: The total distance the ball had rolled after its release was proportional to the square of the integer identifying that equal time interval. The same law related distance and time in free fall motions.

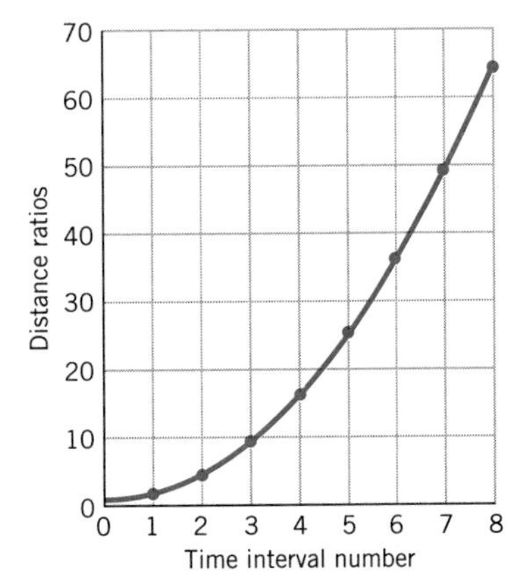

Figure 6: A plot of the total distance traveled in successive time intervals is a parabola.

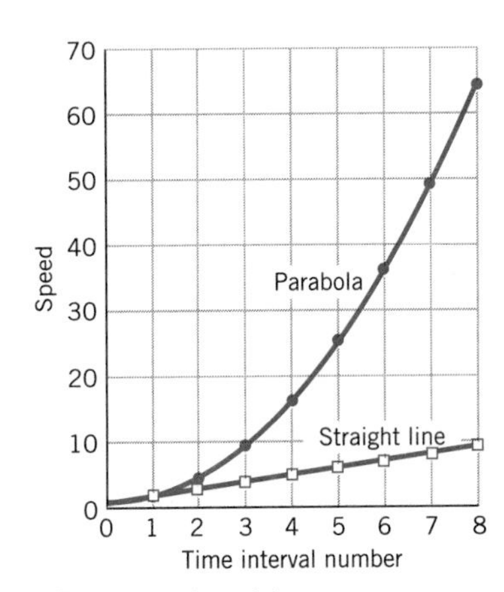

Figure 7: In the plot of the average speed in successive time intervals a parabola or a straight line?

impact, measured, for example, by how much of a depression the object makes in the dirt when it hits. These estimates were misleading. The depth of the impact crater increased like a parabola with the total distance fallen. At first Galileo assumed that the depth represented the object's speed when it hit the ground. The results seemed to confirm this, but eventually he guessed that impact depth is proportional to the square of speed. He later constructed a geometric argument confirming the second choice, showing that speed is proportional to the total time, rather than the distance, of free fall.

8 Formal Publication

Galileo conveyed his new understanding of the proportionality between distance and time squared only in personal correspondence and in teaching students. It was finally incorporated into Galileo's last book, *Two New Sciences*, half a century after legend claims he first dropped weights from the Leaning Tower of Pisa.

By then, Galileo had been discovered by the Roman Inquisition, which had ruled that his work bordered on heresy by entertaining the possibility that the earth moved around the sun. He had been sentenced to life imprisonment (commuted to home incarceration) and prohibited from further publication. *Two New Sciences*, dedicated to a French ambassador who smuggled the manuscript out of Italy, was printed in the Protestant Netherlands by the Elsevier family in 1638.

But Galileo's long process of redesigning experiments and reworking data analyses was not suitable for description in a publication of the time. *Two New Sciences* had to present explanations formally derived from definitions and postulates. The work is in the form of a dialogue inspired by the writings of Plato, and includes discussions of long formal derivations by the characters. The proportionality between distance in free fall and the square of the falling time is proven geometrically as a theorem. The empirical data, through which Galileo first came to identify this law, is not mentioned.

It is easier to understand the classical presentation of *Two New Sciences* than to understand the process by which Galileo derived these results. The former is a treatise similar to the work of Euclid and Plato, while the latter represents a whole new way of thinking about motion. Galileo's actual paths of learning included stops, new starts, interconnections among phenomena and mathematical analogy.

9 Conclusion

While Galileo had described "how" things fall, first by direct measurement and then abstractly with the time-squared rule, he did not aim to explain "why"

Figure 8: Participants in Galileo's *Dialogo sopra i due Massimi Sistemi del mondo Tolemaico e Copernicano* published in Florence in 1632. By permission of the Houghton Library, Harvard University.

things fall. Previous philosophers, in the tradition of Aristotle, regarded their task as finding causes for natural processes. Galileo diverged from this tradition in changing the question from "why do things fall" to "how do things fall". Galileo admits that he does not know the cause of free fall in an exchange between Simplicio (the Aristotelean) and Salviati (the speaker for Galileo) in the *Dialogue* which invoked the Inquisition's condemnation (Figure 8, [4]):

> Simplicio: The cause of this effect is well known; everyone is aware that it is gravity.
>
> Salviati: You are mistaken, Simplicio; what you ought to say is that everyone knows it is called 'gravity'. What I am asking for is not the name of the thing, but its essence, of which essence you know not a bit more than you know about the essence of whatever moves stars around... the name [*gravity*]... has become a familiar household word through the daily experience we have of it. But we do not really understand what principle or what force it is that moves stones downward, any more than we understand what moves them upward after they leave the thrower's hand, or what moves the moon around[9].

By revising experiments to make measurement practical, and analyzing those measurements with mathematics, Galileo composed the first accurate account of how things fall. He came to see numerical patterns in his data that he did not expect to see. These patterns were the first indication that terrestrial motions could be described by mathematics.

The outgrowth of that mathematical metaphor for nature is so tightly knitted into our contemporary physical understanding that it is almost impossible for us to see nature as freshly as he did. We translate physical quantities into algebraic abstractions; Galileo never did. By composing his measurements into ratios, he found a general relationship between distance and time. He worked with a few carefully made measurements; his work was not statistical. Galileo's method of learning from concrete, familiar examples, rather than from abstractions, may have something in common with that of students today.

Without knowing what he was going to find, or what methods might reveal it – and with the opposition of an authoritarian tradition that presumed to dictate how nature worked – Galileo extracted a regularity from falling motions that was new to understanding.

References

[1] *Aristotle's Physics*, W. D. Ross, trans., ed., (Oxford: Clarendon Press) 1966.

[2] L. N. H. Bunt, P. S. Jones, J. D. Bedient, *The Historical Roots of Elementary Mathematics*, (New York NY: Dover Pub., Inc.) 1988.

Elizabeth Cavicchi

[3] Elizabeth Cavicchi, "Halfway Through the Darkness: A Fieldwork Study of Time and Motions", Unpublished manuscript, May 1994.

[4] I. Bernard Cohen, *The Birth of a New Physics*, (New York NY: W. W. Norton & Co.) 1985.

[5] Bern Dibner, *Heralds of Science*, (Cambridge MA: MIT Press) 1969.

[6] Stillman Drake, *Galileo at Work: His Scientific Biography*, (Chicago IL: University of Chicago Press) 1978.

[7] Stillman Drake, *Galileo*, (New York, NY: Hill and Wang) 1980.

[8] Stillman Drake, "Galileo's physical measureents", *American Journal of Physics*, **54** 1985, 302-306.

[9] Stillman Drake, *Galileo: Pioneer Scientist*, (Toronto: University of Toronto Press) 1990.

[10] Howard Eves, *A Survey of Geometry*, (Boston MA: Allyn and Bacon, Inc.) 1972.

[11] Galileo Galilei, *Dialogue Concerning the Two Chief World Systems–Ptolemaic & Copernican*, Stillman Drake, trans., Albert Einstein, forward, (Berkeley CA: University of California Press) 1967.

[12] Galileo Galilei, *Discoveries and Opinions*, Stillman Drake, trans., (New York, NY: Doubleday) 1957.

[13] Galileo Galilei, *Two New Sciences Including Centers of Gravity and Force of Percussion*, Stillman Drake, trans., (Madison WI: University of Wisconsin Press) 1974.

[14] Morris Kline, *Mathematical Thought from Ancient to Modern Times*, (new York NY: Oxford University Press) 1972.

[15] A. Rupert Hall, *The Scientific Revolution 1500-1800: The Formation of the Modern Scientific Attitude*, (Boston MA: Beacon Press) 1966.

[16] David C. Lindberg, *The Beginnings of Western Science: The European Scientific Tradition in Philosophical, Religious, and Institutional Context, 600BC to AD 1450*, (Chicago IL: University of Chicago Press) 1992.

[17] George Sarton, *Ancient Science Through the Golden Age of Greece*, (New York NY: Dover Pub., Inc.) 1980.

[18] Thomas Settle, "Galileo and Early Experimentation" in *Springs of Scientific Creativity: Essays on Founders of Modern Science*, R Aris, H. T. Davis, R. H. Stuewer, eds., (Minneapolis MN: University of Minnesota Press) 1983.

[19] Thomas Settle, "An Experiment in the History of Science", *Science*, Jan. 1961, v. 133, 19-23.

These are the texts which contributed to this reading on free fall. You might find others. This essay closely follows the accounts and approach of Stillman Drake.

Acknowledgements

This essay was written at the request of Judy Clark and Linda Kime, to accompany students' work with the free fall experiment in their course, "Explorations in College Algebra". Linda Kime invited me to observe her teaching this course in fall 1993, and encouraged my interviews of her students. From Linda and her students I began noticing how involved learning can be, and its fascination for students and teacher. Hazel Garland, a student in Linda's class, volunteered to meet me regularly in 1994, to experimentally explore motion and learning. Hazel's spontaneous work and perceptive insights opened me to see the possibilities, uncertainties, questions, and delights that can develop when student and teacher explore their understandings together. My own teachers and advisors, Eleanor Duckworth and Philip Morrison, made possible the questioning and thought about learning and physics that began emerging through my work with Hazel and reading about Galileo. Their responses changed my understandings. Courtney Cazden encouraged this essay, as did Irene Hall, my classmate in Cazden's course. Judy Clark consistently supported my efforts. Phylis Morrison advised the preparation of figures. Alva Couch worked with me on figure design and document preparation and shared my excitement at learning.

Algebra Aerobics Answers

CHAPTER 1

Algebra Aerobics 1.1

1. a. Frequency Count (FC) for Age 1–20 interval is 38% of total: 38% of 137 = 0.38(137) = 52.
   FC for Age 61–80 interval is total FC minus all the others: 137 − (52 + 35 + 28) = 137 − 115 = 22.
   Relative Frequency (RF) for each interval is its FC divided by total:

   RF of (21–40) interval is: $\frac{35}{137} = 0.255 = 26\%$

   RF of (41–60) interval is: $\frac{28}{137} = 0.204 = 20\%$

   RF of (61–80) interval is: $\frac{22}{137} = 0.161 = 16\%$

   To convert decimal to percent, multiply by 100%.

   Table 1.3

Age	Frequency Count	Relative Frequency (%)
1–20	52	38
21–40	35	26
41–60	28	20
61–80	22	16
Total	137	100

   b. 20% + 16% = 36%

2.

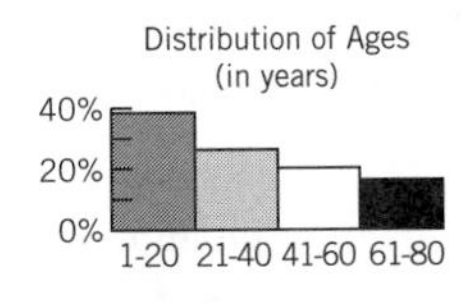

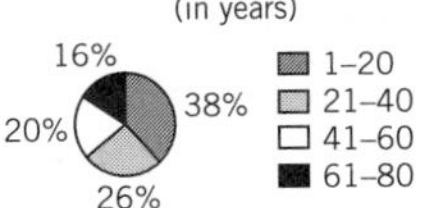

3. Table for the histogram in Figure 1.6

Age	Relative Frequency (%)	Frequency Count
1–20	20%	(0.20)(1352) = 270
21–40	35%	(0.35)(1352) = 473
41–60	30%	(0.30)(1352) = 406
61–80	15%	(0.15)(1352) = 203
Total	100%	1352

4. a. sum = $8750, so mean = $8750/9 = $972.22;
   median = $300

   b. sum = 4.7, so mean = 4.7/8 = 0.59;
   median = (0.4 + 0.5)/2 = 0.45

5. One of the values ($6,000) is much higher than the others which forces a high value for the mean. In cases like this, the median is generally a better choice for measuring central tendency.

6. a. 10% of $200 = $20; $200 − $20 = $180 (Sale price).
   Your discount card reduces that by 10%.
   10% of $180 = $18. → You pay $180 − $18 = $162.

   b. $200 − $162 = $38, so you save $38.

   c. $\frac{\$38}{200} = 0.19 = 19\%$, so the price is reduced by 19%.

Algebra Aerobics 1.2

1. a–d. Compare your topic sentences with someone else.

Algebra Aerobics 1.4

1. a. $36,575; 1989; (1989, $36575)

   b. $35,922; 1993; (1993, $35922)

   c. The median household income slightly increased between 1986 and 1989, then decreased greatly between 1989 and 1993, then continued to rise 1993–1996.

2. a. Square the value of x, then multiply that result by 3, then add that product to the opposite of x (x times -1) and add $+1$.

   b. (0, 1) is the only one of those ordered pairs that is a solution.

   c.

x	0	1	-1	2	-2	3	-3
y	1	3	5	11	15	25	31

3. a. Subtract 1 from the value of x, then square the result.

   b. (0, 1) and (1, 0) are the only ones of those ordered pairs that are solutions.

   c.

x	0	1	-1	2	-2	3	-3
y	1	0	4	1	9	4	16

Algebra Aerobics 1.5

1. a. Yes, it passes the vertical line test.

   b. As x approaches 0, the graph reaches the maximum value of the function.

2. Graphs (b) and (c) represent functions, while (a) and (d) do not represent functions since they fail the vertical line test.

3. Neither is a function of the other. The graph fails the vertical line test, so weight is not a function of height. At height of 51 there are two weights (115 and 200). If we reverse axes, so that weight is on the horizontal axis, and height is on the vertical axis, that graph will also fail the vertical line test for weight. 140 pounds has height of 58 inches and 60 inches.

4. a. D is a function of Y, since each value of Y determines a unique value of D.

   b. Y is not a function of D, since one value of D, \$2.7, yields two values for Y, 1993 and 1997.

Algebra Aerobics 1.6a

1. a. Tip = 15% of the cost of meal, $T = 0.15\text{ M}$
   Independent variable: M (meal price)
   Dependent variable: T (tip)

   The equation is a function since to each value of M there corresponds a unique value of T. The domain and range are all positive numbers.

   b. $T = (0.15)(\$8) = \$1.20$

   c. $T = (0.15)(\$26.42) = \$3.96$. One would probably round that up to \$4.00.

   d. The domain and the range are all the real numbers.

2. a. Since the dosage depends on the weight, the logical choice is W (expressed in kilograms) as the independent, and D (expressed in milligrams) as the dependent variable.

   b. In this formula, each value of W determines a unique dosage D, so D is a function of W.

   c. Values for the domain must be larger than 0 and less than 10 kilograms. So we have

   Domain = all values of W greater than 0 and less than 10 kilograms
   = all values of W (kg) with $0 < W < 10$

   As is often the case in the social sciences, the domain includes some questionable values (we don't expect a child to weigh close to 0 lb), but it spans all the appropriate ones.

   Range = all values of D (mg) greater than 0 to a maximum of up to $10 \cdot 50 = 500$ mg
   = all values of D with $0 < D < 500$

   d. The following table and graph are representations of the function. Note that (0, 0) and (10, 500) are calculated only to help us draw the graph and are not actually included in the model:

Daily Ampicillin Dosage by Weight

W	D
Child's Weight (kg)	Daily Maximum Dosage (mg)
0	0
1	50
2	100
3	150
4	200
5	250
6	300
7	350
8	400
9	450
10	500

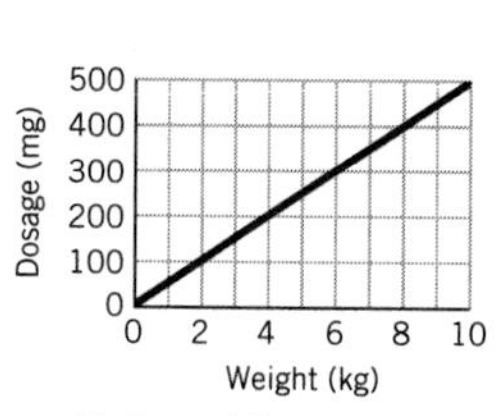

Maximum daily ampicillin dosage as a function of weight

   e. The domain and range are now all the real numbers. The table should now include negative values, and the graph will continue indefinitely in two directions.

3. a. The respiration rate (R) is not a function of the pulse rate (P), since a pulse rate of 75 determines two different respiration rates: 12 and 14.

   b. Pulse rate (P) is a function of respiration (R).

Algebra Aerobics 1.6b

1. $f(0) = (0)^2 - 5(0) + 6 = 6$, so $f(0) = 6$
   $f(1) = (1)^2 - 5(1) + 6 = 1 - 5 + 6 = 2$, so $f(1) = 2$
   $f(-3) = (-3)^2 - 5(-3) + 6 = 9 + 15 + 6 = 30$,
   $f(-3) = 30$

2. $2(x - 1) - 3(y + 5) = 10 \rightarrow 2x - 2 - 3y - 15 = 10 \rightarrow$ $2x - 17 - 3y = 10 \rightarrow 2x - 27 = 3y \rightarrow$ $\frac{1}{3}(2x - 27) = \frac{1}{3}(3y) \rightarrow \frac{2}{3}x - \frac{27}{3} = y$ or $y = \frac{2}{3}x - 9.$

So y is a function of x: $f(x) = \frac{2}{3}x - 9$ and the domain is all real numbers.

3. $x^2 + 2x - 3y + 4 = 0 \rightarrow x^2 + 2x + 4 = 3y \rightarrow$ $y = \frac{1}{3}(x^2 + 2x + 4)$. So y is a function of x: $f(x) = \frac{1}{3}(x^2 + 2x + 4)$ and its domain is all real numbers.

4. $7x - 2y = 5 \rightarrow -2y = -7x + 5 \rightarrow$ $\left(-\frac{1}{2}\right)(-2y) = \left(-\frac{1}{2}\right)(-7x + 5) \rightarrow y = \frac{7}{2}x - \frac{5}{2}$, a function of x, $f(x) = \frac{7}{2}x - \frac{5}{2}$, domain is all real numbers.

5. $2xy = 6 \rightarrow \left(\frac{1}{2x}\right)(2xy) = \left(\frac{1}{2x}\right)6 \rightarrow y = \frac{3}{x}$, a function of x, $f(x) = \frac{3}{x}$, domain is all real numbers except 0.

6. $\frac{x}{2} + \frac{y}{3} = 1 \rightarrow 6\left(\frac{x}{2}\right) + 6\left(\frac{y}{3}\right) = 6(1) \rightarrow 3x + 2y = 6$ $\rightarrow 2y = -3x + 6 \rightarrow \frac{1}{2}(2y) = \left(\frac{1}{2}\right)(-3x) + \frac{1}{2}(6) \rightarrow$ $y = -\frac{3}{2}x + 3$, a function of x, $f(x) = -\frac{3}{2}x + 3$ and its domain is all real numbers.

7. $f(-4) = 2; f(-1) = -1; f(0) = -2; f(3) = 1$
When $x = -2$ or $x = 2$, then $f(x) = 0$

8. $f(0) = 20, f(20) = 0$, if $x = 10$ or 30, $f(x) = 10$. It is a function because every value of x has only one value of $f(x)$.

CHAPTER 2

Algebra Aerobics 2.1

1. (143 − 135)lb/5 yr = 1.6 lb/yr

2. (871 − 581) \$trillion/(1997–1983) $= \frac{\$290 \text{ trillion}}{14 \text{ years}} \rightarrow$ \$20.71 trillion/year

3. a. (105,700 − 114,600) deaths/(1980–1970) $= \frac{-8900 \text{ deaths}}{10 \text{ years}}$ $\rightarrow -\frac{890 \text{ deaths}}{\text{yr}}$ or a decrease of 890 deaths per year.

b. (93,900 − 105,700) deaths/(1996–1980) $= -\frac{11{,}800 \text{ deaths}}{16 \text{ years}}$ $\rightarrow -\frac{737.5 \text{ deaths}}{\text{yr}}$ or a decrease of 737.5 deaths per year.

Algebra Aerobics 2.2

1. a. **World Population**

Year	Total Population (in millions)	Average Rate* of Change
1800	980	n.a.
1850	1260	(1260 − 980) m/(1850–1800) = 280 m/50 yrs → 5.6 m/yr
1900	1650	(1650 − 1260) m/(1900–1850) = 390 m/50 yrs → 7.8 m/yr
1950	2520	(2520 − 1650)m/(1950–1900) = 870 m/50 yrs → 17.4 m/yr
2000	6060	(6060 − 2520) m/(2000–1950) = 3540 m/50 yr → 70.8 m/yr
2050	8910	(8910 − 6060) m/(2050–2000) = 2850 m/50 yr → 57.0 m/yr

*m = million

1. b.

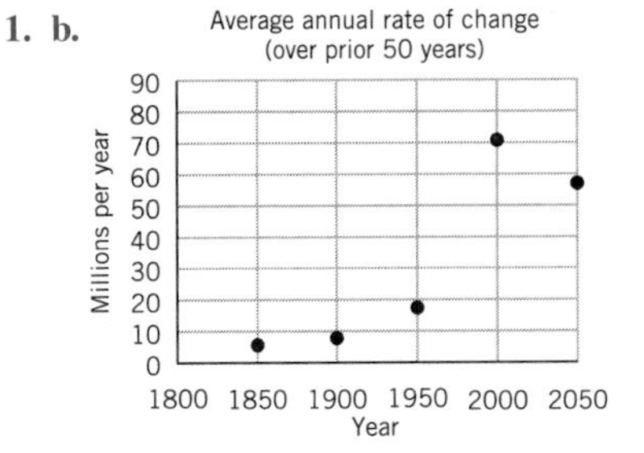

c. 1950–2000, average annual rate of change was 70.8 m/yr.

d. The average rate of change increased in the time interval 1800 to 2000, then despite a projected increase in world population from 2000 to 2050, the rate of change is projected to decrease to 57.0 m/yr.

Algebra Aerobics 2.3

1. a. i

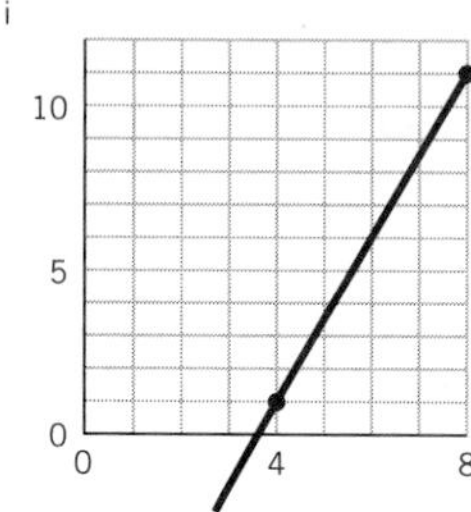

ii

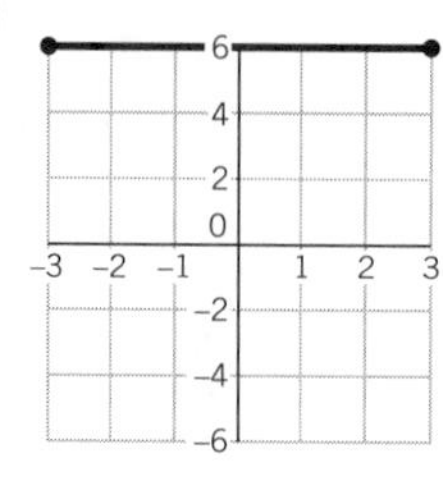

iii

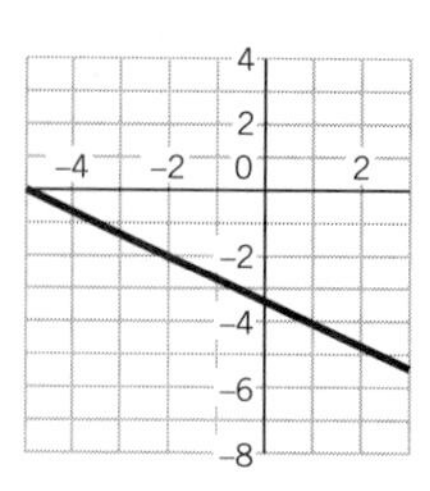

$m = (11 - 1)/(8 - 4) = 10/4$
$= 5/2$

$m = (6 - 6)/[2 - (-3)]$
$= 0$

$m = [-3 - (-1)]/[0 - (-5)]$
$= -2/5 = -\frac{2}{5}$

b. $m = (1 - 11)/(4 - 8) = -10/-4$
$= 5/2$

$m = (6 - 6)/(-3 - 2)$
$= 0$

$m = [-1 - (-3)]/(-5 - 0)$
$= 2/-5 = -\frac{2}{5}$

2. 1986–1987

3. positive: 1987–1990, 1991–1992, 1994–1995
negative: 1992–1994, 1995–1996
zero: 1990–1991

Algebra Aerobics 2.4

1. a. From 1911–1920 to 1921–1930, there was a slight decline in immigration population of 1,629,000.

b. From 1931–1940 to 1981–1990, there was a dramatic increase in immigration population of 6,810,000.

Algebra Aerobics 2.5

1. a. (2, 10) satisfies the equation because $7.0 + 1.5(2) = 10$.
b. (5, 13.5) does not satisfy the equation because $7.0 + 1.5(5) = 7.0 + 7.5 = 14.5$ and does not $= 13.5$.
c. $(-1.3, 5.05)$ satisfies the equation because $7.0 + 1.5(-1.3) = 7.0 - 1.95 = 5.05$.

2. The weight of a 4.5 month old baby girl appears to be ~ 14 lbs. From the equation, the exact weight is $W = 7.0 + 1.5(4.5) = 13.75$ lb. Our estimate is within 0.25 lb of the exact weight.

3. The age appears to be ~ 2.5 months. Solve $11 = 7 + 1.5A \rightarrow 11 - 7 = 1.5A \rightarrow 4 = 1.5A \rightarrow$
$\frac{4}{1.5} = A \rightarrow A = 2.67$ months.

4. One choice of points is: (2, 10) and (4, 13).
$$m = \frac{13 - 10}{4 - 2} = \frac{3}{2} = 1.5$$

5. If $A = -2 \rightarrow W = 7 + 1.5(-2)$
$= 7 - 3$
$= 4 \rightarrow (-2, 4)$ is a solution.
If $A = 100 \rightarrow W = 7 + 1.5(100)$
$= 7 + 150$
$= 157 \rightarrow (100, 157)$ is a solution.

6. Expect slope to be 1.5.
The slope between $(-2, 4)$ and $(100, 157)$ is:
$$m = \frac{157 - 4}{100 - (-2)} = \frac{153}{102} = 1.5$$

7. a. The unit for 15 is dollars/person.
The unit for 10 is dollars.

b. $\text{dollars} = \left(\frac{\text{dollars}}{\text{person}}\right) \text{persons} + \text{dollars}$

8. No answers required.

Algebra Aerobics 2.6a

1. a. $m = 5, b = 3$

b. $m = 3, b = 5$

c. $m = 5, b = 0$

d. $m = 0, b = 3$

e. $m = -1, b = 7.0$

f. $m = -11, b = 10$

g. $m = -\frac{2}{3}, b = 1$

h. $2y + 6 = 10x \rightarrow$
$2y = 10x - 6 \rightarrow y = 5x - 3;$
$m = 5, b = -3$

2. a. $f(x)$ is a linear function because it is represented by an equation of the form $y = mx + b$, where $b = 50$ and $m = -25$.

b. $f(0) = 50 - 25 \cdot (0) = 50 - 0 = 50$
$f(2) = 50 - 25 \cdot (2) = 50 - 50 = 0$

c. Since the line passes through (0, 50) and (2, 0), the slope is $\frac{50-0}{0-2} = \frac{50}{-2} = -25$.

Algebra Aerobics 2.6b

1. $y = 1.2x - 4$

x	y
−6	−11.2
−3	−7.6
0	−4.0
3	−0.4
6	3.2

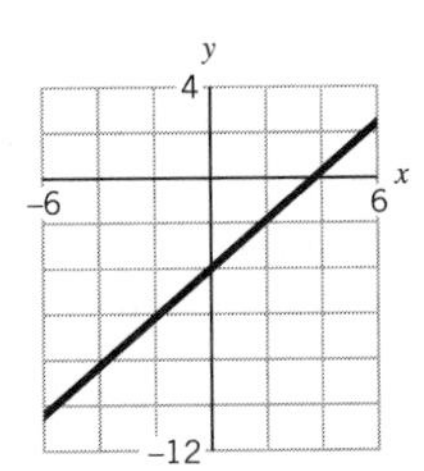

2. $y = 300 - 400x$

x	y
−4	1900
−2	1100
0	300
2	−500
4	−1300

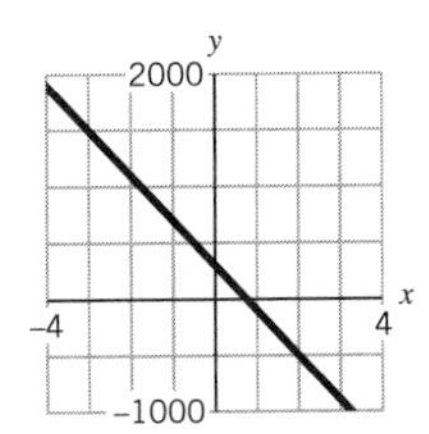

3. The vertical intercept is at (0, 1), so $b = 1$. The line passes through (0, 1) and (3, −5), so $m = \frac{-5-1}{3-0} = \frac{-6}{3} = -2$. So, the equation is: $y = -2x + 1$.

4. a. $S = \$12{,}000 + \$3{,}000x$

b.

Year	$ Salary
0	12,000
1	15,000
3	21,000
5	27,000
7	33,000

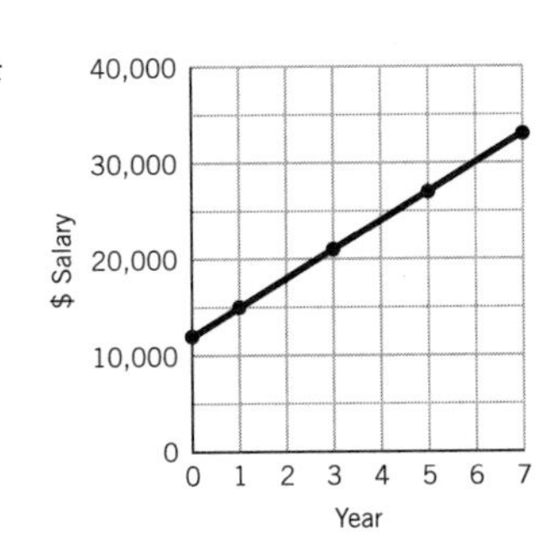

5. a.

15
10
5
0
$ Hourly wage
(13, 13.30)
(10, 8.50)
(8, 5.3)
0 5 10 15
Years of education

b. Yes. Calculate the rate of change between each pair of points:

$\frac{(8.5-5.3)}{(10-8)} = \frac{3.2}{2} \rightarrow 1.6$ $\quad \frac{(13.3-8.5)}{(13-10)} = \frac{4.8}{3} \rightarrow 1.6$

$\frac{(13.3-5.3)}{(13-8)} = \frac{8}{5} \rightarrow 1.6$ $\quad$ So all points lie on line with slope 1.6.

c. Use (8, 5.3) so $x = 8$, $y = 5.3$, $m = 1.6$ in $y = mx + b$, and solve for b.
$5.3 = (1.6)\,8 + b \rightarrow 5.3 = 12.8 + b \rightarrow -7.5 = b$, so equation is $y = 1.6x - 7.5$.

6. No answers required.

Algebra Aerobics 2.7a

1. If (0, 0) is on each line, then $b = 0$, in $y = mx + b$.

a. If $m = -1$, $b = 0$, the equation is: $y = (-1)x + 0 \rightarrow y = -x$.

b. If $m = 0.5$, $b = 0$, the equation is: $y = 0.5x + 0 \rightarrow y = 0.5x$.

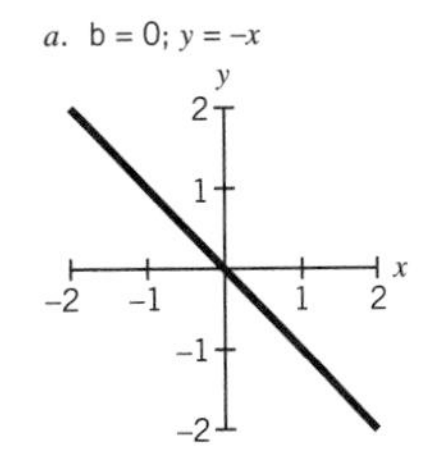

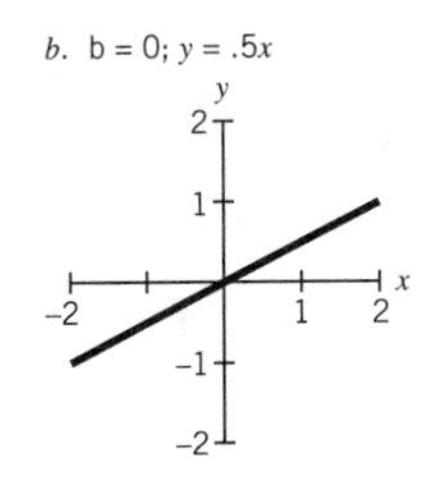

2. a. The variables x and y are directly proportional; the equation is $y = -3x$.

b. The variables x and y are not directly proportional; the equation is $y = 3x + 5$.

3. D–Deutsch marks; U–U.S. dollars

a. $D = mU + b$; $b = 0$, $m = \frac{1.45}{1} \rightarrow D = 1.45\,U$

b. $D = 1.45\,(U - 2.50) \rightarrow D = 1.45U - 3.63$

c. (a) does because the number of Deutschmarks you receive is directly proportional to the number of American dollars you convert.

Algebra Aerobics 2.7b

1. a. $y = -5$; b. $y = -3$; c. $y = 5$; d. $y = 3$

2. a. $x = 3$; b. $x = 5$; c. $x = -3$; d. $x = -5$

3. a. $y = -7$ b. $x = -4.3$

4. Slope is -1, y intercept is 0, so $m = -1, b = 0 \rightarrow y = -x$.

5. $m = 358.9, x = 4, y = 1{,}000$ in $y = mx + b$, to solve for b. $\rightarrow 1000 = (358.9)(4) + b \rightarrow 1000 = 1{,}435.6 + b \rightarrow b = -435.6$. So, the equation is: $y = 358.9x - 435$.

6. **a.** $m = -\dfrac{1}{(-3)} \rightarrow m = 1/3$; **b.** $m = -\dfrac{1}{1} \rightarrow m = -1$;

c. $m = -\dfrac{1}{3.1} \rightarrow (-1) \div (3.1) = -0.32$

d. $m = -\dfrac{1}{(-3/5)} \rightarrow (-1) \div \left(-\dfrac{3}{5}\right) \rightarrow (-1)\left(-\dfrac{5}{3}\right) \rightarrow$ $5/3$. So, $m = 5/3$.

7. **a.** Slope of $y = 2x - 4$ is 2, so line perpendicular to it has slope $-1/2$ or -0.5. Since it passes through $(3, -5)$, $x = 3$ when $y = -5$. So, $y = mx + b$ is: $-5 = (-0.5)(3) + b$. Solve it for $b \rightarrow -5 = -1.5 + b \rightarrow b = -3.5$. So, equation is: $y = -0.5x - 3.5$.

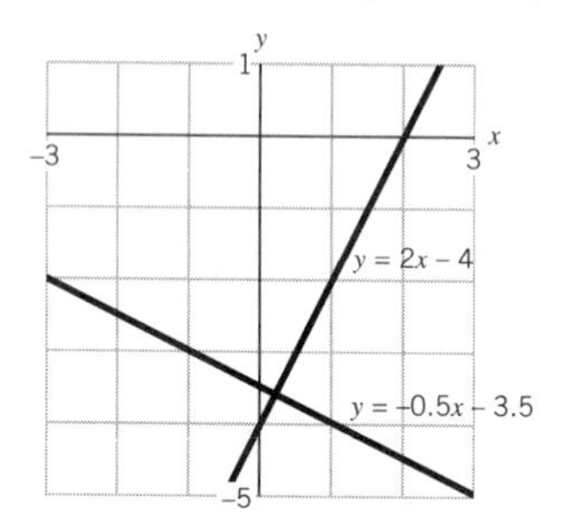

b. Any line parallel to $y = -0.5x - 3.5$, but not equal to it, will be perpendicular to $y = 2x - 4$, but will not pass through $(3, -5)$. They have same m, (-0.5), but different values of b, in $y = mx + b$. Two examples are $y = -0.5x$ or $y = -0.5x - 7.5$.

c. The three lines are parallel.

Algebra Aerobics 2.8

1. **a.**

Population in thousands
3000
2000
1000
0
1980
1990
2000
Year

b. The estimated coordinates of two points on the line are (1980, 1250) and (1995, 2800) where the first coordinate is the year and the second coordinate is the number of students in the thousands. The approximate slope of the line is:

$$\frac{(2800 - 1250)}{(1995 - 1980)} = \frac{1550}{15} \rightarrow \sim 103 \text{ thousand students/yr}$$

c. If x = the number of years *since* 1980 and y = the number (in thousands) of U.S. students 35 or older, then (1980, 1250) becomes the point (0, 1250), the vertical intercept. The point (1995, 2800) becomes (15, 2800).

d. Using the values of x and y from part c, $b = 1250$, $m = 103$, the equation is: $y = 103x + 1250$, where x = number of years since 1980; y = number of students (in thousands) 35 years or older.

e. Each year between 1980 and 1998, approximately 103 thousand (or 103,000) more people 35 or older enrolled in school. Each *decade* 1030 thousand (or 1.03 million) more people 35 or older became students.

CHAPTER 3

Algebra Aerobics 3.1

1. Gas is the cheapest system from approximately 17.5 years of operation to approximately 32.5 years of operation. Solar becomes the cheapest system after approximately 32.5 years of operation.

2. **a.** $(3, -1)$ is a solution since: $5(3) + 2(-1) = 15 - 2 = 13$ and $4(3) + 3(-1) = 12 - 3 = 9$. It is a solution of both equations.

b. $(1, 4)$ is not a solution since: $5(1) + 2(4) = 5 + 8 = 13$ but $4(1) + 3(4) = 4 + 12 = 16 \neq 9$, so it is not a solution of $4x + 3y = 9$.

3. **a.** $12x - 9y = 18$ is equivalent to $4x = 6 + 3y$ because:

given	$12x - 9y = 18$
divide by 3	$4x - 3y = 6$
add $3y$ to each side of equation	$4x = 6 + 3y$

Thus, the equations are equivalent.

b. There are an infinite number of solutions to the system of equations in part (a).

Algebra Aerobics 3.2a

1. **a.** $y = y \rightarrow x + 4 = -2x + 7 \rightarrow 3x = 3 \rightarrow x = 1$; $y = (1) + 4 = 5$
Solution (x, y) is $(1, 5)$.

b. $y = y \rightarrow -1700 + 2100x = 4700 + 1300x \rightarrow$ $800x = 6400 \rightarrow x = \dfrac{6400}{800} = 8$;
$y = 4700 + 1300(8) = 4700 + 10{,}400 = 15{,}100$
Solution (x, y) is $(8, 15100)$.

c. $F = F \rightarrow C = 32 + \left(\dfrac{9}{5}\right)C \rightarrow \dfrac{5}{5}C - \dfrac{9}{5}C = 32 \rightarrow$ $-\dfrac{4}{5}C = 32 \rightarrow \left(-\dfrac{5}{4}\right)\left(-\dfrac{4}{5}\right)C = \left(-\dfrac{5}{4}\right)(32) \rightarrow$
$C = (-5)(8) = -40$; $F = C \rightarrow F = -40$
Solution (C, F) is $(-40, -40)$.

2. a. Substitute $y = x + 3$ into $5y - 2x = 21 \rightarrow$ $5(x + 3) - 2x = 21 \rightarrow 5x + 15 - 2x = 21 \rightarrow$ $3x = 6 \rightarrow x = 2$; $y = (2) + 3 = 5$. Solution (x, y) is $(2, 5)$.

   b. Substitute $z = 3w + 1$ into $9w + 4z = 11 \rightarrow$ $9w + 4(3w + 1) = 11 \rightarrow 9w + 12w + 4 = 11 \rightarrow$ $21w = 7 \rightarrow \frac{21w}{21} = \frac{7}{21} \rightarrow w = \frac{1}{3}$; $z = 3\left(\frac{1}{3}\right) + 1 =$ $1 + 1 = 2$. Solution (w, z) is $\left(\frac{1}{3}, 2\right)$.

   c. Substitute $x = 2y - 5$ into $4y - 3x = 9 \rightarrow$ $4y - 3(2y - 5) = 9 \rightarrow 4y - 6y + 15 = 9 \rightarrow$ $-2y = -6 \rightarrow y = 3$; $x = 2(3) - 5 = 6 - 5 = 1$ Solution (x, y) is $(1, 3)$.

   d. Solve $r - 2S = 5$ for r, and substitute the resulting expression for r into $3r - 10S = 13$. $r - 2S = 5 \rightarrow$ $r = 2S + 5 \rightarrow 3(2S + 5) - 10S = 13 \rightarrow$ $6S + 15 - 10S = 13 \rightarrow -4S = -2 \rightarrow S = 1/2$ $r = 2(1/2) + 5 = 1 + 5 = 6$. Solution (r, S) is $(6, 1/2)$.

Algebra Aerobics 3.2b

1. a. By elimination:
$$\begin{array}{rl} 2y - 5x &= -1 \\ +\,3y + 5x &= 11 \\ \hline 5y &= 10 \end{array}$$
$\rightarrow y = 2$; $3(2) + 5x = 11$ $\rightarrow 6 + 5x = 11 \rightarrow 5x = 5 \rightarrow x = 1$. Solution (x, y) is $(1, 2)$.

   b. Use elimination. Multiply $(3x + 2y = 16)$ by 3; and $(2x - 3y = -11)$ by 2.
$$\begin{array}{rl} 3 \cdot (3x + 2y) = 3 \cdot (16) \rightarrow & 9x + 6y = 48 \\ 2 \cdot (2x - 3y) = 2 \cdot (-11) \rightarrow & 4x - 6y = -22 \\ \hline & 13x = 26 \rightarrow x = 2; \end{array}$$
$3(2) + 2y = 16 \rightarrow 6 + 2y = 16 \rightarrow 2y = 10 \rightarrow y = 5$. Solution (x, y) is $(2, 5)$.

   c. By substitution of $t = 3r - 4$ into $4t + 6 = 7r \rightarrow$ $4(3r - 4) + 6 = 7r \rightarrow 12r - 16 + 6 = 7r \rightarrow$ $-10 = -5r \rightarrow 2 = r$; $t = 3(2) - 4 = 6 - 4 = 2$. Solution (r, t) is $(2, 2)$.

   d. Substitute $z = z \rightarrow 2000 + 0.4[x - 10{,}000] = 800 + 0.2x$ $\rightarrow 2000 + 0.4x - 4{,}000 = 800 + 0.2x \rightarrow$ $-2000 + 0.4x = 800 + 0.2x \rightarrow 0.4x - 0.2x =$ $800 + 2000 \rightarrow 0.2x = 2800 \rightarrow x = \frac{2800}{0.2} = 14{,}000$; $z = 800 + 0.2[14{,}000] = 800 + 2800 = 3600$. Solution (x, z) is $(14000, 3600)$.

2. a. By substitution of $(y = -6x + 4)$ into $(5y + 30x = 20)$ $\rightarrow 5(-6x + 4) + 30x = 20 \rightarrow -30x + 20 + 30x = 20$ $\rightarrow 20 = 20$. So, both equations must be equal. To verify, solve $5y + 30x = 20$ for y: $\rightarrow 5y = -30x + 20 \rightarrow$ $\frac{1}{5}(5y) = \frac{1}{5}(-30x + 20) \rightarrow y = -6x + 4$. Infinitely many solutions since both equations describe same line.

2. a.

2. b.

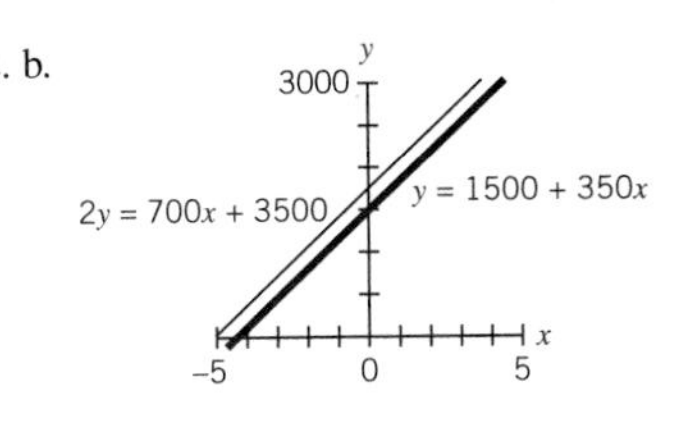

2. c.

   b. $2y = 700x + 3500 \rightarrow y = \frac{1}{2}(700x + 3500)$ $= 350x + 1750$. The slopes are both 350, but the y-intercepts are different (1500 and 1750), so the lines are parallel. There is no solution.

   c. By substitution of $y = y \rightarrow 2x + 4 = -x + 4 \rightarrow 3x = 0$ $\rightarrow x = 0$. $y = -(0) + 4 = 4$. So solution (x, y) is $(0, 4)$.

3. In order for a system of equations to have no solutions, they must produce parallel lines with the same slope, but different y-intercepts. One example is: $y = 5x + 10$; $y = 5x - 3$.

4. Simplify $\frac{x}{2} + \frac{y}{3} = 3$, by multiplying both sides by 6. $6\left(\frac{x}{2}\right) + 6\left(\frac{y}{3}\right) = 6(3) \rightarrow 3x + 2y = 18$; substitute $y = x + 4 \rightarrow 3x + 2(x + 4) = 18 \rightarrow 3x + 2x + 8 = 18 \rightarrow$ $5x = 10 \rightarrow x = 2$; $y = (2) + 4 = 6$. Solution (x, y) is $(2, 6)$.

Algebra Aerobics 3.4a

1. a. i. flat tax ii. flat tax iii. graduated tax

   b. i. $f(\$5000) = 0.15(\$5000) = \$750$
   $g(\$5000) = 0.10(\$5000) = \$500$
   $f(\$5000)$ is \$250 larger than $g(\$5000)$

   ii. $f(\$15{,}000) = 0.15(\$15{,}000) = \$2250$
   $g(\$15{,}000) = \$1000 + 0.20(\$15{,}000 - \$10{,}000)$
   $= \$1000 + 0.20(\$5000)$
   $= \$2000$
   $f(\$15{,}000)$ is \$250 larger than $g(\$15{,}000)$

   iii. $f(\$40{,}000) = 0.15(\$40{,}000) = \$6000$
   $g(\$40{,}000) = \$1000 + 0.20(\$40{,}000 - \$10{,}000)$
   $= \$1000 + 0.20(\$30{,}000)$
   $= \$1000 + \$6000$
   $= \$7000$
   $g(\$40{,}000)$ is \$1000 larger than $f(\$40{,}000)$

2. a. $g(i) = \begin{cases} 0.05i & \text{for } i \le \$50{,}000 \\ 2500 + 0.08(i - 50{,}000) & \text{for } i > \$50{,}000 \end{cases}$
   since 5% of \$50,000 = .05(\$50,000) = \$2500.

   b. $g(i) = \begin{cases} 0.06i & \text{for } i \le \$30{,}000 \\ 1800 + 0.09(i - 30{,}000) & \text{for } i > \$30{,}000 \end{cases}$
   since 6% of \$30,000 = .06(\$30,000) = \$1800.

Algebra Aerobics 3.4b

1. a. The estimated point of intersection from Fig. 3.19 is (58000,3500). At the point of intersection, flat tax = graduated tax. Since \$58,000 income occurs in the second segment, for values of i: $\$50,200 \le i \le \$90,000$, use $G(i) = \$2761 + 0.088\,(i - \$50,200)$.

$$T_F = T_G$$
$$0.0595i = 2761 + 0.088i - 4417.6$$
$$0.0595i = 0.088i - \$1{,}656.6$$
$$-0.0285i = -\$1{,}656.6$$
$$i = 58{,}126.315;\ P\,(\$58{,}126.315) = \$3{,}458.516$$

So: the more accurate point of intersection is calculated by the equations as: (58126.315, 3458.516). An income of \$58,126.32 is taxed by both systems as \$3,458.52.

b. i. The flat tax is slightly higher than graduated tax for incomes less than \$58,126.32, which are to the left of the intersection point.
ii. The flat tax is significantly less than graduated tax for incomes greater than \$58,126.32, which are to the right of the intersection point.

2. a. $f(30{,}000) = 0.0595\,(30{,}000) = \$1785$ $\quad g(30{,}000) = 0.055\,(30{,}000) = \$1650$
Flat tax is \$135 higher for \$30,000 income.

b. $f(60{,}000) = 0.0595\,(60{,}000) = \$3570$

$$g(60{,}000) = 2761 + 0.088\,(60{,}000 - 50{,}200) = 2761 + 0.088\,(9{,}800) = 2761 + 862.40 = \$3623.40$$

Graduated tax is \$53.40 higher for \$60,000 income.

c. $f(120{,}000) = 0.0595(120{,}000) = \$7{,}140$

$$g(120{,}000) = 6263 + 0.098(120{,}000 - 90{,}000) = 6263 + 2940 = \$9{,}203$$

Graduated tax is \$2063 higher for \$120,000 income.

CHAPTER 4

Algebra Aerobics 4.1

1. a. To express 10 billion as a power of ten, look at 1.0, and count the ten place values the decimal must be moved to the right, in order to produce 10 billion.
1.0000000000 = 10,000,000,000 so it is 10^{10}.

b. The decimal point in 1.0 must be moved 14 place values to the left to produce 0.000 000 000 000 01;
000 000 000 000 01, so it is 10^{-14}.

2. a. 10^{-8} = .00000001.0 = 0.00000001

b. 10^{13} = 1.0000000000000 = 10,000,000,000,000

3. a. 10^{-9} or 0.000 000 001 sec b. 10^{-1} or 0.1 m
c. 10^9 or 1,000,000,000 bytes (a byte is a term used to describe a unit of computer memory).

4. a. $7 \cdot 10^{-2}$ or 0.07 m; b. $9 \cdot 10^{-3}$ or 0.009 m;
c. $5 \cdot 10^3$ or 5000 m

5. 602,000,000,000,000,000,000,000

6. $3.84 \cdot 10^8$ m

7. $1 \cdot 10^{-8}$ cm

8. 0.000 000 002 m

9. a. $-705{,}000{,}000$; b. $-0.000\,040\,3$

10. a. $-4.3 \cdot 10^7$; b. $-8.3 \cdot 10^{-6}$

Algebra Aerobics 4.2a

1. a. $10^{5+7} = 10^{12}$
b. $8^{6+14} = 8^{20}$
c. $z^{5+4} = z^9$
d. Cannot be simplified because bases are different.
e. $2 \cdot 7^3$

2. a. $10^{15-7} = 10^8$
b. $8^{6-4} = 8^2$
c. $3^{5-4} = 3^1$ or 3
d. Cannot be simplified because bases are different.

3. a. $10^{4(5)} = 10^{20}$
b. $7^{2(3)} = 7^6$
c. $x^{4(5)} = x^{20}$
d. 2^4x^4 or $16x^4$
e. $2^3a^{4(3)} = 8a^{12}$
f. $(-2)^3a^3 = -8a^3$

4. a. $\left(-\dfrac{4}{5}\right)\left(-\dfrac{4}{5}\right) = \dfrac{16}{25}$

b. $\left(\dfrac{-2x}{4y}\right)\left(\dfrac{-2x}{4y}\right)\left(\dfrac{-2x}{4y}\right) = \dfrac{-8x^3}{64y^3} = \dfrac{-x^3}{8y^3}$

c. $(-5)(-5) = 25$ d. $-(5)(5) = -25$

5. (capacity of hard drive)/(capacity of diskette) = $(2 \cdot 10^9 \text{ bytes})/(1.44 \cdot 10^6 \text{ bytes}) = \dfrac{2}{1.44} \cdot \dfrac{10^9}{10^6} = 1.4 \cdot 10^3 = 1400$ diskettes.

Algebra Aerobics 4.2b

1. a. $(3 \cdot 10^{-4})(4 \cdot 10^7) = 12 \cdot 10^3 = 1.2 \cdot 10^4 \approx 1 \cdot 10^4$

b. $\dfrac{(5 \cdot 10^5)(2 \cdot 10^6)}{4 \cdot 10^{-3}} = \dfrac{5 \cdot 2}{4} \cdot \dfrac{10^5 10^6}{10^{-3}} = \dfrac{10}{4} \cdot \dfrac{10^{11}}{10^{-3}} = 2.5 \cdot 10^{11} \cdot 10^3 = 2.5 \cdot 10^{14} \approx 3 \cdot 10^{14}$

2. Use 3.14 to approximate π.

a. Surface area of Jupiter $= 4\pi r^2 \approx 4\pi(7.14 \cdot 10^4 \text{ km})^2$
$= 4\pi(7.14)^2(10^4)^2 \text{km}^2$
$\approx 4(3.14)(50.98)\ 10^8 \text{ km}^2$
$\approx 641 \cdot 10^8 \text{ km}^2$
$\approx 6.41 \cdot 10^2 \cdot 10^8 \text{ km}^2$
$\approx 6.41 \cdot 10^{10} \text{ km}^2$

b. Volume of Jupiter $= \left(\frac{4}{3}\right)\pi r^3$
$\approx (1.3)(3.14)(7.14 \cdot 10^4 \text{ km})^3$
$\approx (4.18)(7.14)^3(10^4)^3 \text{ km}^3$
$\approx (4.18)(364)10^{12} \text{ km}^3$
$\approx 1522 \cdot 10^{12} \text{ km}^3 \cdot$
$\approx 1.522 \cdot 10^3 \cdot 10^{12} \text{ km}^3$
$\approx 1.52 \cdot 10^{15} \text{ km}^3$

3. According to Example 8, the ratio of people per square mile of farmable land is $\frac{6 \cdot 10^9 \text{ people}}{12 \cdot 10^6 \text{ sq. mi.}}$. If only 3/7 of the farmable land is used, the people/sq. mi. of used farmland is:

$$\frac{6 \cdot 10^9 \text{ people}}{(3/7)(12)10^6 \text{ sq. mi.}} = \left(\frac{7}{3}\right)\frac{6 \cdot 10^9 \text{ people}}{12 \cdot 10^6 \text{ sq. mi.}}$$
$$= \frac{7}{3}(500) \text{ people/sq. mi.}$$
$$\approx 1166.7 \text{ people/sq. mi.}$$

Whenever the denominator of a fraction is decreased, the value of that fraction is increased. So, one expects this ratio to be larger than the ratio of people to farmable land.

Algebra Aerobics 4.3

1. a. $10^{5-7} = 10^{-2} = \frac{1}{10^2}$

b. $11^{6-(-4)} = 11^{6+4} = 11^{10}$

c. $3^{-5-(-4)} = 3^{-5+4} = 3^{-1} = 1/3$

d. Cannot be simplified, different bases

e. $7^{3-3} = 7^0 = 1$

f. $a^{-2-3} = a^{-5} = \frac{1}{a^5}$

2. The time for a TV signal to travel across the United States =
(time to travel 1 kilometer) · (kilometers) =
$(3.3 \cdot 10^{-6})$ sec/km $\cdot (4.3 \cdot 10^3)$ km =
$(3.3 \cdot 4.3) \cdot (10^{-6} \cdot 10^3)$ sec $\approx$
$14 \cdot 10^{-3}$ sec $\approx$
$1.4 \cdot 10 \cdot 10^{-3}$ sec $\approx 1.4 \cdot 10^{-2}$ sec. or 0.014 sec.
So it would take less than 2 hundredths of a second for the signal to cross the United States.

3. $x^{-2}(x^5 + x^{-6}) = x^{-2}x^5 + x^{-2}x^{-6}$
$= x^{(-2+5)} + x^{(-2-6)}$
$= x^3 + x^{-8}$
$= x^3 + \frac{1}{x^8}$

4. a. $10^{(4)(-5)} = 10^{-20} = \frac{1}{10^{20}}$

b. $7^{(-2)(-3)} = 7^6$

c. $\frac{1}{(2a^3)^2} = \frac{1}{4a^6}$ or $2^{-2}a^{3(-2)} = \frac{1}{4}a^{-6}$

d. $\left(\frac{8}{x}\right)^{-2} = \left(\frac{x}{8}\right)^2 = \frac{x^2}{64}$

5. a. $\frac{t^{-3}(1)}{t^{-12}} = t^{-3-(12)} = t^{-3+12} = t^9$

b. $\frac{V^{-3}W^7}{V^{-6}W^{-10}} = V^{-3-(-6)}W^{7-(-10)} = V^{-3+6}W^{7+10}$
$= V^3W^{17}$

Algebra Aerobics 4.4

1. 1 km = 1000 m, so the conversion factor from kilometers to meters is 1000 m/1 km.

Hence $7.8 \cdot 10^8 \text{ km} \cdot \frac{1000 \text{ m}}{1 \text{ km}} = \frac{7.8 \cdot 10^8 \cdot 10^3 \text{ m}}{1}$
$= 7.8 \cdot 10^{11}$ m.

2. From Table 4.1 we have 1 km = 0.62 mi. So the conversion factor to convert from km to mi is 0.62 mi/1 km. Hence
$9.46 \cdot 10^{12} \text{km} \cdot \frac{0.62 \text{ mi}}{1 \text{ km}} \approx (9.46)(0.62)10^{12}\text{m} = 5.87 \cdot 10^{12}$ mi,
which is close to $5.88 \cdot 10^{12}$ mi.

3. 1 m = 100 cm, so the conversion factor for converting from cm to m is 1 m/100 cm.

So 1 Å $= 10^{-8} \text{ cm} \cdot \frac{1 \text{ m}}{100 \text{ cm}} = 10^{-8} \cdot \frac{1 \text{ m}}{10^2} = 10^{-8-2}$ m
$= 10^{-10}$ m.

4. 1 km = 0.62 mi, so the conversion factor for converting from km to miles is $\frac{0.62 \text{ mi}}{1 \text{ km}}$. So, $218 \text{ km} \cdot \frac{0.62 \text{ mi}}{1 \text{ km}} \approx 135$ miles.

5. If a dollar bill is 6 in. long → 2 dollar bills/12 in. = 2 dollars/ft.

The number of dollars needed to reach from Earth to the sun is: $93{,}000{,}000 \text{ miles} \cdot 5{,}280 \frac{\text{ft}}{\text{miles}} \cdot \frac{2 \text{ dollars}}{\text{ft}} =$
$9.3 \cdot 5.3 \cdot 2 \cdot 10^{7+3} \cdot 2$ dollars $\approx (29)(2)10^{7+10}$ dollars $\approx$
$99 \cdot 10^{10}$ dollars $\approx 9.9 \cdot 10 \cdot 10^{10}$ dollars $\approx 9.9 \cdot 10^{11}$ dollars or 990,000,000,000 dollar bills, almost a trillion dollars.

Algebra Aerobics 4.5a

1. a. $\sqrt{81} = 9$ b. $\sqrt{144} = 12$

2. a. $\sqrt{9}\,x^{1/2} = 3x^{1/2}$

b. $\sqrt{\dfrac{x^2}{25}} = \dfrac{\sqrt{x^2}}{\sqrt{25}} = \dfrac{x}{5}$ or $\dfrac{(x^2)^{1/2}}{\sqrt{25}} = \dfrac{x^1}{5} = \dfrac{x}{5}$

3. a. $S = \sqrt{30 \cdot 60} = \sqrt{1800} \approx 42$ mph

b. $S = \sqrt{30 \cdot 200} = \sqrt{6000} \approx 77$ mph

4. a. $\sqrt{25} < \sqrt{29} < \sqrt{36} \rightarrow 5 < \sqrt{29} < 6$; 5 and 6

b. $\sqrt{81} < \sqrt{92} < \sqrt{100} \rightarrow 9 < \sqrt{92} < 10$; 9 and 10

5. a. $\sqrt[3]{27} = 3$ b. $\sqrt[4]{16} = 2$ c. $\dfrac{1}{\sqrt[3]{8}} = \dfrac{1}{2}$

6. a. 6 b. 72

7. a. $(-27)^{1/3} = -3$ since $(-3)^3 = -27$

b. There is no real number solution to the 4th root of a negative number, since a negative or positive number, raised to the 4th power is always positive.

c. $(-1000)^{1/3} = -10$ since $(-10)^3 = -1000$

8. $V = 4/3\,\pi r^3 \rightarrow 3/4\,V = \pi r^3 \rightarrow \dfrac{3V}{4\pi} = r^3 \rightarrow$

$r = \sqrt[3]{\dfrac{3 \cdot 2 \text{ feet}^3}{4\pi}} = \sqrt[3]{\dfrac{3 \text{ feet}^3}{2\pi}} \approx \sqrt[3]{0.478 \text{ feet}^3} \approx 0.78$ feet.

We can express that with a more meaningful figure if we convert it into inches. Since 1 ft = 12 in., we can use a conversion factor of 1 = (12 in.)/(1 foot). The radius of the balloon is:

$(0.78 \text{ ft})\left(\dfrac{12 \text{ in.}}{1 \text{ ft}}\right) = 9.36$ in. or ~ 9.4 in.

Algebra Aerobics 4.5b

1. a. $2^{1/2}2^{1/3} = 2^{1/2+1/3} = 2^{5/6}$

b. $5^{1/2}5^{1/4} = 5^{3/4}$

c. $3^{1/2}9^{1/3} = 3^{1/2}(3^2)^{1/3} = 3^{1/2}3^{2/3} = 3^{7/6}$

2. a. $\dfrac{2^{1/2}}{2^{1/3}} = 2^{1/2-1/3} = 2^{1/6}$

b. $\dfrac{2^1}{2^{1/4}} = 2^{1-1/4} = 2^{3/4}$

c. $\dfrac{5^{1/4}}{5^{1/3}} = 5^{1/4-1/3} = 5^{-1/12}$ or $\dfrac{1}{5^{1/12}}$

3. a. $c = 17.1 \cdot (0.25)^{3/8}$ cm $= 17.1 \cdot (0.59)$ cm $= 10.2$ cm

b. $c = 17.1 \cdot (25)^{3/8}$ cm $= 17.1 \cdot (3.34)$ cm $= 57.2$ cm

Algebra Aerobics 4.6a

1. The Armenian earthquake had tremors one order of magnitude larger than those in Los Angeles.

2. The maximum tremor size of the Hawaiian earthquake of 1983 was 100 times smaller than the maximum tremor size of the largest earthquake.

3. Your salary is \$1,000,000, and Henry's is \$1,000.

4. a. Since $\dfrac{\text{radius of the Milky Way}}{\text{radius of the sun}} = \dfrac{10^{21}}{10^9} = 10^{21-9} = 10^{12}$, the radius of the Milky Way is 12 orders of magnitude larger than the radius of the sun, or equivalently the radius of the sun is 12 orders of magnitude smaller than the radius of the Milky Way.

b. Since $\dfrac{\text{radius of a proton}}{\text{radius of the hydrogen atom}} = \dfrac{10^{-15}}{10^{-11}} = 10^{-15-(-11)} = 10^{-4}$, then the radius of the proton is four orders of magnitude smaller than the radius of the hydrogen atom, or equivalently the radius of the hydrogen atom is four orders of magnitude larger than the radius of a proton.

Algebra Aerobics 4.6b

1. a. $1{,}000{,}000{,}000 \text{ m} = 10^9$ m

b. $0.000\,000\,7 \text{ m} = 7 \cdot 10^{-7} \text{ m} \approx 10 \cdot 10^{-7} \text{ m} = 10^{-6} \text{ m} \rightarrow$ plot it at 10^{-6} m

2. a. Plot at $10^7 \cdot 10^{-2} = 10^5$; b. Average height of humans

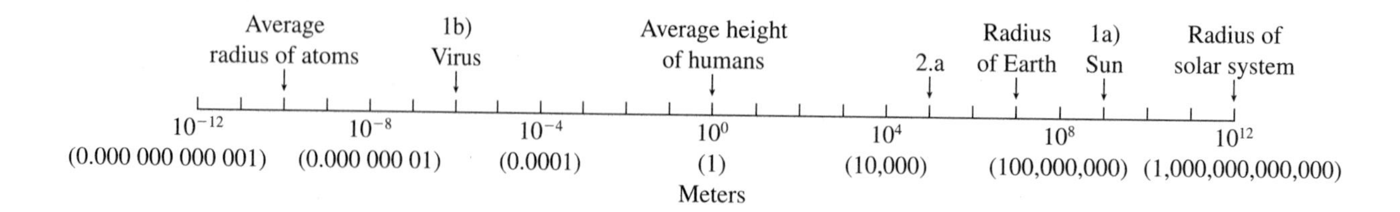

Algebra Aerobics 4.7a

1. a. Since $10{,}000{,}000 = 10^7$, $\log 10{,}000{,}000 = 7$.

   b. Since $0.000\,000\,1 = 10^{-7}$, $\log 0.000\,000\,1 = -7$.

   c. Since $10{,}000 = 10^4$, $\log 10{,}000 = 4$.

   d. Since $0.0001 = 10^{-4}$, $\log 0.0001 = -4$.

   e. Since $1{,}000 = 10^3$, $\log 1{,}000 = 3$.

   f. Since $0.001 = 10^{-3}$, $\log 0.001 = -3$.

   g. Since $1 = 10^0$, $\log 1 = 0$.

2. a. $100{,}000 = 10^5$; c. $10 = 10^1$

   b. $0.000\,000\,01 = 10^{(-8)}$; d. $0.01 = 10^{(-2)}$

Algebra Aerobics 4.7b

1. a. Since $10^{0.4} \approx 2.511886$ and $10^{0.5} \approx 3.162278$, then $\log 3 < 0.5$.
   A calculator gives $\log 3 \approx 0.477121255$.
   So $\log 3 \approx 0.48$

   b. Since $10^{0.7} \approx 5.011872$ and $10^{0.8} \approx 6.30957$, then $\log 6 < 0.8$.
   A calculator gives $\log 6 \approx 0.77815125$.
   So $\log 6 \approx 0.78$

   c. Since $10^{0.8} \approx 6.30957$, then $\log 6.37 \approx 0.8$.
   A calculator gives $\log 6.37 \approx 0.804139432$.

2. a. Write 3,000,000 in scientific notation — $3{,}000{,}000 = 3 \cdot 10^6$
   and substitute $10^{0.48}$ for 3 — $3{,}000{,}000 \approx 10^{0.48} \cdot 10^6$
   then combine powers — $3{,}000{,}000 \approx 10^{6.48}$
   and rewrite as a logarithm — $\log 3{,}000{,}000 \approx 6.48$

   A calculator gives $\log 3{,}000{,}000 \approx 6.477121255$.

   b. Write 0.006 in scientific notation — $0.006 = 6 \cdot 10^{-3}$
   and substitute $10^{0.78}$ for 6 — $0.006 \approx 10^{0.78} \cdot 10^{-3}$
   then combine powrs — $0.006 \approx 10^{0.78-3} \approx 10^{-2.22}$
   and rewrite as a logarithm — $\log 0.006 = -2.22$

   A calculator gives $\log 0.006 \approx -2.22184875$.

3. a. $0.000\,000\,7 = 10^{\log 0.0000007} \approx 10^{-6.1549}$

   b. $780{,}000{,}000 = 10^{\log 780{,}000{,}000} \approx 10^{8.892}$

   c. $0.0042 = 10^{\log 0.0042} \approx 10^{-2.3768}$

   d. $5{,}400{,}000{,}000 = 10^{\log (5{,}400{,}000{,}000)} = 10^{9.732}$

CHAPTER 5

Algebra Aerobics 5.1a

1. a. Initial Population = 350
   Growth Factor = 5

   b. Initial Population = 25000
   Growth Factor = 1.5

   c. Initial Population = 7000
   Growth Factor = 4

2. a. $y = 3000 \cdot 3^t$

   b. $y = (4 \cdot 10^7)(1.3)^t$

   c. $y = 75 \cdot 4^t$

Algebra Aerobics 5.1b

1.

t	$N = 10 + 3t$	$N = 10 \cdot 3^t$
0	10	10
1	13	30
2	16	90
3	19	270
4	22	810

2. For every increase in t, N increases by 3, so there is a constant slope of 3. The vertical intercept is 10.

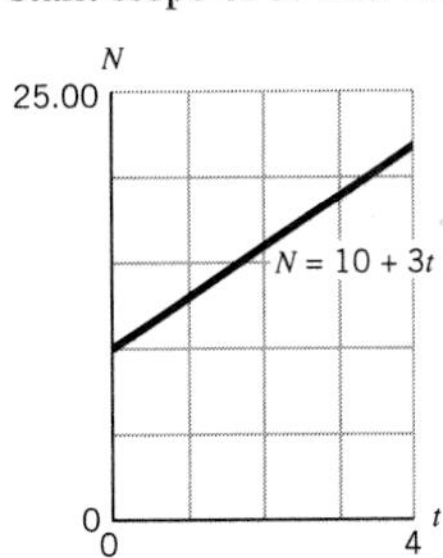

$N = 10 + 3t \qquad 0 \le t \le 4$

3. The slope here is not constant as in the previous problem. For example, the average rate of change between 0 and 1 is $(30 - 10)/1 = 20$ and between 1 and 2 is $(90 - 30)/1 = 60$. But the vertical intercept for both problems is the same, 10.

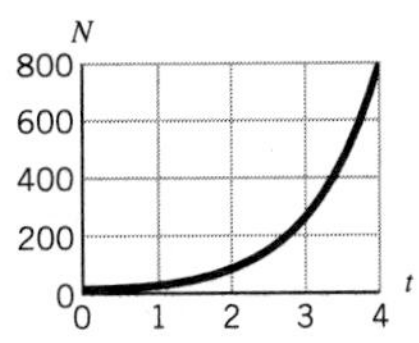

$N = 10 \cdot 3^t \qquad 0 \le t \le 4$

4.

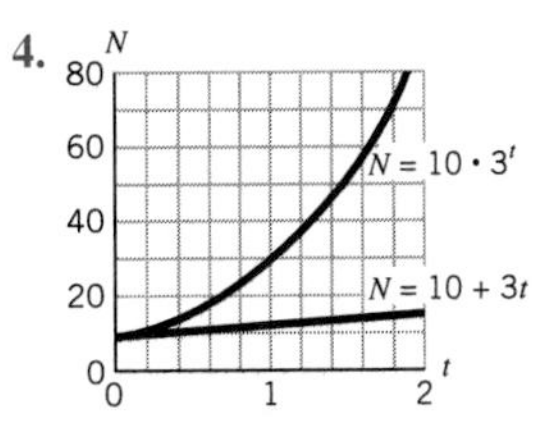

Although both graphs begin at (0, 10), $N = 10 \cdot 3^t$ increases dramatically as x increases, while $N = 10 + 3t$ increases very slightly as x does.

Algebra Aerobics 5.2

1. a. growth c. decay e. decay

   b. decay d. growth f. growth

2. a. $y = 2300\,(1/3)^t$ c. $y = (375)(0.1)^t$

   b. $y = (3 \cdot 10^9)(0.35)^t$

3. $y = 12\,(5)^{-x}$
   $= 12\,(5^{-1})^x$
   $= 12\,(1/5)^x$ so it represents decay.

Algebra Aerobics 5.3

1. a.

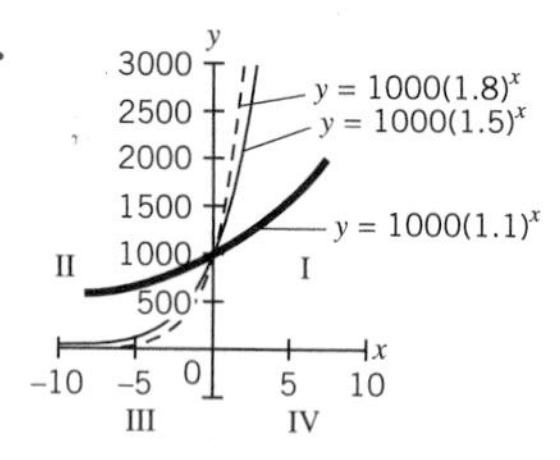

b. Yes, at the y-intercept where $x = 0$, $y = 1000$.

c. In the first quadrant, $y = 1000(1.8)^x$ is on top, $y = 1000(1.5)^x$ is in the middle and $y = 1000(1.1)^x$ is on the bottom.

d. In the second quadrant, $y = 1000(1.1)^x$ is on top, $y = 1000(1.5)^x$ is in the middle and $y = 1000(1.8)^x$ is on the bottom.

2. a.

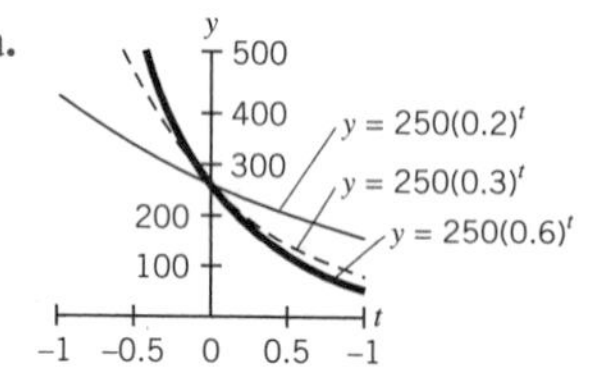

b. Yes, they intersect at (0, 250).

c. In the first quadrant: $y = 250(0.2)^t$ is on top, $y = 250(0.3)^t$ is in the middle, and $y = 250(0.6)^t$ is on the bottom.

d. In the second quadrant: $y = 250(0.6)^t$ is on top, $y = 20(0.3)^t$ is in the middle, and $y = 250(0.2)^t$ is on the bottom.

3. a.

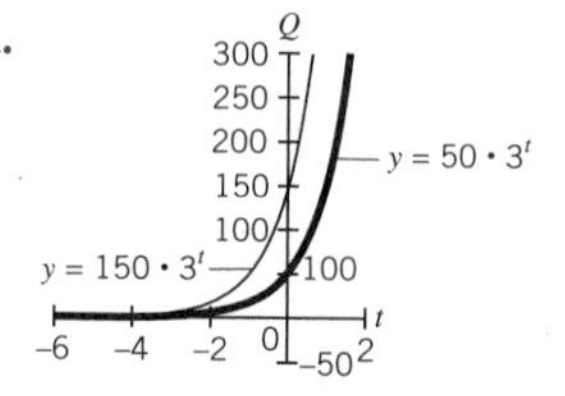

b. No, the curves do not intersect.

c. $y = 150 \cdot 3^t$ is always above $y = 50 \cdot 3^t$.

4. a. $Q = 0$

   b. $g(r) = 0$

   c. No horizontal asymptotes. The graph of this function is a line.

Algebra Aerobics 5.4

1. a. $\dfrac{(4094 - 3912)}{3912} \approx 0.465\,(100) = 4.65\%$

   b. $S = 3912\,(1.0465)^t$
   S is the spending in dollars per person.
   t is the time in years since 1997. So, in the year 2000, $t = 3$.
   Growth rate = 4.65%
   Growth factor = 1.0465

   c. in 2005, $t = 8$
   So, $S = 3912\,(1.0465)^8$
   $S = \$5627.46$ per person

2.

Exponential Function	Initial Value	Growth or Decay?	Growth or Decay Factor	Growth or Decay Rate
$A = 4(1.03)^t$	4	growth	1.03	0.03 or 3%
$A = 10(0.98)^t$	10	decay	0.98	0.02 or 2%
$y = 1000(1.005)^x$	1000	growth	1.005	0.005 or 0.5%
$y = 30(0.96)^x$	30	decay	0.96	0.04 or 4%
$A = 50{,}000(1.0705)^x$	\$50,000	growth	1.0705	0.0705 or 7.05%
$y = 200(0.51)^x$	200 g	decay	0.51	0.49 or 49%

Algebra Aerobics 5.5a

1. a. If the world's population in 1999 was 6 billion, and it grows at a rate of 1.3% per year, then it is only the first year that there is a net addition of 78 million people. The next year, the increase would be 1.3% of 6,078,000,000, which is 79,014,000. It would continue to have an increase that is more than 78 million each year. The increase is a fixed amount only in a linear model.

   b. $P = 6{,}000{,}000{,}000 + 78{,}000{,}000t$, for t the years after 1999, and P the world's population.

c. $P = 6{,}000{,}000{,}000\,(1.013)^t$

d. In 2020, $t = 21$. The linear model predicts the population to be $P = 6.10^9 + 7.8 \cdot 10^7(21)$
$= 6.10^9 + 163.8 \cdot 10^7$
$= 6.10^9 + 1.638 \cdot 10^2 \cdot 10^7$
$= 6.10^9 + 1.638 \cdot 10^9$
$= 7.638 \cdot 10^9$ or 7.638 billion.
(To calculate exponential model use scientific calculator.)
The exponential model predicts: $P = 6(1.013)^{21}$ billion
$= 6(1.31)$ billion
$= 7.86$ billion

In 2050 $t = 51$. The linear model predicts the population to be:
$P = 6 \cdot 10^9 + 7.8 \cdot 10^7(51) = 6 \cdot 10^9 + 397.8 \cdot 10^7$
$= 6 \cdot 10^9 + 3.978 \cdot 10^2 \cdot 10^7 = 6 \cdot 10^9 + 3.978 \cdot 10^9$
$= 9.98 \cdot 10^9$
The exponential model predicts: $P = 6(1.013)^{51}$ billion $\approx$ 7.87 billion $\approx 6(1.31)$ billion $\approx 11.59 \cdot 10^9 = 1.159 \cdot 10^{10}$

2. a. i. $\dfrac{20 - 10}{2 - 0} = \dfrac{10}{2} = 5$
ii. see table
iii. $P = 5t + 10$

t (years)	P	Q
0	10	10
2	20	20
4	30	40
6	40	80
8	50	160

b. i. 2
ii. see table
iii. Since the 2 yr. growth factor is 2, the annual growth factor is $2^{1/2} = \sqrt{2} \approx 1.41$
$a^2 = 2 \rightarrow (a^2)^{1/2} = \sqrt{2}$, so $a \approx 1.41$.
iv. $9 = 10 \cdot 1.41^t$

Algebra Aerobics 5.5b

1. a. $70/2 = 35$ yr

b. $70/0.5 = 140$ months

c. $70/8.1 = 8.64$ yr

d. growth factor $= 1.065 \rightarrow$ growth rate $= 6.5\%$
$70/6.5 = 10.77 \approx 11$ yr

2. a. $70/R = 10 \Rightarrow 70 = 10R \Rightarrow R \approx 70/10 = 7\%/\text{yr}$

b. $R \approx 70/5 = 14\%/\text{min}$

c. $R \approx 70/25 = 2.8\%/\text{sec}$

3. a. Since $a = 0.95$, $r = 0.05$, so this is decay of $R = 5\%$, and half life is $70/5 = 14$ months.

b. Since $a = 0.75$, $r = 0.25$, so this is decay of $R = 25\%$, and half-life is $70/25 = 2.8$ sec.

c. Half-life is $70/35 = 2$ yr.

Algebra Aerobics 5.5c

1. a. $70/3 \approx 23.3$ yr b. $70/5 = 14$ yr c. $70/7 = 10$ yr

2. a. $70/5 = 14\%$ b. $70/10 = 7\%$ c. $70/7 = 10\%$

3. a. $P = \$1000(1.04)^n$ b. $P = \$1000(1.11)^n$
c. $P = \$1000(2.10)^n$

4. If inflation is 10%/month, then what cost 1 cruzeiros this month would cost 1.10 cruzeiros next month. We have 1 cruzeiro $\approx$ 91% of 1.10 cruzeiros (since $1/1.1 \approx 0.91$), so a month later 1 cruzeiro would only be worth 0.91 cruzeiros or 91% of its original value. Thus, the decay factor is 0.91. So the exponential decay function $Q = 100\,(0.91)^n$, gives the purchasing power of 100 of today's cruzeiros at n months in the future.
When $n = 3$ months, then Q, the value of 100 of today's cruzeiros, will be $100(0.91)^3 \approx 100\,(0.75) = 75$ cruzeiros. When $n = 6$ months, then Q, the value of 100 of today's cruzeiros, will be $100(0.91)^6 \approx 100(0.57) = 57$ cruzeiros. When $n = 12$ months or 1 yr, then Q, the value of 100 of today's cruzeiros will be $100(0.91)^{12} \approx 100(0.32) = 32$ cruzeiros.
With a 10% monthly inflation rate, the value of a cruzeiros will shrink by more than two-thirds by the end of a year.

Algebra Aerobics 5.5d

1. a.

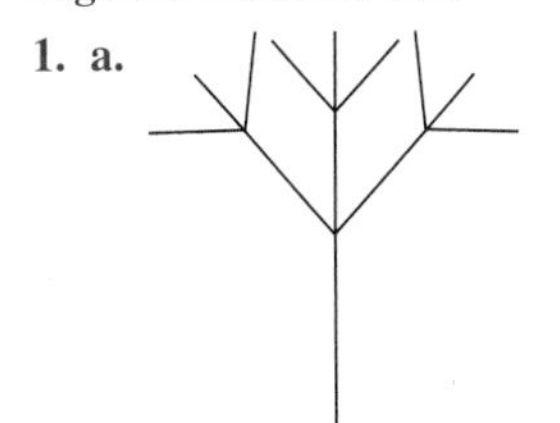

b. $N = 3^L$

2.

Level	# of People Called	Total Called
0	3^0	1
1	3^1	4
2	$3^2 = 9$	13
3	$3^3 = 27$	40
4	$3^4 = 81$	121
5	$3^5 = 243$	364
6	$3^6 = 729$	1,093
7	$3^7 = 2187$	3,280
8	$3^8 = 6561$	9,841

So, it will take 8 levels to reach 8,000 people.

Algebra Aerobics 5.6

1. a. Judging from the graph, the number of *E. coli* bacteria grows by a factor of 10 (for example, from 100 to 1000, or 100,000 to 1,000,000) in a little over 3 time periods.

b. From the equation $N = 100 \cdot 2^t$, we know that every three time periods, the quantity is multiplied by 2^3 or 8.

c. Every four time periods, the quantity is multiplied by 2^4 or 16.

d. The answers are consistent, since somewhere between three and four time periods the quantity should be multiplied by 10 (which is between 8 and 16).

CHAPTER 6

Algebra Aerobics 6.1a

1. a. When $t = 0$, $M = 250(3)^0 = 250(1) = 250$.

   b. When $t = 1$, $M = 250(3)^1 = 250(3) = 750$.

   c. When $t = 2$, $M = 250(3)^2 = 250(9) = 2{,}250$.

   d. When $t = 3$, $M = 250(3)^3 = 250(27) = 6{,}750$.

$M(t) = 250 \cdot 3^t$	(t, m)
$M(0) = 250$	(0, 250)
$M(1) = 750$	(1, 750)
$M(2) = 2250$	(2, 2250)
$M(3) = 6{,}750$	(3, 6750)

2. a. $4 < t < 5$ b. $8 < t < 9$

3. To read each value of x from this graph, draw a horizontal line from the y-axis at given value until it hits curve. Then draw a vertical line to the x-axis to identify the appropriate value of x.

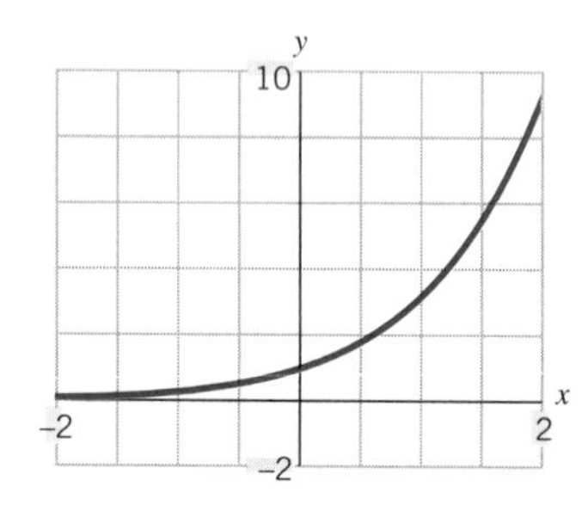

   a. horizontal line $y = 7$, from y-axis to curve, then down to x-axis $\rightarrow x \approx 1.8$.

   b. horizontal line $y = 0.5$, from y-axis to curve, then down to x-axis $\rightarrow x \approx -0.5$.

4. a. $2 < x < 3$ b. $-1 < x < 0$

Algebra Aerobics 6.1b

1. a. $\log(10^5/10^7) = \log 10^{-2} = -2\log 10 = -2(1) = -2$ and $\log 10^5 - \log 10^7 = 5\log 10 - 7\log 10 = 5 - 7 = -2$

   So, $\log(10^5/10^7) = \log 10^5 - \log 10^7$.

   b. $\log(10^5 \cdot (10^7)^3) = \log(10^5 \cdot 10^{21}) = \log 10^{26} = 26\log 10 = 26$ and $\log 10^5 + 3\log 10^7 = 5\log 10 + 3(7)\log 10 = 5 + (3 \cdot 7) = 26$

   So, $\log[10^5 \cdot (10^7)^3] = \log(10^5) + 3\log(10^7)$.

2. $$\log\sqrt{\frac{2x-1}{x+1}} = \log\left(\frac{2x-1}{x+1}\right)^{1/2} = \frac{1}{2}\log\left(\frac{2x-1}{x+1}\right) = \frac{1}{2}[\log(2x-1) - \log(x+1)]$$

3. $$\frac{1}{3}[\log x - \log(x+1)] = \frac{1}{3}\log\frac{x}{x+1} = \log\left(\frac{x}{x+1}\right)^{1/3} = \log\sqrt[3]{\frac{x}{x+1}}$$

4. $$\log(x+12) - \log(2x-5) = 2 \rightarrow \log\frac{x+12}{2x-5} = 2$$
$$\Rightarrow 10^2 = \frac{x+12}{2x-5} \Rightarrow$$
$$100 = \frac{x+12}{2x-5} \rightarrow$$
$$200x - 500 = x + 12 \Rightarrow$$
$$199x = 512 \rightarrow x = \frac{512}{199}$$

5. $\log 10^3 - \log 10^2 = 3\log 10 - 2\log 10 = 3 - 2 = 1$.
$$\frac{\log 10^3}{\log 10^2} = \frac{3\log 10}{2\log 10} = \frac{3}{2} = 1.5.$$
Since $1 \neq 1.5 \rightarrow \log 10^3 - \log 10^2 \neq \dfrac{\log 10^3}{\log 10^2}$

Algebra Aerobics 6.1c

1. a. $60 = 10 \cdot 2^t \rightarrow 6 = 2^t$ (dividing both sides by 10)
   $\log 6 = \log 2^t$ (taking the log of both sides)
   $\log 6 = t\log 2$ (using property 3 of logs)
   $\log 6/\log 2 = t$
   $0.778/0.301 \approx t$ or $t \approx 2.58$

   b. $500(1.06)^t = 2000 \rightarrow (1.06)^t = \dfrac{2000}{500} \rightarrow$
   $(1.06)^t = 4 \rightarrow \log(1.06)^t = \log 4 \rightarrow$
   $t(\log 1.06) = \log 4 \rightarrow t = \dfrac{\log 4}{\log 1.06} = \dfrac{0.6021}{0.0253} \approx 23.8$

   c. $80(0.95)^t = 10 \rightarrow (0.95)^t = 1/8 = 0.125 \rightarrow$
   $\log(0.95)^t = \log 0.125 \rightarrow t\log 0.95 = \log 0.125 \rightarrow$
   $t = \dfrac{\log 0.125}{\log 0.95} \approx \dfrac{-.903}{-.022} \approx 41.05$

2. a. $7000 = 100 \cdot 2^t \Rightarrow 70 = 2^t \Rightarrow \log 70 = \log 2^t \Rightarrow$
   $\log 70 = t\log 2 \Rightarrow t = \log 70/\log 2 \approx \dfrac{1.845}{0.301} \approx 6.13$
   It will take 6.13 time periods or approximately (6.13)(20) min = 122.6 min (or a little more than 2 hours) for the bacteria count to reach 7000.

   b. $12{,}000 = 100 \cdot 2^t \Rightarrow 120 = 2^t \Rightarrow \log 120 = \log 2^t \Rightarrow$ $\log 120 = t\log 2 \Rightarrow t = \log 120/\log 2 \approx 6.907$ time periods or approximately (6.907)(20) min = 138.14 min (or a little more then $2\frac{1}{4}$ hrs.) for the bacteria count to reach 12,000.

3. Using the rule of 70, since $R = 6\%$ per yr, then $70/R = 70/6 \approx 11.7$ yr. More precisely, we have:

$$2000 = 1000(1.06)^t \Rightarrow 2 = 1.06^t \Rightarrow$$
$$\log 2 = \log 1.06^t \Rightarrow \log 2 = t \log 1.06 \Rightarrow$$
$$t = \log 2/\log 1.06 \approx \frac{.301}{.0253} \approx 11.9 \text{ yr}$$

4. $$1 = 100\,(0.9755)^t \rightarrow 0.01 = (0.9755)^t \rightarrow$$
$$\log 0.01 = \log (0.9755)^t \rightarrow \log 0.01 = t \log 0.9755 \rightarrow$$
$$t = \frac{\log 0.01}{\log 0.9755} = \frac{-2}{-0.0108} \approx 185.2 \text{ years,}$$
almost 2 centuries!

Algebra Aerobics 6.2

1. a. $\$1000(1.085) = \$1{,}085$

   b. $\$1000\left(1 + \frac{0.085}{4}\right)^4 = \$1{,}087.75$

   c. $\$1000e^{0.085} = \$1{,}088.72$

2. a. $e^{0.04} = 1.0408 \Rightarrow 4.08\%$ is effective rate

   b. $e^{0.125} = 1.133 \Rightarrow 13.3\%$ is effective rate

   c. $e^{0.18} = 1.197 \Rightarrow 19.7\%$ is effective rate

3.

n	$1/n$	$1 + 1/n$	$(1 + 1/n)^n$
1	1	2	2
100	0.01	1.01	2.704813829
1000	0.001	1.001	2.716923932
1,000,000	0.000001	1.000001	2.718280469
1,000,000,000	0.000000001	1.000000001	2.718281827

The values for $(1 + 1/n)^n$ come closer and closer to the finite irrational number we define as e and are consistent with the value for e in the text of 2.71828.

Algebra Aerobics 6.3

1. a. $\ln e^2 = 2 \ln e = 2(1) = 2$

   b. $\ln 1 = 0$

   c. $\ln \frac{1}{e} = \ln 1 - \ln e = 0 - 1 = -1$

   d. $\ln \frac{1}{e^2} = \ln 1 - \ln e^2 = \ln 1 - 2 \ln e = 0 - 2(1)$
   $= 0 - 2 = -2$

   e. $\ln \sqrt{e} = \ln e^{1/2} = \frac{1}{2} \ln e = \frac{1}{2}(1) = 1/2$

2. $$\ln \frac{\sqrt{x+2}}{x(x-1)} = \ln \frac{(x+2)^{1/2}}{x(x-1)} = \ln (x+2)^{1/2} - \ln x(x-1)$$
$$= \frac{1}{2} \ln (x+2) - \ln x(x-1)$$
$$= \frac{1}{2} \ln (x+2) - [\ln x + \ln (x-1)]$$
$$= \frac{1}{2} \ln (x+2) - \ln x - \ln (x-1)$$

3. $\ln x - 2 \ln (2x - 1) = \ln x - \ln (2x - 1)^2 = \ln \frac{x}{(2x-1)^2}$

4. $e^r = 1 + 0.064 = 1.064 \Rightarrow r = \ln 1.064 = 0.062 = 6.2\%$

5. $50{,}000 = 10{,}000e^{0.078t} \Rightarrow 5 = e^{0.078t} \Rightarrow \ln 5 = 0.078t \Rightarrow$
   $t = \ln 5/0.078 \approx 20.6$ yr

6. a. $\ln e^{x+1} = \ln 10 \Rightarrow (x + 1) \ln e = \ln 10 \rightarrow$
   $(x + 1)(1) \approx 2.30 \rightarrow x \approx -1 + 2.30 \rightarrow x \approx 1.30$

   b. $\ln e^{x-2} = \ln 0.5 \rightarrow (x - 2) \ln e = \ln 0.5 \rightarrow$
   $x - 2 \approx -0.69 \rightarrow x \approx 2 - 0.69 \rightarrow x \approx 1.31$

Algebra Aerobics 6.4

1. Since $\log x^2 = 2 \log x$ (by Property 3 of logarithms), the graphs of $y = \log x^2$ and $y = 2 \log x$ will be identical (assuming $x > 0$).

2. The graph of $y = -\ln x$ will be the mirror image of $y = \ln x$ across the x-axis.

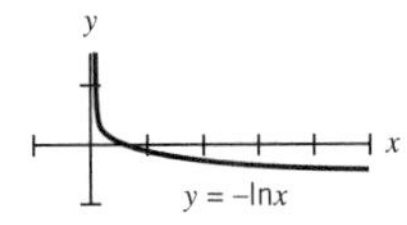

3. Acidic. If $4 = -\log [H^+]$, then $-4 = \log [H^+]$. So $[H^+] = 10^{-4}$. Since the pH is 3 less than pure water's, it will have a hydrogen ion concentration 10^3 or 1000 times higher than pure water's.

4. $$N = 10 \log \left(\frac{1.5 \cdot 10^{-12}}{10^{-16}}\right) = 10 \log(1.5 \cdot 10^4)$$
$$= 10 (\log 1.5 + \log 10^4) \approx 10 (0.176 + 4)$$
$$= 10 (4.176) \approx 42$$

5. Multiplying the intensity by $100 = 10^2$ corresponds to adding 20 to the decibel level.
   Multiplying the intensity by $10{,}000{,}000 = 10^7$ corresponds to adding 70 to the decibel level.

Algebra Aerobics 6.5

1. a. decay; b. decay; c. growth

2. a. $e^x = 1.062 \Rightarrow x = \ln 1.062 \approx 0.060 \Rightarrow y = 1000e^{0.06t}$

b. $e^x = 0.985 \Rightarrow x = \ln 0.985 \approx -0.015 \Rightarrow y = 50e^{-0.015t}$

3. Equation of the form $y = Ce^{rx}$, where x = number of days, C = initial amount of iodine-131, and y = amount after x days. The points (8, 2.40) and (20, 0.88) satisfy the equation, so $2.40 = Ce^{8r}$ and $0.88 = Ce^{20r}$. So

$$\frac{0.88}{2.40} = \frac{Ce^{20r}}{Ce^{8r}} \Rightarrow 0.367 = e^{12r} \Rightarrow$$

$\ln(0.367) = \ln e^{12r} \Rightarrow -1.002 = 12r \Rightarrow r = -0.084$, the decay rate.

Plugging this back in we get,

$2.40 = Ce^{-0.084 \cdot 8} \Rightarrow 2.40 = Ce^{-0.672} \Rightarrow$

$2.40 = C \cdot 0.511 \Rightarrow 2.40/0.511 = C \Rightarrow$

$C = 4.70$ micrograms (μg)

So the exponential decay function is: $y = 4.70e^{-0.084x}$.

Algebra Aerobics 6.6a

1. $f(x) = 2x + 3 \quad g(x) = x^2 - 4$

a. $g(2) = (2)^2 - 4 = 4 - 4 = 0 \rightarrow$
$f(g(2)) = f(0) = 2(0) + 3 = 3$

b. $f(2) = 2(2) + 3 = 4 + 3 = 7 \rightarrow$
$g(f(2)) = g(7) = (7)^2 - 4 = 49 - 4 = 45$

c. $g(3) = (3)^2 - 4 = 9 - 4 = 5 \rightarrow$
$f(g(3)) = f(5) = 2(5) + 3 = 10 + 3 = 13$

d. $f(3) = 2(3) + 3 = 6 + 3 = 9 \rightarrow$
$f(f(3)) = f(9) = 2(9) + 3 = 18 + 3 = 21$

e. $(f \circ g)(x) = f(g(x)) = f(x^2 - 4) = 2(x^2 - 4) + 3 = 2x^2 - 8 + 3 = 2x^2 - 5$

f. $(g \circ f)(x) = g(f(x)) = g(2x + 3) = (2x + 3)^2 - 4 = 4x^2 + 12x + 9 - 4 = 4x^2 + 12x + 5$

2. $F(x) = \dfrac{2}{x - 1}$, $G(x) = 3x - 5$

a. $(F \circ G)(x)) = F(G(x)) = F(3x - 5) = \dfrac{2}{(3x - 5) - 1} = \dfrac{2}{3x - 6}$

b. $(G \circ F)(x) = G(F(x)) = G\left(\dfrac{2}{x - 1}\right) = 3\left(\dfrac{2}{x - 1}\right) - 5 = \dfrac{6}{x - 1} - 5$

Combine these by expressing both terms as fractions with the same denominator $(x - 1)$, then add those resulting fractions by adding their numerators.

$$\frac{6}{x - 1} - \frac{5}{1}\frac{(x - 1)}{(x - 1)} = \frac{6}{x - 1} - \frac{5x - 5}{x - 1} = \frac{6 - (5x - 5)}{x - 1} = \frac{6 - 5x + 5}{x - 1} = \frac{11 - 5x}{x - 1}$$

c. To determine if $\dfrac{2}{3x - 6}$ is equivalent to $\dfrac{11 - 5x}{x - 1}$, evaluate each for the same value of x. If $x = 3 \rightarrow$

$$\frac{2}{3x - 6} = \frac{2}{3(3) - 6} = \frac{2}{9 - 6} = \frac{2}{3}$$

$$\frac{11 - 5x}{x - 1} = \frac{11 - 5(3)}{(3) - 1} = \frac{11 - 15}{2} = \frac{-4}{2} = -2;$$

$\dfrac{2}{3} \neq -2$ so, $(F \circ G)(x) \neq (G \circ F)(x)$

Algebra Aerobics 6.6b

1. a. $f(x) = 2x + 1, \quad g(x) = \dfrac{x - 1}{2}$

$$(f \circ g)(x) = f(g(x)) = f\left(\frac{x - 1}{2}\right) = 2\left(\frac{x - 1}{2}\right) + 1 = \frac{2(x - 1)}{2} + 1 = x - 1 + 1 = x$$

$$(g \circ f)(x) = g(f(x)) = g(2x + 1) = \frac{(2x + 1) - 1}{2} = \frac{2x}{2} = x$$

b. $f(x) = \sqrt[3]{x + 1}, \quad g(x) = x^3 - 1$

$(f \circ g)(x) = f(g(x)) = f(x^3 - 1) = \sqrt[3]{(x^3 - 1) + 1} = \sqrt[3]{x^3} = (x^3)^{1/3} = x^1 = x$

$(g \circ f)(x) = g(f(x)) = g(\sqrt[3]{x + 1}) = (\sqrt[3]{x + 1})^3 - 1 = [(x + 1)^{1/3}]^3 - 1 = (x + 1) - 1 = x$

2. $$f(f^{-1}(y)) = f\left(\frac{1 + y}{y}\right) = \frac{1}{\dfrac{1 + y}{y} - 1} = \frac{y}{y}\left(\frac{1}{\dfrac{1 + y}{y} - 1}\right) = \frac{y}{y\left(\dfrac{1 + y}{y}\right) - 1(y)} = \frac{y}{1 + y - y} = y$$

$$f^{-1}(f(x)) = f^{-1}\left(\frac{1}{x - 1}\right) = \frac{1 + \dfrac{1}{x - 1}}{\dfrac{1}{x - 1}} = \frac{(x - 1)}{(x - 1)}\left(\frac{1 + \dfrac{1}{x - 1}}{\dfrac{1}{x - 1}}\right) = \frac{(x - 1) + \dfrac{(x - 1)}{1}\dfrac{1}{(x - 1)}}{\dfrac{(x - 1)}{1}\dfrac{1}{(x - 1)}} = \frac{x - 1 + 1}{1} = x$$

So, $f(f^{-1}(y)) = y$ and $f^{-1}(f(x)) = x$.

3. f and its inverse f^{-1} along with $y = x$

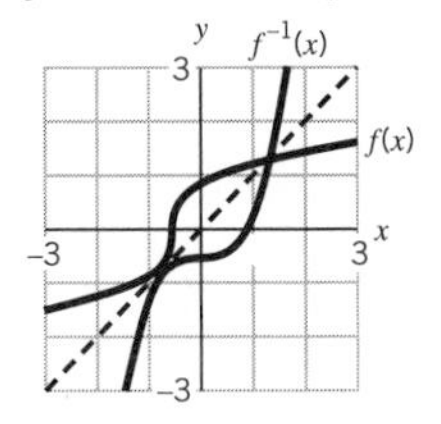

4. a. $f(x) = x^3 = y \to x = \sqrt[3]{y} = f^{-1}(y); f^{-1}(y) = \sqrt[3]{y}$

b. $f(x) = \dfrac{1}{x+1} = y \to$

$(x+1)\left(\dfrac{1}{x+1}\right) = (x+1)y \to 1 = y(x+1) \to$

$1 = yx + y \to 1 - y = yx$

$\to \dfrac{1-y}{y} = x; \to f^{-1}(y) = \dfrac{1-y}{y}; y \neq 0$

c. $f(x) = 150(0.097)^x \to y = 150(0.097)^x \to$

$\dfrac{y}{150} = (0.097)^x \to \log\left(\dfrac{y}{150}\right) = \log(0.097)^x \to$

$\log y - \log 150 = x\log(0.097) \to$

$x = \dfrac{\log y - \log 150}{\log 0.097} \to$

$f^{-1}(y) = \left(\dfrac{\log y - \log 150}{\log 0.097}\right)$

CHAPTER 7

Algebra Aerobics 7.1

1. a. When radius is doubled:

i. $s(2r) = 4\pi(2r)^2 \to 4\pi(4)r^2 = 16\pi r^2$ or $4(4\pi r^2)$ $= 4s(r)$.
So the surface area increases by a factor of 4.

ii. $V(2r) = \dfrac{4}{3}\pi(2r)^3 = \dfrac{4}{3}\pi(8)r^3 = 8\left(\dfrac{4}{3}\pi r^3\right)$ $= 8V(r)$.
So the volume increases by a factor of 8.

b. i. The volume will grow faster than the surface area.

ii. So, the ratio of $\dfrac{\text{surface area}}{\text{volume}}$ will decrease as the radius increases.

Algebra Aerobics 7.2

1.

	Power Function	Independent Variable	Dependent Variable	Constant of Proportionality	Power
a.	yes	r	A	π	2
b.	yes	z	y	1	5
c.	no	—	—	—	—
d.	no	—	—	—	—
e.	yes	x	y	3	5

2. a. y is directly proportional to x^2.

b. y is not directly proportional to x^2.

c. y is directly proportional to x^{-2}.

3. $g(x) = 5x^3$

a. $g(2) = 5(2)^3 = 5(8) = 40$ $g(4) = 5(4)^3 = 5(64) = 320$ So $g(4)$ is eight times larger than $g(2)$

b. $g(5) = 5(5)^3 = 5(125) = 625$ $g(10) = 5(10)^3 = 5(1{,}000) = 5{,}000$ So $g(10)$ is eight times larger than $g(5)$

c. $g(2x) = 5(2x)^3 = 5(8x^3) = 8(5x^3) = 8g(x)$ So $g(2x)$ is eight times larger than $g(x)$

d. $g\left(\dfrac{1}{2}x\right) = 5\left(\dfrac{1}{2}x\right)^3 = 5\left(\dfrac{1}{8}x^3\right) = \dfrac{1}{8}(5x^3) = \dfrac{1}{8}g(x)$ So $g\left(\dfrac{1}{2}x\right)$ is eight times smaller than $g(x)$

4. a. y is equal to 3 times the fifth power of x.

b. y is equal to 2.5 times the cube of x.

c. y is equal to one-fourth times the fifth power of x.

5. a. y is directly proportional to x^5 with a proportionality constant of 3.

b. y is directly proportional to x^3 with a proportionality constant of 2.5.

c. y is directly proportional to x^5 with a proportionality constant of 1/4.

6. a. $P = aR^2 \to \dfrac{P}{a} = \dfrac{aR^2}{a} \to \dfrac{P}{a} = R^2 \to \sqrt{\dfrac{P}{a}} = \sqrt{R^2} \to$

$R = \sqrt{\dfrac{P}{a}}$

b. $V = \left(\dfrac{1}{3}\right)\pi r^2 h \to 3(V) = 3\left(\dfrac{1}{3}\right)\pi r^2 h \to$

$3V = \pi r^2 h \to \dfrac{3V}{\pi r^2} = \dfrac{\pi r^2 h}{\pi r^2} \to h = \dfrac{3V}{\pi r^2}$

c. $3V = \pi r^2 h \rightarrow \frac{3V}{\pi h} = \frac{\pi r^2 h}{\pi h} \rightarrow \frac{3V}{\pi h} = r^2 \rightarrow$

$\sqrt{\frac{3V}{\pi h}} = \sqrt{r^2} \rightarrow r = \sqrt{\frac{3V}{\pi h}}$

d. $Y = Z(a^2 + b^2) \rightarrow \frac{Y}{a^2 + b^2} = \frac{Z(a^2 + b^2)}{a^2 + b^2} \rightarrow$

$\frac{Y}{a^2 + b^2} = Z$

e. $Y = Z(a^2 + b^2) \rightarrow Y = Za^2 + Zb^2 \rightarrow Y - Zb^2$
$= Za^2 \rightarrow \frac{Y - Zb^2}{Z} = \frac{Za^2}{Z} \rightarrow a^2 = \frac{Y - Zb^2}{Z} \rightarrow$
$\sqrt{a^2} = \sqrt{\frac{Y - Zb^2}{Z}} \rightarrow a = \sqrt{\frac{Y - Zb^2}{Z}}$

Algebra Aerobics 7.3a

1. a. $f(2) = 4(2)^3 = 4(8) = 32$
   b. $f(-2) = 4(-2)^3 = 4(-8) = -32$
   c. $f(s) = 4s^3$
   d. $f(3s) = 4(3s)^3 = 4(27)s^3 = 108s^3$

2. a. $g(2) = -4(2)^3 = -4(8) = -32$
   b. $g(-2) = -4(-2)^3 = -4(-8) = 32$
   c. $g\left(\frac{1}{2}t\right) = -4\left(\frac{1}{2}t\right)^3 = -4\left(\frac{1}{8}\right)t^3 = -\frac{1}{2}t^3$
   d. $g(5t) = -4(5t)^3 = -4(125)t^3 = -500t^3$

3. a.

x	$f(x) = 4x^2$	$g(x) = 4x^3$
-4	64	-256
-2	16	-32
0	0	0
2	16	32
4	64	256
$\frac{1}{2}$	1	$\frac{1}{2}$

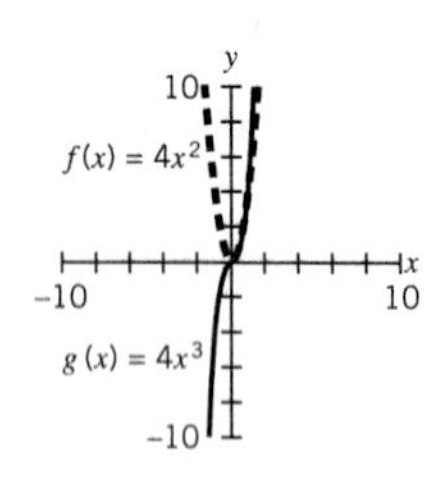

   b. As $x \rightarrow +\infty$, both $f(x)$ and $g(x) \rightarrow +\infty$.
   c. As $x \rightarrow -\infty$, $f(x) \rightarrow +\infty$, but $g(x) \rightarrow -\infty$.
   d. The domain of both functions is all the real numbers, the range of g is also all the real numbers; however, the range of f is only all the nonnegative real numbers.

   e. They intersect at the origin and at the point (1, 4).
   f. All values greater than $x = 1$.

Algebra Aerobics 7.3b

1. a.

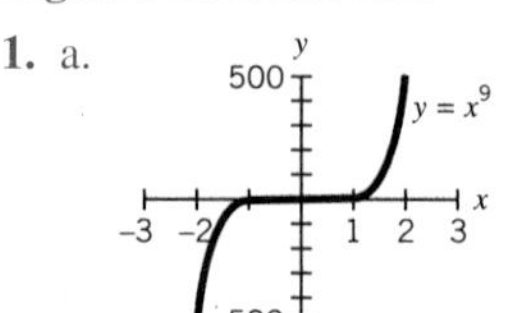

   b.

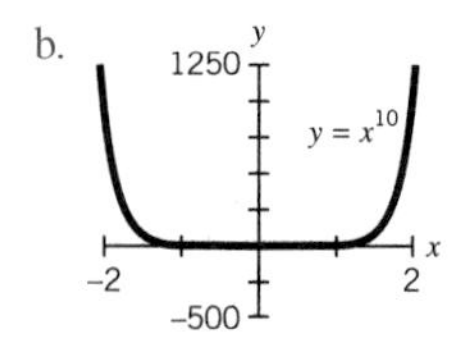

2. a. $y = x$ and $y = 4x$ and $y = -4x$

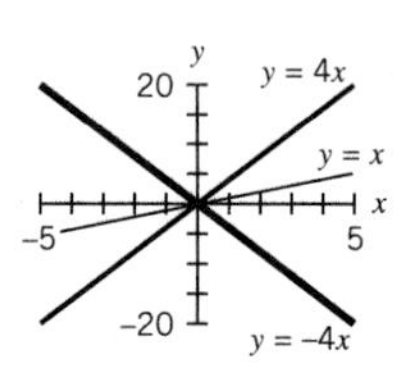

   b. $y = x^4$ and $y = 0.5x^4$ and $y = -0.5x$

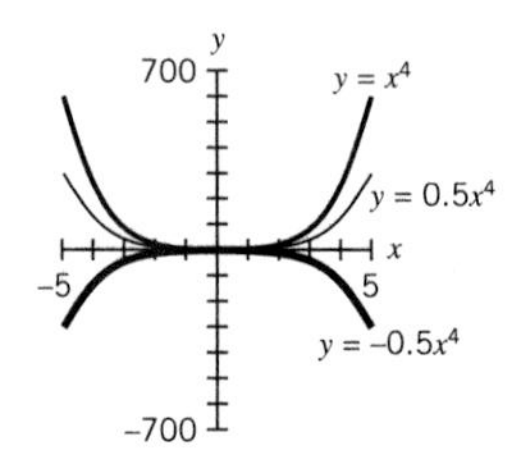

Algebra Aerobics 7.4

1. a.

x	$y = 4^x$	$y = x^3$
0	1	0
1	4	1
2	16	8
3	64	27
4	256	64
5	1024	125
6	4096	216
7	16,382	343

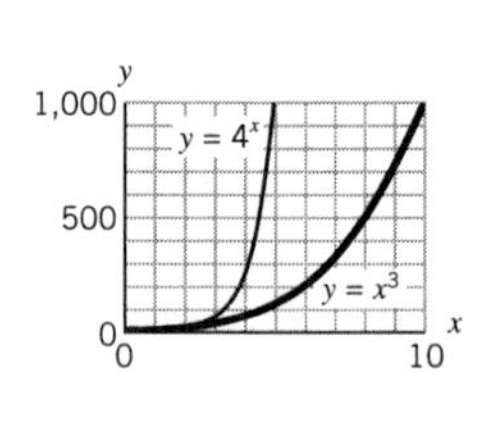

   b. $y = 4^x$ always dominates $y = x^3$.

2. a. $y = 2^x$ dominates eventually for extremely large values of x.
   b. $y = (1.000005)^x$ dominates eventually for extremely large values of x.

Algebra Aerobics 7.5a

1. The volume will increase by a factor of four, since $V = \frac{1}{P}$, and in Table 7.7, the volume at 99 ft is 1/4 of the volume at 0 ft, and P is 4 atm at 99 ft, and 1 atm at 0 ft.

2. a.

x	$r(x)$
0.01	600
0.25	24
0.50	12
1.00	6
2.00	3
5.00	1.2
10.00	0.6

b. The function's value decreases as x increases.

c. The function $\to \infty$ as x approaches zero.

d. No, the function is undefined at $x = 0$.

Algebra Aerobics 7.5b

1. a. i. $\frac{15}{x^3}$ ii. $\frac{-10}{x^4}$ iii. $\frac{3.6}{x}$

b. i. $1.5x^{-2}$ ii. $-6x^{-3}$ iii. $-\frac{2}{3}x^{-2}$

2. $t = kd^{-2}$ or $t = \frac{k}{d^2}$ for some constant k

3. Intensity at 4 ft is $\frac{k}{4^2} = \frac{1}{16}k$, but at 2 ft. intensity is $\frac{k}{(2)^2} = \frac{1}{4}k$. The light is four times as bright, at a distance of 2 ft. away, than it is at 4 ft. away.

4. $g(x) = \frac{3}{x^4}$

a. $g(2x) = \frac{3}{(2x)^4} = \frac{3}{16x^4} = \frac{1}{16}\left(\frac{3}{x^4}\right)$; so,

$g(2x) = \frac{1}{16}g(x)$.

b. $g\left(\frac{x}{2}\right) = \frac{3}{\left(\frac{x}{2}\right)^4} = 3 \div \left(\frac{x}{2}\right)^4 = 3 \div \frac{x^4}{16} = 3 \cdot \frac{16}{x^4}$

$= 16\left(\frac{3}{x^4}\right)$ so, $g\left(\frac{x}{2}\right) = 16\,(g(x))$.

5. a.

x	$f(x) = 1/x^2$	$g(x) = 1/x^3$
0.01	10,000	1,000,000
0.25	16	64
0.50	4	8
1	1	1
2	0.25	0.125
10	0.01	0.001

b. The functions are not defined when x is zero. They are defined for negative values of x.

c. As $x \to +\infty$, both $f(x)$ and $g(x) \to 0$. When $0 < x < 1$ as $x \to 0$, both $f(x)$ and $g(x) \to +\infty$.

Algebra Aerobics 7.6

1. a. $\frac{1}{x^2}$; $\frac{1}{(-2)^2} = \frac{1}{4}$ or 0.25

b. $\frac{1}{x^3}$; $\frac{1}{(-2)^3} = -\frac{1}{8}$ or -0.125

c. $\frac{4}{x^3}$; $\frac{4}{(-2)^3} = -\frac{1}{2}$ or -0.5

d. $\frac{-4}{x^3}$; $\frac{-4}{(-2)^3} = \frac{-4}{-8} = \frac{1}{2}$ or 0.5

2. a. $y = x^{-10}$

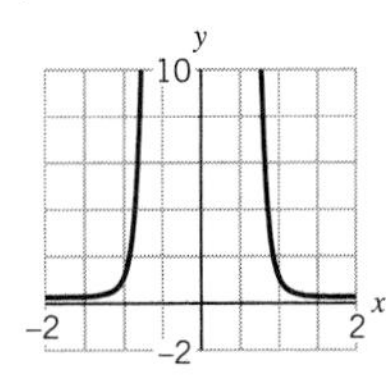

b. $y = x^{-11}$

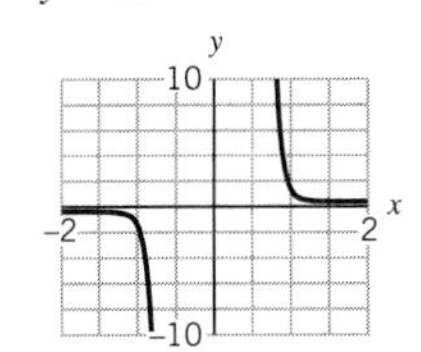

3. a. $y = x^{-2}$ and $y = x^{-3}$

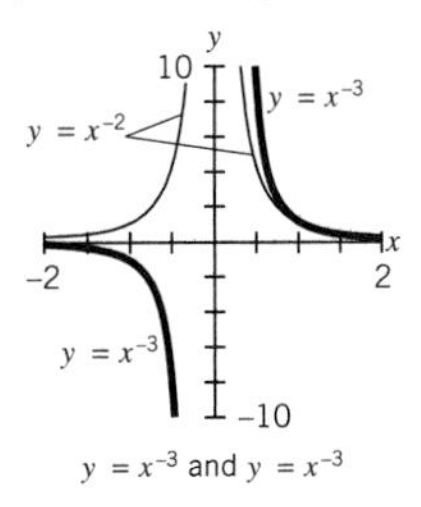

$y = x^{-3}$ and $y = x^{-3}$

The graphs intersect at (1, 1). As $x \to +\infty$, or $\to -\infty$, both graphs approach the x-axis, but $y = x^{-3}$ is closer to the x axis, at each point after $x = 1$. When $x > 0$, as $x \to 0$, both graphs approach $+\infty$, but $y = \frac{1}{x^3}$ is steeper. When $x < 0$, as $x \to 0$, the graph of $y = x^{-2}$ approaches $+\infty$, and the graph of $y = x^{-3}$ approaches $-\infty$.

b. $y = 4x^{-2}$ and $y = 4x^{-3}$

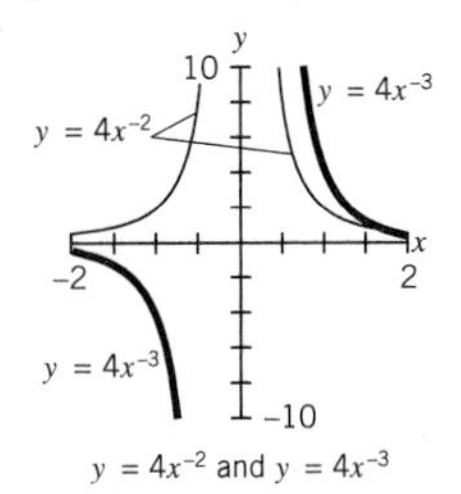

$y = 4x^{-2}$ and $y = 4x^{-3}$

The graphs intersect at (1, 4). As $x \to +\infty$ or $x \to -\infty$, both graphs approach the x-axis, but $y = 4x^{-3}$ approaches it quicker after $x > 1$, than $y = 4x^{-2}$ does. When $x > 0$, as $x \to 0$, both graphs approach $+\infty$. When $x < 0$, as $x \to 0$, the graph of $y = 4x^{-2}$ approaches $+\infty$, and the graph of $y = 4x^{-3}$ approaches $-\infty$.

Algebra Aerobics 7.7a

1. a. $y = 4x^3$ is a power function;

   b. $y = 3x + 4$ is a linear function;

   c. $y = 4 \cdot (3^x)$ is an exponential function.

Function	2. Type of Plot on Which Graph of Function Would Appear as a Straight Line	3. Slope of Straight Line
$y = 3x + 4$	standard linear plot	$m = 3$
$y = 4 \cdot (3^x)$	semi-log	$m = \log 3$
$y = 4 \cdot x^3$	log-log	$m = 3$

Algebra Aerobics 7.7b

1. Note this is a log-log plot, so the slope of a straight line corresponds to the exponent of a power function
   Younger ages: slope $= 1.2 \Rightarrow$ original function is of the form $y = a \cdot x^{1.2}$, where $x =$ body height in cm, and $y =$ arm length in cm.
   Older ages: slope $= 1.0 \Rightarrow$ original function is of the form $y = a \cdot x^1$, where $x =$ body height in cm, and $y =$ arm length in cm.

2. a. exponential

   b. exponential

   c. power

Algebra Aerobics 7.8

1. Estimated surface area is about 20,000 cm^2. Calculate surface area $10(70{,}000)^{2/3}$ with a scientific calculator as: $10 \cdot (70000)^{2/3} = 10 \cdot 1700 = 17{,}000$ cm^2. Since 1 cm = 0.394 in, 1 cm^2 = $(0.394 \text{ in})^2 = 0.155$ in^2. So 17,000 cm^2 = (17000)(0.155) = 2,635 in^2. Since 1 kg = 2.2 lb, a weight of 70 kg translated to pounds is: 70 kg = (70)(2.2) lb = 154 lbs.

2. a. power

   b. $y = ax^{-0.23}$

   c. When body mass increases by a factor of 10, heat rate is multiplied by $10^{-0.23}$ or about 0.589.

3. Estimated rate of heat production is somewhere between 10^3 and 10^4 kilocalories per day—say about 2,000–3,000 Kcal/day.

CHAPTER 8

Algebra Aerobics 8.1a

1. a. Degree is 5: $f(-1) = 11(-1)^5 + 4(-1)^3 - 11 = 11(-1) + 4(-1) - 11 = -11 - 4 - 11 = -26$

   b. Degree is 4: When $x = -1$, $y = -5(-1)^3 + 7(-1)^4 + 1 = -5(-1) + 7(1) + 1 = 5 + 7 + 1 = 13$

   c. Degree is 4: $g(-1) = -2(-1)^4 - 20 = -2(-1) - 20 = -2 - 20 = -22$

   d. Degree is 2: When $x = -1$, $z = -2(-1)^2 + 3(-1) - 4 = -2(1) + 3(-1) - 4 = -2 - 3 - 4 = -9$

2. a. $h(x) = (x^3 - 3x^2) + (2x^2 + 4) = x^3 - x^2 + 4$

   b. $h(0) = 0 - 0 + 4 = 4$
   $h(2) = 8 - 4 + 4 = 8$
   $h(-2) = -8 - 4 + 4 = -8$

Algebra Aerobics 8.1b

1. a. The y-intercept is at -3. The graph only crosses the x-axis once. It happens at about $x = 1.3$.

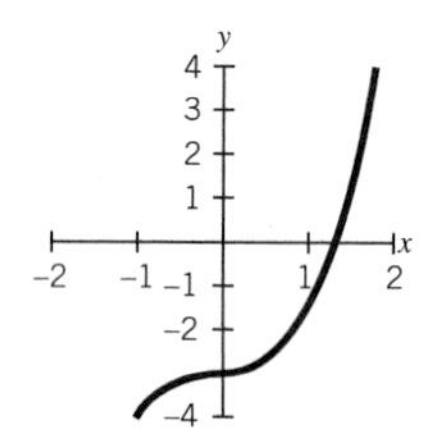

   b. The y-intercept is at 3. The graph does not intersect the x-axis.

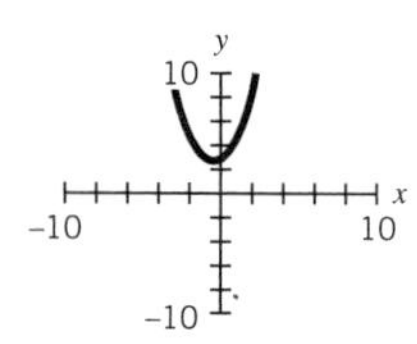

2. a. Degree is 1. x-intercept is -2.
   $3x + 6 = 0 \rightarrow 3x = -6 \rightarrow x = -2$

   b. Degree is 2. x-intercepts are -4 and 1.
   $(x + 4) = 0 \rightarrow x = -4$; $(x - 1) = 0 \rightarrow x = 1$
   To sketch the graph, determine which segment of the function between those x-intercepts is positive (above the x-axis), or negative (below the x-axis).
   Evaluate the function for a value of x between those x-intercepts, x: $-4 < x < 1$. If we use $x = 0$, $f(0) = (0 + 4)(0 - 1) = -4$, so that segment is negative.
   For values of $x > 1$, both factors $(x + 4)$ and $(x - 1)$ are

(Answer is continued on next page.)

positive, so their product $(x + 4)(x - 1)$ is also positive, and the segment to the right of $x = 1$ is positive.
For values of $x < -4$, both factors $(x + 4)$ and $(x - 1)$ are negative, so their product $(x + 4)(x - 1)$ is positive, and the segment to the left of $x = -4$ is positive.

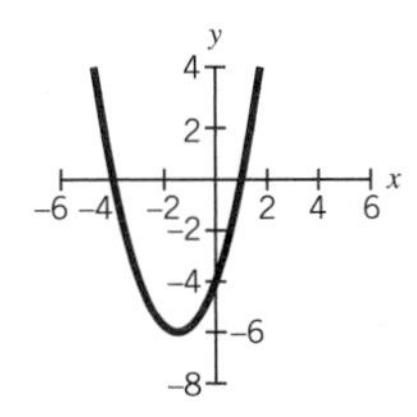

c. Degree is 3. x-intercepts are -5, 3, and $-2\frac{1}{2}$.
$(x + 5) = 0 \rightarrow x = -5$
$(x - 3) = 0 \rightarrow x = 3$
$(2x + 5) = 0 \rightarrow 2x = -5 \rightarrow x = -\frac{5}{2}$ or $-2\frac{1}{2}$
Determine whether the segments between consecutive x-intercepts are positive or negative by evaluating the sign of each factor for a value of x between those x-intercepts. You might use $x = 0$ for x: $-2\frac{1}{2} < x < 3$, to easily determine that segment is negative, and use $x = -3$ for x: $-5 < x < -2\frac{1}{2}$, to easily determine that segment is positive.
For values of $x > 3$, all factors $(x + 5)$, $(x - 3)$ and $(2x + 5)$ are positive, so the product $(x + 5)(x - 3)(2x + 5)$ is positive, as is segment to the right of $x = 3$.
A comparable analysis of values of $x < -5$, shows all those factors are negative, as is their product. So, the segment to the left of $x = -5$ is negative.

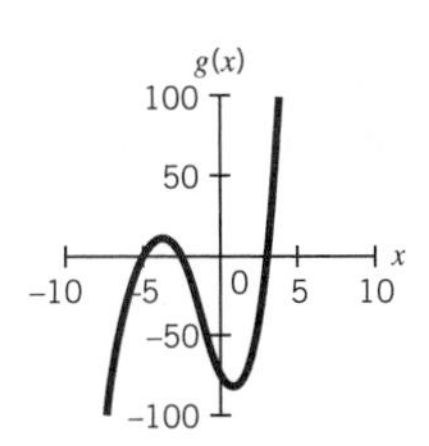

d. Degree is 5. x-intercepts are 5, 3, 0, -3 and -5.

$(x - 5) = 0 \rightarrow x = 5$
$(x - 3) = 0 \rightarrow x = 3$
$x = 0$
$(x + 3) = 0 \rightarrow x = -3$
$(x + 5) = 0 \rightarrow x = -5$

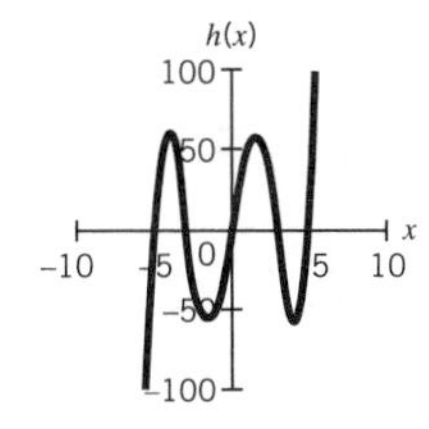

3. There are infinitely many examples of such functions. To have those four x-intercepts, they are of the form:

$f(x) = ax^n(x + 3)^m(x - 5)^p(x - 7)^r$, where a is a real number, and n, m, p and r are natural numbers.

(i) $f(x) = x(x + 3)(x - 5)(x - 7)$
$= (x^2 + 3x)(x^2 - 12x + 35)$
$= x^4 - 9x^3 - x^2 + 105x$

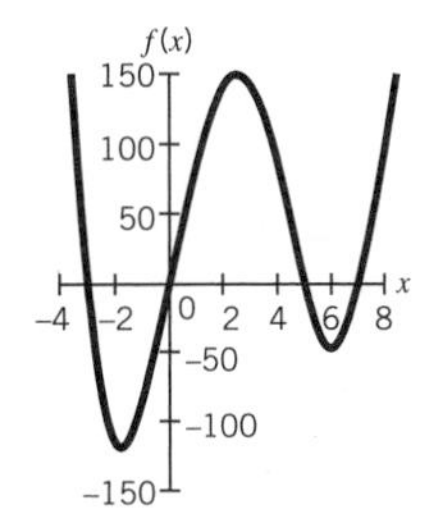

$f(x) = x(x + 3)(x - 5)(x - 7)$

(ii) $h(x) = 2x^2(x + 3)(x - 5)(x - 7)$
$= (2x^3 + 6x^2)(x^2 - 12x + 35)$
$= 2x^5 - 18x^4 - 2x^3 + 210x^2$
Note: $h(x) = 2xf(x)$

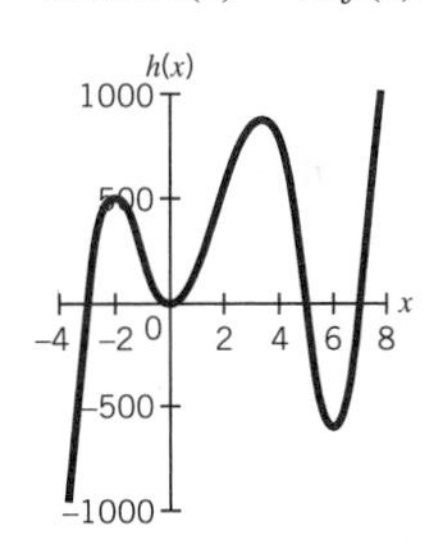

$h(x) = 2x^2 (x + 3)(x - 5)(x - 7)$

(iii) $g(x) = -20x^2(x + 3)(x - 5)(x - 7)$
$= -20x^5 + 180x^4 + 20x^3 - 2100x^2$
Note: $g(x) = -20xf(x)$

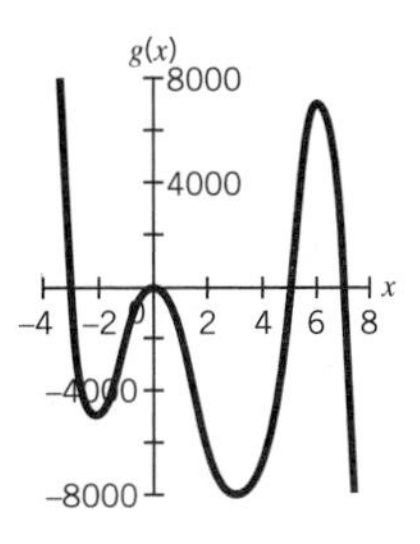

$g(x) = -20x^2(x + 3)(x - 5)(x - 7)$

Algebra Aerobics 8.2

1. a. vertex: (0, 2); axis of symmetry is $x = 0$; concave down; two x-intercepts; the y-intercept is (0, 2)

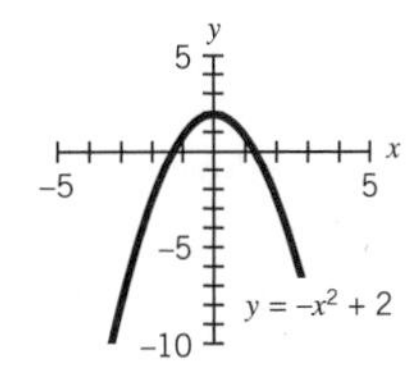

b. vertex: $(-1, 0)$; axis of symmetry is $x = -1$; concave up; one x-intercept; the y-intercept is $(0, 1)$.

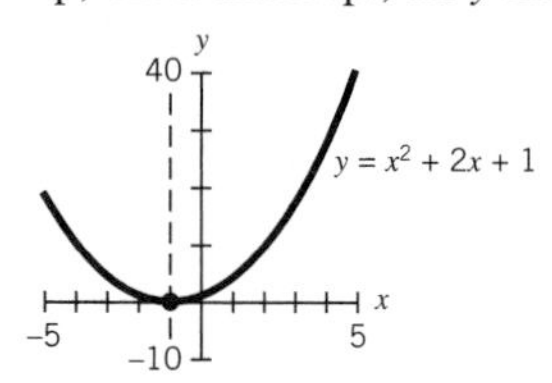

c. vertex: $(2, 1)$; axis of symmetry is $x = 2$; concave up; no x-intercepts; the y-intercept is $(0, 3)$.

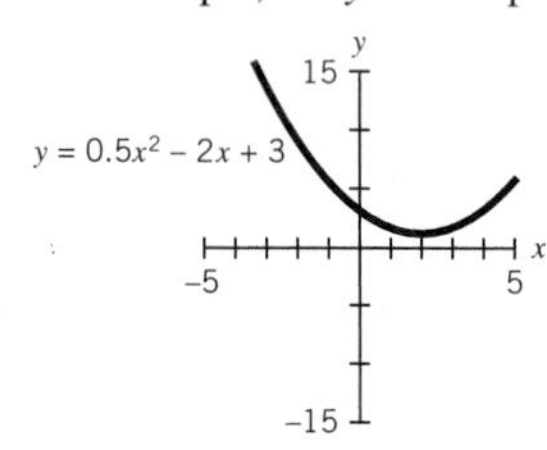

2. $y = ax^2 + bx + c$ compared to $y = x^2$ $(a = 1)$

a. $y = 2x^2 - 5 \rightarrow a = 2$;
i. $a > 0 \rightarrow$ minimum at the vertex
ii. $|a| > 1$, so parabola is narrower than $y = x^2$.

b. $y = 0.5x^2 + 2x - 10 \rightarrow a = 0.5$;
i. $a > 0 \rightarrow$ minimum at the vertex
ii. $|a| < 1$, so parabola is broader than $y = x^2$.

c. $y = 3 + x - 4x^2 \rightarrow a = -4$;
i. $a < 0 \rightarrow$ maximum at the vertex
ii. $|a| > 1$, so parabola is narrower than $y = x^2$.

d. $y = -0.2x^2 + 11x + 8 \rightarrow a = -0.2$;
i. $a < 0 \rightarrow$ maximum at the vertex
ii. $|a| < 1$, so parabola is broader than $y = x^2$.

In 3–7 references to a, b, c are from $y = ax^2 + bx + c$.

3. $y = 2x^2 + 2$ is narrower than $y = x^2 + 2$ because $|(a = 2)| > |(a = 1)|$. Both have vertical intercept at $(0, 2)$ since $c = 2$ in both equations.

4. They have the same shape since both have $a = 1$. $g(x)$ is 6 units higher than $f(x)$ because $(c = 8)$ is 6 more than $(c = 2)$.

5. $d = -t^2 = 5$ is concave down since $a = -1$ is negative, $d = t^2 + 5$ is concave up since $a = 1$ is positive. They have the same shape because $|a| = 1$ in both. They both cross vertical axis at 5 since $c = 5$ in both equations.

6. Both are concave down because both have $a < 0$. Both have vertex at $(0, 0)$ since b and $c = 0$ in both equations. $g(z)$ is broader than $f(z)$ because $|a = -0.5| < |a = -5|$.

7. They have the same shape and are concave down since $a = -3$ is negative in both equations. $h = -3t^2 + t - 2$ is 3 units higher than $h = -3t^2 + t - 5$ since $(c = -2)$ is 3 more than $(c = -5)$.

Algebra Aerobics 8.3a

1. a. The vertex is above x-axis and opens up, so there are no x-intercepts. The function has 2 imaginary zeros.

b. The vertex is on the x-axis, so there is one x-intercept. The function has one real zero.

c. The vertex is above the x-axis and opens down, so there are two x-intercepts. The function has two real zeros.

2. Discriminant is $b^2 - 4ac$, from $ax^2 + bx + c = y$

a. $a = -5, b = -1, c = 4 \rightarrow$ discriminant $= 1 - 4(-5)(4) = 81 > 0 \Rightarrow$ two x-intercepts. $\sqrt{81} = 9$, so roots are rational. The function has two real zeros at:

$$x = \frac{1 \pm 9}{-10} \rightarrow$$

$$x = \frac{1 + 9}{-10} = \frac{10}{-10} = -1 \quad \text{and} \quad x = \frac{1 - 9}{-10} = \frac{-8}{-10} = \frac{4}{5}$$

So the x-intercepts are $(-1, 0)$ and $(4/5, 0)$.

b. $a = 4, b = -28, c = 49 \rightarrow$ discriminant $= 784 - 4(4)(49) = 784 - 784 = 0 \Rightarrow$ one x-intercept. $\sqrt{0} = 0$, so root is rational. The x-intercept is at:
$$x = \frac{28 \pm 0}{8} = \frac{7}{2}; \left(3\frac{1}{2}, 0\right).$$

c. $a = 2, b = 5, c = 4 \rightarrow$ discriminant $= 25 - 4(2)(4) = 25 - 32 = -7$ which is negative $\Rightarrow$ no x-intercepts. The function has two imaginary zeros at
$$x = \frac{-5 \pm \sqrt{-7}}{4} = \frac{-5 \pm \sqrt{7}i}{4}.$$

3. a. $h = -4.9t^2 + 50t + 80$. The vertical intercept is the initial height (at 0 seconds), which is 80 m; $(0, 80)$. $a = -4.9, b = 50, c = 80$. So, the horizontal intercepts are:

$$t = \frac{-50 \pm \sqrt{2500 - 4(4.9)(80)}}{2(-4.9)} = \frac{-50 \pm \sqrt{2500 + 1568}}{-9.8} = \frac{-50 \pm \sqrt{4068}}{-9.8} \rightarrow$$

$$t = \frac{-50 + 63.8}{-9.8} = \frac{13.8}{-9.8} = -1.41 \quad \text{and} \quad t = \frac{-50 - 63.8}{-9.8} = \frac{-113.8}{-9.8} = 11.61 \text{ seconds.}$$

Negative values of t have no meaning in a height equation, so the horizontal intercept at $(11.61, 0)$ means that the object hits the ground after 11.61 seconds.

b. $h = 150 - 80t - 490t^2$. The vertical intercept is (0, 150) which means that the initial height is 150 cm. $a = -490$, $b = -80$, $c = 150 \rightarrow$
The horizontal intercepts are:

$$t = \frac{80 \pm \sqrt{6400 - 4(-490)(150)}}{2(-490)}$$
$$= \frac{80 \pm \sqrt{6400 + 294{,}000}}{-980}$$
$$= \frac{80 \pm \sqrt{300{,}400}}{-980} = \frac{80 \pm 548}{-980} \rightarrow$$
$$t = \frac{80 + 548}{-980} \quad \text{and} \quad t = \frac{80 - 548}{-980}$$
$$= \frac{628}{-980} \qquad = \frac{-468}{-980}$$
$$= -0.64 \qquad = 0.48 \text{ seconds.}$$

Discard negative solution. The object hits the ground after 0.48 seconds.

Algebra Aerobics 8.3b

1. **a.** $y = -16t^2 + 50t \rightarrow y = -2t(8t - 25)$; $y = 0 \rightarrow t = 0$, or $(8t - 25) = 0 \rightarrow t = 3\frac{1}{8}$. So, the horizontal intercepts are (0, 0) and $\left(3\frac{1}{8}, 0\right)$.

   b. $y = x^2 + x - 6 \rightarrow y = (x + 3)(x - 2)$; $y = 0 \rightarrow$ $(x + 3) = 0 \rightarrow x = -3$, or $(x - 2) = 0 \rightarrow x = 2$. So the horizontal intercepts are $(-3, 0)$ and (2, 0).

   c. $y = 2x^2 + x - 5$ does not factor, so use quadratic formula. $\rightarrow a = 2, b = 1, c = -5 \rightarrow$. The discriminant is: $1 - 4(2)(-5) = 1 + 40 = 41$, which is not a perfect square, so, the function does not factor. The zeros of the function given by the quadratic formula occur at $\frac{-1 \pm \sqrt{41}}{4}$. Since these are real numbers, they also represent the x-intercepts.

$$\sqrt{41} \approx 6.4 \rightarrow x \approx \frac{-1 + 6.4}{4} \quad \text{and} \quad x \approx \frac{-1 - 6.4}{4}$$
$$\approx \frac{5.4}{4} \qquad \approx \frac{-7.4}{4}$$
$$\approx 1.4 \qquad \approx -1.9$$

   So the x-intercepts are approximately at: (1.4, 0) and $(-1.9, 0)$.

   d. $h(t) = 69 - 9t^2 \rightarrow h(t) = 3(23 - 3t^2)$, but does not factor easily into two linear terms. To find the x-intercepts, we solve $0 = 3(23 - 3t^2)$, $\rightarrow 23 - 3t^2 = 0$ or $t^2 = \frac{23}{3} \rightarrow t = \pm\sqrt{\frac{23}{3}} \approx \pm\sqrt{7.7} \approx \pm 2.77$. So the x-intercepts are approximately at: (2.8, 0) and $(-2.8, 0)$.

   e. $f(x) = -2x^2 + 12x + 54 \rightarrow f(x) = -2(x^2 - 6x - 27)$ $\rightarrow -2(x - 9)(x + 3)$. Setting $f(x) = 0$, we get $x = 9$ or $x = -3$. The x-intercepts are (9, 0) and $(-3, 0)$.

   f. $g(x) = 64x^2 + 16x + 4 \rightarrow g(x) = 4(16x^2 + 4x + 1)$ which does not factor to linear factors. $a = 16$, $b = 4$, $c = 1 \rightarrow$ the discriminant is: $16 - 4(16)(1) = -48$, which is negative, so $g(x) = 16x^2 + 4x + 1 = 0$ has no real roots and hence no x-intercepts.

2. **a.** When $x = 0$, $y = 2(0) - 3(0) = 0$, so y-intercept is (0, 0).

   b. When $x = 0$, $y = 5 - (0) - 4(0) = 5$, so y-intercept is (0, 5).

   c. When $x = 0$, $y = \frac{2}{3}(0) + 6(0) - \frac{11}{3} = -\frac{11}{3}$, so y-intercept is $\left(0, -3\frac{2}{3}\right)$.

Algebra Aerobics 8.4a

1. **a.** $f(0) = -4$, so vertex is $(0, -4)$.

   b. $g(0) = 6$, so vertex is (0, 6).

   c. $t = 0 \rightarrow w = 1$, so vertex is (0, 1).

2. **a.** vertex is (0, 3)

   b. vertex is (0, 3)

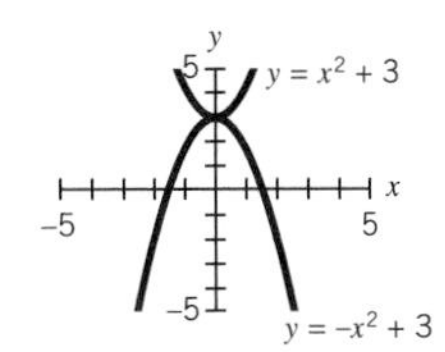

3. **a.** Vertex is $f(0) = -3(0)^2 = 0 \rightarrow (0, 0)$. Opens down since $a = -3$, which is negative. There is one x-intercept (0, 0), since vertex is on the x-axis.

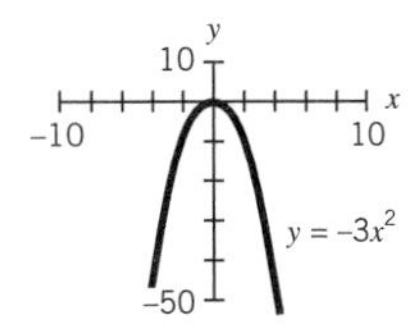

   b. Vertex is $f(0) = -2(0)^2 - 5 = -5 \rightarrow (0, -5)$, which is below x-axis. Opens down since $a = -2$ which is negative. So, there are no x-intercepts.

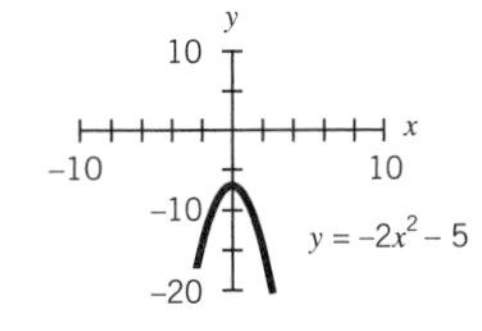

In c and d, use formula $\frac{-b}{2a}$ as horizontal coordinate of the vertex $(f(x) = ax^2 + bx + c)$.

c. $a = 1, b = 4 \rightarrow \frac{-b}{2a} = \frac{-4}{2(1)} = -2,$
$f(-2) = (-2)^2 + 4(-2) - 7 = 4 - 8 - 7 = -11.$
Vertex is $(-2, -11)$. Opens up since a = 1 which is positive. There are two x-intercepts. They are approximately $(-5, 0)$ and $(1, 0)$.

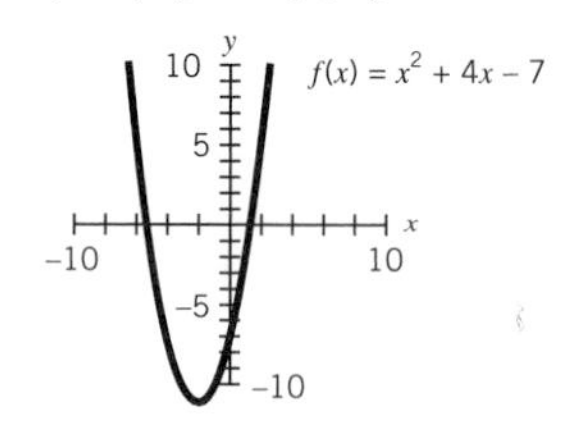

d. $f(x) = 4 - x - 2x^2 = -2x^2 - x + 4$; so $a = -2$, $b = -1$. $\frac{-b}{2a} = \frac{-(-1)}{2(-2)} = \frac{1}{-4} = -\frac{1}{4}$;
$f\left(-\frac{1}{4}\right) = 4 - \left(-\frac{1}{4}\right) - 2\left(-\frac{1}{4}\right)^2 = 4 + \frac{1}{4} - 2\left(\frac{1}{16}\right) =$
$4 + \frac{1}{4} - \frac{1}{8} = 4\frac{1}{8}$. So, vertex is $\left(-\frac{1}{4}, 4\frac{1}{8}\right)$. Opens down since $a = -2$, which is negative. There are two x-intercepts. They are approximately $(-2, 0)$ and $(1, 0)$.

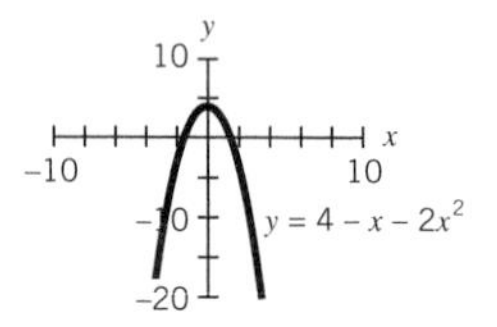

4. a. $a = 1, b = 3$; so horizontal coordinate of vertex is:
$\frac{-b}{2a} = \frac{-3}{2} = -1\frac{1}{2}$. If $x = -\frac{3}{2}, \rightarrow$
$y = \left(-\frac{3}{2}\right)^2 + 3\left(-\frac{3}{2}\right) + 2 = \frac{9}{4} - \frac{9}{2} + 2 =$
$2\frac{1}{4} - 4\frac{2}{4} + 2 = -\frac{1}{4}$. So, vertex is $\left(-1\frac{1}{2}, -\frac{1}{4}\right)$.
To find horizontal intercepts, set $y = 0$, $0 = x^2 + 3x + 2$
$\rightarrow 0 = (x + 2)(x + 1)$. $(x + 2) = 0 \rightarrow x = -2$;
$(x + 1) = 0 \rightarrow x = -1$, so horizontal intercepts are -2 and -1, $(-2, 0)$ and $(-1, 0)$.

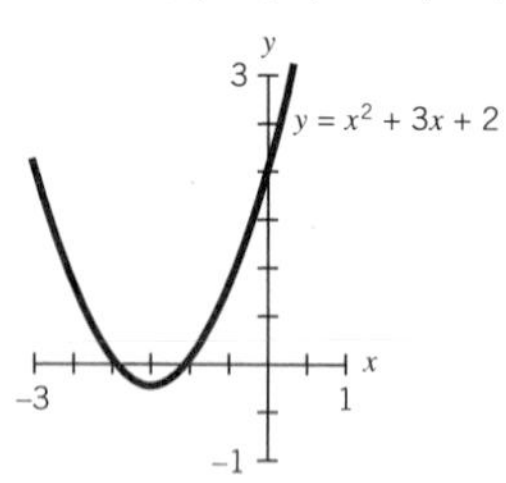

b. $a = 2, b = -4$; so horizontal coordinate of vertex is:
$\frac{-b}{2a} = \frac{-(-4)}{2(2)} = 1$. $f(1) = 2 - 4 + 5 = 3$, so vertex is $(1, 3)$. Since vertex is above x-axis, and $a = 2$ which is positive, it opens up, it does not cross x-axis, so no horizontal intercepts.

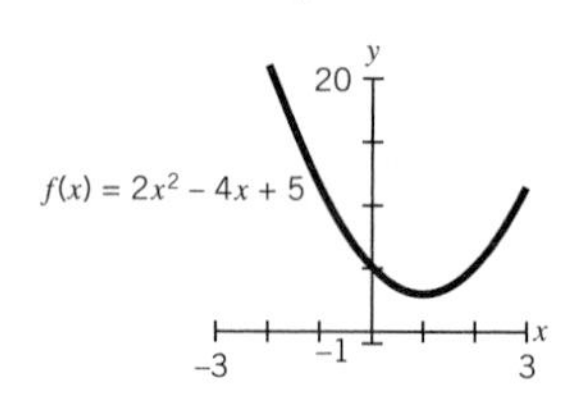

c. $a = -1, b = -4; \rightarrow \frac{-b}{2a} = \frac{-(-4)}{2(-1)} = \frac{4}{-2} = -2.$
$g(-2) = -(-2)^2 - 4(-2) - 7 = -4 + 8 - 7 = -3$, so vertex is $(-2, -3)$. Since vertex is below t-axis, and $a = -1$ which is negative, means it opens down, it does not cross t-axis, so no horizontal intercepts.

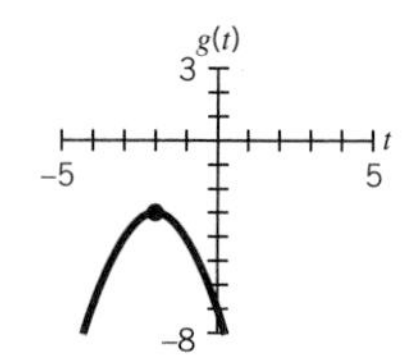

Algebra Aerobics 8.4b

1. a. vertex is $(0, 0)$; **b.** vertex is $(-3, 0)$;

c. vertex is $(2, 0)$.

All vertices lie on the x-axis. Vertex of (b) is 3 units to the left of the vertex of (a). Vertex of (c) is 2 units to the right of vertex of (a). All have same shape and open upward.

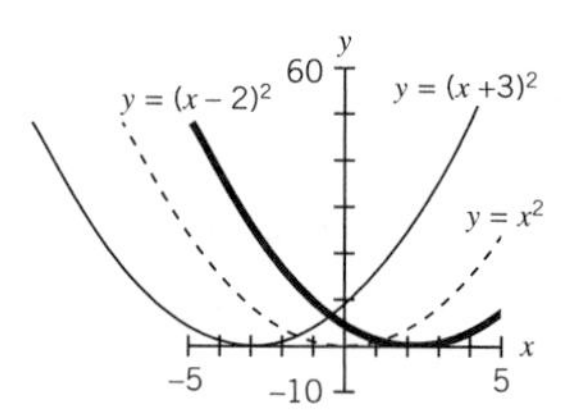

2. a. vertex is $(0, 0)$; **b.** vertex is $(1, 0)$;

c. vertex is $(-4, 0)$.

All vertices lie on the x-axis. The vertex of (b) is 1 unit to the right of the vertex of (a). The vertex of (c) is 4 units to the left of vertex of (a). All have same shape and open upward.

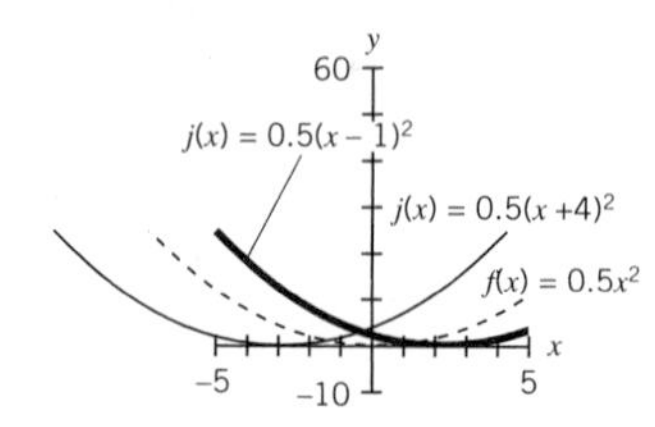

3. a. vertex is (0, 0); b. vertex is (− 1.2, 0);

c. vertex is (0.9, 0).

All of the vertices lie on the horizontal axis. The vertex of (b) is 1.2 units to the left of the vertex of (a). The vertex of (c) is 0.9 units to the right of the vertex of (a). All have same shape and open downward.

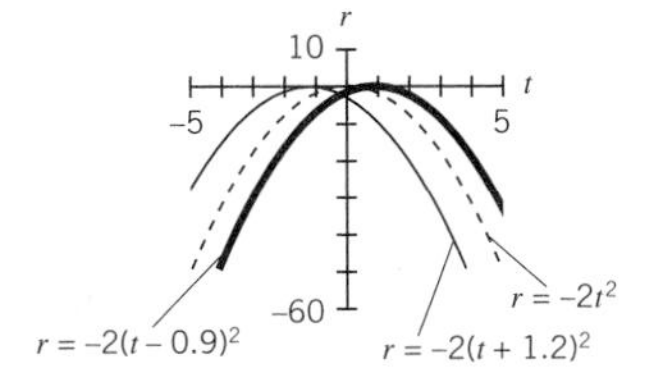

4. $y = a(x - h)^2 + k$. The value of $h = 2$ in (b), (c) and (d), so the vertices of those parabolas are 2 units to the right of vertex of (a), where $h = 0$. All open up with the same shape since $a = 1$ in all four equations. (c) is 4 units above (a) and (b); (d) is 3 units below (a) and (b); since $k = 0$ in (a) and (b), $k = 4$ in (c) and $k = -3$ in (d).

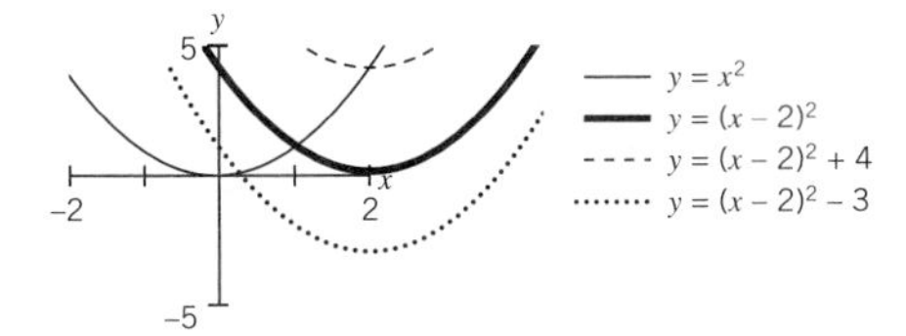

5. $y = a(x - h) + k$. All open down with the same shape since $a = -1$ in all four equations. (b), (c) and (d) have vertices 3 units to the left of vertex of (a). Since $h = -3$ in those equations, and $h = 0$ in (a), $k = 0$ in (a) and (b), but $k = -1$ in (c) so (c) is 1 unit below (a) and (b). $k = 4$ in (d), so (d) is 4 units above (a) and (b).

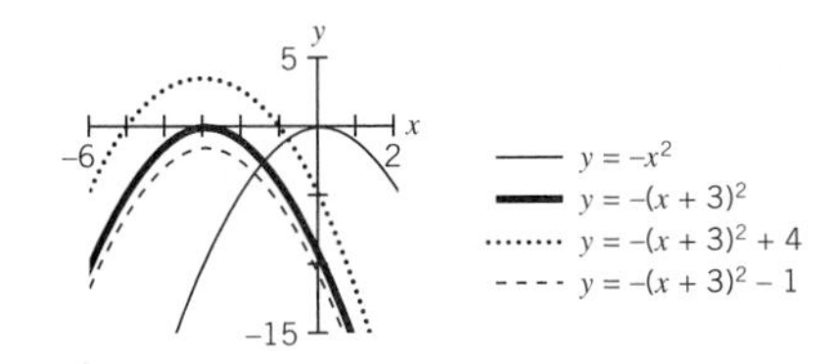

6. a. vertex is (1, 5), above x-axis, $a = 3$ is positive so it opens up; so, there are no x-intercepts.

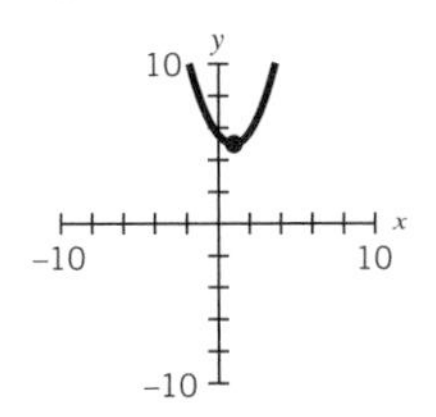

b. vertex is (− 4, − 1), below x-axis, $a = -2$ is negative so it opens down; so, there are no x-intercepts.

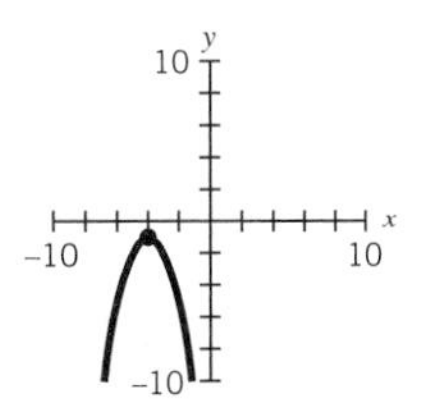

c. vertex is (− 3, 0), on x-axis, $a = -5$ is negative so it opens down; so, there is one x-intercept.

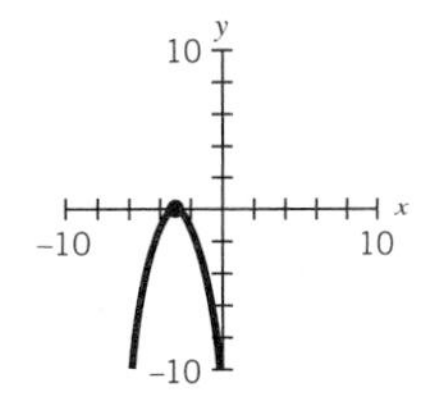

d. vertex is (1, − 2), below x-axis, $a = 3$ is positive so it opens up; so, there are two x-intercepts.

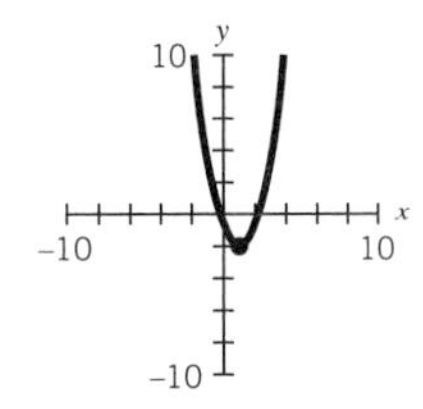

Algebra Aerobics 8.4c

1. a. $f(x) = x^2 + 2x - 1 = x^2 + 2x + (1 - 1) - 1$
$= (x^2 + 2x + 1) + (-1 - 1) \rightarrow$
$f(x) = (x + 1)^2 - 2$

b. $j(z) = 4z^2 - 8z - 6 = 4(z^2 - 2z) - 6$
$= 4(z^2 - 2z + 1) - 4(1) - 6 \rightarrow$
$j(z) = 4(z - 1)^2 - 10$

c. $h(x) = -3x^2 - 12x = -3(x^2 + 4x)$
$= -3(x^2 + 4x + 4) + 3(4) \rightarrow$
$h(x) = -3(x + 2)^2 + 12$

2. a. $y = 2(x - \frac{1}{2})^2 + 5 = 2(x - \frac{1}{2})(x - \frac{1}{2}) + 5$
$= 2(x^2 - x + \frac{1}{4}) + 5 = 2x^2 - 2x + \frac{1}{2} + 5 \rightarrow$
$y = 2x^2 - 2x + 5\frac{1}{2}$

b. $y = -\frac{1}{3}(x + 2)^2 + 4 = -\frac{1}{3}(x + 2)(x + 2) + 4$
$= -\frac{1}{3}(x^2 + 4x + 4) + 4$
$= -\frac{1}{3}x^2 - \frac{4}{3}x - \frac{4}{3} + 4 \rightarrow$
$y = -\frac{1}{3}x^2 - \frac{4}{3}x + 2\frac{2}{3}$

3. a. $y = x^2 + 6x + 7 = x^2 + 6x + 9 - 9 + 7 \rightarrow$
$y = (x + 3)^2 - 2$

b. $y = 2x^2 + 4x - 11 = 2(x^2 + 2x) - 11$
$= 2(x^2 + 2x + 1) - 2(1) - 11 \rightarrow$
$y = 2(x + 1)^2 - 13$

4. a. $y = x^2 + 8x + 11 \rightarrow y = x^2 + 8x + 16 - 16 + 11 \rightarrow$
$y = (x + 4)^2 - 5$
So, vertex is $(-4, -5)$. y-intercept at $x = 0 \rightarrow$
$y = 0 + 0 + 11 \rightarrow y = 11$ is $(0, 11)$. To solve for x-intercepts use: $y = x^2 + 8x + 11 \rightarrow a = 1, b = 8, c = 11 \rightarrow$

$$y = 0 = x^2 + 8x + 11 \rightarrow x = \frac{-8 \pm \sqrt{64 - 4(1)(11)}}{2(1)} \rightarrow$$

$$\frac{-8 \pm \sqrt{64 - 44}}{2} \rightarrow \frac{-8 \pm \sqrt{20}}{2} \rightarrow \frac{-8 \pm 2\sqrt{5}}{2} \rightarrow$$

$$\frac{2(-4 \pm \sqrt{5})}{2} \rightarrow x = -4 \pm \sqrt{5}:$$

$\sqrt{5} \approx 2.2 \rightarrow x = -4 + 2.2 = -1.8$ and
$x = -4 - 2.2 = -6.2$
So, x-intercepts are approximately at $(-1.8, 0)$ and $(-6.2, 0)$.

b. $y = 3x^2 + 4x - 2$ when $x = 0 \rightarrow y = 0 + 0 - 2 = -2$
y-intercept is $(0, -2)$. $a = 3, b = 4, c = -2 \rightarrow$

$$\frac{-b}{2a} = \frac{-4}{6} = -\frac{2}{3};$$

$$f\left(-\frac{2}{3}\right) = 3\left(-\frac{2}{3}\right)^2 + 4\left(\frac{-2}{3}\right) - 2 \rightarrow$$

$$f\left(-\frac{2}{3}\right) = 3\left(\frac{4}{9}\right) - \frac{8}{3} - 2 = \frac{4}{3} - \frac{8}{3} - 2 =$$

$$-\frac{4}{3} - 2 = -1\frac{1}{3} - 2 = -3\frac{1}{3}, \text{ so vertex is}$$

$\left(-\frac{2}{3}, -3\frac{1}{3}\right)$. To solve for x-intercepts $\rightarrow y = 0 =$

$$3x^2 + 4x - 2 \rightarrow x = \frac{-4 \pm \sqrt{16 - 4(3)(-2)}}{2(3)} =$$

$$\frac{-4 \pm \sqrt{16 + 24}}{6} = \frac{-4 \pm \sqrt{40}}{6} = \frac{-4 \pm 2\sqrt{10}}{6} =$$

$$\frac{2(-2 \pm \sqrt{10})}{2 \cdot 3}$$

$$x = \frac{-2 \pm \sqrt{10}}{3}. \quad \left(\sqrt{10} \approx 3.16\right)$$

$$x = \frac{-2 + 3.16}{3} = .387;$$

$$x = \frac{-2 - 3.16}{3} = -1.72$$

So x-intercepts are approximately $(0.4, 0)$ and $(-1.7, 0)$.

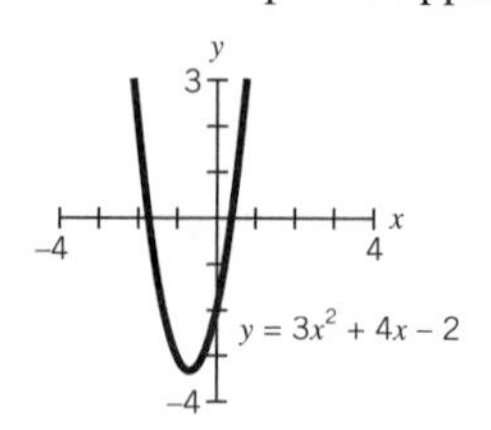

5. a. Vertex is $(-5, -11)$. y-intercept at $x = 0 \rightarrow$
$y = 0.1(5)^2 - 11 = 0.1(25) - 11 = 2.5 - 11 = -8.5$. So, y-intercept is -8.5. To solve for x-intercepts $\rightarrow y = 0 \rightarrow$
$0 = 0.1(x + 5)^2 - 11 \rightarrow 11 = (0.1)(x + 5)^2 \rightarrow 110 =$
$(x + 5)^2 \rightarrow \pm\sqrt{110} = x + 5 \rightarrow x = -5 \pm \sqrt{110}$.
$(\sqrt{110} \approx 10.5) \rightarrow x \approx -5 + 10.5 \approx 5.5$ and $x \approx$
$-5 - 10.5 \approx -15.5$ so, x-intercepts are approximately $(5.5, 0)$ and $(-15.5, 0)$.

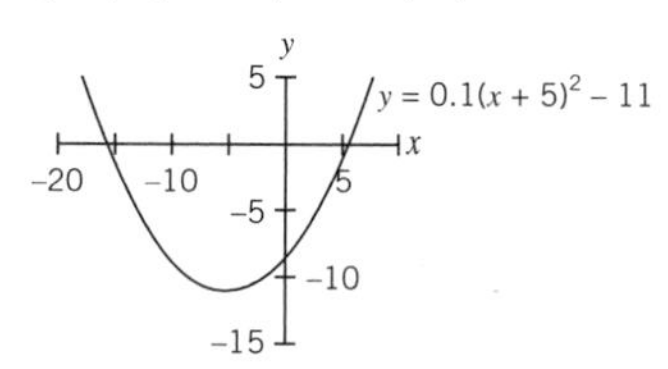

b. Vertex is $(1, 4)$. y-intercept at $x = 0 \rightarrow$
$y = -2(-1)^2 + 4 = -2 + 4 = 2$. So y-intercept is 2.
To solve for x-intercept $\rightarrow y = 0 \rightarrow$
$0 = -2(x - 1)^2 + 4 \rightarrow 2(x - 1)^2 = 4 \rightarrow$
$(x - 1)^2 = 2 \rightarrow (x - 1) = \pm\sqrt{2} \rightarrow$
$x = 1 \pm \sqrt{2}$. $(\sqrt{2} \approx 1.4) \rightarrow$
$x = 1 + \sqrt{2}$ and $x = 1 - \sqrt{2}$
$\approx 1 + 1.4 \qquad \approx 1 - 1.4$
$\approx 2.4 \qquad \approx -0.4$
So x-intercepts are approximately $(2.4, 0)$ and $(-0.4, 0)$.

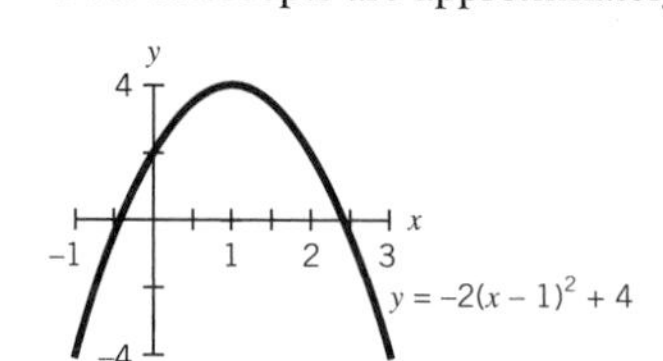

Brief Solutions to Odd-Numbered Problems

CHAPTER 1

Exercises for Section 1.1

1. **a.** Mean = 386/7 = 55.14; median = 46.

   b. Changing any entry in the list that is greater than the median to something still higher will not change the median of the list.

3. Mean = \$24,700; median = \$18,000. The mean is heavily weighted by the two high salaries. The mean salary is more attractive but not likely to be an accurate indicator.

5. None is given here. (In general, when answers can vary quite a bit, either a typical answer or none is given.)

7. He is correct, provided the person leaving state A has an I.Q. that is below the average I.Q. of people in state A and above the average I.Q. of the people in state B.

9. The mean net worth of a group, e.g., American families, is heavily biased upwards by the very high incomes of a relatively small subset of the group. The median net worth of such groups is not as biased.

11. **a.** $\left(\sum_{i=1}^{5} x_i\right)/5$ **b.** $\left(\sum_{i=1}^{n} t_i\right)/n$ **c.** $\sum_{k=1}^{5} 2k = 30$

13. From 35 to 35.5 years, depending on one's selection of group midpoints.

15. **a.** 44 **b.** Quantitative; measured in units of \$1000.

    c. 22.7%

    d.

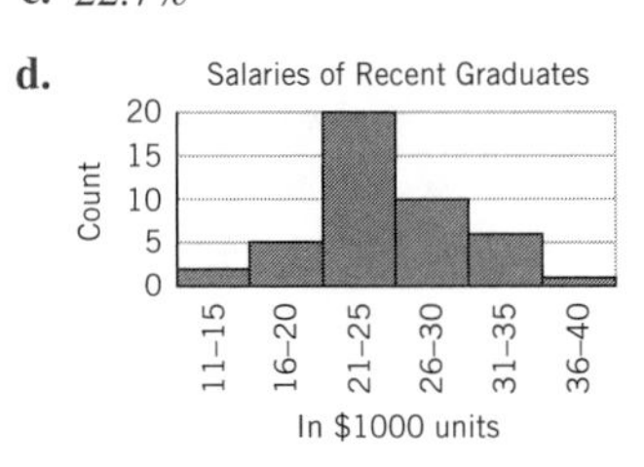

17. **a.** Housing; 41.4% **b.** \$5220.00

    c. "Where does the average American household's money go?"

19. Mean = \$3.24; median = \$3

Exercises for Section 1.2

21. Answers will vary. One very noticeable trend is the increase in 2050 of the percentages of persons in each 5 year age spread, starting at 45 and going up.

23. **a.** Northeast: mean = 204.9/9 = 22.77; median = 9.20
    Midwest: mean = 238.9/12 = 19.91; median = 18.05

    b. The means and medians are for occupancy rates in NE and MW and the great discrepancy between these are for personnel in both sections.

25. The population of China is substantially younger than that of the U.S.

27. **a.** Sex is a qualitative variable; age and weight are quantitative.

    b. Below is a table with the computed data–all entries are measured in pounds.

1993 data	Weights of Wolves in the Northwest	
	Mean	Median
Male	103.3	98
Female	68.5	72

    c. The median is preferred to the mean.

    d. If one counts a pup's age as 0.5 years and takes the midpoint when a range is given, then the male mean age = 3.63 years, the female mean = 1.96 years, and the mean age overall is 2.33 years.

29. a. i. 11.4 million

ii. 6.1 million

iii. males: 7.1 million; females: 12 million

iv. 53.4 million

b. By 2050 there will be a huge increase in the 85+ age range.

31. This is open to interpretation. By 2050 the mean age may be greater than the median for the United States, as it is more affected than the median by outlier values (e.g., by those living longer because of medical advances). On the other hand, the pyramid seems as if it will be weighted downward and this could bring the mean value down.

Exercises for Section 1.3

33. a. 86–87 and 89–93

b. 87–89 and 93–94

c. \$12 billion in 1993

d. \$7.5 billion minimum in sales in 1989

35. a. Approximately 37.4 °C.

b. Approximately 36.3 °C.

c. His temperature was fairly the same from noon to 6 PM and then started going down until it reached a low around 1 AM, then slowly rose until 7 AM and finally crept up to where it was around noon on the previous day.

37. a. From $2.4 \cdot 10^5$ to $5.8 \cdot 10^5$; from $1.8 \cdot 10^5$ to a high of $3.8 \cdot 10^5$.

b. 1900 to 1930 **c.** 1930 to 1990

d. Sometime around 1938

Exercises for Section 1.4

39. a. Add 1 to the value of x; divide the result by what one gets by subtracting 1 from the value of x.

b. (5, 1.5) **c.** (2, 3)

d. No, the formula is not defined if $x = 1$.

41. a. $x = 0$ implies $y = 0$ **b.** If $x > 0$ then $y < 0$.

c. If $x < 0$ then $y < 0$. **d.** No

43. a. Only $(-1, 3)$ solves $y = 2x + 5$.

b. (1, 0) and (2, 3) solve $y = x^2 - 1$.

c. $(-1, 3)$ and (2, 3) solve $y = x^2 - x + 1$.

d. Only (1, 2) solves $y = 4/(x + 1)$.

Exercises for Section 1.5

45. a. $S_1 = 0.90 \cdot P$; \$90 **b.** $S_2 = (0.90)^2 \cdot P$; \$81

c. $S_3 = (0.90)^3 \cdot P$; \$72.90

d. $S_5 = (0.90)^5 \cdot P$; \$59.05; 40.95%

47. Yes; it passes the vertical line test.

49. No; an input of 120 results in two different outputs. Weight, however, is a function of height since there are no repeated heights in the second column.

51. a. The graph labeled (i).

b. It alone passes the vertical line test.

53. a. True; each weight input has only one cost output.

b. False; the cost, e.g., to mail a 3 oz. parcel can be \$3.59 or \$3.90, depending on the zone. Weight, however, is a function of cost if one sticks to a single zone.

c. True; the cost for all zones is the same.

d. True; the same weight can be mapped to different zones.

e. False, knowing just the zone does not tell the cost; knowing just the weight also does not tell the cost.

55. a. The formulae are: $y = x + 5$; $y = x^2 + 1$; $y = 3$

b. All 3 represent y as a function of x; each input of x has only one output y.

57. a. Year is the independent variable and the ppm of carbon monoxide is the dependent variable.

b. It is a function; each Year is associated with a unique level of carbon monoxide.

c. The carbon monoxide content varies according to the year.

d.

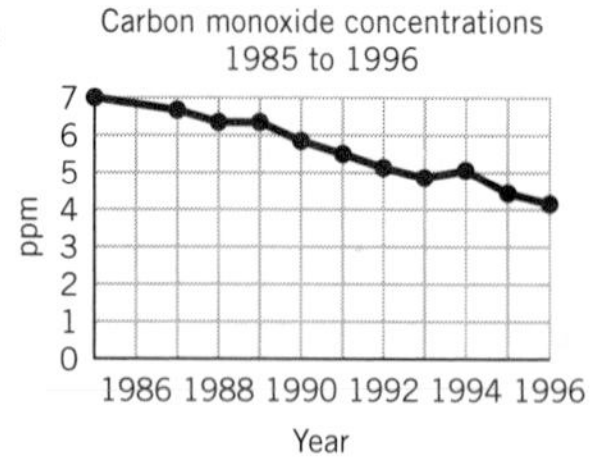

Carbon monoxide concentrations
1985 to 1996
ppm: 0 1 2 3 4 5 6 7
Year: 1986 1988 1990 1992 1994 1996

e. Function is decreasing from 1985 to 1993 and is increasing from 1993 to 1994.

f. maximum: 6.97 ppm; minimum: 4.20 ppm

59. a. P is a function of Y since the inputs are all distinct.

b. The domain is the set of years from 1990 to 1995 inclusive; the range is the set of corresponding money values, namely $\{-0.5, 0, 1.2, 1.4, 2.3\}$

c. The maximum P value is \$2.3; it occurs in 1991.

d. P is increasing from 1990 to 1991 and from 1993 to 1994.

e. Y is not a function of P since an input of \$1.4 has two different outputs.

61. a. $C = 1.24 \cdot G$, where C is cost in dollars and G = number of gallons purchased.

b. G is the independent variable and C the dependent one.

c. For each input of G, there is one and only one value for C; thus it is a function.

d. A suitable domain would be from 0 to the maximum capacity in gallons of a gas tank. We can play it safe and say 50 gallons. The corresponding range is then from \$0 to \$62. Clearly if there are larger tanks on the market, the domain and range will adjust accordingly.

e. Table of values and corresponding graph:

Gas (gallons)	Cost($)
10	12.40
20	24.80
30	37.20
40	49.60
50	62.00

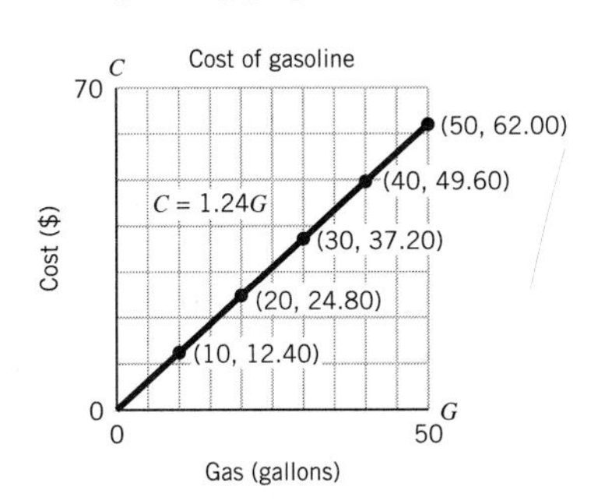

63. The equation is $C = 2.00 + 0.32M$. It represents a function. The independent variable is M. The dependent variable is C. Its graph is below. Here is a table of values:

Miles	Cost	Miles	Cost
0	2.00	30	11.60
10	5.20	40	14.80
20	8.40	50	18.00

Some of these values and others are marked in the graph.

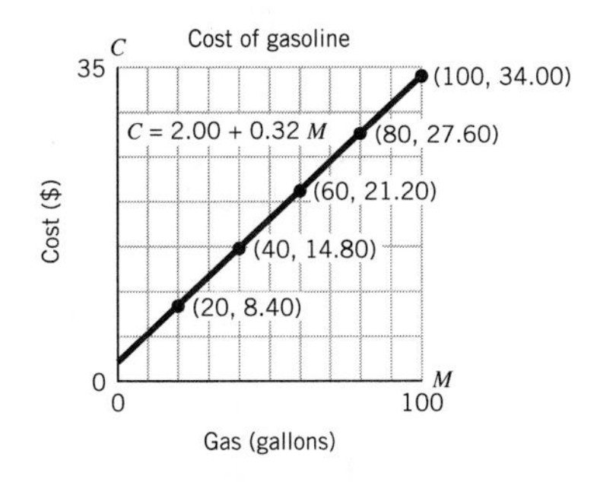

65. a. $f(2) = 4$ **b.** $f(-1) = 4$

c. $f(0) = 2$ **d.** $f(-5) = 32$

67. a. $p(-4) = 0.063, p(5) = 32$ and $p(1) = 2$.

b. $n = 1$ only.

69. a. $f(-2) = 5, f(-1) = 0, f(0) = -3$ and $f(1) = -4$.

b. $f(x) = -3$ if and only if $x = 0$ or 2.

c. The range of f is from -4 to ∞ since we may assume that its arms extend out indefinitely.

71. $f(0) = 1, f(1) = 1$ and $f(-2) = 25$

73. a. 2 **b.** -8 **c.** 10 **d.** -5

CHAPTER 2

Exercises for Section 2.1

1. a. -1.59 points per year **b.** 0.00 points per year

3. a. 570,607.07 computers/year

b. -4.33 students per computer per year.

5. a. For *whites:* 0.99; *blacks:* 1.18; *Asian/Pacific Islanders:* 1.08; *all:* 1.01.

b. For *whites:* 85.68; *blacks:* 78.36; *Asian/Pacific Islanders:* 87.16; *all:* 84.82.

c. There was a great increase in the percent of each group of persons over the past 58 years.

d. For *whites:* mid 2015; *blacks:* early in 2019; *Asian/Pacific Islanders:* near the end of 2012; *all:* early in 2016.

7. a. Highest: in 1900 and in 2000: white females. Lowest: in 1900 and in 2000: black males.

b. white males: 0.276; white females: 0.318; black males: 0.321; black females: 0.412; *black females* had the highest average rate of change in life expectancy from 1900 to 2000.

c. With only one exception, each group mentioned has enjoyed great increases in life expectancy during the past century from one decade to the next. Possible reasons for the one exception (black males during the '60's) could include urban poverty and overrepresentation among the infantry in Vietnam.

Exercises for Section 2.2

9. a.

Year	Salary ($ millions)	Rate of Change Annual Average
1990	0.60	n.a.
1991	0.85	0.25
1992	1.03	0.18
1993	1.08	0.05
1994	1.17	0.09
1995	1.11	− 0.06
1996	1.12	0.01

b. The average rate was smallest in the 1995–1996 period. Its line segment slope is the least steep.

c. The average rate was greatest in the 1990–1991 period. The graph is steepest there.

d. In general, salaries increased at a shrinking rate; they have leveled off since 1994.

11. a. From 1850 to 1950; 0.113 years of age per year.

b. 0.080; 0.146; 0.11; 0.048–all measured in years per calendar year.

c. Better health care, better delivery of food and better nutrition or a shrinking birth rate caused, in part, by a rise in the use of birth control; increase in median age could imply that people are living longer.

d. The units in the 3rd column are years per calendar year.

Med. Age of US Population 1850–2000		
Year	Median Age	Annual Average Rates of Change Over Prior Decade
1850	18.9	n.a.
1860	19.4	0.050
1870	20.2	0.080
1880	20.9	0.070
1890	22.0	0.110
1900	22.9	0.090
1910	24.1	0.120
1920	25.3	0.120
1930	26.4	0.110
1940	29.0	0.260
1950	30.2	0.120
1960	29.5	− 0.070
1970	28.0	− 0.150
1980	30.0	0.200
1990	32.8	0.280
2000	35.7	0.290

e. See labeled plot of the first and third columns.

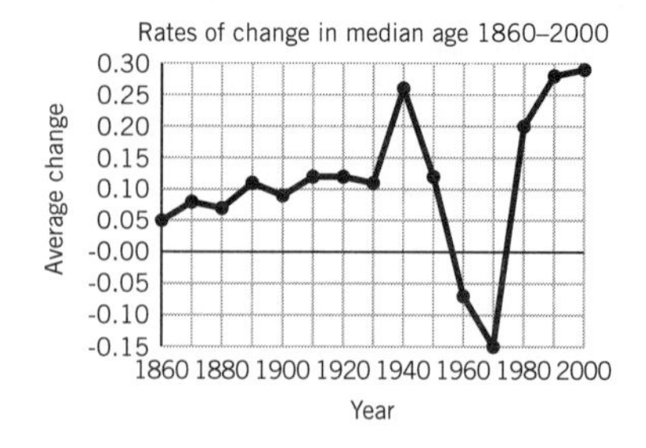

f. Negative rates of change show up in the 1950-to-1960 and 1960-to-1970 periods. During those two periods there were two major wars involving the US. that took the lives of many men. Other causes might be increase in birth rate and immigration of young people.

13. a. In 1920 there were 0.26 papers printed per person. In 1990 there were 0.25 papers printed per person.

b.

Year	Avg. Annual Change in TV Stations	Avg. Annual Change in Newsp. Publ.
1950	n.a.	n.a.
1955	62.6	− 2.4
1960	20.8	0.6
1965	10.8	− 2.4
1970	21.6	− 0.6
1975	5.8	1.6
1980	5.6	− 2.2
1985	29.8	− 13.8
1990	41.8	− 13
1995	88.0	− 5.6

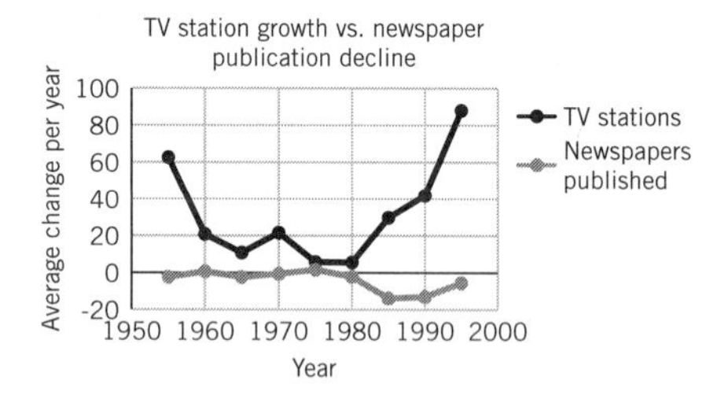

c. 2412 TV stations in 2005; it seems too large

d. Fewer newspapers are surviving (while the readership of these fewer newspapers goes up). The number of on air TV stations has soared. Our intuition and the evidence seem to tell us that more news is received by watching TV. Yet the data alone do not necessarily indicate a change in the way most people get their news. One has no specific information here as to whether TV watchers are flocking to CNN or to the Home Shopping Network or some other programming.

Exercises for Section 2.3

15. a. $t = -19$ **b.** $t = 9.5$

17. The point sequences are collinear if the slopes of the line segments between each successive pair have the same value.

a. These three points ARE collinear since the slopes of the line segments joining various pairs is always 2.

b. The slopes of the line segments between different pairs is different: 3/5 and 4/5. Thus these three points are NOT collinear.

19. a. Positive over $B < x < F$; negative over $F < x \leq H$; zero over $A \leq x \leq B$ and at $x = F$.

b. Positive over $B < x < C$ and $D < x < E$; negative over $E < x < G$; zero over $A \leq x \leq B$, $C < x < D$ and $G < x \leq H$. (Some dividing points are corners and at these the slope is not uniquely defined.)

Exercises for Section 2.4

23. a. Encouraging fact: the number of cases reported went down from 1991 to 1992 by 385 cases and again from 1993 to 1994 by 2341 cases—the drop being more than 6 times as great.

b. The most discouraging fact is the sharp rise in the 1992–1993 period, more than 15 times as great as the rise from 1990 to 1992.

c. What appears to be a sudden increase in the spread of AIDS could be due to increased spending in a new, free, and well-advertised testing program, resulting in an increased number of diagnoses of existing cases.

25. The horizontal axis of time is the same in each of the four graphs. The differences in appearance are due to the changing of the vertical scales. The scales used take advantage of the power of an image to convey sizes. That power is much stronger than the numbers themselves (which do not vary) in causing impressions to be formed for most persons.

Exercises for Section 2.5

29. a. $E(0) = 5000$, $E(1) = 6000$, and $E(20) = 25000$

b. (0, 5000), (1, 6000) and (20, 25000)

31. a. (5000, 0) is not a solution to either equation.

b. (15, 24000) is a solution to the second equation but not the first.

c. (35, 40000) is a solution to the first equation and the second.

33. $\text{Dollars} = \text{dollars} + \dfrac{\text{dollars}}{\text{year}} \cdot \text{number of years}$

35. a. hours **b.** miles/gallon **c.** calories/gram of fat

Exercises for Section 2.6

37. a. Slope = 0.4, vertical intercept = -20.

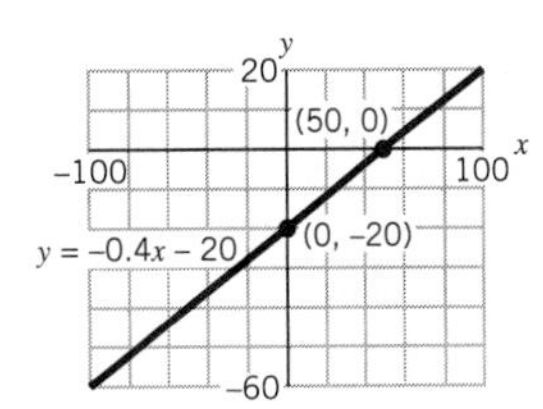

b. Slope = -200, vertical intercept = 4000.

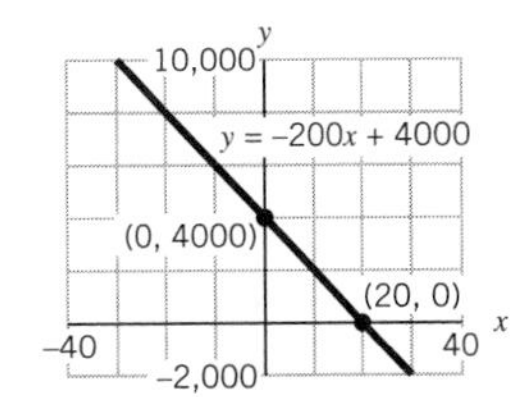

39.

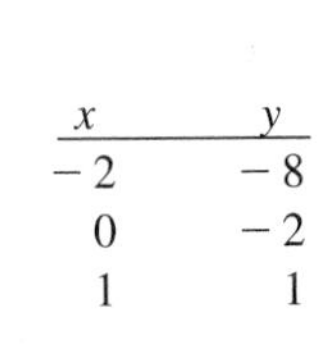

x	y
-2	-8
0	-2
1	1

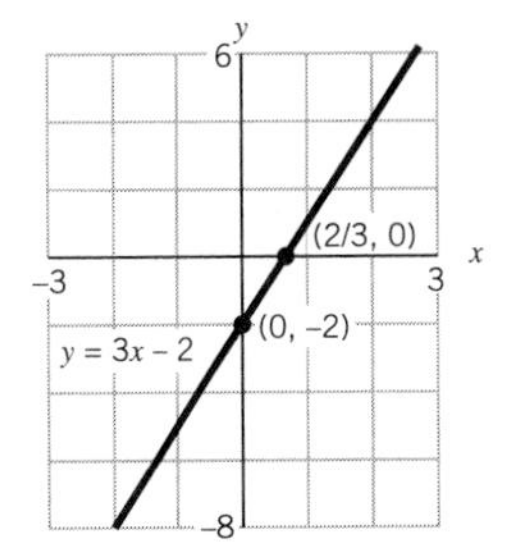

41. The equation is $y = 0x + 1.5$.

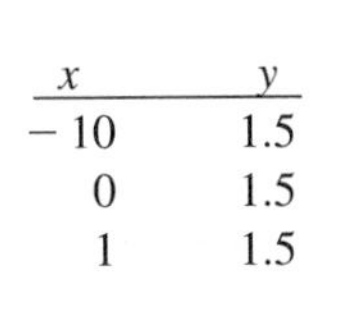

x	y
-10	1.5
0	1.5
1	1.5

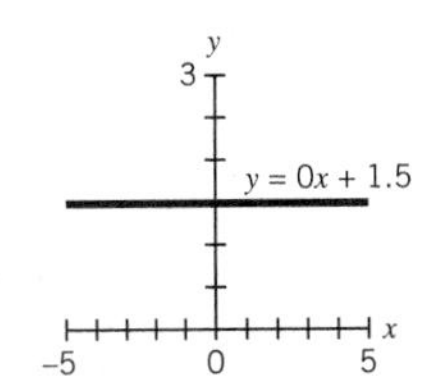

43. a. $y = 5x + 13$ **b.** $y = -0.75x + 1.5$

c. $y = 3$

45. a. $m = -1.25$ and $y - 7.6 = -1.25(x - 2)$ or $y = -1.25x + 10.1$

b. $m = 2$ and $W - 12 = 2(A - 5)$ or $W = 2A + 2$

47. Cost = 150 + 120 · credits where credits = total number of credits registered for in all classes and 150 is measured in dollars and 120 in dollars per credit.

49. $C(n) = 2.50 + 0.10n$ where n = number of checks cashed that month.

51. a. Annual increase = (32000 − 26000)/4 = \$1500

b. $S(n) = 26000 + 1500n$, where $S(n)$ is measured in dollars and n in years from start of employment.

c. Here $0 \le n \le 20$ since the contract is for 20 years.

53. a. Slope = 4 and vertical intercept = − 160.

b. Units for K and − 160 are number of chirps per minute; units for F are degrees Fahrenheit; units for 4 are chirps per minute per degree Fahrenheit.

c. A reasonable domain of this model is from $F = 40$ to, say, $F = 80$.

d. A small table is given

K	F
40	0
60	80
80	160

e. 4; yes, it is what was expected because for a linear equation the rate of change between any two points is the same.

f. Cricket chirps as a function of temperature.

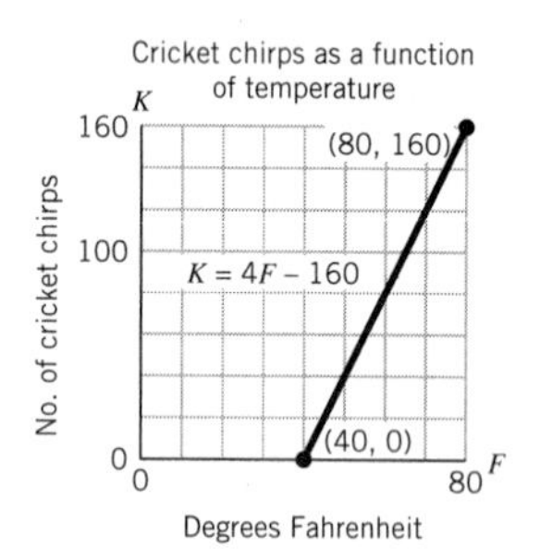

55. a. $P = 285 - 15t$. **b.** 16.33 years

57. a. $V(t) = 20{,}000 + 8846.15t$, where t measures years since 1973.

b. \$347,307.55

c. 54.26 years since 1973. If now means 1999, then in 1973 she was 31. In 54.26 years from 1973 she will be 85.26 years old.

59. The equation of the line illustrated is $y = (8/3)x - 4$. If the horizontal scale were tighter or the vertical scale more spread out, the line would appear steeper. If the horizontal scale were more spread out or the vertical scale tighter, the line would seem less steep.

61. a. Mean Salary for Women vs. Years of Education

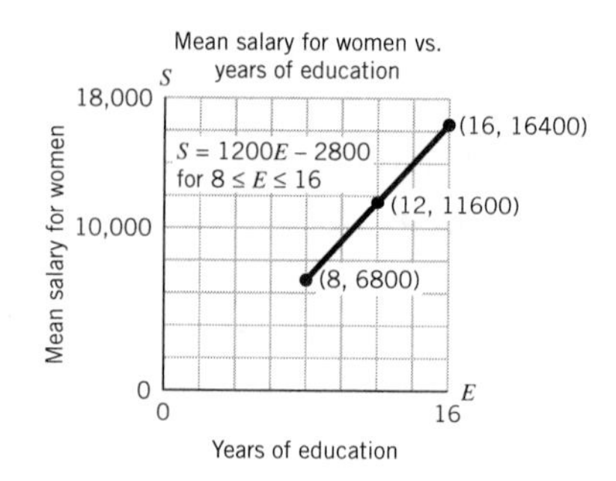

b. The two variables are linearly related. The common slope between pairs of points is \$1200/years of education

c. $S = 1200E - 2800$, if $8 \le E \le 16$.

63. The relationship is linear. The average rate of change in salinity per degree Celsius is a constant: − 0.054. Since P for 0 salinity is 0 degrees Celsius, we have that $P = -0.054S$, where S is the salinity measured in ppt and P, the freezing point, is measured in degrees Celsius.

Exercises for Section 2.7

65. a. The independent variable is the price P; the dependent variable is the sales tax T. The equation is $T = 0.065P$.

b. Independent variable is S the amount of sunlight received; dependent variable is the height of the tree H. The equation is $H = kS$, where k is a constant.

c. Time t in years since 1985 is the independent variable and salary S in dollars is the dependent variable. The equation is $S = 25000 + 1300t$.

67. $d = 5t$; yes, d is directly proportional to t; it is more likely to be the person jogging; the rate is too slow for a car.

69. $D(t) = 15000$, where t is measured in years from 1983 and $D(t)$ is measured in the number of flexible disk drives.

71. a. horizontal: $y = -4$; vertical: $x = 1$; slope 2: $y + 4 = 2(x - 1)$ or $y = 2x - 6$

b. horizontal: $y = 0$; vertical: $x = 2$; slope 2: $y - 0 = 2(x - 2)$ or $y = 2x - 4$

c. horizontal : $y = 50$; vertical: $x = 8$; slope 2: $y - 50 = 2(x - 8)$ or $y = 2x + 34$

73. a. Assuming a steady rate, it must be 10 lbs. per month.

b. $w(t) = 175$ where t = months until fall training and 175 is measured in lbs. The graph is a horizontal line.

c. If the independent variable were the days from end of spring training instead of months, then the data extreme values would probably be less apart from each other but the oscillations would probably occur more frequently.

75. a. $y - 7 = -1(x - 3)$ or $y = -x + 10$.

b. $y - 7 = 1(x - 3)$ or $y = x + 4$

77. a. For graph A: both slopes are positive; same y-intercept.

b. For graph B: one slope is positive, one negative; same y-intercept.

c. For graph C: lines are parallel; different y-intercepts.

d. For graph D: one slope is positive, one negative; different y-intercepts.

79. a. $y = (-A/B)x + (C/B)$, $B \neq 0$

b. The slope is $-A/B$, $B \neq 0$

c. The slope is $-A/B$, $B \neq 0$

d. The slope is B/A, $A \neq 0$.

Exercises for Section 2.8

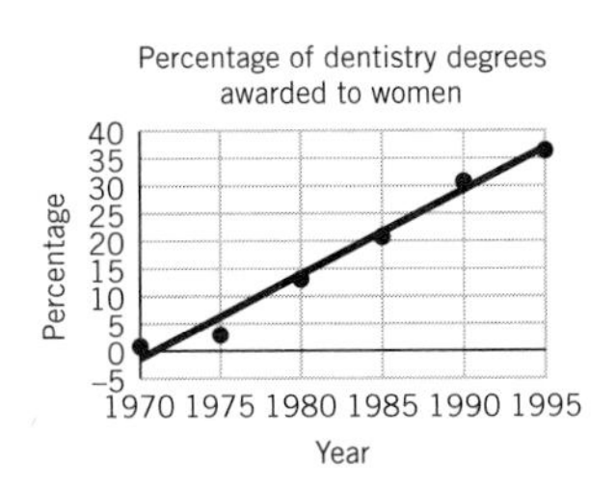

81. a. The line shown in the diagram has the equation $y = -1.6 + 1.5x$. Here x = number of years since 1970 and y = percent of dental degrees given to women. The rage of change of percentage of degrees awarded to women per year is approximately 1.5%.

b. 100% of dentist degrees would be awarded to women in 2036–37.

c. Typical answers could be: growth will slow down; the percent will most likely level off to a ceiling.

85. a.

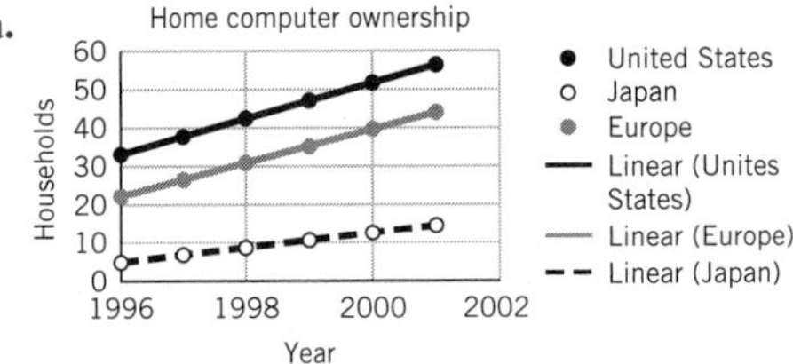

b. The data do suggests that the growth in home computers is roughly linear for each area. Using the first and last points in each table, we get the following linear models:

United States:	$y = 33.1 + 4.62x$
Japan:	$y = 5.4 + 1.88x$
Europe:	$y = 22.6 + 4.34x$

where x = number of years since 1996 and y = number (in millions) of households with computers. (The graphs of the linear models are also in the diagram given.) The slopes represent the overall growth rate of home computers per year. The US dominates both in numbers and growth rate. In 2005 the US would have 74.68 million home computers in use, Japan would have 22.32 million, and Europe would have 61.66 million.

c. The data and linear models are given here:

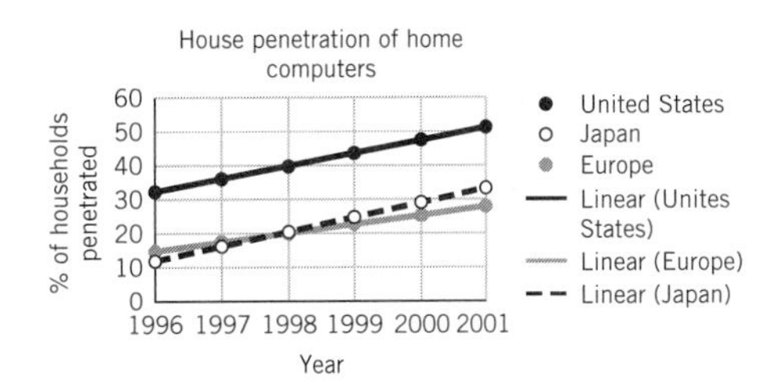

d. The growth seems linear. Again, using the first and last data points, one gets the following linear models:

United States:	$y = 32 + 3.8x$
Japan:	$y = 13 + 4.2x$
Europe:	$y = 15 + 2.6x$

The slopes of these lines represent the overall rate of growth per year in percent of households penetrated by home computers. In 2005, the US would have a 66.2% penetration, Japan would have a 50.8% penetration and Europe would have a 38.4% penetration.

e. There is growth in all three areas and utter dominance by the US in the overall numbers of home computers and percent penetration, but Japan is growing in percent penetration faster than the US and Europe.

87. a. Answers will vary. One eyeballed linear model is $y = 4.9 + 0.24t$, where t measures years since 1960.

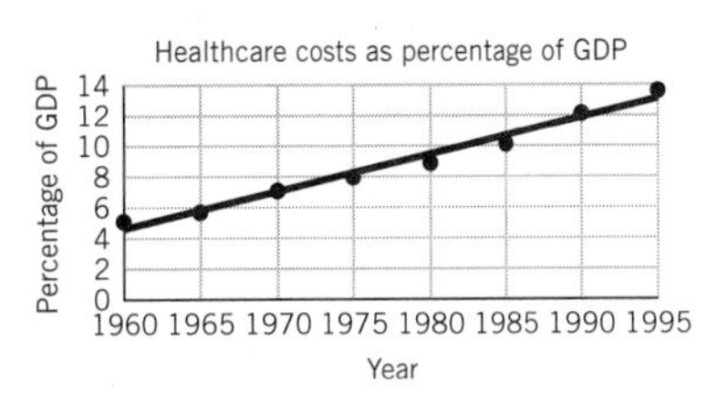

b. 14.5%

c. One expected reason is that the GDP has grown very large and thus the percentage for health care is not growing as much as the health care costs themselves.

EXTENDED EXPLORATION: LINKS BETWEEN EDUCATION AND INCOME

1. a. 0.65, 0.68, 0.07, 0.70 **b.** 0.07, 0.65, 0.68, 0.70

3. a. Slope = 4210; vertical intercept = − 16.520; correlation coefficient = 0.88.

b. It means that for each additional year of education it is expected that the person's mean personal wages will increase by $4210.

c. $4210; $42,100

5. a. 4370

b. It means that the rate of change of mean personal income with respect to years of education is $4370 per year.

c. $4370 for 1 year; $43,700 for 10 years.

d. Individual values and outliers are not well described by the regression line.

7. a. 2100 in this equation is the rate of change of personal total income with respect to years of education; its units are dollars per year of education.

b. (4, 5000), (8, 13400) and (10, 17600). The slope = 2100.

c. This slope is the same as that given in part a.

d.

Personal total income vs. years of education

P = −3400 + 2100Y

(10, 17600)

(8, 13400)

(4, 5000)

Personal total income

Years of education

e. Income increases much more rapidly for males.

9. a. 0.516 is the rate of change of the mean height of a son in inches for each increase of 1 inch in height of the father.

b. For fathers 64 in. tall: 66.754 in.; for fathers 73 in. tall : 71.398 in.

c. Approximately 70 in.

d. There are only 17 different heights for fathers given.

11. These answers are from using technology. Eyeball answers will differ.

a and b. Best fit formulae:
private: $C = 6453 + 819y$ and public: $C = 1258 + 169y$
where y = years from 1985 and C = cost in dollars

Graphs of best linear fits:
Private Colleges

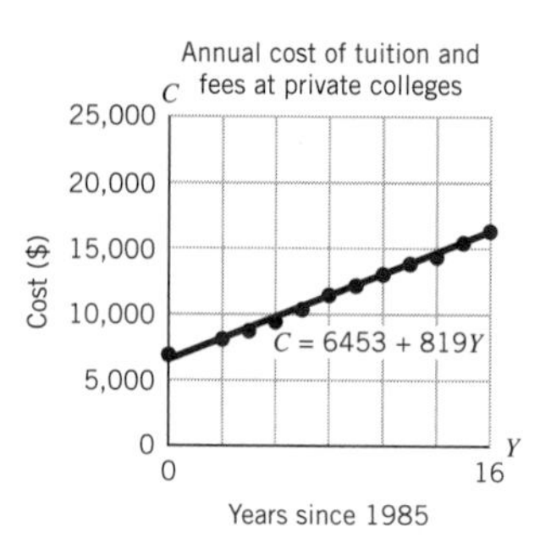

Public Colleges

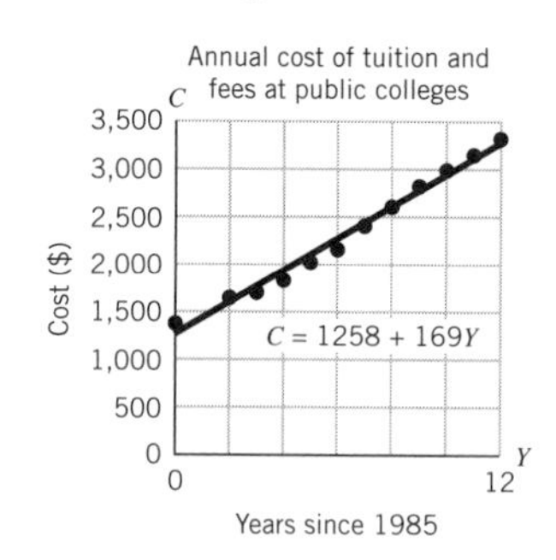

c. For private: $26,928 and for public: $5483; reasonable for public colleges; way too low for private colleges.

13. a.

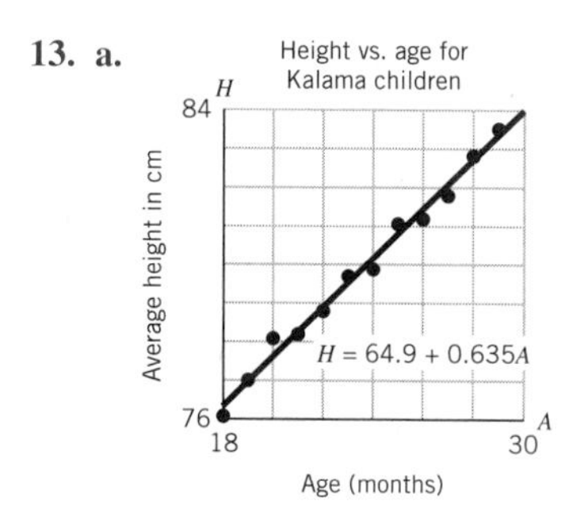

The equation of the best fit regression line is: $H = 64.9 + 0.635A$.

b. A represents age in months; H the average height in centimeters; 64.9 is the vertical intercept and its represents the (extrapolated) height of a Kalama child at birth; the slope is 0.635 and represents the number of centimeters on average that a Kalama child would grow in one month; the correlation coefficient is 0.994 and thus the line is a good fit. A reasonable domain, built on the data, is from 18 to 29 months.

c. 81.7 cm.

15. a. Scatter plot and regression line graph:

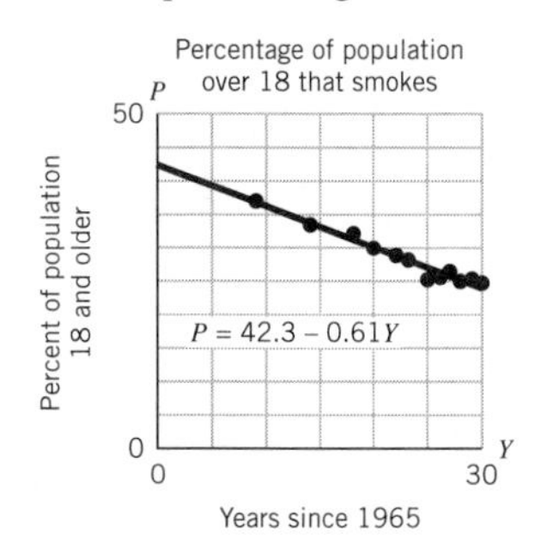

Here where y = number of years since 1965 and P = % of population 18 and older that smokes.

i. − 0.59; its units are change in % of those who smoke per year.

ii. − 0.16; its units are the same as in i.

b. and **c.** The plot of the regression line along with the data is given in the chart. The equation of the regression line is $p = 42.35 - 0.61y$ where p and y are defined as in **a.** The average rate of change from this line is − 0.61% change per year, i.e., the % of smokers is decreasing by 0.61 each year on average. Its correlation coefficient is − 0.991825. Student hand-generated rates will vary.

d. For males: $p = 50.33 - 0.83y$ and for females: $p = 35.34 - 0.42y$; the correlation coefficient is − 0.988 for males and − 0.963 for females.

e. Given the constant negative information being provided about the dangers of smoking, the downward trend will probably continue.

17. One way to check to see if there is hope of a linear relationship is to look at correlation coefficients of each possible pair of data columns. The closer to 1 in absolute value the higher the linear correlation.
From the table those that show the highest linear correlation are:

Life expectancy of males compared to that for females has a cc. of 0.947.
Life expectancy of males and infant mortality rates has a cc. of − 0.910.
Life expectancy of males and literacy % has a cc. of 0.724.
Life expectancy of females and infant mortality rates has a cc. of − 0.965.
Life expectancy of females and literacy % has a cc. of 0.809.
Infant mortality and literacy % have a cc. of − 0.855.

19. a. and **b.** $F = -213.21y + 33458.71$ is the equation of the regression line for farm population where y stands for years since 1880; the cc. is − 0.75. This not a very good fit. The graph for the regression line for the farm population as a percentage of the whole population is given in the diagram. Its equation is: $P = 46.99 - 0.44y$ where P stands for percentage of total U.S. population and y stands for years since 1880. Its cc. is − 0.989. This is a much better linear fit.

Graph for # 19 a. Farm Population since 1880

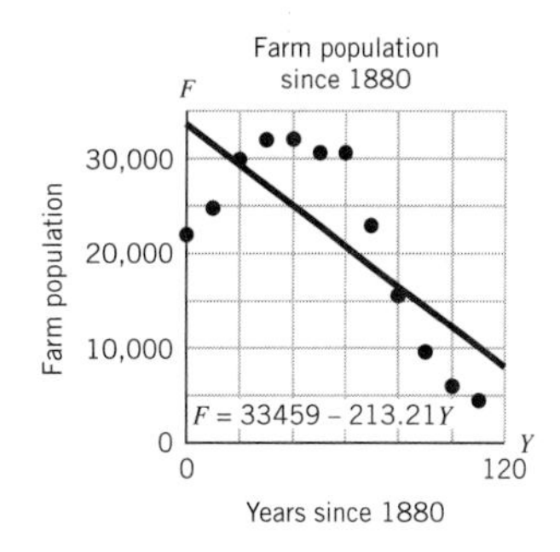

Graph for # 19 a. Percentage of Total Population since 1880

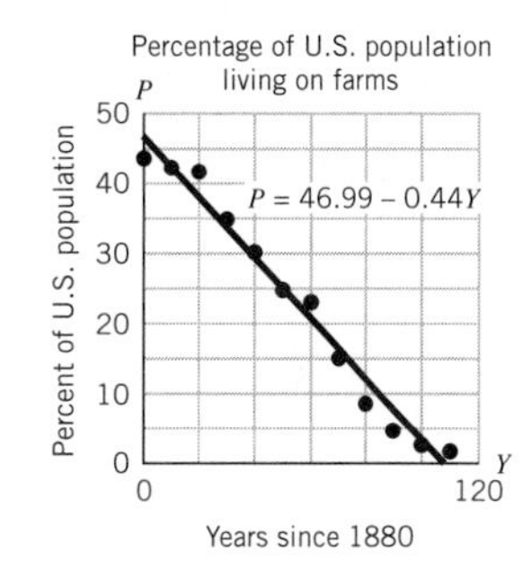

	Correlation Coefficients for Nations Data Columns								
	GDP	GDP pc	Pers/TV	LifexpM	LifexpF	Pers/Be	Pers/dr	Infmor	Litr %
GDP pc	0.425								
Pers/TV	− 0.085	− 0.268							
LifexpM	0.203	0.607	− 0.492						
LifexpF	0.220	0.662	− 0.499	0.947					
Pers/Bed	− 0.105	− 0.333	0.304	− 0.466	− 0.519				
Pers/Dr	− 0.128	− 0.383	0.581	− 0.623	− 0.636	0.286			
Infmor	− 0.213	− 0.664	0.498	− 0.910	− 0.965	0.494	0.662		
Litr %	0.219	0.591	− 0.319	0.724	0.809	− 0.479	− 0.625	− 0.855	

c. If one measures the farm population in absolute numbers from 1950, one gets a much better linear fit as is seen in the graph for this part. The equation is:
$F = -464.98y + 21107$; the cc. is -0.97.

Graph for #19 c. Farm Population since 1950

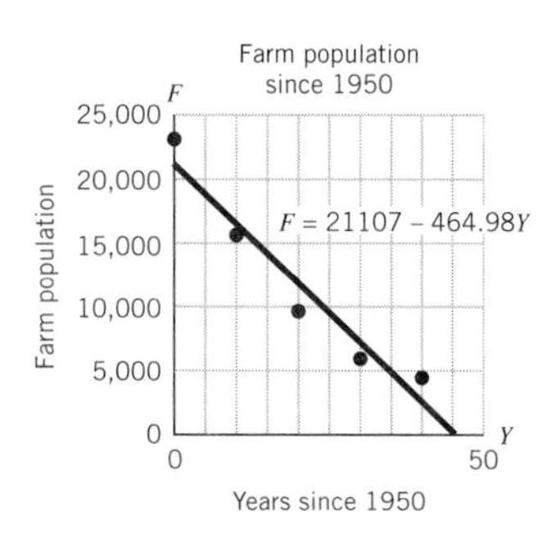

21. a. Its equation is $T = 169.055 - 1.176Y$; the cc. is -0.79. The data and the line are graphed with two domains.

Women's Marathon Times 1972 to 2000

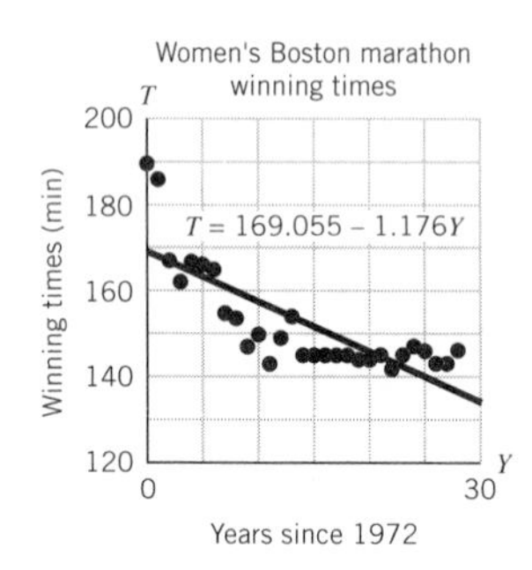

b. 124.372 minutes

Women's Marathon Times 1981 to 2000

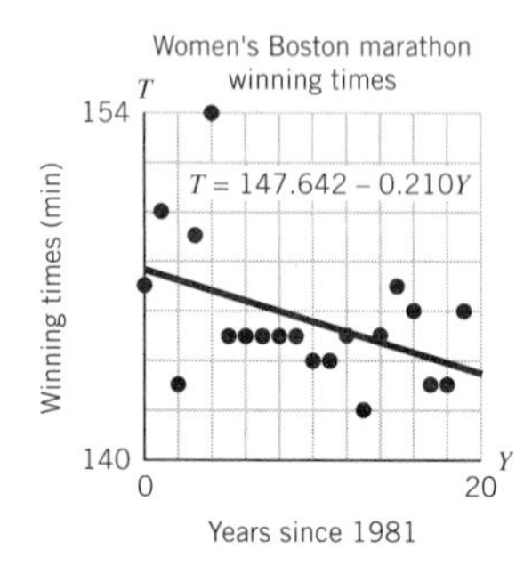

c. The best linear fit going from 1981 to 2000 is $T = 147.64 - 0.210Y$, where Y is measured in years since 1981 and T is measured in minutes. The cc is -0.447. The predicted time for 2010 is 141.543 minutes; it is much more realistic than the estimate given in part (**b**).

23. a. $C = 99.91 - 0.00098H$ or when suitably rounded off: $C = 100 - 0.001H$, where H is measured in feet above sea level and C is measured in degrees Celsius; the cc. is -0.9999.

b. (i) $F = 212 - .002H$, where H is measured in feet above sea level and F is measured in degrees Fahrenheit; the cc. is -0.9998.
(ii) Answers will vary.

c. On Mt. McKinley it boils at 171.36 °F; in Death Valley it boils at 212.57 °F.

d. 56,700 ft. **e.** 100,000 ft.

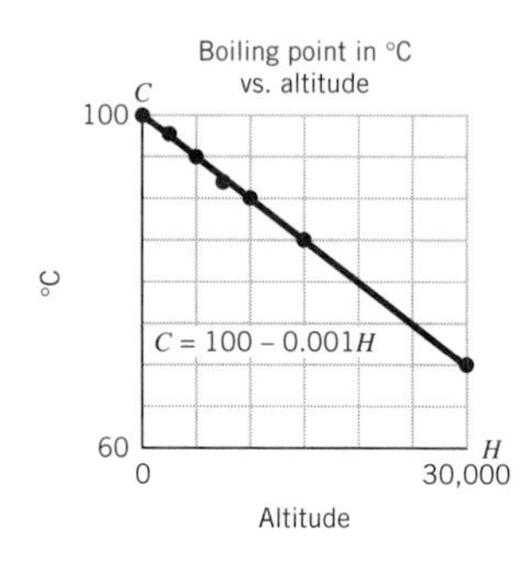

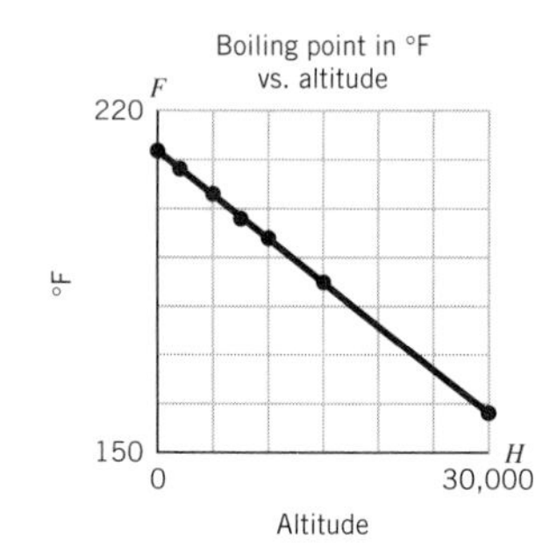

25. a. The best linear approximation equation is: $R = 26.81 + 3.53y$, where y = years since 1945 and R = number of motor vehicles in millions registered in year y.

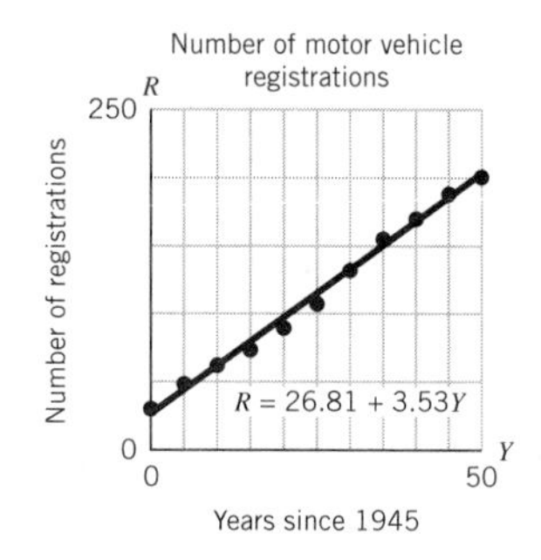

b. Each year, on average the number of motor vehicle registrations increases by 3.53 million.

c. $R(65) = 256.26$ million

27. A high correlation between two factors does not mean that one causes the other. In the case of smoking and lung cancer, however, there is much other evidence to link them causally.

29. In 1993 the Republican candidate received more machine votes than the Democratic candidate in a senatorial election in Philadelphia. But the Democratic candidate received so many absentee ballot votes that the combined number of votes (machine and absentee) put him ahead. There were protests that the Democratic candidate unduly influenced absentee ballot votes. The courts threw out the absentee ballots and announced that they would award the election to the Republican if they could be persuaded that his vote count would not have been surpassed by the absentee ballot votes had they been regarded as legitimate.

The Princeton economics professor used regression analysis to show this. He plotted a scatter diagram. The x-axis measured differences between Republican and Democratic machine votes in the last 22 senatorial elections in Philadelphia. The y-axis measured the corresponding differences in absentee ballots. He showed that the points in this scatter diagram tended to fall around a line "representing an ideal correlation" between the machine and absentee tallies. (The newspaper was referring to the regression line of this data.) This ideal situation suggests an agreement between machine and absentee ballot voting. Moreover, the graph shows that the point representing the 1993 election falls well outside the area where 95% of the results would be expected to fall. In fact he showed that the probability of this reversal being due to chance was less than 1%. Based on his arguments, the court awarded the election to the Republican.

CHAPTER 3

Exercises for Section 3.1

1. "A solution to a system of equations" is an ordered pair of numbers (or set of ordered pairs of numbers) that satisfies *all* of the equations in the system.

3. **a.** $(5, -10)$ does *not* solve the given system.

   b. The coordinates work for the 1st equation but not the second.

5. **a.** (40, 3.5)

   b. To the left of the point of intersection, the population of Palm City is greater than the population of Johnsonville; the opposite is true to the right of that point.

7. **a.**

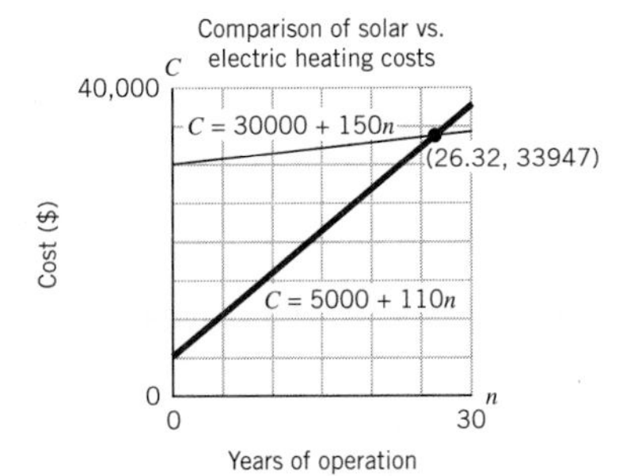

   b. 1100 and 150 represent the slopes of two lines. In terms of heating costs; 1100 represents the rate of change of the total cost in dollars per year for electric heating; 150 represents the total cost in dollars per year for solar heating.

   c. 5000 is the initial cost in dollars of installing the electric heat; 30000 is the initial cost (in dollars) of installing solar heat. It cost a lot more initially to install gas than to install electricity.

   d. The point of intersection is approximately: (26, 34000)

   e. (26.32, 33947)

   f. Assuming simultaneous installation of both heating systems, the total cost of solar heat was higher than the total cost of electric heat up to year 23 (plus nearly 4 months); after that the total cost of electric heat will be greater than that of solar heat.

Exercises for Section 3.2

9. **a.** For $n = 10$ we have $S_A = S_B = 45{,}000$

   b.

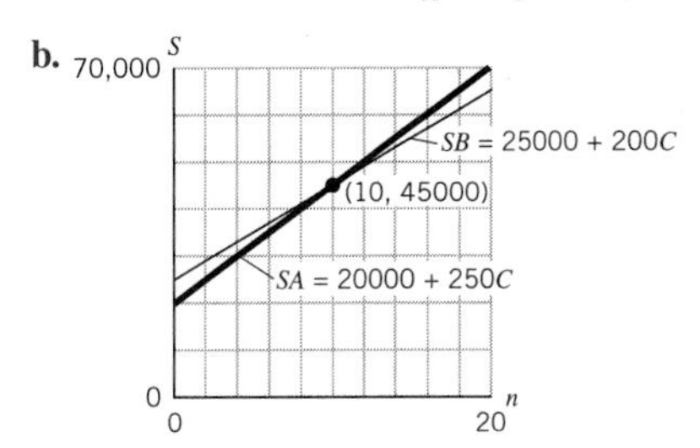

   The intersection occurs when $n = 10$ and when the common salary value is 45,000.

11. **a.** The solution is at $(-3, 3)$.

   b.

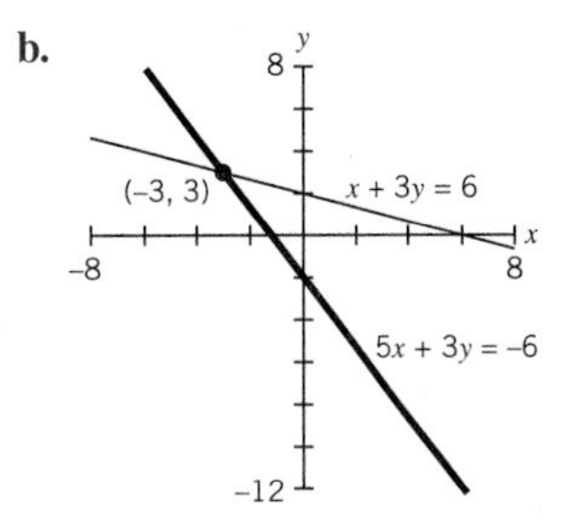

13. $x = \$1500$ and $y = \$500$. [*Check:* $1500 + 500 = 2000$ and $.04 \cdot 1500 + .08 \cdot 500 = 100$.]

15. **a.** $x = 2$ and $y = 2 - 4/3 = 2/3$. [*Check:* $2 - 2/3 = 4/3$ and $2/3 + 1/3 = 1$.]

   b. $x = 12$ and $y = 6$. [*Check:* $12/4 + 6 = 9$.]

17. **a.** $m = -3$ and $b = 4$.

   b. $m = 5$ and $b = -12$

19. Two equations are called equivalent if they have exactly the same solutions or the same graph. An example is: $3x = 12$ and $2x - 8 = 0$.

21. **a.** $11x + 7y = 68$ (4) **b.** $9x + 7y = 62$ (5)

   c. $x = 3$ and $y = 5$ solve (4) and (5)

   d. $z = 10$ **e.** $x = 3, y = 5, z = 10$

23. a. $y = x + 5$ and $y = x + 6$

b. $y = x + 5$ and $y = -x + 5$

c. Algebraically: setting $x + 5 = -x + 5$ gives $2x = 0$ or $x = 0$ and thus $y = 5$. The graph confirms this.

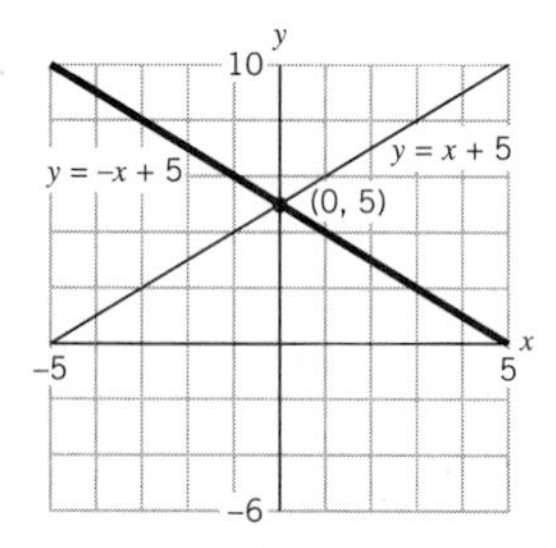

25. This system has no solution if the graphs of these two equations are distinct parallel lines. This occurs when $m_1 = m_2$ (parallel means the same slope) but $b_1 \neq b_2$ (different vertical intercepts).

27. a. The equations of this pair of lines are:

$$y = -4x + 13 \quad \text{and} \quad y = 3.5x - 2$$

b.

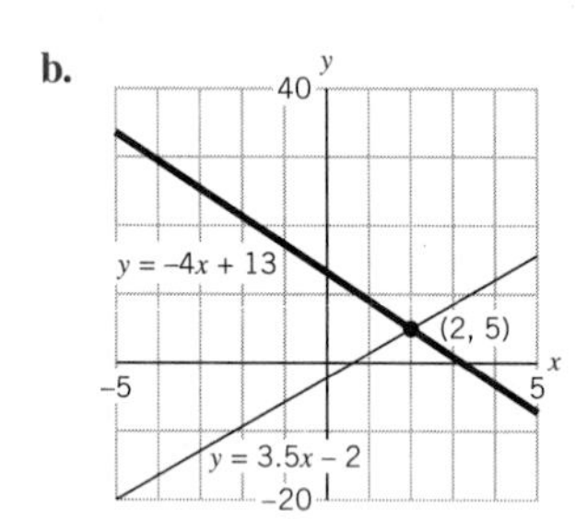

29. a. It will take $T = Q/(A + B)$ hours to lay $A + B$ bricks.

b. Alpha works 7 hours.

c.

A Time (hrs.)	A Bricks Laid	A Cost ($)	B Time (hrs.)	B Bricks Laid	B Cost ($)
1	50	42	1	70	45
2	100	72	2	140	90
3	150	102	3	210	135
4	200	132	4	280	180
5	250	162	5	350	225
6	300	192	6	420	270
7	350	222	7	490	315
8	400	252	8	560	360

The equation for the cost of Alpha laying bricks vs. the number of bricks laid is: $C = 0.60x + 12$. The equation for the cost of Beta laying bricks vs. the number of bricks laid is $C = (9/14)x$. Alpha and Beta have laid the same number of bricks and have been paid the same amount of money when $x = 280$ bricks and $C = \$180$. It is cheaper to hire Alpha when there are more than 280 bricks to be laid.

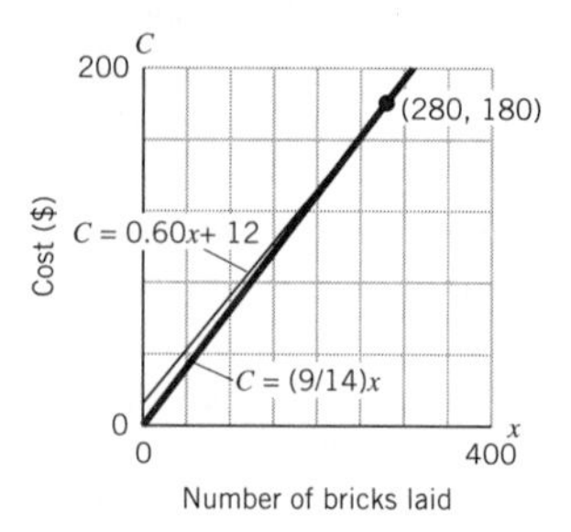

d. Alpha lays 50 bricks an hour and thus $A_{\text{bricks}} = 50T_\alpha$, where T_α is measured in hours and A_{bricks} in bricks. Similarly Beta lays 70 bricks an hour and thus $B_{\text{bricks}} = 70T_\beta$. Also, $A_{\text{cost}} = 12 + 30\ T_\alpha$ is cost of Alpha working T_α hours; $B_{\text{cost}} = 45T_\beta$ is the cost of Beta working T_β hours.

Alpha and Beta lay the same number of bricks and get paid the same amount when the number of bricks is 280 and the payment is $180. But they have worked a different number of hours: Alpha lays 280 bricks in 5.6 hours or 5 hours and 36 minutes, and Beta lays 280 bricks in 4 hours.

31. a. $B_y = 30$, $A_y = 0.625A_x$. **b.** $x = 48$ and $y = 30$.

c. Plane A will arrive at this point nearly 6 minutes after plane B. It is a safe situation.

Exercises for Section 3.3

33. a.

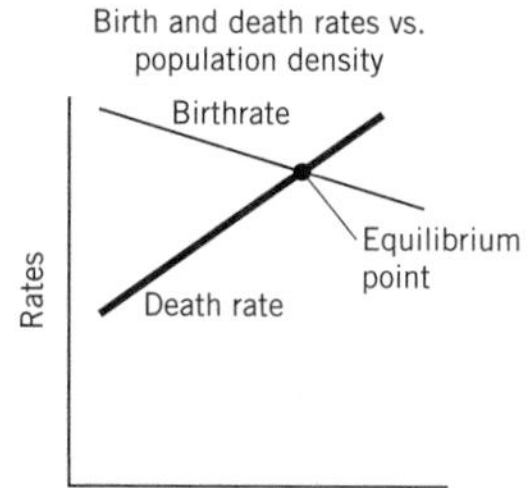

b. The growth rate of population at the equilibrium point is zero.

c. If the death rate decreases (and the birth rate stays the same) the equilibrium point tends to shift down the right.

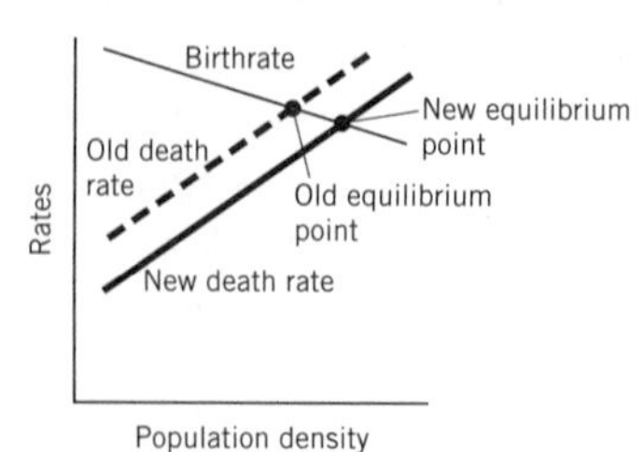

d. If the death rate increases (and the birth rate stays the same) the equilibrium point tends to shift left and go up. (See the appropriately labeled graph.)

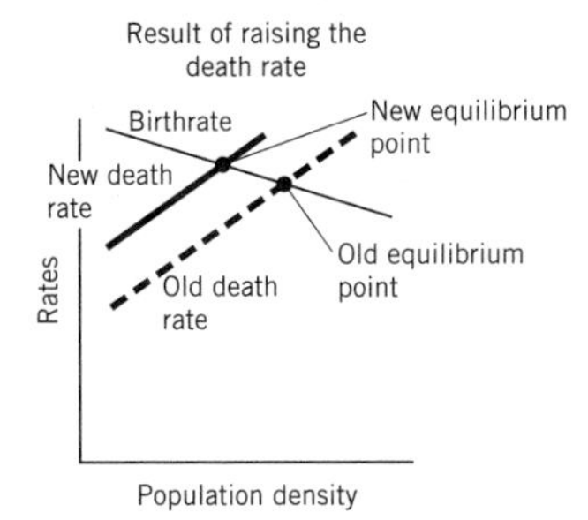

Exercises for Section 3.4

35. Answers will vary from state to state. Check student answers against the local tax form itself.

37. a. $y = 1$ for $0 \le x \le 1$
$= x$ for $x > 1$

b. $y = 1 - x$ for $0 \le x \le 1$
$= 1.5x - 1.5$ for $x > 1$

39. If $f(x) = 2x + 1$ for $x \le 0$ and $= 3x$ if $x > 0$ then $f(-10) = -19$, $f(-2) = -3$, $f(0) = 1$, $f(2) = 6$ and $f(4) = 12$.

41. a. Table for f whose domain is all real x

x	$f(x)$
0	5
3	5
8	5
10	5
15	15
20	25

b. Table for g whose domain is $-10 \le t < 10$

t	$g(t)$
−10	11
−5	6
1	0
3	3
5	5
10	10

Graph of f

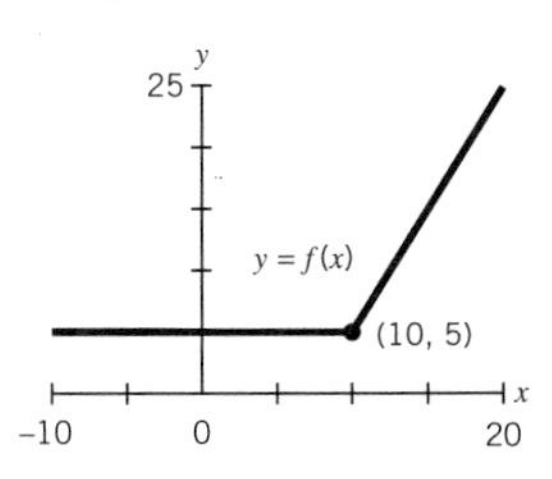

Graph of g (note gap at $t = 1$)

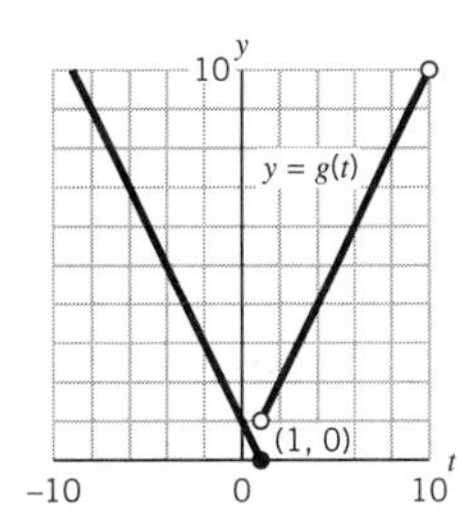

43. a. flat tax: $f(x) = 0.10 \cdot x$ if $x > 0$
grad tax $g(x) = 1000$ if $0 < x \le 20000$;
$= 1000 + 0.20(x - 20000) = 0.2x - 3000$ if $x > 20000$.

b.

Income	Flat	Graduated
0	0	0
10000	1000	1000
20000	2000	1000
30000	3000	3000
40000	4000	5000

c.

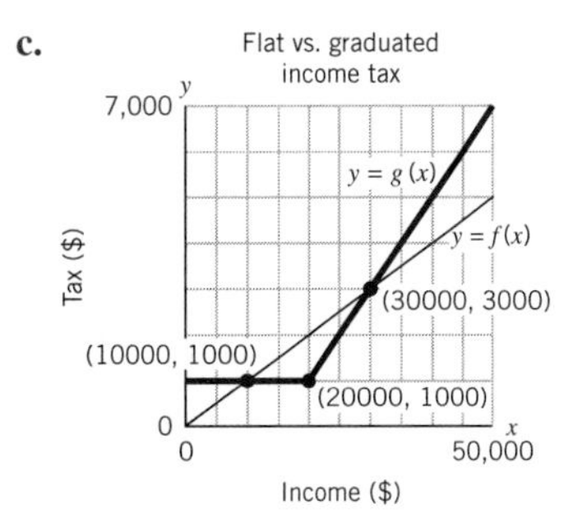

d. $x = 10{,}000$ and $x = 30{,}000$

e. The flat tax is more than the graduated tax for $10000 < x < 30000$. The graduated tax is more for $0 < x < 10000$ and $x > 30000$.

f. If $0 < x < 1000$, then the tax under the graduated plan would be \$1000; thus there would be negative income.

45. a and **b.**

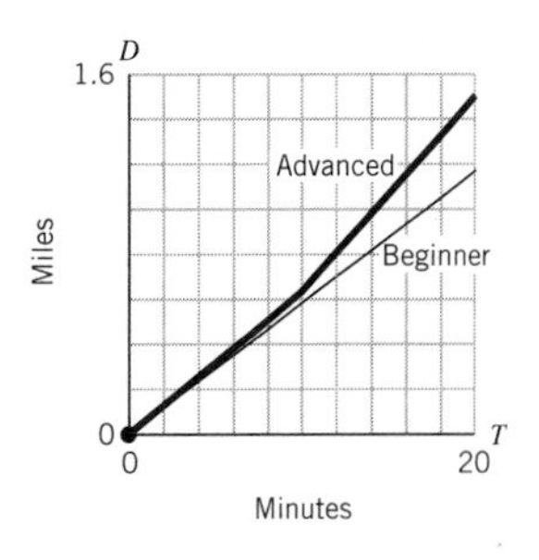

For $0 \le T \le 20$: $D_{\text{beginner}} = (3.5/60)T$ since there is 1/60 of an hour in a minute; note that T is measured in minutes and $D_{beginner}$ is measured in miles. The table is found below.

b. For $0 \le T \le 10$: $D_{advanced} = (3.75/60)T$ and for $10 < T \le 20$: $D_{advanced} = 0.0875T - 0.25$

c. The beginner and advanced function graphs intersect for $T = 0$ only, at the start of the race. The advanced is always ahead if $T > 0$. The table given below confirms this.

T	$D_{beginner}$	$D_{advanced}$	T	$D_{beginner}$	$D_{advanced}$
0	0.0000	0.0000	11	0.6417	0.7125
1	0.0583	0.0625	12	0.7000	0.8000
2	0.1167	0.1250	13	0.7583	0.8875
3	0.1750	0.1875	14	0.8167	0.9750
4	0.2333	0.2500	15	0.8750	1.0625
5	0.2917	0.3125	16	0.9333	1.1500
6	0.3500	0.3750	17	0.9917	1.2375
7	0.4083	0.4375	18	1.0500	1.3250
8	0.4667	0.5000	19	1.1083	1.4125
9	0.5250	0.5625	20	1.1667	1.5000
10	0.5833	0.6250			

47. a. $C = 7$ if $N \le 100$; $C = 3.5$ if $N > 100$.

b. $C = 7$ if $0 < N \le 100$; $= 3.5$ if $N > 100$
Cost is measured in cents.

c. If you have from 1 to 100 photocopies of a page made, then the charge is 7 cents per copy, but if you have more than 100 copies made, the charge is 3.5 cents per copy.

d. $B(N) = 7$ if $0 < N \le 5000$;
$= 3.5$ if $N > 5000$

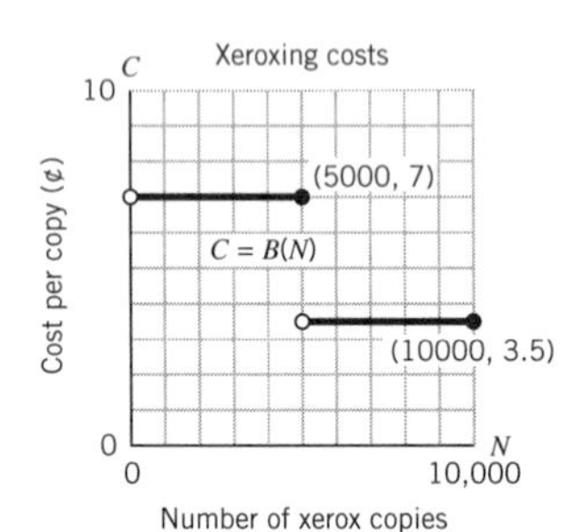

e. $TC(N) = 0.07\,N$ if $0 < N \le 100$
$= 7 + 0.035(N - 100)$ if $N > 100$
with the cost now measured in dollars and N representing the number of pages photocopied.

f. $TB(N) = 0.07N$ if $0 < N \le 5000$;
$= 350 + 0.035(N - 5000)$ if $N > 5000$
with the cost measured in dollars and N representing the number of pages photocopied.

CHAPTER 4

Exercises for Section 4.1

1. a. 10^6 **b.** 10^{-5} **c.** 10^9 **d.** 10^{-3}

3. a. 10^{-2} m **b.** $4 \cdot 10^3$ m **c.** $3 \cdot 10^{12}$ m **d.** $6 \cdot 10^{-9}$ m

5. a. $2.9 \cdot 10^{-4}$ **b.** $6.54456 \cdot 10^2$ **c.** $7.2 \cdot 10^5$ **d.** $1.0 \cdot 10^{-11}$

7. a. 723,000 **b.** 0.000526 **c.** 0.001 **d.** 1,500,000

9. a. 9 **b.** 9 **c.** 1000 **d.** − 1000

11. a.

x	$\lvert x \rvert$
− 2	2
− 1	1
0	0
1	1
2	2

b. The graphs of both $y = |x|$ and $y = |x - 2|$ are displayed to allow for comparison.

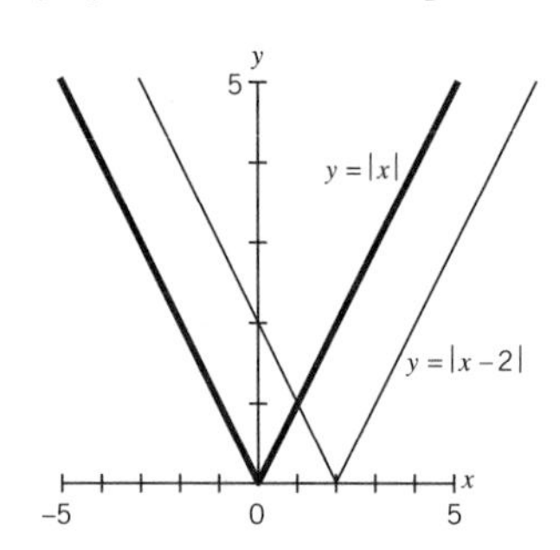

Exercises for Section 4.2

13. a. 10^7 **b.** $1.1 \cdot 10^4$ **c.** $2 \cdot 10^3$ **d.** x^{15}

e. x^{50}

f. $1.6409 \cdot 10^4$ (scientific notation was chosen; other forms are also allowed).

g. z^5 **h.** 1 or as is **i.** 3^{-1} **j.** 4^{11}

15. a. $16a^4$ **b.** $-2a^4$ **c.** $-x^{15}$ **d.** $-8a^3b^6$

e. $32x^{20}$ **f.** $18x^6$ **g.** $2500a^{20}$

h. as is—nothing is simpler

17. a. $-\left(\dfrac{5}{8}\right)^2 = -0.625^2 = \dfrac{-25}{64}$

b. $\left(\dfrac{3x^3}{5y^2}\right)^3 = \dfrac{3^3x^9}{5^3y^6} = \dfrac{27x^9}{125y^6}$

c. $\left(\dfrac{-10x^5}{2b^2}\right)^4 = \dfrac{10^4x^{20}}{4^2b^8} = \dfrac{10{,}000x^{20}}{16b^8} = 625\dfrac{x^{20}}{b^8}$

d. $\left(\dfrac{-x^5}{x^2}\right)^3 = -x^9$

19. a. $10^9/10^6 = 10^3 = 1000$ **b.** $1000/10 = 10^2 = 100$

c. $1000/0.001 = 10^6 = 1{,}000{,}000$

d. $10^{-6}/10^{-9} = 10^3 = 1000$

21. a. 826.89 persons per square mile

b. 75.30 persons per square mile

c. The population density of Japan is almost 11 times as great as that for the USA.

23. a. and **b.** If $a = 0$, then $(-a)^n = 0$. If $a \neq 0$, then, when n is an even integer, $(-a)^n$ is positive and when n is odd, then $(-a)^n$ has the opposite sign to a.

25. Three cases are distinguished:
If $n = 0$, then $(ab)^0 = 1$ and $a^0 \cdot b^0 = 1 \cdot 1 = 1$
If $n > 0$, then $a^n = a \cdot \cdots \cdot a$ (n factors) and $b^n = b \cdot \cdots \cdot b$ (n factors) and thus $a^n \cdot b^n = (a \cdot \cdots \cdot a) \cdot (b \cdot \cdots \cdot b) = (ab) \cdot \cdots \cdot (ab)$ (n factors), after rearrangement, and this product is what is meant by $(ab)^n$ when $n > 0$.
If $n < 0$, then $a^n = (1/a) \cdot \cdots \cdot (1/a)$ ($-n$ factors) and $b^n = (1/b) \cdot \cdots \cdot (1/b)$ ($-n$ factors) and thus $a^n \cdot b^n = (1/a) \cdot \cdots \cdot (1/a) \cdot (1/b) \cdot \cdots \cdot (1/b) = (1/ab) \cdot \cdots \cdot (1/ab)$ ($-n$ factors), after rearrangement, and this product is what is meant by $(ab)^n$ when $n < 0$.

27. a. $1.71 \cdot 10^2$ kwhs per person; $1.069 \cdot 10^5$ kwhs per square mile.

b. $2.059 \cdot 10^2$ kwhs per person; $1.5129 \cdot 10^4$ kwhs per square mile.

c. UK is generating per square mile 7 times the kwhs that the US is.

d. US is generating 1.2 times the number of kwhs (or 20% more) per person that the UK is generating.

Exercises for Section 4.3

29. a. $\dfrac{1}{2^2x^4} = \dfrac{1}{4x^4}$ **b.** xy^{12} **c.** $\dfrac{1}{x^3y^2}$

d. $(x + y)^{11}$ **e.** $\dfrac{ab^7}{c^6}$

31. a. $4.6 \cdot 10^{10}$ **b.** $4.07 \cdot 10^3$ **c.** $1.525 \cdot 10^{11}$

d. $5.1669 \cdot 10^{-8}$ **e.** $2.3833 \cdot 10^{158}$

f. $2.601 \cdot 10^{-21}$

33. 200 times longer or $1.6 \cdot 10^{-2}$ seconds.

35. a. $\left(\dfrac{1}{x^2} - \dfrac{1}{y}\right) \cdot (xy^2) = \left(\dfrac{y - x^2}{x^2y}\right) xy^2 = (y - x^2) \cdot \dfrac{y}{x} = \dfrac{y^2 - x^2y}{x}$

b. $\dfrac{x^4y^6}{5^2} = \dfrac{x^4y^6}{25}$

Exercises for Section 4.4

37. a. Estimates will vary. There are approximately 30.77 cm. per foot.

b. Estimates will vary. 1 ft = (1/3.28) = 0.305 m.

39. 9.8 mps = 32.144 ftps

41. a. 186,000 miles per second

b. $9.4608 \cdot 10^{15}$ meters per year

43. 500 seconds; 500 seconds or 8 minutes and 20 seconds from now.

45. Light travels 0.98208 ft per nanosecond.

47. a. Clinton's BMI $\approx$ 27.79 kg/m^2; overweight.

b. For Clinton we get: $704.5 \cdot 216/74^2 = 27.79$

c. Answers will vary but note that $0.45 \approx 1/2.2$ and $.0254 \approx 1/39.37$

d. A kilogram, more precisely, is 2.2046 lbs. and $39.37^2/2.2046 \approx 703.07$ and thus Brent Kigner is correct.

49. a. 10^{-35} m $= 10^{-38}$ km

b. 10^{-35} m $= 6.2 \cdot 10^{-39}$ miles. **c.** $x = 3.33 \cdot 10^{-28}$

Exercises for Section 4.5

51. a. $\sqrt{\dfrac{a^2b^4}{c^6}} = \dfrac{ab^2}{c^3}$ **b.** $\sqrt{36x^4y} = 6x^2\sqrt{y}$

c. $\sqrt{\dfrac{49x}{y^6}} = \dfrac{7\sqrt{x}}{y^3}$ **d.** $\sqrt{\dfrac{x^4y^2}{100z^6}} = \dfrac{x^2y}{10z^3}$

53. $r = \sqrt{\dfrac{20}{4\pi}} = 1.262$ meters

55. a. 0.1 **b.** 1/5 **c.** 4/3 **d.** 0.1

57. **a.** 4 **b.** -4 **c.** $\sqrt{2} \approx 1.414$

d. not defined **e.** $2\sqrt{2} \approx 2.8284$ **f.** 1

59. **a.** 9 **b.** 8 **c.** $\frac{1}{125}$ **d.** $(1/3)^3 = 1/27$

61. $V = (4/3)\pi r^3$ and thus if $V = 4$ then $r^3 = 3/\pi$. If one uses 3 as a crude estimate of π then r is approximately 1 foot (using a calculator, one gets 0.985 ft.)

63. **a.** 3673.01 lbs. **b.** 6344.62 lbs.

Exercises for Section 4.6

65. **a.** $4.7304 \cdot 10^{20}$ = number of meters in 50,000 light years. Hence, the Milky Way is $9.4608 \cdot 10^{24}$ times larger or nearly 25 orders of magnitude larger than the first living organism on Earth.

b. Pleiades is 3 orders of magnitude older than *homo sapiens.*

67. **a.** $2.0 \cdot 10^{-5}$ in. **b.** $2.0 \cdot 10^{-5}/39 = 5.128 \cdot 10^{-7}$ meters

c. The name is a bit off—by 2 orders of magnitude (looking at meters).

d. The tweezers would have to be made able to grasp things 2 orders of magnitude smaller than they can grasp now.

69. **a.** i. moon: $1.76 \cdot 10^6$ meters; ii. Earth: $6.4 \cdot 10^6$ meters; iii. sun: $6.95 \cdot 10^8$ meters.

b. i. The surface area of Earth is one order of magnitude bigger.
ii. The volume of the sun is 7 orders of magnitude bigger than the volume of the moon.

71. Scale a is additive, while scale b appears to be logarithmic.

73. **a.** At -5 on the log scale.

b. At -11 on the log scale.

75. The point would be about 85% of the way between 10^2 and 10^3 along the x-axis and about 40% of the way between 10^1 and 10^2 along the y-axis.

77. **a.** 10^{-7} cm.

b. A radio wave is 10^5 cm. long on average.

c. 13 orders of magnitude (X-rays, on average, are 10^{-8} cm. long.)

d. From 10 Å to 10^3 Å **e.** From 10^4 Å to 10^7 Å

Exercises for Section 4.7

79. **a.** $\log(10{,}000) = 4$ **b.** $\log(0.01) = -2$

c. $\log(1) = 0$ **d.** $\log(0.00001) = -5$

81. Since $\log(375) = 2.574$ we see that $375 = 10^{2.574}$.

83. **a.** $\log_{10}(100) = 2$ **b.** $\log_{10}(10{,}000{,}000) = 7$

c. $\log_{10}(0.01) = -2$ **d.** $10^1 = 10$

e. $10^4 = 10{,}000$ **f.** $10^{-4} = 0.0001$

85. **a.** $1 < \log 11 < 2$ **b.** $4 < \log 12{,}000 < 5$

c. $-1 < \log 0.125 < 0$

87. **a.** Multiplying by 10^{-3} **b.** Multiplying by $\sqrt{10}$

c. Multiplying by 10^2 **d.** Multiplying by 10^{10}

89. **a.** 7 **b.** 2.85 **c.** $3.16 \cdot 10^{-12}$

d. A higher pH means a lower hydrogen ion concentration because $y = -\log(x)$ is a decreasing function.

e. Pure water is neutral, orange juice is acidic, and ammonia is basic. In plotting, one uses the top numbers to find the right spots. Thus water would be placed at the 7 mark, orange juice 85% of the way between the 2 and 3 marks, and ammonia halfway between the 11 and 12 marks.

CHAPTER 5

Exercises for Section 5.1

1. **a.** Initial population = 275; growth factor = 3

b. Initial population = 15000, growth factor = 1.04

c. Initial population = $6 \cdot 10^8$; growth factor = 5

3. **a.** $G(t) = 5 \cdot 10^3 \cdot 1.185t$ cells/ml.

b. $G(8) = 5 \cdot 10^3 \cdot 1.185^8 = 19440.92$ cells/ml.

5. a and b.

t	$Q = 5 + 1.5t$	Average Rate of Change (between $t-1$ and t)	$Q = 5 \cdot 1.5^t$	Average Rate of Change (between $t-1$ and t)
1	6.5	1.5	7.5	2.5
2	8	1.5	11.25	3.75
3	9.5	1.5	16.875	5.625
4	11	1.5	25.313	8.4375

c. The graphs of the two functions are in the accompanying diagram. They intersect only at $t = 0$ and $Q = 5$. From the shape of the curves, for $t \geq 0$, the linear function will never lie above the exponential. The exponential will grow rapidly for values of $t > 4$ and stay much bigger in value than the linear function.

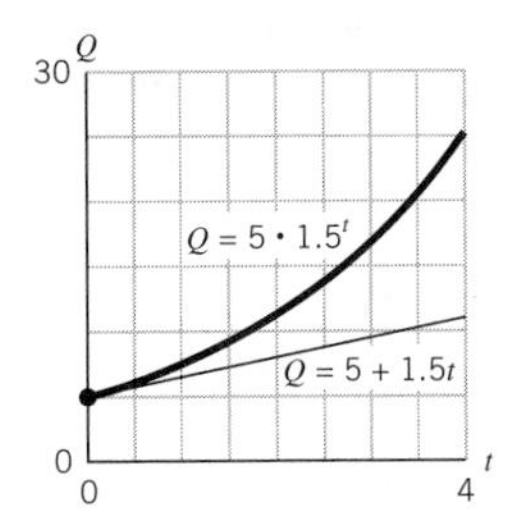

d. The plots of the two average rate of change functions are given in the accompanying diagram. The rate of change for the exponential is ever increasing while that for the linear function is a constant.

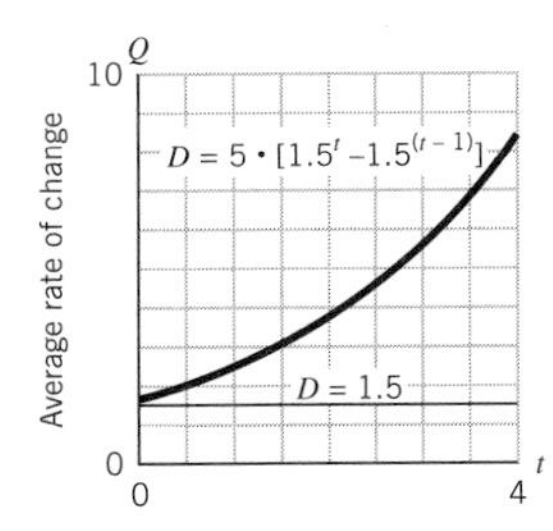

Exercises for Section 5.2

7. a. $f(t) = 10{,}000 \cdot 0.4^t$ **b.** $g(T) = 2.7 \cdot 10^{13} \cdot 0.27^T$

c. $h(x) = 219 \cdot 0.1^x$

9. a. This is not exponential since the base is the variable.

b. This is exponential; the decay factor is 0.5. The vertical intercept is 100.

c. This is exponential; the decay factor is 0.999. The vertical intercept is 1000.

11. a. $P(t) = 100 \cdot b^t$. $99.2 = 100 \cdot b$ or $b = 0.992$ and therefore $P(t) = 100 \cdot 0.992^t$.

b. $P(50) = 100 \cdot 0.992^{50} = 66.9$ grams; $P(500) = 100 \cdot 0.992^{500} = 1.8$ grams

Exercises for Section 5.3

13. Let $y_1 = 2^x$, $y_2 = 5^x$ and $y_3 = 10^x$.

a. $C = 1$ for each case; $a = 2$ for y_1, $a = 5$ for y_2 and $a = 10$ for y_3.

b. Each represents growth; y_1's value doubles, y_2's is multiplied by 5, y_3's by 10.

c. They all intersect at $x = 0$, $y = 1$.

d. In first quadrant, the graph of y_3 will be on top, the graph of y_2 will be in the middle, and the graph of y_1 will be on the bottom.

e. All have the graph of $y = 0$ as their horizontal asymptote.

f. Small table for each:

x	y_1	y_2	y_3
0	1	1	1
1	2	5	10
2	4	25	100

g. The graphs of the three functions are in the accompanying diagram. They indeed confirm the answers given to questions **a** through **f**.

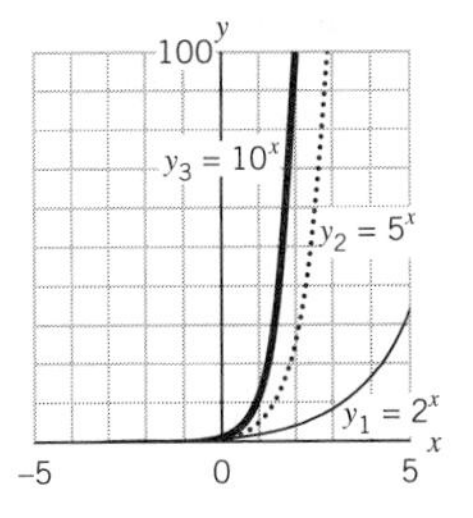

15. Let $y_1 = 3^x$, $y_2 = (1/3)^x$ and $y_3 = 3 \cdot (1/3)^x$.

a. For y_1 and y_2, $C = 1$ and for y_2, $C = 3$. For y_1, $a = 3$; for y_2 and y_3, $a = 1/3$.

b. y_1 represents growth; y_2 and y_3 decay.

c. y_1 and y_3 intersect at $x = 0.5$, $y = 30.5$; y_1 and y_2 intersect at $x = 0$, $y = 1$.

d. The graph of y_1 is on top; the graph of y_3 is in the middle and the graph of y_2 is on the bottom.

e. All graphs have the graph of $y = 0$ as their horizontal asymptote.

f. A small table for each is given by:

x	y_1	y_2	y_3
0	1	1	3
1	3	1/3	1
2	9	1/9	1/3

g. The graphs are given in the diagram in the accompanying diagram. They verify the answers to the questions in **a** through **f**.

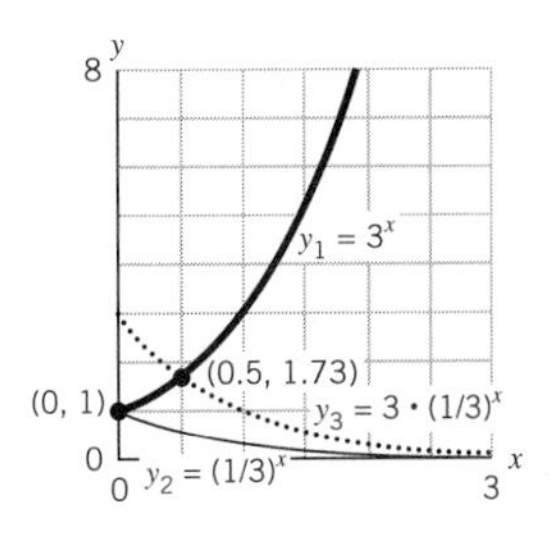

17. The function P goes with graph C; the function Q goes with graph A; the function R goes with graph B and the function S goes with graph D.

Exercises for Section 5.4

19. a. $Q(T) = 1000 \cdot 3^T$ **b.** $Q(T) = 1000 \cdot 1.30^T$

c. $Q(T) = 1000 \cdot 1.03^T$ **d.** $Q(T) = 1000 \cdot 1.003^T$

21. a. $P(t) = 150 \cdot 3^t$ **b.** $P(t) = 150 - 12t$

c. $P(t) = 150 \cdot 0.93^t$ **d.** $P(t) = 150 + t$

23. In 2020, 350 million; in 2050, 490 million. The general formula is $P(t) = 250 \cdot 1.4t$, where t represents the number of 30 year periods from 1990; $y = 1000$ when $t = 123.6$ years.

25. a. $A(n) = A_0 \cdot 0.75^n$, where n measures the number of years from the original dumping of the pollutant.

b. $y = 0.01$ when $n = 16.008$ years.

27. a. 8.7% per year. **b.** 0.4% per month

Exercises for Section 5.5

29. a. The best exponential fit of this data is given by $P(t) = 4.96 \cdot 1.02^t$, where $P(t)$ measures the population of the US in millions and t years since 1780. The annual growth factor 1.02 and the annual growth rate is 2% and the estimated initial population is 4.96 million (about 1.27 times its actual value).

b.

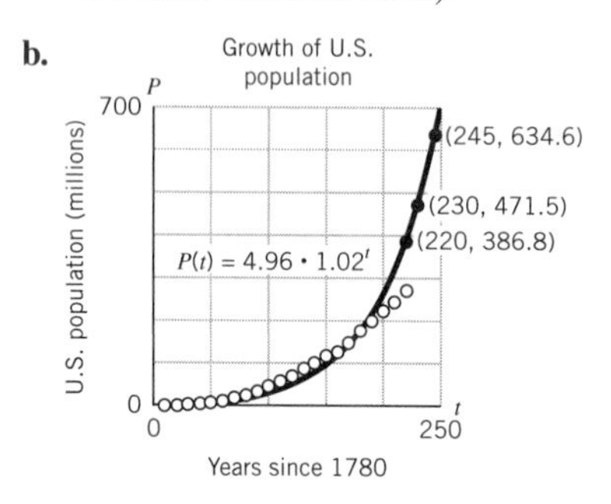

For the first 50 years, the estimated population is only slightly larger than the actual one; then for the next 100 years the real population is larger and thereafter the estimated population grows much more quickly than the actual one. For example, it predicts 386.8 million in 2000.

c. In 2010 : 471.5 million and in 2025 : 634.6 million.

31. a. The data are plotted in the accompanying figure along with the best straight line and exponential fitting curves. It is clear from inspection of the graphs and the data that the exponential graph fits the data much better than the linear graph.

b. Between 1995 and 1999, the number of internet users increased by a factor of $119.2/18 = 6.62$. So for an individual year the growth factor is $(6.62)^{(1/4)} \approx 1.604$. so the annual growth factor is 1.604 and the annual growth rate is 60.4%.

c. The best exponential fit (rounded to two decimal places) is: $U(x) = 18.1 \cdot 1.61^x$ where x = number of years since 1995 and $U(x)$ = number of users in millions; the annual growth factor is 1.61 and the annual growth rate is 61%. The growth factor and rate are slightly higher than what comes from using just the first and last data points.

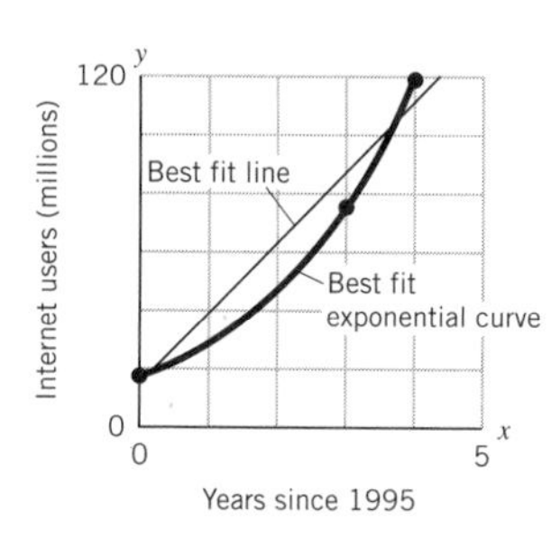

d. In 2010, 22,912 million internet users; in 2025, 29,003,797 million users.

e. Interpolation gives that the US population was approximately 262 million and 76 million is 21.7% of that population. In 1999 interpolation gives 272.1 million in the US and 119.2 million is approximately 31.5% of that population.

f. Using the figures given in **29(c)** and in **31(d)** we see that in both 2010 and 2025 the predicted number of Internet users far exceeds the predicted population of the US. The two models are not very compatible that far into the future.

33. a. Here $A(t)$ represents the salary that would be given by Aerospace Engineering Corp. and where $B(t)$ is the salary that would be given by Bennington Corp.

t	$A(t)$	$B(t)$	t	$A(t)$	$B(t)$
0	50,000	35,000	5	60,000	56,368
1	52,000	38,500	6	62,000	62,005
2	54,000	42,350	7	64,000	68,205
3	56,000	46,585	8	66,000	75,026
4	58,000	51,244			

b. The formulae would be: $A(t) = 50000 + 2000t$ and $B(t) = 35{,}000 \cdot 1.1^t$, where t measures years since hiring and $A(t)$ and $B(t)$ represent the salaries in years after t years of employment.

c. The intersection point of the graphs is marked in the accompanying diagram.

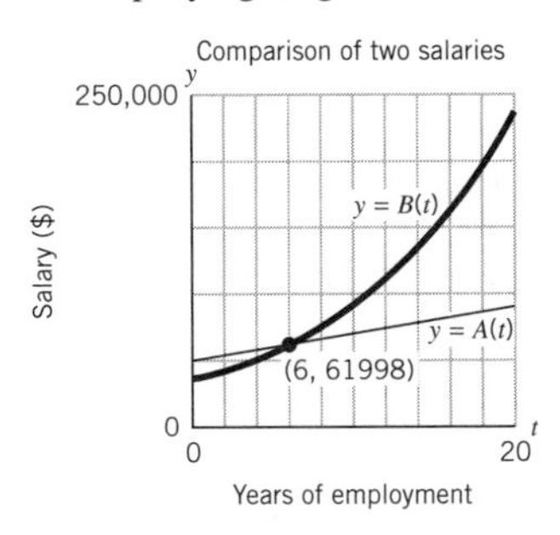

d. The salary from Bennington would exceed the salary from AeroSpace at the start of year 6.

e. $A(10) = 70.000$ and $B(10) = 90{,}780.99$; $A(20) = 90{,}000$ and $B(20) = 235{,}462.50$

35. a. Linear model: $LS(t) = 100 - 10t$, where t is measured in minutes and $LS(t)$ is measured in percent chance of survival. $LS(t) \leq 50$ if $100 - 10t \leq 50$ or $5 \leq t$. Thus, using this model, one's chances are less than 50% after 5 minutes.

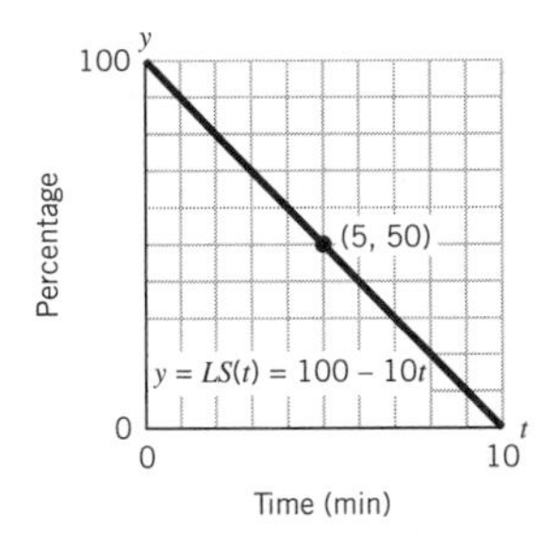

b. $ES(t) = 100(0.90)^t$, where $ES(t)$ and t are respectively measured in the same units as $LS(t)$ and t above. $ES(t) \leq 50$ if $100(0.90)^t \leq 50$ or $0.90t \leq 0.5$ or $t \cdot \log(0.90) \leq \log(0.5)$ or $-0.4576 \cdot t \leq -0.3010$ or $t \geq 6.57$ minutes. Thus, one's chances of survival are less than 50% in the exponential model after 6.57 minutes. The graph for this model is given in the accompanying diagram.

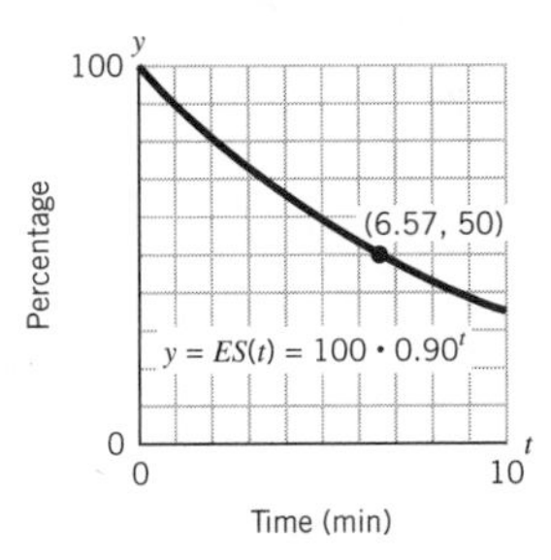

37. The graph goes through the points (0, 100) and (12.3, 50), where the first coordinate is measured in years since the tritium weighed 100 kg. and the second is measured in kilograms. Using the graphing utility gives $A(t) = 100 \cdot 0.945^t$ as best exponential fit.

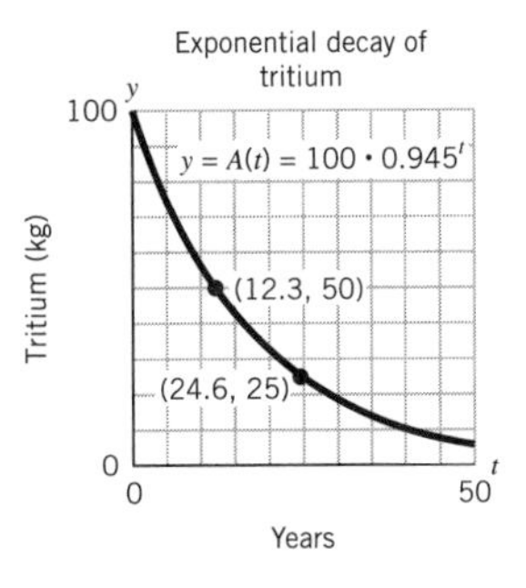

39. a. 1/32nd or 3.125% of the original dosage is left.

b. i. Approximately 2 hrs. ii. $A(t) = 100(1/2)^{t/2}$

iii. 10 hours and 3.125 milligrams

iv. Student answers will vary. They should mention the half life and present a graph to make the drug's behavior more clear to the prospective buyer.

41. a. $R = 70/5730 = 0.012$ percent

b. $R = 70/11460 = 0.0061$ percent

c. $R = 70/(5/31{,}536{,}000) = 441{,}504{,}000$ percent

d. $R = 220{,}752{,}000$ percent

43. b. $P = 817.17 \cdot 1.0092^t$ is the best fitting exponential equation (when rounded off to 2 significant decimal places); where t is measured in years since 1800 and P is measured in millions of people. Its graph is given in the accompanying diagram.

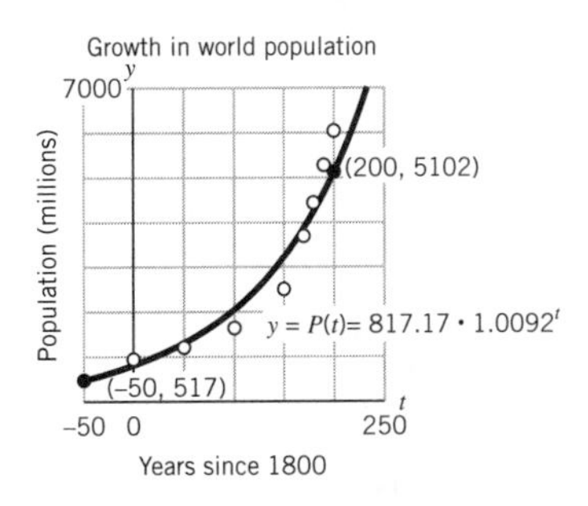

c. The 817.17 is the model's value for the world's population size in 1800 ($t = 0$), measured in millions. (Note that it is approximately 163 million smaller than the actual size.) P's domain should not go much farther back than 1800 nor much past 1980 since (as can be seen from inspecting the graph above) the actual growth is faster after 1980 than the model predicts. The range in that domain is from 817.17 million to 4248.31 million. (If one carries it to 2000, the range goes up to 5102.24 million in the model.)

d. The base 1.0092 means that the growth rate for this model is 0.92%. This rate is measured in millions of people per year.

e. i. Using a calculator gives the $P(t)$ values in the table on the right.

t	$P(t)$
−50	516.95
120	2452.34
225	6414.93
250	8065.34

Eyeball estimates should be close to these.

ii. 1 billion was reached during early 1823; 3.2 billion during 1950; 4 billion half way through 1974; and 8 billion will be reached during early 2050.

iii. It takes approximately 75.69 years for the population to double.

45. a. 10 years gives a 99.9% reduction.

b. $A(2) = 0.25A_0$, $A(3) = 0.125A_0$, $A(4) = 0.0625A_0$. After n half lives the amount left is $A(n) = 0.5^n \cdot A_0$.

47. a. Let $M(t)$ = size of the tumor in grams after t months. Then $M(3) = 20$, $M(6) = 40$, $M(9) = 80$ and $M(12) = 160$. In general, $M(t) = 10 \cdot 2^{1/3} = 10 \cdot (2^{1/3})^t = 10 \cdot 1.2599^t$.

b. Solving $M(t) = 2000$ gives $t = 22.93$ months or 7.64 or 8 quarters or 2 years.

c. $2000/10 = 200$ and thus at 2000 grams it is 20,000% of its original size.

49. a. i. $A(t) = 5000 \cdot (1.035)^t$

ii. $B(t) = 5000 \cdot (1.0675)^t$

iii. $C(t) = 5000 \cdot (1.125)^t$

b. $A(40) = 19{,}796.30$
$B(40) = 68{,}184.45$
$C(40) = 555{,}995.02$

51. a. $V(n) = 100 \cdot (1.06)^n$. Solving $V(n) = 200$ for n graphically gives $n = 11.90$ years.

b. $W(n) = 200 \cdot (1.03)^n$. Solving $W(n) = 400$ for n graphically gives $n = 23.45$ years.

c. The two graphs intersect each other at approximately $n = 24.13$ years, when either investment amounts to \$408.28.

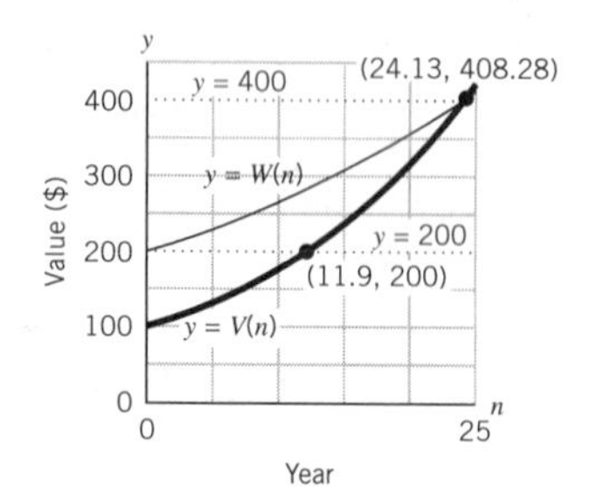

53. Since $1.00/1.06 = 0.943$, we have that the value of a dollar in t years is $V(t) = (0.943)^t$. Solving $V(t) = 0.50$ for t gives $t = 11.8$ years.

55. a. and **b.** It is an exponential growth function. The formula for the best exponential fit (after 2 place roundoff) is: $D = 334.71 \cdot 1.73^x$, where x measures 5 year periods from 1970 and D measures billions of dollars.

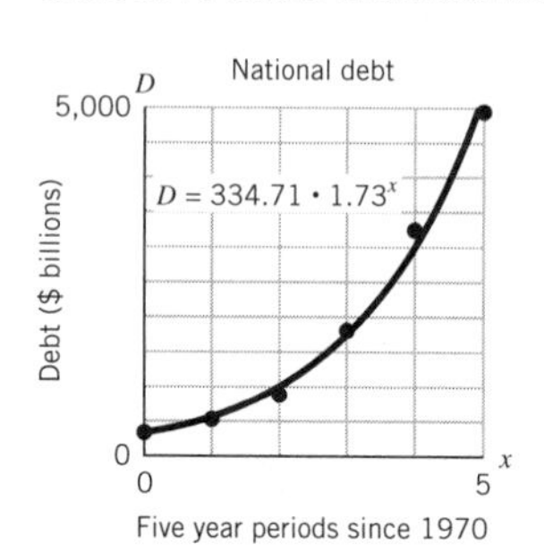

c.

Period	Debt Ratio
70 to 75	1.499
75 to 80	1.680
80 to 85	2.010
85 to 90	1.788
90 to 95	1.519

The average of these ratios is 1.699. Thus $G = 1.699$ is the average growth factor.

d.

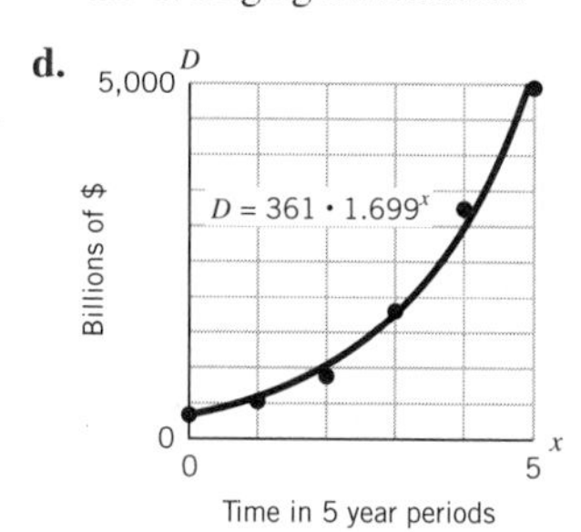

If $4961 = 361 \cdot G^5$, then $G = 1.689$ (when rounded to 3 decimal places).

f.

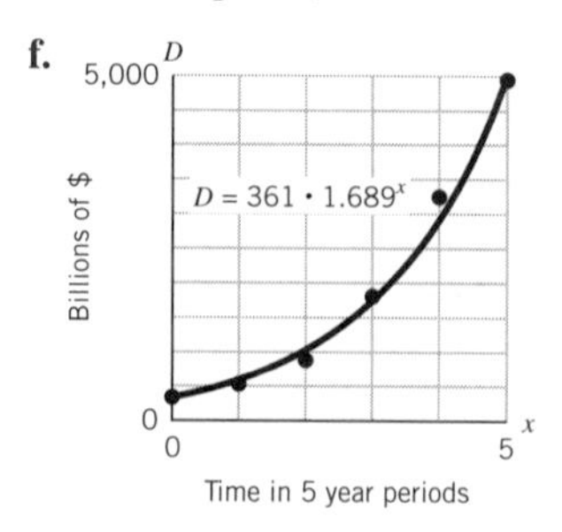

Answers as to which is the better fit will vary. The original best fit model along with the other two predict the following values (rounded off to two decimal places) for 2000 and 2005:

Model	2000	2005
Best fit model	8,973.16	15,523.56
Model from **d**	8,682.95	14,752.34
Model from **e**	8,380.79	14,155.16

g. The national debt seems to be growing exponentially.

57. a. $M_{new}(n) = 10^n$, where n measures the number of rounds and $M(n)$ measures the number of people participating in the nth round of recruiting.

b. $M_{Total}(n) = 1 + 10 + \cdots + 10^n$

c. $M_{Total}(5) = 1 + 10 + \cdots + 10^5 = 11{,}111$ but only 11,110 of these stem from the originator. After 10 rounds the number recruited (not including the first person) would be 11,111,111,110.

d. Note how fast the number grows and how the amounts expected are not quite what one would have thought from the advertisements.

59. a.

	Females		Males	
t	Female Births	Total Females	Male Births	Total Males
0		100		100
1	200	300	200	300
2	600	900	600	900
3	1800	2700	1800	2700
4	5400	8100	5400	8100
5	16200	24300	16200	24300
6	48600	72900	48600	72900
7	145800	218700	145800	218700
8	437400	656100	437400	656100

b. $F(t) = 100 \cdot 3^t$ and $M(t) = 100 \cdot 3^t$

c. There would be 11,809,800 cats altogether. The cat population has grown by 5 orders of magnitude over 5 years.

d. There would be 2,952,450 cats in all.

61. a. Semilog plot of the white blood cell counts over 40 days:

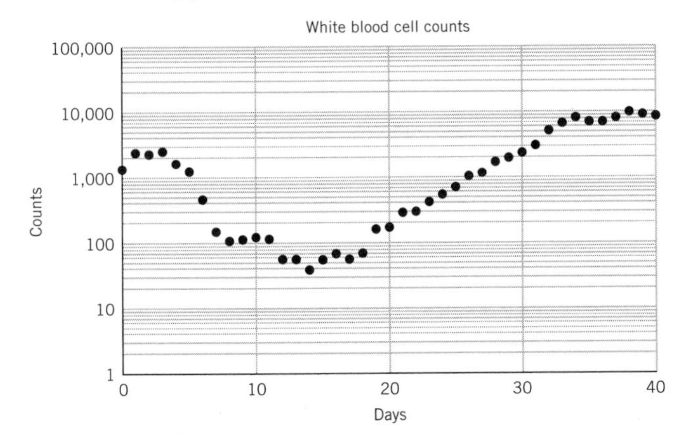

The data from Oct 17th to Oct. 30th seems to be exponential since the data seems to fall on a straight line. The progress from Sept. 30th to Oct 5th could be considered exponential decay but the period is rather short.

b. Semilog plot of the E-coli counts:

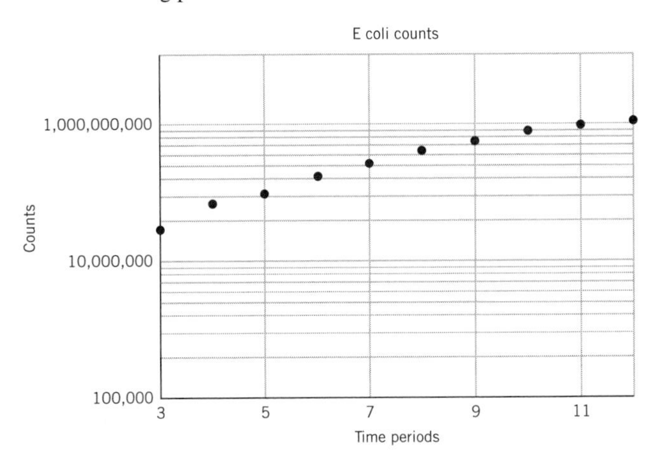

Since the plot looks somewhat linear and is increasing with time, the data represents exponential growth.

63. Note that all numbers in the answer to this question are eyeball estimates.

a. This is a semi-log plot.

b. In 1929 it was 400; in 1933 it was 40; this was the Depression era.

c. 1000 first in 1966; 3000 first in 1994; 6000 first in 1997.

d. A straight line on semi-log graph means that the growth is exponential.

e. In 1980 the DJA is 900 and in 1997 it is 7000. Thus if x is measured in years since 1980, we have that $7000 = 900a^{17}$ and thus $a = 1.128$. Thus the annual rate of growth over that period is 12.8%.

f. Similarly, $7000 = 40a^{64}$ and thus $a = 1.084$. Thus the annual growth rate over that period is 8.4%.

CHAPTER 6

Exercises for Section 6.1

1. a. Approximately 37 years

b. Approximately 22 years

c. Approximately 15 years

3. Eyeball estimates will vary. One set of guesses is **a.** 1.3 hours **b.** 2.4 hrs. **c.** 4.3 hrs.

5. a. 0.001 **b.** 10^6 **c.** $10^{1/3}$

d. 1 **e.** 10 **f.** 0.1

7. If $w = \log(A)$ and $z = \log(B)$ then $10^w = A$ and $10^z = B$. Thus $A/B = 10^w/10^z = 10^{w-z}$ and therefore $\log(A/B) = \log(10^{w-z}) = w - z = \log(A) - \log(B)$.

9. a. $\log\left(\dfrac{K^3}{(K+3)^2}\right)$ **b.** $\log\left(\dfrac{(3+n)^5}{m}\right)$

11. Let $w = \log(A)$. Then $10^w = A$ and thus $A^p = 10^{wp}$ and then $\log(A^p) = w \cdot p = p \cdot w = p \cdot \log(A)$

13. a. 1000 **b.** $x = 999$ **c.** $10^{5/3} \approx 46.416$

d. 0.5 **e.** 1/9 **f.** none

15. a. 37.17 years **b.** 22.51 years **c.** 16.24 years

17. a. 1.71 20 minute time periods or 34.19 min.

b. 5.68 20-minute time periods or 113.58 min.

19. a. $B(t) = B_o(0.5)^{0.05t}$ where t is measured in minutes.

b. 12.5% is left. **c.** Its quarter life is 40 minutes.

d. 66.439 minutes.

21. a. $S(W) = 300(0.9)^W$ if $0 \le W \le 10$, where W is measured in weeks and $S(W)$ in dollars.

b. 6.58 weeks; $104.60.

Exercises for Section 6.2

23. $A_k(n) = 10000 \cdot (1 + 0.085/k)^{kn}$ gives the value of the $10,000 after n years if interest is compounded k times per year and $A_c(n) = 10000 \cdot e^{0.085n}$ gives that value if the interest is compounded continuously.

a. annually: 8.50% **b.** semi-annually: 8.68%

c. quarterly: 8.77% **d.** continuously: 8.87%

25. a. $U(x) = 10 \cdot (1/2)^{x/5}$ where x is measured in billions of years.

b. 16.6096 billion years.

27. a. 6.12 years **b.** 5.86 years **c.** 5.78 years

29. a. 3.37% **b.** 4.50%

Exercises for Section 6.3

31. a. $10^n = 35$ **b.** $N = N_0 e^{-kt}$

33. Let $w = \ln(A)$ and $z = \ln(B)$ and thus $e^w = A$ and $e^z = B$. Therefore $A \cdot B = e^{w+z}$ and $\ln(A \cdot B) = w + z = \ln(A) + \ln(B)$.

35. a. $\ln\sqrt[4]{(x+1)(x-3)}$ **b.** $\ln\left(\dfrac{R^3}{\sqrt{P}}\right)$

c. $\ln\left(\dfrac{N}{N_0^2}\right)$

37. a. $x = \ln(10) \approx 2.303$ **b.** $x = \log(3) \approx 0.477$

c. $x = \log(5)/\log(4) \approx 1.161$ **d.** $x = e^5 \approx 148.413$

e. $x = e^3 - 1 \approx 19.086$ **f.** no solution (ln increases)

Exercises for Section 6.4

39. a. At $B = B_0$.

b.

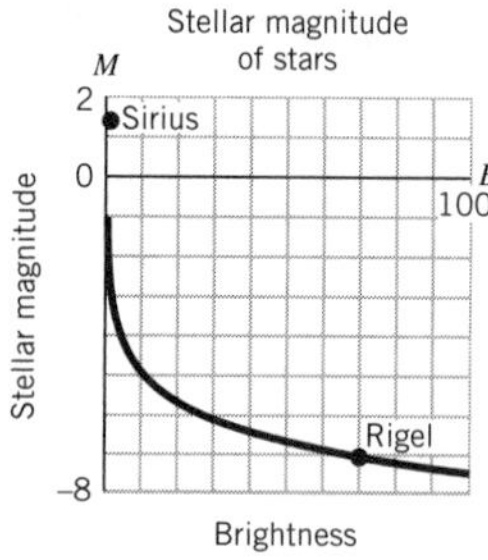

c. As the brightness increases the magnitude decreases. Thus, a 6th magnitude star is less bright than a first magnitude star.

d. The magnitude becomes $2.5 \cdot \log(5) \approx 1.75$ units less in size.

41. 28 dB has $I = 10^{-13.2}$ watts/cm^2 and 92 dB has $I = 10^{-6.8}$ watts/cm^2.

43. Quintuplets crying are about 7 decibels louder than 1 baby crying.

Exercises for Section 6.5

45. a. $N = 10e^{0.044017t}$ **b.** $Q = 5 \cdot 10^{-7} \cdot e^{-2.631089A}$

47. a. decay **b.** decay **c.** growth

49. 10.430 km.

51. Half life is $\ln(1/2)/(-r) = \ln(2)/r = 100 \cdot \ln(2)/R \approx 69.3147/R$ which is approximately $70/R$.

53. Draw a curve that goes through roughly the middle of each cluster. Without the data table it is not possible to give the best fit exponential with any precision.

55. $T = 13.817$, $n_0 = 9.95$ or 10 neutrons

57. a.

Year	1965	1970	1975	1980	1985	1990	1995
Cost in $ per person	214	358	606	1088	1799	2805	3770

b. and **c.** The data suggest that the growth is exponential. The curve's equation is: $C(x) = 25.43 \cdot 1.115^x$, where x measures years since 1960 and $C(x)$ measures the total medicine expenditures in year x in billions of dollars. The model predicts that in 1995 the total expenditure will be $1148 billion. Actual costs were $156 billion less.

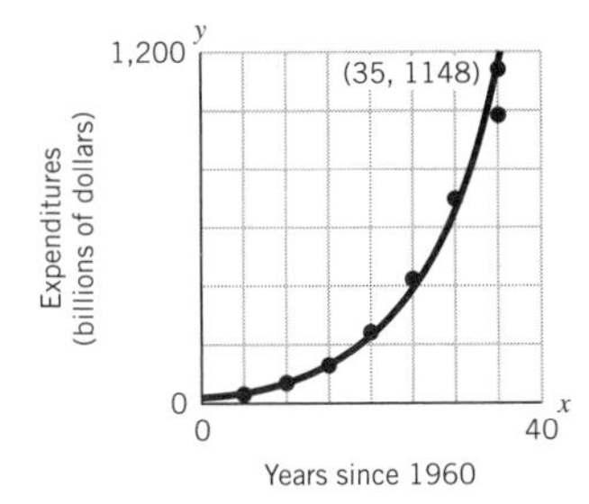

Exercises for Section 6.6

59. a. $F(G(1)) = F(0) = 1$ **b.** $G(F(-2)) = G(-3) = 4$

c. $F(G(2)) = F(0.25) = 1.5$

d. $F(F(0)) = F(1) = 3$

e. $F \circ G(x) = F(G(x)) = F\left(\dfrac{x-1}{x+2}\right) = 2\left(\dfrac{x-1}{x+2}\right) + 1 = \dfrac{3x}{x+2}$

f. $G \circ F(x) = G(2x+1) = \dfrac{(2x+1)-1}{(2x+1)+2} = \dfrac{2x}{2x+3}$

61. a. $f(g(1)) = f(0) = 2$ **b.** $g(f(1)) = g(1) = 0$

c. $f(g(0)) = f(1) = 1$ **d.** $g(f(0)) = g(2) = 3$

e. $f(f(2)) = f(3) = 0$

63. a. $g(f(2)) = g(0) = 1$ **b.** $f(g(-1)) = f(2) = 0$

c. $g(f(0)) = g(4) = -3$ **d.** $g(f(1)) = g(3) = -2$

65. a. $V = (4/3)\pi(5+p)^3$ **b.** $V = g(f(p))$

67. a. $T = 32 - 5s$ **b.** $S(x) = \left[1 - \dfrac{1}{2}\cdot\left(\dfrac{x}{20}\right)^2\right]S_d$

c. At the middle of the road $x = 0$ and $S(0) = S_d$. At the edge of the road $x = 20$ and $S(20) = 0.5S_d$.

d. $T(x) = 32 - 5\left[1 - \dfrac{1}{2}\cdot\left(\dfrac{x}{20}\right)^2\right]S_d$

e. $T(0) = 32 - 5S_d$ and $T(20) = 32 - 2.5S_d$.

69. a. $f(g(x)) = f\left(\dfrac{x+1}{2}\right) = 2\left(\dfrac{x+1}{2}\right) - 1 = x + 1 - 1 = x$ and $g(f(x)) = g(2x-1) = \dfrac{[2x-1]+1}{2} = \dfrac{2x}{2} = x$

b. $f(g(x)) = f\left(\dfrac{x^3-5}{4}\right) = \sqrt[3]{4\left(\dfrac{x^3-5}{4}\right)+5} = \sqrt[3]{x^3} = x$ and $g(f(x)) = g(\sqrt[3]{4x+5}) = \dfrac{(\sqrt[3]{4x+5})^3 - 5}{4} = \dfrac{(4x+5)-5}{4} = x$

71. They are inverse functions by definition, i.e. $e^{\ln(x)} = x$ and $\ln(e^x) = x$. See the tables of properties.

73. a.

x	$f^{-1}(x)$
5	−2
1	−1
2	0
4	1

b.

x	$f^{-1}(x)$
5	0
3	1
2	−2
−7	4

75.

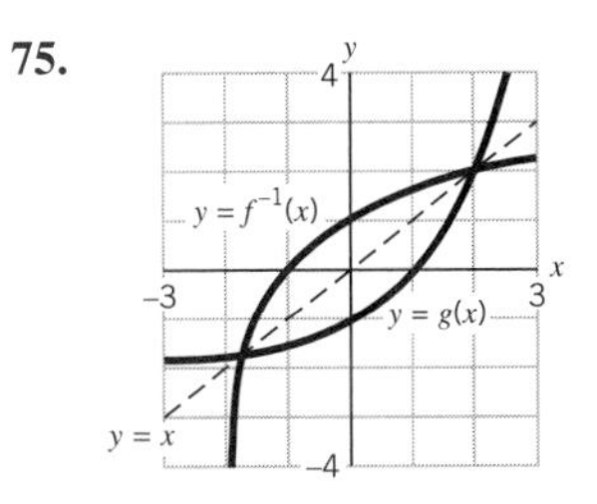

77. a. $x = 4y - 7$ or $4y = x + 7$ or $y = (x+7)/4$ Thus $f^{-1}(x) = (x+7)/4$

b. $x = 8y^3 - 5$ or $x + 5 = y^3$ or $y = (x+5)^{1/3}$ Thus $g^{-1}(x) = (x+5)^{1/3}$

c. $x = (y+1)/2y$ or $2xy = y + 1$ or $(2x-1)y = 1$ Thus $h^{-1}(x) = 1/(2x-1)$

79. a. The form is $T = A + Ce^{-kt}$. The data are: $A = 70$ °F; at $t = 0$, $T = 375$ °F and at $t = 30$, $T = 220$ °F. Hence $375 = 70 + C$ and thus $C = 305$. Now we also have $220 = 70 + 305e^{-30k}$ and thus $150/305 = e^{-30k}$ or $k = \ln(150/305)/(-30) \approx 0.237$ Thus $T = 70 + 305e^{-0.237t}$ is the law of cooling in this setup.

b., c. and **d.** $t = \dfrac{1}{0.237}\ln\left(\dfrac{305}{T-70}\right)$ is the inverse function. At $T = 370$ we have $t = 0.0697$ and at $T = 374$ we have $t = 0.0139$ and at $T = 375$ we have $t = 0$.

e. The graph of T:

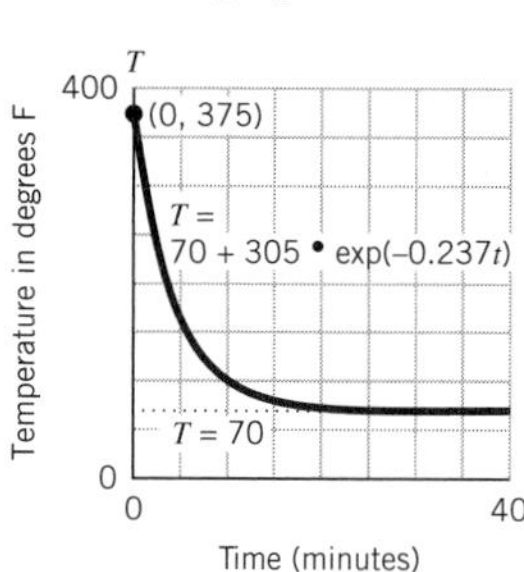

f. Graph of inverse:

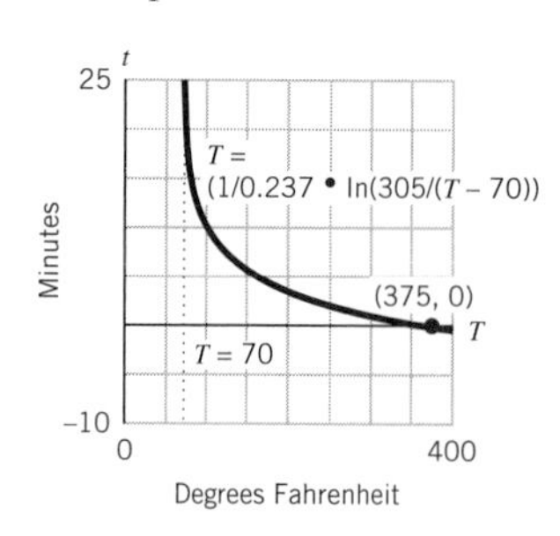

17. a. and **b.** The graphs of D_4 and D_{safe}

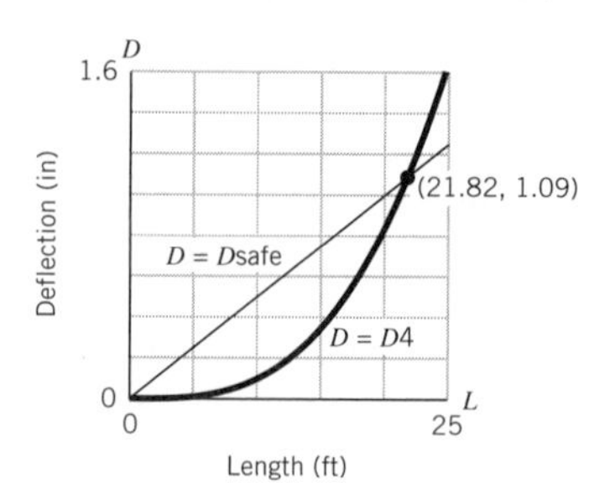

c. One notes that the safety deflections are well above the actual deflections for all values of L between 0 and 20. It would cease to be safe if the plank were longer than 21.82 ft. (the L value where the graphs of D_{safe} and D_4 meet–this can be solved for graphically or by setting the two equations equal to each other).

CHAPTER 7

Exercises for Section 7.1

1. a. $S = 1.26 \cdot 10^{-19}\ \text{m}^2$ **b.** $V = 4.19 \cdot 10^{-30}\ \text{m}^3$

c. $S/V = 3.0 \cdot 10^{10}\ \text{m}^{-1}$

3. a. S is multiplied by 16; V is multiplied by 64.

b. S is multiplied by n^2; V is multiplied by n^3.

c. S is divided by 9; V is divided by 27.

d. S is divided by n^2; V is divided by n^3.

5. a. The volume doubles if the height is doubled.

b. The volume is multiplied by 4 if the radius is doubled.

7.

radius	C	A	Circumference ratio	Area Ratio
R	2πR	πR^2	1 to 1	1 to 1
2R	4πR	4πR^2	2 to 1	4 to 1
3R	6πR	9πR^2	3 to 1	9 to 1
0.5R	πR	0.25πR^2	1 to 2	1 to 4

When you multiply the radius by k, you multiply the circumference by k.
When you multiply the radius by k, you multiply the area by k^2.

Exercises for Section 7.2

9. a. 20, 20 **b.** 40, – 40 **c.** – 20, – 20

d. – 40, 40

11. a. $Y = kX^3$ **b.** $k = 1.25$

c. Increased by a factor of 125

d. Divided by 8 **e.** $X = \sqrt[3]{\dfrac{Y}{k}}$

13. a. L is multiplied by 32. **b.** M is multiplied by 2^p.

15. $k = 144$; I(8) = 2.25 watts per sq. meter and $I(100) =$ 0.0144 watts per sq. meter.

19. In all the formulae below, k is the constant of proportionality.

a. $d = \text{k} \cdot t^2$ **b.** $E = \text{k} \cdot m \cdot c^2$. **c.** $A = \text{k} \cdot b \cdot h$

d. $R = \text{k}[O_2] \cdot [NO]^2$ **e.** $v = \text{k} \cdot r^2$

Exercises for Section 7.3

21. The graphs of the 5 functions:

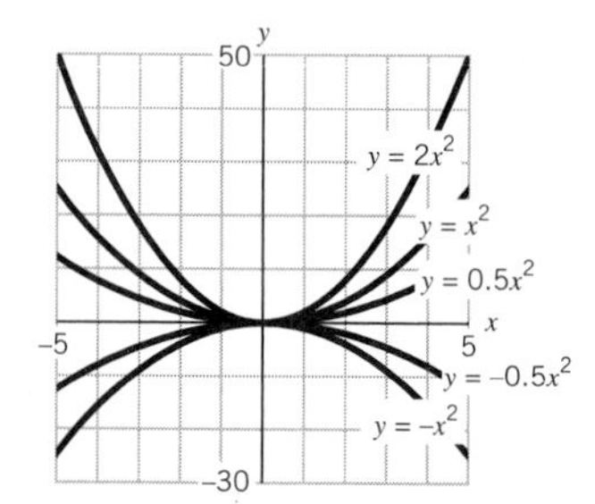

Each function has basically the same shaped graph. They are all of the form $y = ax^2$, for various values of a. The differences among the graphs are caused by this constant multiplier. If $a > 0$, then the graph faces up. If $a < 0$, then the graph faces down. The larger the absolute value of a, the narrower the graph.

23. Graphs of first four functions

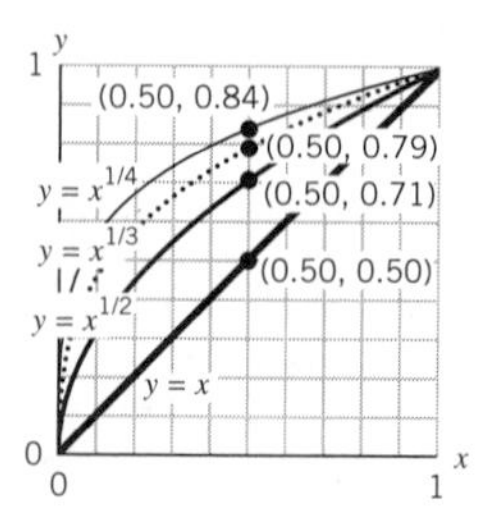

Graphs of the first and the last 3 functions

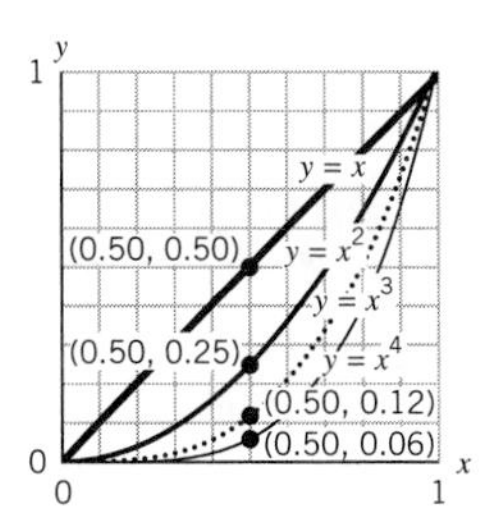

All seven graphs go through (0, 0) and (1, 1). The points where $x = 0.5$ on each are marked with their y values for comparison. One notes that the smaller the fractional power the bigger the y value here. One may also note that the fractional powers all have y values above or on the graph of $y = x$ and the whole number powers all have y values below or on the graph of the graph of $y = x$ over the interval [0, 1].

Exercises for Section 7.4

25. **a.** The two graphs intersect at (2, 16) and (4, 256). Thus, the functions are equal for $x = 2$ and $x = 4$. (See the graphs below.)

b. As x increases both functions grow. For $0 \le x < 2$ we see that $x^4 < 4^x$; from $2 < x < 4$ we have $4^x < x^4$; for $x > 4$ we have that $x^4 < 4^x$.

c. Eventually (for $x > 4$) the graph of $y = 4^x$ dominates: see the graph below.

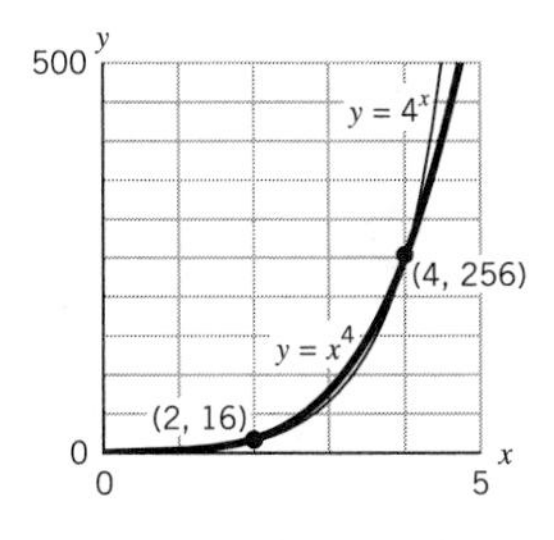

27. As the graphs of $y = 2^x$ and $y = x^2$ show, there are intersections at $x = 2$, and $x = 4$ and $y = 2^x$ is above the graph of $y = x^2$ for $0 \le x < 2$ and for $x > 4$.

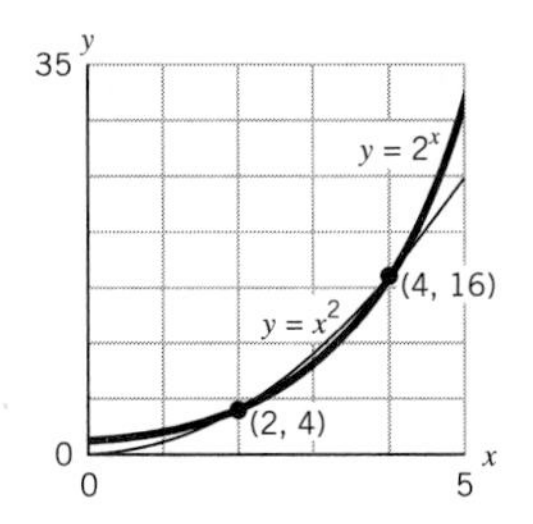

The graphs of $y = x^3$ and $y = 3^x$ show one point of intersection (at $x = 3$, $y = 27$) and that for $x > 0$ we have $x^3 < 3^x$ except at $x = 3$.

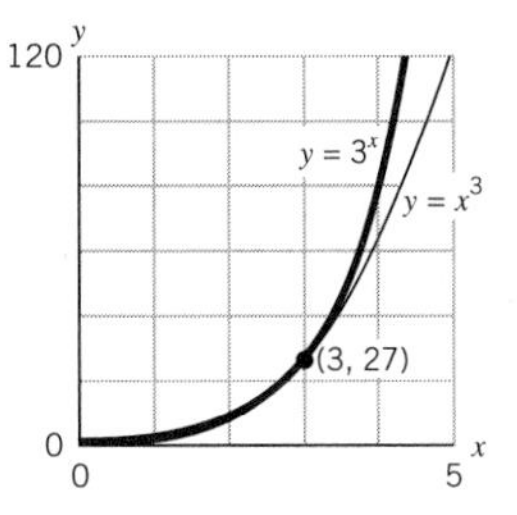

29. $3 \cdot 2^x > 3 \cdot x^2$ if $0 < x < 2$ and if $x > 4$. Also, $3 \cdot 2^x < 3 \cdot x^2$ if $2 < x < 4$.

Exercises for Section 7.5

31. $I(d) = k/d^2$ and thus $I(4)/I(7) = 49/16 = 3.06$–it will be 3.06 times as great.

33. **a.** The volume becomes 1/3 of what it was.

b. It becomes $1/n$ of what it was.

c. The volume is doubled.

d. It becomes n times what it was.

35. **a.** Letting L = the wavelength of a wave, v = the velocity of a wave, and t = the time between waves, we have that $t = L/(k\sqrt{L}) = (1/k) \cdot \sqrt{L}$.

b. 4 times as far apart.

37. **a.** $x = k \cdot \dfrac{y}{z}$ **b.** Solving $4 = k \cdot \dfrac{16}{32}$ for k gives $k = 8$

c. $x = 8 \cdot 25/5 = 40$

39. **a.** $H = k/P$ is suggested, where k is a proportionality constant.

b. The software gives $H = 87.19 \cdot P^{-0.99}$ as a best fit, which is very close.

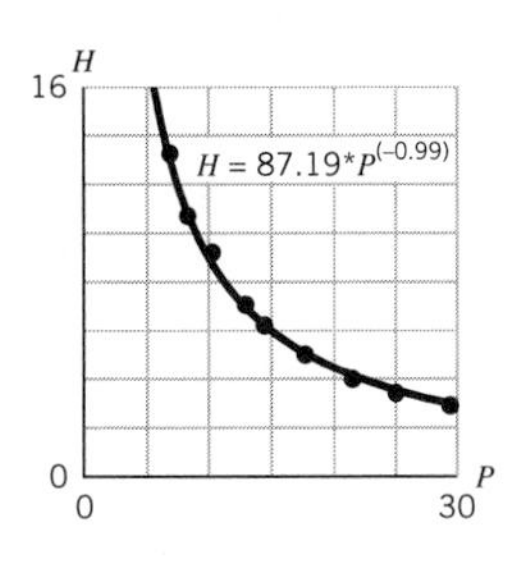

Exercises for Section 7.6

41.

x	$r(x)$
1	6
2	3
3	2
6	1
−1	−6
−2	−3
−3	−2
−6	−1

The domain of the abstract function is $x \neq 0$. For $x > 0$, as $x \to 0$ we have $r(x) \to +\infty$ and for $x < 0$, as $x \to 0$ $r(x) \to -\infty$.

43. a.

x	$5x$	x	$x/5$	x	$1/x$	x	$5/x$
−2	−10	−2	−0.4	−2	−1/2	−2	−5/2
2	10	2	0.4	2	1/2	2	5/2
−1	−5	−1	−0.2	−1	−1	−1	−5
1	5	1	0.2	1	1	1	5
0	0	0	0.0	0.5	2	0.5	10

b. The graphs of g and h are both straight lines through the origin with positive slope. In absolute value, the y values of g's graph are 25 times those of h. As for the graphs f and t: both have the x and y axes as asymptotes, i.e., they approach but never touch these axes—and both are confined to the first and third quadrants. The y values of the graph of f are 5 times the values of the graph.

Graphs of $g(x) = 5x$ and $h(x) = x/5$

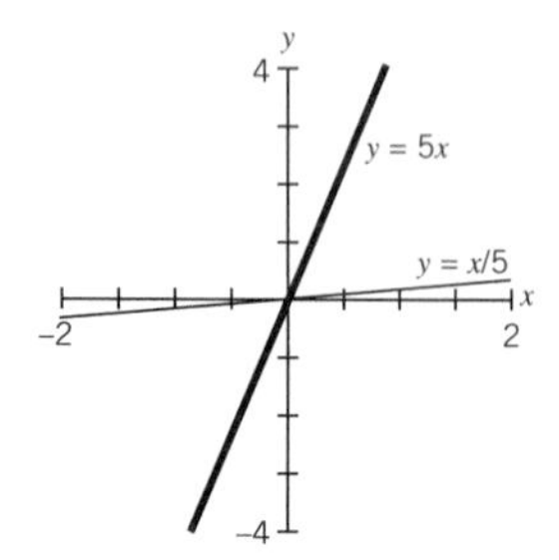

Graphs of $t(x) = 1/x$ and $f(x) = 5/x$

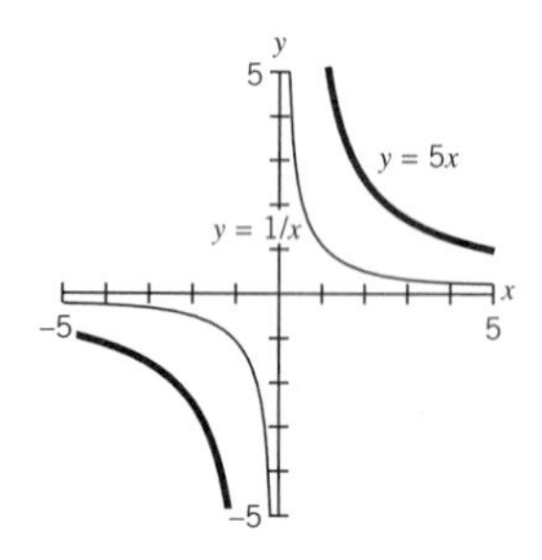

45. a. $y = x^2$ decreases for $x < 0$ and increases for $x > 0$; $y = x^{-2}$ increases for $x < 0$ and decreases for $x > 0$.

b. The two graphs intersect at $(-1, 1)$ and $(1, 1)$.

c. As x approaches $\pm\infty$, the graph of $y = x^2$ approaches ∞. As x approaches $\pm\infty$, the graph of $y = x^{-2}$ approaches 0.

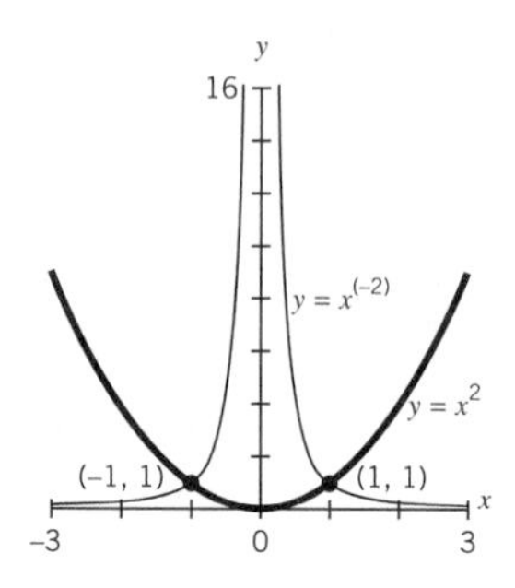

47. a. intersect at $x = 0$ and at $x = 1/2$

b. intersect at $x = 0$ and at $x = 1$

c. intersect at $x = (1/4)^{1/4} = \sqrt{(1/2)}$ and $-\sqrt{(1/2)}$

d. intersect at $x = 1$ **e.** intersect at $x = 1$.

Exercises for Section 7.7

49. The graph given in the text suggests an exponential decay function. It is nearly a straight line graph and the vertical scale is logarithmic.

51. a.

Species	Adult Mass in Grams	Egg Mass in Grams	Egg/Adult Ratio
ostrich	113,380.0	1700.0	0.015
goose	4,536.0	165.4	0.036
duck	3,629.0	94.5	0.026
pheasant	1,020.0	34.0	0.033
pigeon	283.0	14.0	0.049
hummingbird	3.6	0.6	0.167

Notable is the fact that the egg/adult mass ratio for the hummingbird is very high and it is very low for the ostrich. The other ratios are not far apart from each other.

b. In the three scatter plots given below, x stands for adult bird mass in grams and y stands for egg mass in grams.

i. linear x vs. linear y plot

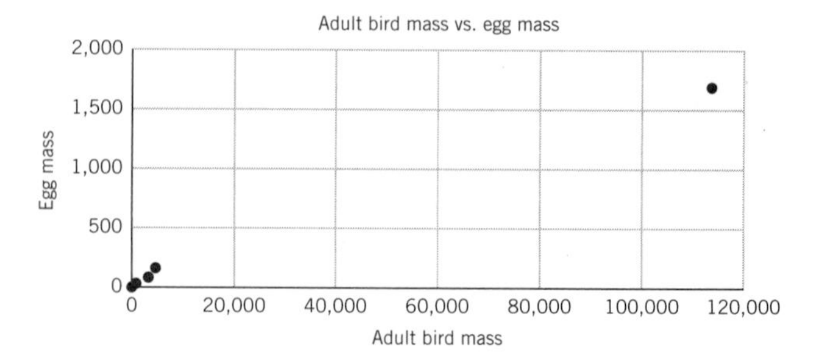

ii. linear x *vs.* log y plot

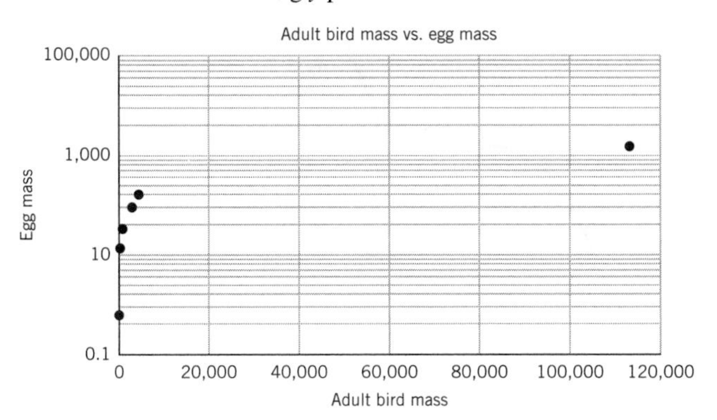

iii. log x *vs.* log y plot

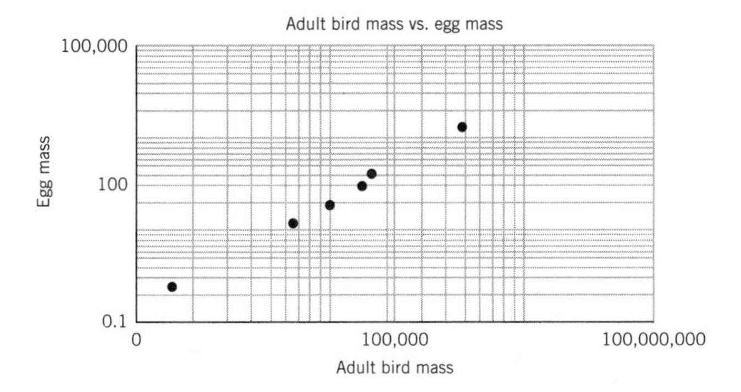

c. The log *vs.* log scatter diagram is, of course, the most linear. The best straight-line fit equation for this is: $\log(y) = \log(0.1918) + 0.7719 \cdot \log(x)$ or, in the more ordinary linear equation form: $Y = -0.717 + 0.7719X$

d. Regrouping, using the laws of logarithms, we get: $\log(y) = \log(0.1918 \cdot x^{0.7719})$ and thus $y = 0.1918 \cdot x^{0.7719}$ where x = mass of the adult and y is the corresponding mass of the egg, both measured in grams.

e. 1.36 grams.

f. 20.848 grams.

53. a. Since ℓ_1 and ℓ_2 are parallel lines they have the same slope but different vertical intercepts. Thus their equations can be written as $\log(y) = mx + \log(b_1)$ and $\log(y) = mx + \log(b_2)$, with $\log(b_1) \neq \log(b_2)$. Solving each for y in terms of x we have: $y = b_1 \cdot 10^{mx}$ and $y = b_2 \cdot 10^{mx}$ with $b_1 \neq b_2$. Thus they have the same power of 10 but differ in their y-intercepts.

b. The equation of ℓ_3 is $\log(y) = m \cdot \log(x) + \log(b_3)$ and the equation of ℓ_4 is $\log(y) = m \cdot \log(x) + \log(b_4)$, with $\log(b_3) \neq \log(b_4)$. Thus they have different $\log(y)$ intercepts. Solving for y in terms of x in each gives: $y = b_3 x^m$ and $y = b_4 x^m$ with $b_3 \neq b_4$. Thus they have the same power of x but differ in their constant multipliers.

Exercises for Section 7.8

55. a. The population density is directly proportional to the -2.25 power of the length of the organism. Let x be the length of the organism measured in meters and let p be the population density of that organism, i.e., the number of individual organisms per square kilometer. Then, since the scale on both axes of the graph is logarithmic and the graph is a straight line, the relationship between population density and length of an organism is: $\log(p) = k - 2.25 \cdot \log(x)$. Since we can write $k = \log(c)$ for some positive c, we can combine and get $\log(p) = \log(c \cdot x^{-2.25})$ or $p = c \cdot x^{-2.25}$.

b. The population density of the organism is directly proportional to the -0.75 power of the body mass of the organism. Its derivation is similar to that done in part (**a**).

c. From (**a**) and (**b**) we have that $d \cdot m^{-0.75} = c \cdot x^{-2.25}$ or that $m = (d/c)^{4/3} \cdot x^3$. Thus, mass is directly proportional to the cube of the length.

CHAPTER 8

Exercises for Section 8.1

1. a. 1/8, − 1/8 **b.** 1/2, − 1/2

c. − 1/2, 1/2 **d.** − 32, 32

3. a. $j(x) = 3x^5 + x^2 + x - 1$; $k(x) = 3x^5 - x^2 + x + 1$; $l(x) = 3x^7 - 3x^5 + x^3 - x$

b. $j(2) = 101$; $k(3) = 724$; $l(-1) = 0$

5. a. $y = 2x - 3$ goes with Graph 1: linear, positive slope, y-intercept at -3.

b. $y = 2 - x$ goes with Graph 4: linear, negative slope, y-intercept at 2.

c. $y = 3(2^x)$ goes with Graph 3: exponential; increasing, y-intercept at 3.

d. $y = (x^2 + 1)(x^2 - 4)$ goes with Graph 2: 2 zeros at ± 2, 4th degree, etc.

7. a.

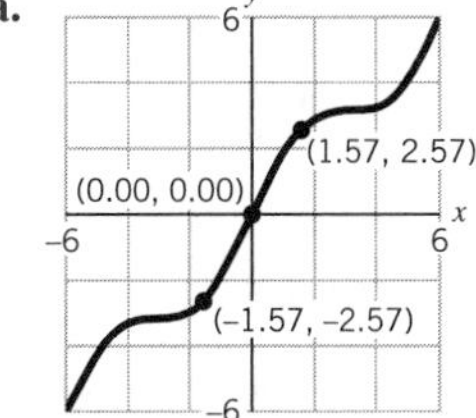

x	y_1	y_2	$y_1 + y_2$
0.0	0.0	0.0	0.0
1.6	1.6	1.0	2.6
− 1.6	− 1.6	− 1.0	− 2.6
...	...	...	...

b. Graph of $y = -x^2 + x^3$

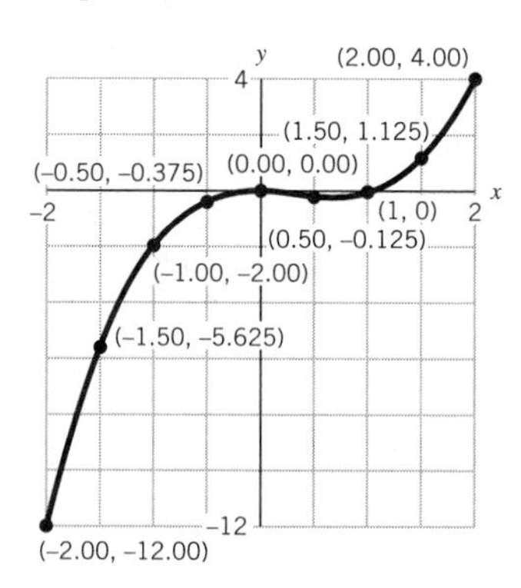

c. Graph of $y = 2^x - x^2$

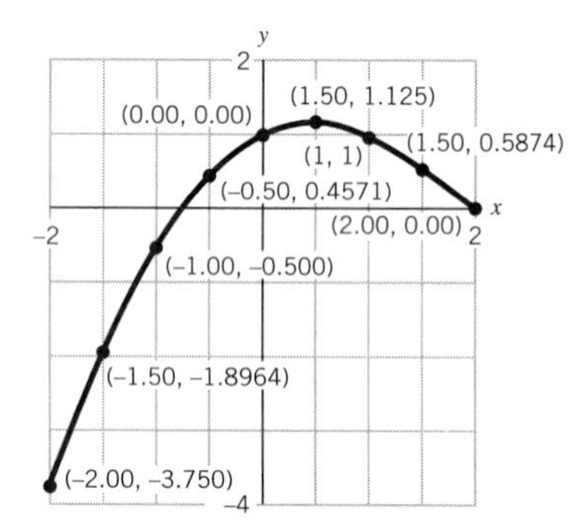

9. a. $8701.5 = 8n^3 + 7n^2 + 0n + 1n^0 + 5n^{-1}$

b. $239 = 2n^2 + 3n + 9n^0$; if $n = 2$, this polynomial would have the value 23.

c. The number written in base 2 as 11001 and equals 25 when written in base 10 notation since $1 \cdot 2^4 + 1 \cdot 2^3 + 0 \cdot 2^2 + 0 \cdot 2^1 + 1 \cdot 2^0 = 25$.

d. $35 = 1 \cdot 2^5 + 0 \cdot 2^4 + 0 \cdot 2^3 + 0 \cdot 2^2 + 1 \cdot 2^1 + 1 \cdot 2^0$ or 100011 in base 2.

11.

Function	x-Intercepts	Degree
$y = 2x + 1$	-0.5	1
$y = x^2 - 3x - 4$	$-1, 4$	2
$y = x^3 - 5x^2 + 3x + 5$	$-0.709, 1.806, 3.903$	3
$y = 0.5x^4 + x^3 - 6x^2 + x + 3$	$-4.627, -0.610, 0.916, 2.320$	4

One might be tempted to conclude that a polynomial of degree n has n distinct, real zeros.

13. a. $y = 3x^3 - 2x^2 - 3$ has only one real zero at $x \approx 1.28$.

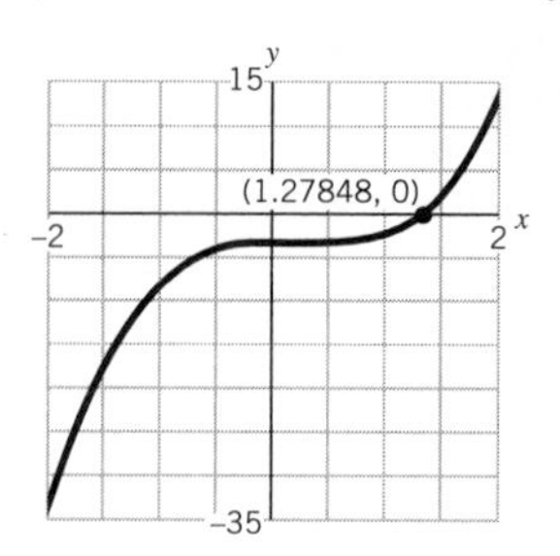

b. $y = x^2 + 5x + 3$ has two real zeros: $x = -2.5 \pm 0.5\sqrt{13} \approx -4.30$ or -0.70

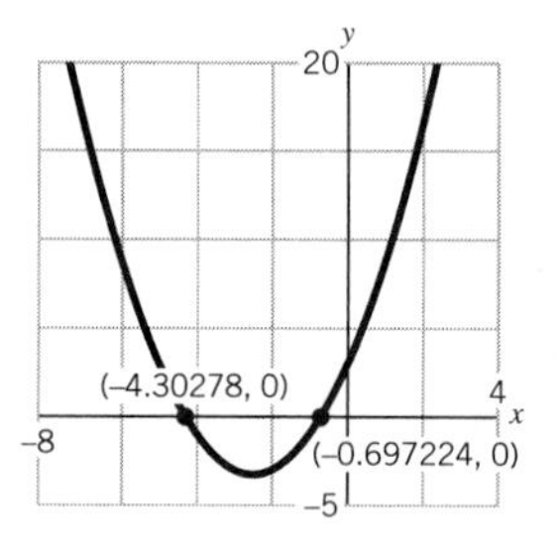

15. a. When the degree of a polynomial is odd, the "tails" of its graph head in opposite directions; i.e., as x increases or decreases without bound, the y values approach ∞ on one side and $-\infty$ on the other. Since all polynomials have graphs that are one piece, the graph of a polynomial of odd degree must cross the x-axis at least once.

b. i. $y = x^2 + 1$ is an example of a polynomial of degree 2 that has no real zeros. See the accompanying graph.

ii. $y = -1 - x^4$ is an example of a polynomial of degree 4 that has no real zeros. See the accompanying graph.

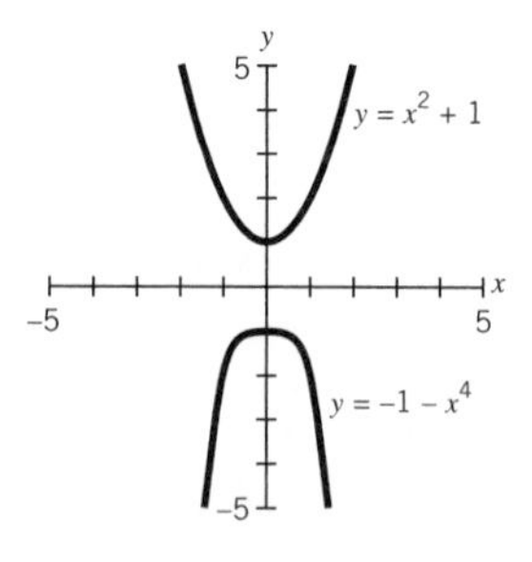

c. i. $y = x^3$ is an example of a polynomial of degree 3 that has exactly one real zero. See the accompanying graph.

ii. $y = (x + 1)(x - 3)(x - 5)$ is an example of a polynomial of degree 3 that has exactly 3 real zeros. See the accompanying graph.

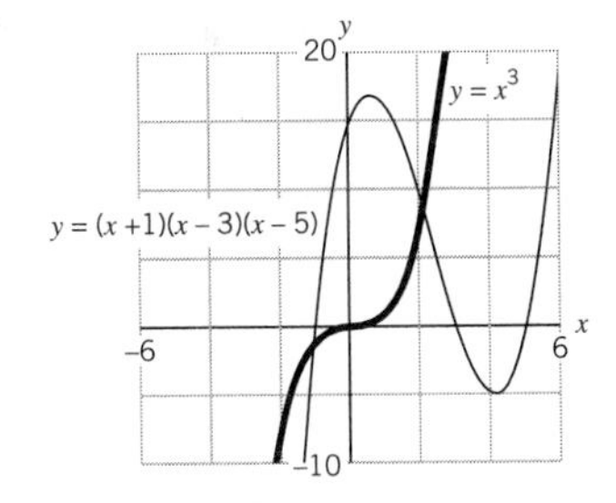

d. The graph of $y = (x^4 - 1)$ has exactly two real zeros: 1 and -1. See the accompanying graph.

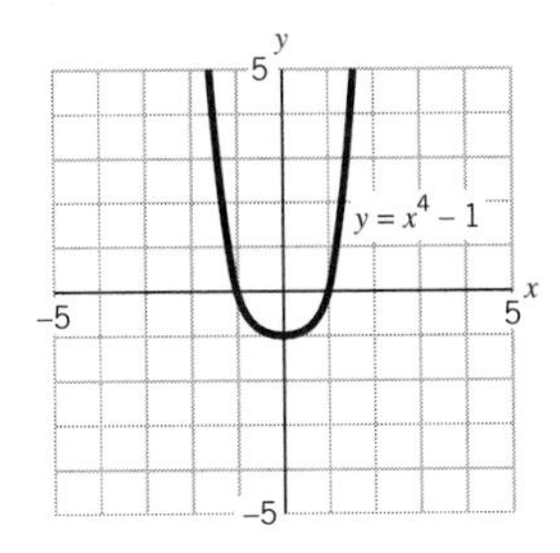

17. a. If $f(x) = a \cdot x^{2k}$, then $f(-x) = a(-x)^{2k} = ax^{2k} = f(x)$

b. If $f(x) = a \cdot x^{2k+1}$ then $f(-x) = a(-x)^{2k+1} = -ax^{2k} = -f(x)$.

c. Even functions have graphs that are symmetric about the y-axis. Odd functions have graphs that are symmetric about the origin. An inspection of the graphs of $y = x^3$ and of x^2, for example, will show this.

d. i. $f(-x) = (-x)^4 + (-x)^2 = x^4 + x^2 = f(x)$–this is an even function.

ii. $u(-x) = (-x)^5 + (-x)^3 = (-x^5) + (-x^3) = -(x^5 + x^3) = -u(x)$–this is an odd function.

iii. $h(-x) = (-x)^4 + (-x)^3 = x^4 - x^3 \neq h(x)$ and $\neq -h(x)$–neither even nor odd.

iv. $g(-x) = 10 \cdot 3^{-x} \neq g(x)$ and $\neq -g(x)$–neither even nor odd.

e. The graphs of even functions are symmetric with respect to the y-axis and the graphs of odd functions are symmetric with respect to the origin as the graphs of the functions in **d. i** and **d. ii** will show.

Exercises for Section 8.2

19. a. The graphs of these are all parabolas facing up and with vertex at (0, 0). But the larger the coefficient, the faster the y values grow as x moves away from the origin in either direction. (See their graphs.)

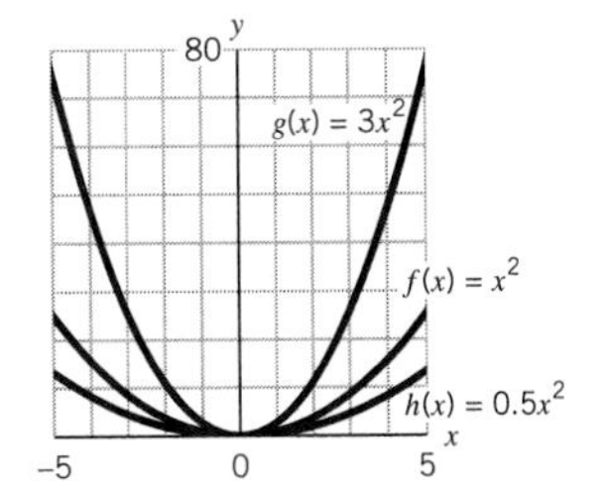

b. The graphs of these functions are all parabolas. Two have their vertex at (0, 4) and one has the vertex at (0, -4). The first and last are mirror images of each other with respect to the x-axis. The first and 2nd are mirror images of each other with respect to the graph of $y = 4$. (See their graphs.)

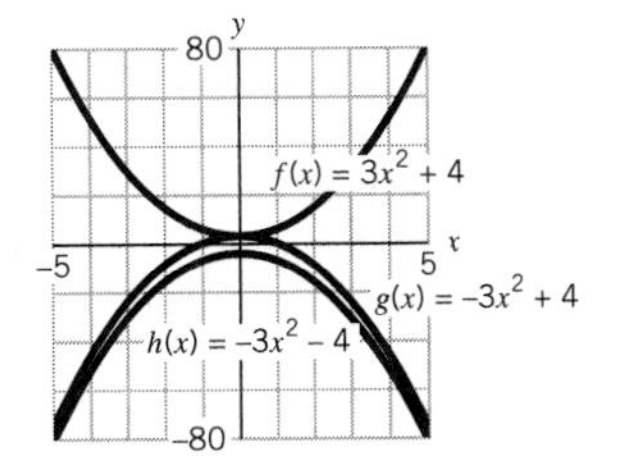

c. The y values of all three increase as x increases when $x > 0$. Their relative rates of increase vary but eventually the growth of $y = 2x$ becomes the slowest, that of $y = 2^x$ becomes the quickest and the growth of $y = x^2$ stays between the others. (See their graphs.)

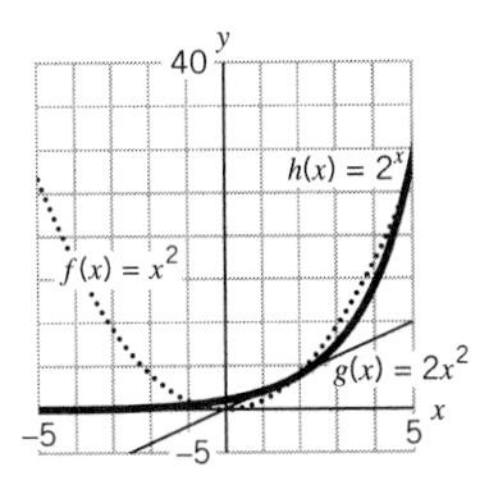

21.

23. Any quadratic of the form $y = ax^2 + bx + 10$ will have a y-intercept of 10. If a is positive then the graph will turn up and if a is negative the graph will turn down. The value of b is not relevant here. The graph of $y = x^2 - 7x + 10$ is given here; it is turned up and has two x-intercepts, one at 5 and the other at 2, while the graph of $y = 10 - x^2$ will turn down and has two zeros, namely $= \pm\sqrt{10}$.

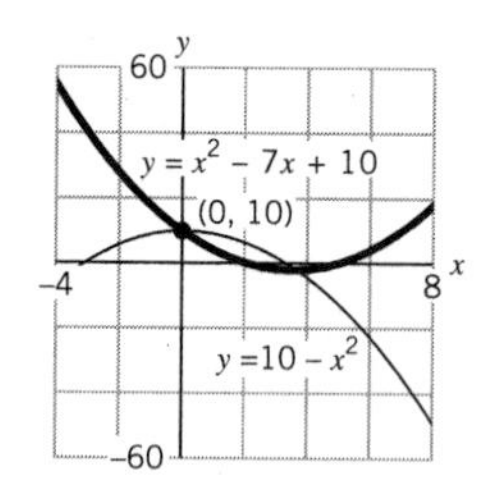

25. **a.** $A(x) = x^2 + (20 - x)^2$

b. Its domain is $0 < x < 20$

27. **a.** 50 meters by 50 meters

b. $P/4$ meters by $P/4$ meters

29. $V(h) = (20 - h)\cdot(10 - h)\cdot h$, with $0 < h < 10$, is the volume formula. An inspection of its graph shows that the maximum volume occurs at $h \approx 4.227$ feet. The corresponding volume is 384.900 cu.ft. (Note that the dimensions of the box are approximately 4.227 ft. by 5.773 ft. by 15.773 ft. That is a large box!)

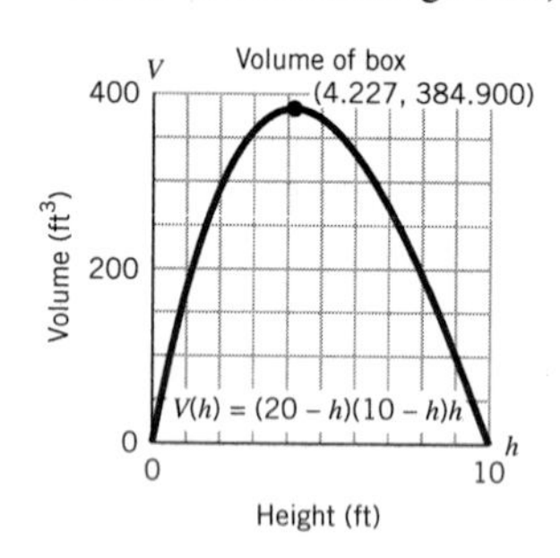

31. **a.** $W = 1600\cdot 4\cdot 100/120 = 5333.33$ lbs.

b. $W = 1600\cdot 10\cdot 16/120 = 2133.33$ lbs. It has been reduced to 40% of what it should be.

c. $W = 2666.67$ if L is doubled, W is halved.

d. $1600\cdot 4\cdot d^2/192 = 1600\cdot 4\cdot 100/120$ gives $d^2 = 160$ or $d = 12.65$ in.

Exercises for Section 8.3

33. **a.** $t = 5/3$ or $-1/2$ **b.** $x = 2/3$

c. $z \approx 1.91$ or -1.57 **d.** $x \approx$ -1.59 or -4.41

35. **a.** $y = x^2 - 5x + 6$ has 2 and 3 as roots.

b. $y = 3x^2 - 2x + 5$ has no roots at all

c. $y = 3x^2 - 12x + 12$ has one double root at 2.

37. **a.** y-intercept is (0, 1); x-intercepts at (1/3, 0) and (− 1, 0).

b. y-intercept is (0, 11); x-intercepts at approximately (1.42, 0) and (2.58, 0).

39. **a.** $0 = x^2 - 9 = (x - 3)(x + 3)$; thus $x = 3$ or -3

b. $0 = x^2 - 4x = x(x - 4)$; thus $x = 0$ or 4.

c. $0 = 3x(3x - 25)$; thus $x = 0$ or 25/3

d. $0 = x^2 + x - 20 = (x + 5)(x - 4)$; $x = -5$ or 4.

e. $0 = 4x^2 - 12x + 9 = (2x - 3)^2$; $x = 3/2$

f. $0 = 3x^2 - 13x - 10 = (3x + 2)(x - 5)$; $x = 5, -2/3$

g. $0 = x^2 + 4x + 4 = (x + 2)^2$; $x = -2$

h. $0 = 2x^2 - 5x - 3 = (2x + 1)(x - 3)$; $x = -0.5, 3$

41. **a.** $y = (x + 4)(x + 2)$

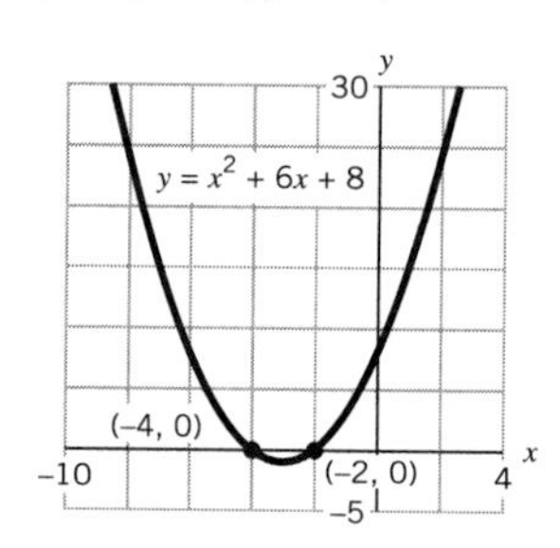

b. $z = 3(x - 3)(x + 1)$

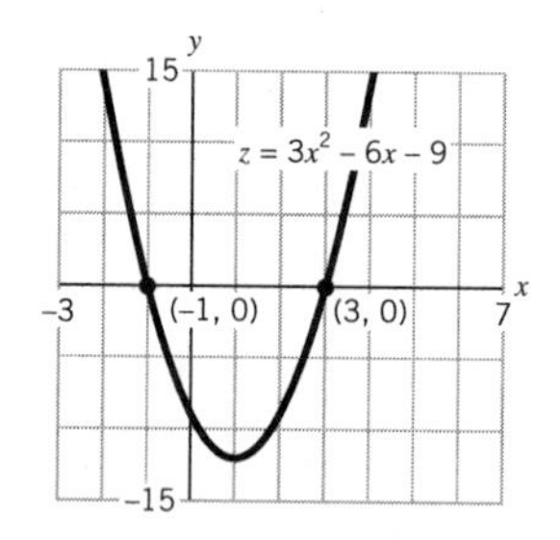

c. $f(x) = (x - 5)(x + 2)$

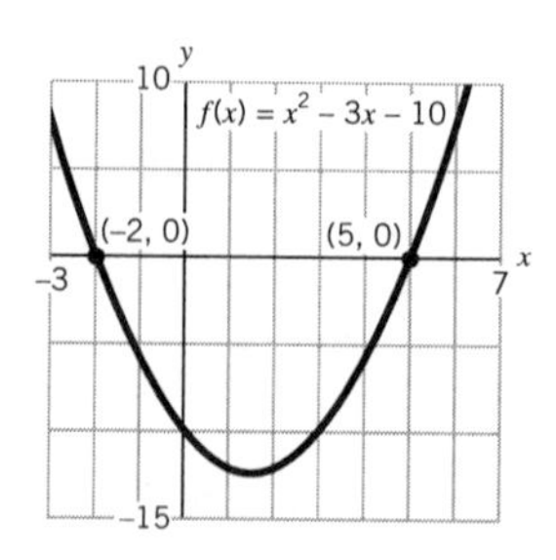

d. $w = (t - 5)(t + 5)$

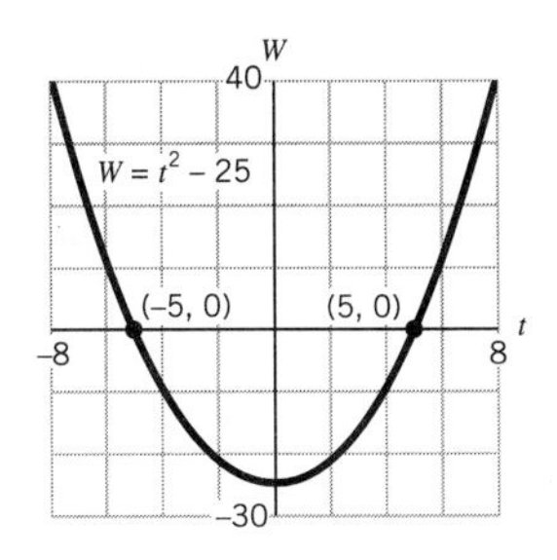

e. $r = 4(s - 5)(s + 5)$

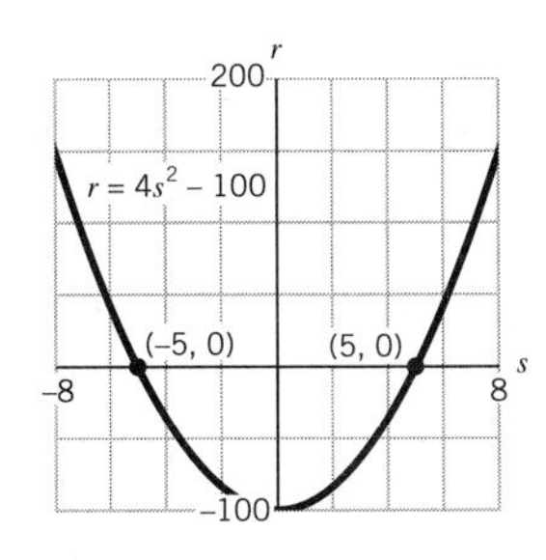

f. $g(x) = (x + 1)(3x - 4)$

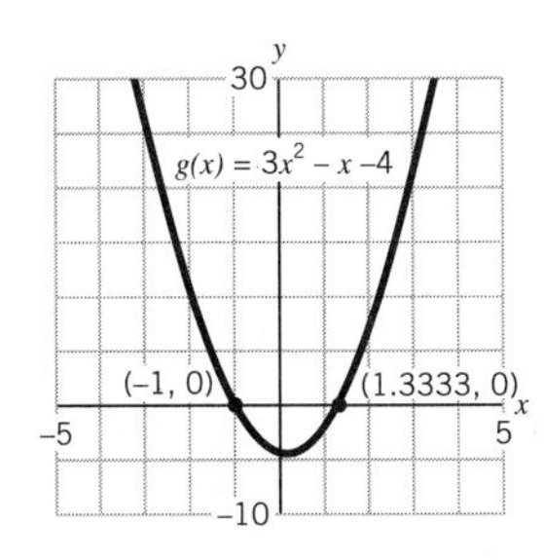

43. a. Algebraically one finds the values of x that satisfy the two equations:

$2x^2 - 3x + 5.1 = -4.3x + 10$ or

$2x^2 + 1.3x - 4.9 = 0$ or

$x = [-1.3 \pm \sqrt{[1.3^2 - 4(2)(-4.9)]}]/4$ or

$= [-1.3 \pm \sqrt{40.89}]/4$ or

$= [-1.3 \pm 6.39]/4$ or

$x = 1.27 \text{ or } -1.92$

b.

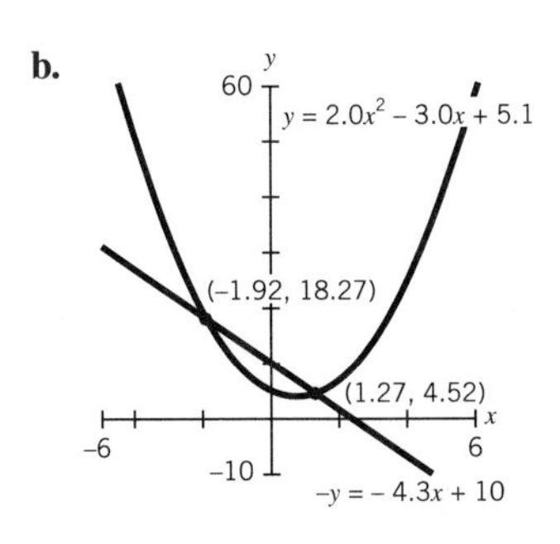

Exercises for Section 8.4

45. a.

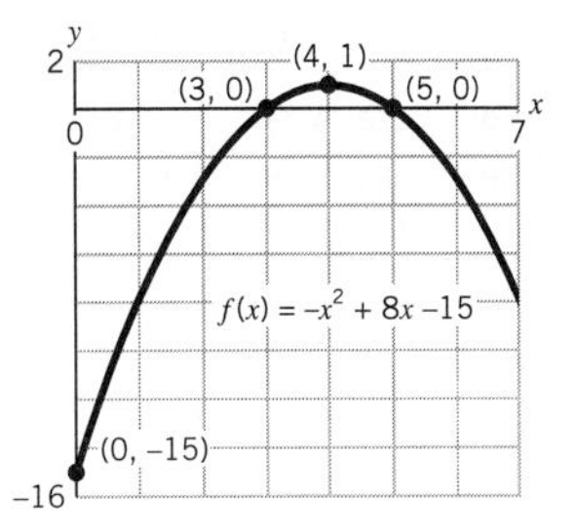

The formula can be factored into $y = -(x - 3)(x - 5)$ and thus the zeros are 3 and 5.

b. $f(0) = -15$ gives the y-intercept.

c. The vertex is at $x = -b/(2a) = -8/(-2) = 4$

47. An expenditure of $x = 20$ yields the maximum profit, which is 430 thousand dollars.

49. The maximum occurs when $x = 8$ computers and the revenue will be 192 million dollars.

51. a. At $t = 0$, $h = 4$ ft.

b. $h = 0$ when $t = 3.20$ or -0.08 and the latter is rejected because h is not defined for negative values of time.

c. $h = 30$ ft. at $t = 0.659$ sec. and at $t = 2.466$ sec. (via quadratic formula); it is never 90 feet high since the discriminant is negative for $h = 90$.

d. It reaches its maximum height at $t = -50/(-32) = 1.56$; and the height at that moment is 43.06 ft.

53. We are given that $2W + L = 1$. Thus $L = 1 - 2W$ and thus the area $A = W(1 - 2W) = W - 2W^2$. Note this is at its maximum at the vertex, i.e., when $W = -1/[2\cdot(-2)] = 1.4$. $L = 1/2$. Thus, Dido was correct.

55. a.

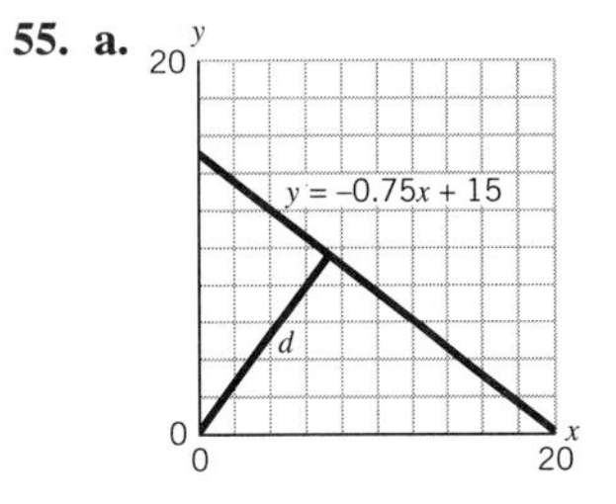

The graph show x and y marked off in miles, and with a typical point (x, y) marked on the highway along with the straight line to that point.

b. The highway goes through the points (0, 15) and (20, 0) and thus has the equation $y = -0.75x + 15$.

c. and **d.** The distance squared $D = x^2 + y^2 = x^2 + (15 - .75x)^2 = 1.5625x^2 - 22.5x + 225$

e. The minimum occurs at the vertex, i.e., where $x = 22.5/(2 \cdot 1.5625) = 7.2$. The minimum for D is 144.

f. Thus the minimum distance $\sqrt{D} = \sqrt{144} = 12$ miles. The coordinates of the point of shortest distance from (0, 0) are (7.2, 9.6).

57. a. 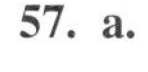

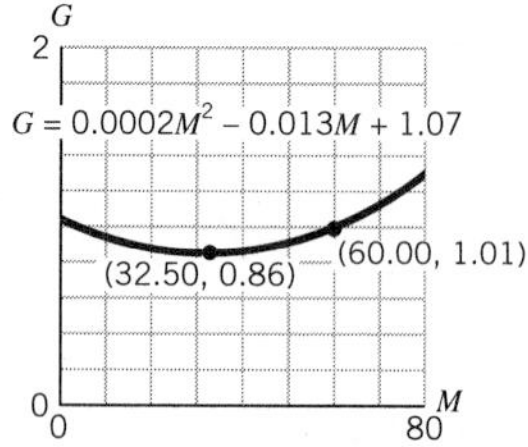

The minimum gas consumption rate suggested by the graph occurs when $M = 32.5$ mph and it is approximately 0.86 gph.

b. Computation on a calculator gives the same M but the gas consumption rate is 0.86 when rounded off.

c. In 2 hours, 1.72 gallons will be used and one will have traveled 65 miles.

d. It takes 1 hour and 5 minutes to travel 65 miles at 60 mph and one will have used 1.094 gallons.

e. Clearly traveling at the speed that is supposed to minimize the gas consumption rate does not conserve fuel if the trip lasts only 2 hrs.

f. and g.

mph	gph	gpm	mpg
0	1.07	—	0.0
10	0.96	0.09600	10.4
20	0.89	0.04450	22.5
30	0.86	0.02867	34.9
40	0.87	0.02175	46.0
50	0.92	0.01840	54.3
60	1.01	0.01684	59.4
70	1.14	0.01629	61.4
80	1.31	0.01638	61.1

f. $y = G/M$

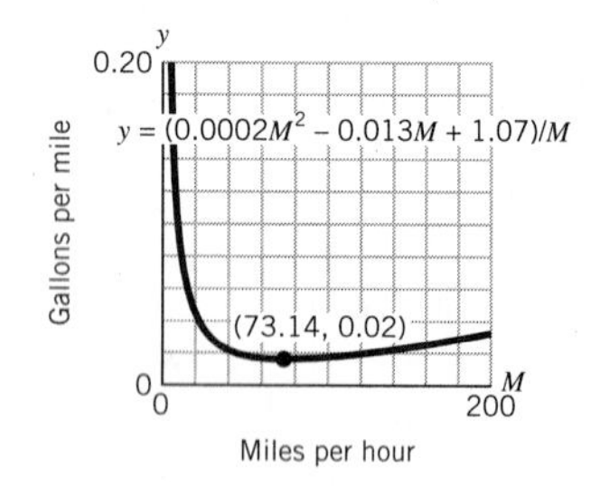

g. $y = M/G$

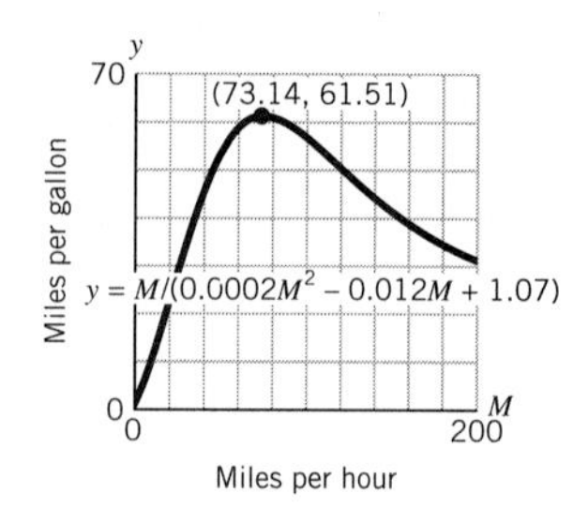

For **f**: $G/M = (0.0002M^2 - 0.013M + 1.07)/M$; eyeballing its graph gives the minimum y at $M \approx 73$ mpg. For **g**: $M/G = M/(0.0002M^2 - 0.013M + 1.07)$; eyeballing its graph gives the maximum y at the same M. This is expected since max = 1/min.

59. In general: The a-h-k form is the easier to use to find the vertex and it is easier to find the y-intercept from the a-b-c form.

a. a-b-c form: $y_1 = 2x^2 - 3x - 20$
a-h-k form: $y_1 = 2(x - 0.75)^2 - 21.125$

a-b-c form: $y_2 = -2x^2 + 4x - 5$
a-h-k form: $y_2 = -2(x - 1)^2 - 3$

a-b-c form: $y_3 = 3x^2 + 6x + 3$
a-h-k form: $y_3 = 3(x + 1)^2 + 0$

a-b-c form: $y_4 = -2x^2 + 2x + 12$
a-h-k form: $y_4 = -2(x - 0.5)^2 + 12.5$

b. The vertex for y_1 is at $(0.75, -21.125)$; its y-intercept is -20; its x-intercepts are at $x = -2.5$ and $x = 4$.

Graph of y_1

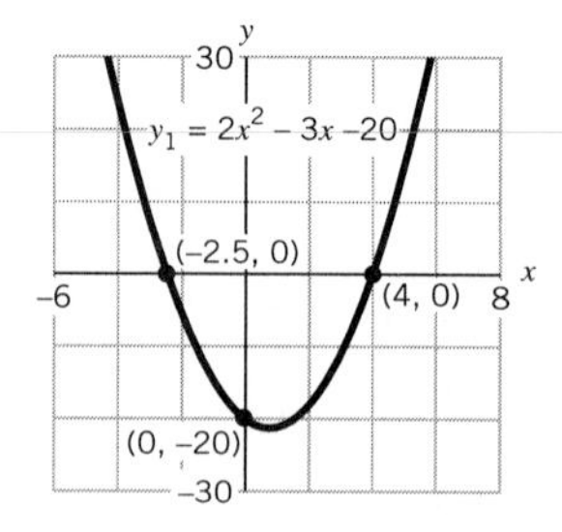

The vertex for y_2 is at $(1, -3)$; its y-intercept is -5; it has no x-intercepts since $(x - 1)^2$ cannot be equal to $-3/2$ and the discriminant is -24. (See the graph.)

Graph of y_2

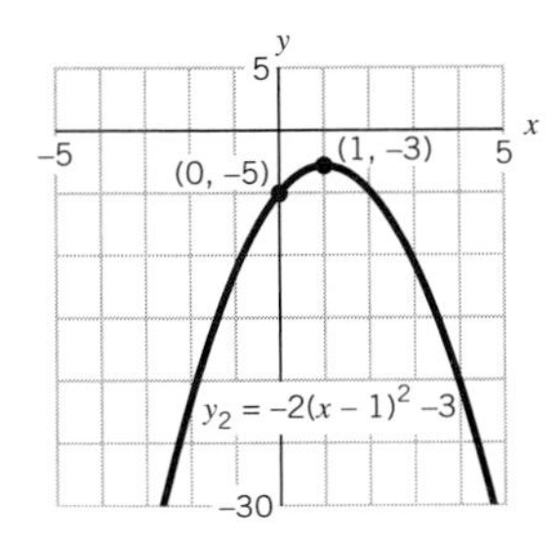

The vertex for y_3 is at $(-1, 0)$; its y-intercept is 3 and its x-intercept is at $x = -1$ only.

Graph of y_3

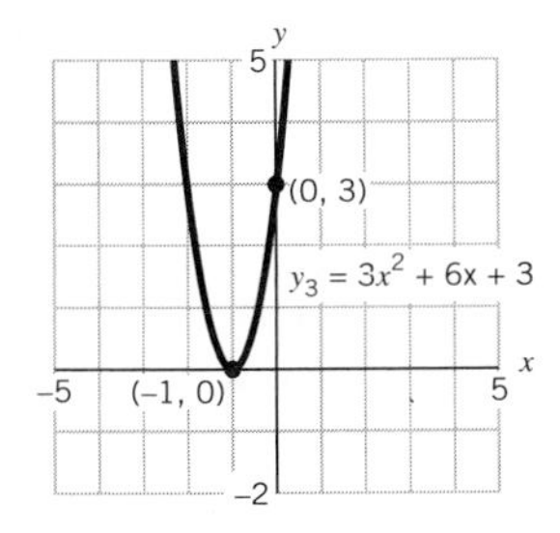

The vertex for y_4 is at (0.5, 12.5); its y intercept is 12; its x-intercepts are at $x = -2.$ and $x = 3$.

Graph of y_4

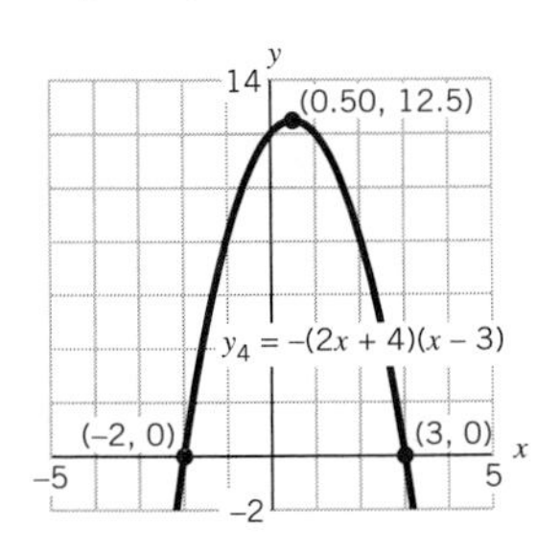

61. a. $y = a(x - 2)^2 + 4$ and $7 = a(1 - 2)^2 + 4$ gives us $a =$ 3 and thus $y = 3(x - 2)^2 + 4 = 3x^2 - 12x + 16$

b. $y = a(x - 2)^2 - 3$ is concave up if $a > 0$ and concave down if $a < 0$.

63.

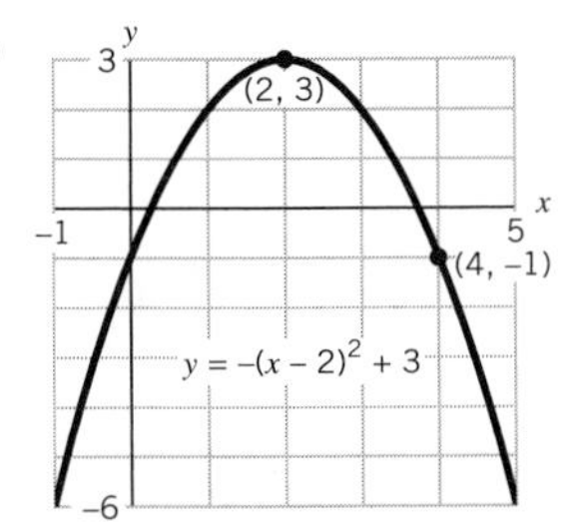

Derivation: from the data we have: $y = a(x - 2)^2 + 3$; and thus $-1 = a(4 - 2)^2 + 3$. This implies that $a = -1$. Thus $y = -(x - 2)^2 + 3$ is its equation.

65.

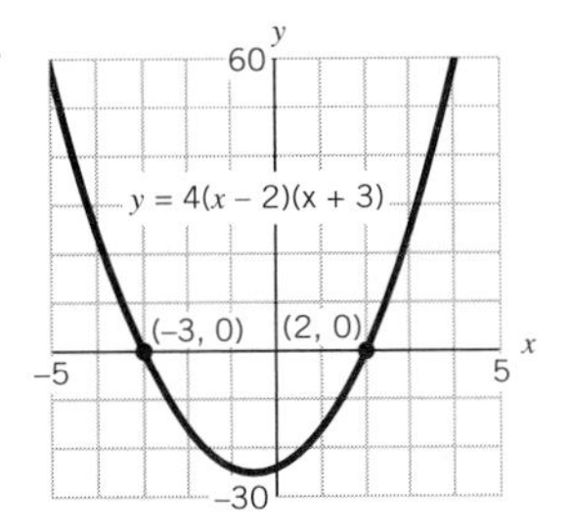

The equation is $y = 4(x - 2)(x + 3)$. Reasons: The coefficient of x^2 is always a constant multiplier of the factored form and the two roots are given. Multiplying it out gives $y = 4x^2 + 4x - 24$.

67. a. $\left(\dfrac{-b}{2a}, c + \dfrac{1 - b^2}{4a}\right)$

b. i. $(-3/2, 0)$ and 1/4 ii. (1, 3.125) and 0.125

c. Graph of $y = x^2 + 3x + 2$ showing its focus

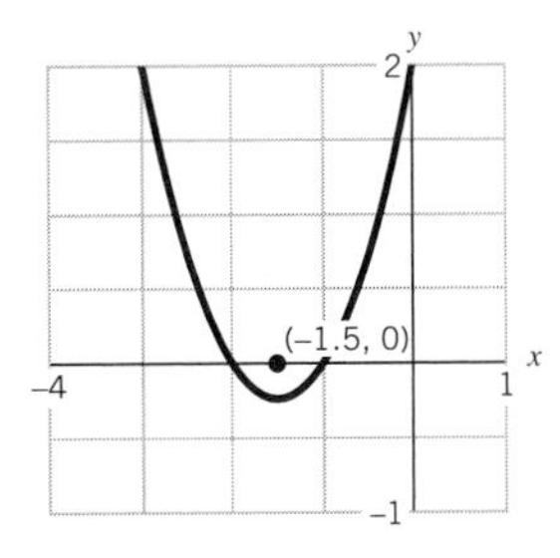

Graph of $y = 2x^2 - 4x + 5$ showing it focus

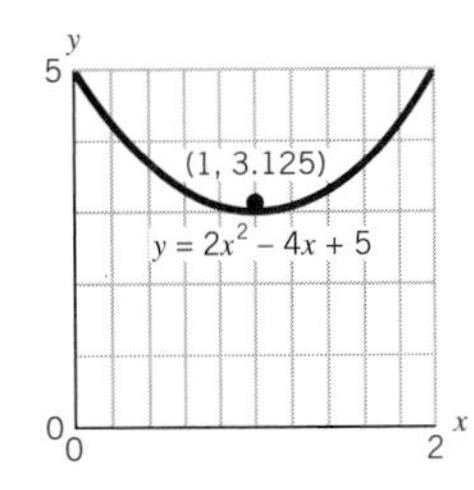

d. The smaller in absolute value is a, the more open is the parabola and thus the larger is the focal length since the focal length is $|1/(4a)|$

69. In these answers, the origin is chosen as the location of the vertex.

a. $y = ax^2$ where x and y are in feet; since for $x = \pm 2.5$ feet we have $y = 1.25$ ft., we get $1.25 = a \cdot (2.5)^2$ or $a =$ 0.2; thus the equation is $y = 0.2x^2$

b. Focal length $= |1/(4a)| = |1/(4 \cdot 0.2)| = 1.25$ feet.

c. The horizontal line through the focal point at (0, 1.25) intersects the parabola at $x = \pm 2.5$ ft. Hence the diameter there is 2.5 ft.

Exercises for Section 8.5

71. From the table you can see the average rate of change has a constant slope of -2 and thus the function is linear, namely, $y = -2x$.

x	y	Average Rate of Change of y wrt x	Average Rate of Change of Average Rate of Change
−3	−3	n.a.	n.a.
−2	1	4	n.a.
−1	3	2	−2
0	3	0	−2
1	1	−2	−2
2	−3	−4	−2
3	−9	−6	−2

Graph of average rate of change vs. x

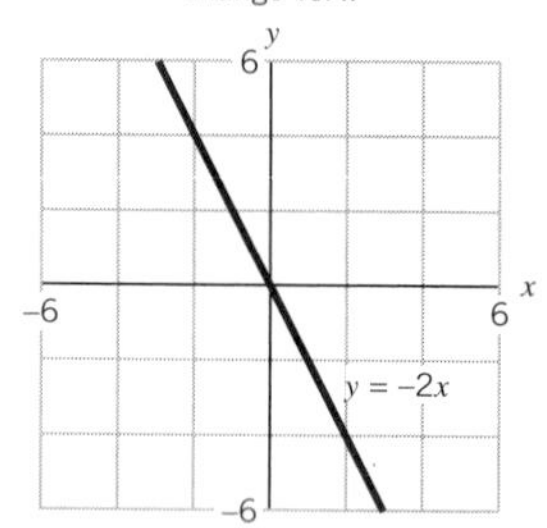

EXTENDED EXPLORATION: THE MATHEMATICS OF MOTION

1.

Time (sec)	Distance (cm)	Avg Vel over previous 1/30th sec. (cm/sec)
0.0000	0.00	n.a.
0.0333	3.75	113
0.0667	8.67	147
0.1000	14.71	181
0.1333	21.77	212
0.1667	29.90	243

The average velocity over each 1/30th of a second increases rapidly as time progresses.

3. For $d = 490t^2 + 50t$:

a. 50 is measured in cm/sec; it is the initial velocity of the object falling; 490 is measured in (cm/sec)/sec and is half the acceleration due to gravity when measured in these units.

b.

t	d
0.0	0.0
0.1	9.9
0.2	29.6
0.3	59.1

c.

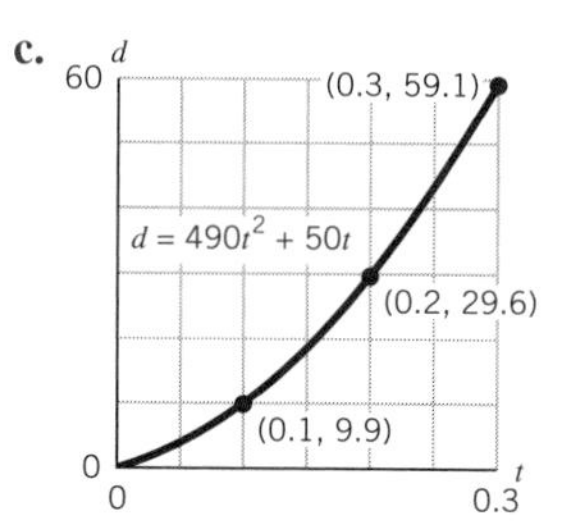

5. For $d = 4.9t^2 + 1.7t$

a. 1.7 is the initial velocity of the object falling; it is measured in meters per sec.; 4.9 is half the gravitational constant when this is measured in (meters/sec)/sec.

b.

t	d
0.0	0.000
0.1	0.219
0.2	0.536
0.3	0.951

c.

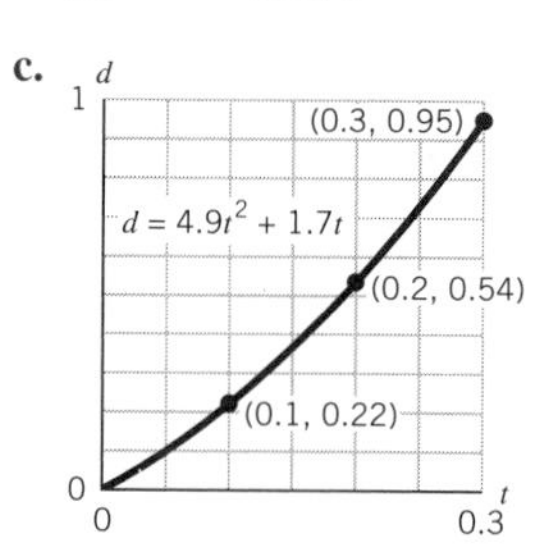

7. $(\text{m/sec}^2) \cdot \text{sec}^2 + (\text{m/sec}) \cdot \text{sec} = \text{m}$.

9. a. $D(t) = 490t^2 + 50t$ $\quad V(t) = 980t + 50$

b. $D(1) = 540$ cm, $V(1) = 1030$ cm/sec.; $D(2.5) = 3187.5$ cm, $V(2.5) = 2500$ cm/sec.

11. a. $d = 16t^2 + 12t$

b.

t	d
0	0
1	28
2	88
3	180
4	304
5	460

c.

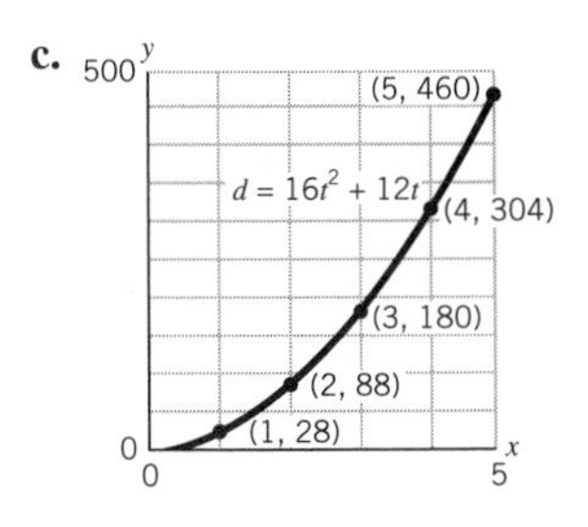

13. The distance is measured in meters if the time is measured in seconds. The use of 4.9 for 1/2 of the gravity constant is the indicator.

15. a. Answers will vary considerably.

b. Since the velocity is changing at a constant rate, a straight line should be a good fit. The graph of this line is a representation of velocity.

17. a. The coefficient of t^2 is one half the gravity constant. When distance is measured in centimeters and time in seconds, it is measured in cm/sec^2. The coefficient of t is an initial velocity and is measured in cm/sec.

b. $d(0.05) = 1.59$ cm. $d(0.10) = 5.617$ cm.
$d(0.30) = 45.993$ cm.

19. It represents an initial velocity of the object measured in meters per second.

21. a. When $t = 0.1$, $h = 19.951$ m; when $t = 2$, $h = 380.4$ m; when $t = 10$, $h = 1510$ m.

b.

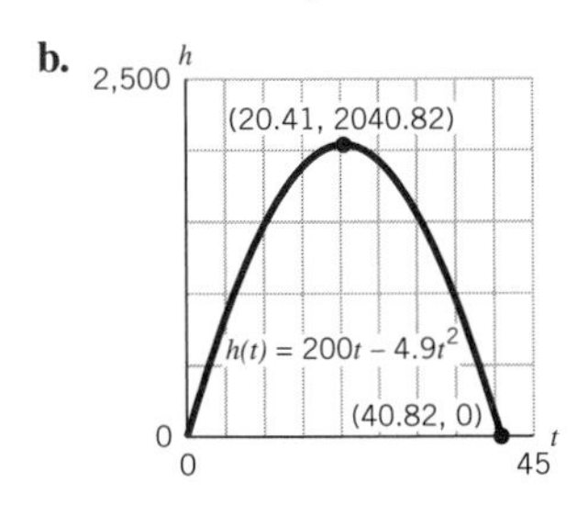

c. The object reaches a maximum height of 2040.82 meters after 20.41 seconds. It reaches the ground after 40.82 seconds of flight.

23. a. 85 is the height in centimeters of the falling object at the start; -490 is half the gravitational constant when measured in (cm/sec.)/sec.; it is negative in value since h measures height above the ground and the gravitational constant is connected with pulling objects down. This will mean subtraction from the original height.

b. 0 cm/sec.

c.

t	h
0.0	85.0
0.1	80.1
0.2	65.4

d.

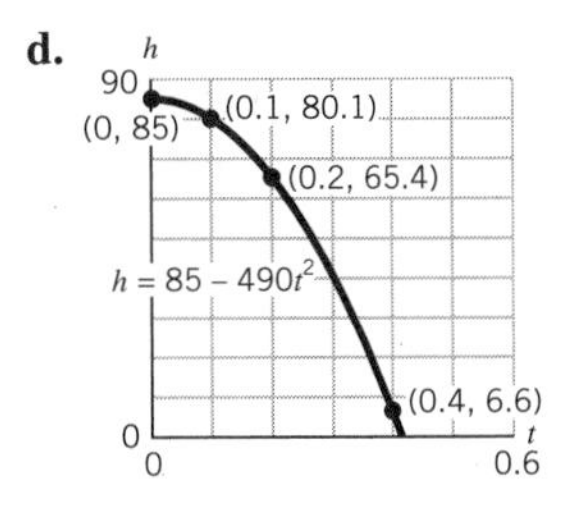

25. a. The initial velocity is positive since we are measuring height above ground and the object is going up at the start.

b. The equation of motion is $h = 50 + 10t - 16t^2$, where h is measured in feet and t in seconds.

27. a. 980 cm/sec^2 since we are measuring in cm and in sec.

b.

t	v
0	-66
1	-1046
2	-2026
3	-3006
4	-3986

c. The object is traveling faster and faster towards the ground. The increase in downward velocity is at a constant rate, as we can see from the constant slope of the graph. This constant acceleration, of course, is due to gravity.

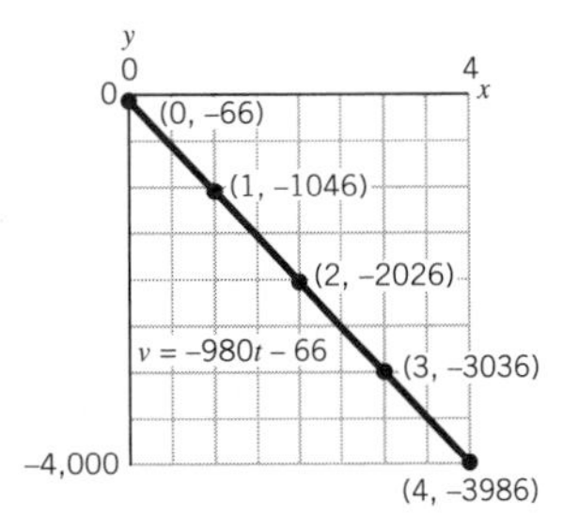

d. Ordinarily, if $t = 0$ corresponds to the actual start of the flight, then the initial condition given would indicate that the object was thrown downwards at a speed of 66 meters per second. This interpretation comes from the negative sign given to the initial velocity. But this would contradict the statement in the problem to the effect that it is a "freely falling body." In this context, another interpretation is suggested by the laboratory experiment, namely that the object started being timed at a point along its downward flight.

29. **a.** Negative, since the object is falling; h is measured in cm above the ground; t is measured in seconds.

b. $h = 150 - 25t - 490t^2$; for $0 \leq t \leq 0.528$ (the second value being the approximate time in seconds when the object hits the ground).

c. Average velocity is -270 cm/sec; the initial velocity is -25 cm/sec; the average velocity in the first half second is 10.8 times as great in magnitude as the initial velocity.

31. Forming $\dfrac{d}{t} = \dfrac{v_0 + (v_0 + at)}{2}$ and solving for d, we get

$$d = \frac{2v_0t + at^2}{2} = v_0t + \frac{1}{2}at^2$$

This is very similar in form to the falling body formula. The acceleration factor increases the velocity in a manner proportional to the square of the time traveled and the initial velocity increases the distance in a manner proportional to the time.

33. **a.** 110 cm/sec; 660 cm/sec; $60 + 10t$ cm/sec.

b. 85 cm/sec.

35. **a.** $V(t) = 200 + 60t$ meters/sec.

b. $D(t) = 200t + 30t^2$ meters

37. **a.** The units used are feet and seconds and thus the equation governing the water spout is $d = -16t^2 + v_0t$, where d is measured in feet and t in seconds and where v_0 is the sought after initial velocity. We are given that the maximum height reached is 120 ft. The maximum height is achieved at the vertex; i.e., when $t = -v_0/-32$. Solving the equation: $120 = -16\left(\dfrac{v_0}{32}\right)^2 + v_0\left(\dfrac{v_0}{32}\right) = -\dfrac{v_0^2}{64} + 2\dfrac{v_0^2}{64} = \dfrac{v_0^2}{64}$ for v_0 one obtains that $v_0^2 = 7680$ or $v_0 \approx 87.64$ ft. per second.

b. $t = v_0/32 = 87.64/32 \approx 2.75$ sec.

39. **a.** $d_c = v_ct + a_ct^2/2$; $d_p = a_pt^2/2$. One wants to solve for the t at which $d_c = d_p$, i.e., when $0 = v_ct + a_ct^2/2 - a_pt^2/2 = t(v_c + a_ct/2 - a_pt/2) = t(v_c + [a_c/2 - a_p/2]t)$ This occurs at $t = 0$ (when the criminal passes by the police car) and again when $t = 2v_c/(a_p - a_c)$

b. Now $v_c = a_ct + v_c$ and $v_p = a_pt$. One wants to solve for the t at which $v_c = v_p$, i.e., when $a_ct + v_c = a_pt$ or for $t = (a_p - a_c)/v_c$. This does not mean that the police have caught up to the criminal but rather that the police are at that moment going as fast as the criminal and are starting to go faster.

Index

Page references followed by italic *n* indicate material in footnotes.